Geographies of Development

PEARSON
Education

Geographies of Development

An Introduction to Development Studies

Third Edition

Robert B. Potter
Tony Binns
Jennifer A. Elliott
David Smith

Harlow, England • London • New York • Boston • San Francisco • Toronto • Sydney • Singapore • Hong Kong
Tokyo • Seoul • Taipei • New Delhi • Cape Town • Madrid • Mexico City • Amsterdam • Munich • Paris • Milan

Pearson Education Limited
Edinburgh Gate
Harlow
Essex CM20 2JE
England

and Associated Companies throughout the world

Visit us on the World Wide Web at:
www.pearsoned.co.uk

First published 1999
Second edition published 2004
Third edition published 2008

ISBN: 978-0-13-222823-7

British Library Cataloguing-in-Publication Data
A catalogue record for this book is available from the British Library

Library of Congress Cataloging-in-Publication Data
Geographies of development : an introduction to development studies / Robert B. Potter . . . [*et al.*]. – 3rd ed.
p. cm.
Includes bibliographical references and index.
ISBN-13: 978-0-13-222823-7
1. Economic development. 2. Economic geography. 3. Human geography.
I. Potter, Robert B.
HD82.G387 2008
338.9–dc22

2008009454

10 9 8 7 6 5 4 3 2 1
11 10 09 08

Typeset in 9.75/13pt Minion by 35
Printed and bound by Ashford Colour Press, Gosport

The publisher's policy is to use paper manufactured from sustainable forests.

Brief contents

Contents

Part II Development in practice: components of development

5 People in the development process 183

6 Resources and the environment 229

7 Institutions of development 275

Part III Spaces of development: places and development

8 Movements and flows 327

9 Urban spaces 381

List of boxes

List of case studies

List of key ideas

List of key thinkers

List of critical reflections

Preface to the Third Edition

From its first publication in 1999, the intention of *Geographies of Development* was to provide an up-to-date and innovative approach to teaching and learning in the broad interdisciplinary fields of development geography and development studies. From the outset, we were keen to get away from the sector-by-sector approach that had been so typical of earlier texts, together often with a distinctly regional orientation. The First and Second Editions attempted this by means of a threefold structure, broadly dealing respectively with: (i) conceptualising development, (ii) development in practice and (iii) spaces of development.

We have, of course, been delighted that both the First and Second Editions have been welcomed in both critical and commercial terms, and that the general tenor of the comments we have received have been positive, whether in the form of written reviews or general comments and reactions received from those who are using the book. It seems therefore that, as intended, *Geographies of Development* has generally been well received as an innovative and comprehensive text for undergraduates, as well as for some taught postgraduates, who are studying development in a variety of fields, not just geography.

As well as those reviews appearing in journals, running up to the Third Edition, the publishers commissioned a number of detailed reviews of the Second Edition. We should like to thank those involved in this process for their constructive and generally highly positive responses, as these greatly helped us in shaping this Third Edition. A couple of reviewers who had not previously been using the book stated that they had not done so because the word 'geographies' appears in the title, but would do so now having actually looked at the contents! We hope, therefore, that the addition of the subtitle 'An introduction to development studies' in respect of this Edition will provide just a little more encouragement to those concerned with development issues but who are working in other disciplinary fields.

In embarking on the Third Edition of *Geographies of Development*, once again we did not feel that the structure of the book needed to be changed in any significant fashion. Inevitably, it was clear that the text should be improved by means of general and specific updates and revisions, and this is exactly what we have done. And this time round, the publishers were enthusiastic about upgrading the overall presentation of the book. Thus, this Third Edition of *Geographies of Development* appears in a larger format and employs a modified layout. In addition, a second colour has been used throughout.

The main difference between the First and Second Editions, other than minor changes and updates, was that the Second Edition was substantially longer than the First – making a total of 509 pages, as opposed to 312 in the case of the First Edition. As a result of this and other background changes, the publishers were keen that as authors we should make every effort to provide more entry points into the text. We have responded to this by increasing the number of sections and subsections throughout the book.

In order to further aid the reader in accessing the text, short statements concerning the aims and content are provided right at the start of each chapter and these are then fleshed out by means of more detailed bullet-point summaries. In addition, a listing of key points is provided at the end of each chapter.

Further, in this Third Edition, a new 'hierarchy' of boxed materials has been introduced to support the text. Thus, the substantive boxed *Case studies* presented in the first two editions are still to be found, but an innovation is the inclusion of *Key idea* and *Key thinker* boxes where these are likely to inform and further assist the reader. Last, but by no means least, this Edition includes *Critical reflections*, which seek to engage the reader with key issues and debating points. It is our intention that groups in a classroom or tutorial setting can use these just as easily as the individual reader.

As with the Second Edition published in 2004, David Smith remains as a listed author, despite his untimely death in December 1999. Once again all three of us as continuing authors were unanimous that David should remain as a joint author of the Third Edition. Quite simply, David's intellectual input to the First Edition was clear, and it follows that it is no less recognisable in subsequent editions.

As with the First and Second Editions, we look forward to receiving the reactions of students, lecturers and general readers who use this Third Edition, in the form of reviews, the passing of comments as mentioned previously and, of course, as is more likely these days, via e-mail messages sent to us via our respective institutions. All of these will help us to shape the next edition of *Geographies of Development*.

Finally we are extremely grateful to Andrew Taylor at Pearson, who from the outset showed genuine and sustained enthusiasm for the Third Edition to be produced in a timely fashion. No publisher could have shown more interest in the project or provided more support: thank you Andrew from us all. A little further into the process, Sarah Busby helped substantially in all manner of ways and we extend our warm thanks to her for this support. Philippa Fiszzon proved to be a most efficient and very supportive editor, for which we are most grateful. Joan Dale Lace did a very thorough job copy-editing the manuscript, as did Margaret Binns in compiling the index.

Rob Potter, Tony Binns and Jenny Elliott
September 2007

Guided tour

Chapter outline
tells you what the chapter is about and introduces the topics to be covered.

Boxed case studies
include a range of international examples and illustrations to add a real world relevance to topics discussed.

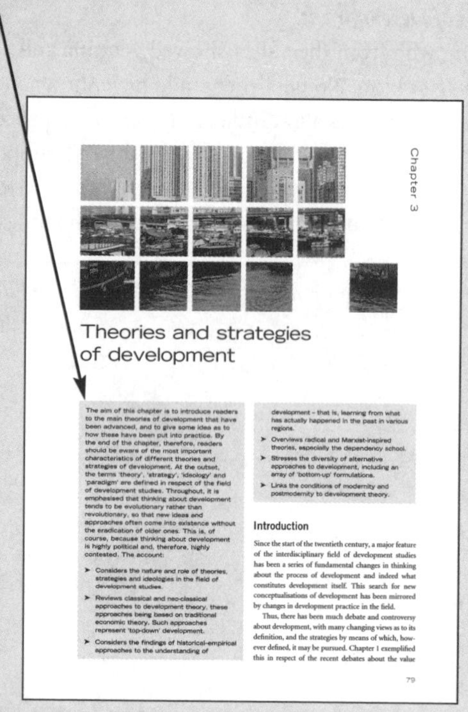

Chapter 3

Theories and strategies of development

The aim of this chapter is to introduce readers to the main theories of development that have been advanced, and to give some idea as to how these have been put into practice. By the end of the chapter, therefore, readers should be aware of the most important characteristics of different theories and strategies of development. At the outset, the terms 'theory', 'strategy', 'ideology' and 'paradigm' are defined in respect of the field of development studies. Throughout, it is emphasised that thinking about development tends to be evolutionary rather than revolutionary, so that new ideas and approaches often come into existence without the eradication of older ones. This is, of course, because thinking about development is highly political and, therefore, highly contested. The account:

- Considers the nature and role of theories, strategies and ideologies in the field of development studies.
- Reviews classical and neo-classical approaches to development theory, these approaches being based on traditional economic theory. Such approaches represent 'top-down' development.
- Considers the findings of historical–empirical approaches to the understanding of development – that is, learning from what has actually happened in the past in various regions.
- Overviews radical and Marxist-inspired theories, especially the dependency school.
- Stresses the diversity of alternative approaches to development, including an array of 'bottom-up' formulations.
- Links the conditions of modernity and postmodernity to development theory.

Introduction

Since the start of the twentieth century, a major feature of the interdisciplinary field of development studies has been a series of fundamental changes in thinking about the process of development and indeed what constitutes development itself. This search for new conceptualisations of development has been mirrored by changes in development practice in the field.

Thus, there has been much debate and controversy about development, with many changing views as to its definition, and the strategies by means of which, however defined, it may be pursued. Chapter 1 exemplified this in respect of the recent debates about the value

79

Chapter 3 Theories and strategies of development

Case study 3.1

Industrialisation and development: the case of Singapore

Singapore is a city state and in 1996 it had a total population of 3,044,000 persons. Its per capita income stood at US$32,810 in 2000, higher than for many European countries. Further, in 1996, the country recorded a 7.3 per cent growth in its GDP. Indeed, Singapore is frequently held up as a nation which has created a strong 'Third World' economy in a relatively short period. Yet when Singapore became an independent republic in 1965 the prospects for growth did not seem much better than for many newly emerging nations. As noted by Drakakis-Smith (2000), Singapore is a very clean and green city which has led by example with respect to its environmental policies. But, as the same author notes, in social terms there has been a price to pay for this continued growth.

Singapore is a good example of a state that has grown by early industrialisation. The programme which was embarked upon in 1968 focused on both light industrialisation and some forms of heavy industry, such as oil refining, iron and steel, ship building and repairing. Thus, the contribution of manufacturing to GDP grew from 11.9 per cent in 1960, the average figure for developing nations, to 29.1 per cent in 1980. By 1995, this figure had increased to over one-quarter of GDP, standing at 26.5 per cent. The Government of Singapore was one of the first in Asia to realise fully the limitations of growth via low technology industrial development. Accordingly, throughout the 1980s it sought to transform the economy by focusing on high-tech, high-value-added industries. It was fully intended that this 'second industrial revolution' would transform Singapore into the 'Switzerland' of Asia. Singapore's major trading partners are now the USA, Malaysia, Hong Kong, Japan and the European Union. The city-state has also become an important centre for financial services, with 149 commercial banks and 79 merchant banks in 2000 (Whitaker's, 1999).

The World Bank frequently holds Singapore up as a model of what can be achieved by free market policies and industrial development. But several analysts have noted that there are local factors and that these are unlikely to be repeated elsewhere. For instance, Drakakis-Smith (2000) argues that Singapore's success has been strongly predicated on its people, and that the degree of human resource management has been intensive and ultimately authoritarian. Specifically, as in many states in the region, there have been very tough controls on labour unions, with the general introduction of factory unions rather than occupational unions. Many argue that this has also served to reduce social class solidarity. Japanese-style company loyalty has been the desired outcome. In addition, since the 1960s there has been strict state population control, both in respect of migration and also directed education programmes, with children being allocated to 'hand' and 'brain' streams at an early stage of their education. Drakakis-Smith (2000) also notes that ethnic disparities in wealth are prominent in Singapore, with Malays forming the most disadvantaged group, but within a general societal context where middle-class consumerism dominates.

Rostow envisaged that there were five stages through which all countries have to pass in the development process: the traditional society, preconditions to the take-off phase, take-off, the drive to maturity, and the age of mass consumption, as depicted in Figure 3.6. Rostow's stage model encapsulates faith in the capitalist system, as expressed by the subtitle of the work: a non-communist manifesto for economic growth. For Rostow, the critical point of take-off can occur where the net investment and savings as a ratio to national income grows from 5 to 10 per cent, thereby facilitating industrialisation.

Although Rostow's framework can in many respects be regarded as a derivation of Keynesian economics, its real significance lies in the simple fact that it seemed to offer every country an equal chance to develop (Preston, 1996). In particular, saliently, the 'take-off' period was calibrated at 20 or so years, long enough to be conceivable but short enough not to seem unattainable. The importance of the Rostowian framework was that it purported to explain the advantages of the Western development model. Further, in the words of Preston (1996: 178), the 'theory of modernisation follows on from growth theory but is heavily influenced by the

90

Chapter 3 Theories and strategies of development

stage of the development debate at particular points in time.

However, each approach still retains currency in certain quarters. Hence, in the realm of development theory and academic writing, left-of-centre socialist views may well be more popular than classical–traditional and neo-classical formulations. But in the area of practical development strategies, the period since the 1980s has seen the implementation of neo-liberal interpretation of classical theory, stressing the liberalisation of trade, along with public sector cut-backs, as a part of structural adjustment programmes (SAPs), aimed at reducing the involvement of the state in the economy and promoting the free market.

But even so, the account which follows uses these four divisions to overview the leading theories, strategies and ideologies that have been used to explain and promote the development process.

Classical–traditional approaches: early views from the developed world

Introduction

The traditional approach to the study of development derives from classical and neo-classical economics and has generally dominated policy thinking at the global scale. Classical economic theory, dating from before 1914, was strongly based on the writings of Adam Smith (1723–1790) and David Ricardo (1772–1823). Both Smith and Ricardo equated economic development with the growth of world trade and the law of comparative advantage (Sapsford, 2008) (see Key idea box). Neo-classical theories, those having generally been produced since 1945 (although some date back to the 1870s), take an essentially similar worldview, stressing the importance of liberating world trade as the essential path to growth and development.

Essentially traditional approaches regard developing countries as being characterised by a dualistic structure. Hettne (1995) notes the strong role of dichotomous thinking in early anthropology, where comparisons were made between what were referred to as 'backward' and 'advanced' societies, the 'barbarian' and the 'civilised', and the 'traditional' and the 'modern'.

The fundamental dualism exists between what is seen as a traditional, indigenous, underdeveloped sector on the one hand, and a modern, developed and Westernised one on the other. It follows that the global development problem is seen as a scaled-up version of this basic dichotomy. Seminal works include those of Hirschman (1958), Meier and Baldwin (1957), Myrdal (1957), Perloff and Wingo (1961), Perroux (1950) and Schultz (1953).

The basic framework: the contribution of A.O. Hirschman

In this framework, underdevelopment is an initial state beyond which the West has managed to progress (Rapley, 1996). It also envisages that the experience of the West can assist other countries in catching-up by sharing both capital and know-how. The avowed intention, therefore, is to bring developing countries to the modern age of capitalism and liberal democracy (Rapley, 1996).

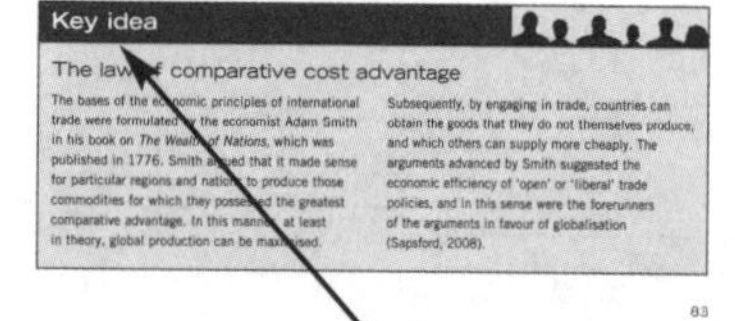

Key idea

The law of comparative cost advantage

The bases of the economic principles of international trade were formulated by the economist Adam Smith in his book on *The Wealth of Nations*, which was published in 1776. Smith argued that it made sense for particular regions and nations to produce those commodities for which they possessed the greatest comparative advantage. In this manner, at least in theory, global production can be maximised. Subsequently, by engaging in trade, countries can obtain the goods that they do not themselves produce, and which others can supply more cheaply. The arguments advanced by Smith suggested the economic efficiency of 'open' or 'liberal' trade policies, and in this sense were the forerunners of the arguments in favour of globalisation (Sapsford, 2008).

83

Chapter 3 Theories and strategies of development

system, a major development that is detailed in Chapter 4.

The top-down paradigm of development and the 'Western world view'

All of these approaches, involving unequal and uneven growth, modernisation, urban industrialisation, the diffusion of innovations and hierarchic patterns of change and growth poles may be grouped together and regarded as constituting the 'top-down' paradigm of development (Stöhr and Taylor, 1981). Such an approach advocates the establishment of strong urban-industrial nodes as the basis of self-sustained growth and is premised on the occurrence of strong trickle-down effects, by means of which, through time, it is believed that modernisation will inexorably be spread from urban to rural areas (Figures 3.3 and 3.5). This gives rise to the concept of the planned growth pole. Case study 3.1 presents the case of Singapore, where industrial development has formed an important component of development since independence in 1965.

As with modernism, all such approaches 'had a great appeal to a wider public due to the paternalistic attitude toward non-European cultures' (Hettne, 1995: 64). These approaches, together with modernisation, reflected the desire of the USA to order the post-war world, and were used to substantiate the logic of 'authoritative intervention', as noted in Chapter 1 (Preston, 1996). As Mehmet (1999: 1) stated, 'a Western worldview is the distinctive feature of the mainstream theories of economic development, old and new'. This world view has been predominantly 'bipolar', stressing a strict belief in Western rationality, science and technology.

Rostow's Stage Model of Economic Growth

Such models, including Rostow's (1960) classic *The Stages of Economic Growth*, see urban-industrial nodes as engines of growth and development. Rostow's work can be seen as the pre-eminent theory of modernisation to appear in the early 1960s (Preston, 1996). Rostow's position was avowedly right-wing politically (see Key thinker box).

Key thinker

The contribution of Walt Rostow

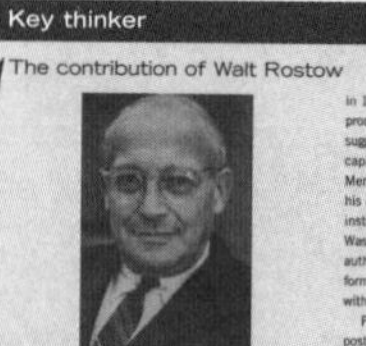
Plate 3.1 Walt Rostow
Source: Getty Images/Time & Life Pictures

Walt Whitman Rostow's (1916–2003) classic work *The Stages of Economic Growth* carried the subtitle *A Non-Communist Manifesto*, bearing testimony to its highly political orientation. Rostow was fiercely anti-communist. The book, which was published in 1960 at the height of the Cold War, offered the prospect of automatic or almost formulaic growth, suggesting that by following a few simple rules the capitalist Western model could be re-enacted. As Menzel (2006) notes, Rostow was fully aware that his development theory could be employed as an instrument in East–West relations, stressing the Washington path to development. As the same author notes, Rostow's theory was a very simple formulation, which was presented and recommended with 'missionary-zeal'.

Following a series of academic and governmental posts, when John Kennedy became President of the United States Rostow became a full-time staff member and was successful in promoting development policy as US foreign policy. After the assassination of President Kennedy Rostow continued to work under the new President Lyndon B. Johnson, and did so up until Richard Nixon became President. Above all else, Rostow's work shows the strongly political nature of development theory.

89

Key idea and key thinker boxes
provide an in-depth focus on a key idea or thinker in development studies.

Critical reflection
asks you to engage in a challenging development issue in more detail and take a view. Ideal for discussion and debate.

Key points summaries
recap and reinforce the key points you take away from the chapter. They also provide a useful revision tool

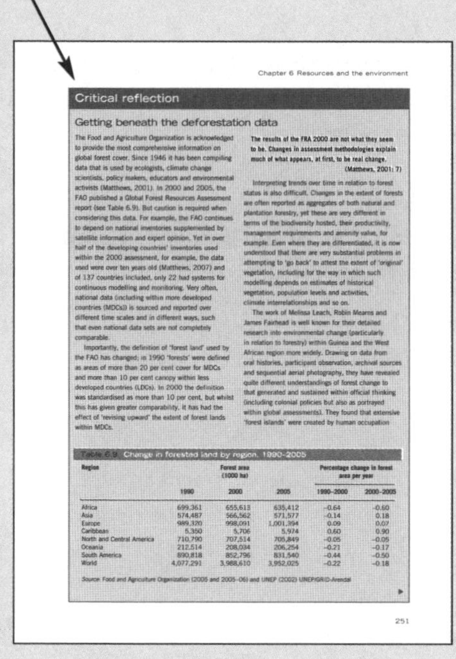

Critical reflection

Getting beneath the deforestation data

The Food and Agriculture Organization is acknowledged to provide the most comprehensive information on global forest cover. Since 1946 it has been compiling data that is used by ecologists, climate change scientists, policy makers, educators and environmental activists (Matthews, 2001). In 2000 and 2005, the FAO published a Global Forest Resources Assessment report (see Table 6.9). But caution is required when considering this data. For example, the FAO continues to depend on national inventories supplemented by satellite information and expert opinion. Yet in over half of the developing countries' inventories used within the 2000 assessment, for example, the data used were over ten years old (Matthews, 2007) and of 137 countries included, only 22 had systems for continuous modelling and monitoring. Very often, national data (including within more developed countries (MDCs)) is sourced and reported over different time scales and in different ways, such that even national data sets are not completely comparable.

Importantly, the definition of 'forest land' used by the FAO has changed; in 1990 'forests' were defined as areas of more than 20 per cent cover for MDCs and more than 10 per cent canopy within less developed countries (LDCs). In 2000 the definition was standardised as more than 10 per cent, but whilst this has given greater comparability, it has had the effect of 'revising upward' the extent of forest lands within MDCs.

The results of the FRA 2000 are not what they seem to be. Changes in assessment methodologies explain much of what appears, at first, to be real change.
(Matthews, 2001: 7)

Interpreting trends over time in relation to forest status is also difficult. Changes in the extent of forests are often reported as aggregates of both natural and plantation forestry, yet these are very different in terms of the biodiversity hosted, their productivity, management requirements and amenity value, for example. Even where they are differentiated, it is now understood that there are very substantial problems in attempting to 'go back' to attest the extent of 'original' vegetation, including for the way in which such modelling depends on estimates of historical vegetation, population levels and activities, climate interrelationships and so on.

The work of Melissa Leach, Robin Mearns and James Fairhead is well known for their detailed research into environmental change (particularly in relation to forestry) within Guinea and the West African region more widely. Drawing on data from oral histories, participant observation, archival sources and sequential aerial photography, they have revealed quite different understandings of forest change to that generated and sustained within official thinking (including colonial policies but also as portrayed within global assessments). They found that extensive 'forest islands' were created by human occupation

Table 6.9 Change in forested land by region, 1990–2005

Region	Forest area (1000 ha)			Percentage change in forest area per year	
	1990	2000	2005	1990–2000	2000–2005
Africa	699,361	655,613	635,412	−0.64	−0.60
Asia	574,487	566,562	571,577	−0.14	0.18
Europe	989,320	998,091	1,001,394	0.09	0.07
Caribbean	5,350	5,706	5,974	0.60	0.90
North and Central America	710,790	707,514	705,849	−0.05	−0.05
Oceania	212,514	208,034	206,254	−0.21	−0.17
South America	890,818	852,796	831,540	−0.44	−0.50
World	4,077,291	3,988,610	3,952,025	−0.22	−0.18

Source: Food and Agriculture Organization (2005 and 2005–06) and UNEP (2002) UNEP/GRID-Arendal

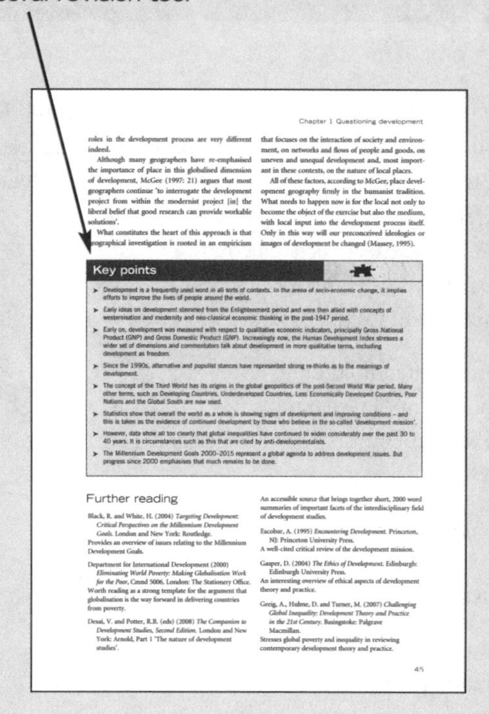

roles in the development process are very different indeed.

Although many geographers have re-emphasised the importance of place in this globalised dimension of development, McGee (1997: 21) argues that most geographers continue 'to interrogate the development project from within the modernist project [in] the liberal belief that good research can provide workable solutions'.

What constitutes the heart of this approach is that geographical investigation is rooted in an empiricism that focuses on the interaction of society and environment, on networks and flows of people and goods, on uneven and unequal development and, most important in these contexts, on the nature of local places.

All of these factors, according to McGee, place development geography firmly in the humanist tradition. What needs to happen now is for the local not only to become the object of the exercise but also the medium, with local input into the development process itself. Only in this way will our preconceived ideologies or images of development be changed (Massey, 1995).

Key points

- Development is a frequently used word in all sorts of contexts. In the arena of socio-economic change, it implies efforts to improve the lives of people around the world.
- Early ideas on development stemmed from the Enlightenment period and were then allied with concepts of westernisation and modernity and neo-classical economic thinking in the post-1947 period.
- Early on, development was measured with respect to qualitative economic indicators, principally Gross National Product (GNP) and Gross Domestic Product (GNP). Increasingly now, the Human Development Index stresses a wider set of dimensions and commentators talk about development in more qualitative terms, including development as freedom.
- Since the 1990s, alternative and populist stances have represented strong re-thinks as to the meanings of development.
- The concept of the Third World has its origins in the global geopolitics of the post-Second World War period. Many other terms, such as Developing Countries, Underdeveloped Countries, Less Economically Developed Countries, Poor Nations and the Global South are now used.
- Statistics show that overall the world as a whole is showing signs of development and improving conditions – and this is taken as the evidence of continued development by those who believe in the so-called 'development mission'.
- However, data show all too clearly that global inequalities have continued to widen considerably over the past 30 to 40 years. It is circumstances such as this that are cited by anti-developmentalists.
- The Millennium Development Goals 2000–2015 represent a global agenda to address development issues. But progress since 2000 emphasises that much remains to be done.

Further reading

Black, R. and White, H. (2004) *Targeting Development: Critical Perspectives on the Millennium Development Goals*. London and New York: Routledge.
Provides an overview of issues relating to the Millennium Development Goals.

Department for International Development (2000) *Eliminating World Poverty: Making Globalisation Work for the Poor*, Cmnd 5006. London: The Stationery Office.
Worth reading as a strong template for the argument that globalisation is the way forward in delivering countries from poverty.

Desai, V. and Potter, R.B. (eds) (2008) *The Companion to Development Studies, Second Edition*. London and New York: Arnold, Part 1 'The nature of development studies'.
An accessible source that brings together short, 2000 word summaries of important facets of the interdisciplinary field of development studies.

Escobar, A. (1995) *Encountering Development*. Princeton, NJ: Princeton University Press.
A well-cited critical review of the development mission.

Gasper, D. (2004) *The Ethics of Development*. Edinburgh: Edinburgh University Press.
An interesting overview of ethical aspects of development theory and practice.

Greig, A., Hulme, D. and Turner, M. (2007) *Challenging Global Inequality: Development Theory and Practice in the 21st Century*. Basingstoke: Palgrave Macmillan.
Stresses global poverty and inequality in reviewing contemporary development theory and practice.

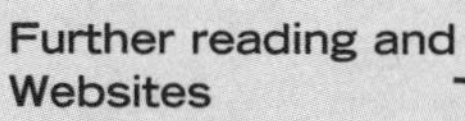

Further reading and Websites
provide guidance and ideas about where to find useful further paper/online resources and information.

Key points

- Development theories and strategies have been many and varied, with new approaches generally being added alongside existing ones.
- Classical and neo-classical economic approaches generally stress the need for unrestrained, polarised growth and of letting the market decide for itself.
- Neo-liberalism as a generic development paradigm stems from the New Right and emphasises what is seen as the continuing need for market liberalisation and for the economy to be market- and performance-driven.
- Historical models give a normative impression of the degree to which in the past, since mercantilism and colonialism, development has been highly uneven and spatially polarised.
- Both dependency (radical) approaches and alternative/another development can be seen as direct critiques of modernisation theory. Thus the economic growth paradigm of the 1950s was challenged by socialist and environmentally oriented paradigms in the 1960s and 1970s respectively.
- Development thinking reflects political views – views on how economies should work and how societies should be structured. Postmodernity is leading to less monumental approaches to development.

Further reading

Cowan, M.P. and Shenton, R.W. (1996) *Doctrines of Development*. London: Routledge.
A comprehensive text dealing with principles of development.

Desai, V. and Potter, R.B. (eds) (2008) *The Companion to Development Studies*, 2nd edn. Part 2: Theories and strategies of development. London: Hodder-Arnold.
Contains a range of short essays covering the most important aspects of development theories and strategies.

Greig, A., Hulme, D. and Turner, M. (2007) *Challenging Global Inequality: Development Theory and Practice in the 21st Century*. Basingstoke: Palgrave Macmillan.
Stresses global poverty and inequality in reviewing contemporary development theory and practice.

Hettne, B. (1995) *Development Theory and the Three Worlds*, 2nd edn. London: Longman.
Although produced in the mid-1990s, this book still affords a very clear introduction to the principal paradigms of development.

Preston, P.W. (1996) *Development Theory: An Introduction*. Oxford: Blackwell.
Another book dealing with the theory and practice of development.

Simon, D. (ed.) (2006) *Fifty Key Thinkers on Development*. London: Routledge.
A useful source which brings together short essays on those who are deemed to have had a noticeable impact on studies of development.

Websites

www.iedconline.org
The website of the International Economic Development Council based in Washington DC. IEDC is a non-profit membership organisation dedicated to assisting 'economic developers' to do their job more effectively. With over 4500 members worldwide, IEDC offers support for professional development, advisory and legal services and regular conferences on development topics.

www.ids.ac.uk
The Institute of Development Studies (IDS) at the University of Sussex offers an extensive website giving access to the online catalogue of the IDS Library, with good holdings on development strategies and ideologies.

www.id21.org
Also out of IDS, and known as 'Communicating Development Research', this is a reporting scheme, bringing a selection of the 'latest and best' UK-based development research.

Discussion topics

- Assess the extent to which 'modern' development theories have been discredited by the rise of so-called 'alternative' and 'postmodern' approaches.
- Examine the extent to which Vance's mercantile model can be seen as a graphical representation of classical dependency theory.
- 'Old development theories never die. In fact, they don't even seem to fade away.' Discuss.
- For one developing nation, outline and assess the national development strategies employed since 1947.

Discussion topics
highlight debates, controversies and ideas from the chapter that could be thought about individually or discussed in a wider group.

Acknowledgements

The publishers would like to thank Rob, Tony and Jenny for their considerable investment of time, expertise and goodwill to ensure this new edition is of the highest quality and has been a pleasure to produce. We would also like to thank the following reviewers for their valuable comments on the book:

Sylvi Endresen, Associate Professor
Department of Human Geography
University of Oslo

Professor Ragnhild Lund
Department of Geography
Norwegian University of Science and Technology (NTNU)

Dr Arjan Verschoor
School of Development Studies
University of East Anglia

Professor Sören Eriksson
Jönköping University, Sweden

Dr Alan Terry
Geography and Environmental Management
University of the West of England

Professor Sylvia Chant
Geography and Environment
London School of Economics

Aina Tollefsen
Department of Social and Economic Geography
Umeå University, Sweden

Dr Dina Abbott
Senior Lecturer, Geographical Sciences
University of Derby

Paul Hoebink
Associate Professor, Development Studies
Catholic University Nijmegen, The Netherlands

Faiza Zafar
Graduate Student
Department of Geography
University of Illinois at Urbana-Champaign

Dr Shelagh Waddington
Department of Geography
National University of Ireland, Maynooth

We are grateful to the following for permission to reproduce copyright material:

Table 1.1 adapted from, *Southeast Asia: The Human Landscape of Modernisation and Development*, Routledge, (Rigg, J. 1997) by permission of Cengage Learning Services Limited; Figure 1.1 adapted from *Development Theory: An Introduction*, reproduced by permission of Blackwell Publishers, (Preston, P.W. 1996); Figure 1.2 from *Human Development Report 1990: Concept and measurement of human development*, Oxford University Press, (United Nations Development Programme, 1990). By permission of Oxford University Press, Inc; Table 1.3 adapted from *The Geography of Economic Development: Regional Changes, Global Challenges*, 2e, McGraw Hill, (Fik, T.J. 2000) © 2000 The McGraw-Hill Companies, Inc., *Global Issues: An Introduction*, 2e, Blackwell Publishing, (Seitz, J.L. 2002) and *Human Development Report 1998*, Oxford University Press (United Nations Development Programme, 1998). By permission of Oxford University Press, Inc; Table 1.4 adapted from 'The Millennium Development Goals' (Rigg, J.) in *The Companion to Development Studies*, 2e, Hodder-Arnold, (Desai, V. and Potter, R.B., eds, 2008), based on www.un.org/millenniumgoals/index.html; Figure 1.5 adapted from 'Profiles of the Third World' in *Pacific Viewpoint* 5(2), Blackwell Publishing, (Buchanan, K. 1964); Figure 1.6, adapted from 'What is a socialist developing country?' in *Geography*, 72(4), Geographical Association, (Drakakis-Smith, D., Doherty, J. and Thrift, N. 1987) www.geography.org.uk; Figure 1.7 adapted from Willy Brandt, *North-South: A Program for Survival*, figure "Models on the 1980s: North and South; core, periphery and semi-periphery", © 1980 The

Independent Bureau on International Development Issues, by permission of the MIT Press and also from *North–South: A Programme For Survival*, Pan, (Brandt, W. 1980). Copyright © W. Brandt, 1980, with the permission of Pan Macmillan; Figures 1.8, 1.9 and 4.7 adapted from *Human Development Report, 2001: Promoting Linkages*, Oxford University Press (United Nations Development Programme, 2001). By permission of Oxford University Press, Inc.; Figure 1.12 from *Millennium Development Project*, UNEP/GRID-Arendal, (United Nations Environmental Programme, 2005); Figure 2.1 adapted from *Development Theory: An Introduction* reproduced by permission of Blackwell Publishers (Preston, P. W. 1996); Figures 2.2 and 2.3 adapted from *Political Geography*, Taylor, P., Pearson Education Ltd © 1985; Figure 2.4 adapted from *Inside Third World Cities*, Croom Helm (Lowder, S. 1986) by permission of Cengage Learning Services Limited; Figure 2.5 adapted from *The African Inheritance*, Griffiths, I.L. Copyright (© 1995) Routledge. Reproduced by permission of Taylor and Francis Books UK.; Table 3.1 after 'Development from below: the bottom-up and periphery-inward development paradigm', in Stöhr, W.B. and Taylor, D.R.F. (eds) *Development from Above or Below?* pp 39–72, copyright 1981, © John Wiley & Sons Limited, reproduced with permission Chichester: John Wiley, (Stöhr, W.B. 1981); Table 3.2 adapted from *The Condition of Postmodernity*. Blackwell Publishing (Harvey, D. 1989), adapted with the kind permission of David Harvey; Figure 3.7, Figure 8.1 and Figure 10.4 Material on pages 95, 342 and 485 from Oxfam's campaign against IMF policies including SAPs from Oxfam IMF campaign poster and adaptation of Map of Ethiopia from *Behind the Weather: Lessons to be Learned. Drought and Famine in Ethiopia* and the framework from Neefjes, K. *Environments and Livelihoods: Strategies for Sustainability* pp. 82 (2000) Oxfam Publishing is reproduced with the permission of Oxfam GB, Oxfam House, John Smith Drive, Cowley, Oxford OX4 2JY, UK www.oxfam.org.uk. Oxfam GB does not necessarily endorse any text or activities that accompany the materials, nor has it approved the adapted text.; Figure 3.8 Friedmann, John, *Regional Development Policy: A Case Study of Venezuela*, 1 figure: "Core- periphery model", © 1966 Massachusetts Institute of Technology by permission of the MIT Press; Figure 3.9(b) Cluff cartoon Trickle down: from rich to poor within a single country reprinted with kind permission of *Private Eye*; Figure 3.11 and Figure 3.17 adapted from *Urbanisation in the Third World*, by permission of Oxford University Press (Potter, R.B. 1992); Figure 3.13 from *Latin America and the Caribbean: A Systematic and Regional Survey*, (Blouet, B.W. and Blouet, O.M.), © 2002 John Wiley & Sons Inc. Reprinted with permission of John Wiley & Sons, Inc.; Figures 3.14 and 9.12 adapted from 'Urbanisation and development in the Caribbean' in *Geography*, 80, Geographical Association, (Potter, R.B. 1995), www.geography.org.uk; Figure 3.15 adapted from *Development Theory: An Introduction* reproduced by permission of Blackwell Publishers (Preston, P. W. 1996); Figure 3.16 Courtesy of *New Internationalist Magazine*/reproduced by permission of Paul Fitzgerald; Table 4.1 adapted from Keeling, D.J. 'Transport and the world city paradigm' in Knox, P.L. and Taylor, P.J. (eds) *World Cities in a World-System*, (1995) Cambridge University Press; Figure 4.1a from Hellman, © Pressdram Limited 2002, reproduced by permission; Figure 4.1b Artwork appears courtesy of David Simonds/*The Guardian* Copyright Guardian News & Media Ltd 1999; Table 4.2 from www.internetworldstats.com 2007; Table 4.3 from *Industrialization and Development in the Third World*, Routledge, Cengage Learning Services Limited, (Chandra, R. 1992); Figures 4.2, 4.10, 4.11 and 4.14 Reproduced by permission of SAGE Publications, London, Los Angeles, New Delhi and Singapore, from Dicken, *Global Shift: Transforming the World Economy*, 3e, Paul Chapman Publishing, Copyright © Sage Publications, 1998; Figure 4.3 adapted from 'Annihilating space? The speed-up of communications' (Leyshon, A. 1995) in Allen, J. and Hamnett, C. (eds) *A Shrinking World?* Oxford University Press and the Open University. By permission of the Oxford University Press, Inc.; Table 4.4 Reproduced by permission of SAGE Publications, London, Los Angeles, New Delhi and Singapore, from Dicken, P., *Global Shift: The Internationalization of Economic Activity*, 2e, Copyright (©Paul Chapman, 1992); Figure 4.4 adapted from Keeling, D.J. 'Transport and the world city paradigm' in Knox, P.L. and Taylor, P.J. (eds) *World Cities in a World-System* (1995) Cambridge University Press and by permission of Dr David Keeling; Table 4.5 from *World Development Indicators 2005* (Table 4.3) © International Bank for Reconstruction and Development/The World Bank; Figure 4.5 from AND Cartographic Publishers Ltd, 1997, © RM Education plc (2008) All Rights Reserved. Helicon Publishing is a division of RM Education plc; Table 4.8 Reprinted with permission from Yue-Man Yeung (Ed.) *Global Change and the Commonwealth*, Hong Kong: Hong Kong Institute of Asia-Pacific Studies, The Chinese University of Hong Kong, 1996, pp 111–112 and also with permission from John Hopkins University; Figures 4.8, 4.9 and 4.12 Reproduced by permission of SAGE Publications, London, Los Angeles, New Delhi and

Singapore, adapted from Dicken, P., *Global Shift: Mapping the Changing Contours of the World Economy*, 5e, Copyright (© Sage Publications Ltd, 2007); Figure 4.18 adapted from Friedmann, J. 'The world city hypothesis' *Development and Change*, 17, Wiley-Blackwell Publishing Ltd, 1986; Figure 4.20 adapted from 'Third World urbanisation in a global context' *Geography Review*, 10 (3), Philip Allan Updates, (Potter, R.B. 1997), by permission of Hodder Education; Figure 4.21 reproduced by permission of *New Internationalist*; Figure 4.22 Peter Schrank cartoon from *The Independent on Sunday*, 5 December 1999, © 1999 *The Independent*; Figure 4.23 and Table 6.1 from 'The crazy logic of the continental food swap' in *The Independent*, 25 March (Lucas, C. 2001), © 2001 *The Independent*; Figure 4.24 from *A Shrinking World*, pp 233–254, by permission of Oxford University Press (Allen, J. and Hamnett, C. 1995); Table 5.1 from Bongaarts, J. 1995. Global and regional population projections to 2025. In *Population and Food in the Early Twenty-First Century: Meeting Future Food Demand of an Increasing Population*, ed. N. Islam. IFPRI Occasional Paper 30, Tables 2.1 & 2.2. Washington, D.C.: International Food Policy Research Institute. (The sources of these tables are: Merrick, T. 1989. World population in transition. *Population Bulletin* 41 (2). [for Table 2.1], Bos, E., M.T. Vu, A. Levin, and R. Bulatao. 1992. World population projections, 1992–93 edition. Baltimore: Johns Hopkins University Press for the World Bank. [for Table 2.2] and United Nations. 1993. World population prospects: The 1992 revision. New York. [for Table 2.2]; Table 5.2 and Table 5.3 from UNICEF, The State of the World's Children 2007, UNICEF, New York, pp. 102–5 and 122–5; Figure 5.2 adapted from the following sources: Reproduced by permission of SAGE Publications, London, Los Angeles, New Delhi and Singapore, from Jones, G. and Hollier, G., *Resources, Society and Environmental Management*, Copyright (© Paul Chapman, 1997) and from *A Dictionary of Geography*, Oxford University Press, (Mayhew, S. 1997). By permission of Oxford University Press, Inc.; Figure 5.3a adapted from *The Independent* 12 Jan 1998 © reproduced by permission of *The Independent*; Table 5.4: Column 1 from WHO *Global InfoBase Online*, www.who.int/ncd_surveillance/infobase/web/InfoBaseCommon; correspondence on health expenditure, May, Geneva, Column 2 from World Health Statistics 2006, Geneva www.who.int/whosis/whostat2006/en/index.html whostat2006_healthsystems.pdf with the permission of the World Health Organization; Figure 5.4 adapted from 'Ghana: West Africa's latest success story?' in *Teaching Geography* 19 (4), (Binns, T. 1994), Geographical Association, www.geography.org.uk; Table 5.5 updated from http://www.unaids.org/en/KnowledgeCentre/HIVData/EpiUpdate/EpiUpdArchive/2007default.asp. using Slide 3 from http://data.unaids.org/pub/EPISlides/2007/071118_epicore2007_slides_en.pdf Reproduced with kind permission of UNAIDS www.unaids.org; Table 5.6 from *Epidemiological fact sheets on HIV/AIDS and sexually transmitted infections*, www.who.int/globalatlas/default.asp (World Health Organization, 2007); Figure 5.6b Adapted by permission from Macmillan Publishers Ltd: *European Journal of Clinical Nutrition* 54, (3), p. 250, R Martorell, L Kettel Khan, M L Hughes, L M Grummer-Strawn, 'Obesity in women from developing countries', copyright 2000.; Table 5.7 created from material from the following sources: **HIV prevalence and HDI rank:** United Nations Development Program, Human Development Report 2006, (2006) Palgrave Macmillan, reproduced with permission of Palgrave Macmillan, and **Deaths to AIDS:** *Epidemiological fact sheets on HIV/AIDS and sexually transmitted infections*, www.who.int/globalatlas/default.asp, WHO (2007), with permission of the World Health Organization; Figure 5.7 from Figure 4.1 'Impact of AIDS on life expectancy in five African countries, 1970–2010', AIDS epidemic update, *2006 Global Report*, UNAIDS. Reproduced by kind permission of UNAIDS, www.unaids.org; Table 5.8 from *State of the World's Children 2007*, www.unicef.org/sowc07/docs/sowc07.pdf UNICEF (2006), New York; Figure 6.1 from *Global Environment Outlook 3*, Earthscan, (UNEP/GRID-Arendal 2002); Table 6.3 from *Tears of the Crocodile: From Rio to Reality in the Developing World*, Pluto Press, (Middleton, N., O'Keefe, P. and Moyo, S. 1993); Figure 6.3 from *Dams and Development: A New Framework for Decision-Making*, The Report of the World Commission on Dams, Earthscan, (World Commission on Dams, 2000); Table 6.4 Compiled from 'Comparative study of cut roses for the British market produced in Kenya and the Netherlands', Précis Report for World Flowers, Cranfield University, (Williams, S. 2007) by kind permission of Ian Finlayson, World Flowers, www.world-flowers.co.uk; Figure 6.4 compiled using material from *Human Development Report 2006*, Palgrave Macmillan, (UNDP, 2006), reproduced with permission of Palgrave Macmillan; Figure 6.5 from 'Energy' (Holdern, J., & Pachauri, R.K.,) in *An Agenda of Science for Environment and Development into the 21st Century*, Cambridge University Press, (Dooge, (ed) 1992); Table 6.6 from Water and sanitation data, www.wssinfo.org/en/233_wat_africaS.html Copyright © WHO and UNICEF Joint Monitoring Programme. All rights reserved.; Table 6.7 from *World Development Report, 1992*, Tables 2.2 and 2.3 on p. 49, (World Bank, 1992), © International Bank for Reconstruction and Development/The World Bank; Figure 6.6 cartoon, reproduced by permission of David Hughes; Figure 6.7 adapted from *Humid Tropical Environments*, Blackwell Publishing, (Reading, A.J., Thompson, R.D. and Millington, A.C. 1995); Table 6.8 adapted from *Humid Tropical Environments*, Blackwell Publishing, (Reading, A.J., Thompson, R.D. and Millington, A.C. 1995) and also by kind permission of Andrew Millington; Figure 6.8 from *World Resources 1998–99: Environment and Health*, Oxford University Press (World Resources Institute, 1998). Reproduced by permission of World Resources Institute, Washington DC; Table 6.9 compiled from *Statistical Yearbook 2005–06*, Food and Agriculture Organization of the United Nations and *Global Environment Outlook 3*, Earthscan, (UNEP/GRID-Arendal, 2002); Figure 6.9 from from *Human Development Report 2006*, Palgrave Macmillan, (UNDP, 2006) reproduced with permission of Palgrave Macmillan; Table 6.10 from *Climate Change 2007: The Physical Science Basis, Summary for Policymakers*, (IPCC, 2007), The Cambridge University Press; Table 6.11 compiled from *Human Development Report, 2000*, Oxford University Press, (United Nations Development Programme, 2000). By permission of Oxford University Press, and *Human Development Report 2006*, Palgrave Macmillan, (UNDP, 2006), reproduced with permission of Palgrave Macmillan; Figure 6.11 www.who.int/heli/risks/urban/urbanenv/en/index.html accessed 9 May 2007; Table 6.12 from *The State of the World, 2006: The Challenge of Global Sustainability*, Earthscan, (Worldwatch Institute, 2006); Table 6.14 from *World Development Indicators 2007*, Table 3.13 'Air Pollution' on pp 174–5, (World Bank, 2007), © International Bank for Reconstruction and Development/The World Bank; Figure 7.3 adapted from 'Information about the World Bank Group', *Annual Report 2005*, (World Bank, 2005), © International Bank for Reconstruction and Development/The World Bank; Table 7.4 from www.worldbank.org 'About Us: Organization: Executive Directors: Voting Powers: IBRD: Votes and Subscriptions', © International Bank for Reconstruction and Development/The World Bank; Figure 7.4 adapted from 'Regional lending by theme and sector', *Annual Report 2006* (World Bank, 2006) © International Bank for Reconstruction and Development/The World Bank; Table 7.5 from World Bank *World Bank and the Environment: Fiscal 1993*, pp 3–4, (World Bank, 1994) © International Bank for Reconstruction and Development/ The World Bank; Figure 7.5 from *Global Development Finance*, Figure 2.1, p. 35, (World Bank 2007), © International Bank for Reconstruction and Development/The World Bank; Table 7.6 from 'Heaven or hubris: reflections on the "New Poverty Agenda"' (Maxwell, S.) in *Targeting Development: Critical Perspectives on the Millennium Development Goals*, pp 25–46, Routledge, (Black, R. and White, H. (eds) 2004) by permission of Cengage Learning Services Limited; Figure 7.6 adapted from *Environment Matters at the World Bank*, ((a) and (c) from *Good Governance and Environmental Management: Reporting on Environmental Sustainability*, (b) from Appendix 10 pp 131–2), (World Bank, 2006), © International Bank for Reconstruction and Development/The World Bank; Figure 7.7 from *Making Sustainable Commitments: An Environment Strategy for the World Bank, Summary*, December, Fig 1 'What's new in the Environment Strategy', p xxvii, (World Bank, 2001), © International Bank for Reconstruction and Development/The World Bank; Table 7.8 from *Structural Adjustment: Theory, Practice and Impacts*, Routledge, (Mohan, G., Brown, E., Milward, B. and Zack-Williams, A.B. 2000), by permission of Cengage Learning Services Limited; Figure 7.8 from *Making Sustainable Commitments: An Environment Strategy for the World Bank, Summary*, December, Chapter 2 'Lessons from the World Bank Experience', Fig 2.2, (World Bank, 2001), © International Bank for Reconstruction and Development/The World Bank; Table 7.9 from *The Companion to Development Studies*, (Desai, V. and Potter, R.B. (eds)) 'The World Bank and NGOs', (Nelson, P.J.), Edward Arnold (Publishers) Ltd, (© 2002). Reproduced by permission of Edward Arnold (Publishers) Ltd; Figure 7.9 adapted from *Fairer Global Trade: The Challenge for the WTO*, Understanding global issues 96/6. European Schoolbooks (Buckley, R. (ed.) 1996) © Understanding Global Issues Ltd; Figure 7.10 from *Business and Human Rights: A Geography of Corporate Risk*, IBLF/Amnesty International, (International Business Leaders Forum, 2002), © Amnesty International/IBLF 2002; Table 7.12 from *Global Citizen Action* edited by Michael Edwards and John Gaventa. Copyright © 2001 by Lynne Rienner Publishers. Used with permission and also with the permission of Earthscan; Table 8.1 from World Tourism Highlights: 2006 edition, www.unwto.org/facts/menu.html, World Tourism Organization (2006), © UNWTO, 9284400508; Table 8.2 from

Cashin, P.C., Liang, H. and McDermott, C.J., 'Do commodity price shocks last too long for stabilization schemes to work?' in *Finance and Development*, 36(3), International Monetary Fund, (1999); Figure 8.2 adapted from *The West Pacific Rim*, copyright John Wiley & Sons Ltd, reproduced with permission (Hodder, R. 1992); Table 8.3 from Table 1.1, *Debt Relief for the Poorest: An Evaluation Update of the HIPC Initiative*, Chapter 1, p. 4, World Bank Independent Evaluation Group (WB-IEG, 2006), © International Bank for Reconstruction and Development/ The World Bank; Figure 8.5 from *Development Cooperation Report, 2005*, Organisation for Economic Cooperation and Development. Copyright OECD, 2006; Table 9.1 from *Urbanisation in the Third World* by permission of Oxford University Press (Potter, R. B. 1992); Figure 9.1, Figure 9.17 and Figure 9.27 adapted from *The Third World City*, 2e, Drakakis-Smith, D., Copyright (© 2000), Routledge. Reproduced by permission of Taylor & Francis Books UK.; Figure 9.4 adapted from *The Urban Transformation of the Developing World*, by permission of Oxford University Press (Gugler, J. 1996); Table 9.6 compiled from *A Geography of the Third World*, 2nd Edn., reproduced by permission of Routledge (Dickenson, J., Gould, B., Clarke, C., Mather, C., Prothero, M., Siddle, D., Smith, C. and Thomas-Hope, E. 1996) and *World Development Report, 1994*, Table 31 'Urbanization', p 222, (World Bank, 1994) © International Bank for Reconstruction and Development/The World Bank; Table 9.7 from *The City in the Developing World*, Longman, (Potter, R.B. and Lloyd-Evans, S. 1998), by permission of Pearson Education Ltd; Figure 9.7, Figure 9.8 and Figure 9.25 adapted from *Urbanisation in the Third World* by permission of Oxford University Press (Potter, R. B. 1992); Table 9.8 based on 'On rank-size distributions of cities: an ecological approach' in *Economic Development and Cultural Change*, 17, The University of Chicago Press, (Vapnarsky, C.A. 1969); Table 9.11 after 'National urban development strategies in developing countries', *Urban Studies*, Vol. 18, pp 267–283, reproduced by permission of Taylor & Francis, www.tandf.co.uk/journals (Richardson, H.W. 1981); Figure 9.13 after 'Rural–urban interaction in Barbados and the southern Caribbean', in Potter, R.B. and Unwin, T. (eds) *Urban–Rural Interaction in Developing Countries*, reproduced by permission of Routledge (Potter, R.B. 1989) and © *The Urban Caribbean in an Era of Global Change*, R.B. Potter, 2000, Ashgate Publishing; Table 9.14 reprinted from *Habitat International*, Vol. 19, No. 4, C. Rakodi, 'Poverty lines or household strategies?' pp 407–426, copyright 1995, reprinted with permission from Elsevier Science; Table 9.15 based on 'Towards Environmental Strategies for Cities' (Bartone, C. *et al.* 1994), in *Strategic Options for Managing the Urban Environment*, Paper 18, Chapter 1, Figure 1.1 'Spatial scale of urban environmental problems' , p. 15, World Bank, © International Bank for Reconstruction and Development/The World Bank; Figure 9.15 and Figure 9.16 adapted from 'The emergence of desakota regions in Asia: expanding a hypothesis', (McGee, T.G.) in *The Extended Metropolis: Settlement Transition in Asia*, University of Hawaii Press, (Ginsburg, N., Koppell, B. and McGee, T.G. (eds) 1991); Figure 9.18 adapted from 'International transport and communications interactions between Pacific Asia's world cities', (Rimmer, P.J.), in *Emerging World Cities in Asia*, United Nations University Press, (Lo, F.-C. and Yeung, Y.-M. (eds) 1991). Reproduced with the permission of United Nations University Press; Figure 9.20 Reprinted with permission from *Urban Geography*, Vol. 16, No. 6, pp. 521–554. © Bellwether Publishing, Ltd, 8640 Guilford Road, Suite 200, Columbia, MD 21046. All rights reserved.; Figure 9.21 and Figure 9.24 adapted from *The City in the Developing World*, Longman (Potter, R.B. and Lloyd-Evans, S. 1998) by permission of Pearson Education Ltd; Figure 9.26 adapted from 'Legitimising informal housing: accommodating low income groups in Alexandria, Egypt' *Environment and Urbanisation*, 1, (Soliman, A.M. 1996); Figure 9.28 adapted from *Ecology and Development in the Third World*, Methuen (Gupta, A. 1988) by permission of Cengage Learning Services Limited; Table 10.1 reprinted from *World Development*, 34(1), Rigg, J., 'Land, farming, livelihoods, and poverty: rethinking the links in the Rural South', pp 180–202. Copyright 2006, with permission from Elsevier; Figure 10.1 from *Rural Livelihoods and Diversity in Developing Countries*, Oxford University Press, (Ellis, F. 2000) By permission of Oxford University Press, Inc.; Table 10.2, 2004 figures from *FAO Statistical Yearbook 2005–06*, Food and Agricultural Organization of the United Nations; Figure 10.2 adapted from 'The demise of the moral economy: food and famine in a Sudano-Sahelian region in historical perspective', (Watts, M.) in *Life Before the Drought*, pp 124–48, Allen & Unwin, (Scott, E. (ed.) 1984) by permission of Cengage Learning Services Limited; Table 10.3 adapted from *World Development Report 2000/2001*, Table 4 'Poverty', (World Bank, 2001), © International Bank for Reconstruction and Development/The World Bank; Figure 10.3 adapted from Geheb, K. and Binns, T. '"Fishing farmers" or "farming fishermen"? The quest for household income and nutritional security on the Kenyan shores of Lake Victoria', in *African Affairs*, 1997, 96, pp 73–93, Copyright © 1997 The Royal

African Society, by permission of The Royal African Society, http://afraf.oxfordjournals.org/; Table 10.4 2004 figures from 'Meeting the MDG drinking water and sanitation target – the urban and rural challenge of the decade' (http://who.int/water_sanitation_health/monitoring/jmpfinal.pdf) World Health Organization; Table 10.6 This table adapted from *Whose Reality Counts?*, Intermediate Technology Publications, (Chambers, R. 1997), with the kind permission of Practical Action Publishing; Table 10.7 from 'Farming in the public interest', in *World Institute State of the World 2002*, Earthscan, (Halweil, B. 2002); Table 10.8 from *Southeast Asia*, Routledge, (Rigg, J. 1997) by permission of Cengage Learning Services Limited; Table 10.9 from *An Introduction to Agricultural Geography*, 2e, Routledge (Grigg, D. 1995) by permission of Cengage Learning Services Limited; Table 10.10 after 'Diversification and risk management amongst East African herders', *Development and Change*, Vol. 32, pp 401–433, reproduced by permission of Blackwell Publishers (Little, P.D. *et al.* 2001); Table 10.11 from *World Resources, 2005: The Wealth of the Poor*, World Resources Institute (2005); Table 10.12 reprinted from *World Development*, Vol. 28, No. 11, W. Cavendish, 'Empirical regularities in the poverty–environment relationship of rural households: evidence from Zimbabwe', pp 1979–2000, copyright 2000, reprinted with permission from Elsevier Science; Table 10.13 compiled from 'Dryland forestry: manufacturing forests and farming trees in Nigeria', (Cline-Cole, R.), in *The Lie of the Land*, James Currey Publishers, (Leach, M. and Mearns, R. (eds) 1996); Table 10.14 from *Agroforestry: Realities, Possibilities and Potentials*, Martinus Nijhoff, (Gholz, H.L. (ed.) 1987), with kind permission of Springer Science and Business Media; Table 10.15 *Women and Development in the Third World*, Routledge, (Momsen, J.H. 1991) by permission of Cengage Learning Services Limited; Table 10.16 from *FAO Statistical Yearbook 2005–06*, Food and Agricultural Organization of the United Nations.

Plate 2.2 from 'Tents for the Colonies' from *Notes on Outfit*, circa early 1900s, reproduced by permission of the Royal Geographical Society; Plate 3.3 (Sean Sprague), Plate 3.4, Plate 6.3 and Plate 10.3 (Ron Giling), Plate 3.9 (Eric Millex), Plate 5.5 (Bjorn Omar Evju), Plate 6.5 (Jeremy Hartley), Plate 8.2(Howard J Davies), Plate 10.2 (Philip Wolmuth) © Panos Pictures; Plate 4.1, Plate 6.7, Plate 8.5 and Plate 10.6 (Mark Edwards), Plate 4.3 (Jorgen Schytte), Plate 4.4 (Ron Giling), Plate 4.6 (Harmutt Schwarzbach), Plate 7.1 (Teit Hornbak) © Still Pictures; Plate 4.9 Barbados: 'just beyond your imagination' © 1998 Barbados Tourism Authority with kind permission of Barbados Tourism Authority.

Exhibit 6.1 created from The Stern Review Report © Crown copyright 2006, *The Economics of Climate Change: The Stern Review* © Cambridge University Press 2007 and from *Climate Change 2007: The Physical Science Basis, Summary for Policymakers*, Cambridge University Press, (IPCC, 2007); Exhibit 6.2 *Climate Change 2007: The Physical Science Basis, Summary for Policymakers*, Cambridge University Press, (IPCC, 2007); Exhibit 6.3 from 'Biodiversity and ethics' (Dolman, P.) in *Environmental Science for Environmental Management*, Pearson Education Ltd, (O'Riordan, T. (ed.) 2000); Exhibit 7.1 from *World Resources, 2005: The Wealth of the Poor*, World Resources Institute (2005); Exhibit 7.2 adapted from *World Development Report: The State in a Changing World*, Box 1.1 'State and government: Some concepts', p. 20, (World Bank, 1997), © International Bank for Reconstruction and Development/The World Bank; Exhibit 10.1 from *Rural Livelihoods and Diversity in Developing Countries*, Oxford University Press, (Ellis, F. 2000) By permission of Oxford University Press, Inc.

We are grateful to the following for permission to reproduce the following texts:

Box 7.2 from *Global Governance: A Review of Multilateralism and International Organizations*, vol. 6, 11. Copyright © by Lynne Rienner Publishers. Used with permission.; Critical Reflection Box, Chapter 8 created using material from 'Achieving grassroots transformation in post-apartheid South Africa' in *International Journal of Development Issues*, 5(2), Emerald Group Publishing Limited, (Bek, D., Binns, T., Nel, E. and Ellison, B. 2006)); Box 8.5 from *Trade for Life: Making Trade Work for Poor People*, Christian Aid, (Curtis, M. 2001); Extract on page 368 © The Economist Newspaper Limited, London (7 December 2006); Quotes on pages 347–8, 352, 353, 369 and 370, extracts from '**Africa: Make or Break. Action for recovery**' Oxfam 1993 are reproduced with the permission of Oxfam GB, Oxfam House, John Smith Drive, Cowley, Oxford OX4 2JY UK www.oxfam.org.uk. Oxfam GB does not necessarily endorse any text or activities that accompany the materials.

In some instances we have been unable to trace the owners of copyright material, and we would appreciate any information that would enable us to do so.

The publisher would like to thank the following for their kind permission to reproduce their photographs:
(Key: b-bottom; c-centre; l-left; r-right; t-top)

Alamy Images: Andrew Hasson 135; Dionodia Images 15; H Mark Weidman Photography 271; Popperfoto 64; **Andre Gunder Frank:** 109b; **Corbis:** Guenter Rossenbach/Zefa 172; Kevin R Morris/Bohemian Nomad Picturemakers 266; Tibor Bognar 54; Xiaoyang Liu 236; **Getty Images:** AFP 12, 170; Hulton Archive 185; News 317; Time & Life Pictures 89; **PA Photos:** 109t; **Rob Potter:** Dept of Geography, Emeritis, T McGee 407

Picture Research by: Alison Prior.
Every effort has been made to trace the copyright holders and we apologise in advance for any unintentional omissions. We would be pleased to insert the appropriate acknowledgement in any subsequent edition of this publication.

Introduction

This Third Edition of *Geographies of Development* aims to build on the contribution made by the previous two editions in providing a comprehensive introductory textbook for students, primarily those taking courses in the field of development geography and the interdisciplinary area of development studies. The feedback the authors received concerning the First and Second Editions showed that although the text is mainly directed at the second-year undergraduate market, given the global importance of the subject matter dealt with, the book is just as appropriate for first-year students taking broader courses, along with those reading for more specialist options in the third year of their degree programmes. Indeed, we are directly aware that the book is also recommended as a key text on a number of taught Masters programmes.

At the outset, the distinctive aim of *Geographies of Development* was to move away from what had at that time become the traditional structure of geography and development textbooks, which all too frequently started with definitions of the Third World and colonialism, and then proceeded to consider, step by step, topics such as population and demography, agriculture and rural landscapes, mining, manufacturing, transport, urbanisation, development planning and so on. Having provided detailed accounts on such topics, many texts unfortunately terminated at that juncture, but those that endeavoured to provide a broader picture generally went on to present a selection of country- or region-based case studies.

Geographies of Development endeavours to break this mould of development-oriented textbooks in a manner that reflects the rapidly changing concerns of development itself. In this sense, its *raison d'être* is to provide a text for learning and teaching about development in the early twenty-first century. As such, the structure of this Third Edition remains broadly the same as the first two editions, with a division into three relatively equal parts, dealing respectively with conceptualising development (Part I), development in practice (Part II) and the spaces of development (Part III). This structure is shown diagrammatically in the figure.

Part I (Chapters 1, 2, 3 and 4) provides a detailed overview of the concepts, ideas and ideologies that have underpinned writings about the nature of development as well as pragmatic attempts to promote development in the global arena. It also addresses how 'development' has been conceptualised and measured, and gives detailed consideration to important topics such as the histories, meanings and strategies of development, the emergence of the Third World, the nature of imperialism and colonialism and its various stages of mercantile, industrial and late colonialism, together with key concepts such as the new international division of labour and the new international economic order. Part I also provides thorough reviews of relevant and related topics such as modernity,

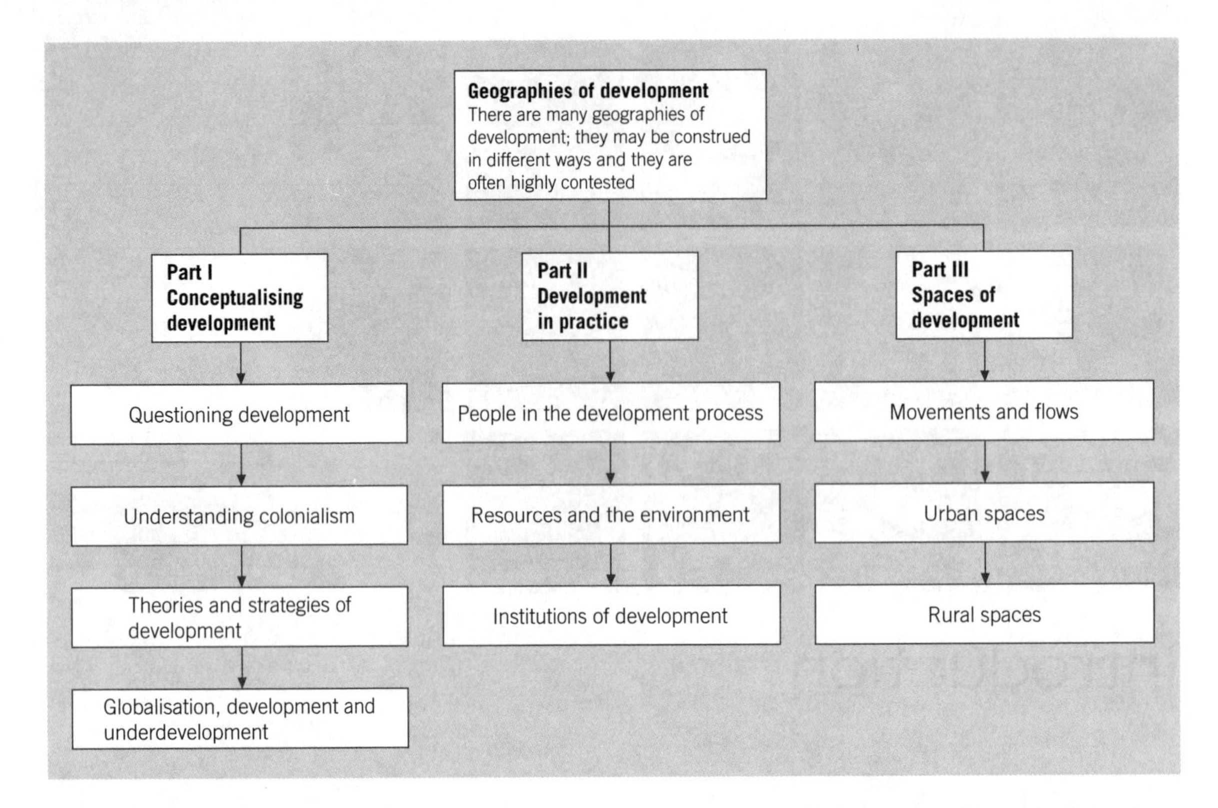

enlightenment thinking, the relevance of postmodernity to the Developing or Poor Worlds, anti-developmentalism, the socialist Third World, responsibility to distant others, globalisation, global shifts and time–space convergence. Updated sections emphasise important topics such as anti-development and anti-capitalism, global poverty and inequalities, the digital divide, global shifts, progress with the Millennium Development Goals (MDGs), decolonisation, the legacies of colonialism, post-colonialism and participatory and 'bottom-up' development strategies.

As with the other parts of the book, these early chapters exemplify the title and the overarching theme of the volume. Part I makes it clear that ideas concerning development have been many and varied, and have been highly contested through time. Thus definitions of, and approaches to, development have varied from place to place, from time to time, from country to country, region to region, and group to group within the general populace. It is essentially this plural nature of development that *Geographies of Development* seeks both to exemplify and to illustrate. Furthermore, this part of the book demonstrates that current global processes are not leading to the homogenisation of the world's regions. Far from it, the evidence shows all too clearly that contemporary global processes are leading to increasing differences between places and regions, and thus to the generation of progressively more unequal patterns of development and change. This is nowhere more the case than in relation to the so-called 'digital divide'. Hence, the emphasis is on multiple geographies of development.

Part II (Chapters 5, 6 and 7) covers what may be regarded as the basic components of the development equation – people, environments, resources, institutions and communities – together with the complex and multifaceted interconnections that exist between them. New sections have been included on the effects of the HIV/AIDS pandemic, the position of children in conflict situations and the effects of ageing populations on development processes. In considering resources and environment, further attention is now given to issues of resource scarcity (such as of water and oil), to 'new resources' such as Genetically Modified Organisms, and to the 'curse' that natural resources can provide to development (through the association with conflict and inequalities, for example). Chapter 6 draws on new data concerning the causes and impacts of climate change and examines the operation of new markets in carbon. The inclusion of a chapter specifically dealing with institutions and communities as the primary decision makers involved in the development process serves to exemplify the utility of the overall approach adopted in *Geographies of Development*. The decision makers considered extend from the agents of global governance – the United Nations, World Bank, International Monetary Fund and World Trade

Organization – via the country level, involving the role of the state, down to civil society, community participation and the empowerment of the individual, embracing non-governmental and community-based organisations. This account serves to stress the plurality of decision makers affecting geographies of development, just as the detailed expositions on population, resources, environment and development exemplify the diversity of views on salient topics of the moment. These include the character of sustainable development, people–resource relations, the concept of limits to growth, the environmental impacts of development, biodiversity loss, land degradation, pollution and global warming, health, education and human rights.

Part III (Chapters 8, 9 and 10) focuses on what development means in relation to particular places and people. This is achieved by consideration of the flows and movements that occur between geographically separate locales, and in terms of the distinctive issues raised by development and change in both urban and rural spaces. Once again, notwithstanding the difference in focus, the theme is the diversity and complexities of the movements and flows of people, capital and innovations, along with the diverse realities of transport and communications. Pressing topics of current significance, such as patterns of aid, international tourism, the realities of world trade and the debt crisis, receive detailed attention in this part of *Geographies of Development*. The nature and scale of urbanisation in the contemporary Third World, evolving urban systems and the incidence of unequal development, the need for urban and regional planning, the salience of basic needs and human rights and the quest for sustainable cities in relation to the 'brown agenda' are the prominent topics reviewed in relation to urban spaces and development imperatives. Consideration of the importance of urban–rural relations is an additional feature. Rural spaces are analysed with particular reference to diverse rural livelihood systems (particularly the rising significance of non-farm incomes), plus the examination of the multiple meanings and outcomes of approaches to rural development, such as land reform, the green revolution, irrigation and the promotion of non-farm activities. Forming the last major part of the book, these chapters draw heavily on earlier accounts presented in Parts I and II, and they make frequent reference to the realities of globalisation, convergence, divergence, urban bias, industrialisation and sustainable development, as well as other topics.

The thematic structure and orientation of *Geographies of Development* means that important contemporary development issues such as social capital, civil society, NGOs, anti-development, anti-capitalism, postmodernity, globalisation, gender and development, structural adjustment, poverty reduction programmes, climate change, sustainable development, environmental degradation, human rights, basic needs, empowerment and participatory democracy are not dealt with in standalone chapters, but rather are treated as appropriate at various points in the text, and sometimes from a variety of different perspectives. This approach reflects the complexity of these issues in the context of multiple geographies of development. A case in point is the relationship between tourism and development. This is first covered in detail in Part I in considering processes of globalisation. International tourism then reappears when Chapter 8 in Part III considers global movements and flows.

Geographies of Development focuses on the processes that are leading to change, whether for better or worse. In this sense, the book follows Brookfield's (1975) simple and straightforward definition of development as change, whether positive or negative. Thus, although the primary remit of the book is the so-called 'Third', 'Developing', or 'Poor' World, the focus of the book is very much on development as change, regardless of where or how it is occurring. We can take the case of tourism once more, and the first major example of the use of mass tourism as a strategy of development is provided by Spain, a European colonial power. This is presented as an example of the early stage of internationalisation in relation to development in Part I. Naturally, throughout the hope is that development can lead to positive change that is reflected in the marked betterment of the lives of people around the world.

As in the First and Second Editions, every effort has been made in the Third Edition of *Geographies of Development* to elucidate clear and cogent examples of the issues under discussion, in the form of diagrams, maps, tables, photographs, boxed materials and critical reflections. Many new illustrations are included in this edition, and updated and entirely new boxed case studies and examples are presented throughout the chapters. These seek either to extend definitions of basic concepts or to provide detailed illustrations of the generic topics under consideration or to promote critical reflection and discussion. These are as diverse as Aboriginal town camps in Alice Springs, Australia, globalisation and the production of athletic footwear, the environmental impacts of tourism, and China's one-child population policy.

In Part I of *Geographies of Development*, the nature and definition of the Third World is the subject of detailed discussion. Although much debate surrounds the term, it can generally be argued that the expression *Third World* still serves to link those countries that, all bar a very few, are characterised by a colonial past and relative poverty in the current global context. It is for this reason that the term *Third World* is still employed as a shorthand collective noun for what must be appreciated as a diverse set of nations. When referring to nations and areas that make up this broadly defined category, the expressions 'developing countries', 'developing areas', 'poor regions and areas' are used interchangeably, as befits the overarching title *Geographies of Development*. But the title of the book implicitly recognises that these specific terms can be used just as readily in relation to the former socialist states of Eastern Europe, to southern Spain, to Aboriginal Australia or indeed to disadvantaged areas within urban regions. Some might suggest that in the contemporary context the term 'poor countries' is a more indicative and more useful one, reflecting the need to implement progressive and effective poverty reduction strategies.

As authors we have embarked on this Third Edition with the firm belief that teaching, learning and researching about territories other than the ones in which we live, and of which we have direct experience, are demanding, but vitally important tasks (Potter and Unwin, 1988, 1992; Unwin and Potter, 1992). The amount of media attention given to development issues in poor countries seems to have declined steadily in recent years. Potter (2003) has recently cited the results of a survey carried out by the Third World and Environment Broadcasting Trust (3WE), funded by Oxfam, Christian Aid, Comic Relief and other charities. The survey provided a detailed analysis of programming on British television during 2001, revealing that only four programmes dealing with the politics of developing countries were shown during that year. Further, three of the five major channels broadcast no programmes at all in this category. Not only was it found that the serious international documentary is virtually dead, but when the developing world was depicted on television it was usually in the context of travel programmes, or in providing 'exotic' backgrounds for holiday 'challenges', reality television and 'docusoaps' featuring celebrities.

We believe that the post-war development of geography as a discipline has pivoted too strongly around a UK/Europe/North America 'core' focus, leading to a relative neglect of the study of distant places, and also the existence of little empathy among the broad academic community for the relatively few colleagues who have directed their research activities towards an investigation of patterns and processes in the Third World 'periphery'. Such issues have been the subject of a lively debate in the pages of academic geography journals such as *Area* (Potter, 2001a, 2002a; A. Smith, 2002).

We would advocate a reshaped vision of geography, in which both theories and empirical studies travel in all directions, recognising the porosity of boundaries in this era of increasing trannationality and globalisation. Furthermore, it seems important that geography and geographers should show greater responsibility to distant 'others' at a time when increasing interdependence is occurring alongside progressively greater inequality between the world's 'haves' and the 'have-nots' (Smith, 1994). It is the ultimate aim of *Geographies of Development* to assist students and teachers alike in structuring their observations and discussions of the multiple meanings of development in the increasingly complex and interdependent contemporary world.

PART I

Conceptualising development: meanings of development

Chapter 1

Questioning development

Having outlined the overall aims and the structure of *Geographies of Development* in the Introduction, the present chapter provides a background context for understanding the nature and meanings of development. In this way, the account also provides an overarching context for the chapters that make up the rest of this introductory book on development studies. This initial chapter is about the ways in which actors in the development process think about development; how they seek to define it, determine its components and conceptualise its purpose. It is also about understanding fundamental critiques of development, or so called 'anti-development'. In the second half of the chapter, the spatial expression of development in the form of the Third World, Developing World, Global South and Poor Countries is considered in the light of current patterns and processes of development. More specifically the chapter:

- Overviews how development has been, and can be defined and conceptualised for academic and policy-related purposes;
- Explores how development has been measured, from quantitative counts of relative wealth per person such as Gross Domestic Product/Gross National Product (GDP/GNP), to the Human Development Index (HDI) and the qualitative conception of development as 'freedom';
- Seeks to make readers aware of recent critiques of development such as those presented by anti-development, post-development and beyond-development;
- Reviews and assesses the genesis and nature of spatial categorisations of development such as the 'Third World', 'Developing Countries', the 'Global South' and 'Poor Countries';
- Stresses that while general indicators show that the developing world has witnessed substantial socio-economic improvements as a whole since the 1970s, during that same period the world has become twice as unequal. Little has changed in the case of sub-Saharan Africa;
- Introduces the Millennium Development Goals (MDG) as an agreed set of global development targets adopted in 2000 and reviews progress in reaching these by 2015;
- Finishes by linking geography and development via a concern with what we may refer to as 'distant others' – people who live far away from us.

Introduction: from 'development' to 'anti-development'

This chapter firstly looks at the ways in which development has been defined and characterised. This proceeds from the simple consideration of the general use of the word 'development' in everyday life. Following on, a summary overview of the multifarious approaches that have been adopted to implement development in practice is presented.

The closely related argument that such development initiatives have not worked effectively in the past, and indeed, by definition, that the types of development attempted could never ultimately be successful, is considered. This line of argument is referred to as 'anti-development', 'post-development' or 'beyond development', and is associated with what has been referred to as the 'impasse in development studies' (Schuurman, 2008; Power, 2003).

In the second half of the chapter, spatial aspects of development and development initiatives are considered in detail. Such an approach involves interrogating the utility of terms such as the 'Third World', 'Developing World', 'Global South', 'Poor Countries' and the like. Globally speaking, to which spaces do these sorts of terms apply? Are they helpful labels? Which terms have the widest currency at the present time?

The current state of the gap existing between the poorer and richer nations of the world is also considered in this chapter, with emphasis being placed on whether conditions are improving or worsening, that is 'converging' (getting more similar) or 'diverging' (getting more varied), at the international scale.

As part of this discussion, efforts to improve conditions in developing countries are considered, specifically in respect of what were known as the International Development Targets or, more commonly now, the Millennium Development Goals.

The chapter finishes with a brief discussion of the changing relationships between geography and development. It is the express aim of this chapter to set out a number of major themes that will have pertinence at many points in the rest of the book.

The meaning of the word 'development'

The *Concise Oxford Dictionary of Current English* defines the word 'development' as '[g]radual unfolding, fuller working out; growth; evolution . . . ; well-grown state, stage of advancement; product; more elaborate form . . . ; Development area, one suffering from or liable to severe unemployment'.

As this dictionary definition suggests all too clearly, 'development' is a word that is almost ubiquitous within the English language. People talk about the 'development of the child' and the 'development of the self'. Many firms have 'research and development' divisions in which the creation and evolution of new products, from sports trainers and car exhausts to laptop computers and mobile phones, is the specific focus of attention.

Turning to the level of the state, 'physical development (land use) plans' are produced; so too are 'national economic development plans', dealing with the economy as a whole. These sorts of plans are expressly designed to guide the process of development and change in the sense of unfolding and working out how things should be in the future. In this sense, development has a close connection with planning. Planning itself may be defined as foreseeing and guiding change (Hall, 1982; Potter, 1985; Pugh and Potter, 2005).

Thus, in the arena of development policy, development processes are influenced by development planning, and most plans are in turn shaped by development theories that ultimately reflect the way in which development is perceived; in other words, by what we may refer to as the ideology of development.

However, the development process is affected by many factors other than ideologies (Tordoff, 1992), although ideologies often condition state and institutional reactions to these. The precise nature of development theories, development strategies and development ideologies forms the subject of the major review of development theories and strategies that is provided in Chapter 3.

In the context of this book, the major use of the term 'development' is at the global scale. At this level, one of the main divisions of the world is between so-called 'developed nations' and 'developing nations', in a manner that is frequently understood to involve stages of advancement and evolution, as in the dictionary definition provided at the beginning of this section. At the simplest level, developed countries are seen as assisting the developing countries by means of development aid, in an effort to reduce unemployment and other indicators of 'underdevelopment'.

In practical terms though, what exactly is meant by *development?* (See Critical reflection on development.) And when it comes right down to it, do individuals, firms, states and global institutions understand the word 'development' to mean much the same thing?

Critical reflection

The nature of development

In considering the ethics of development, Gasper (2004), citing Thomas (2001), recognises a number of different usages of the word 'development' in the development studies literature. These are worth noting here as they effectively expand upon the simple dictionary definition of 'development' given at the outset of this section:

1. Development as fundamental or structural change – for example, an increase in income;
2. Development as intervention and action, aimed at improvement, regardless of whether betterment is, in fact, actually achieved;
3. Development as improvement, with good as the outcome;
4. Development as the platform for improvement – encompassing changes that will facilitate development in the future.

These sub-definitions start to make us think that development may not always lead to an overall improvement, only a partial one. For example, income per head may go up, but inequality might increase rapidly at the same time. And if when incomes increase more people can afford cars, and more large cars, then road congestion, increased journey times, parking problems and pollution are likely to follow soon after.

Critical reflection

It is quite often observed that those with higher incomes may not always be the most contented when asked to evaluate their level of satisfaction with different aspects of their lives. Why might this be the case – can't money help to buy happiness? Looking at points 1–4 above, what other factors might be involved and what other things may people be looking for in their lives? And can the same sorts of arguments be scaled up and applied at the level of nations? Are the richest countries likely to be those within which the population is, on average, the most satisfied? Are you aware of data that support or refute any such broad association between income levels and social satisfactions?

Table 1.1 lists some 'good' and 'bad' outcomes that are frequently associated with the process of development. On the plus side is the idea that development brings economic growth and national progress, and should involve other positive outcomes such as the provision of basic daily needs (food, clothing, housing, basic education and health care), better forms of governance and a move towards patterns of growth that are more sustainable in the long term.

In respect of the negative consequences of development, the occurrence of inequalities between rich and poor regions, countries and groups of people is often referred to, along with the perpetuation of relative poverty. Another line of criticism suggests that so-called development is associated with the dependency of poor countries on richer nations, and the maintenance of forms of economic, social, political and cultural subordination.

Table 1.1 Alternative interpretations of development

Good	Bad
Development brings economic growth	Development is a dependent and subordinate process
Development brings overall national progress	Development is a process creating and widening spatial inequalities
Development brings modernisation along Western lines	Development undermines local cultures and values
Development improves the provision of basic needs	Development perpetuates poverty and poor working and living conditions
Development can help create sustainable growth	Development is often environmentally unsustainable
Development brings improved governance	Development infringes human rights and undermines democracy

Source: Adapted from Rigg (1997).

For the most part in this chapter, and indeed in the book as a whole, the concept and practice of development are discussed in relation to the experiences of what are frequently referred to as 'developing countries' or 'poor countries'. But it should be borne in mind that development relates to all parts of the world at every level, from the individual to the global. Thus, development relates just as much to poor areas in cities, and relatively poor regions in rich nations (see Potter, 2000, 2001b).

However, development has become most often linked with the so-called 'Third World', itself a value-laden term, the emergence of which has been closely associated with the rapid evolution of the concept of development in the second half of the twentieth century. The second part of this chapter (p. 22 onwards) will therefore examine the emergence, use and persistence of what some now regard as an outmoded terminology, and will associate this with thinking about development itself.

In conclusion, the working definition of development assumed by this text at the outset is that initially provided by Brookfield (1975), namely that development is change, either for the better or for the worse. Specifically in this text it is assumed that progressive and effective development represents change that is intended to lead to the betterment of people and places around the globe.

Thinking about development

Histories of development: the enlightenment, modernity, neo-colonialism, trusteeship and the like

Most people writing about both development and what has come to be referred to recently as 'anti-development' (Escobar, 1995; Preston, 1996; Sachs, 1992; Power, 2003; Simon, 2007) situate the origins of the modern process of development in the late 1940s. More precisely, they link the *modern era of development* to a speech made by President Truman in 1949 in which he employed the term 'underdeveloped areas' to describe what was soon to be known as the *Third World.* Truman also set out what he saw as the duty of the West to bring 'development' to such relatively underdeveloped countries.

If *colonialism* is defined as the direct political control and administration of an overseas territory by a foreign state, then effectively Truman was establishing a *new colonial*, or *neo-colonial* role for the USA within the newly independent countries that were emerging from the process of decolonisation. He was encouraging the so-called 'underdeveloped nations' to recognise their condition and to turn to the USA for long-term assistance.

Modernism may be defined as the belief that development is all about transforming 'traditional' countries into *modern, westernised nations.* Viewed in this light, it is undoubtedly true that the genesis of much modern(ist) development theory and practice lay in the period between 1945 and 1955.

For many Western governments, particularly former colonial powers, such views represented a continuation of the late colonial mission to develop colonial peoples within the concept of *trusteeship* (Cowen and Shenton, 1995; Chapter 2).

Trusteeship can be defined as the holding of property on behalf of another person or group, with the belief that the latter will better be able to look after it themselves at some time in the future. There was little recognition that many traditional societies might in fact have been content with the ways of life they already led. Indeed, development strategists often tried to persuade them otherwise. Thus Rigg (1997: 33) cites the American advisers to the Thai government of the 1950s as trying to prevent the monks from preaching the virtues of contentedness, which was seen as retarding modernisation.

Many other writers, however, recognise that the origins of modern development lay in an earlier period. Specifically, such a move was closely linked with the rise of rationalism and humanism in the eighteenth and nineteenth centuries, respectively. During this period, the simple definition of development as change became transformed into what was seen as a more directed and logical form of evolution.

Collectively, the period when these changes took place is known as the 'Enlightenment'. The Enlightenment generally refers to a period of European intellectual history that continued through most of the eighteenth century (Power, 2008).

In broad terms, Enlightenment thinking stressed the belief that science and rational thinking could progress human groups from barbarism to civilisation. It was the period during which it came to be increasingly believed that by applying rational, scientific thought to the world, change would become more ordered, more predictable and more valuable.

The new approach challenged the power of the clergy and largely represented the rise of a secular (that is a non-religious) intelligentsia. Hall and Gieben (1992) list a number of threads which made up Enlightenment thinking: the primacy of reason/rationalism; the belief in empiricism (gaining knowledge through observation); the concept of universal science and reason; the idea of orderly progress; the championing of new freedoms; the ethic of secularism; and the notion that all human beings are essentially the same (cited in Power, 2008).

Those who could not adapt to such views came to be thought of as 'traditional' and 'backward'. As an example of this, the Australian Aborigines were denied any rights to the land they occupied by the invading British in 1788 because they did not organise and farm it in a systematic, rational way, that is in what was construed as a 'Western' manner.

It was in this fashion, and at this juncture, that the whole idea of development became directly associated with Western values and ideologies. Thus, Power (2002: 67) notes that the 'emergence of an idea of "the West" was also important to the Enlightenment . . . it was a very European affair which put Europe and European intellectuals at the very pinnacle of human achievement'. Thus, development was seen as being directly linked to Western religion, science, rationality and principles of justice. This theme is extensively explored in Chapters 3 and 4.

In the nineteenth century, Darwinism began to associate development with evolution; that is, a change towards something more appropriate for future survival (Esteva, 1992). When combined with the rationality of Enlightenment thinking, the result became a narrower but 'correct' way of development, one based on Western social theory.

During the Industrial Revolution this became heavily economic in nature. But by the late nineteenth century a clear distinction seems to have emerged between the notion of 'progress', which was held to be typified by the unregulated chaos of pure capitalist industrialisation, and 'development', which was representative of Christian order, modernisation and responsibility (Cowen and Shenton, 1996; Preston, 1996).

It is this latter notion of development that, as Chapter 2 discusses, began to permeate the colonial mission from the 1920s onwards, firmly equating development in these lands with an ordered progress towards a set of standards laid down by the West; or as Esteva views it, 'robbing people of different cultures of the opportunity to define the terms of their social life' (Esteva, 1992: 9).

Little recognition was given to the fact that 'traditional' societies had always been responsive to new and more productive types of development. Indeed, had they not done so they would not have survived. Furthermore, the continued economic exploitation of the colonies made it virtually impossible for such development towards Western standards and values to be achieved. In this sense, underdevelopment was the creation of development, an argument that is considered in several of the following chapters of this text, but especially in Chapter 3, in relation to what is known as 'dependency theory'.

Conventional development: 'authoritative intervention'

Chapter 3 discusses in detail the theories and strategies by means of which development is portrayed as a materialist process of change. This present section overviews the process and its impact on the revision of 'development' as a set of ideas about change.

President Truman, in his speech of 1949, noted how the underdeveloped world's poverty is 'a handicap and threat both to them and more prosperous areas . . . greater production is the key to prosperity and peace. And the key to greater production is a wider and more vigorous application of modern scientific and technical knowledge' (Porter, 1995: 78).

Enlightenment values were thus combined with nineteenth-century humanism to justify the new trusteeship of the neo-colonial mission, a mission that was to be accomplished by 'authoritative intervention', primarily through the provision of advice and aid programmes (Preston, 1996).

Salient aspects of this approach are summarised in Figure 1.1. Clearly, this 'modern notion of development' (Corbridge, 1995: 1) had long and well-established antecedents. Figure 1.1 sees the origins of growth theory and authoritative intervention in the three strands of Keynesianism, the rise of the political agenda of the USA, along with nationalist developmentalism.

These forces were then articulated via economic growth models, planning systems and aid mechanisms, giving rise to the ultimate goal of recapitulating the historical experience of the First World in the Third World. It is, therefore, perhaps not too surprising that,

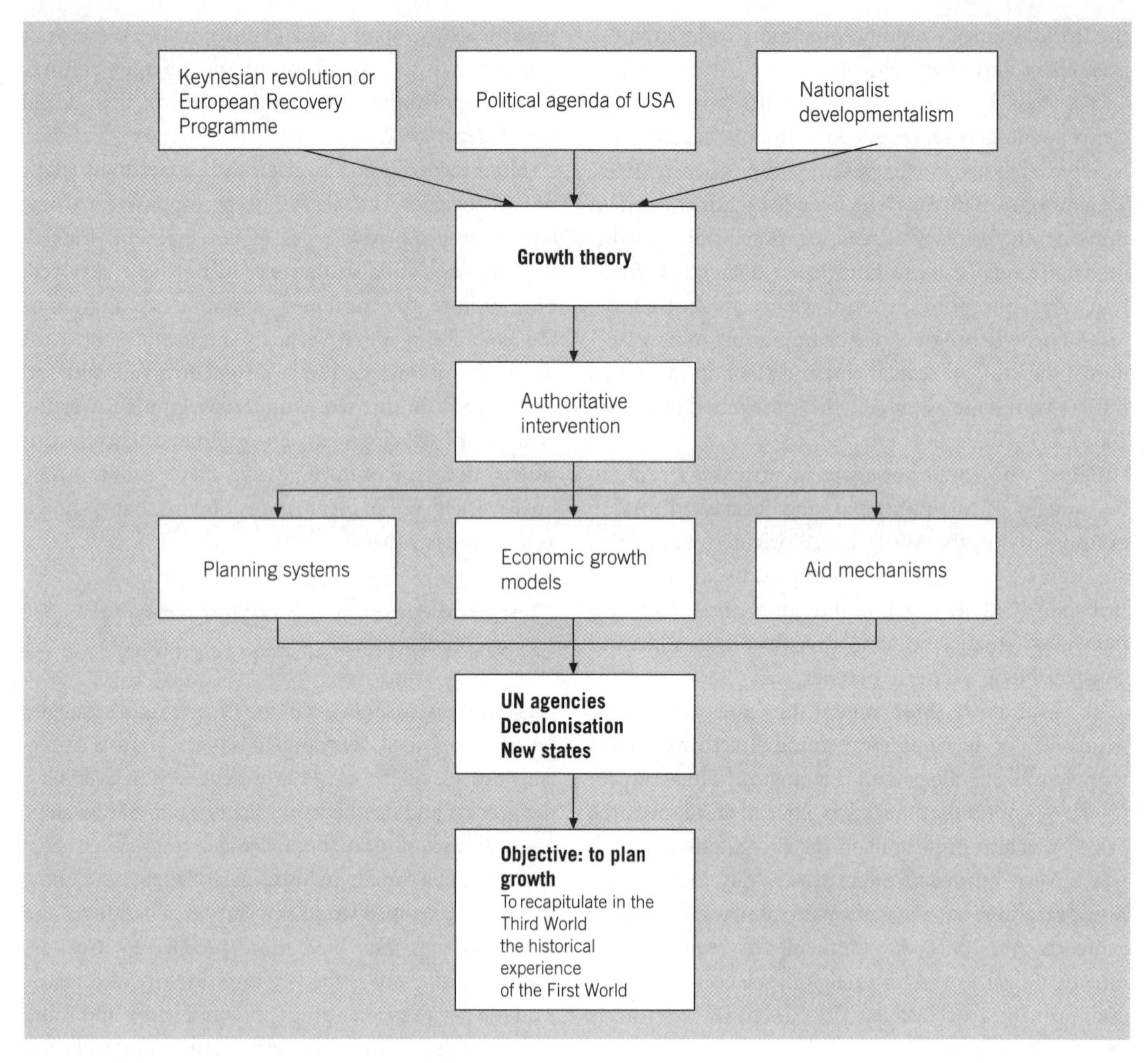

Figure 1.1 Post-colonial growth theory
Source: Adapted from Preston (1996)

in its earliest manifestation in the 1950s, development became synonymous with economic growth.

One of the principal 'gurus' of this approach, Arthur Lewis, was uncompromising in his interpretation of the modernising mission: 'it should be noted that our subject matter is growth, and not distribution' (Esteva, 1992: 12). In other words, increasing incomes and material wealth were seen as being of far more importance than making sure that such income was fairly or equitably spread within society.

During the second half of the twentieth century, therefore, debates about development were dominated by economists. This is not to say that other aspects of development have not contributed, often crucially, to the debate. This is particularly true of sociologists and geographers in respect of the social and spatial unevenness of development, but the dominant influence in both theory and practice has been economics.

The prominence and influence of development economics in the 1950s and 1960s have clear repercussions on other terminologies related to development, most notably the way in which underdeveloped countries were identified and described, a point which is elaborated in the second half of this chapter.

The earliest and for many still the most convenient way of quantifying underdevelopment has been through the level of *Gross National Product* (*GNP*) per capita pertaining to a nation or territory. GDP is defined in the Key idea box, but can broadly be seen as measuring income per head of the population. As Michael Watts (1996a) has noted, this is still a principal way in which the poverty of the Third World and the failure of development are blandly laid out in the statistical sections of World Bank and United Nations development reports.

However, some analysts preferred a classification linked to resource potential, the bases of which were equally shaky. For example, in the 1950s the eminent economist Myint (1964) argued that small, overpopulated states – a category in which he included Hong Kong, Singapore, South Korea and Taiwan – faced the direst prospect for the future; little did he know!

Key idea

Measuring development: from GNP to the HDI

By the end of the first United Nations Development Decade, not only was concern arising over the interpretation of development as economic growth, there was also considerable criticism of GNP per capita as the indicator of such growth.

GNP per capita is measured by the total domestic and foreign value added of a nation divided by its total population. The real problem with this measure is that it gives no indication of the distribution of national wealth between different groups within the population.

Nevertheless, as Seers (1972: 34) points out, to argue that GNP per capita is a totally inappropriate measure of a nation's development is to weaken the significance of the growing GNP per capita gap between rich and poor nations, a gap which is dealt with extensively later in this chapter.

In other words, the serious criticisms that one can make of development statistics do not deny them some use in the analysis of the development process, particularly its unevenness.

Seers himself, with his egalitarian leaning, suggested the use of three criteria to measure comparative development: poverty, unemployment and inequality. He accepted that the statistical difficulties involved in doing so were considerable, but argued that they produced data that were no less reliable than GNP per capita, and were a far better reflection of the distribution of the benefits of growth. Although Seers considered them to be economic criteria, they clearly include social dimensions; indeed, Seers suggested social surrogates for their measurement.

The 1970s and 1980s were conspicuous for the appearance of a whole series of social indicators of development, such as those relating to health, education or nutrition, which were produced either as tables attached to major annual reviews, for example the annual *World Development Report* (produced by the World Bank), or less frequently as maps that accompanied attempts to identify the developing world per se.

Eventually these social indicators were broadened further still to incorporate measures of gender inequality, environmental quality and political and human rights.

As with all statistical measures, these data are open to a variety of criticisms, some technical, some interpretational. For example, how does one measure human rights when cultural interpretations are not consistent, as successive East–West disputes have indicated (Drakakis-Smith, 1997).

Moreover, by the late 1980s a plethora of economic, social and other indicators were being produced on an annual basis. These were not always consistent with one another and could be manipulated to show that some 'development' had occurred almost anywhere.

The consequence was, not surprisingly, that as indicators multiplied so there emerged a renewed enthusiasm for the single composite measure. Such measures did not always produce results that matched the GNP-based categories of development that have graced the pages of the *World Bank Development Report* for so long.

In Richard Estes' (1984) Index of Social Progress, the USA was ranked well below countries such as Cuba, Colombia and Romania. As usual, one can always prove a point with statistics. Other measures were even more complex in an effort to be all-embracing.

Tata and Schultz (1988) constructed a human welfare index from ten variables using factor analysis. The final scores, however, were more or less arbitrarily divided into three sets, producing a table and map little different from those of the three worlds (First, Second and Third World) in vogue at that time.

However, almost inevitably, single measures, usually in conjunction with multiple tables of individual indicators, remain popular as easily digestible summaries of world development trends. One of the

▶

Key idea (continued)

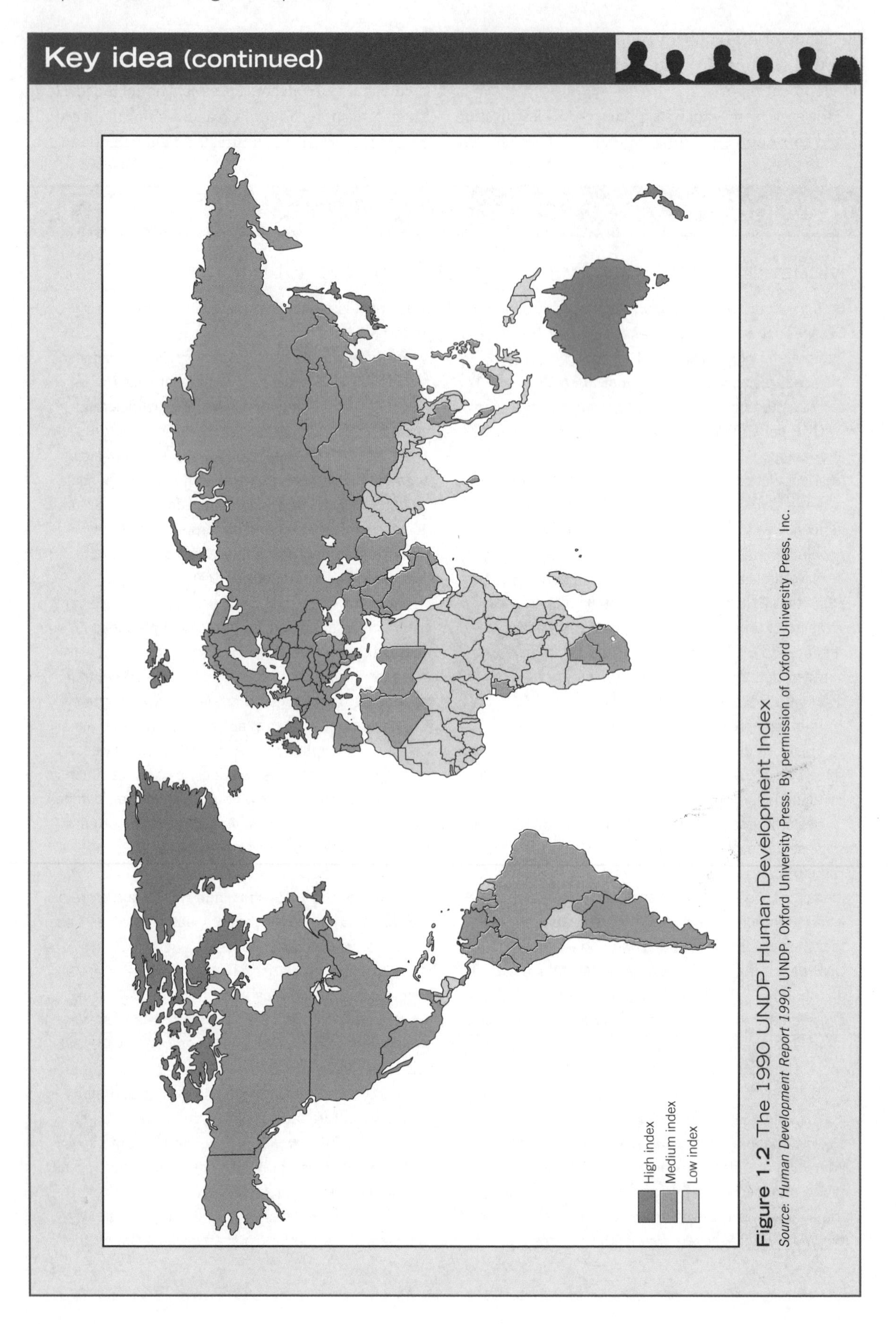

Figure 1.2 The 1990 UNDP Human Development Index

Source: Human Development Report 1990, UNDP, Oxford University Press. By permission of Oxford University Press, Inc.

Key idea (continued)

most widely used is the *Human Development Index* introduced and developed by the United Nations. In the words of the 2001 Human Development Report, the 'HDI measures the overall achievements in a country in three basic dimensions of human development – longevity, knowledge and a decent standard of living' (UN, 2001: 14). Thus, the HDI is measured by life expectancy, educational attainment (adult literacy and combined primary, secondary and tertiary enrolment), plus adjusted income per capita in purchasing power parity (PPP) US dollars.

The manner in which the basic index is calculated is shown in Figure 1.3. The three basic dimensions are translated into a series of indicators, and these are summed to give a single Human Development Index. This summary measure has come to be used in a wide variety of contexts. For example, the Government of Barbados has recently used the HDI in a number of promotional contexts, basically to show that Barbados is 'the most highly developed of developing nations'; indeed, one Barbadian administration went so far as to announce that its express aim was to make Barbados a 'First World' nation within the foreseeable future (Potter, 2000).

In respect of the HDI of those nations classified as characterised by high human development, the majority are First World or 'developed countries', such as the USA, Canada, Sweden, Japan, Switzerland, United Kingdom and New Zealand. But a number of those recording high scores are 'Third World' nations, such as Hong Kong, Singapore, the Republic of Korea, Barbados, Chile, the Bahamas and the United Arab Emirates. Medium-level human development nations include Trinidad and Tobago, Venezuela, Romania, Peru, Sri Lanka, Jamaica, China, Egypt and Namibia. Low HDI scores are returned by Pakistan, Haiti, Tanzania, Senegal, Guinea, Rwanda, Niger, Sierra Leone and Burundi.

It needs to be stressed that the HDI is a summary, and not a comprehensive measure of development. As such, over the years since its introduction, various methodological refinements have been tried by the United Nations. Such refinements include Human Poverty Indexes 1 and 2, the Gender-related Development Index and the Gender Empowerment Measure. These are all variations on the basic Human Development Index. In each instance, additional variables are brought in to reflect the revised index. For example, for the Human Poverty Index 2, a measure of social exclusion is included in the calculation, measured by the long-term unemployment rate. For the gender-related index variables such as female life expectancy, literacy and estimated earnings are factored into the calculation of the HDI.

As Esteva (1992) notes, human development is thus translated into a linear process indicated by measuring levels of deprivation, or how far countries depart from the Western ideal. Moreover, if one chooses other similar variables in the same categories, quite different overall indices can be obtained. And yet we cannot dispose of development indicators too readily, for above all they indicate trends over time and even the anti-development critics (defined shortly below) use collated statistics of this nature in order to consolidate their starting point that 'development' has been a myth. Indeed, Ronald Horvath (1988) has conceded that in endeavouring to measure development, he was 'measuring a metaphor'.

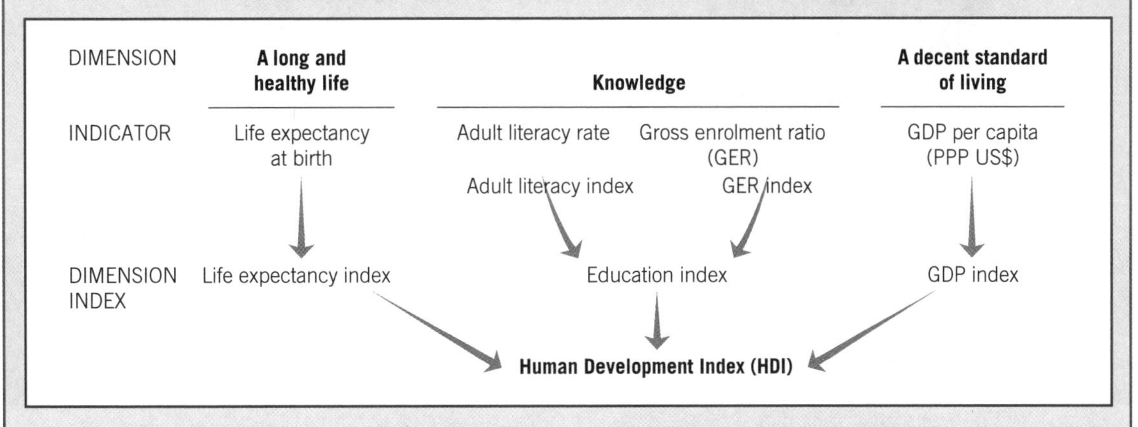

Figure 1.3 How the Human Development Index (HDI) is calculated

To be sure, the notion of development as economic growth has broadened over the years to incorporate social indicators and political freedoms. This has recently been more fully explored by Amartya Sen (2000), in his book *Development as Freedom*. The approach outlined by Sen is summarised in the Key thinker box.

Yet the most recent of the composite indicators which enables international comparisons to be made, the *Human Development Index of the United Nations* (see Key idea box for a definition), is still basically underpinned by economic growth and is finally expressed as a single statistical measure.

In a similar vein, the model for economic development is still seen to be that of the capitalist West, as this quote from Nobel prizewinner Douglas North (cited in Mehmet, 1995: 12) seeks to confirm:

> **The modern western world provides abundant evidence of markets that work and even approximate the neo-classical ideal . . . Third World countries are poor because the institutional constraints define a set of payrolls to political/economic activity that do not encourage productive activity.**

Critiques of development: Eurocentrism, populist stances, anti-development and postmodernity

Critiques of development: introduction

Criticism of development as conceptualised and practised in the ways described above has been continuous since the 1960s, and has clearly influenced theory and strategy, as will be outlined in Chapter 3.

Even the narrow focus on economic growth by the agents of development failed to produce a convergence of income indicators. Far from it, there is evidence that inequality between and within countries has increased substantially (Griffin, 1980).

This is referred to as the 'convergence debate', and we will return to it in detail later in the chapter. Trickle-down economics had not worked and the call came for a more diversified and broader interpretation of development (Dwyer, 1977).

Key thinker

Amartya Sen and *Development as Freedom*

Plate 1.1 Amartya Sen
(photo: Getty Images/AFP)

The Nobel Laureate Amartya Sen published a book in 2000 under the title *Development as Freedom*. Sen was awarded the Nobel Memorial Prize for Economic Science in 1998. Over the years he has written widely on many aspects of development economics, including poverty, famine, capabilities, inequality, democracy, and issues of public policy in developing economies.

Sen (2000) argues that development consists of the removal of various types of unfreedoms that leave people with little choice and little opportunity for exercising their reasoned agency.

One of the vital points is that one human freedom tends to promote freedoms of other kinds: 'There are very many different interconnections between distinct instrumental freedoms' (Sen, 2000: 43).

For example, Sen argues that there is strong evidence that economic and political freedoms help reinforce one another, although some argue the reverse. But in a less contested manner, it can be argued that social opportunities in the fields of health care and education, which generally require public action, complement individual opportunities for economic and political participation. Such linkages emphasise the intrinsic importance of human freedoms.

Key thinker (continued)

Sen's emphasis has to be on substantive freedoms. It makes little sense to celebrate the freedom to pollute, torture or employ child labourers, as noted in a recent review of the strengths and weaknesses of Sen's approach to development as freedom by Corbridge (2002a).

Sen's focus is very much upon 'instrumental freedoms', that is those which allow us to live lives free from starvation, undernourishment, escapable morbidity, premature mortality, illiteracy and innumeracy. Being able to enjoy political participation and free speech are further vital freedoms. How much more difficult it is likely to be if people cannot read?

It is clear that this list maps out political freedoms, such as the right to vote, but also relates to the existence of economic opportunities, social facilities, transparency within society (trust and openness), as well as a measure of protective security.

A particular issue that Sen deals with is gender discrimination. As noted by Corbridge (2002a), Sen writes movingly in *Development as Freedom* about the estimated 100 million women who are missing from the world today as a consequence of sex-selective abortion. Sen interprets this in terms of women not enjoying the same substantive freedoms as men. They suffer from unfair food sharing and health care within households, and have little voice.

It is clear that in Sen's work the differences that matter the most are those that define us as individual human beings (Corbridge, 2002a). Centrally, Sen's approach serves to emphasise that development needs to be measured by means other than GDP. The approach celebrates individual freedom.

On the other hand, using instances mainly drawn from China and India, Corbridge (2002a) argues that some examples that Sen uses suggest that the curtailment of individual freedoms can have beneficial impacts on the poor, or on a certain, prescribed social grouping. And there are further points: individual freedoms may be enhanced by social mobilisations, some of which may be anything but liberal and democratic. These do not invalidate Sen's argument, but they do limit the agency of individual freedoms.

Critical reflection

What are the relative freedoms that you most enjoy day to day and in what areas of your life do you consider that you experience unfreedoms, or a relative lack of freedom? How do you feel these compare with somebody living in a developing region of the world? Thinking about the freedoms you most enjoy, do you feel that some of them might be bought at the expense of someone else's freedoms?

Explanations were sought and offered for the failure of the *modernisation* project (Brookfield, 1978) and new strategies were devised. But in most of this discussion, development as a linear and universal process was seldom questioned or addressed. What was debated was the variable and erratic nature of development, and explanations were sought in relation to both its chronological and spatial unevenness.

Eurocentricity and development

For some critics the answer was straightforward; it was, and still is, the Eurocentricity (European orientation) of economic development theories that had distorted patterns and processes of development, especially through their pseudoscientific rationale (Table 1.2). Mehmet (1995), in particular, has been virulent in his criticism:

Table 1.2 Eurocentricity: some principal points of criticism

Denigration of other people and places
Ideological biases
Lack of sensitivity to cultural variation
Setting of ethical norms
Stereotyping of other people and places
Tendency towards deterministic formulations
Tendency towards empiricism in analysis
Tendency towards male orientation (sexism)
Tendency towards reductionism
Tendency towards the building of grand theories
Underlying tones of racial superiority
Unilinearity
Universalism

As a logical system, Western economics is a closed system . . . in which assumptions are substituted for reality, and gender, environment and the Third World are all equally dismissed as irrelevant . . . [However], mainstream economics is neither value-free nor tolerant of non-western cultures.

Of course, Eurocentrism is a criticism that can be levied at more than mainstream economics and its associated modernisation strategy (Hettne, 1990; McGee, 1995). Indeed, as Chapter 3 indicates, almost all of the major strategies for development have been Eurocentric in origin and in bias, from modernisation through neo-Marxism to the neo-liberal 'counter-revolution' of the 1980s.

Moreover, it can be argued that all such approaches tend to equate 'development' with capitalism (Harriss and Harriss, 1979). Certainly, all were universalist in their assumptions that development is a big issue that needs to be understood through grand theories, or so-called 'metanarratives'.

It is not surprising that such arrogant approaches began to be criticised. After all, these were approaches in which the developed nations devised the parameters of development, set the objectives and shaped the strategies.

Not only was this 'westernised' development not working for most Third World countries, but the West itself was continuing to be the 'beneficiary' of the distorted development that it produced.

Since 1960, the start of the first *United Nations Development Decade*, disparities of global wealth distribution doubled, so that by the mid 1990s the wealthiest quintile of the world's population controlled 83 per cent of global income, compared to less than 2 per cent for the lowest quintile (UNDP, 1996).

However, despite the extensive criticism that began to appear, we must also recognise the fact that some societies were able to absorb selectively from this imposed development to their own advantage. The Asian industrialising societies, for example, provided ample evidence of this.

Two principal sets of voices began to be heard in the widespread criticism of the general situation. The first was characterised by a stance that recommended greater input into defining development and its problems from those most affected by it – 'development from below', as Stöhr and Taylor (1981) expressed it, or 'putting the last first', as Robert Chambers (1983) memorably termed it. This is often seen as having given rise to distinctly alternative and populist approaches to development and change. These are fully aired in Chapter 3.

The second set of views exhibited similar values, but its supporters were not prepared to work within what they regarded as an unfair and heavily manipulated dialogue of development in which the West, through the medium of international development agencies and 'national governments', assigns to itself the ability to speak and write with authority about development (Corbridge, 1995: 9). This group has become known, therefore, for its 'anti-development' stance; perhaps a somewhat misleading description, as we will see later.

Some of the values of this group cut across the opinions of what might be termed 'postmodern development' with its rejection of metatheory (large-scale, all-embracing theories) and its embracing of meso- or micro-approaches to development problems, which would include gender and environmental issues. Stated simply, postmodern development is development that rejects the tenets of modernity and the Enlightenment, a theme considered in further detail in Chapters 3 and 4.

Both of these approaches can be interpreted as forming part of what came to be referred to as the 'impasse in development studies' (see Booth, 1985; Schuurman, 1993; 2008). Seen as affecting development studies from the mid 1980s, the impasse was attributed to the failure of development itself, along with a growing postmodern critique of the social sciences, and the rise of globalisation.

In respect of the latter, what was regarded as the lessening importance of the role of the state was increasingly seen as cutting right across existing theories and conceptualisations of development and change. But at this juncture, the alternative/populist reactions and the anti-development school are now reviewed in turn.

Alternative and populist approaches to development

Alternative or 'other' forms of development, which were much discussed in the 1990s, are not necessarily recent phenomena. Indeed, many had taken up the earlier lead provided by the Indian activist Mahatma Gandhi (see Key thinker box).

Even in the 1960s there were reactions against the idea that development could be narrowly defined and superimposed upon a variety of situations across the Third World. More locally oriented views began to emerge; for example, in the Dag Hammarskjöld Foundation's concept of 'another development', one that was more human centred (see also Chapter 3).

These approaches were soon co-opted into official development policies, underpinning the 'basic needs' strategies of the 1970s, which fragmented the monolithic targets for development into what were seen as more locally and socially oriented goals.

Key thinker

Mahatma Gandhi

Plate 1.2 Mahatma Gandhi
(photo: Alamy Images/Dionodia Images)

For many, the Indian activist Mahatma Gandhi has come to symbolise the call for peaceful principles of locally driven development and change. Born in western India in 1869, and having studied law in England, Gandhi lived in South Africa for over 20 years. He opposed the 'pass laws' and all forms of racial discrimination (see Singh, 2005).

The time that Gandhi spent in South Africa greatly influenced his views, and on returning to India in 1914 he became a leading figure in the rise of the Indian nationalist and development movements. He referred to what we would today call the process of development as 'progress' (Singh, 2005). Above all, Gandhi proposed a philosophy of non-violent agitation, with the intention of creating mass awareness and cultural unity.

Linked to this, Gandhi stressed that every human has the right to feed, clothe and house themselves. To this end, villages should become self-sufficient, on a local, 'bottom-up' basis. It was maintained that the locus of power should firmly reside with the village or neighbourhood, and the aim should be an equitable distribution of resources. Gandhi was a great advocate of the development of small-scale rural-based industries.

Regrettably, in 1948 while he was conducting a prayer meeting in Delhi, Gandhi was shot dead by a fanatic (Singh, 2005). But by then Gandhi had become the doyen of rural-based, bottom-up development, based on principles of peaceful action and socio-economic change.

Unfortunately, these worthy objectives relating to shelter, education or health not only competed with one another for funding but were also tackled with the same universalist approaches that had bedevilled earlier development strategies. The theme of participation is picked up in detail in Chapter 3 in relation to the evolution of alternative development theories.

However, the concept of locally oriented, endogenous development was by now firmly established and was given a considerable boost by Robert Chambers (1983) with his 'development from below' philosophy.

Although initially discussed with reference to peasants and rural development in general, the philosophy of community participation has been widely adopted as an interpretation of development that is people oriented (Plate 1.1; Friedmann, 1992). But, as Munslow and Ekoko (1995) have suggested, empowerment of the poor has been stronger on rhetoric than in reality; in particular, the widening of political participation has been very slow compared to the improvement in social and economic rights that has occurred. Nevertheless, facilitating 'people's participation' now has a place on the agenda of the major development institutions, as witnessed by the extensive review of participation and democracy which was provided by the UN (1993).

A major facilitator in this process of empowering the poor through participation in the development process has been non-governmental organisations (NGOs) (for the discussion of the role of NGOs see also Chapter 7).

The blanket term 'NGO' covers a wide variety of community-based organisations (CBOs). The largest have operating budgets greater than those of some developing countries, whereas the smallest struggle on with little official encouragement or funding, blending almost imperceptibly with social movements.

The role of NGOs has been scrutinised intensively over recent years, with some seeking to promote linkages away from purely local, community-level projects. In these circumstances, NGOs become involved with more comprehensive larger-scale planning projects, building stronger bridges with the state (Korten, 1990a, 1990b).

Plate 1.3 Rural hawker and child in Guyana: ultimately development is about improving the life chances of people (photo: Rob Potter)

Others, however, already see NGOs, particularly the larger, Western NGOs, as extensions of the state, helping to maintain existing power relations and legitimising the existing political system (Botes, 1996). Indeed, within the changes that have accompanied structural adjustment programmes (that is economic recovery programmes imposed by international agencies), NGOs may be seen as facilitators in the process of the privatisation of welfare functions, freeing the state from its social obligations within the development arena.

Many other criticisms have been levelled at NGOs and the role of outsiders in community participation; most of them have been lucidly summarised by Botes (1996). These cover the paternalistic actions of development experts who see their role as transferring knowledge to those who know less, disempowering them in the process; selective participation of local partners, often bypassing the less articulate or visible groups; favouring 'hard' issues, such as technological matters, over the more difficult and time-consuming 'soft' issues, such as decision-making procedures or community involvement; promoting 'gatekeeping' by local elites; and, particularly important, accentuating product at the expense of process.

Development from below *can* be qualitatively different from conventional development as envisaged by modernisationists, but this process must be realistic rather than romantic in its praxis. Societies, even at the local scale, can be heterogeneous, divided and fractious; and grassroots development, keen to encourage participatory development, must take this into account. Of course, neo-liberal development strategists would argue that their recommendations encourage the empowerment of individuals through greater freedom of choice within an open market economy.

This approach is criticised by Munslow and Ekoko (1995: 175) as the 'fallacy of empowerment' and the 'mirage of power to the people'. In reality, they argue, 'participatory democracy is really about a transfer of power and resources, [if] not to people directly, [then] to NGOs and other representatives at grassroots level'. This is a theme taken up by another group of developmental thinkers, the *anti-developmentalists.*

Before we look at anti-development, it is necessary to stress that ideas about what development is, and how it may be achieved, have changed significantly since the 1980s. As revealed in this text so far, up to the late 1970s development was almost universally seen as being concerned with increasing incomes and overall national levels of economic growth (Figure 1.4). In the main, so-called 'modernisation' was seen as the vehicle by means of which this improvement would occur. Reflecting the times, this could be directly monitored and assessed in quantitative terms (measured, of course, in US dollars, or the volume of goods, numbers of cars and so on).

However, the 2000/2001 *Human Development Report* (UNDP, 2001) of the United Nations stated that 'human development is about much more than the rise or fall of national incomes' (UNDP, 2001: 9). The passage goes on to note that development should be about creating an environment in which people can develop their full potential in order that they should be enabled to lead productive and creative lives, in accord with their needs and interests. Thus, development is about 'expanding the choices people have to lead lives that they value' (UNDP, 2001: 9).

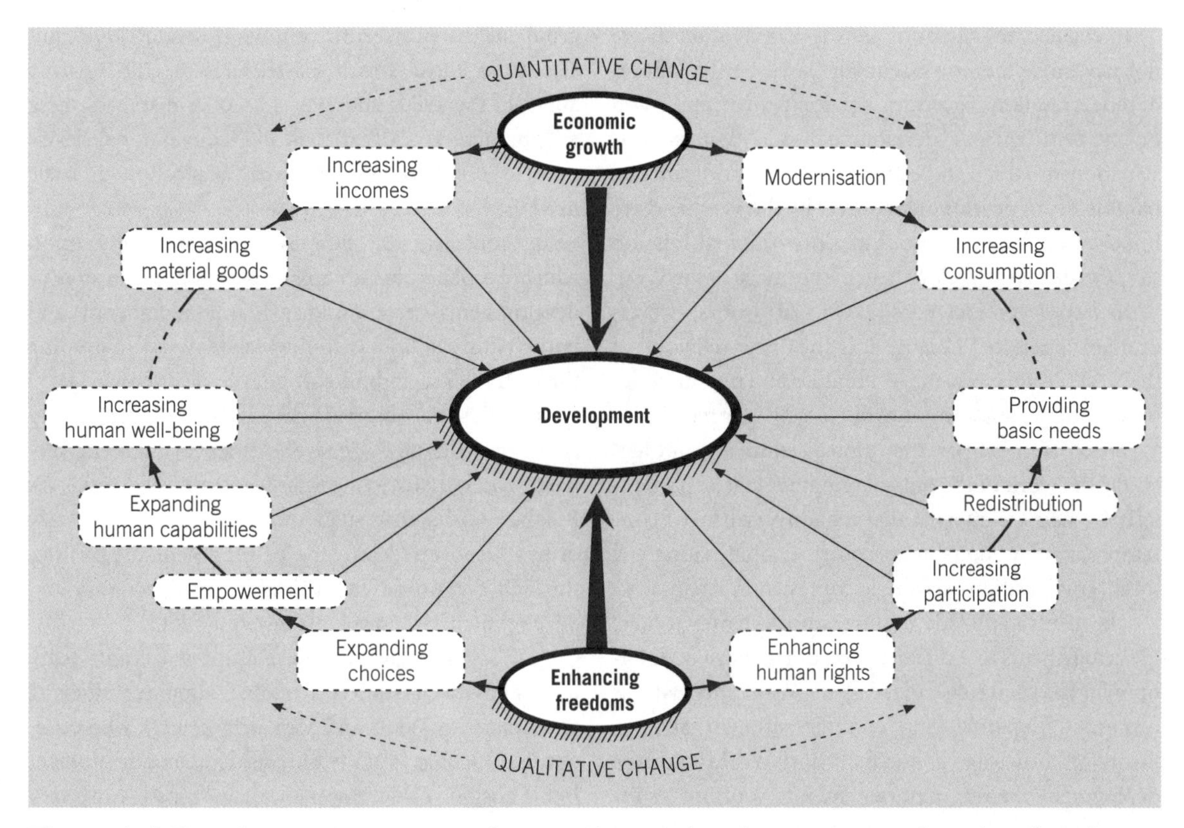

Figure 1.4 Development as economic growth and development as enhancing freedoms

Such a perspective promotes the idea that development witnesses the building of human capabilities. Some of the arguments which follow from this are summarised in Figure 1.4. The approach is directly linked to the primary indicators that are used to calculate the United Nation's Human Development Index.

The most basic of human capabilities are to lead long and healthy lives, to be knowledgeable, and to have access to the resources which are needed to achieve a decent standard of living, and to be able to participate in the life of the community. Approached in this light, the promotion of human well-being can be regarded as the ultimate purpose or end-purpose of development, so that development shares a common vision with the enhancement of human rights.

The *Human Development Report* concludes its argument on the definition of development by equating it directly with the promotion of human freedom: 'The goal is human freedom . . . People must be free to exercise their choices and to participate in decision-making that affects their lives' (UNDP, 2001: 9).

Again, several of the strands concerning the nature and definition of development are summarised in Figure 1.4. These approaches stress the qualitative dimensions of the development equation.

Anti-development, post-development and beyond development

There is considerable overlap between populist interpretations of development and the anti-developmentalists who have emerged in recent years to challenge the notion of development as a whole.

However, as Corbridge (1995) argues, there are long antecedents to anti-(Western) developmentalism stretching back to the nineteenth century. Anti-development is sometimes also referred to as post-development and beyond development (Blaikie, 2000; Corbridge, 1997; Nederveen Pieterse, 2000; Parfitt, 2002; Schuurman, 2000, 2002; Sidaway, 2008; Simon, 2007).

It is also claimed that the failures of neo-Marxism 'to provide practical assistance to those on the front lines of development' (Watts and McCarthy, 1997: 79) have turned disillusioned radicals towards anti-developmentalism (Booth, 1993). Thus, Nederveen Pieterse (2000: 175) comments that 'along with "anti-development" and "beyond development", post-development is a radical reaction to the dilemmas of development'.

In essence, the theses of anti-developmentalism are not new since they are essentially based on the failures of modernisation. Thus, anti-developmentalism is based on the criticism that development is a Western construction in which the economic, social and political parameters of development are set by the West and are imposed on other countries in a neo-colonial mission to normalise and develop them in the image of the West.

In Nederveen Pieterse's (2000: 175) words, 'Development is rejected because it is the "new religion" of the west'. In this way, the local values and potentialities of 'traditional' communities are largely ignored.

There is much of the 'globalisation steamroller' about the work of the anti-developmentalists, particularly in their assumption that the universalism of contemporary development discourse is obliterating the local. This position is strongly contested in Chapter 4.

The central thread holding anti-developmentalist ideas together is that the discourse or language of development has been constructed by the West, and that this promotes a specific kind of intervention 'that links forms of knowledge about the Third World with the deployment of terms of power and intervention, resulting in the mapping and production of Third World societies' (Escobar, 1995: 212).

Escobar argues that development has 'created abnormalities' such as poverty, underdevelopment, backwardness, landlessness and has proceeded to address them through a normalisation programme that denies value or initiative to local cultures.

There are, in these arguments, many similarities to Said's orientalism, similarities that are both implicit and explicit in the views of the anti-development school of thought, which sees both the 'problems' of the Third World and their 'solutions' as the creations of Western development discourse and practice. Of course, there is a recognition that they are not static but change according to contemporary power structures.

However, a consistent factor within the anti-developmental discourse is the role of the Third World state in facilitating the 'Westernisation' of the so-called 'development mission'. It follows, therefore, that the restructuring of development must come from below. Here the anti-developmentalists in general, and Escobar in particular, place enormous emphasis not just on grassroots participation but more specifically on new social movements as the medium of change (Case study 1.1).

The nature of these new social movements is allegedly quite different not only from the class-based group of the nineteenth century (Preston, 1996) but also from those which Castells (1978, 1983) wrote about in the 1970s and 1980s. Escobar dismisses these as 'pursuing goals that look like conventional development objectives (chiefly, the satisfaction of basic needs)' (Escobar, 1995: 219).

In contrast, the new social movements upon which Escobar pins so many of his hopes are anti-developmental, promoting egalitarian, democratic and participatory politics within which they seek autonomy through the use and pursuit of everyday knowledge.

Indeed, some observers have gone even further, and claim that the new social movements 'transcend any narrow materialist concerns' (Preston, 1996: 305–6). Escobar warns that such movements must be wary of being subverted into the developmentalist mission through compromised projects such as 'women and development' or 'grassroots development'.

Not surprisingly, anti-developmentalism in these terms has been subject to some stinging criticism, particularly by Watts and McCarthy (1997), who point out that Escobar is guilty of considerable reductionism in his critique of development, painting a picture very much resembling the dependency theories of the 1970s, in which a monolithic capitalism, particularly in the guise of the World Bank, monopolises development within a largely complacent Third World.

As Rigg (1997) observes, Escobar is very selective in his evidence, with little discussion of those Asia-Pacific countries that might contradict his polemic. Corbridge (1995) also argues that Escobar ignores the many positive changes that Western-shaped development has brought about in terms of improved health, education and the like, no matter how uneven this has been.

Escobar attributes to Third World people his own mistrust of development, a view they may not share. Indeed, some would argue that the intellectual tradition in many Asian universities is to support the state and its policies rather than to criticise them (Rigg, 1997).

Assumptions of widespread anti-developmentalism are therefore as arrogant as assumptions of widespread approval of the modernist-development project. Indeed, as many observers have noted, the poor of the Third World simply get on with the business of survival; holding views on development is a luxury of the privileged.

It can be argued that many poor people in the Third World are inherently conservative and resent imposed or introduced change of any kind, despite the fact they are often very innovative and adaptive in their own coping mechanisms and survival strategies.

Case study 1.1

Urban social movements in Australia

Over the years there has been considerable debate as to what constitutes an urban social movement and Drakakis-Smith (1989) considered it in the Australian context elaborated in this case study. We can define a social movement as a collective, territorially based action, operating outside the formal political system, with the objective of defending or challenging the provision of urban services against the interests and values of the dominant groups in society.

Although urban social movements (USMs) are essentially local and non-political in origin, their effectiveness in improving the quality of life is strongly influenced by the broader social, political and economic contexts in which they are situated, not only at the urban level but at the regional, national and international levels as well.

This case study illustrates how the success of USMs can vary, even in apparently similar situations, in relation to the way in which local circumstances can be manipulated by more broadly based processes. It relates the attempts by Aboriginal-based USMs in the towns of Darwin and Alice Springs to obtain improved housing conditions. Both towns are located in the Northern Territory of Australia but the outcome of the respective USMs was quite different. To understand this situation, it is necessary to discuss briefly both the general history of Aborigines in the development of Australia and the specific circumstances in each town.

When the British arrived in Australia in 1788 they simply took possession of the land in the name of the crown. There were no negotiations with the Aborigines and no treaty was signed. Exploitation of Aboriginal land for agriculture, pastoral industries and mining was accompanied by exploitation of Aboriginal labour as stockmen and domestic servants. More recently, Aboriginal culture has been exploited by the tourist industry. Indigenous Australians were effectively ignored by the Australian government throughout this period, being largely confined to missions/reserves where their labour was reproduced.

The breaking of the link between the Aboriginal people and their land devastated their culture and spiritual basis, and by the mid twentieth century those who had drifted into towns tended to live in small shanty towns in communities ravaged by alcoholism (Rowley, 1978). Even the incorporation of the Aboriginal population into the state welfare system in the 1960s failed to halt this situation, but it did create thousands of jobs for white Australians 'servicing' the Aboriginal community.

One of the areas in which the Aboriginal community was poorly served was housing. Excluded from the private sector by their poverty, they were also excluded from state housing by virtue of their inability to meet rental payments and by their 'inability to cope' with a state house. This was certainly the situation in Alice Springs, where by the 1970s around 80 per cent of the housing stock was occupied by white families, whereas most Aboriginals lived in some 30 camps outside the town (see Plate 1.4).

This unequal situation was the end product of a set of ideological forces (Aboriginal people deserved no rights), ethnic prejudice and class antagonism. This situation was changed in Alice Springs not just by new economic circumstances, but by the translation of broader, global civil rights movements into national and local situations. First, the federal state, under external and internal pressure, made it possible for Aborigines to claim lease rights to the land they occupied around Alice Springs. Despite opposition from entrenched white interest groups, Aborigines were successful in their bids for leaseholds largely because they formed a collective of camp leaders called the Tangatjira Association. This took on the role of adviser to various camp groups and later became a facilitator for the introduction of appropriate housing technologies and building maintenance, all of which also created employment for local Aborigines.

Although the federal state, through the Department of Aboriginal Affairs, certainly facilitated this process, the success in obtaining tenure security and improved shelter undoubtedly came from the

▶

Case study 1.1 (continued)

Plate 1.4 Aboriginal town camp on the outskirts of Alice Springs
(photo: David Smith)

activities of a very focused urban social movement. The situation in Darwin, however, in similar global and national circumstances, was quite different. There, Aboriginal groups failed to mobilise into a social movement to challenge the system, despite the fact they were even more marginalised than in Alice Springs. In many ways this is the consequence of local political and economic circumstances.

As Darwin is the territory capital, it has a large percentage of both territorial and federal employees, most of whom received large incentives, usually in the form of housing subsidies. The housing rental market, both private and public, is therefore marked by high rents and those not favoured with subsidies have to share in order to obtain adequate shelter. The low-income groups in Darwin who cannot afford rents live in a variety of accommodation from camps to caravan parks or boats. However, in contrast to Alice Springs, there is no ethnic unity as many of the poor comprise white males who have 'dropped out' of conventional Australian society. Compared with Alice Springs, Aboriginal groups therefore comprise a smaller proportion not only of Darwin as a whole, but of the low-income group. Because of this no urban grouping has coalesced around a social issue such as housing. Moreover, as Darwin is the state capital, individuals with the potential for group leadership have tended to be co-opted into the political system, thus reducing the danger of social mobilisation.

These two examples illustrate the fragile distinction between success and failure in urban social movements, even in similar situations, and they emphasise the fact that local circumstances can impede as well as facilitate mobilisation.

Corbridge (1995) suggests that anti-developmentalism romanticises and universalises the lifestyles of indigenous peoples. Do the actions of the poor and vulnerable really constitute a resistance to development or are they simply seeking to manipulate development to improve their access to basic resources and to justice?

Certainly, the anti-developmentalists have reinforced our sense of the local in the face of what appears to be an overwhelming process of globalisation. Indeed, the alleged retreat of the state and return to democracy that have occurred in some parts of the world, have opened up new spaces in which social movements can seize the initiative.

Yet, as Watts and McCarthy (1997: 84) note, 'a central weakness of the social movements-as-alternative approach is precisely that greater claims are made for the movements than the movements themselves seem to offer'. Moreover, what is wrong with social movements having modest, self-centred aims which focus on basic needs if it results in improvements in the quality of life for a group that subsequently disbands? Who are we to castigate this as mere self-seeking satisfaction of conventional development appetites?

Much of the problem with anti-developmentalism is that it 'overestimates the importance of discourse as power . . . when the political economy of changing material relations between capital and native remain central to the process of capital accumulation' (Watts and McCarthy, 1997: 89). In short, the intention to develop is confused with the process itself. Although they can and do overlap, there is much more to the unsatisfactory nature of development than its intellectual discourse.

However, the anti-development movement *has* brought about a re-emphasis of the importance of the local in the development process, as well as the important skills and values that exist at this level; it also reminds us what can be achieved at the local level in the face of the 'global steamroller', although few such successes are free of modernist goals or external influences (see Case study 1.1).

The postmodernist stance

Of all the recent changes within developmental thinking, perhaps the most successful and least heralded has been the shift away from large-scale theory to meso-conceptualisations that focus on specific issues or dimensions of development in an attempt not merely to separate out a slice of development for scrutiny, but to see how it relates to the development process as a whole and to local situations. A good illustration of this might be the fusion of gender and shelter debates that has made a strong impact on theory and policy in the 1990s (Chant, 1996).

Some might claim that this illustrates the influence of postmodernism in development studies (Corbridge, 1992), involving a liberation of thought, a recognition of a local 'otherness' and support for small-scale development. However, one could argue that such approaches have been part of development geography for some time, reflecting its empirical traditions. As McGee (1997) points out, the accumulated experiences (histories) of empirical studies are invaluable in bringing out a sense of the local within the development process.

On the other hand, postmodernism has also been interpreted as merely 'the cultural logic of late capitalism, effectively representing the new conservatism . . . preoccupied with commodification, commercialisation and cheap commercial developments' (Potter and Dann, 1994: 99). Under this latter formulation, most of the new meso- and micro-narratives of development thinking have little in common with postmodernism.

Indeed, the rise of post-development is seen as a part of the rise of 'postmodernism' or 'post-structuralism' (see Parfitt, 2002; Simon, 1998). Thus, postmodern theory would deny that if history is examined a process of progression to 'higher' levels of civilisation can be identified. Rather, postmodernism sees history as a contingent succession of events, so that it is difficult to think in terms of goals – including development goals (Parfitt, 2002).

Following this brief introduction to the links between postmodernism and development, the topic will be pursued in further detail in Chapters 3 and 4.

Critical reflection

Postmodernity

It is often joked that, architecturally speaking, postmodernity is signified by a triangle or a pillar on a new building! Any form of contemporary ornamentation or embellishment that has resonances of the past is regarded as possibly postmodern, along with features that emphasise style rather than function. This is relatively easy to discern when looking at architecture and buildings.

But what other signs of postmodernity are you aware of around you? Are there wider aspects to be discerned in the environment in which you live, or places that you have recently visited? What about the ways in which change and development have recently been thought about and enacted – does this seem to be in any way postmodern to you? The topic of postmodernity is dealt with extensively in Chapter 3, as well as in outline in this chapter.

Reviewing development

So, what are we left with after all this discussion of the nature of development? For the anti-developmentalists, development has become 'an amoeba-like concept, shapeless but ineradicable [which] spreads everywhere because it connotes the best of intentions [creating] a common ground in which right and left, elites and grass roots fight their battles' (Sachs, 1992: 4). But in its naivety, to some critical commentators, the anti-development alternative of 'cosmopolitan localism' based on regeneration, unilateral self-restraint and a dialogue of civilisations unfortunately seems no more than a Utopia for New Age travellers.

Despite its eighteenth- and nineteenth-century origins, 'development has never been a scientific concept, it has always been ideology' (Friedmann, 1992: v) – quite simply, development is always political. Thus, development can mean all things to all people; poor squatters are highly likely to have a completely different view of what constitutes change for the better in their lifestyle as opposed to a senior politician or national planner.

This is clearly evident in discussions of the 'brown agenda', in which Satterthwaite (1997) and others have pointed out that many of the concerns of the international agencies and national planners with global warming and ozone layers reflect a 'Northern' agenda that is far away from the clean water needed by most squatter households. As Hettne (1990: 2) notes, 'there can be no fixed and final definition of development, only suggestions of what development should imply in particular contexts'.

The debates over the definition of what development is and how people think about it are not simply academic, although for some this is the limit of their interest. Thoughts and views about the development condition underlie policy formulation and subsequent implementation.

A common example of conflict is that which often occurs between the national government with its economic impetus and external linkages, and NGOs or CBOs which tend to emphasise democratisation, political involvement and the local, immediate needs of these disadvantaged groups (Thomas, 1996).

Development is an historical process of change which occurs over a very long period but it can be, and usually is, manipulated by human agency. It is often forgotten that culture (particularly religion) can play an important role in characterising national and local development strategies. Many of the industrialising states in Asia claim to have followed an 'Asian way' to development, although this in itself has been criticised for its reductionism and selectivity (Rigg, 1997). It is the nature of these manipulations and the associated goals that will be discussed in Chapter 3.

However, what we have now established is that development is not unidirectional. Improvement in the human condition has many different dimensions and the speed of change may vary enormously for any individual or community. While fair and balanced development may be a desirable goal, for most of the world's population it is far from being a realistic one (Friedmann, 1992). It is to the definition of this proportion of the world's population that this chapter now turns.

Spatialising development: the Third World/Developing World/Global South/Poor Countries

The term that has most commonly been employed to refer to spatial contrasts in types of development, different levels of development and different patterns of development is examined in this section, along with the other terms that have come to be employed alongside it, or instead of it.

Specifically, the evolution of the term *Third World* will be traced, along with its proponents, its critics and the alternatives they have posed, such as the *Global South*, the *Developing World* and *Poor Nations* (Dodds, 2008).

Almost inevitably there will be a degree of overlap with the first half of the chapter, since in many ways we are examining the public lexicon concerning the more theoretical issues discussed previously. Indeed, the wider currency that the terms *development* and *underdevelopment* allegedly experienced as a result of President Truman's inaugural address of 1949 (Esteva, 1992; Sachs, 1992) was to a certain extent clarified by the new 'three world' terminology that was also emerging at the same time.

Thus, the First World was promoting development, the Second World was opposing it and the Third World was the express object of the exercise. In the rigidities and absurdities of the Cold War politics of the 1950s this seemed to make good sense to at least some. However, over the years the association between the notion of three worlds and the development process has changed considerably. It will be useful to examine this association within a broad chronological framework.

The emergence of the Third World in the 1950s and 1960s

As with *development*, the antecedents of the term Third World go back beyond 1949, although not much further.

Moreover, in contrast to the current largely economic interpretation of the Third World (Milner-Smith and Potter, 1995), particularly in terms of poverty, the origins of the term were political, largely centring around the search for a 'third force' or 'third way' as an alternative to the Communist–Fascist extremes that dominated Europe in the 1930s.

In the Cold War politics of the immediate post-war years, this notion of a third way was revived initially by the French Left, which was seeking a non-aligned path between Moscow and Washington (Pletsch, 1981; Wolfe-Phillips, 1987; Worsley, 1979).

It is this concept of non-alignment that was seized upon by the newly independent states in the 1950s, led in particular by India, Yugoslavia and Egypt, and culminating in the first major conference of non-aligned nations held in Bandung in Indonesia in 1955. Indeed, at one point 'Bandungia' appeared to be a possibility for their collective title. Friedmann (1992: iii) claims that as a result of this meeting 'the Third World was an invention of the non-western world', in spirit if not etymologically.

The sociologist Peter Worsley (1964) played a major role in the popularisation of the term *Third World*, principally via his book under this title. For Worsley the term was essentially political, labelling a group of nations with a colonial heritage from which they had recently escaped, and to which they had no desire to return within the ambit of new forms of colonialism, or 'neo-colonialism'.

Nation building was therefore at the heart of the project and it is no coincidence that the loudest voices came from those states with the most charismatic leaders. For India, Yugoslavia and Egypt, therefore, read Nehru, Tito and Nasser.

But the emerging Third World was not quite the same as it is today; many countries had still to gain their independence and Latin American countries were not present in Bandung. Moreover, both the Bandung group and Worsley's Third World 'excluded the communist countries' (Worsley, 1964: ix) (see Box 1.1).

Nevertheless, for a while in the 1950s and 1960s, this Afro-Asian bloc did attempt to pursue a middle way in international relations. In economic terms, however, it was a different story.

Almost all newly independent states lacked the capital to sustain their colonial economies, let alone expand or diversify. Most remained trapped in the production of one or two primary commodities, the prices of which were steadily falling in real terms, unable to expand or improve infrastructure and their human resources.

Once Worsley had identified the common political origins of the Third World in anti-colonialism and non-alignment, he cemented this collectively through the assertion that its current bond was poverty (Plate 1.5).

The same feature had also been noted by Keith Buchanan (1964) in the first substantial geographical contribution to the debate. Buchanan's diagrammatic representation of the Third World, shown here as Figure 1.5, bears a close resemblance to that of the

Plate 1.5 People making a living by a variety of means, Old Delhi
(photo: Rob Potter)

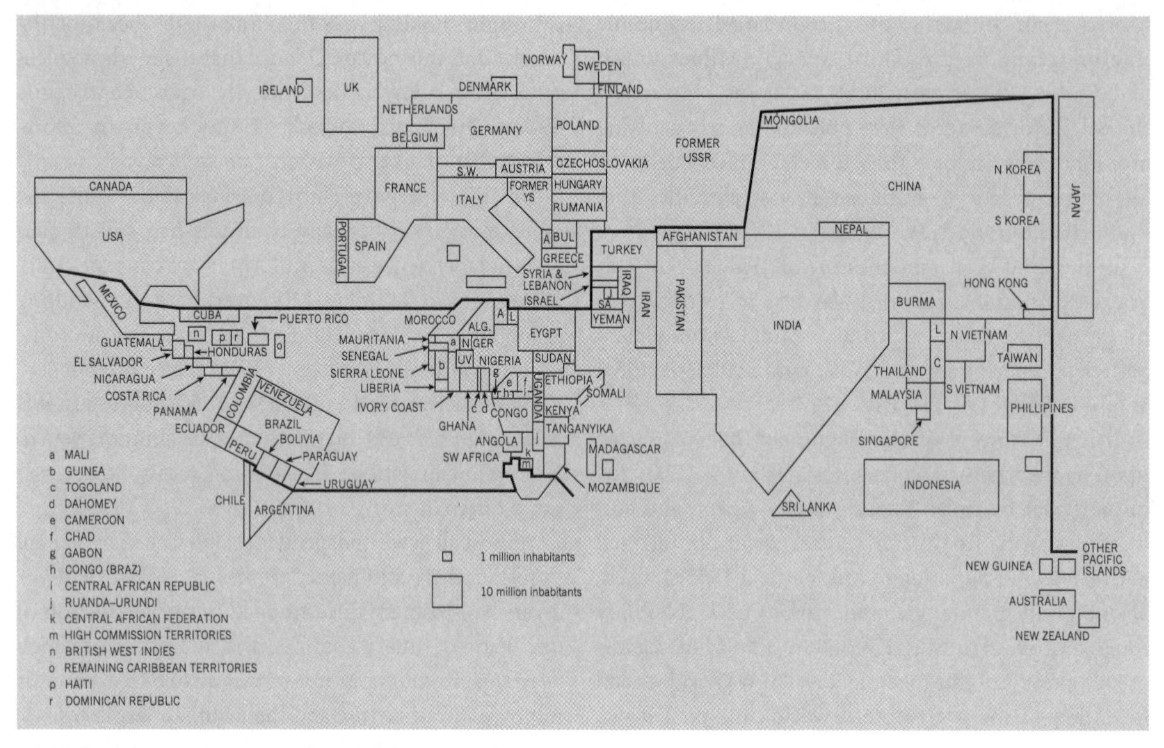

Figure 1.5 Buchanan's Third World in the 1950s
Source: Adapted from Buchanan (1964)

Brandt Commission 20 years later, but makes somewhat more geographical sense.

In particular, Buchanan's diagram is helpful in showing the population size of countries. Hence, the overwhelming contribution of the populations of China and India within the Third World is one of the most prominent features of the map.

The 1960s witnessed a major shift in interest on the part of several social science disciplines towards the nature of development and underdevelopment, prompted largely by the failure of modernisation strategies (see Chapter 3 for a detailed view) to bring predicted growth to what was increasingly becoming called the Third World.

It is important to note, however, that much of this economic debate was predicated on deeper political concerns: the fear that widespread and persistent poverty would lead to insurrection and a further round of Communist coups. In Asia, the puppet regimes of South Vietnam and South Korea were looking somewhat shaky, whereas the continuing strength of Castro's revolution in Cuba raised fears of a Caribbean domino effect and the spread of communism.

The principal concern, particularly among development economists, was to find out what had gone wrong and where the problems were located. In geography this was the era of the quantitative revolution, and from both disciplines there arose a series of measurements designed to rank Third World nations in terms of needs with the usual signifier being Gross National Product (GNP) per capita (see Key idea box on pp. 9–11).

Within some of the individual states, 'modernisation surfaces' were produced which indicated spatially uneven development by means of multiple indices of development 'attributes' (schools, hospitals, roads, street lighting, etc), most of which closely mirrored the spatial imprint of colonialism itself (Gould, 1970; Soja, 1968). This approach is considered in detail in Chapter 3.

Despite the largely uninformative nature of these academic developments, the term *Third World* was by this stage in widespread use, even by its constituent states in forums such as the United Nations (Wolfe-Phillips, 1987).

Conceptually, therefore, by this juncture, the world was firmly divided into three clusters, namely the West, the Communist bloc and the Third World: but the terms being used were etymologically inconsistent.

The first is an abstract geographical term (west of where?), the second is a political epithet, and the last is numerical – hardly an example of consistent logic, but one which had by the 1970s obtained a significant measure of popular acceptance.

The 1970s: critiques of the Third World

By the early 1970s, the rather loose combination of political and economic features that constituted the Third World had already come in for criticism. The French Socialist Debray (1974: 35) argued that it was a term imposed from without rather than within, although more developing nations were beginning to use it.

Anti-developmentalists consider the 1970s to be a critical point in the development process, a time when the Third World was beginning to recognise its own underdevelopment, adopting Western evaluations of its condition.

Many other critics, however, also felt the term was derogatory since it implied that developing countries occupied third place in the hierarchy of the three worlds (Merriam, 1988). An even more valid criticism was that users of the term had still failed to situate the socialist developing states in the three-world terminology (Box 1.1).

The main cause of the doubts that emerged during the 1970s was related to the growing political and economic fragmentation of the Third World as it 'postmodernised' from a 'meta-region' into a plethora of sub-groupings.

Ironically, perhaps the biggest impetus to the break-up of the Group of 77 non-aligned nations came from within, when the Organization of Petroleum Exporting Countries (OPEC) nations raised the price of their oil massively in 1973–74, with a second wave in 1979 following the fundamentalist revolution in Iran.

Initially conceived as a political weapon against the West for its support of Israel, the price rise had a much greater effect on the non-oil-producing countries of the developing world, many of which were following oil-led industrial and transport development programmes. The result was a widening income gap between different developing countries.

This was further reinforced by the new international division of labour in the 1970s, in which capital investment via multinational corporations and financial institutions poured out of Europe and North America in search of industrial investment opportunities in developing countries.

Most of this overseas investment was highly selective and cheap labour alone was not sufficient to attract investment: good infrastructure, an educated and adaptable workforce, local investment funds, docile trade unions were also important factors.

The outcome, of course, was that investment focused on a handful of developing countries (specifically, the four Asian tigers plus Mexico and Brazil) where GNP per capita began to rise rapidly, further stretching relative economic and social contrasts within the Third World. This relative 'global shift' is the subject of detailed attention in Chapter 4.

BOX 1.1

The socialist Third World

Within the evolution of the term *Third World*, the place of the socialist developing countries always seemed to pose difficult conceptual problems. Although most writers seemed to have no problem in distinguishing between developed and developing capitalist states, the same did not hold true for the centrally planned economies. Peter Worsley solved this problem by not discussing the issue; others have added on various developing states to the main Socialist bloc, effectively shifting them from the Third World to the Second World at the stroke of a pen.

Part of the problem has emanated from the difficulty of defining exactly what a socialist state is, particularly as socialists and Marxists often disagree fundamentally about this. In the mid 1980s, before the collapse of the Second World, Thrift and Forbes (1986) listed the attributes of a socialist government as follows:

- one-party rule;
- egalitarian goals;
- high or increasing degree of state ownership of industry and agriculture;
- collectivisation of agriculture;
- centralised economic control.

Using these criteria in a non-doctrinaire way, Drakakis-Smith *et al.* (1987) identified a surprisingly extensive map of socialist Third World states (see Figure 1.6). Of course, such a map cannot be static

▶

BOX 1.1 (continued)

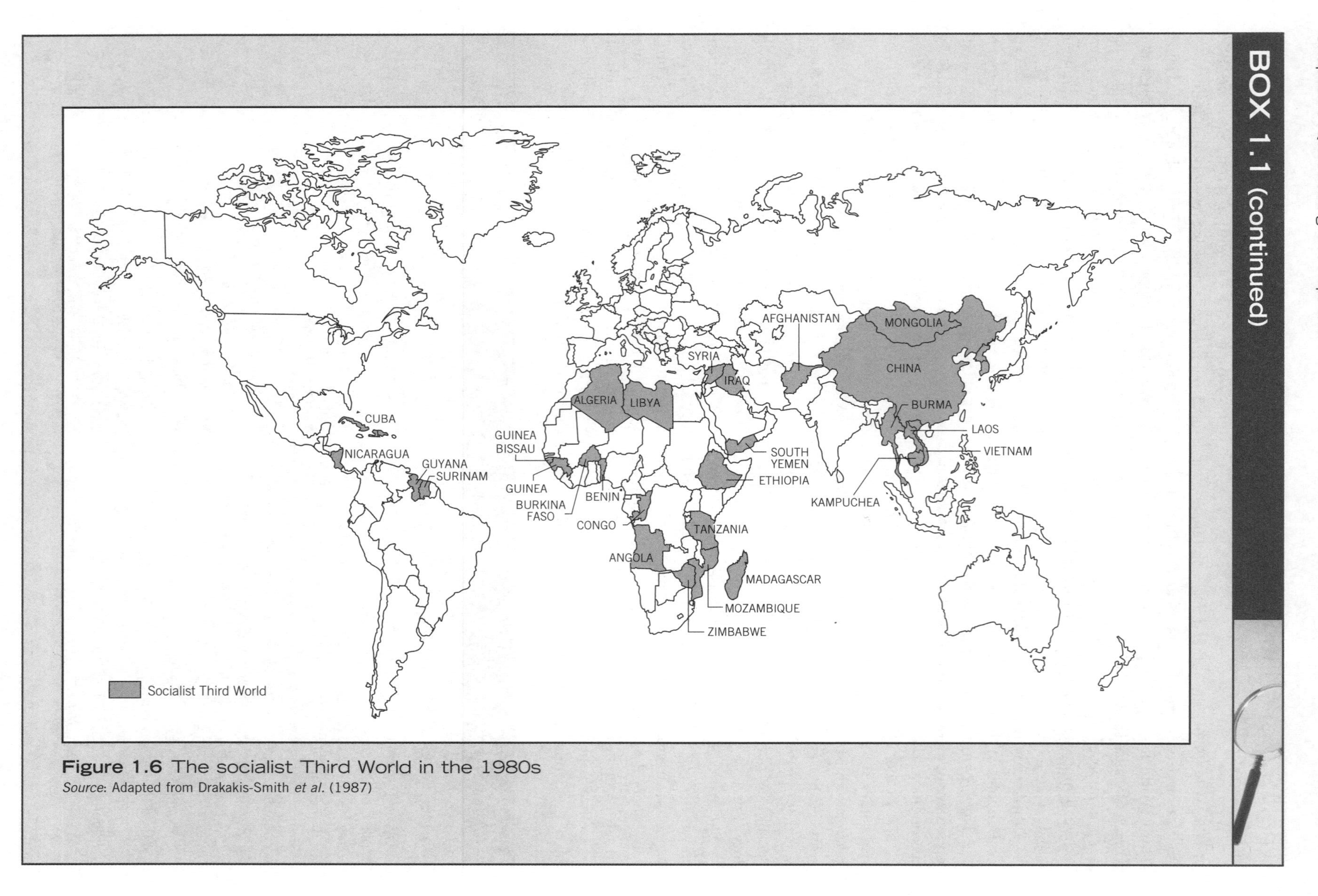

Figure 1.6 The socialist Third World in the 1980s
Source: Adapted from Drakakis-Smith *et al.* (1987)

BOX 1.1 (continued)

as governments come and go and, moreover, there are immense political, economic and social differences between the socialist states. In some, such as Tanzania, socialism has clear rural origins, often from an anti-colonial struggle, leading to persistent suspicion of urbanites and reflected in anti-urban policies. However, few if any Third World socialist states reflect the nature and roots of European socialism in the class contradictions of industrial societies.

Perhaps the most important distinction between the socialist Third World states is the nature of their political structure – put crudely, the distinction between authoritarian and democratic socialism. The massive transformations in Eastern Europe were bound to have an impact on the socialist Third World, but it is difficult to separate this from other major events that have affected the Third World in general since the mid 1980s, particularly structural adjustment. It is too easy and arrogant to read into the events in Eastern Europe and the former Soviet Union the 'end of history' in the rest of the world.

Perhaps surprisingly, socialist states persisted throughout the Third World in quite considerable numbers, but often with substantially modified forms of socialism. The region that has seen most change is sub-Saharan Africa, where change of government has been as much the result of structural adjustment policies, with their insistence on 'good' (meaning Western-style democratic) governance as one of the conditions for renewed loans, as the breakdown of support from the former Soviet Union. In several states, such as Mozambique, there has been a shift towards a more democratic form of socialism, at least as indicated by multiparty elections. In some states, however, this has released forces that the socialist government has struggled to suppress. This has been very evident in some Muslim countries, where Islamic fundamentalism has challenged the socialist state – successfully in Afghanistan and substantially in Algeria.

Elsewhere, it is the *rapprochement* with capitalism that has been more noticeable, particularly in Pacific-Asia, where economies were opened to investment in the 1990s, with immediate and substantial impact on both economic and social life (Drakakis-Smith and Dixon, 1997). However, despite such changes, all these states are still nominally socialist, particularly in terms of their being one-party states. In these countries, such as China, the struggle for democracy within socialism is still ongoing and is linked more with the individual economic freedoms associated with capitalism rather than fundamental changes in belief at government level. This has resulted in a heated debate on the nature of democracy between the West and many Asia-Pacific states, with both socialist and capitalist states in the region arguing for democracy through collective responsibilities rather than individual freedoms.

Recent changes in socialist states were reviewed by Sutton and Zaimeche (2002). These authors chart the relative shrinkage of socialist Third World states, and the movement of several towards market economies. For example, this has been true of Mozambique, Tanzania, Nicaragua, Benin, Mongolia, Burkina Faso, Guinea-Bissau, Afghanistan, Albania and Madagascar. The main current socialist states include China, Cuba, North Korea, Laos and Vietnam. A number of other countries, such as Libya, Syria, Egypt, Guyana, Tanzania and Venezuela retain clear constitutional references to socialism. But fundamental changes have occurred in the nature of the internal structures and organisation of continuing socialist states. Above all, there has been a steady, but often contested, shift towards more democratic forms of socialism, although such shifts are under threat from indigenous reactionary forces and also from the inherent inequalities of the free market economy.

The widening differences began to exercise academic minds. Journals such as *Area* and the *Third World Quarterly* published articles about the merits and demerits of the term 'Third World' as a descriptive concept (Auty, 1979; Mountjoy, 1976; O'Connor, 1976; Worsley, 1979).

The debate soon spread to some of the 'serious' journals of the popular press, where various ways of regrouping the developed and developing countries were suggested. For example, *Newsweek* identified four worlds; the Third World comprised those developing countries with significant economic potential and the Fourth World consisted of the 'hardship cases'.

Not to be outdone, *Time* magazine subsequently put forward a five-world classification in which the Third World contained those states with important natural resources, the Fourth World were the newly industrialising countries (NICs) and the Fifth World comprised what were clearly regarded as the 'basket cases'.

Many academics joined in this semantic debate. Goldthorpe (in Worsley, 1979) produced a list of nine worlds; at the lower end of which came 'the better-off poor', 'the middling poor', 'the poor' and 'the poorest' – indefinable refinements of poverty that were of little conceptual value and even less comfort to those under such scrutiny.

To cause even greater confusion, the term 'Fourth World' was also coming into general use to describe underdeveloped regions within developed nations, particularly where this referred to the exploitation of indigenous peoples, as in the cases of the Canadian Inuit or Australian Aborigines (see the special edition of *Antipode* 13(1) in 1981).

The changes were reflected to a certain extent in the classification system employed by the World Bank in its annual Development Reports. In the early 1980s, the developed countries were classified by their dominant mode of industrial production. Thus, the socialist states of Eastern Europe and the former Soviet Union were politically identified as 'centrally planned'. The non-oil-exporting developing countries were divided on the basis of wealth into low- and middle-income states.

Subsequently, following the apparently worldwide demise of socialism, the classification has regressed to an entirely income-based one. After 30 years of constant criticism, GNP per capita still rules as a development indicator with the World Bank.

The 1980s: the 'lost decade' for development in the Third World

Despite the regression at the World Bank to an economically based stratification of the Third World, the 1980s in general saw considerable widening of the range of indicators used for classifying the various nations of the developing world and soon they were being amalgamated into composite indices of well-being or quality of life as previously reviewed in Box 1.1. However, such indices did little to address the debate on the concept of the Third World per se.

Nevertheless, during the 1980s a growing critique of the term began to emerge from the new right-wing development strategists who argued that the Third World is merely the result of Western guilt about colonialism, a guilt which is exploited by the developing countries through the politics of aid.

Economist Lord Bauer (1975: 87), one of the leading exponents of this view, expressed it like this: 'The Third World [is] the collection of countries whose governments, with the odd exception, demand and receive foreign aid from the West . . . the Third World is the creation of foreign aid; without foreign aid there is no Third World.'

In the eyes of the New Right, virtually all developing countries are tainted with socialism and their groupings have invariably been anti-Western and therefore anti-capitalist, a view which has effectively been taken to task by John Toye (1987). Ironically, many Marxists also found it difficult to accept the term 'Third World' because they regarded the majority of its constituent countries as underdeveloped capitalist states linked to advanced capitalism.

Thus, in the eyes of Marxists there were only two worlds, those of capitalism and Marxian socialism, with Marxian socialism subordinate to capitalism. Unfortunately, there was little agreement among Marxists as to what constituted the socialist Third World (Box 1.1).

The notion of two worlds perhaps represented the most concerted challenge to the three-world viewpoint and, indeed, most of the semantic alternatives that we currently use are structured around this dichotomy, namely rich and poor, developed and underdeveloped (or less developed), North and South. Indeed, this perspective leads to dualism, a concept which is extensively reviewed in Chapter 3.

The North–South labelling, in particular, received an enormous boost in popularity with the publication of the Brandt Report (1980). As many critics have noted, the Brandt Report set out a rather naive and impractical set of recommendations for overcoming the problems of underdevelopment, relying as it did on the governments of the South to pass on the recommended financial support from the governments of the 'North' (Singer, 1980; Potter and Lloyd-Evans, 2008).

Moreover, much of the impetus behind the new 'concern' for development was fuelled by the economic crises of the developed countries and their search for new markets in the rest of the world (Frank, 1980). The heads of state assembled at Cancun in the early 1980s to discuss the Brandt Report and the plight of the world and, having been publicly seen to be concerned, duly dispersed affirming their faith in market forces rather than Willy Brandt.

From a developmental perspective, one of the Brandt Report's major defects was its simplistic subdivision of the world into two parts based on an inadequate conceptualisation of rich and poor (Figure 1.7). Some critics have claimed that this is spatial reductionism of the worst kind, apparently undertaken specifically to divide the world into a wealthy, developed top half and a poor, underdeveloped bottom half – North and South, them and us – although the terms did no more than rename pre-existing spatial concepts (see Figure 1.7).

However, the labels *North* and *South* do seem to be used with disturbing geographical looseness, since the South includes many states in the northern hemisphere, such as China and Mongolia, whereas Australasia comprises part of the North (Figure 1.7).

No definitions were discussed in the report, and the contorted dividing line that separates the two halves of the world stretches credulity more than a little as it is bent around Australia and New Zealand, totally ignoring the many small island states of the Pacific, generously, but erroneously, giving them developed status.

One problem with the North–South division of the Brandt Report was that it lacked explanatory power and compares unfavourably with another dichotomous model that also became popular in the 1980s. This is the core and periphery model (Wallerstein, 1979). The Brandt line does not allow for, or explain, the immense variety that exists both in the core and periphery, nor does it incorporate change over time, whether growth or decline.

In order to accommodate this, a semi-periphery was introduced; this is a category of countries allegedly incorporating features of both the periphery and the core. The core, semi-periphery and periphery classification of the world is also shown in Figure 1.7. The outcome is referred to as the 'world systems approach' (for a summary see Klak, 2008), which is fully reviewed in Chapter 3.

Effectively, the semi-periphery gives us another division of the world into three sections, but although apparently based on very different principles from those identified earlier, the various components are still bound together by the overarching operations of capitalism. The detailed and accurate map of core, semi-periphery and periphery after the 1990s, following the break-up of the former Soviet Union, clearly looks quite different over much of Central Asia and Eastern Europe (Klak, 2008).

As the 1980s wore on, however, the old Truman goal of development towards the Western capitalist model began to fade. The finishing line had in any case been moving away from most Third World countries faster than they were moving towards it.

The unified social and economic objectives of the second *United Nations Development Decade* began to look limp in the face of worsening world recession and a harder attitude towards a set of nations that were now being looked upon as a drag on world development through their incessant demand for aid and their growing debt defaulting.

The dualism of North and South thus took on a much harsher complexion as the World Bank, the IMF and the regional banks began to impose what are referred to as 'structural adjustment programmes' (SAPs) on the Third World. SAPs are economic austerity packages that were made conditional on countries wanting financial loans and aid. Such programmes are looked at in greater detail in Chapters 3 and 7 (see also Simon, 2008).

The growing confusion over whether the world should be divided into two or three components, both in conceptual and policy terms, was further accentuated in the 1980s by a feeling among some commentators that the original universalism of the United Nations had somehow been lost and that we should return to thinking of the world as a single entity.

Allan Merriam (1988) notes that views on the unity of humanity are long established, and he cites the seventeenth-century Czech educator Comenius, who stated that 'we are all citizens of one world, we are all of the same blood . . . let us have but one end in view, the welfare of humanity'. Such sentiments have featured frequently in the speeches of some Third World leaders, such as Indira Gandhi and Julius Nyerere, although often as rhetoric rather than reality.

Much of this growth in one-worldism was sustained by a belief that development in the Third World is characterised by a *convergence* along those paths experienced by the West towards the current lifestyles and political–economic structure of developed countries (Armstrong and McGee, 1985; Potter, 1990, 2008a).

For many, the thought of such convergence is alarming, since pursuit of the same economic ends by the same means will simply lead to a faster use of the Earth's finite resources and will only exacerbate its environmental problems. However, many of these concerns spring from self-interest in that ultimately it is

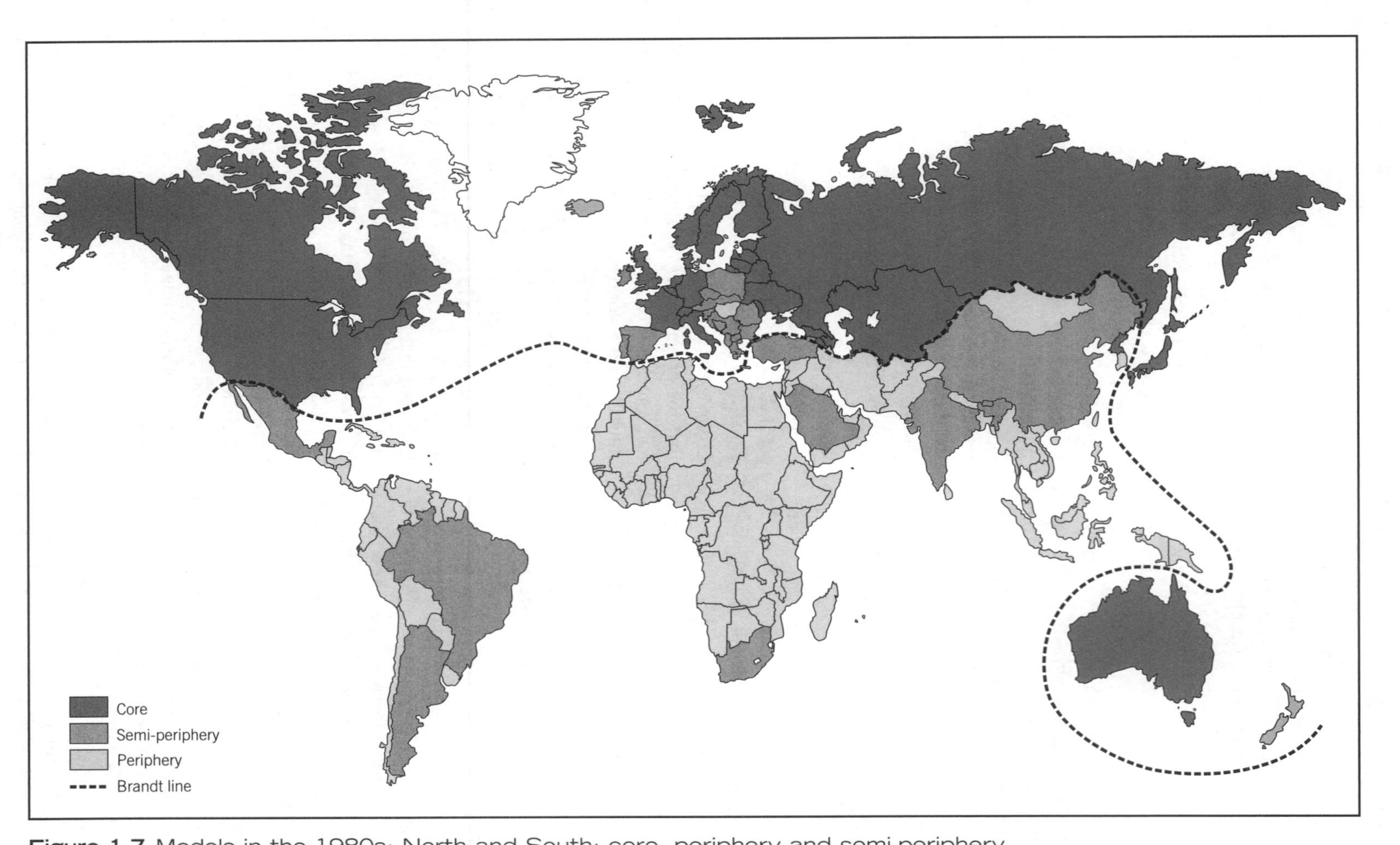

Figure 1.7 Models in the 1980s: North and South; core, periphery and semi-periphery

Source: Adapted from Willy Brandt, *North-South: A Program for Survival*, figure 'Models on the 1980s: North and South; core, periphery and semi-periphery', © 1980 The Independent Bureau on International Development Issues, by permission of the MIT Press and also from *North-South: A Programme for Survival*, Pan, (Brandt, W. 1980). Copyright © W. Brandt, 1980, by permission of Pan Macmillan.

the Western way of life that may be threatened. The people of the Third World are therefore being asked to make sacrifices 'for the greater good of humankind', sacrifices that those in the West have never made.

Sachs (1992) sees the convergence theories of the ecologists as yet another example of universalism, perpetuating the single goal, single strategy of the Truman doctrine and denying any role or opportunity for diversity. He claims that the 'one world or no world' warnings of environmental scientists suggest that preservation of our fragile global ecosystem demands that everyone has a responsible and specific role to play. 'Can one imagine a more powerful motive for forcing the world into line than that of saving the planet?' he asks (Sachs, 1992: 103).

As the Third World poor have been conveniently found to be the worst offenders in resource destruction, so their re-education could usefully be combined with scaled-down poverty reduction programmes through 'sustainable development'. The West can now give less aid and still feel good about it!

However, perhaps the basic premise of this concern is unfounded. There may be convergence towards the Western model, but this is very selective and uneven, an issue which is fully addressed in Chapter 4.

Moreover, even the high-flyers among the NICs still have a long way to go to match the levels of economic and social well-being attained in the West. Although South Korea may have almost as many fax machines per 100 business telephones as the USA, GNP per capita is still only one-third of the Organisation for Economic Co-operation and Development (OECD) average; and despite its spectacular urban centres the country still has a persistent informal sector, squatter housing and a large external debt, all of which have contributed to instability in the past.

Although attention in recent years has focused on these growing contrasts within the Third World itself (and this is a valid concern), it has also masked the more important fact that global contrasts too are continuing to widen.

In particular, there has been much concern that a large number of countries, particularly in Africa, have not only failed to exhibit any signs of development but have actually deteriorated, saddled as they now are with the spiralling debts of poverty and harsh structural adjustment programmes.

In this context, convergence theory could be seen as a myth. Indeed, it is arrogant to assume that the process of economic and cultural transfer is one-way. The West has not merely exported capitalism to the developing world, capitalism itself was built up from resources transferred to the West from those same countries.

Similarly, acculturation is not simply the spread of Gucci and McDonald's around the world. In almost every developed country, clothes, music and cuisine, together with many other aspects of day-to-day living, are permeated with influences from Asia, Africa, Latin America and the Caribbean such as bamboo furniture, curries and salsa music.

The Third World since the 1990s

The extension of the world recession into the 1990s meant that fragmentation of interests continued, and weaker communities at both local and global levels faced increasing difficulties.

One response to this was the emergence of regional economic blocs in the image of the European Union, such as NAFTA (North Atlantic Free Trade Agreement) and APEC (Asia-Pacific Economic Cooperation), all of which were designed to protect their member states and which cut across the traditional boundaries of the three worlds. Of course, this conceptualisation has suffered an even greater blow from the apparent demise of the Second World with the break-up of the Soviet Empire and the, admittedly uneven, democratisation and capitalisation of Eastern Europe.

If the Second World no longer exists, can there be a Third World? In this etymological sense, there is little justification for retaining the term, particularly since the early commonality of non-alignment and poverty has also long been fragmented.

Many commentators in the 1990s, particularly those who form part of the anti-development school, suggested that it is time for the term Third World to be abandoned. Sachs (1992: 3), inelegantly but forcefully, stated that, 'the scrapyard of history now awaits the category "Third World" to be dumped'. Corbridge (1986: 112) too joined 'with others in questioning the current validity of the term the Third World'. Friedmann (1992) also rejected the term in favour of a focus on people rather than places, preferring to identify and build policy around the disempowered. And yet, despite such strong condemnation, the term persists in common usage, even by some of those who have criticised its overall validity.

Critical reflection

Different worlds – different words?

How do you view the term 'Third World' – do you use it? If not, when you are talking about poorer countries within the world context how do you refer to such nations? As we have seen, people use a very wide array of terms, including 'Developing Countries', 'Less Economically Developed Countries' ('LEDCs'), 'Underdeveloped Countries', the 'Global South', 'Poor Countries', 'Former Colonies', etc.

What do you regard as the merits and limitations of each of these descriptors? Might it be better to always refer to broad continental divisions, thereby suggesting that African, Asian, Latin American and Caribbean countries should be seen as possessing their own distinct characteristics, and that these outweigh any broad commonality?

So why does the term persist in this way when the Second World has all but gone and the developing countries continue to fragment in their interests, among themselves and within themselves?

Perhaps, as Norwine and Gonzalez (1988) have remarked, some regions are best defined or distinguished by their diversity. An analogous situation in the biophysical world is the tropical rainforest: 'More diverse in flora and fauna than any other terrestrial biogeographic type, a rainforest is nonetheless one organic whole, consisting of many disparate parts, yet far greater than the sum of them' (Norwine and Gonzalez, 1988: 2). In other words, although highly diverse, it remains a recognisable entity.

Despite the variations in the nature of the Third World that we have noted in this section, most people in most developing countries continue to live in grinding poverty with little real chance of escape. This is the unity that binds the diversity of the casual labourer in India, the squatter resident in Soweto, or the street hawker in Lima. All are victims of the unequal distribution of resources that the world exhibits.

Moreover, this unity is not merely one of pattern or distribution, but of fundamental processes that are linked to the past, present and probable future roles of these states within the world economy, as exploited suppliers of resources – human as well as physical. It is these countries that have faced structural adjustment programmes, and now are the focus of poverty reduction strategies. It is among these countries that the debt crisis and massive levels of poverty and preventable death loom large.

Further, it still holds true that there is a unity provided by colonisation, decolonisation and antipathy, but a lack of resistance to imperialism (socialist as well as capitalist), something noted by Mao Tse Tung, Peter Worsley and John Toye. The same sort of view gives rise to the argument that 'the Third World is **SIC**', that is it is the outcome of the forces of **S**lavery, **I**mperialism and **C**olonialism.

It is for exactly these sorts of reasons that some commentators still approve of the use of the term the 'Third World', in that it stresses the historical–political and strategic commonalities of relatively poor, primarily ex-colonial countries. Thus, virtually all Third World nations, save for Thailand, Iran, parts of Arabia, China and Afghanistan, share a history of colonial rule and external domination. Thus, a case can certainly be made, on the grounds of history, for the continued use of the collective noun the 'Third World'.

In this sense, the concept of the Third World is an 'extremely useful figment of the human imagination . . . The Third World exists whatever we choose to call it. The more difficult question is how can we understand it' (Norwine and Gonzalez, 1988: 2–3), and change it according to priorities set out by its own inhabitants.

Most of the students of development who continue to use the term the *Third World* must realise, therefore, that it is not simply a semantic or geographical device (Killick, 1990), but a concept that refers to a persistent process of exploitation through which contrasts at the global, regional and national levels are growing wider.

No matter what abstract conceptualisations are used to structure development debates – three worlds, two worlds, nation states, cities or whatever – we must not forget that we are discussing human beings. Their welfare and how to improve it must be the focus for our debates, rather than the sterile question of what label is politically correct.

Rich and poor worlds: relative poverty and inequalities at the global scale

The salience of inequalities and relative poverty

Descriptive phrases and terms such as the 'Third World', are just that: they are descriptors of ongoing dynamic processes of change. Viewed in this light, it is unrealistic to expect any one term to describe the global pattern of development over time.

The present volume focuses attention on the process of development and underdevelopment, wherever they occur, be it in the former colonial world, poor regions in former colonial powers, in what has in the past been referred to as the Third World, or wherever. In this text, the terms 'Third World', 'Developing Nations' and the Global South are used almost interchangeably.

But at the same time, as recognised above, some commentators emphasise that the real commonality between the countries that we study is their relative poverty. They are the countries that are generally poorer than other nations.

It is worth recalling Worsley's identification of the commonalities of Third World countries in colonial domination, non-alignment and poverty. But per capita income levels, literacy levels and educational enrolment still represent major dimensions in the United Nations Human Development Index.

Indeed, the argument that inequalities within society are more important than the overall average level of income or wealth is a telling one. It can be argued that the gap between the rich worlds and the poor worlds is most significant in determining the kind of world in which we live.

Progress in development from the 1970s to the 2000s

As noted earlier in this chapter, one of the reactions to anti-developmentalist thinking is the argument that, in overall terms, impressive gains have been made to conditions in developing nations over the past 30 years or so. Indeed, the United Nations *Human Development Report* (UNDP, 2001) argued that this is recognised by all too few people, and that the progress made thus far demonstrates the possibility of eradicating poverty in the future.

Some of the key aspects of change cited by the United Nations are summarised in Figure 1.8. Notably, the key variables which the United Nations focuses upon are the ones that relate most closely to those employed in the Human Development Index.

The *Human Development Report* (UNDP, 2001: 10) noted that a child born today can on average expect to live eight years longer than one who was born only 30 years ago. During the same time period, the level of adult literacy has increased considerably, from 64 per cent in 1970 to 79 per cent in 1999.

Average incomes in developing countries almost doubled in real terms between 1975 and 1998, from US$1300 to US$2500. The same report also argued that the basic conditions for achieving human freedoms were also transformed in the ten years between 1990 and 2000, as more than 100 developing and transitional countries ended military or one-party rule, thereby opening up political choices.

In addition, from 1990, formal commitment to international standards in human rights has spread dramatically, as indicated in Figure 1.8, which charts the number of countries that have ratified six major human rights conventions and covenants.

Yet even the United Nations noted that 'behind this record of overall progress lies a more complex picture of diverse experiences across countries, regions, groups of people' (UNDP, 2001: 10). Thus, as a corollary of the general improvement in socio-economic conditions indicated in the foregoing account, the United Nations Development Programme (2001: 10) noted that 'across the word we see unacceptable levels of deprivation in people's lives'.

Thus, of the 4.6 billion people living in developing nations at that time, more than 850 million were functionally illiterate, while an estimated 1 billion lacked access to improved water supplies, and 2.4 billion lacked access to basic sanitation. Some 325 million children did not attend school (UNDP, 2001). A staggering 11 million children under the age of five die each year from preventable causes, representing a massive 30,000 children a day (UNDP, 2001). Finally, approximately 1.2 billion people were living on less than US$1 a day, and 2.8 billion on less than US$2 a day (UNDP, 2001). It remains only too obvious that much needs to be done in the future in respect of reducing global inequalities.

An unequal world

A large number of people around the world still face conditions that can only be described as far from acceptable. But how big is the gap between those who have much and those who have little? A first impression of the order of this disparity may be gained if we look at trends in average incomes for different regions of the world. In Figure 1.9, such trends in average incomes are set out for the major global regions for the 38-year period between 1960 and 1998.

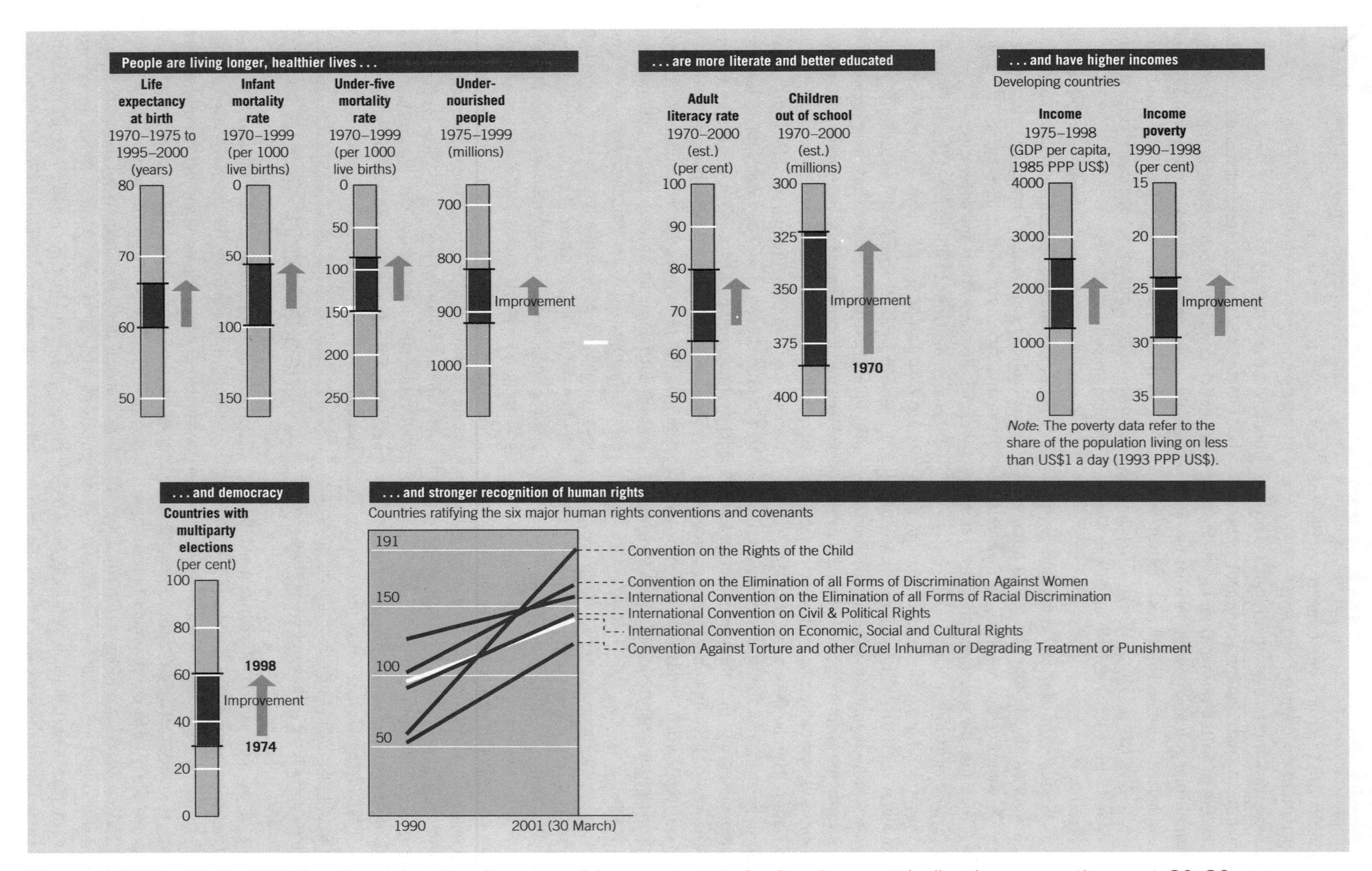

Figure 1.8 The United Nations' graphs showing overall improvements in development indications over the past 20–30 years

Source: Adapted from *Human Development Report, 2001*, UNDP, Oxford University Press. By permission of Oxford University Press, Inc.

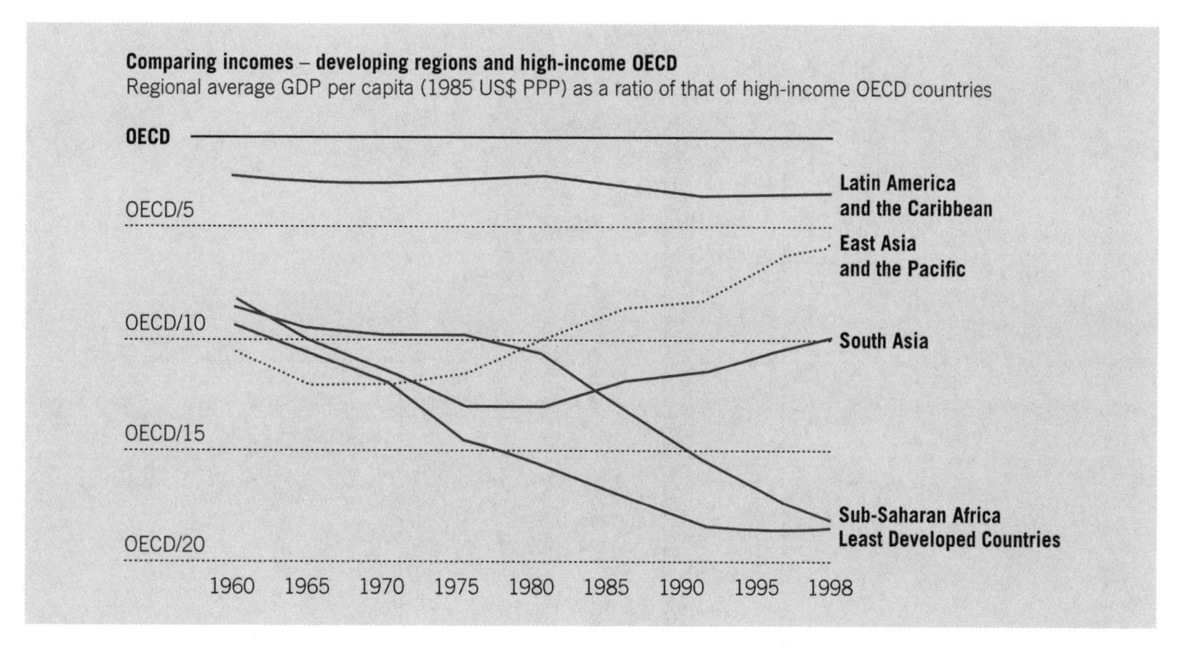

Figure 1.9 Regional variations in relative incomes 1960–1998
Source: Adapted from *Human Development Report, 2001*, UNDP, Oxford University Press. By permission of Oxford University Press, Inc.

In 1960, East Asia and the Pacific, South Asia, sub-Saharan Africa and the least developed countries all had average per capita incomes that amounted to around one-ninth to one-tenth of those of the high-income countries of the OECD (consisting of 30 of the wealthiest nations, including the USA, UK, Sweden, Japan).

Latin America and the Caribbean fared a little better, showing average incomes of around one-third to one-half those recorded by the OECD nations.

Between 1960 and 1998, East Asia experienced a major improvement in incomes, increasing over the period to nearly one-fifth of OECD levels (Figure 1.9). During the 1980s and 1990s, the average income pertaining to South Asia improved significantly, having shown relative declines earlier, staying at an average income level of around one-tenth that of the OECD nations.

The Latin America and Caribbean region also remained at about the same level during this period (Figure 1.9). But, tellingly, for sub-Saharan Africa, and for the least developed countries, relative income levels fell quite dramatically between 1960 and 1998, as displayed in Figure 1.9.

In Figure 1.10 we see affirmation of the growing heterogeneity of the 'Third World'. Thus, while the average incomes of East Asian and Pacific nations have shown *convergence* on those of the rich nations, *divergence* in average incomes has been true of sub-Saharan Africa and the least developed nations.

In fact, UNDP (2001: 16) states quite candidly that, 'despite a reduction in the relative differences between many countries, absolute gaps in per capita income have increased' during the period 1960–1998.

This is cogently summarised in Figure 1.10. Even for the fastest growing region of East Asia and the Pacific, the absolute difference in income against the high-income OECD nations widened from about US$6000 in 1960 to more than US$16,000 by 1998. In absolute terms, South Asia fell behind, as did sub-Saharan Africa and the least developed countries (Figure 1.10; see also Morrissey, 2001; Rapley, 2001).

Stated simply, world inequality is very high. In a study conducted in the mid 1990s, and reported in UNDP (2001), it was estimated that the richest 1 per cent of the world's population received as much as the poorest 57 per cent. And approximately 25 per cent of the world's people receive 75 per cent of total income.

Not only that, as we have already seen via the study of average incomes by major world region, the relatively rich and the relatively poor are getting further apart at their extremes. This is clearly demonstrated in Table 1.3, which is adapted from Fik (2000) and Seitz (2002), and once again is based on United Nations' data.

In 1960 the ratio of the income of the world's richest 20 per cent of the population to the income of the world's poorest 20 per cent stood at 30:1. By 1980 this ratio had widened considerably to 45:1. The United

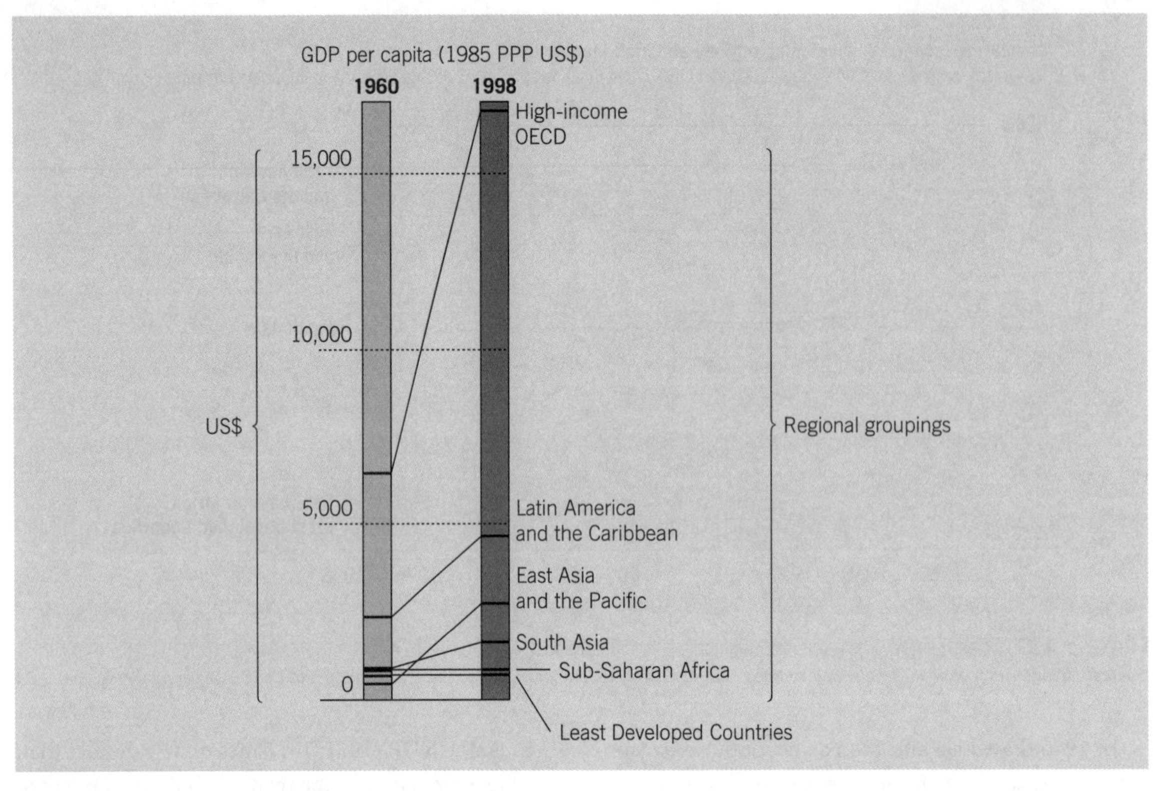

Figure 1.10 The widening gap in absolute incomes between world regions 1960–1998
Source: Adapted from United Nations (2001)

Table 1.3 Income ratios between the richest and poorest nations, 1820–2000

Year	Income of richest 20% divided by income of poorest 20%
1820	3:1
1913	11:1
1960	30:1
1970	32:1
1980	45:1
1990	59:1
2000	70:1

Sources: Fik (2000), Seitz (2000), UNDP (1998)

Nations estimated that for the year 2000 this income disparity had widened dramatically, rendering a ratio as high as 70:1 by the turn of the century (see Table 1.3). The data show that this ratio of disparity was only 3:1 in the nineteenth century.

In the light of these summary statistics concerning current global trends, it is tempting to return briefly to some of our earlier discussions concerning the nature of development and the developing world. For example, as we have already noted, the widening gap in incomes between rich and poor nations is taken by the anti-development school as a direct affirmation that development is not working. However, as was also noted in an earlier section, taken on its own this argument falls into the trap of equating development with income.

On the other hand, the absolute progress in incomes that has been made in Asia and the Pacific, South Asia, Latin America and the Caribbean is cited by those who wish to argue that development is working, at least in this narrow sense.

In so far as the gap between the richest and the poorest is continuing to widen, some do indeed argue the case for the retention of titles such as the 'Third World'. As noted previously, Norwine and Gonzalez (1988) argue that heterogeneity need not destroy a common identity, and sometimes can even serve to strengthen it. They argue that the Third World is a useful figure of the imagination, and that it exists whatever we choose to call it.

Once again, it is tempting to argue that something like a 'Third World' does exist, even if it is referred to as the 'Global South' or 'Poor Nations'. This grouping of nations remains united by their colonial past and continued subordinate role within the global economy (Drakakis-Smith, 2000). This view is pictorially played out in Figure 1.11.

Figure 1.11 Depiction of the question 'Are we all in the same boat?'
Source: Drakakis-Smith (2000)

The Millennium Development Goals (MDGs) 2000–2015: poverty, development and the future

What can be done to work towards a more equal world at the beginning of the twenty-first century? Reflecting the enormous magnitude of the inequalities that characterise the contemporary world order, the intention to do something about it exists in the form of an agreed international set of development targets – what are now referred to as the Millennium Development Goals, to which over 190 world leaders committed the support of their governments at the very start of the new Millennium.

The MDGs were formally adopted at the General Assembly of the United Nations held in New York on 18 September 2000, which was referred to as the UN Millennium Summit (Rigg, 2008). The MDGs consist of eight overarching goals, each to be achieved by 2015:

- Eradicate extreme hunger and poverty;
- Achieve universal primary education;
- Promote gender equality and empower women;
- Reduce child mortality;
- Improve maternal health;
- Combat HIV/AIDS, malaria and other diseases;
- Ensure environmental sustainability;
- Develop a global partnership for development.

These eight specific goals are reflected in 18 specific targets. In turn, the targets are linked to 48 detailed indicators, although some of these are not defined in very precise terms (Black and White, 2004). Full details of the goals, targets and indicators are listed in Table 1.4, based on Rigg (2008) in Desai and Potter (2008). Together they can be regarded as making up a comprehensive agenda for global development in the twenty-first century.

In fact, the targets were first enumerated by the OECD in a document under the title *Shaping the Twenty-First Century* (OECD, 1996), and were strongly supported in the UK by the Department for International Development (DFID) as the International Development Targets (see, for example, the UK White Paper, DFID, 2000a).

The whole idea is that the MDGs should amount to realistic and reachable targets and goals. Reflecting

Table 1.4 The Millennium Development Goals, Targets and Indicators

Goals	Targets	Indicators
1. Eradicate extreme hunger and poverty	1. Reduce by half the proportion of people living on less than a dollar a day	1. Proportion of population below $1 (PPP) per day 2. Poverty gap ratio, $1 per day 3. Share of poorest quintile in national income or consumption
	2. Reduce by half the proportion of people who suffer from hunger	4. Prevalence of underweight children under five years of age 5. Proportion of the population below minimum level of dietary energy consumption
2. Achieve universal primary education	3. Ensure that all boys and girls complete a full course of primary schooling	6. Net enrolment ratio in primary education 7. Proportion of pupils starting grade 1 who reach grade 5 8. Literacy rate of 15- to 24-year-olds
3. Promote gender equality and empower women	4. Eliminate gender disparity in primary and secondary education, preferably by 2005, and at all levels by 2015	9. Ratio of girls to boys in primary, secondary and tertiary education 10. Ratio of literate women to men 15–24 years old 11. Share of women in wage employment in the non-agricultural sector 12. Proportion of seats held by women in national parliaments
4. Reduce child mortality	5. Reduce by two-thirds the mortality rate among children under five	13. Under-five mortality rate 14. Infant mortality rate 15. Proportion of one-year-old children immunised against measles
5. Improve maternal health	6. Reduce by three-quarters the maternal mortality ratio	16. Maternal mortality ratio 17. Proportion of births attended by skilled health personnel
6. Combat HIV/AIDS, malaria and other diseases	7. Halt and begin to reverse the spread of HIV/AIDS	18. HIV prevalence among 15- to 24-year-old pregnant women 19. Condom use rate of the contraceptive prevalence rate and population aged 15–24 years with comprehensive correct knowledge of HIV/AIDS 20. Ratio of school attendance of orphans to school attendance of non-orphans aged 10–14 years
	8. Halt and begin to reverse the incidence of malaria and other major diseases	21. Prevalence and death rates associated with malaria 22. Proportion of population in malaria risk areas using effective malaria prevention and treatment measures 23. Prevalence and death rates associated with tuberculosis 24. Proportion of tuberculosis cases detected and cured under directly observed treatment short courses
7. Ensure environmental sustainability	9. Integrate the principles of sustainable development into country policies and programmes; reverse loss of environmental resources	25. Forested land as percentage of land area 26. Ratio of area protected to maintain biological diversity to surface area 27. Energy supply (apparent consumption; kg oil equivalent) per $1000 (PPP) GDP 28. Carbon dioxide emissions (per capita) and consumption of ozone-depleting CFCs
	10. Reduce by half the proportion of people without sustainable access to safe drinking water	29. Proportion of the population with sustainable access to and improved water source 30. Proportion of the population with access to improved sanitation
	11. Achieve significant improvement in lives of at least 100 million slum dwellers, by 2020	31. Slum population as percentage of urban population (secure tenure index)

Table 1.4 (continued)

Goals	Targets	Indicators
8. Develop a global partnership for development	12. Develop further an open trading and financial system that is rule-based, predictable and non-discriminatory, includes a commitment to good governance, development and poverty reduction – nationally and internationally 13. Address the least developed countries' special needs. This includes tariff- and quota-free access for their exports; enhanced debt relief for heavily indebted poor countries; cancellation of official bilateral debt; and more generous official development assistance for countries committed to poverty reduction 14. Address the special needs of landlocked and small island developing states 15. Deal comprehensively with developing countries' debt problems through national and international measures to make debt sustainable in the long term 16. In cooperation with the developing countries, develop decent and productive work for youth 17. In cooperation with pharmaceutical companies, provide access to affordable, essential drugs in developing countries 18. In cooperation with the private sector, make available the benefits of new technologies – especially information and communications technologies	*Official development assistance* (ODA) 32. Net ODA as percentage of OECD/DAC donors' gross national product (targets of 0.7% in total and 0.15% for LDCs) 33. Proportion of ODA to basic social services (basic education, primary health care, nutrition, safe water and sanitation) 34. Proportion of ODA that is untied 35. Proportion of ODA for environment in small island developing states 36. Proportion of ODA for transport sector in landlocked countries *Market access* 37. Proportion of exports (by value and excluding arms) admitted free of duties and quotas 38. Average tariffs and quotas on agricultural products and textiles and clothing 39. Domestic and export agricultural subsidies in OECD countries 40. Proportion of ODA provided to help build trade capacity *Debt sustainability* 41. Proportion of official bilateral HIPC debt cancelled 42. Total number of countries that have reached their HIPC decision points and number that have reached their completion points (cumulative) (HIPC) 43. Debt service as a percentage of exports of goods and services 44. Debt relief committed under HIPC initiative 45. Unemployment of 15- to 24-year-olds, each sex and total 46. Proportion of population with access to affordable, essential drugs on a sustainable basis 47. Telephone lines and cellular subscribers per 100 population 48. Personal computers in use and Internet users per 100 population

Source: Rigg 2008 based on http://www.un.org/millenniumgoals/index.html

their earlier origins, statistics measuring progress with the MDGs are frequently presented from 1990, although the MDGs proper relate to the period 2000–2015. Thus, 2007–2008 represents the mid-point toward the MDGs. So, what has been achieved so far, and on the present trends, which of the goals are likely to be achieved by 2015?

As Rigg (2008) concludes, the unequivocal answer has to be that while there has been good progress with some of the goals in some parts of the world, overall progress is best described as 'patchy'.

This is well exemplified if we look at one of the major targets relating to the 'eradication' of poverty and hunger – that of reducing by half the proportion of people living on less than US$1 a day. Table 1.5 shows the relevant data for 1990 and 2004, derived from the United Nations. Overall, this measure of poverty has fallen since 1990, and if the current rate remains on track, this global poverty indicator will fall to 12.5 per cent in 2015, less than half the level recorded in 1990.

But it is clear that progress has been uneven when viewed by major continental division (Table 1.5). While very substantial progress has been made in the case of Southern Asia, Eastern Asia, and South-East Asia, little progress has been made in the case of sub-Saharan Africa. In sub-Saharan Africa, over 40 per cent

Table 1.5 Percentage of the total population living on less than $1 per day, 1990 and 2004

Continental Region	1990	2004
Sub-Saharan Africa	46.8	41.1
Southern Asia	41.1	29.5
Eastern Asia	33.0	9.9
Latin America and the Caribbean	10.3	8.7
South-Eastern Asia	20.8	6.8
Western Asia	1.6	3.8
North Africa	2.6	1.4
Trans. Cs of SE Europe	<0.1	0.7
Comm. Ind. States S Asia	0.5	0.6
Developing Countries	*31.6*	*19.2*

Source: Adapted from United Nations (2007c), p.6

of the total population still live on less than US$1 per day.

We can look at the likelihood that the MDG targets can be met in the various world regions by 2015. This has been analysed graphically by the Millennium Project (2005), which is supported by the United Nations Development Group.

The 16 targets that can be quantified were assessed for the ten regions of the global system, as shown in Figure 1.12. Of the total of 160 thereby identified, in 16 cases the target has already been met or is very close to being met. In some 40 other cases, the target is expected to be met by 2015. However, in the case of 34 targets, no progress has been made or deterioration and reversal have occurred. Finally, in a further 66 cases, it is not expected that the MDG target will be met.

Thus, in respect of the majority of targets viewed by continental division (some 100 out of 160), the outcome looks set to be negative by 2015, judging by present progress. Figure 1.12 shows where the biggest number of missed targets are likely to be encountered – namely in sub-Saharan Africa, followed by Oceania, Commonwealth of Independent States (CIS) Asia and Western Asia.

The box pertaining to sub-Saharan Africa is by far the most frightening, with all 16 constituent boxes showing either no progress or that the target is not expected to be met. Overall, therefore, there clearly remains a great deal to be done. And as Rigg (2008) asks, are the mechanisms in place to achieve the MDGs other than by exhortation and moral persuasion. For example, forms of global taxation, such as a so-called Tobin-style tax, could raise the annual monies needed to establish universal primary education in one go, rather than moving slowly towards the target over 15 years. This argument is covered in detail in Chapter 4.

Once more we are reaching the same general conclusions regarding the current nature and disposition of global development patterns as we did in the last major section on trends from 1970. While progress is being made in certain regions, and in particular respects, there remains an enormous amount to do if gross inequalities are to be meaningfully reduced.

Development and anti-development are extremely important concepts, for they exist in a global context where differences in wealth, opportunity and choice appear to be widening (diverging) at their extremes rather than narrowing (converging).

Thus, while some poorer nations are showing enhanced incomes in relation to the rich nations of the West, the majority are continuing to fall yet further behind.

The politico-strategic salience of this widening gap seems likely to increase in the future. Talking in the aftermath of the events of 11 September 2001, Bill Clinton (in *The Richard Dimbleby Lecture*, 2001) argued that 'we in the wealthy countries have to spread the benefits of the twenty-first century world and reduce the risks so we can make more partners and fewer terrorists in the future' (Clinton, 2001: 2).

In the same speech, Bill Clinton attributed the events of 11 September to increased liberalisation and globalisation: 'we have built a world where we tore down barriers, collapsed distances and spread information. And the UK and America have benefited richly' (Clinton, 2001: 2) (see also Potter, 2003).

But in addressing what he regards as the burdens of the twenty-first century, Clinton observes that over half the world are excluded from the benefits of the new global economy, and he asks, 'what kind of economy leaves half the people behind?' (Clinton, 2001: 3).

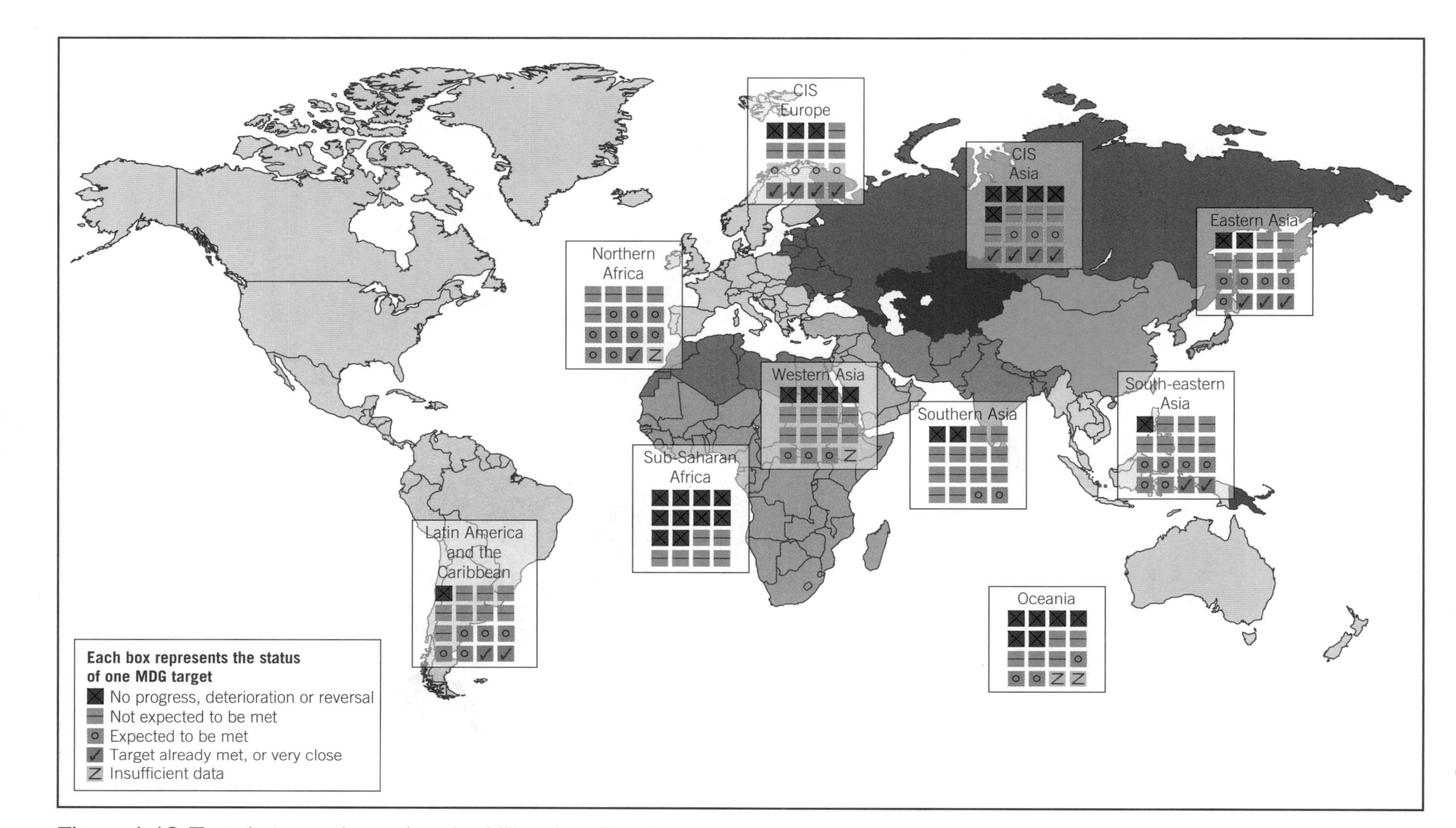

Figure 1.12 Trends toward meeting the Millennium Development Goal targets by 2015
Source: Based on UNEP (2005c). UNEP/GRID-Arendal

Plate 1.6a Part of a recycling system close to the Yumuna River in Delhi, India
(photo Rob Potter)

Plate 1.6b Refuse is sorted into various types
(photo Rob Potter)

Plate 1.6c In an adjacent 'Dhobi' or clothes washer settlement, recycled fabrics and clothes are washed using old tyres and trainers as fuel
(photo Rob Potter)

Plate 1.6d Cleaned and recycled fabrics are dried before distributed from the Dhobi settlement
(photo Rob Potter)

Concluding issues: geography, development and 'distant others'

Terry McGee (1997) has drawn attention to some of the implications of the shifts into postmodern development, particularly those that emphasise globalisation as 'a variable geometry of production or consumption, labour, capital management and information – a geometry that denies the specific meaning of place outside its position in a network whose shape changes relentlessly' (Castells, in McGee, 1997: 8).

For some time, since Alvin Toffler's *Future Shock* (1970), commentators have been arguing for the end of geography. Toffler himself based his arguments on increased flows of people, goods and information that serve to dissolve difference and distinctions.

Apart from geographical differences still being very evident in the world (Chapter 4), Toffler also ignored the fact that linkages and flows between places are largely the province of the geographer. Similarly, Richard O'Brien (1991) also claimed an end to geography on the basis that location matters much less for economic development than it has done in the past.

Although the recent development of technologies does 'challenge conventional notions of distance, boundaries and movement . . . geography matters . . . because global relations construct unevenness in their wake *and* operate through the pattern of uneven development laid down' (Allen and Hamnett, 1995: 235).

Thus, the choice for new investment is often conditioned by the facilities and resources that are already there, thus perpetuating inequalities. Not surprisingly, therefore, many would argue (Massey and Jess, 1995) that place is more fundamental than ever, since the realities of development within the Third World are represented by an unevenness and by a constantly shifting fusion and conflict between the global and local, usually filtered through national or regional agency. Our review of current development patterns and global inequalities in the last section has shown as much.

Thus, it is appropriate to emphasise in this opening chapter that the relationships between the realities of development and geography remain very strong. If development is regarded as the improvement of lives and opportunities, then the challenge is generally about improving matters for other groups in different – and often far-off – regions.

Indeed, Potter (1993a, 2001a, 2002a) has argued that questions of development have been somewhat neglected in the academic discipline of geography, in the sense of faraway places and peoples receiving a surprisingly small share of the attention of academic geographers, relative to their salience in the world at large.

In general, a disproportionate amount of attention is focused on the European and North American realms. Potter (2001a) argues that an increasing responsibility to what may be described as 'distant geographies' is urgently required. This concept is an embellishment of the more general 'responsibility to distant others' or 'distant strangers' (see Smith, 1994, 2008).

The issue of our responsibility to distant others basically asks how likely we are to be beneficient to people who are worse off than ourselves, but who live far away. Beneficience is the process of active kindness, and may be distinguished from benevolence, which involves charitable feelings and the desire to do good (Smith, 2002).

The central question posed, therefore, is 'how spatially extensive beneficence can be justified by moral argument, given what might appear to be the natural human tendency to favour our nearest and dearest over more needy strangers farther away' (Smith, 2000: 132).

In many ways, this moral issue is at the heart of the complex relations that exist between contemporary geography and development. The rapidly ongoing processes of globalisation make this pressure much more intense than in the past (see Chapter 4 on this). For greater responsibility to be shown to distant others, a greater commitment to the mobilisation of resources at the international level is required (Smith, 2000). Thus, the issue underpins giving to charitable foundations and the provision of aid at the global level.

It is true that at one level the processes of globalisation are creating super-regions, such as the European Union, NAFTA and APEC, all with varying degrees of cohesion, together with their global and regional mega-cities, such as Paris, London or Tokyo.

But at other levels and in other localities in the interstices of the global network, neglect, ignorance and even resistance combine to produce patterns of development that are strongly geared to place and history, and which must be studied as such in order for development to be fully understood (Chapter 4).

Such places can even occur within the mega-cities themselves, as migration throws together people whose roles in the development process are very different indeed.

Although many geographers have re-emphasised the importance of place in this globalised dimension of development, McGee (1997: 21) argues that most geographers continue 'to interrogate the development project from within the modernist project [in] the liberal belief that good research can provide workable solutions'.

What constitutes the heart of this approach is that geographical investigation is rooted in an empiricism that focuses on the interaction of society and environment, on networks and flows of people and goods, on uneven and unequal development and, most important in these contexts, on the nature of local places.

All of these factors, according to McGee, place development geography firmly in the humanist tradition. What needs to happen now is for the local not only to become the object of the exercise but also the medium, with local input into the development process itself. Only in this way will our preconceived ideologies or images of development be changed (Massey, 1995).

Key points

- Development is a frequently used word in all sorts of contexts. In the arena of socio-economic change, it implies efforts to improve the lives of people around the world.
- Early ideas on development stemmed from the Enlightenment period and were then allied with concepts of westernisation and modernity and neo-classical economic thinking in the post-1947 period.
- Early on, development was measured with respect to qualitative economic indicators, principally Gross National Product (GNP) and Gross Domestic Product (GNP). Increasingly now, the Human Development Index stresses a wider set of dimensions and commentators talk about development in more qualitative terms, including development as freedom.
- Since the 1990s, alternative and populist stances have represented strong re-thinks as to the meanings of development.
- The concept of the Third World has its origins in the global geopolitics of the post-Second World War period. Many other terms, such as Developing Countries, Underdeveloped Countries, Less Economically Developed Countries, Poor Nations and the Global South are now used.
- Statistics show that overall the world as a whole is showing signs of development and improving conditions – and this is taken as the evidence of continued development by those who believe in the so-called 'development mission'.
- However, data show all too clearly that global inequalities have continued to widen considerably over the past 30 to 40 years. It is circumstances such as this that are cited by anti-developmentalists.
- The Millennium Development Goals 2000–2015 represent a global agenda to address development issues. But progress since 2000 emphasises that much remains to be done.

Further reading

Black, R. and White, H. (2004) *Targeting Development: Critical Perspectives on the Millennium Development Goals*. London and New York: Routledge.
Provides an overview of issues relating to the Millennium Development Goals.

Department for International Development (2000) *Eliminating World Poverty: Making Globalisation Work for the Poor*, Cmnd 5006. London: The Stationery Office.
Worth reading as a strong template for the argument that globalisation is the way forward in delivering countries from poverty.

Desai, V. and Potter, R.B. (eds) (2008) *The Companion to Development Studies, Second Edition*. London and New York: Arnold, Part 1 'The nature of development studies'.
An accessible source that brings together short, 2000 word summaries of important facets of the interdisciplinary field of development studies.

Escobar, A. (1995) *Encountering Development*. Princeton, NJ: Princeton University Press.
A well-cited critical review of the development mission.

Gasper, D. (2004) *The Ethics of Development*. Edinburgh: Edinburgh University Press.
An interesting overview of ethical aspects of development theory and practice.

Greig, A., Hulme, D. and Turner, M. (2007) *Challenging Global Inequality: Development Theory and Practice in the 21st Century*. Basingstoke: Palgrave Macmillan.
Stresses global poverty and inequality in reviewing contemporary development theory and practice.

Kothari, U. (ed.) (2005) *A Radical History of Development Studies: Individuals, Institutions and Ideologies*. Cape Town: David Phillip; London and New York: Zed Books.
An interesting set of specially commissioned chapters that consider the nature and development of development studies.

Simon, D. (ed.) (2005) *Fifty Key Thinkers on Development*, London and New York: Routledge.
A useful source which brings together short essays on those who are deemed to have had a noticeable impact on studies of development.

Websites

www.un.org/millenniumgoals/
A useful site, which provided access to salient documents like the Millennium Development Goals Report 2007.

www.undp.org
Site of the United Nations Development Programme (UNDP) giving direct access to the United Nations Human Development Reports.

www.dfid.gov.uk
The website of the Department for International Development (DFID) of the UK Government. Provides details of, and access to, DFID publications.

www.eldis.org
Described as the gateway to development information, providing access by subject area and country, plus news, etc.

Discussion topics

- Examine the argument that development should be about reducing 'unfreedoms' rather than promoting economic development.
- Elaborate the view that measures of relative development must include non-economic variables.
- Assess the current applicability of the assertion that the 'Third World' exists whatever we choose to call it.
- To what extent is it helpful to think in terms of poor countries rather than developing nations?

Chapter 2

Understanding colonialism

This chapter examines the development and demise of colonialism and its impact on developing countries. After considering definitions of colonialism and imperialism, the issue of post-colonialism is critically reviewed. With reference to case study material, three phases of colonialism are then discussed; mercantile colonialism, industrial colonialism and late colonialism. This is followed by an examination of the process of decolonisation and a consideration of some of the most significant legacies of colonialism.

This chapter:

- Examines different perceptions of colonialism and imperialism;
- Reviews the debate about post-colonialism and the effects of colonisation on local people;
- Considers the key features of the three phases of colonialism;
- Examines the increasing pressure for independence and the process of decolonisation;
- Assesses some of the beneficial and detrimental aspects of colonialism;
- Evaluates the concepts of a New International Division of Labour and a New International Economic Order.

Introduction: colonialism and imperialism

The literature on colonialism is as plentiful, and at times as opaque, as the literature on development. There is an unfortunate tendency to equate colonialism with the expansion of capitalism in the nineteenth and twentieth centuries, implying that it primarily comprises an economic process (Key idea, Colonialism and imperialism). Clearly, there is an essential, and at times overwhelming, economic impetus to colonialism, but to construct a framework of analysis based on such a simple equation would be unhelpful. Colonialism is essentially a political process, and the establishment of colonies long pre-dates the genesis and subsequent globalisation of European capitalism. Right through this period, colonies were acquired for motives other than the economic imperative for material resources, labour or markets. As in Roman times, otherwise barren or unpromising territory was annexed for strategic reasons: to protect the periphery of pre-existing colonies, to control important military routes or simply to prevent the expansion of rival European powers. Although some of these lands eventually proved to have some economic value, the original motivation was often quite different.

Much of the development literature conceptualises the global expansion of capitalism as imperialism rather than colonialism, although even here there is a debate about when this process began. For some neo-Marxists, imperialism begins with the division of Africa at the Treaty of Berlin in 1885, reinforced by the assertion that the term *imperialism* was first coined in the nineteenth century by Napoleon III. Most, however, believe that imperialism began in the late fifteenth and early sixteenth centuries with the rise to prominence of the European nation state, although the commercial underpinning to the feudal system had also stimulated some colonialism before this period (Blaut, 1993).

At some point in the eighteenth or nineteenth century, depending on whom you read, development itself becomes an identifiable process in its own right, underpinning and fusing with imperialism and the expansion of capitalism (Dixon and Heffernan, 1991). Throughout this period, colonies continued to be founded for a variety of complex motives and in many different forms. The particular nature of colonialism varied not only with the motives, but also with contemporary political economies and cultures of both the metropolitan power and colonised territory (Box 2.1). Indeed, in recent years there has been a tendency to define colonialism as much by the means it employed as by the impetus behind it. Thus 'colonialism is often defined as a system of government which seeks to defend an unequal system of commodity exchange' (Corbridge, 1993a: 177), whereas Said (1979, 1993) maintains that colonialism existed in order to impose the superiority of the European way of life on that of the Oriental, a colonisation of minds and bodies as much as of space and economies and 'much harder to transcend or throw off' (Corbridge, 1993a: 178).

Key idea

Colonialism and imperialism

Colonialism and imperialism are not interchangeable ideas, although they overlap considerably. Anthony King (1976: 324), in the index of his seminal text on colonial urbanisation, states simply, 'see colonialism' under the entry for imperialism. Consequently, his definition of colonialism is only partial: 'the establishment and maintenance for an extended time, of rule over an alien people that is separate and subordinate to the ruling power'. The concept of one state establishing political control over another is a recurring theme in definitions of colonialism, for example, 'the policy or practice of acquiring political control over another country, occupying it with settlers, and exploiting it economically' (Concise *Oxford Dictionary*, 1999: 282).

Blauner (quoted in Wolpe, 1975: 231) is more comprehensive than King in his approach, defining colonialism as, 'the establishment of domination of a geographically extended political unit, most often inhabited by people of a different race and culture, where this domination is political and economic and the colony exists subordinated to and dependent on the mother country'. However, once again the specific mode of domination and exploitation is left unidentified and, as with so many analyses of colonialism, the focus is on internal processes at the expense of outward linkages.

A useful working definition of imperialism is, 'a policy of extending a country's power and influence through colonization, use of military force, or other means' (Concise *Oxford Dictionary*, 1999: 711). But there are some serious inconsistencies in the various uses of the term 'imperialism'. One of the most obvious is that the term is employed in two distinct ways: 'a technical sense – to define the latest stage in the evolution of capitalism – and a colloquial sense – to describe the relationships between metropolitan countries and underdeveloped countries' (Bell, 1980: 49). These need not be incompatible, although difficulties of reconciliation between the two approaches have certainly led to contradictions in the chronology of imperialism. Marxist (Leninist) analysts believe that this monopoly stage of capitalism only began around the start of the twentieth century, at least that is what Bell (1980) argues. Barratt-Brown (1974), on the other hand, has extended consideration of imperialism to roughly the last 400 years.

Barratt-Brown's chronology is preferable, largely because it permits a broader definition of imperialism to be used, referring to 'both formal colonies and privileged positions in markets, protected sources of materials and extended opportunities for profitable employment of labour' (Barratt-Brown, 1974: 22). This permits us to examine the way in which the expansion of imperialism affected, and was affected by, the parallel colonisation process in the Americas, Asia and Africa from the early sixteenth century to the decolonisation phases of the 1950s and 1960s.

BOX 2.1

Politics, society and trade in pre-colonial sub-Saharan Africa

Before the arrival of European traders from the sixteenth century onwards, there were many substantial and sophisticated communities throughout Asia, Africa and Latin America. Although we must recognise the achievements of these societies, given the distorted picture communicated by early European trader-colonists (Blaut, 1993), we must not romanticise 'traditional' societies. Many such societies were not repositories of simple communism and were probably rather unpleasant places for most of their inhabitants, with a substantial slave trade in which indigenous chiefs willingly participated. Slavery was widespread before the contact with Europeans. As Hopkins observes in West Africa, 'Slaves were employed as domestic servants, they acted as carriers, they maintained oases and cut rock salt from the desert, they laboured to build towns, construct roads and clear paths, they were drafted as front line troops, and they were common in all types of agricultural work' (Hopkins, 1973: 24). Change was sometimes for the better. In sub-Saharan Africa, the 'traditional' indigenous crops, sorghum and millet, were nutritionally inferior to the manioc (cassava) and maize introduced from the Americas by European traders. According to Iliffe, 'In moist savanna regions maize produces nearly twice as many calories per hectare as millet and 50 per cent more than sorghum. Cassava produces 150 per cent more calories than maize and is less vulnerable to drought' (Iliffe, 1995: 138).

Organisationally, most societies in the immediate pre-European period were (semi) subsistent, but this did not preclude trade or the emergence of large, powerful states. At various times in pre-colonial West Africa a series of empires and kingdoms existed, for example, Ghana, Mali, Songhai in the savanna–sahel region, and Ashanti, Oyo and Benin in the forest region. Much of the power and wealth of these states was based on a flourishing north–south trade in commodities such as gold, ivory and kola nuts from the forest region, salt from the Sahara, and manufactured goods from Europe and north Africa. Busy trade routes developed across the Sahara, linking the north African coast and Europe with sub-Saharan Africa. Certain West African towns grew rapidly as a result of this profitable commerce, most notably Djenné (now in Mali), founded in the ninth century, Gao (Mali), and Kano (Nigeria), all dating from the late tenth century, and Zaria (Nigeria) and Timbuctu (Mali) founded in the eleventh century. Timbuctu became an important religious and educational centre with the first university in West Africa, while Kano, now in northern Nigeria, had some 75,000 inhabitants by the sixteenth century (Binns, 1994a).

Essentially, these societies were structured on two levels. At the local level was a patriarchal agrarian community, where land was allocated and used on a more or less equitable basis and in which the redistribution of surplus took place on a social rather than a market basis. Reciprocity and obligation brought status. At the elite level, status was inherited or was enhanced by wealth accumulated by raids, confiscation or conquest outside the community. Trade in valuable items such as gold or salt was dominated by a small group of elites. Pre-European, pre-capitalist societies were therefore ravaged by constant wars, fought on behalf of the elites in order to control the production or trade of valuable items or to control people. Before the arrival of European capitalism, life was not comfortable for most Africans. Unfortunately, it did not change for the better with the arrival of the market economy and the ensuing revaluation of commodities.

Perceptions of the non-European world

This myriad of possibilities does not mean to say that some common ground cannot be discerned or that broad phases of colonial development cannot be identified. Indeed, in order to make some sense of the colonial discourse, we need a framework within which we can address it, identify its principal processes and set out the legacies that persist within current society, in both developed and developing countries. Preston (1996: 140) has identified several dimensions to what he terms 'the process of absorption and reconstruction of other peoples' (Figure 2.1). However, he does not

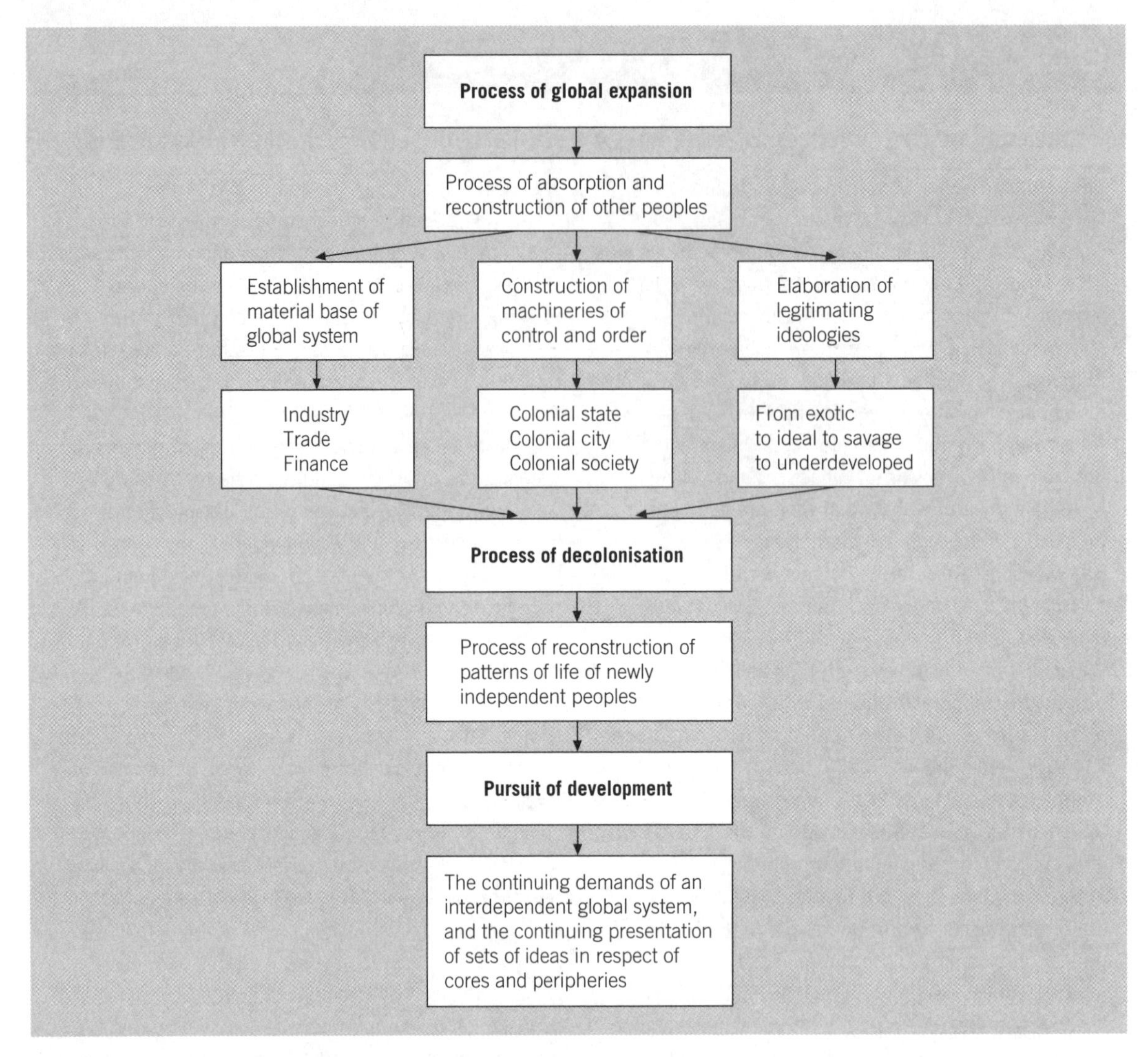

Figure 2.1 Principal processes of colonialism
Source: Adapted from Preston (1996)

elaborate on the links between parallel processes and suggests a rather simplistic sequence of ways in which Europeans represented the non-European world: first as exotic cultural equals, then as representatives of innocence and noble savagery during the Age of Enlightenment, subsequently as the uncivilised savages of the nineteenth century who had to be controlled, then improved and eventually guided to independence. Although this sequence is not untrue, it suggests a set of ideologies that were uniform over space and through time. The reality was, of course, very different, particularly over the long period from the early sixteenth to the early nineteenth century. The Enlightenment, as we noted in Chapter 1, is a period in which the concept of development and the role of the 'enlightened' within this was crystallised. The romantic notion of the noble savage was applied rather sparingly during this period and was very much influenced by the nature of the non-European society encountered. Even Cook on the same journey around the Pacific could both admire the Polynesians and despise Australian Aborigines according to a particular set of British values.

'Waves' of colonialism

The chronological sequence suggested by Preston, despite its oversimplicity, presents an approach that has been used by many other analysts of colonialism, especially by world system advocates such as Wallerstein (1979). Taylor (1985), in particular, has set out very clearly a sequence of waves or phases in which imperialism and colonialism combine to produce a series of long and short waves or cycles of development. The long waves coincide with major economic systems: feudalism, mercantilism and industrial capitalism. The shorter waves, often termed Kondratieff waves after their founder, are said to fit into the long waves in roughly 50-year cycles. All the waves are characterised by phases of growth and stagnation; during the stagnation phases economic restructuring occurs in order to re-establish economic strength. Such restructuring can involve one or more of a variety of actions, from the development of new technologies, through social change to new sources of raw materials or cheap labour. It is alleged that the acquisition of colonies formed part of this restructuring process by giving access to materials, food and labour. The long waves of mercantilism and industrial colonialism can be seen, therefore, to coincide with the rise and fall of major cycles of colonialism (Figure 2.2).

Despite its rigidity, the chronological sequences of world system theory do present a useful framework from which to examine phases of colonial–imperial development and within which the legitimating ideologies, material base and machineries of control and order may be examined. However, as many writers have noted (Kabbani, 1986; Said, 1993), the narratives from which we draw our material to interpret or read colonialism are themselves subject to deeply embedded prejudices that have found expression in both development thinking and development practice, albeit varying through time (Key idea, Post-colonialism). In similar fashion, we need to be cognisant of the fact that the power exercised within colonialism is not homogeneous; it is often diffuse, fragmented, local and, above all, highly personalised.

In this account, *colonialism* is used to refer to the period from the beginning of the sixteenth century onwards in which economic and political motivations fused together to give spatial expression to the accelerating globalisation of capitalism. The phases of

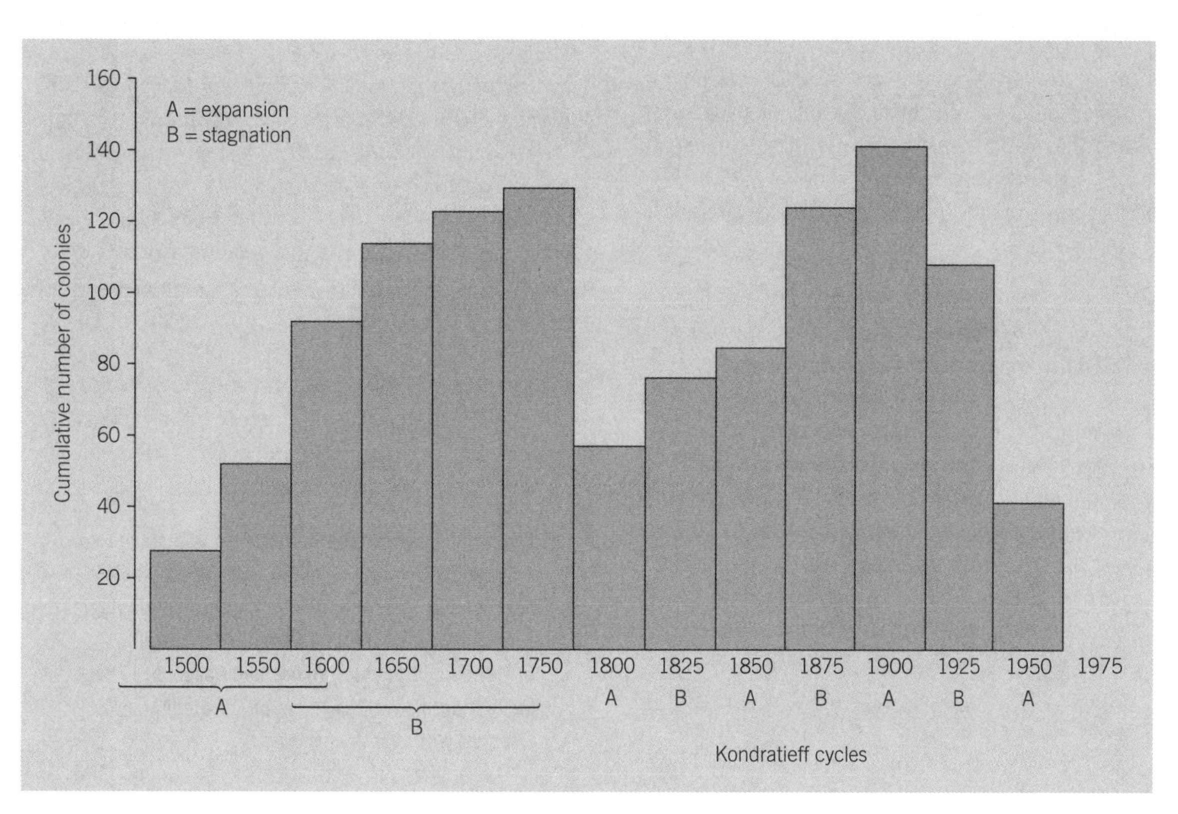

Figure 2.2 Long and short waves of colonialism
Source: Adapted from *Political Geography*, Taylor, P., Pearson Education Ltd © 1985

Key idea

Post-colonialism

In the last two decades there has been much debate across the humanities and social sciences about the nature of colonialism and the perceptions of both the colonisers and those who have been colonised. Much of the debate has surrounded what are referred to as 'post-colonial' perspectives or 'post-colonialism', which is concerned with the effects of the process of colonisation on cultures and societies. These terms are confusing, since in one sense they refer to the post-colonial period, which was 'after' the period of decolonisation and the winning of independence, but in another sense, since the 1980s, they have been more widely used to refer to the political, linguistic and cultural experience of societies that were former European colonies (Ashcroft *et al.*, 1998). Writers on post-colonialism, who adopt a broadly anti-colonial standpoint, critique the popular discourses relating to the period of European colonialism, which they argue are 'unconsciously ethnocentric, rooted in European cultures and reflective of a dominant Western worldview' (McEwan, 2002: 127).

Post-colonial studies attempt to dissect the many attributes of the colonial experience, '(revealing) the historical and geographical diversity of colonialism and the need to ground such critiques in material and specific contexts' (Blunt and Wills, 2000: 170). Importantly, post-colonial studies attempt to appreciate the viewpoints and empathise with those who were marginalised and oppressed.

'Subaltern' studies

Gayatri Spivak, an Indian literary critic and a key figure in post-colonial studies, refers to such marginalised people as 'subalterns', 'men and women among the illiterate peasantry, the tribals, the lowest strata of the urban subproletariat' (Spivak, 1993), for example, lower caste Indian women under British colonial rule. 'Subaltern' is a military term relating to lower ranking officers in the British army, below the rank of captain. Spivak argues that histories of the colonial period, and accounts of the modern world, are usually written by the powerful, invariably men, and the voices of the 'subalterns' are rarely heard. Spivak has played a key role in gathering together a group of Indian historians to form the 'Subaltern Studies collective' who aim to write critical histories of South Asia, 'which followed the traditions neither of imperialist histories of conquest, nor of nationalist histories that charted a singular and linear development of nationalist consciousness.

Subaltern histories have focused on the lives, agency and resistance of those people who had been silenced and erased from both imperialist and nationalist accounts of the past' (Blunt and Wills, 2000: 190). As Ranajit Guha, another member of the Subaltern Studies collective comments, 'The historiography of Indian nationalism has for a long time been dominated by elitism – colonialist elitism and bourgeois-nationalist elitism . . . sharing the prejudice that the making of the Indian nation was exclusively or predominantly an elite achievement' (Guha, 1982: 1).

Edward Said: East and West

A particularly important book, *Orientalism* by Edward Said, first published in 1978, is 'commonly regarded as the catalyst and reference point for postcolonialism' (Gandhi, 1998: 64). Said's book, which examines how the West imagined the East, has a strong cultural focus and is concerned with understandings and images. He shows how Western ideas about the world as a whole are still informed by ideas that were widespread during the colonial period, and he considers how visual and textual representations can shape knowledge about a place, and condition behaviour in relation to that place. Said believes that the West was able to manage the East as colonial, dependent territory because 'orientals' were seen as being in need of Western guidance and guardianship. As he says,

> **Neither imperialism nor colonialism is a simple act of accumulation and acquisition. Both are supported and perhaps even impelled by impressive ideological formations which include notions that certain territories and people require and beseech domination, as well as forms of knowledge affiliated with that domination.**
>
> **(Said, 1993: 8)**

Said is concerned with the views of both the coloniser and the colonised, and how dichotomies developed

Key idea (continued)

and were perpetuated, for example, if the East was static then the West was dynamic; if the East was savage, the West was civilised; if the East was despotic, the West was enlightened.

Homi Bhabha: hybridity, mimicry and ambivalence

Another writer, Homi Bhabha (1994), is concerned with the place of colonised people in these discourses. His work is complex, but in essence he deals with with three concepts: hybridity, mimicry and ambivalence.

- Hybridity is concerned with the fact that Europeans who took their culture with them to the colonies had their beliefs, values and practices affected by the culture of the indigenous people they encountered and vice versa. The end product of the cultural encounter, Bhabha suggests, is neither a fixed and pure European identity, nor a pre-existing Asian, African or Latin American identity.
- Linked to hybridity is the practice of mimicry, where the British and the Indians, Africans and others, adopted aspects of each other's cultures. For example, the colonial encounter between the Indians and the English led to influential Indians emulating the British upper classes with their European clothing, Christianity, private education and sports such as cricket and polo, while Britons acquired a desire to eat Indian food, to build extravagant Indian-style homes, such as Brighton's Royal Pavilion (Plate 2.1), and terms such as bungalow and jodhpur became part of the English vocabulary.
- Ambivalence is reflected in the way that colonial discourse was grounded in an innate assumption of European control and superiority, while at the same time resting on somewhat insecure foundations and anxiety. In relation to individual colonisers there was disgust about the savagery and backwardness of those being colonised, while at the same time a desire to be more like the colonised and to have sex with them (Bhabha, 1994).

As Blunt and Wills suggest,

> **Rather than represent the colonized subject as simply either complicit or opposed to the colonizer, Bhabha suggests the coexistence of complicity and resistance. The hegemonic authority of colonial power is made uncertain and unstable because the ambivalent relationships between colonizers and colonized are complex and contradictory.**
>
> **(Blunt and Wills, 2000: 187)**

Post-colonial geographies

During the 1990s, much interest developed in so-called 'post-colonial geographies'. As Livingstone has suggested, 'Geography was the science of imperialism par excellence' . . . [because] exploration, topographic and social survey, cartographic representation, and regional inventory . . . were entirely suited to the colonial project' (Livingstone, 1993: 160, 170). From its foundation in 1830, the Royal Geographical Society played an important role in shedding light on the 'dark continent' by supporting and reporting on major expeditions to Africa (Plate 2.2). As Binns comments, '[this] fascination with distant and different peoples and environments, together with a burning desire to expand the British Empire and develop world trade (often under the guise of eradicating slavery and spreading the gospel), were the driving forces behind many expeditions' (Binns, 1995a: 310). A strong justification for enhancing the position of geographical education in schools in the late nineteenth century was to educate young people about the Empire. As a leading geographer of the time, Halford Mackinder, wrote, 'We should aim at educating the citizens of the many parts of the British Empire to sympathize with one another and to understand Imperial problems by teaching geography visually, not only from the point of view of the Homeland, but also of the Empire' (Mackinder, 1911–12: 86, quoted in Binns, 1995a). In 1906, the journal *The Geographical Teacher*, aimed at informing school teachers about new developments in geography, contained a fascinating, yet somewhat arrogant and jingoistic, report on recent impressions of northern Nigeria by Louis La Chard,

▶

Key idea (continued)

Plate 2.1 Brighton Royal Pavilion
Source: Corbis/Tibor Bognar.

> **Perhaps in no part of our world-wide Empire has that indefatigable energy and cheerful indifference to depressing circumstances, so characteristic of the Anglo-Saxon race, been better displayed than in the various colonies and protectorates which form our West African possessions. In the very face of disease and death the white pioneer has marched forward and has built, almost in defiance of nature herself, a firm and comparatively healthy basis for the construction of the western cornerstone of our Empire in the Dark Continent.**
>
> **(La Chard, 1906: 191, quoted in Binns, 1995a)**

More recently there has been a call for 'critical, contextual histories of geography that examine the culture of imperialism' (Driver, 1992). Meanwhile, there is also concern from geographers and others that perspectives and writing about the colonial experience have been strongly gendered, 'embodied in exclusively masculine terms of virility and bravery' (Blunt and Wills, 2000: 196), with little attention given to the voices of women. Those women who did travel overseas encountered what today might be regarded as sexist and patronising advice in early travel books about the importance of the appearance and behaviour of the traveller herself (Blunt, 1994). As Driver suggests, 'contemporary writings on geography were infused with assumptions about gender, as well as empire' (Driver, 1992: 28).

Post-colonialism has undoubtedly been an important and thought-provoking perspective on the colonial experience, not least because 'It demonstrates how the production of Western knowledge forms is inseparable from the exercise of Western power. It also attempts to loosen the power of Western knowledge and reassert the value of alternative experiences and ways of knowing'

Key idea (continued)

TENTS

FOR THE COLONIES.

Fitted with VERANDAH, BATHROOM, &c.

As used by most eminent Travellers, and supplied to H.M. Government for East, West, Central, and South Africa, &c.

SPECIAL TENTS FOR EXPLORERS & MOUNTAINEERING

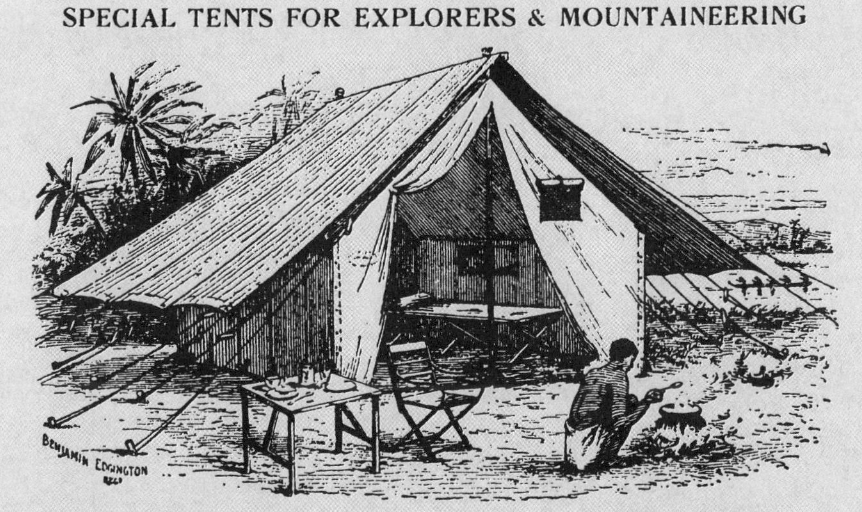

COMPLETE EQUIPMENT.

CAMP FURNITURE WITH LATEST IMPROVEMENTS.
AIR AND WATERTIGHT TRUNKS.
UNIFORMS AND CLOTHING OF ALL KINDS.

Plate 2.2 Tents for the colonies.
Source: Allen (1979).

(McEwan, 2002: 130). The works of Bhabha, Said and Spivak are not always easy to follow, and some would criticise their lack of clarity and their need to explain their ideas with empirical evidence. Meanwhile, Marxists would argue that post-colonialism has too strong a focus on culture at the expense of class, while others would even question the appropriateness of the term 'post-colonial' when the world continues to experience other forms of colonialism.

colonialism identified in Figure 2.3 were common to most parts of the non-European world, but the chronology, rationale and reactions involved varied enormously. In Clapham's (1985: 13) words, there were 'the Americas, both rich and easy to control; Asia, rich but difficult to control; and Africa, for the most part poor and so scarcely worth controlling'. Latin America and sub-Saharan Africa therefore experienced more intensive plundering activities than either Asia or North America, but during very different historical periods (Figure 2.4). Indeed, most of both North and South America passed through these phases into independence,

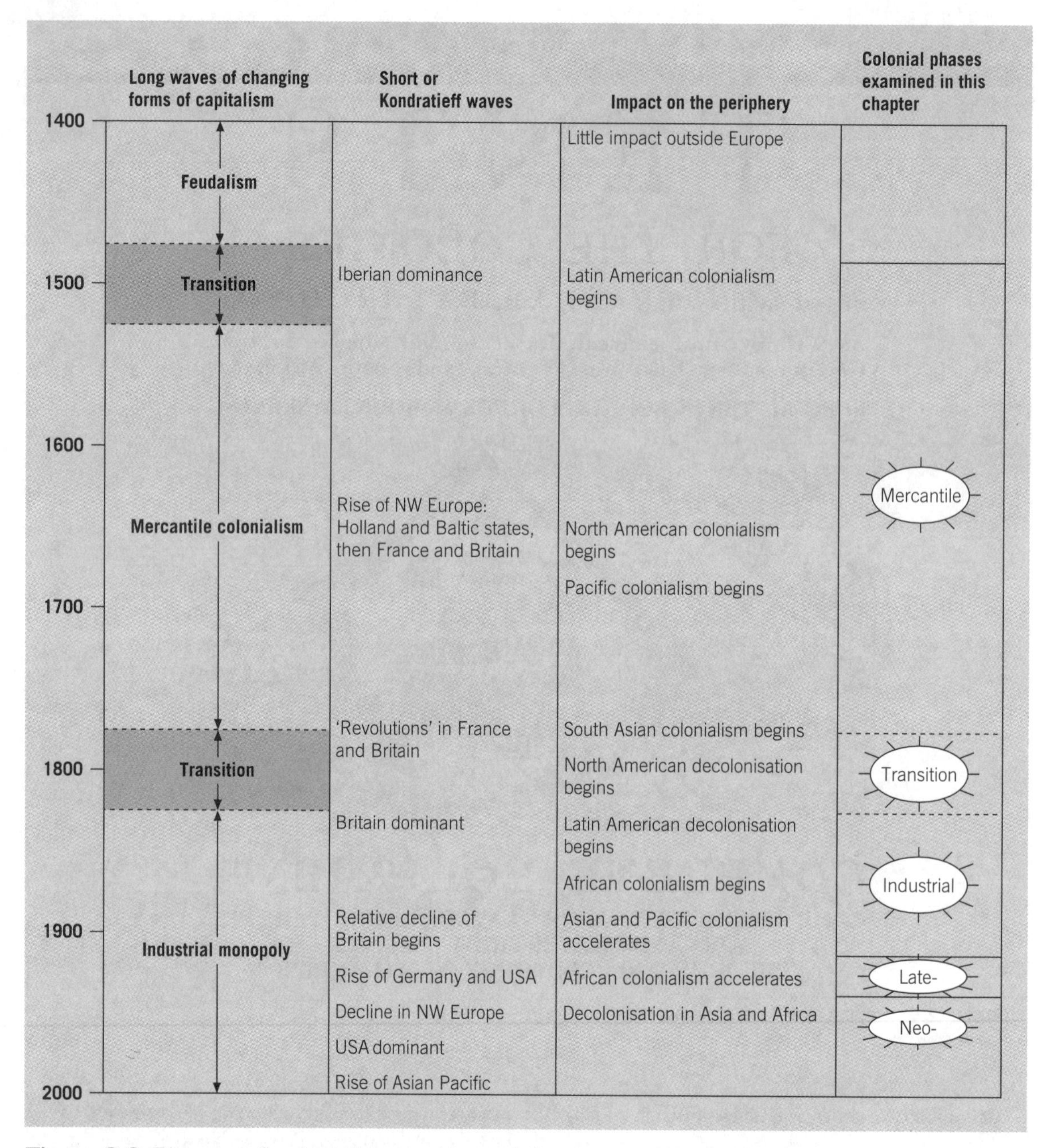

Figure 2.3 Phases of colonialism and imperialism
Source: Adapted from *Political Geography*, Taylor, P., Pearson Education Ltd © 1985, and Bernstein (1992a)

and in the case of South America into neo-colonialism, before the intensive phase of the colonial project began in Asia or Africa. In most of sub-Saharan Africa, formal colonies were a relatively short-lived political process, lasting only around 80 years from the 1880s to the 1960s, although exploitation was present before and after this period. The phases identified here are therefore indicative rather than definitive of the major changes that occurred in the expansion of colonial capital.

Phases of colonialism

Mercantile colonialism

The predominant features of this first phase of colonialism were commerce and trade, although in the earliest stages of contact in Latin America it was plunder and conquest which motivated the conquistadors. Within North America and the Caribbean, trade and commerce were underpinned by production

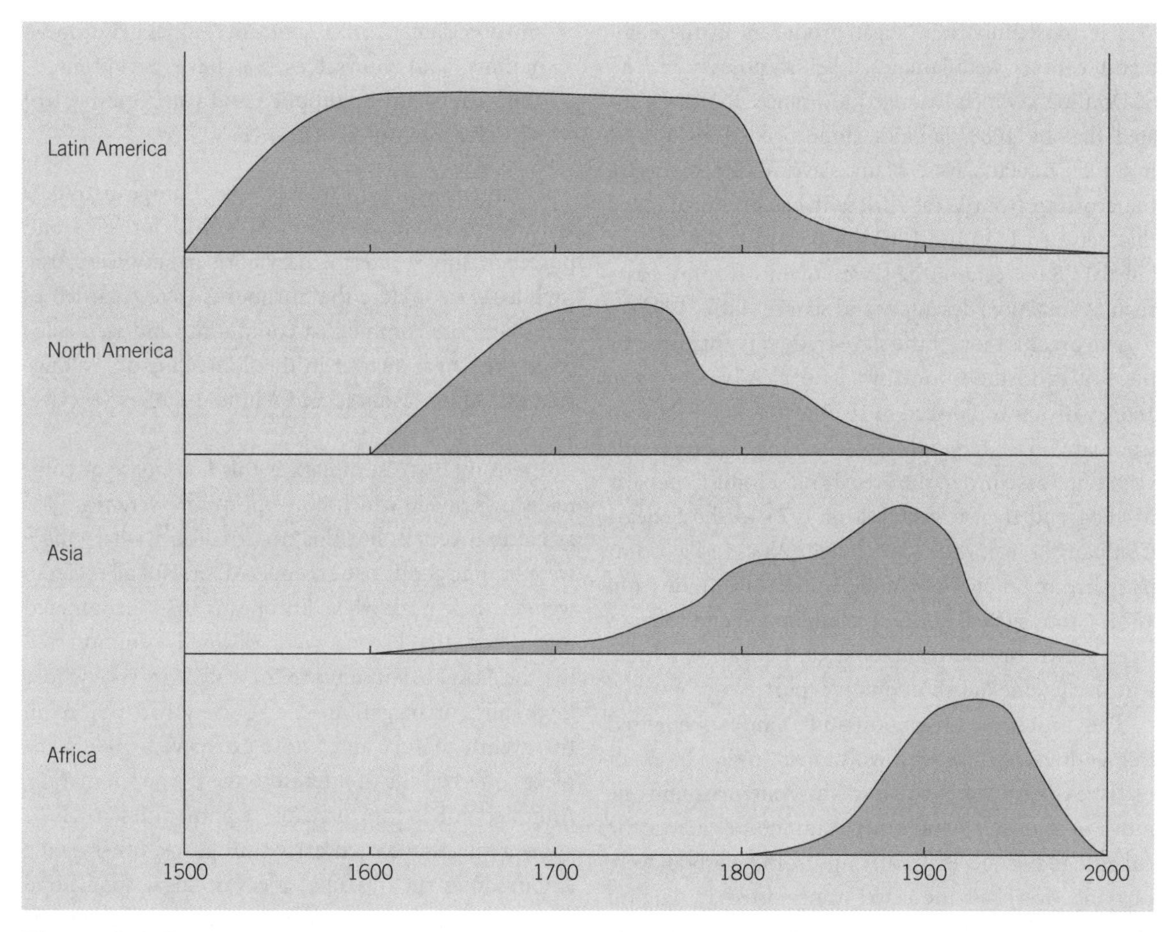

Figure 2.4 Regional colonialism: a chronology of the rise and fall in the numbers of colonies
Source: Adapted from Lowder (1986)

within the plantation system using slave labour (see Chapter 3).

In Africa, and more especially Asia, the initial contact was structured much more around commodity exchange. In this early period, the impact of mercantile colonialism was determined by a variety of factors, including the type of European involvement, the nature of the commodities sought by the Europeans and the strength, culture and organisation of the non-European state.

The plantation system and slave labour

The plantation system and the massive demand for slave labour had a very significant impact on economy and society in Africa and the Americas from the late sixteenth and seventeenth centuries. The plantation system originated in the offshore islands of West Africa during the fifteenth and sixteenth centuries. It was from the Canary Islands that the Spanish took the system to the Caribbean, and from the Cape Verde Islands and the island of San Thomé in the Gulf of Guinea that the Portuguese introduced it into Brazil.

During the seventeenth century, intense competition between the Dutch, English and French in the Caribbean region, and the growing European popularity of sugar in Europe, led to a massive demand for labour to work the plantations, resulting in the development of the slave trade between West Africa and the Americas. A decree issued by Louis XIV of France in 1670 read, 'There is nothing which contributes more to the development of the colonies and the cultivation of their soil than the laborious toil of the Negroes'. Whereas before 1600, an estimated 900,000 slaves had been taken from West Africa to the Americas, in the seventeenth century this trade, initially led by the Dutch, increased to 2.75 million, and in the eighteenth and nineteenth centuries rose further to a massive 7 million and 4 million, respectively (Oliver and Fage, 1966).

The most important sugar producers in the eighteenth century were Jamaica, a British possession, and St Domingo, which belonged to France. It was estimated that by 1688, Jamaica alone needed an annual input of 10,000 slaves. Many slaves did not survive the crossing from West Africa. It has been estimated that between 1630 and 1803 the average Dutch voyage killed 14.8 per cent of the slaves, mainly from diseases such as smallpox, dysentery and scurvy (Iliffe, 1995).

The organisation of the slave trade was entrusted by the British to the Company of Royal Adventurers in 1663, which was replaced in 1672 by the Royal African Company. The monopoly of the French slave trade was at first assigned to the French West India Company in 1664 and then transferred in 1673 to the Senegal Company. Meanwhile, the Dutch West India Company had the monopoly of the Dutch slave trade from 1621. Crops such as coffee, cotton, indigo and tobacco were grown on the plantations, but sugar was overwhelmingly the most important export.

The cities of Liverpool and Nantes benefited tremendously from their involvement in the shipment of slaves from West Africa to the Americas, and the movement of sugar and other commodities across the Atlantic to Europe. By 1750, ships from Liverpool were carrying over half the slaves transported in English vessels and the Liverpool slavers had acquired a reputation for ruthless efficiency compared with their rivals in cities such as Bristol (Hopkins, 1973). There is much debate among historians concerning the total number of slaves transported from West Africa during the entire period of the Atlantic slave trade, but a figure somewhere between 8 million and 10.5 million seems appropriate, thus representing one of the greatest migrations of all time (Hopkins, 1973).

The intensification of trading links

The organisation of the non-European state varied enormously and European traders found themselves in contact with societies whose ways of life were in material, administrative and spiritual terms often far superior to their own. China, for example, thought little of the European goods brought in trade, and in 1793 Emperor Chen Lung condescendingly informed George III's emissary Earl Macartney that,

> **our celestial empire possesses all things in prolific abundance and lacks no product within its borders. There is therefore no need to import the manufactures of outside barbarians. . . . But as tea, silk and porcelain . . . are absolute necessities to European nations, and yourselves, we have permitted . . . your wants [to] be supplied and your country [to] participate in our beneficence.**

For many years up to this date, European traders were forced to exchange their goods for silks and porcelain only at intermediary ports in Southeast Asia, such as Macao. Here the Europeans were regarded as just one more commercial community and were allocated their own quarter in the flourishing port alongside the Arab, Javanese and Chinese traders (McGee, 1967).

Preston (1996) summarises this first phase of colonialism as one in which non-Europeans were regarded as cultural equals, but this occurred only where there were trading goods to be competed for. Not all societies were at their peak when Europeans first encountered them, and the French were distinctly unimpressed by the Angkorian empire of the Khmers, whose 'hydraulic' or irrigation-based society had peaked in the twelfth century and whose extensive temple complexes offered little of interest to the European market. Although inherently precious commodities, such as gold or silver, attracted early Europeans, other exotic commodities such as silk, spices or sugar soon lured many adventurers to particular parts of the world. Initially, trade with these 'distant others' was a high-risk enterprise into which vast sums were invested and from which huge profits were realised. To varying degrees, these trading adventurers were accompanied by other kinds of Europeans, such as missionaries, emissaries or even scientists, curious about the non-European world.

The mercantile phase of colonialism in Asia and Africa lasted for some considerable time without extensive European settlement and with no uniform sign of the dominant–subordinate relationship which was to come later. In Africa,

> **For centuries Europeans knew the coastline of Africa but not the interior. Climate, tropical diseases, Islam and resistant Africans deterred exploration. There was also a lack of interest, partly because trade at the coast was adequate and partly because there was little spirit of curiosity.**
>
> **(Griffiths, 1995: 30)**

But this was to change in the late eighteenth century. In the Americas, the situation was quite different, with intensification of trade in the seventeenth and eighteenth centuries accompanied by much more extensive settlement from Iberia, France and Britain. Moreover, in North America and the Caribbean, the colonisers were heavily involved in the production process, something which did not occur in Asia and Africa until much later. However, as trade with these two continents grew in both volume and value, so it became more organised in its structure, usually within the context of the trading company. In Asia, it was the Dutch that began this trend in the seventeenth century, and soon the other European nations had their own East India companies too.

The acceleration in the scale and organisation of mercantile colonialism not only expanded profits, but also involved increased European commitment to a physical presence in the trading region where commodities were to be assembled, stored and protected (Plate 2.3). In order to acquire both commodities and protection, Europeans involved themselves increasingly in local politics, making alliances and inciting conflicts, all of which had enormous repercussions. In much of Africa the slave trade dramatically increased the power of locally based traders and indigenous chiefs often living well away from the coastal area, intensifying conflicts because of the rewards that could be achieved through the sale of prisoners into slavery.

Although at this time the Europeans had only a relatively small physical presence in much of the non-European world outside the Americas, this varied enormously. Re-victualling posts were, however, established across the globe as part of a vital network for supporting exploration and commerce. For example, European settlement in South Africa was initiated in 1652, when the powerful Dutch East India Company ordered that a supply station for passing ships be established by a group of Company employees in the sheltered harbour of what is now Cape Town. In that year, Jan van Riebeeck, the commander of the outpost, organised the setting up of facilities for growing fresh fruits and vegetables in the so-called 'Company gardens', while relations were developed with the local Khoikhoi people to supply cattle for meat in exchange for European products (Lester *et al.*, 2000; Plate 2.4).

But even with limited European settlement, change had occurred on an extensive scale by the end of the eighteenth century. The extended trading networks had increasingly drawn many parts and peoples of the non-European world into the capitalist system. European goods, particularly weaponry, European values and ideas, religious and secular, had penetrated most regions. Even where direct impact was still relatively limited, change occurred; for example, in Siam where the present Chakri dynasty was established in the mid eighteenth century through a series of reformist, modernising monarchs who sought to resist the Europeans by becoming more like them. For the great mass of peasants in Asia or Africa, however, life seemed to continue as it had for thousands of years,

Plate 2.3 Macao: remnants of Portuguese presence during the mercantile colonial period
(photo: David Smith)

Plate 2.4 Cape Dutch style home in Graaff-Reinet, South Africa
(photo: Tony Binns)

but their activities, whether subsistence or market-oriented, had over the long mercantile colonial period been subtly linked to a fledgling world economy, the core of which lay in Europe.

The transition to industrial colonialism

The mercantile colonial period merged into the era of industrial colonialism in a highly differentiated transition period. In North America, the USA had decolonised itself with the declaration of independence in 1776 and was preparing to become an enthusiastic and powerful metropolitan power in its own right. In Latin America and the Caribbean, colonial production and trade were beginning to be challenged from within, if not by indigenous peoples. In Asia, the East India companies were going bankrupt, as their shift into commodity production, in order to ensure quantity and quality of supplies, had escalated their costs of administration and protection. In Europe itself, political revolutions and continental-scale war consumed state resources and attracted the individual adventurers who had underpinned much of the mercantile colonialism. But, above all, Europe offered new and lucrative profits for the reinvestment of accumulated merchant capital in its accelerating industrial transformation. Even so, the nineteenth century witnessed the establishment of colonial concessions related to trade and commerce, in parallel with the broader changes wrought by state-structured colonialism. Thus, the trading islands of Penang (1786), Singapore (1819) and Hong Kong (1841) were acquired, respectively, by Francis Light, Thomas Stamford Raffles and Charles Elliot on behalf of the crown rather than their companies, as were the treaty ports in China (Guangzhou, Xiamen, Fuzhou, Ningbo and Shanghai) following the Treaty of Nanking signed by Britain and China in 1842.

But if mercantile colonialism had begun to fade, its impact was already fuelling the Industrial Revolution and the renewed burst of colonialism that began in the nineteenth century. The fortunes that had been made from plunder, from commodity trade and, particularly, from the triangular trade between West Africa, the Caribbean and Europe, were underpinning the accelerating industrial age. As Blaut (1993) argues, the point is not just that profits had been made, but that they were in the hands of a new breed of entrepreneur rather than the old elite. The mercantile colonial period not only created new money, but it was accompanied by a social and political revolution, the combination of which gave Britain and other European powers a strong platform from which to launch into a more spatially extended and economically intensive form of colonialism. In short, by the late eighteenth and early nineteenth century, 'capitalism arose as a world-scale process: as a world system. Capitalism became concentrated in Europe because colonialism gave Europeans the power both to develop their own society and to prevent development from occurring elsewhere' (Blaut, 1993: 206).

Industrial colonialism

Certain changes characterised the colonialism of the nineteenth and twentieth centuries. The first was related to the dynamics of capitalism itself. Although commerce and trade still made money for the merchants of Liverpool, Bristol and London, the manufacturers themselves were eager to find methods of expanding production, or at least stabilising costs and extending their profits. Two obvious ways were to seek expanded and/or cheaper sources of raw materials and to find new markets overseas. A further development was to expand the production of cheap food overseas, thus lowering the costs of labour production in Europe by keeping wages down. Although markets took a while to develop, all of these and more were made available in the restructured colonies of the nineteenth century. These colonies were established and organised by the state rather than the company, although business and the state worked together through their representatives to transform production, consumption and cultures. The key to this process was territorial acquisition. Before 1870, annexation and occupation tended to follow resource exploitation, whereas after 1870 they tended to precede it.

Although the needs of capitalism may have been the driving force behind the industrial colonialism of the nineteenth century, the rationale for the colonial project itself was provided by a consolidation of the ideology of justifiable intervention and occupation of what had become either 'uncivilised savages' or traditional groups whose history was ignored and whose societies and activities were seen as either static or disintegrating (Box 2.2).

Science, reason and, above all, organisation for most nineteenth-century thinkers elevated Europeans to their 'superior' position and placed them above the brutality and poverty of the peoples in their occupied lands. For Porter (1995) there were 'master metaphors' provided by physics (stability, equilibrium) and biology (constituent parts functioning for the whole) that shaped the ideologies of both colonialism and development. These gave rise to a modernist theme, a universal process of change which is clear and predetermined (Porter, 1995). Sympathetic motives could therefore be written into this process underpinned by a parent–child metaphor (Manzo, 1995), often expressed vividly by the image of the 'mother country' and her fledgling colonies.

BOX 2.2

The scramble for Africa

Although colonialism expanded rapidly throughout the nineteenth century, the speed of Africa's partition was new. The 'scramble for Africa' usually refers to the 30-year period between 1884 and 1914 when most of Africa was partitioned among the European powers. The key event in this period was when Bismarck called the European powers to Berlin in 1884 to draw up rules to regulate the partition of Africa. The General Act of the Conference of Berlin was signed in February 1885, which was concerned with 'the development of trade and civilization in Africa; the free navigation of the Rivers Congo, Niger, etc; the suppression of the slave trade by sea and land; the occupation of territory on the African coasts' (Griffiths, 1995: 38). However, even before the Berlin Conference, some of the continent had already passed under the control of the imperial powers, through various treaties with African leaders. One of the key factors motivating the conference was undoubtedly the speed of developments in South Africa, where diamonds were discovered in 1868 just across the Cape Colony's northern border in Griqualand West, and in the following year rich reserves were found in Kimberley. Britain annexed the tribal territory to prevent costly warfare generated by settler land-grabbing and security concerns on the Cape's borders. Then, in the year of the Berlin Conference, 1884, vast quantities of gold were discovered on the Witwatersrand, leading to the rapid expansion of mining activities and the mushrooming of the city of Johannesburg. European, and particularly British, investment in South African mining was massive, and between 1887 and 1898

▶

BOX 2.2 (continued)

gold production increased in value from £80,000 to £16 million, representing one-quarter of the world total (Lester *et al.*, 2000).

We can summarise the processes, both within Africa and outside, which gave rise to the scramble for territory. Some of these processes were chronologically or spatially specific.

External processes

1. Between the first and second industrial revolutions (i.e. the shift from coal and iron to oil, electricity and steel), Europe went through a deep recession. As rates of profit fell, European firms began to seek new material sources, new markets and new investment opportunities on an extensive scale, partly to forestall other European rivals.
2. The last quarter of the nineteenth century saw the newly united countries of Italy and Germany using colonialism to sidestep internal tensions. As Africa was at the time the largest uncolonised area, it became the focus of a national scramble for territory and prestige, with France seeking compensation for its defeat (1871)

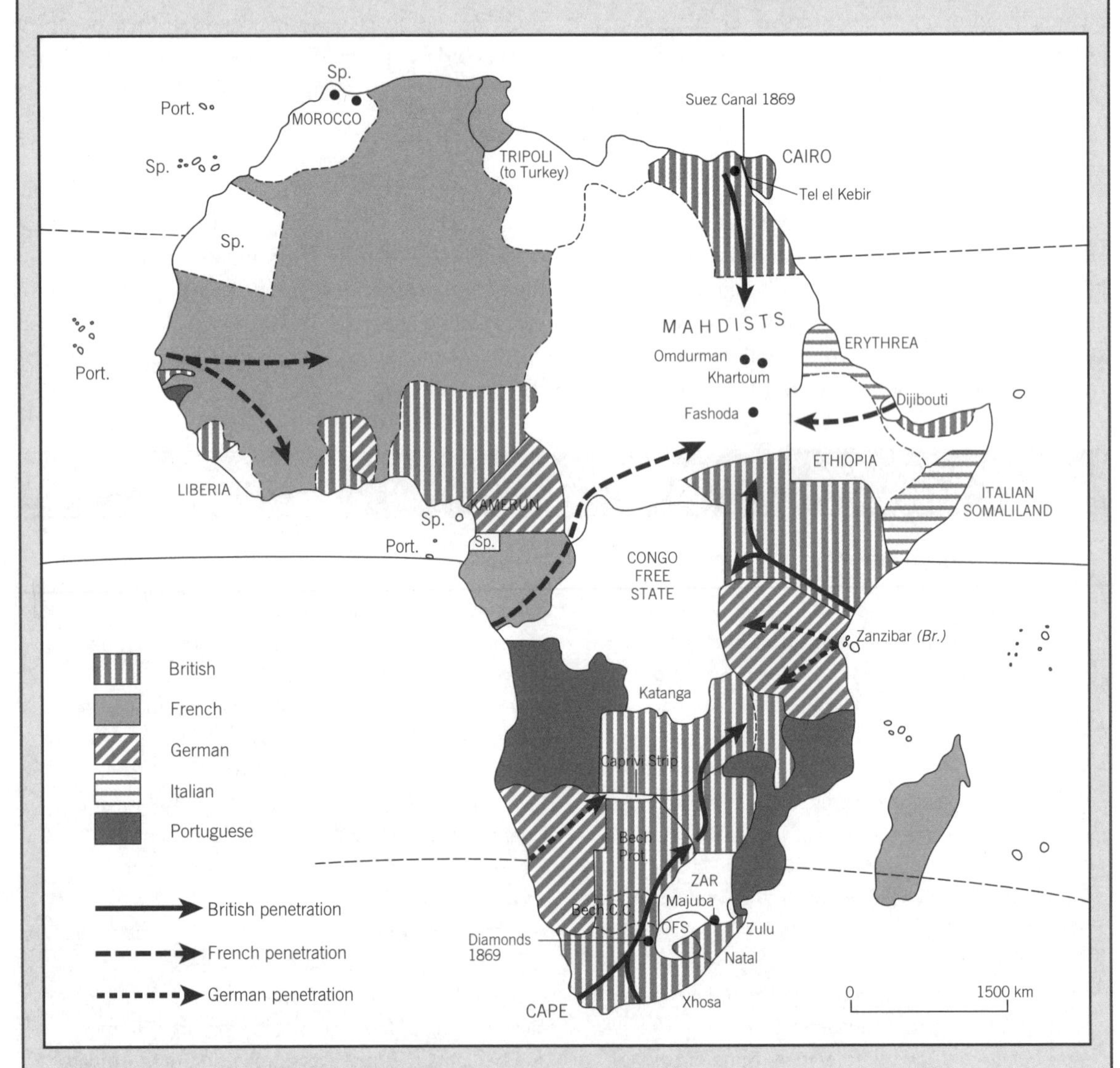

Figure 2.5 The scramble for Africa

Source: Adapted from *The African Inheritance*, Griffiths, I.L. Copyright (© 1995) Routledge. Reproduced by permission of Taylor & Francis Books UK.

BOX 2.2 (continued)

by Germany within Europe. In this process, governments were supported by a popular imperialism created and sustained by a jingoistic media boom in newspapers, journals and books.

3. These processes were facilitated by a technological revolution, particularly in transport, where steamships, railways and telegraphic links accelerated both decision making and physical advances into Africa. New armaments, such as machine-guns, facilitated this process by smaller and smaller European forces.
4. Once acquired, many colonies were also seen as healthy places for the surplus European population that technological advances in medicine and hygiene were beginning to produce. The Kenyan highlands, for example, had a pleasant climate and proved to be a particularly attractive place for white settlers (mainly British), accessing the area by the railway that was constructed from Mombasa on the coast, through Nairobi to Lake Victoria. The settlers eventually took some 18 per cent of the colony's best agricultural land. Further south in Southern Rhodesia (now Zimbabwe), white settlers 'seized one-sixth of its land during the 1890s, mostly on the central highveld and including almost the entire Ndebele Kingdom, together with most of its cattle' (Iliffe, 1995: 205). In North Africa, by 1914, Europeans (mainly French) owned 920,000 hectares of land in Tunisia, and had also settled in large numbers in Algeria and Morocco. There was also considerable expansion of white settlement in South Africa during the nineteenth and early twentieth centuries.

Internal processes

1. The acceleration and intensification of European capitalism brought about a breakdown of existing relationships between traders and African societies, goading many of the African societies into reaction and providing excuses for further European invasion.
2. Sub-imperialism occurred when European settlement became extensive and decision making was wrested from the metropolitan centre by ambitious local individuals or groups, forcing retrospective recognition of highly personalised adventurism. Cecil Rhodes provided the most blatant example of such actions, a visionary entrepreneur who amassed great wealth from mining and business deals, and dreamed of a 10,000-km railway from the Cape to Cairo to be built, with imperial as well as commercial considerations, across territories coloured red on the map (Griffiths, 1995) (see Key thinker, Cecil Rhodes; Plate 2.5).
3. Some African groups facilitated and accelerated the colonial process by 'inviting' Europeans to 'collaborate' against other groups. Often, however, such invitations were manufactured. Taken individually, no single reason explains the sudden scramble for Africa, but the conjuncture of many factors in the 1870s and 1880s gave rise to a spiral of European ambition and nationalism that, once started, proved difficult to stop.

The drive for profit and prestige

We should be careful not to overemphasise the power of ideology and discourse within this process; colonialism had at its heart the economic drive for profit. Every Sir Alfred Milner had his Cecil Rhodes whispering or bellowing into his ear about the returns to investment that will follow annexation and control of yet another piece of territory occupied by traditional people not using it to its full potential. As the explorer David Livingstone wrote to Sir Roderick Murchison, President of the Royal Geographical Society, in 1855, justifying the funding of future expeditions, 'The future of the African continent will be of great importance to England in the way of producing the raw materials of her manufactures as well as an extensive market for the articles of her industry' (quoted in Pachai, 1973: 30). To be sure, there was also prestige for the 'mother country', annexation would be one in the eye for other European rivals and would promote pastures new for grazing missionaries of the true church; but the underlying impetus was usually greed.

Key thinker

Cecil Rhodes (1853–1902)

Cecil Rhodes was born the son of a vicar in Bishop's Stortford, Hertfordshire, England in 1853. He first visited South Africa in 1870, where his older brother, Herbert, had a cotton farm in Natal. In 1871 Rhodes left the colony to go west to Kimberley where diamonds had been discovered. He returned to study at Oxford in 1873, but his studies were repeatedly interrupted by frequent visits to South Africa. In 1880 he launched the De Beers Mining Company and by 1888 he had secured a monopoly of the Kimberley diamond production and had amassed an enormous personal fortune. In 1888 he tricked Lobengula, king of the Ndebele of Matabeleland into an agreement by which Rhodes secured important mining concessions in Matabeleland and Mashonaland. He exploited these concessions through his British South Africa Company, which was given a charter by the British Government in 1889, and soon established complete control of the territory.

From 1877 until his death in 1902 Rhodes represented the constituency of Barkly West in the Cape House of Assembly and was Prime Minister of the Cape Colony from 1890 until 1895. Rhodes has been accused of being racist, and during his time as Prime Minister he introduced legislation to push black people from their lands and to restrict the franchise to literate persons, thus reducing the African vote.

Rhodes played a key role in British imperial policies in southern Africa. He was deeply committed to the expansion of the British Empire and yearned for a railway from the Cape to Cairo constructed entirely on land belonging to Britain. In his will he said of the British, 'I contend that we are the finest race in the world and that the more of the world we inhabit the better it is for the human race'. Through close collaboration with agents of the British Government in South Africa, his imperial ambitions and capital investment progressed simultaneously. By 1894, the British South Africa Company controlled an area of 1.1 million sq km between Lake Tanganyika in the north and the Limpopo River on South Africa's northern border. The following year, the name of this territory was changed from 'Zambesia' to 'Rhodesia', which later became Northern and Southern Rhodesia and after independence, respectively, Zambia and Zimbabwe.

Rhodes died in the Cape in 1902 and was buried in the Matobo Hills in present-day Zimbabwe. He left over £6 million and at the time was one of the richest men in the world. His will provided for the establishment of the Rhodes Scholarships, to enable students from territories under British rule, formerly under British rule, or from Germany, to study at Oxford University. More than 80 scholarships are now awarded annually to both men and women from the former British colonies, the USA and Germany.

Plate 2.5 Cecil Rhodes
(photo: Alamy Images/Popperfoto)

Strengthening control over the colonies

Whatever their nature, ideologies need to be translated into action and for colonialism this was through the elaboration and enablement of, first, its material base, i.e. production, trade and finance; and, second, the establishment of the administrative machinery of control and order. There is no necessary sequence in this process, annexation and the provision of administrative structures could follow economic interests, as in the Transvaal (South Africa), or could precede them, as in the French occupation of Indo-China. There was, however, clearly an expanded role for the state vis-à-vis the trading company in the administrative system.

In contrast to the mercantile period, the main medium of exploitation was not the trading concessions, although they continued to be squeezed out of 'independent' states, but rather the acquisition of land on which to organise the mechanics of production. The colonial state then established the infrastructure of legal, transport, administrative and police systems through which the pursuit of wealth and order could be controlled. It is no coincidence that Sir Harry Johnston (Box 2.3) in his address to the Royal Geographical Society in 1895, following a tour of duty as Commissioner in British Central Africa (now Malawi), attributed the transformation of Mlanje District to the fact that the natives 'above all, are trained to respect and to value settled and civilised government' (Crush, 1995a: 2).

But if the colonial state was an administrative state, it was usually a productive state too, since it was regarded as right and proper for the metropolitan state, metropolitan companies and metropolitan individuals to secure a profitable return on their investment. Ensuring such a return from colonising Africa was for some a daunting challenge. As Lord Lugard, Governor-General of Nigeria admitted,

> **Neither the Foreign Office nor the Colonial Office had any experience of Central African conditions and administration, when, at the close of the nineteenth century, the summons for effective occupation compelled this country to administer the hinterlands of the West African colonies, and to assume control of vast areas on the Nile, the Niger, the Zambezi, and the great lakes in the heart of Africa.**
>
> **(Lugard, 1965: 607)**

The production of export commodities

The spatial expression of economic exploitation was experienced for the most part in the rural areas in which the export commodities were produced. This varied substantially according to the nature of the commodity, local customs and the metropolitan power involved. In some areas agricultural restructuring occurred through the creation of large-scale plantations, whereas in others local producers were encouraged to amalgamate their holdings. Both processes resulted in widespread landlessness, creating labour pools for the new commercial holdings. Lonsdale and Berman (1979) report how the Kenyan landscape was transformed by the colonisers:

> **In the 1880s the inland areas of Kenya comprised a web of subsistence economies which exploited complementary ecological niches suited either to predominantly pastoral or predominantly agricultural forms of production. Between cattlemen and cultivators there was a symbiotic exchange of commodities and intermittent adjustment of populations . . . Three decades later the economic and political structures of the region had been subject to profound transformation, under the sway of a state apparatus linking them to the capitalist world economy. Maasailand was now the core of the White Highlands.**
>
> **(Lonsdale and Berman, 1979: 494–5)**

Others were shifted into agricultural or mining industries by new taxes that often forced farmers to migrate into wage labour to meet these demands, often ruining prosperous and well-organised indigenous systems. In many parts of Africa, poll taxes and hut taxes were introduced and, 'tax evasion was brutally discouraged and could lead to harsh punishment and forced labour' (Binns, 1994a: 10). Where local labour proved to be 'inadequate' for commercial agriculture, workers were often imported from elsewhere in the country (as in Vietnam, where the French shifted workers from north to south), or from overseas (as in Malaya where the British brought in workers from India). However, labour also moved 'voluntarily', recruited through family or kinship systems (e.g. from south China to the Malayan tin mines).

The new agricultural systems often meant that, over large areas, the range of crops produced was narrowed to those commodities required by metropolitan industries, such as cocoa, coffee, cotton, groundnuts, palm oil, rubber, sisal, sugar and tea. Colonies thus became associated with the production of one or two items, being forced to import whatever else was needed; 'Economically, colonialism programmed [African] countries to consume what they do not produce and to produce what they do not consume' (Binns, 1994a: 5). Needless to say, metropolitan firms were in control of both directions of trade. Although some of these commodities were new introductions to the colonies, such as rubber or coffee in various Southeast Asian countries, more traditional crops continued to play an important role, for example, coconuts and groundnuts. Local food crops too became an important export crop.

BOX 2.3

The nineteenth-century logic of colonialism

The discourse of colonialism which first justified and then ratified colonial intervention is well expressed by Jonathan Crush (1995b) in his edited book *Power of Development*, where he caricatures the transformation of Mlanje District in British Central Africa (now Malawi) through the eyes of Sir Harry Johnston, its Commissioner (Crush, 1995a: 1–2). Johnston's first description is of Mlanje in 1895 before colonialism extends its benign hand to that unfortunate land:

> **In the Mlanje District there was practically chaos . . . the few European planters were menaced in their lives and property, and the only mission station had to be abandoned . . . throughout all this country there was absolutely no security for life and property for natives, and not over-much for the Europeans . . . Everything had got to be commenced.**

This picture of scorned opportunities was blamed on disinterested local tribes and evil-minded slave traders and was contrasted by Sir Harry Johnston with the scene after just three years of British rule as a placid paradise where:

> **The natives who pass along are clothed in white calico . . . A bell is ringing to call the children to the mission school. A planter gallops past on horseback . . . long rows of native carriers pass in Indian file, carrying loads of European goods. You will see a post office, a court of justice, and possibly a prison, the occupants of which, however, will be out mending roads under the superintendence of some very businesslike policeman of their own colour . . . The most interesting feature in the neighbourhood of these settlements at the present time is the coffee plantation, which, to a great extent, is the cause and support of our prosperity.**
>
> **(Crush, 1995a: 1–2).**

The influential colonial statesman Lord Frederick Lugard had an interesting perspective on the objectives of colonialism. He asserted that there was a need to fulfil what he called a 'Dual Mandate' delivering mutual benefits for coloniser and colonised. In his book of the same title he argued:

> **Let it be admitted at the outset that European brains, capital and energy have not been, and never will be, expended in developing the resources of Africa from motives of pure philanthropy; that Europe is in Africa for the mutual benefit of her own industrial classes, and of the native races in their progress to a higher plane; that the benefit can be made reciprocal, and that it is the aim and desire of civilized administration to fulfil this dual mandate.**
>
> **(Lugard, 1965: 617)**

Siamese rice, for example, was exported to many other Asian countries, largely through British firms, where it helped lower the cost of labour reproduction, particularly in the cities (Dixon, 1998).

Expanding markets

As a result of the drastic economic, social and demographic changes of industrial colonialism, the last quarter of the nineteenth century also witnessed the acceleration of market potential for Western manufactured products. Indeed, in Pacific Asia it was the purchasing power of the Chinese and Japanese markets that was as important as access to their products in encouraging the Western powers in their almost frantic attempts to gain trading concessions. In the colonies themselves, the initial markets for Western goods were confined to wealthy expatriate and indigenous elites. But the quality and price of these goods and the demonstration effect of purchases by the wealthy soon resulted in imported commodities dominating the expenditure pattern of all social groups, even those in rural subsistence, thus further destroying the indigenous artisan economy and increasing dependency on the West.

Particularly poignant in these circumstances was the re-export of cheap food to the growing markets among the urban and rural poor. Their diet of flour, sugar and tea often had colonial origins, but was processed (and value-added) in Europe, thus facilitating a double exploitation of the colonial poor: first, through their labour in growing the crops, and then through their subsequent purchase of it at exorbitant prices.

Critical reflection

Why was manufacturing industry discouraged in the colonies?

In many areas in the colonies where indigenous manufacturing posed a real threat to imported manufactured goods from Europe, local industries were quickly suppressed, as in the Indian textile towns (Blaut, 1993). The corollary of this situation is that manufacturing was relatively limited during the industrial colonial phase. Any manufacturing that existed was largely concerned with the preliminary processing of primary products, such as rice milling or tin smelting. Most of the more sophisticated processing, and the creation of profits, occurred within the booming industrial areas of the metropolitan country. This is the era during which the big dockside manufacturing plants for tobacco and sugar proliferated in Liverpool, Glasgow and London.

With reference to specific examples, consider the nature of manufacturing industry today in former colonies. To what extent is this still dominated by the preliminary processing of primary products which was a feature of the industrial colonialism phase?

It would not be correct to assume that colonial cities were simply points of control and administration. Although few were centres of production, commercial activity, ranging from the manufacture of small consumer goods to the retailing of imported products, was very extensive. Much of this activity was in the hands of non-Europeans. This is not the same as saying that they were the prerogative of local entrepreneurs, because almost all of the colonial powers in East Africa and in Pacific Asia, other than the Japanese, made a point of encouraging or permitting immigrant groups, usually Chinese or Indian, to infiltrate and monopolise local commerce. In this way, a convenient demographic, cultural and economic buffer was placed between the colonised and the colonialists. Discontent on the part of indigenous populations with the cost of living was therefore often directed against those who were immediately available, rather than those who were ultimately responsible.

The period from 1850 to 1920 saw a massive restructuring of urban systems (Drakakis-Smith, 1991). Colonial production may have been based in the countryside, but colonial political and economic control was firmly centred on the city. Usually just one or two centres were selected for development, giving rise to the urban primacy which still characterises many developing countries today (see Chapters 3 and 9). In some cases completely new cities were built, such as Kaduna in Northern Nigeria, which served as the colonial administrative capital of the region. Within these cities, despite the numerical dominance of the indigenous populations, most of the land space was given over to European activities. Spacious residential and working areas were paralleled by extensive military cantonments, all physically separated from the usually cramped, crowded indigenous city by railway lines, parks or gardens. Little face-to-face contact took place between the colonisers and the colonised, except within a dominant–subordinate relationship. It was a situation that seemed as though it would go on for ever, but the First World War intervened and widespread changes ensued. Within a generation, the political world order of 1914 was totally undermined and the sun began to set rapidly over the colonial empires.

Late colonialism

A fundamental change occurred in the ethos of colonialism after 1920. Put simply, the 'heroic' age of creating empires gave way to a more prosaic phase of imperial governance. The key to this change was the concept of 'trusteeship', which had permeated the formation of the League of Nations and which elevated to a high priority the well-being and development of colonial peoples. In practice, this did not necessarily mean indigenous colonial peoples. Indeed, prevailing anthropological theory conveniently explained that such progress was impossible for 'backward' and 'traditional' societies which did not hold in proper esteem social values such as democracy or the business ethic. Not until after 1945 did metropolitan governments seriously consider fairer representation for indigenous interests, but this was too little, too late.

Styles of colonial rule varied (Box 2.4), but between the wars colonial government was dominated by bureaucrats, both in metropolitan capitals and overseas, striving on behalf of the colonies, with little appreciation either of indigenous aspirations or of the changing world economy in which they were situated.

BOX 2.4

Styles of colonialism in Africa

The European powers had different ideas about colonialism and the way their African colonies should be ruled. The general philosophy was to establish and maintain a level of order, if necessary with swift military imposition, and to do this as cheaply as possible, while at the same time generating as much funding as possible from taxes and exports to support the metropolitan power.

British policy was to adopt a pragmatic and decentralised approach to governing its African colonies. The strategy of 'Indirect Rule', introduced by Lord Frederick Lugard into Northern Nigeria, 'became the accepted ideal of British colonial administration' (Fage, 1995: 394). Britain feared the problems which might follow the possible demise of indigenous cultures, and so maintained traditional institutions and African rulers, wherever possible working through these leaders who were delegated the difficult task of keeping order over the mass of the people. If necessary, military forces were used to defeat ruling emirs without destroying their administrations. Although the sort of strong indigenous institutions that existed in Northern Nigeria were not necessarily replicated elsewhere in Africa, the policy of Indirect Rule was nevertheless widely introduced – for example in 1925 in the former German colony of Tanganyika, which came under British control following the First World War. Subsequently, in the 1930s, Indirect Rule was also introduced into Nyasaland (Malawi) and Northern Rhodesia (Zambia), and then into Basutoland (Lesotho), Bechuanaland (Botswana) and Swaziland. As Iliffe comments,

> **The policy's conservative thrust was strong. In Sudan, for example, the Egyptian and Sudanese elites initially employed for their anti-Mahdist sympathies were abandoned after 1924 when an army mutiny revealed the first glimpses of Sudanese nationalism. Instead the British adopted 'Indirect Rule' and rehabilitated 'tribal chiefs' in a policy described by the governor as 'making the Sudan safe for autocracy'.**
>
> **(Iliffe, 1995: 201)**

Although Indirect Rule was not adopted in Southern Rhodesia (Zimbabwe) and Kenya, where the white settlers felt it restricted labour supply and made the local chiefs too powerful, other colonial powers sometimes used similar policies in administering their territories.

The French approach to managing its colonial territories was very different from the British. 'Whereas British colonialism was designed to create Africans with British characteristics, French policy was designed to create black Frenchmen' (Binns, 1994a: 9). The French regarded their colonies as part of France, 'Overseas France', with the aim of assimilating the colonies and their people into France and the French way of life. In fact, the process of assimilation was a driving force in the French colonial design. But to acquire full French citizenship was a lengthy procedure, requiring 'education in French schools, performing military service and a minimum of civilian French employment, and agreeing to be monogamous and to foreswear traditional or Islamic law and custom' (Fage, 1995: 411). By 1939, only 80,000 of the 15 million Africans in French West Africa had actually gained full citizenship. The vast majority of those inhabitants of the colonies that did not achieve this distinction could only be 'associated' with France. White settlement was common in Francophone colonies, particularly in the countries bordering the Mediterranean (notably, Algeria, Morocco and Tunisia), but there were also quite sizeable white minorities in Dakar (Senegal) and in the plantation areas of Cameroon, Guinea, Ivory Coast and Madagascar.

The French established a centralised, bureaucratic, authoritarian and hierarchical state and there was little attempt to work through traditional institutions or leaders as in the British territories. Early French administration in West Africa was preoccupied more with military than with commercial motivation. In 1904, France's eight West African colonies were amalgamated into a single federation with its capital in Dakar (Senegal), where the Governor-General was based, taking his directives from the Colonial Ministry in Paris, and himself disseminating policy through a hierarchy of governors in each colony, and down to their provincial commissioners and *commandants de cercle*, the officers in charge of each district.

Belgium controlled the vast territory known initially as the Congo Independent State from the establishment of its boundaries after the Treaty of

BOX 2.4 (continued)

Berlin in 1885, and later from 1908 it became the Belgian Congo (now Democratic Republic of Congo, formerly Zaire). Congo was initially a personal possession of King Leopold, who seized the territory when others in his country were apparently showing little interest in colonisation. Leopold ruled in a dictatorial manner and decreed that all land in Congo apart from villages and surrounding cultivated areas belonged to him as Head of State. There were strict controls on what produce Africans could buy and sell, and inhabitants were heavily taxed. Leopold established several concessionary trading companies with exclusive trading rights over defined areas and he took a 50 per cent stake in each. The so-called 'Congo System', which heavily exploited both people and resources, 'attracted much contemporary criticism and opprobrium, largely because of the way in which the private companies, backed by the official "army" of the state, brutally abused and exploited Africans' (Griffiths, 1995: 53). This criticism, much of it directed at the King himself, eventually led to the renaming of the colony as 'Belgian Congo' from 1908, when it formally became a colony of Belgium. However, the exploitative Congo System persisted and there was considerable international surprise when at the Treaty of Versailles in 1919, Belgium was given the trust territory of Ruanda-Urundi (now Rwanda and Burundi) under League of Nations mandate.

Portugal was the other major colonial power in Africa, and as early as 1482 had established fortified trading posts on the west coast from Senegal to the Gold Coast. Despite its long-standing contacts with Africa, Portugal was less successful than Britain and France in gaining territory during the competitive 'scramble'. By the early twentieth century, Portugal controlled the small West African territory of Portuguese Guinea (now Guinea-Bissau), and the two much larger colonies of Angola and Mozambique. Like the French, Portugal adopted a policy of assimilation, with all territories being regarded as part of the Portuguese Union. Substantial numbers of white settlers moved to the Portuguese colonies, as in the French territories. According to Rodney, the Portuguese colonial regime had an appalling record of slavery-like practice in its African colonies.

> **One peculiar characteristic of Portuguese colonialism was the provision of forced labour, not only for its own citizens, but also for capitalists outside the boundaries of Portuguese colonies. Angolans and Mozambicans were exported to the South African mines to work for subsistence, while the capitalists in South Africa paid the Portuguese government a certain sum for each labourer supplied.**
>
> **(Rodney, 1972: 167)**

The world wars and the intervening depression severely disrupted colonial economies, with investment from Europe being limited and commodity prices falling steadily. Much of the capital sustaining growth in this inter-war period was American, or, in Asia, from the overseas Chinese community. Thus, in the Dutch East Indies, 80 per cent of all domestic trade was controlled by Nanyang Chinese, whereas the largest rubber plantation was owned by a US tyre company. The declining profitability of commodity exports helped cause a shift in the nature of government investment during this period, with increasing amounts of capital being invested in infrastructure – roads, utilities or railways.

During the economic recession of the 1920s, official metropolitan and colonial ties grew closer. Thus, by 1930 some 44 per cent of British trade was with the empire. This was reinforced by demographic changes, with increasing numbers migrating away from the European recession to the perceived opportunities of the colonies, often encouraged enthusiastically by their governments. Although most British migrants went to the settler colonies, such as Canada or Australia, most Dutch and French migrants moved to already heavily populated colonies and were forced into a variety of urban occupations rather than farming as in earlier decades. Many unqualified migrants took up relatively low paid work in retail or office locations.

The effect of this growing European presence was complex. For the small, educated indigenous group, it made personal advancement even more difficult, sending many off to Europe for further education and to sharpen their political and organisational skills. The huge peasant population, however, suffering dreadfully from the recession, were not organised enough to threaten more than the local representatives of the system. But the growing urban indigenous population posed more of a problem – not an overtly political problem in the late colonial period, rather one associated more with social issues.

The inter-war years

For the majority of European colonists, the inter-war years seemed like 'a golden age', despite the economic vicissitudes. Even those with more modest incomes had status and privilege relative to the indigenous population, which compensated somewhat. This was the era from which most contemporary images of colonialism are drawn in the media, particularly in the cinema, for example *Out of Africa*. The luxurious way of life was facilitated by the shift in the balance of administrative power from the metropolitan centre to the colonies, certainly in the British Empire, so that salaries, privileges and jobs overseas remained secure and rewarding, despite the economic situation at home. This shift in power also enabled the administrators to more or less ignore trusteeship as far as indigenous populations were concerned, favouring the colonial settlers in numerous ways.

Urban planning became a distorted version of European concepts, so the garden city movement in Africa and Asia was largely used to segregate European and indigenous populations further by swathes of greenery, particularly recreational areas such as golf courses and racecourses. Despite the health risks, burgeoning indigenous populations were often crammed into crumbling 'old quarters'. In 1911 in Delhi, on the eve of the emergence of the new imperial capital, the old Mughal city of Shahjahanabad contained almost a quarter of a million people in its 2.5 square miles (King, 1976).

The Second World War destroyed the myth of European invincibility, particularly in Asia as a result of Japanese victories, although ironically the eventual success of the European allies owed not a little to the colonial or imperial forces that served under the metropolitan flag and to the resources that the colonies continued to supply. Indeed, it has been estimated that some 200,000 Africans fought alongside the British and the French during the war, and, as Fage comments,

> **Since the direct or indirect result of their endeavours had been to free lands like Ethiopia, Burma, India and even France from foreign rule . . . these men . . . were naturally inclined to ask why this benefit should be withheld from them. All these people, the ex-soldiers especially, looked for a better life now that the war was over.**
>
> **(Fage, 1995: 476)**

Post-war decolonisation and independence

In the post-war period the movement towards decolonisation gathered momentum. There were several reasons why this occurred.

First, France and Britain were exhausted after the Second World War, their economies were weak and they were dependent on the USA for financial support. Two new superpowers had emerged, the USA and former USSR, both of them committed to the end of overt colonialism. In fact, decolonisation was strongly encouraged by the USA, partly because it wished to extend its own sphere of economic influence through a kind of informal imperialism, which might be more appropriately termed either 'global capitalism' or 'neo-colonialism'. With regard to the former USSR, Kwame Nkrumah, the first leader of independent Ghana (formerly Gold Coast), apparently commented, 'Had it not been for Russia, the African liberation movement would have suffered the most brutal persecution' (Nkrumah, quoted in Iliffe, 1995: 246).

A second motivation for decolonisation came from the Africans and Indians who had received a Western education, and who used their skills of oratory and writing to campaign and mobilise the masses. The Indian Congress Movement, founded as early as 1885, was led by such men, who initially campaigned for more respect and greater elected representation from the British rulers. From 1917, however, the party waged a battle of varied tactics, including periods of intense nationalist agitation, with Gandhi playing a pre-eminent role. The Congress was able steadily to raise the financial and moral price which the British would have to pay if they wished to remain in power. In 1947 India won its independence and Congress helped to form the interim government, in power until 1951, when Jawaharlal Nehru as Prime Minister also assumed the Congress presidency. The independence of India had a very significant impact on nationalist movements in other colonies across the world.

Further momentum was given to the decolonisation process by the signing of the Atlantic Charter during the Second World War, the tone and sentiments of which were taken up later in the UN Charter. Indeed, the UN set timetables for independence in Libya (1951) and Somalia (1960), and also decreed the federation of Italy's former colony, Eritrea, with Ethiopia, while the former USSR supported full independence for Eritrea.

Finally, the increasing mobility of European and US capital also facilitated the decolonisation process, since with the incorporation of the colonies into the global capitalist economy, Western companies could continue to move capital and other resources into and out of these territories irrespective of whether or not they actually governed them. It was suggested that independent governments would continue to permit such transactions, not least because their governing elites would require access to outside capital and technology.

Not all the metropolitan powers responded to the call for decolonisation and to the call of the colonial peoples for independence. Britain found it easier than most to withdraw because of its less formal and more decentralised administrative systems, although there were one or two places such as Southern Rhodesia where settler resistance saw the colonial sun set rather more slowly than elsewhere. Indeed, Southern Rhodesia did not become independent Zimbabwe until 1980, over 20 years after Ghana had been the first British colony in Africa to be given self-rule after the Second World War (1957), following the earlier granting of independence to South Africa (1910) and Egypt (1922). In Kenya, the Mau Mau uprising took place during the 1950s, in which predominantly dispossessed Kikuyu protested over the alienation of land for European settler farms. The raiding of settler farms and indiscriminate killings proved to be a threat to stability and British troops were needed to restore control. This eventually led to the colonial government putting pressure on the white settlers to accept a movement towards independence in 1963, when Jomo Kenyatta, who had been detained during the uprising, was elected the first president.

For France and the Netherlands, however, decolonisation was an altogether more complex and bloody affair, with settlers resentful of being abandoned and unwilling to return to a war-ravaged Europe putting up fierce resistance. Brutal decolonisation wars resulted in Algeria, the East Indies and Indo-China. In North Africa, both Morocco and Tunisia gained independence from France in 1956, whereas in Algeria young militants started a guerrilla war in 1954. In the next eight years, over half a million French troops virtually defeated the *Front de Libération Nationale* within Algeria, but it was kept alive in neighbouring Morocco and Tunisia. President de Gaulle was forced to agree to Algerian independence in 1962, and there followed an exodus of some 85 per cent of the European settlers. Portugal abandoned its African colonies abruptly in 1975 following the *coup d'état* in Portugal in 1974. Colonial administrators were withdrawn quickly with very little time to hand over the trappings of power and bureaucracy to the local people. Portuguese businessmen, industrial managers and technicians also departed. Portugal had been struggling against liberation wars in both Angola and Mozambique, and after the departure of the colonial power both countries disintegrated into many years of civil war, which at various times was fuelled by the intervention of other countries.

Eventually, however, independence came to all except a few small territories, though their development paths were strongly affected, not only by the neo-colonial forces that quickly moved into the political and economic vacuum, but also by the considerable legacies that colonisation had bequeathed to these nascent states.

Legacies of colonialism

Politically, the national units that emerged between the 1940s and 1970s essentially comprised the territorial divisions of the colonial era and often had limited correlation with environmental geographies, pre-colonial structures or with contemporary cultural and ethnic patterns. As Clapham (1985: 20) has remarked, 'There is still no more striking, even shocking, reminder of the impact of colonialism in Africa than to cross an entirely artificial frontier and witness the instant change of language . . . that results'. Moving from Zambia (English) to Angola (Portuguese) illustrates this comment very vividly.

The Malay world of Southeast Asia was divided between three countries, Malaysia, Indonesia and the Philippines, in accordance with the territorial spread of the British, Dutch and Spanish-American empires. Even non-Malay areas were incorporated into these territories if they had been part of the colonial territory. Thus Indonesia has held on to Melanesian Irian Jaya, despite the cultural and ethnic contradictions, and is attempting to flood the area with Javanese migrants as part of its accumulation process.

There is a lively and ongoing debate among historians, political scientists and others as to whether colonialism was beneficial or detrimental to the economies and societies of those countries that were colonised. In the early twenty-first century, it is a fact that many African countries and former colonies are worse off now than they were at independence in the 1960s. Some historians would argue that colonialism was too varied and, in some instances, too brief to judge whether it has left a beneficial or detrimental legacy. Certainly, the local textures of colonialism were immensely complex, but as a component of a broader, changing global process, there were immense overall repercussions.

Two much-quoted and contrasting viewpoints are provided by Bauer (1976) and Rodney (1972). Bauer in his influential book *Dissent on Development* asserts:

> **It is untrue to say that colonial status is incompatible with material progress, and that its removal is a necessary condition of economic development. Some of the richest countries were colonies in their earlier history, notably the United States, Canada, Australia and New Zealand; and these countries were already prosperous while they were still colonies. Nor has colonial status precluded the material advance, from extremely primitive conditions, of the African and Asian territories which became colonies in the nineteenth century. Many of these territories made rapid economic progress between the second half of the nineteenth century, when they became colonies, and the middle of the twentieth century, when most of them became independent.**
>
> **(Bauer, 1976: 148)**

In his important and thought-provoking text *How Europe Underdeveloped Africa*, Walter Rodney (1972) takes a very different view from Bauer on the merits and problems of colonialism:

> **The colonisation of Africa lasted for just over seventy years in most parts of the continent. This is an extremely short period within the context of universal historical development. Yet, it was precisely in those years that in other parts of the world the rate of change was greater than ever before . . . The decisiveness of the short period of colonialism and its negative consequences for Africa spring mainly from the fact that Africa lost power. Power is the ultimate determinant in human society, being basic to the relations within any group and between groups. It implies the ability to defend one's interests and if necessary to impose one's will by any means available.**
>
> **(Rodney, 1972: 224)**

Bauer and Rodney, among many others, provide what might seem to be credible, or at least seriously debatable perspectives. The debate will undoubtedly go on, but some of the legacies of the colonial era will be considered to inform and illuminate the different perspectives in the debate.

Demographic effects of colonialism

The demographic legacy of colonialism was considerable. During the late colonial period the rate of indigenous population growth began to accelerate, partly due to transfers of medical technology, improvements in health care and hygiene and wider food security. Not that they were provided on either a widespread or enthusiastic basis by the colonial authorities, but improvements did occur and fertility did rise to the extent that the newly independent nations inherited an accelerating population growth.

Most colonial powers adopted a top-down hierarchical model for health care provision in their colonies, with an emphasis on curative rather than preventative medicine. In this model, patients were referred to successively higher levels which, in theory at least, possess greater skills and technical resources. In practice, the highest level in the hierarchy, the specialist and/or teaching hospital, was invariably located in large urban centres and in some cases only in the capital cities.

Such a health care system inevitably led to greater expenditure and better provision in the cities, while the poorer and remoter rural areas had (if any) only basic provision. Indeed, much rural health care was, and still is, provided by missionary bodies and non-governmental organisations (Phillips, 1990). Many people in both rural areas and cities depended on traditional medicine, which was (and indeed still is) widely available and usually less expensive than Western medicine.

In the 1970s, Ghana, for example, had one teaching hospital in Accra, the capital city, eight regional hospitals and 32 district hospitals. Although only 18 per cent of the population was concentrated in towns larger than 20,000 persons, over 66 per cent of the country's doctors were based in such towns. Furthermore, in terms of expenditure on health, a specialist teaching hospital benefiting only 1 per cent of the population received 40 per cent of the national health budget, whereas primary health care, serving some 90 per cent of the population, was allocated only 15 per cent of total health spending. Such inequalities were also found in India, where large cities and port towns were given priority, 'Local populations were cut off from modern medicine, but ironically, indigenous systems were often discouraged and certainly neglected' (Phillips, 1990: 114).

Former colonies have had to come to terms with these 'alien' medical systems since independence, but in reality relatively few countries have had either the foresight or the funding to restructure health care delivery completely. A much-quoted exception to this is Tanzania, where under the country's first president, Julius Nyerere, there was a stated commitment to redressing the inherited colonial imbalance of health expenditure and provision, and extending the provision of health care to rural areas through trained community health workers, as in the Chinese 'barefoot doctor' model (see Chapter 5). Much research was done by the colonial powers into the causes of ill health in tropical regions, for example, the incidence and prevention of diseases transmitted by mosquitoes and tsetse flies. Medical research stations were established, such as the British Medical Research Council's hospital at Fajara in The Gambia, while other research stations were concerned with improving the health of livestock.

Colonial governments also introduced emergency programmes when epidemics occurred. For example, when the plague came to Bombay in 1896, the health authorities began to disinfect houses to kill the plague-carrying fleas that infested rats in the city.

A further demographic effect of colonialism was the mixing of populations that occurred as a result of labour movements, whether forced, contracted or voluntary. Add to this the multiple ethnic groups that were already in the artificial colonial states which became independent territories, and the consequence has been an ethnic melting pot that has simmered and bubbled almost everywhere during the post-colonial period, at best considerably hampering the development process, at worst resulting in appalling acts of expulsion or genocide, such as the expulsion by President Idi Amin of 80,000 Asians from Uganda from 1972.

Education

Some would argue that, along with health, the provision of formal education systems was another benefit gained from the colonial experience. In many colonies a respectable group of educated indigenous people developed, who spoke the colonial language and absorbed themselves in the alien culture. As Gould comments,

> **The respectful, educated professional or civil servant of the British colonies, as personified by E.M. Forster's Dr Aziz in *A Passage to India*, or a figure of fun like Joyce Cary's *Mister Johnson* in Nigeria, aspired to the colonial cultural values, and provided an implicit model for others.**
>
> **(Gould, 1993: 13)**

In Francophone countries, there were similar groups of educated indigenes, for example, Léopold Senghor (who became the first president of Senegal) and Aimé Césaire of Martinique. They published poetry and essays in the 1930s which celebrated 'négritude' (blackness), in which they reaffirmed African culture by writing in impressively fluent French. Césaire's poetic account *Return to my Native Land* tells of a journey back to Martinique after years of schooling in France, in which he presents a critique of colonial rule whilst also giving beautiful images of blackness and the survival of African traditions. Both Senghor and Césaire showed equally close attachments to African tradition and to metropolitan French culture.

As part of the administrative machinery of colonialism, the colonisers required a small, educated local workforce. As Rodney comments,

> **In practice, it was not necessary to educate the masses because only a minority of the . . . population entered the colonial economy in such a way that their performance could be enhanced by education. Indeed, the French concentrated on selecting a small minority, who would be thoroughly subjected to French cultural imperialism.**
>
> **(Rodney, 1972: 257)**

The missionaries were also keen to promote reading and writing so that indigenous people could read the Bible. In Muslim areas, the Koranic schools played an important role before the establishment of the Western-style school system and continued alongside their newer counterpart. Like the colonial health systems, the education systems were generally centralised, hierarchical and bureaucratic (Watson, 1982). There was a strong emphasis on achieving academic excellence, and syllabuses and assessment methods were frequently transported with little, if any, modification from the metropolitan country to the colonies.

In terms of educational provision, universities, where they existed, were usually located in the large cities, notably the capitals, and received a disproportionately large amount of funding. In West Africa, for example, provision for higher education was very limited. In fact, in 1951 the whole of predominantly Muslim Northern Nigeria had only a single graduate – a Christian (Iliffe, 1995). Rodney reports that, 'In 1874, when Fourah Bay College (Freetown, Sierra Leone) sought and obtained affiliation with Durham University, *The Times* newspaper declared that Durham should next affiliate with the London Zoo' (Rodney, 1972: 141)

Meanwhile, rural education facilities were rudimentary and concentrated on very little more than 'the three Rs'. In some colonies the provision of secondary and higher education was negligible, as in the Belgian Congo, where although in the early 1950s a higher proportion of the population (one in 12) was attending primary school than almost anywhere else in tropical Africa, the provision of secondary education for the African population was virtually non-existent, with only one secondary school for every 870 primary schools. At the same time, the ratio of secondary schools to primary schools was 1:25 in French West Africa, and 1:70 in both Gold Coast (Ghana) and Nigeria.

Transport

In material terms, too, the legacies of colonialism have been substantial, particularly in the form of transport and communications links. The basic road and rail networks of many contemporary states owe their origins to the colonial era. Unfortunately, most of these lines of communication reflect the economic needs of the colonial powers and do not necessarily coincide with the contemporary needs of independent states. In Southern Rhodesia (now Zimbabwe), for example, the principal road and rail links connected the main areas of European settlement and economic activity on the high veld, and exited through what was British South Africa.

Transport and communication links were, in a way, subordinate to the urbanisation process which, as we have seen, was either established for the first time or completely restructured during the colonial period to favour one or two major ports that functioned as crucial connecting points between the colonial and global economies. Many instances of exaggerated urban primacy in the contemporary world have their origins in the colonial economy (see also Chapter 9). Within these cities, too, there is often a considerable physical legacy of colonial triumphalism in architectural form, planning layout and infrastructure provision (Plates 2.4 and 2.6; King, 1990).

Administrative, legal and judicial systems

Colonial cities were points of administration, rather than production, and some would argue that the most useful legacy of colonialism is the administrative, legal and judicial systems that were established by the metropolitan powers. There is certainly some truth in this as far as orderly and efficient administration was concerned. But the elitism inherent within these systems still continues in their successors, particularly the use of European languages in the highest echelons of the state.

In Hong Kong, the colonial experience ended as recently as 1 July 1997. By the 1980s, Hong Kong's civil service had developed a relatively high degree of autonomy and institutional integrity and had taken effective measures to prevent other groups and institutions from challenging its authority. In 1958, financial and budgetary autonomy was granted to Hong Kong by the UK.

For most of the colonial era, the colonial government banned political parties and a mainly appointed legislature was kept weak and largely ineffective, being answerable only to the Governor. The bureaucracy in colonial Hong Kong managed to remain neutral in relation to business and political interests, and government positions were denied to party activists. However, certain colonial practices undermined morale in the civil service, for example, hiring expatriates to do jobs that could easily be done by local people.

Earlier in the colonial period many of Hong Kong's administrative officers had middle class origins and were recruited from Oxford and Cambridge. Although there was a move towards employing local staff from 1984, even as late as 1997 some 23 per cent of the top 1130 positions in the civil service directorate were held by expatriates (Burns, 1999). The fact that all government business was conducted in English rather than Chinese (Cantonese) also added to both inefficiency and criticism. Only in 1974 did the government adopt both English and Chinese as official languages, though publication of laws in both languages did not occur until 1989.

The common law legal system defined the relationship between state and society and between the state and its employees, and all civil servants held office 'at

Plate 2.6 Grand French colonial architecture in Dakar, Senegal
(photo: Tony Binns)

the pleasure of the Crown'. Elsewhere, former colonies have attempted to modernise their legal systems (e.g. Senegal and Madagascar), but in many countries present-day legal systems still bear a strong resemblance to those inherited from the former colonial power. Fairness was rarely a characteristic of colonial administrative and legal systems, and here again some unfortunate legacies have persisted. Thus Singapore, like many other countries, retained and uses legislation on detention without trial. In similar fashion, building regulations and standards retained from the colonial period have long prevented more effective housing policies being pursued within contemporary developing world cities.

Economic activities

What underpinned almost all these consequences of colonialism were its economic activities and here the direct legacy has also been substantial. The narrowing of production into one or two commodities persisted in the great majority of countries, even to the point of the same commodities being grown. This posed enormous problems for those countries, the great majority of which watched commodity prices continue to fall and failed to diversify their economies. And the spatial concentrations of these activities continues to cause problems of regional inequality and imbalance, often reinforced by post-colonial urbanisation trends. As Blaine Harden (1993) cynically sums up the African colonial experience,

> **Africans were not asked whether they wanted to be guinea pigs. They were bullied into it. Europeans overwhelmed the continent in the last quarter of the nineteenth century, looking for loot. They carved it up into weirdly shaped money-making colonies, many of them landlocked, all of them administered from the top down. The colonies bore little or no relation to existing geographical or tribal boundaries. Total conquest took all of about twenty-five years. Then, after sixty years or so – the shortest introduction to so-called civilization that any so-called primitive people have ever had – the Europeans turned their authoritarian creation over to the Africans.**
>
> **(Harden, 1993: 16–17)**

For some 25 years after the end of the Second World War the economies of the newly independent states of Asia and Africa remained virtually unchanged under the development strategies of neo-classical advisers such as Lewis or Rostow (see Chapter 3). These economies were still reliant on the export of a narrow range of primary commodities, possibly with some diversification into import substitution industries. For the most part, these economies were still controlled from the outside through the medium of tied aid or the activities of transnational corporations (TNCs). Although metropolitan powers continued to be linked in this way with their former colonies, new international players were equally dominant, particularly the USA.

Neo-colonialism

Some would argue that colonialism has given way to 'neo-colonialism', in which powerful states such as the USA, the former USSR, Japan and, collectively, the member states of the European Union, exercise economic and political control over the economies and societies of the underdeveloped world. Although the colonies might have gained their independence, their economic and political systems are still controlled from outside, notably through aid, trade and political relationships. It might be suggested, for example, that the Lomé Conventions linking the EU and some 60 African, Caribbean and Pacific states are a new form of colonialism, or 'neo-colonialism', in laying down guidelines for aid, trade and investment agreements. Similarly, since the 1980s the structural adjustment programmes of the World Bank and International Monetary Fund have imposed liberal economic policies and a wide range of conditions on the poor countries that have sought financial help (see Chapters 3 and 7).

New International Division of Labour (NIDL)

From the 1970s, there occurred a more sustained and substantial outflow of investment funds from Europe and North America as industrial profits began to decline in those areas. The various reasons for this are often related to the rising costs of labour and environmental protection. New financial systems, particularly artificial international currencies such as Eurodollars, and the windfall profits from the OPEC oil price rise facilitated the investment of money overseas in new manufacturing plants in the developing world. The usual term to describe this post-colonial phenomenon is the New International Division of Labour (NIDL), in which the low-cost labour-intensive parts of the manufacturing process are siphoned off to the developing world where costs are lower. A detailed critique of this fragmentation of the production process and the role played by TNCs can be found later in Chapter 4. However, at this point and as a link to Chapter 4, it might be useful to examine the concept of NIDL as it has become almost synonymous with the post-colonial period.

NIDL is perhaps the third international division of labour:

- The first comprised the production and extraction of primary commodities in the colonies and their manufacture in the metropolitan countries.
- The second involved the shift of some industry into the newly independent countries under import substitution policies.
- The third began to expand in the 1970s and involves the fragmentation of the manufacturing process and the shift of a large proportion of this to developing countries, largely through the medium of TNCs.

It is important to realise that all three international divisions of labour are currently in operation. In particular, large amounts of TNC investment continue to sustain primary resource production in countries such as Papua New Guinea, almost all of which is exported to advanced capitalist economies.

The latest international division of labour came about through a conjuncture of changes: reduced profitability in Europe, largely through increased production costs, cheap production costs in developing countries, encouragement given to urban-industrial growth in the developing world by international development agencies, facilitating developments in communications technology and the parallel increasing mobility and flexibility of financial services. The result was new levels of extraction of surplus value created by super-exploitation of poor country labour, which accrues few skills and with limited backward linkages into the local economy (see Chapter 3).

The problem with NIDL is that it overemphasises the inevitability of this exploitation and fails to credit peripheral social formations with any autonomy to manipulate foreign investment to maximise local advantage. The concept is also undermined in its global applicability by the fact that NIDL was initially very selective, with just six countries, usually identified as the four Asian tigers (Hong Kong, Singapore, South Korea, Taiwan), together with Mexico and Brazil, receiving the majority of TNC investment, largely because TNCs rely on more than just cheap labour for efficient and profitable production.

It can be argued that the range of countries which are industrialising through the medium of foreign investment is widening. In particular, analysts point to the rapid rise of countries such as Malaysia, Thailand and Indonesia. But these countries receive much, if not most, of their investment from within the Asia-Pacific region, notably from the four Asian tigers who used NIDL to retain surplus value and generate regional investment.

It is clear in this context that national governments have not been overwhelmed or displaced by Western TNCs, but act in concert with indigenous equivalents to promote joint interests overseas. What has emerged, therefore, is a regional division of labour (RDL) operating both within and in conflict with NIDL. Nascent RDLs exist in other areas of the developing world too, such as the Middle East or South Asia or even southern Africa, but the extended world recession of the 1980s and 1990s will have slowed their emergence considerably.

New International Economic Order (NIEO)

Often confused with NIDL, largely because it first surfaced at about the same time, was the New International Economic Order (NIEO). This concept mainly derived from the United Nations and was underpinned by a mounting concern with the failure of modernisation strategies (Chapter 3) to achieve much for most people in poor countries.

Mounting poverty, growing national debts and the pessimistic environmental predictions of the Club of Rome all resulted in the declaration by the UN in 1994 of its intention to establish an NIEO, although we must also be aware of the self-interest of Western nations threatened by recession, oil price hikes and inflation. The five principal areas of concern identified were trade reform, monetary reforms, debt relief, technology transfer and regional cooperation.

In fact, very little was achieved for the lasting benefit of the developing nations, not least because of their growing diversity of interests (Chapter 1), and eventually the NIEO became suffused into the structural adjustment programmes of the New Right (Chapter 3). In effect, the NIEO became a justification for TNC investment in the developing world in the name of development, and as such was a factor in the internationalisation of the division of labour which paralleled it.

Conclusion

Colonialism has undoubtedly done much to shape the world in which we live today. The configuration and character of particular developing countries was in many cases determined by the European powers dividing territory and drawing boundaries at some distance from the colonies. In the post-independence phase, many former colonies have retained the language, education, legal and health systems of their colonial masters, whilst trading networks and communications links often bear strong similarity to those of the colonial period. The process, experience and legacies of colonialism remain highly controversial and hotly contested issues. The extent to which the European powers benefited from exploitation of their colonies, and whether particular African, Asian and Latin American countries would actually have been better off today without a colonial history, are just two elements of the ongoing debate.

Key points

- The particular nature of colonialism varied according to the motives and also with the political economies and cultures of both the metropolitan power and the colonised territory.
- Post-colonialism has provided an important perspective on the colonial experience in demonstrating how the production of Western knowledge is inseparable from the exercise of Western power.
- Commerce and trade dominated the first phase of colonialism – mercantile colonialism. Trading companies were heavily involved in both production and trade during this phase. The plantation system and slave labour were key features of this phase.
- The second phase of colonialism – industrial colonialism – involved the expansion of overseas markets for European manufactured goods and the overseas production of raw materials and food for European countries. Colonies were increasingly established and organised by the state rather than by trading companies, although business and state often worked together.
- During the phase of late colonialism, the two world wars and the intervening depression severely disrupted colonial economies, with a reduction in investment from Europe and falling commodity prices. The period was characterised by increasing European migration to the colonies and by a shift in the balance of administrative power from the metropolitan centres to the colonies. The post-Second World War period saw a move towards decolonisation and the granting of independence to colonial states.
- There is an ongoing debate about whether colonialism was on balance beneficial or detrimental to the economies and societies of those countries that were colonised.

Further reading

Blunt, A. and McEwan, C. (eds) (2002) *Postcolonial Geographies*. London: Continuum.
A valuable collection of papers which help to de-mystify the concept of post-colonialism.

Blunt, A. and Wills, J. (2000) *Dissident Geographies*. London: Prentice Hall.
This book contains some useful ideas on post-colonial studies.

Crush, J. (ed.) (1995) *The Power of Development*. London: Routledge.
A much-cited text which examines a variety of perspectives on the concept of 'development', written by some leading scholars in the field.

Duncan, J.S., Johnson, N.C. and Schein, R.H. (eds) (2004) *A Companion to Cultural Geography*. Oxford: Blackwell.
A very useful reference book on issues relating to cultural geography.

Griffiths, I.L. (1995) *The African Inheritance*. London: Routledge.
An excellent text on the colonisation and decolonisation of Africa.

McEwan, C. (2002) Postcolonialism, in Desai, V. and Potter, R.B. (eds), *The Companion to Development Studies*. London: Arnold, 127–31.
A particularly clear and concise introductory essay on post-colonialism.

McLeod, J. (2000) *Beginning Postcolonialism*. Manchester: Manchester University Press.
A helpful text which explores the concept of post-colonialism.

Rodney, W. (1972) *How Europe Underdeveloped Africa*. Washington, DC: Howard University Press.
A classic and thought-provoking text which examines how, in the views of the author, Africa suffered as a result of European colonisation.

Said, E. (1978) *Orientalism*. New York: Pantheon Books.
A seminal book which has played an influential role in understanding relationships between colonisers and colonised.

Sartre, J.P. (1964, 2001) *Colonialism and Neo-colonialism*. English translation (2001). London: Routledge.
An important text from a key French thinker on colonialism.

Williams, P. and Chrisman, L. (eds) (1993) *Colonial Discourse and Postcolonial Theory*. London: Prentice Hall.
A collection of essays examining colonisation, colonial perceptions and post-colonialism.

Websites

http://www.zmag.org/zmag/articles/barsaid.htm
Speech by Edward Said on 'Culture and Imperialism', given at York University, Toronto, 10 February, 1993.
An interesting speech from a leading thinker and the author of the influential book, *Orientalism*, first published in 1978.

http://www.mkgandhi.org/
Useful wide-ranging website on Mahatma Gandhi, who played a key role in the drive for independence in India, which in turn had a significant impact on the wider decolonisation process.

DISCUSSION TOPICS

- Select a country that was colonised and examine what you believe are the positive and negative legacies of colonialism.
- Identify some of the key elements in writings on post-colonialism and suggest how these have enhanced understanding of the colonial experience.
- Examine the ideological and practical differences between the phases of mercantile, industrial and late colonialism.

Chapter 3

Theories and strategies of development

The aim of this chapter is to introduce readers to the main theories of development that have been advanced, and to give some idea as to how these have been put into practice. By the end of the chapter, therefore, readers should be aware of the most important characteristics of different theories and strategies of development. At the outset, the terms 'theory', 'strategy', 'ideology' and 'paradigm' are defined in respect of the field of development studies. Throughout, it is emphasised that thinking about development tends to be evolutionary rather than revolutionary, so that new ideas and approaches often come into existence without the eradication of older ones. This is, of course, because thinking about development is highly political and, therefore, highly contested. The account:

- Considers the nature and role of theories, strategies and ideologies in the field of development studies.
- Reviews classical and neo-classical approaches to development theory, these approaches being based on traditional economic theory. Such approaches represent 'top-down' development.
- Considers the findings of historical–empirical approaches to the understanding of development – that is, learning from what has actually happened in the past in various regions.
- Overviews radical and Marxist-inspired theories, especially the dependency school.
- Stresses the diversity of alternative approaches to development, including an array of 'bottom-up' formulations.
- Links the conditions of modernity and postmodernity to development theory.

Introduction

Since the start of the twentieth century, a major feature of the interdisciplinary field of development studies has been a series of fundamental changes in thinking about the process of development and indeed what constitutes development itself. This search for new conceptualisations of development has been mirrored by changes in development practice in the field.

Thus, there has been much debate and controversy about development, with many changing views as to its definition, and the strategies by means of which, however defined, it may be pursued. Chapter 1 exemplified this in respect of the recent debates about the value of the concept of development itself, in terms of anti-development and post-development. This was also demonstrated by the variety of approaches to development that have been adopted, ranging from development as economic growth to development as promoting human rights and freedoms.

In short, the period since the 1950s has seen the promotion and application of many varied geographies of development and there is much vitality in the field. Such diversity is demonstrated by the number and variety of major books and other items published on 'development' (see, for example, Apter, 1987; Brohman, 1996; Clark, 2006; Corbridge, 1986; Cowen and Shenton, 1996; Crush, 1995b; Desai and Potter, 2008; Escobar, 1995; Greig *et al.*, 2007; Hettne, 1995; Kothari, 2005; Leys, 1996; Mehmet, 1999; O'Tuathail, 1994; Panayiotopoulos and Capps, 2001; Power, 2003; Preston, 1987, 1996; Rapley, 1996; Schuurman, 1993; Sen, 2000; Simon, 2006; Slater, 1992a, 1992b, 1993; Streeten, 1995; and Thirlwall, 1999).

The purpose of this chapter is to provide an introduction to the different approaches to development that have been proposed and followed, principally since around 1940, both in theory and in practice. As noted above, these different approaches reflect the changing paradigms of development that were outlined in Chapter 1. Many of these paths to development will be further elaborated in Parts II and III of this text.

A major theme in the present account will be to illustrate that ideas about development have long been highly controversial and contested. This is because thinking about development paths and states is essentially political. This argument is further highlighted in the conclusion to the chapter, which sets out briefly some of the relationships between development theory and the societal conditions of modernity and postmodernity. The linked contention that development theory and development studies have currently reached an impasse or deadlock is also considered as part of this account, an argument that draws on the positions of anti- and post-development introduced in Chapter 1.

Theories, strategies and ideologies of development

Introductory definitions

The view provided of development thinking in this chapter is very wide and eclectic. Accordingly, a broad definition of paths to development is also adopted at the outset. To use Hettne's (1995) nomenclature, the chapter reviews selected aspects of development theories, development strategies and development ideologies. Before doing so, these three basic terms must be defined:

1. *Development theories* may be regarded as sets of apparently logical propositions which purport to explain how development has occurred in the past, and/or should occur in the future. Development theories can either be normative, when they generalise about what should be the case in an ideal world, or positive in the sense of dealing with what has actually been the case. (This distinction is shown in Figure 3.2 below.) The arena of development theory is primarily, although by no means exclusively, to be encountered in the academic literature.
2. On the other hand, *development strategies* can be defined as the practical paths to development that may be pursued by international agencies, states around the world, non-government organisations and community-based organisations, in an effort to stimulate change within particular nations, regions and continents.
3. Different development agendas will reflect different goals and objectives. These goals will reflect social, economic, political, cultural, ethical, moral and even religious influences. Thus, what may be referred to as different *development ideologies* may be recognised. Chapter 1 stressed how, both in theory and in practice, early perspectives on development were almost exclusively concerned with promoting economic growth. Subsequently, however, the predominant ideology within the academic literature changed to emphasise political, social, ethnic, cultural, ecological and other dimensions of the wider process of development and change. Sometimes the term *paradigm* is used to refer to broad sets of ideas about development (see Key idea box).

A sensible approach is to follow Hettne (1995) and to employ the idea of *development thinking* in the body of this chapter. The expression 'development thinking' may be used as a catch-all phrase indicating the sum total of ideas about development, including pertinent aspects of development theory, strategy and ideology. Thus, the present chapter takes a very broad remit in presenting an overview of development strategies.

Key idea

What is a paradigm?

Paradigms are generally defined as supra-models – that is, broad sets of ideas that come to dominate particular groups of scholars and/or particular disciplines. Once accepted, these form the agreed or consensus view of the discipline or scholarly group, and are generally defended until the evidence is so overwhelming that they have to be replaced by a new paradigm or supra-model. In the context of development, major theories of how economies have developed and how best economies should be developed, can be regarded as giving rise to paradigms. It is the evolution and nature of these paradigms in the field of development studies that forms the focus of this chapter.

The contested nature of development thinking

Such an all-encompassing definition is necessary due to the nature of thinking about development itself. Development thinking has shown many sharp twists and turns during the twentieth century. Thus the various theories that have been produced have not commanded attention in a strictly sequential temporal manner. In other words, as a new set of ideas about development has come into favour, earlier theories and strategies have not been totally discarded and replaced. Rather, theories and strategies have tended to stack up upon one another, coexisting in what can sometimes be described as a very convoluted manner. Thus, in discussing development theory, Hettne (1995: 64) has drawn attention to the 'tendency of social science paradigms to accumulate rather than fade away'.

This characteristic of development thinking as a distinct field of enquiry can be considered in more detail, using Thomas Kuhn's ideas on the structure of scientific revolutions (Figure 3.1). Kuhn (1962) argued that scientific disciplines are dominated at particular points in time by communities of researchers and their associated methods, and they define the subjects and the issues deemed to be of importance within them.

Kuhn referred to these as 'invisible colleges', and he noted that they serve to define and perpetuate research which confirms the validity of the existing paradigm; he called this 'normal science'. Kuhn noted that a fundamental change occurs only when the number of observations and questions confronting the status quo of normal science becomes too large to be dealt with by means of small changes. However, if the proposed changes are major, and a new paradigm is adopted, a scientific revolution can be said to have occurred, linked to a period of so-called extraordinary research.

In this model, therefore, scientific disciplines basically advance by means of revolutions in which the prevailing normal science is replaced by extraordinary science, and ultimately a new form of normal science develops.

In dealing with social scientific discourses, it is perhaps inevitable that the field of development theory is

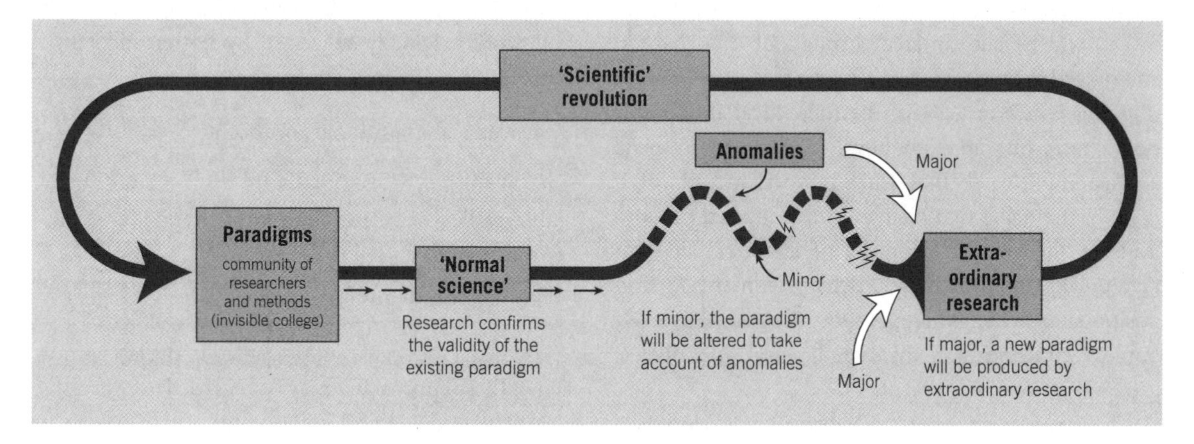

Figure 3.1 Scientific revolutions: picturing Kuhn's model of their structure

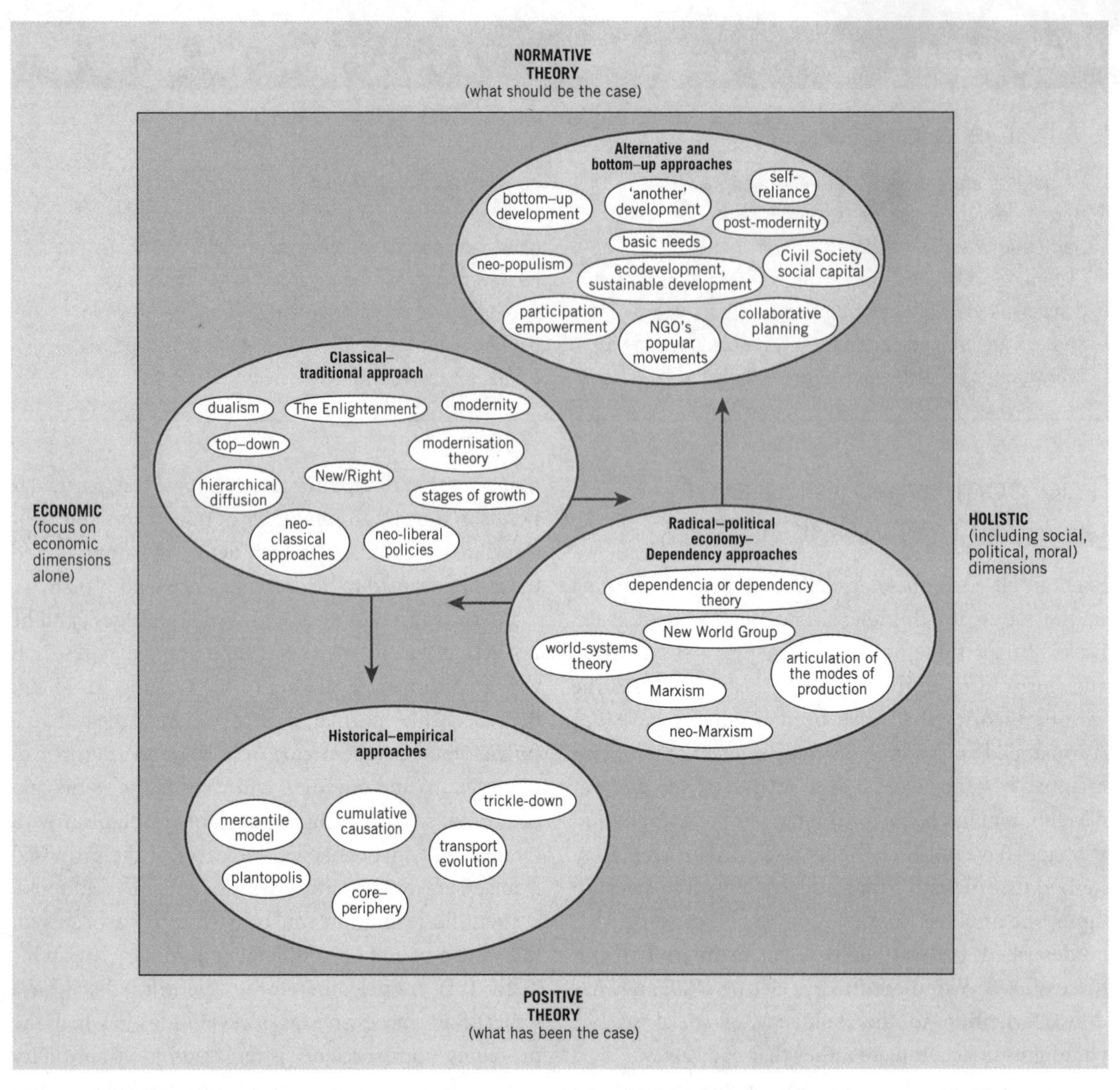

Figure 3.2 Development theory: a framework for this chapter

characterised by evolutionary rather than revolutionary change. Evidence of the persistence of ideas in some quarters, years after they have been discarded elsewhere, will be encountered throughout this chapter, and indeed through this book. Given that development thinking is not just about the theoretical interpretation of facts, but rather about values, aspirations, social goals, and ultimately that which is moral and ethical, it is understandable that change in development theory leads to the parallel evolution of ideas, rather than revolution. Hence conflict, debate, contention and positionality are all inherent in the discussion of development strategies and associated plural and diverse geographies of development.

There are many ways to categorise development thinking through time. Broadly speaking, it is suggested here that four major approaches to the examination of development theory can be recognised, and these are shown in Figure 3.2. This categorisation follows the framework originally suggested by Potter and Lloyd-Evans (1998). The four approaches are:

1 the classical–traditional approach;
2 the historical–empirical approach;
3 the radical–political economy–dependency approach; and
4 alternative and bottom-up approaches.

Following the argument presented in the last section, each of these approaches may be regarded as expressing a particular ideological standpoint, and can also be identified by virtue of having occupied the centre stage of the development debate at particular points in time.

However, each approach still retains currency in certain quarters. Hence, in the realm of development theory and academic writing, left-of-centre socialist views may well be more popular than classical–traditional and neo-classical formulations. But in the area of practical development strategies, the period since the 1980s has seen the implementation of neo-liberal interpretation of classical theory, stressing the liberalisation of trade, along with public sector cutbacks, as a part of structural adjustment programmes (SAPs), aimed at reducing the involvement of the state in the economy and promoting the free market.

But even so, the account which follows uses these four divisions to overview the leading theories, strategies and ideologies that have been used to explain and promote the development process.

Classical–traditional approaches: early views from the developed world

Introduction

The traditional approach to the study of development derives from classical and neo-classical economics and has generally dominated policy thinking at the global scale. Classical economic theory, dating from before 1914, was strongly based on the writings of Adam Smith (1723–1790) and David Ricardo (1772–1823). Both Smith and Ricardo equated economic development with the growth of world trade and the law of comparative advantage (Sapsford, 2008) (see Key idea box). Neo-classical theories, those having generally been produced since 1945 (although some date back to the 1870s), take an essentially similar worldview, stressing the importance of liberating world trade as the essential path to growth and development.

Essentially traditional approaches regard developing countries as being characterised by a dualistic structure. Hettne (1995) notes the strong role of dichotomous thinking in early anthropology, where comparisons were made between what were referred to as 'backward' and 'advanced' societies, the 'barbarian' and the 'civilised', and the 'traditional' and the 'modern'.

The fundamental dualism exists between what is seen as a traditional, indigenous, underdeveloped sector on the one hand, and a modern, developed and Westernised one on the other. It follows that the global development problem is seen as a scaled-up version of this basic dichotomy. Seminal works include those of Hirschman (1958), Meier and Baldwin (1957), Myrdal (1957), Perloff and Wingo (1961), Perroux (1950) and Schultz (1953).

The basic framework: the contribution of A.O. Hirschman

In this framework, underdevelopment is an initial state beyond which the West has managed to progress (Rapley, 1996). It also envisages that the experience of the West can assist other countries in catching-up by sharing both capital and know-how. The avowed intention, therefore, is to bring developing countries to the modern age of capitalism and liberal democracy (Rapley, 1996).

Key idea

The law of comparative cost advantage

The bases of the economic principles of international trade were formulated by the economist Adam Smith in his book on *The Wealth of Nations*, which was published in 1776. Smith argued that it made sense for particular regions and nations to produce those commodities for which they possessed the greatest comparative advantage. In this manner, at least in theory, global production can be maximised. Subsequently, by engaging in trade, countries can obtain the goods that they do not themselves produce, and which others can supply more cheaply. The arguments advanced by Smith suggested the economic efficiency of 'open' or 'liberal' trade policies, and in this sense were the forerunners of the arguments in favour of globalisation (Sapsford, 2008).

The general economic development model of the American economist A.O. Hirschman forms a convenient starting point for discussion of the traditional approach. Hirschman (1958), in his *The Strategy of Economic Development*, advanced a notably optimistic view in presenting the neo-classical position (Hansen, 1981). Specifically, Hirschman argued that polarisation should be viewed as an inevitable characteristic of the early stages of economic development. This represents the direct advocacy of a basically unbalanced economic growth strategy, whereby investment is concentrated in a few key sectors of the economy. It is envisaged that the growth of these sectors will create demand for the other sectors of the economy, so that a chain of disequilibria will lead to growth. The corollary of sectorally unbalanced growth is geographically uneven development, and Hirschman specifically cited Perroux's (1955) idea of the natural growth pole.

The forces of concentration were collectively referred to by Hirschman as polarisation. The crucial argument, however, was that eventually development in the core will lead to the 'trickling down' of growth-inducing tendencies to backward regions. These trickle-down effects were seen by Hirschman as an inevitable and spontaneous process. Thus, the clear policy implication of Hirschman's thesis is that governments should not intervene to reduce inequalities, for at some juncture in the future the search for profits will promote the spontaneous spin-off of growth-inducing industries to backward regions. Hirschman's approach is therefore set in the traditional liberal model of letting the market decide. The process whereby spatial polarisation gives way to spatial dispersion out from the core to the backward regions has subsequently come to be referred to as the point of 'polarisation reversal' (Richardson, 1977, 1980).

The doctrine of unequal growth

The full significance of these ideas concerning polarised development extends far beyond their use as a basis for understanding the historical processes of urban-industrial change, for in the 1950s and 1960s they came to represent an explicit framework for regional development policy (Friedmann and Weaver, 1979). Thus, the doctrine of unequal growth gained both positive and normative currency in the first post-war decade and the path to growth was actively pursued via urban-based industrial growth. The policies of non-intervention, enhancing natural growth centres, and creating new induced sub-cores became the order of the day. As Friedmann and Weaver (1979: 93) observe, the 'argument boiled down to this: inequality was efficient for growth, equality was inefficient', so that, 'given these assumptions about economic growth, the expansion of manufacturing was regarded as the major propulsive force'.

The elaboration of modernisation theory

Hirschman's ideas can be seen as part of a wider modernisation theory, which was in vogue during the 1950s and 1960s. The paradigm was grounded on the view that the gaps in development which exist between the developed and developing countries can gradually be overcome on an imitative basis. The emphasis was placed on a simple dichotomy between development and underdevelopment. Thereby developing countries would inexorably come to resemble developed countries, and 'in practice, modernization was thus very much the same as Westernization' (Hettne, 1995: 52). The modernisation thesis was largely developed in the field of political science, but was taken up from an essentially spatial viewpoint by a group of geographers in the late 1960s (Gould, 1970; Riddell, 1970; Soja, 1968, 1974), although sociologists also spent some considerable time working along these lines.

In such works, sets of indices which were held to reflect modernisation were mapped and/or subjected to multivariate statistical analysis to reveal the 'modernisation surface'. For example, using such an approach, Gould (1970) examined what he regarded as the modernisation surface of Tanzania (Box 3.1).

One of the classic papers written in this mould was by Leinbach (1972), who investigated the modernisation surface in Malaya between 1895 and 1969, using indicators such as the number of hospitals and schools per head of the population, together with the incidence of postal and telegraph facilities and road and rail densities. This modernisation approach served to emphasise that core urban areas of Malaya and the transport corridors running between them were the focus of dynamic change (Leinbach, 1972). In 1895, early growth was almost exclusively related to the west coast, specifically centering on Kuala Lumpur, with a clear inland island around Ipoh. By 1955, the so-called modernisation surface had penetrated to the east of the nation along two 'corridors' (Figure 3.3).

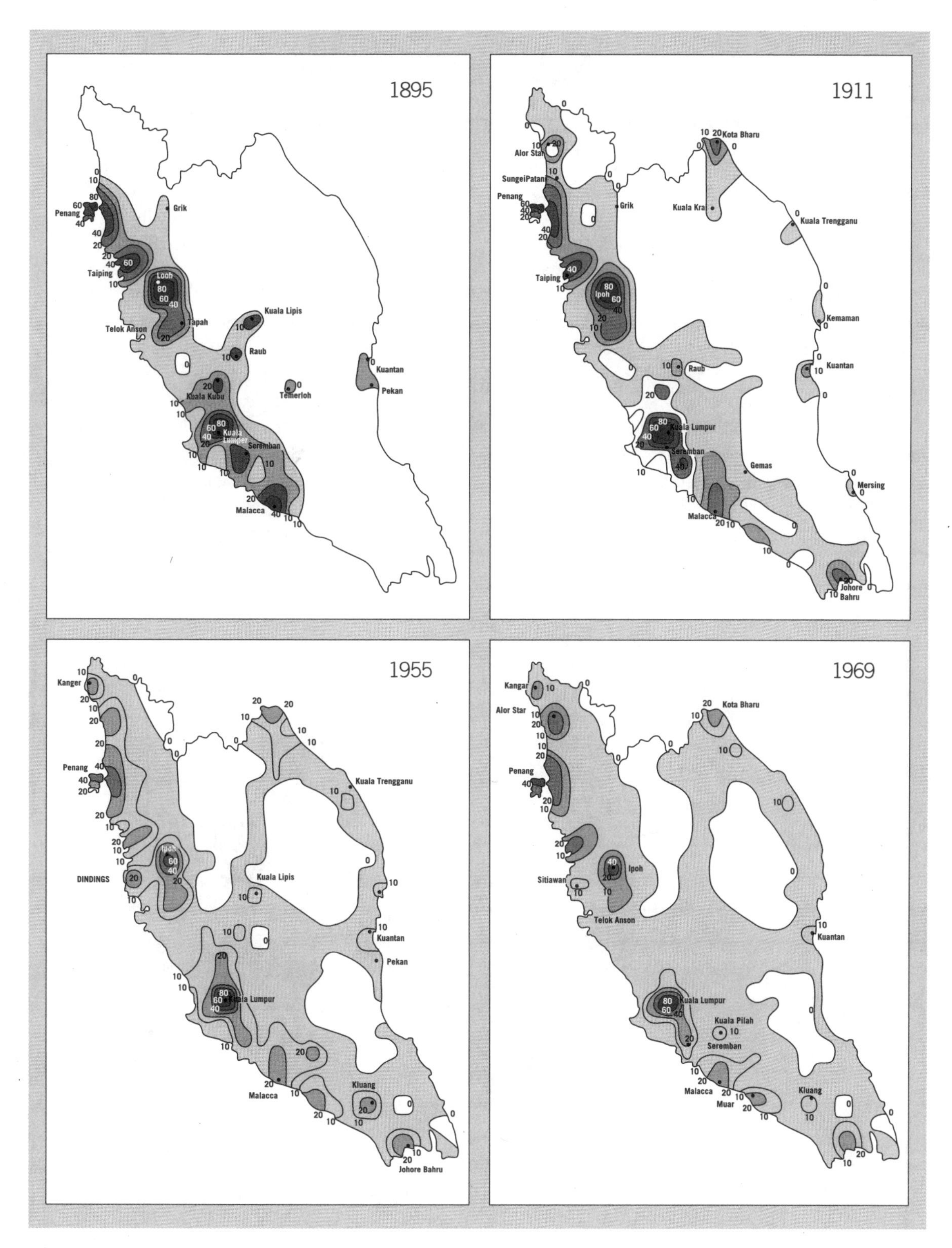

Figure 3.3 The modernisation surface for Malaya, 1895–1969
Source: Adapted from Leinbach (1972)

BOX 3.1

Modernisation and development in Tanzania

Tanzania became independent in 1961 after a British and German colonial history (Hoyle, 1979). The area was occupied by Germans in the 1880s, and after the First World War it became British-administered Tanganyika. Like many former colonies, the population was very concentrated along the narrow coastal region (see Figure 3.4). The other major urban nodes such as Morogoro, Iringa and Mbeya formed a corridor running in a south-westerly direction from Dar-es-Salaam on the Indian Ocean coast.

During the era when modernisation thinking was in vogue, 'islands' of development linked by major transport lines were recognised by geographers such as Gould (1969, 1970), Hoyle (1979) and Safier (1969). Traditionally, the settlement pattern had comprised dispersed villages, although strong urban concentration around Dar-es-Salaam occurred during the colonial period, with Hoyle (1979) referring to it as an 'hypertropic cityport' (O'Connor, 1983).

Lundqvist (1981) identified four main phases of development planning in Tanzania between 1961 and 1980. The period from 1961 to 1966 was indeed seen as the legacy of the colonial era, during which such planning as was carried out was sectoral rather than regional in scope, as a result of which infrastructure remained concentrated in the principal towns and urban–rural disparities were maintained. Thus, one could talk about a highly polarised 'modernised–non-modernised' development surface which largely reflected colonial penetration.

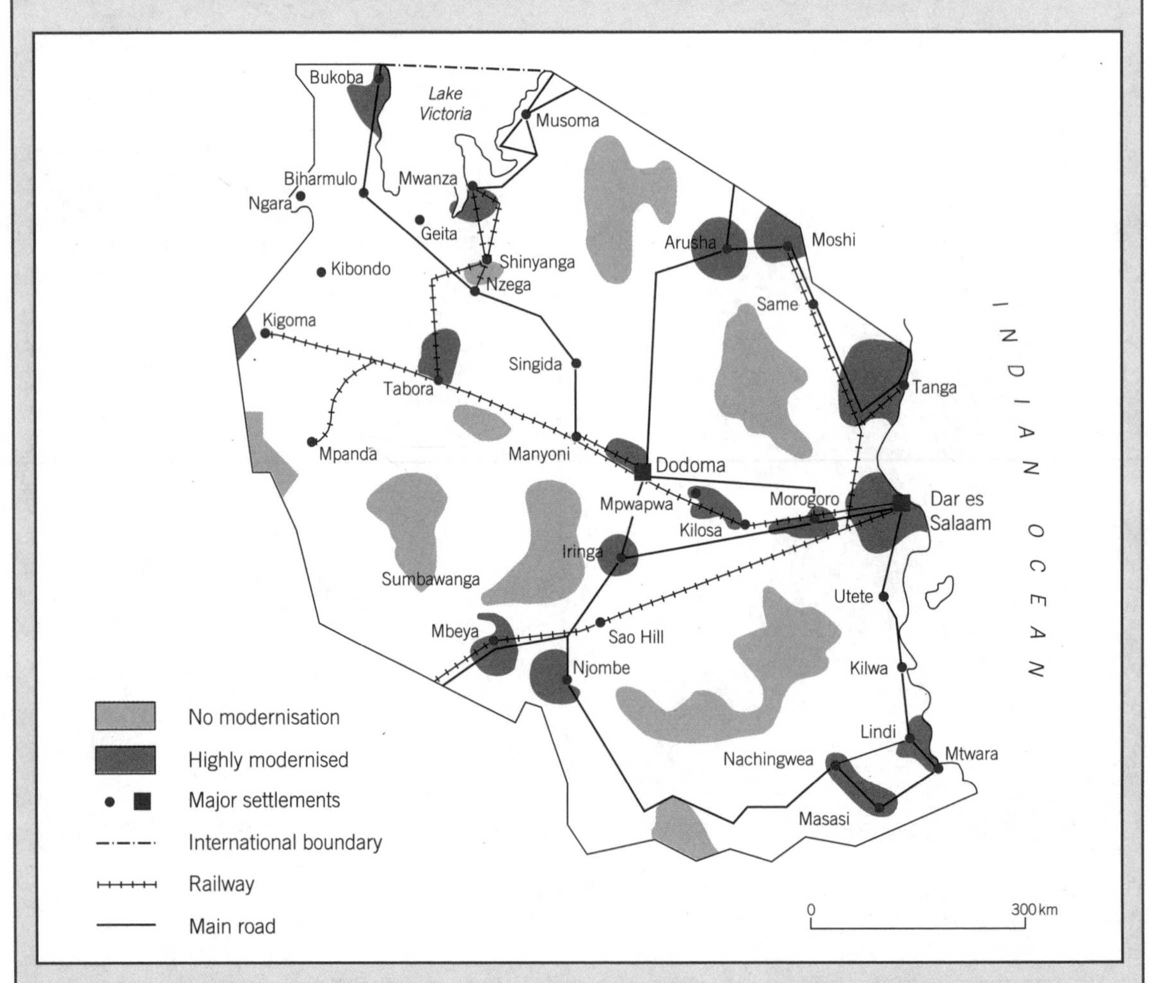

Figure 3.4 Tanzania in the 1970s: settlements and the modernisation surface
Source: Adapted from Gould and White (1974)

BOX 3.1 (continued)

However, subsequently, development in Tanzania has been far more complex than this simplistic overview implies. Thus, the principal policy efforts to reduce urban–rural differences can be identified as giving rise to the second and most important development phase, lasting from 1967 to 1972, and witnessing the emergence of a strong commitment to rural-based development, linked to strong principles of traditional African socialism.

These policies were based on the Arusha Declaration of 1967, which attacked privilege and sought to place strong emphasis on the principles of equality, cooperation, self-reliance and nationalism. Such ideas were put into practice in the second five-year plan, 1969–1974. The major policy imperative was *ujamaa* villagisation, which was regarded as the expression of 'modern traditionalism', i.e. a twentieth-century version of traditional African village life. The word *ujamaa* is Swahili for familyhood. The intention was to concentrate scattered rural populations and, by this process of villagisation, to provide the services required for viable settlements. Reducing rural-to-urban migration was a major goal, along with lessening the dependence on major cities such as Dar-es-Salaam. Ujamaa villages were envisaged as cooperative ventures by means of which initiative and self-reliance would be fostered, along socialist lines. In addition, efforts were also made to spread urban development away from Dar-es-Salaam towards nine selected regional growth centres. In overall terms, President Nyerere regarded these policies as a distinct move away from a slavish imitation of Western-style planning and development, based on uncritical 'modernisation'.

Despite having received much praise from certain quarters, the policies adopted in Tanzania have been viewed with considerable scepticism by others, especially those from a committed Marxist perspective. During the third phase, from 1973 to 1986, enforced movement to development villages occurred. Furthermore, by the fourth stage, starting in 1978, industry and urban development were once again being upgraded at the expense of ujamaa villages and rural progress. Thus, the fourth five-year plan, 1982/83 to 1985/86, gave priority to industrial development, and by this juncture the ujamaa concept appeared to have all but fallen from the consciousness of both planners and politicians alike.

Critical reflection

The word 'modernisation' is frequently used in the media and by politicians. For example, the former Prime Minister of the UK, Tony Blair, talked about 'modernisation' frequently. For a limited period, keep a note of all instances that you encounter the word – in written accounts or via broadcasts. In such references, what is it being suggested needs to be modernised? Are there clear implications as to the essential nature of modernising processes? How often is the term used in relation to processes in developing as opposed to developed nations or regions?

Empirical and conceptual elaborations of modernisation theory

The process involved in the Malaysian case is shown as an ideal-typical sequence in the four boxed diagrams depicted in the lower half of Figure 3.5. In the lower figure, T1 to T4 refer to successive time-periods. The figure essentially represents the diffusion downwards of 'development' from the largest to the smallest settlement, as shown at the top of Figure 3.5. Thus, from a critical perspective, Friedmann and Weaver (1979: 120) argued that the approach only succeeded in 'mapping the penetration of neo-colonial capitalism'.

The hallmark of this work was that it posited that modernisation is basically a temporal–spatial process. In such a vision, underdevelopment is seen as something which can be overcome, principally by the spatial diffusion of modernity. A number of studies argued that growth occurs within the settlement system from the largest urban places to the smallest in a basically hierarchical sequence.

This is shown in the upper part of Figure 3.5. Foremost among the proponents of such a view was Hudson (1989), who applied the ideas of Hagerstrand (1953), concerning spatial diffusion, to the settlement or central place system. Hudson argued that, first, innovations can travel through the settlement system by a process of contagious spread, where there is a neighbourhood or regional effect of clustered growth. This was close to Schumpeter's (1911) general economic theory, in which he argued that the essence of

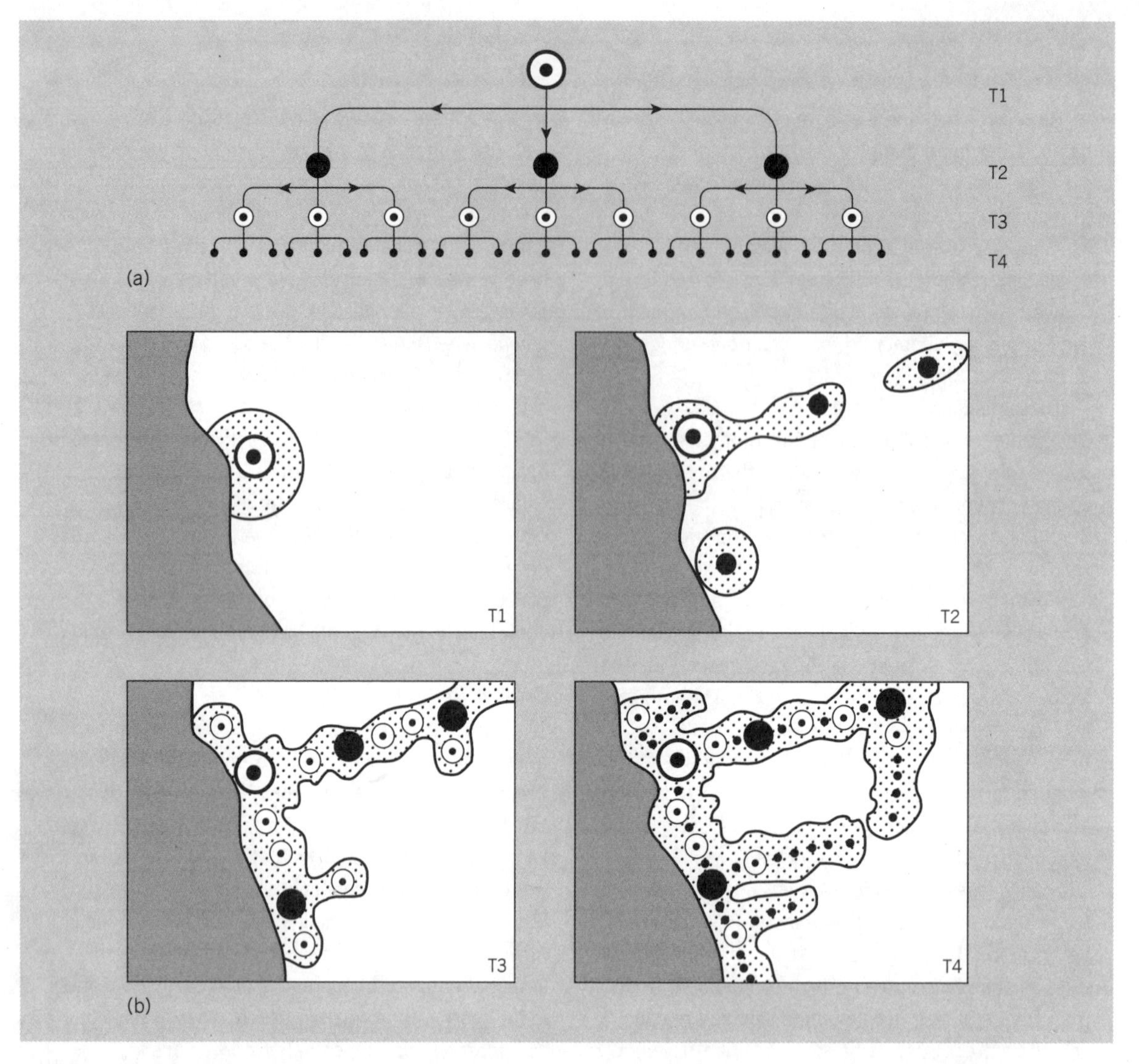

Figure 3.5 The spread of modernisation: hypothetical examples (a) down through the settlement system from the largest places to the smallest places and (b) over the national territory

development is a volume of innovations. Opportunities tend to occur in waves which surge after an initial innovation. Thus, Schumpeter argued that development tends to be 'jerky' and to appear in 'swarms', forming natural, spontaneous growth poles.

Second, Hudson noted that diffusion can occur downwards through the settlement system in a progressive manner, the point of introduction being the largest city. Pedersen (1970) argued the case for a strictly hierarchical process of innovation diffusion, an assertion which seemed to be borne out by some historical–empirical studies carried out in advanced capitalist societies such as the USA (Borchert, 1967) and England and Wales (Robson, 1973). Pedersen drew a very important distinction between domestic and entrepreneurial innovations; entrepreneurial innovations were the instrument of urban growth, not domestic innovations. In another frequently cited paper of the time, Berry (1972) also argued strongly in favour of a hierarchical diffusion process of growth-inducing innovations; this was seen as the result of the sequential market-searching procedures of firms, along with imitation effects.

But, notably, Berry's analysis was based on the diffusion of domestic as opposed to entrepreneurial innovations, namely of television receivers. In other words, it dealt with what was happening to consumption rather than production. Furthermore, the critique of modernisation has to accept that even larger firms are currently coming to dominate the world capitalist system, a major development that is detailed in Chapter 4.

The top-down paradigm of development and the 'Western world view'

All of these approaches, involving unequal and uneven growth, modernisation, urban industrialisation, the diffusion of innovations and hierarchic patterns of change and growth poles may be grouped together and regarded as constituting the 'top-down' paradigm of development (Stöhr and Taylor, 1981). Such an approach advocates the establishment of strong urban-industrial nodes as the basis of self-sustained growth and is premised on the occurrence of strong trickle-down effects, by means of which, through time, it is believed that modernisation will inexorably be spread from urban to rural areas (Figures 3.3 and 3.5). This gives rise to the concept of the planned growth pole. Case study 3.1 presents the case of Singapore, where industrial development has formed an important component of development since independence in 1965.

As with modernism, all such approaches 'had a great appeal to a wider public due to the paternalistic attitude toward non-European cultures' (Hettne, 1995: 64). These approaches, together with modernisation, reflected the desire of the USA to order the post-war world, and were used to substantiate the logic of 'authoritative intervention', as noted in Chapter 1 (Preston, 1996). As Mehmet (1999: 1) stated, 'a Western worldview is the distinctive feature of the mainstream theories of economic development, old and new'. This world view has been predominantly 'bipolar', stressing a strict belief in Western rationality, science and technology.

Rostow's Stage Model of Economic Growth

Such models, including Rostow's (1960) classic *The Stages of Economic Growth*, see urban-industrial nodes as engines of growth and development. Rostow's work can be seen as the pre-eminent theory of modernisation to appear in the early 1960s (Preston, 1996). Rostow's position was avowedly right-wing politically (see Key thinker box).

Key thinker

The contribution of Walt Rostow

Plate 3.1 Walt Rostow
Source: Getty Images/Time & Life Pictures

Walt Whitman Rostow's (1916–2003) classic work *The Stages of Economic Growth* carried the subtitle *A Non-Communist Manifesto*, bearing testimony to its highly political orientation. Rostow was fiercely anti-communist. The book, which was published in 1960 at the height of the Cold War, offered the prospect of automatic or almost formulaic growth, suggesting that by following a few simple rules the capitalist Western model could be re-enacted. As Menzel (2006) notes, Rostow was fully aware that his development theory could be employed as an instrument in East–West relations, stressing the Washington path to development. As the same author notes, Rostow's theory was a very simple formulation, which was presented and recommended with 'missionary-zeal'.

Following a series of academic and governmental posts, when John Kennedy became President of the United States Rostow became a full-time staff member and was successful in promoting development policy as US foreign policy. After the assassination of President Kennedy Rostow continued to work under the new President Lyndon B. Johnson, and did so up until Richard Nixon became President. Above all else, Rostow's work shows the strongly political nature of development theory.

Case study 3.1

Industrialisation and development: the case of Singapore

Singapore is a city state and in 1996 it had a total population of 3,044,000 persons. Its per capita income stood at US$32,810 in 2000, higher than for many European countries. Further, in 1996, the country recorded a 7.3 per cent growth in its GDP. Indeed, Singapore is frequently held up as a nation which has created a strong 'Third World' economy in a relatively short period. Yet when Singapore became an independent republic in 1965 the prospects for growth did not seem much better than for many newly emerging nations. As noted by Drakakis-Smith (2000), Singapore is a very clean and green city which has led by example with respect to its environmental policies. But, as the same author notes, in social terms there has been a price to pay for this continued growth.

Singapore is a good example of a state that has grown by early industrialisation. The programme which was embarked upon in 1968 focused on both light industrialisation and some forms of heavy industry, such as oil refining, iron and steel, ship building and repairing. Thus, the contribution of manufacturing to GDP grew from 11.9 per cent in 1960, the average figure for developing nations, to 29.1 per cent in 1980. By 1995, this figure had increased to over one-quarter of GDP, standing at 26.5 per cent. The Government of Singapore was one of the first in Asia to realise fully the limitations of growth via low technology industrial development. Accordingly, throughout the 1980s it sought to transform the economy by focusing on high-tech, high-value-added industries. It was fully intended that this 'second industrial revolution' would transform Singapore into the 'Switzerland' of Asia. Singapore's major trading partners are now the USA, Malaysia, Hong Kong, Japan and the European Union. The city-state has also become an important centre for financial services, with 149 commercial banks and 79 merchant banks in 2000 (Whitaker's, 1999).

The World Bank frequently holds Singapore up as a model of what can be achieved by free market policies and industrial development. But several analysts have noted that there are local factors and that these are unlikely to be repeated elsewhere. For instance, Drakakis-Smith (2000) argues that Singapore's success has been strongly predicated on its people, and that the degree of human resource management has been intensive and ultimately authoritarian. Specifically, as in many states in the region, there have been very tough controls on labour unions, with the general introduction of factory unions rather than occupational unions. Many argue that this has also served to reduce social class solidarity. Japanese-style company loyalty has been the desired outcome. In addition, since the 1960s there has been strict state population control, both in respect of migration and also directed education programmes, with children being allocated to 'hand' and 'brain' streams at an early stage of their education. Drakakis-Smith (2000) also notes that ethnic disparities in wealth are prominent in Singapore, with Malays forming the most disadvantaged group, but within a general societal context where middle-class consumerism dominates.

Rostow envisaged that there were five stages through which all countries have to pass in the development process: the traditional society, preconditions to the take-off phase, take-off, the drive to maturity, and the age of mass consumption, as depicted in Figure 3.6. Rostow's stage model encapsulates faith in the capitalist system, as expressed by the subtitle of the work: a non-communist manifesto for economic growth. For Rostow, the critical point of take-off can occur where the net investment and savings as a ratio to national income grows from 5 to 10 per cent, thereby facilitating industrialisation.

Although Rostow's framework can in many respects be regarded as a derivation of Keynesian economics, its real significance lies in the simple fact that it seemed to offer every country an equal chance to develop (Preston, 1996). In particular, saliently, the 'take-off' period was calibrated at 20 or so years, long enough to be conceivable but short enough not to seem unattainable. The importance of the Rostowian framework was that it purported to explain the advantages of the Western development model. Further, in the words of Preston (1996: 178), the 'theory of modernisation follows on from growth theory but is heavily influenced by the

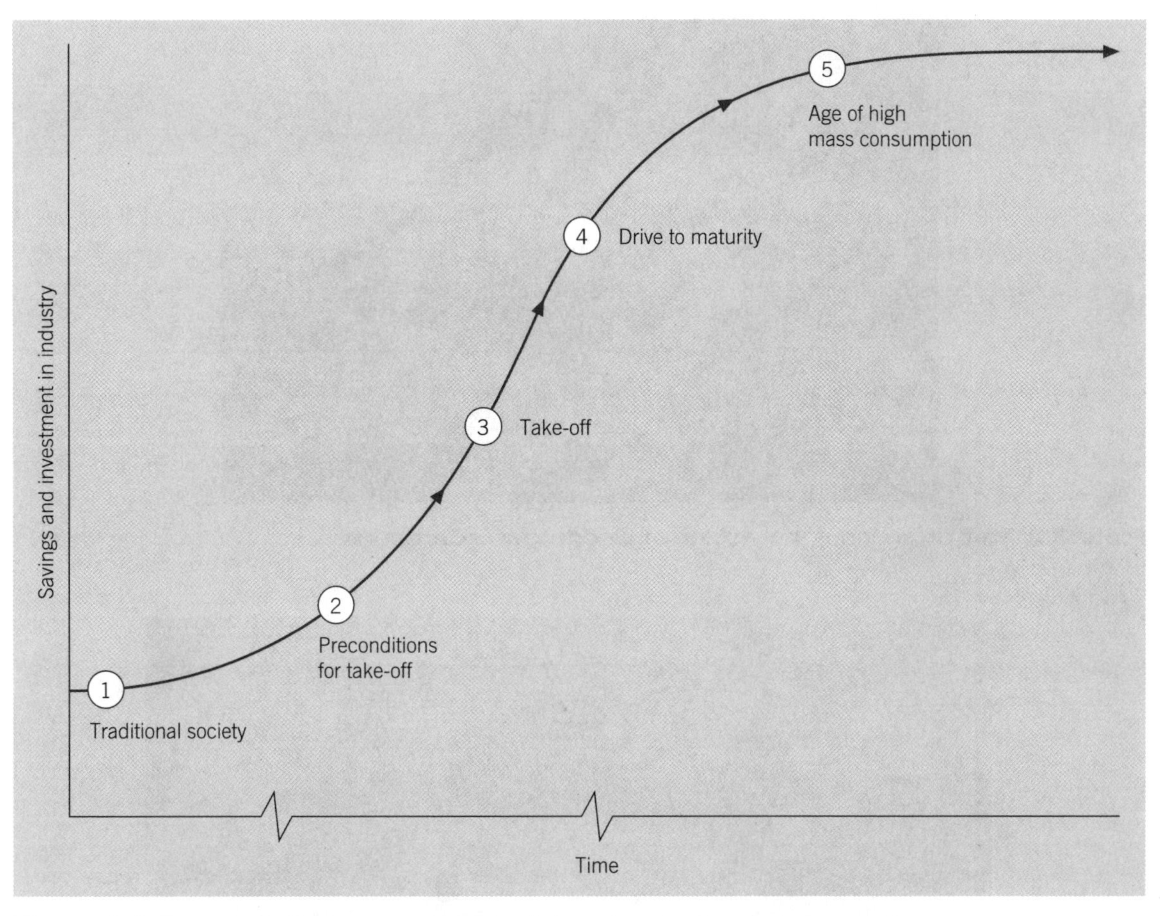

Figure 3.6 Rostow's five-stage model of development

desire of the USA to combat the influence of the USSR in the Third World'.

The central argument was that developing nations needed to industrialise in order to develop. The various approaches that could be followed in pursuing this aim are reviewed in Chapter 4, and include import substitute industrialisation (ISI), industrialisation by invitation (I by I) as well as big-push industrial programmes.

For a number of reasons, it was the industrialisation by invitation model that received the most attention. One reason for this was the influential work of Lewis (1950, 1955), an economist of West Indian origin who was working at the University of Manchester at the time. Lewis set out the foundations of modernisation theory when he maintained that the juxtaposition of a backward traditional sector with an advanced modern sector meant that an 'unlimited supply of labour' existed for development. This duality means that industry can expand rapidly if industrialisation is financed by foreign capital the argument ran. This led to the so-called policy of industrialisation by invitation (Plates 3.2 and 3.3). It was ironic that a St Lucian economist should use the metaphor of a snowball, arguing that once the process started to move it would develop its own self-sustaining momentum, like a snowball rolling downhill. This is, of course, an essentially similar argument to that of Walt Rostow.

The evaluation of modernisation and top-down approaches

Indeed, all such formulations place absolute faith in the existence of a linear and rational path to development, based on Western positivism and science, and the possibility that all nations can follow this in an unconstrained manner. All such thinking was directly related to the 'enlightenment' (see Power, 2008). Modernism was very much an urban phenomenon from 1850 onwards (Harvey, 1989). What is often referred to as 'universal' or 'high' modernism became hegemonic after 1945. Thus, the top-down approach was strongly associated with the 1950s, through to the early 1970s.

Plate 3.2 Part of an industrial estate in Bridgetown, Barbados
(photo: Rob Potter)

Plate 3.3 US-owned baseball factory in Port au Prince, Haiti
(photo: Sean Sprague, Panos Pictures)

Taken together, many writers refer to these theories as representing 'Eurocentric development thinking', i.e. development theories and models rooted in Western European history and experience (Hettne, 1995; Mehmet, 1999; Slater, 1992a, 1992b). Via such approaches, during the 1950s and 1960s development was seen as a strengthening of the material base of society, principally by means of industrialisation (Plate 3.4). Inevitably, the history of the first industrial state was taken as the model which should be followed, not only by the rest of Europe but ultimately by the rest of the world, for 'it is quite natural that the original recipe for development given by the developed countries should emanate from their own experiences and prejudices' (Hettne, 1995: 37). It is notable that all the early theories of development were authored by men, and that virtually all of them were of Anglo-European origin.

It would be wrong, however, to give the impression that the focus on top-down, urban-industrial growth, linked to the quest for modernisation, was associated with a single and unified path. Four more or less distinct development strategies making up the early Western tradition can be identified (Hettne, 1995):

Plate 3.4 A heavily polluted town with the Pannex refinery in the background, Mexico
(photo: Ron Giling, Panos Pictures)

1 *The liberal model*, the strategy implicitly discussed through much of the above account, stresses the importance of the free market and largely accepts the norm based on English development experience during the Industrial Revolution. In the 1980s and 1990s such views gained fresh currency in the form of structural adjustment programmes (SAPs) enforced by the International Monetary Fund (IMF), United States Agency for International Development (USAID) and the World Bank as the condition of loans given for so-called 'economic restructuring'. However, there have been major outcries against the impact of SAPs, principally in respect of the harm that they do to the poor, and especially to women and children (see Chapter 7). This is fully exemplified by the Oxfam campaign that likened SAPs to medicines which need to be withdrawn from the marketplace, due to their risk to the health of vulnerable groups (see Figure 3.7). Noticeably, by 2000 all reference to SAPs had been expunged from the World Bank website (Simon, 2008). More positive sounding programmes focusing on poverty reduction strategies (PRSs) are now the order of the day. There seems little doubt that anti-capitalist demonstrations since Seattle in 1999 have also been influential in this respect. In these recent developments the battle between the left and the right, and its outcome in outpourings of rhetoric and labelling can be witnessed within the liberal and neo-liberal paths. Neo-liberal policies will receive further attention in several sections later in this book; meanwhile its essential nature is reviewed in the Key idea box.

2 *Keynesianism* departs from the liberal tradition by virtue of arguing that the free-market system does not self-regulate effectively and efficiently, thereby necessitating the intervention of the state in order to promote growth in capitalist systems. Since the 1930s, Keynesianism has been a prominent development ideology in the industrialised capitalist world, especially in countries with a social democratic tendency (Hettne, 1995).

3 *State capitalist strategies* refer to an early phase of industrial development in continental Europe, principally tsarist Russsia and Germany. The approach advocated the development of enforced industrialisation based primarily on agrarian economies in order to promote nationalism and for reasons pertaining to national security.

4 *The Soviet model* represents a radical state-oriented strategy inspired by Stalin's five-year mandatory economic development plans. The approach regarded modernisation as the goal, to be achieved by means of the transfer of resources from agriculture to industry. The agricultural sector was collectivised, and heavy industry was given the highest priority. The state completely replaced the market mechanism.

Key idea

Neo-liberalism

Liberalism, as the belief in free markets and the abolition of government intervention in the economy, dates back to the English economist Adam Smith in his *Wealth of Nations* (1776). In contrast to the view of Smith, during the Great Economic Depression of the 1930s the renowned economist John Maynard Keynes argued that governments needed to be involved in creating employment in order to steer economies out of recession. However, with the rise of what is referred to as the 'New Right' in the 1980s, there was a return to calls for a strongly market-driven approach, which is referred to as *Neo-liberalism*, that is new forms of liberal free-market policy.

The chief proponents of the approach were the politicians Ronald Reagan, President of the USA, and Margaret Thatcher, the Prime Minister of the United Kingdom. Their new political project argued that the state should be progressively removed from the economy, with this ideology frequently being referred to as 'the rolling back of the state', and that measures should be taken to deregulate the economy. Since then, the tenets of neo-liberalism have become the policy orthodoxy of international development agencies such as the World Bank and the International Monetary Fund (Power, 2003).

As a critique, Conway and Heynen (2006) argue, citing Bourdieu (1998), that neo-liberalism is more than a belief in free trade. They suggest that the neo-liberal doctrine is based on what they refer to as the structured violence of unemployment, job insecurity and the threatened layoff from work – in other words, that neo-liberalism is a coercive economic system. It is a short step to the argument that the system stresses the supremacy of economic entrepreneurs over the subordination of nation states and that it has exacerbated global inequalities and the global divide reviewed in Chapter 1. The same authors argue that neo-liberalism is serving to hit the Third World once more, just as slavery and colonialism did in the past. Neo-liberalism is giving rise to accumulation and self-interest over communal obligations and social obligations to neighbours and the wider community.

Critical reflection

How do you respond to these arguments? Thinking of a sector of the economy, be it manufacturing, agriculture, education, health or indeed any other, outline the major changes that you know have occurred to it over the past say 15–20 years. How many of these fundamental shifts can be seen as the direct outcome of neo-liberal policies? In the case of many nations, the university sector is an interesting example to consider from this perspective, as there have been many changes.

The common denominator linking these four approaches is an unswerving faith in the efficacy of urban-based industrial growth (Plate 3.5), although some approaches in the early modernisation phase also emphasised resource-based development strategies. Notwithstanding the variations noted above, Hettne (1995) comments on the fact that through time the role of the state has generally been central to Western development strategies. This was certainly true of the Keynesian, state-capitalist and Soviet models. However, the recent orthodoxy has seen the dismantling of the welfare state and the reduced scope of government in the form of neo-liberalism. Arguing that the modern welfare state together with trades unions and state bureaucracies have destroyed the market system, Friedman (1962) and others argued against Keynesianism and promoted the New Right.

In this way, the New Right neo-liberal theorists have been seen as having given rise to a counter-revolution, celebrating the unrestrained power of the unregulated free market, arguing both for its economic efficiency and its role in liberating the social choices of the individual. There can be little doubt that the general pro-market position of the New Right has served to inform the policies of the World Bank, the IMF and the United States government since the 1980s.

Thus, as detailed in many of the chapters of this book, but particularly Chapter 7, the World Bank and the IMF have pressed for economic liberalisation, the elimination of market imperfections and market-inhibiting social institutions, plus the redefinition of planning regulations in Third World countries. Some sections of the New Right have even gone so far as to argue that the 'Third World only exists as a figment of

Figure 3.7 Oxfam's campaign against IMF policies including SAPs

the guilty imaginations of First World scholars and politicians' (Preston, 1996: 260).

In Britain, neo-liberalism was witnessed in the form of Thatcher's popular capitalism in the 1980s, and in America in the guise of Reaganism. Both Reagan and Thatcher saw the extension of the market into new fields such as hospitals, schools, universities and other public sector establishments, often in the form of performance-related league tables and ratings. On the global scale, the 1980s also saw 'liberalisation' being advocated in the Third World, especially by the monetarist school, which advanced an extreme laissez-faire approach, as in the so-called 'global Reaganomics'. Using the example of the newly industrialising countries (NICs), particularly those of Asia, countries were advised to liberalise their economies, encourage entrepreneurship and to seek comparative cost advantage. From the 1980s, monetarism has become the firm policy of the World Bank and the International Monetary Fund.

Plate 3.5 'Urban-based modernisation: the CBD of Johannesburg, South Africa
(photo: Tony Binns)

However, Preston (1996: 260), among others, has argued that 'the schedule of reforms inaugurated by the New Right have not generally proved to be successful'. It is also true to say that the rhetorically important attempt to annex the development experiences of Pacific Asia to the position of the New Right has been widely ridiculed by development specialists. One of the reasons for this is the 'Krugman thesis'. This argues that the so-called miracles of Asian development can largely be attributed to once-in-a-lifetime changes which, having once been enacted, cannot by definition be repeated. This has included massive increases in female participation in the labour force (Watters and McGee, 1997; see also Case study 3.1). Others have pointed to the lack of unionisation, poor working conditions and authoritarian government as factors promoting high productivity, but in ways which are not acceptable elsewhere (see also Case study 3.1).

A further example is to be found in the case of the Caribbean, where the World Bank is wont to point to the low wage levels and high productivity of the Asian tiger economies. Local Caribbean economists and policy makers respond by saying that the policies which underlie Asian development are just not feasible in the Caribbean context. This dialogue can be followed in the World Bank publication edited by Wen and Sengupta (1991).

Historical approaches: empirical perspectives on change and development

The nature of historical approaches

Another way in which scholars and practitioners can seek to generalise about development is by empirical or real-world observations through time. By definition, this approach will give rise to descriptive–positive models of development (see Figure 3.2), and some feel these frameworks have a key role to play in the discussion of development, specifically for grounding theory in the historical realities of developing nations. Although such approaches deal primarily with the colonial and pre-independence periods, it can be argued that they may still afford insights regarding contemporary patterns and processes of development and change.

Gunnar Myrdal and cumulative causation

In contrast to Hirshman, the Swedish economist Gunnar Myrdal (1957), although writing at much the same time, took a noticeably more pessimistic view, maintaining that capitalist development is inevitably marked by deepening regional and personal income and welfare inequalities. Myrdal followed the arguments of the vicious circle of poverty in presenting his theory of 'cumulative causation'. Thereby, it was argued that once differential growth has occurred, internal and external economies of scale will serve to perpetuate the pattern.

Such a situation is the outcome of the 'backwash' effect, whereby population migrations, trade and capital movements all come to focus on the key growth points of the economy. Increasing demand, associated with multiplier effects, and the existence of social facilities also serve to enhance the core region. Although 'spread' effects will undoubtedly occur, principally via the increased market for the agricultural products and raw materials of the periphery, Myrdal concluded that, given unrestrained free-market forces, these spread effects would in no way match the backwash effects. Myrdal's thesis leads to the advocacy of strong state policy in order to counteract what is seen as the normal tendency of the capitalist system to foster increasing regional inequalities.

Core–periphery and the work of John Friedmann

The view that, without intervention, development is likely to become increasingly polarised in transitional societies was taken up and developed by a number of scholars towards the end of the 1960s and the beginning of the 1970s. As such, they ran counter to the conventional wisdom of the time. These works were based mainly on empirical studies which encompassed an historical dimension. Undoubtedly, the best-known example is provided by American planner John Friedmann's (1966) core–periphery model. From a purely theoretical perspective, Friedmann's central contention was that 'where economic growth is sustained over long time periods, its incidence works toward a progressive integration of the space economy' (Friedmann, 1966: 35). This process is made clear in the much-reproduced four-stage ideal-typical sequence of development shown here in Figure 3.8.

The first stage, independent local centres with no hierarchy, represents the pre-colonial stage and is associated with a series of isolated self-sufficient local economies. There is no surplus production to be concentrated in space, and an even and essentially stable pattern of settlement and socio-economic development is the result.

In the second stage, a single strong centre, it is posited that as the result of some form of 'external disruption' – a euphemism for colonialism – the former stability is replaced by dynamic change. Growth is envisaged to occur rapidly in one main region and urban primacy is the spatial outcome. What may be referred to as 'social surplus product' is strongly concentrated – essentially this is a concentrated surplus of production over need. The centre (C) feeds on the rest of the nation, and the extensive periphery (P) is drained. Advantage tends to accrue to a small elite of urban consumers, who are located at the centre. However, Friedmann regarded this stage as inherently unstable.

The outcome of this instability is the development of a single national centre with strong peripheral sub-centres. Over time, the simple centre–periphery pattern is progressively transformed to a multi-nuclear one. Sub-cores develop (SC1, SC2), leaving a series of intermetropolitan peripheries (P1 to P4). This is the graphical representation of the point of polarisation reversal when development starts to be concentrated in parts of the former periphery, albeit on a highly concentrated basis.

The fourth and final stage, which sees the development of a functionally interdependent system of cities, was described by Friedmann as 'organised complexity' and is one where progressive national integration continues, eventually witnessing the total absorption of the intermetropolitan peripheries. A smooth progression of cities by size and a smooth process of national development are envisaged as the outcomes.

An evaluation of core–periphery models and frameworks

The first two stages of the core–periphery model describe directly the history of the majority of developing countries. Indeed, it is often not appreciated that, in the first stage, the line along which the small independent communities are drawn represents the coastline. The occurrence of uneven growth and urban concentration in the early stages of growth is seen as being the direct outcome of exogenic forces. Thus, Friedmann commented that the core–periphery relationship is essentially a colonial one, his work having been based on the history of regional development in Venezuela.

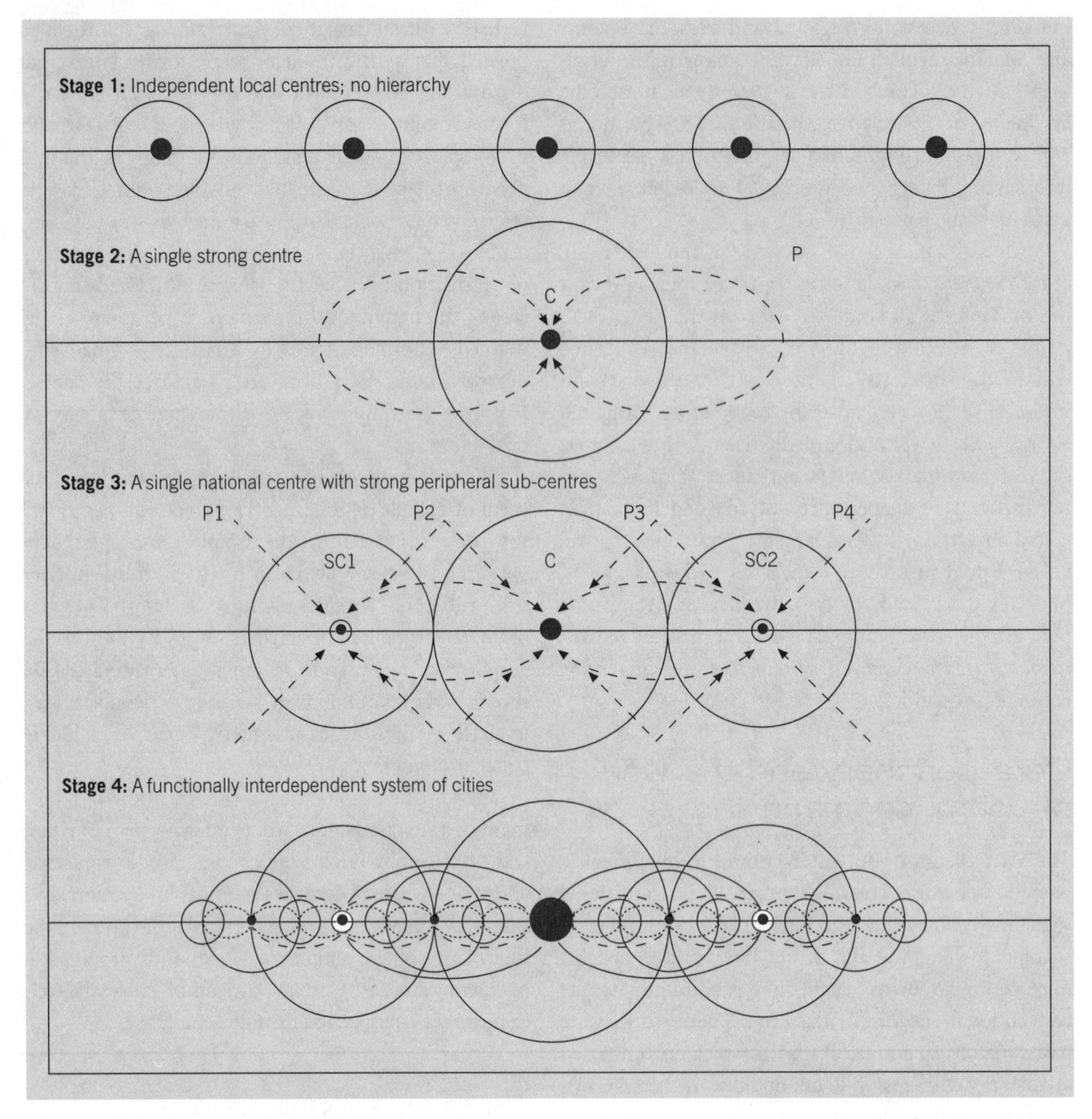

Figure 3.8 An overview of Friedmann's core–periphery model
Source: Adapted from Friedmann (1966). © 1966 Massachusetts Institute of Technology, by permission of the MIT Press

The principal idea behind the centre–periphery framework is that, early on, factors of production will be displaced from the periphery to the centre, where marginal productivities are higher. Thus, at an early stage of development nothing succeeds like success. However, the crucial change is the transition between the second and third stages, where the system tends towards equilibrium and equalisation. Friedmann's model is one which suggests that, in theory, economic development will ultimately lead to the convergence of regional incomes and welfare differentials.

But at the very same time as he was presenting the simplified model as a template, Friedmann observed that in reality there was evidence of persistent disequilibrium. Thus, in a statement which appeared right alongside the model, Friedmann (1966: 14) observed that there was 'a major difficulty with the equilibrium model: historical evidence does not support it'. Despite this damning caveat, many authors have represented the model as a statement of invariant truth, ignoring Friedmann's warning that 'disequilibrium is built into transitional societies from the start' (Friedmann, 1966: 14).

Effectively, Friedmann was maintaining that, without state intervention, the transition from the second stage to the third stage will not occur in developing

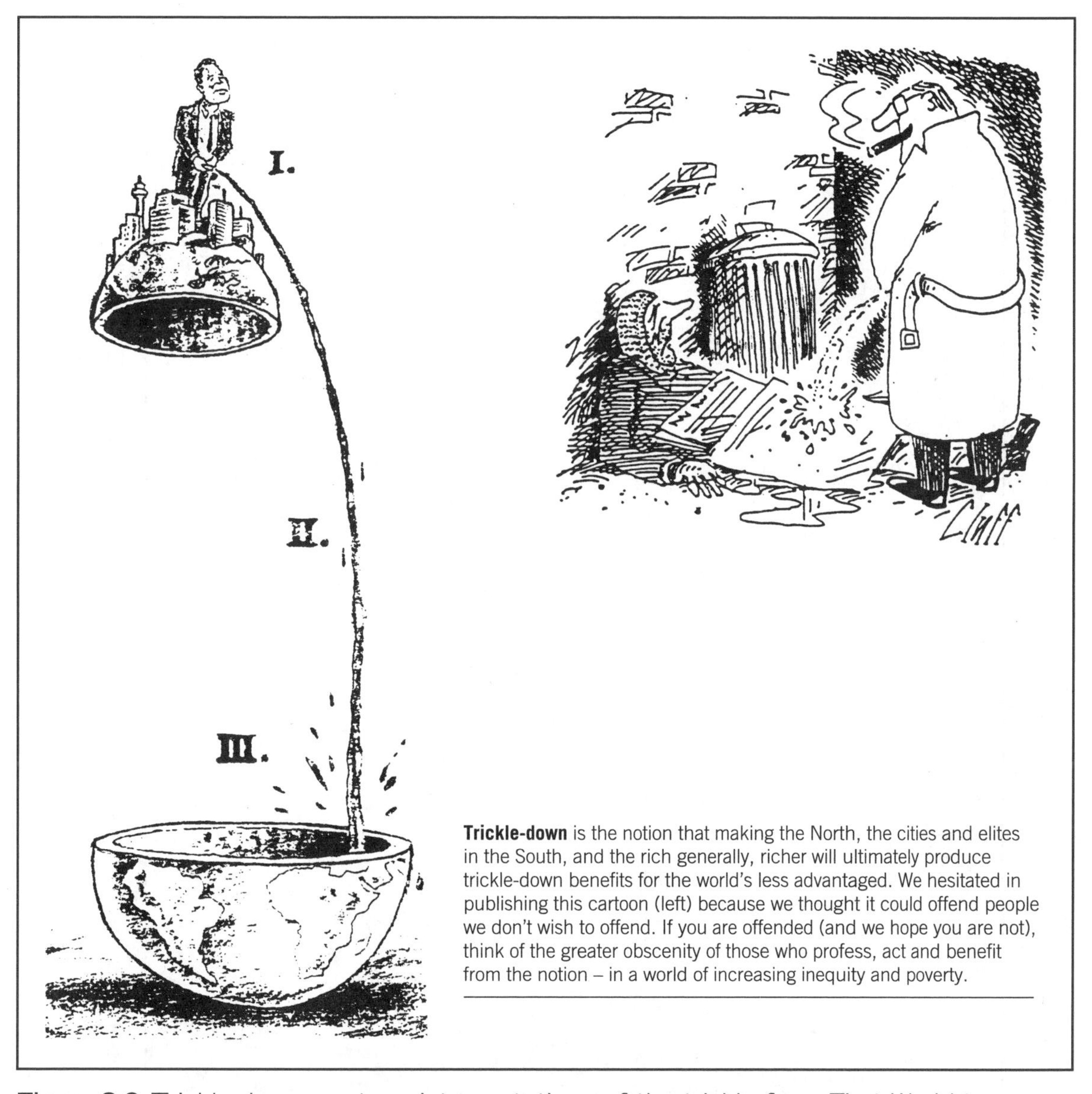

Figure 3.9 Trickle-down: cartoon interpretations of the trickle from First World to Third World (left), and From rich to poor within a single country (right)
Source: Cartoon on left from *Te Amokura, Na Lawedua, The New Zealand and Pacific Islands Development Digest*; cartoon on right reprinted with kind permission from *Private Eye*

societies; in this respect he was in agreement with Myrdal's prescription that development will become ever more concentrated in space, with polarisation always tending to exceed the so-called trickle-down effect.

Figure 3.9 depicts a somewhat irreverent view of this argument, stressing a lack of faith in spontaneous trickle-down, and the view that such forces will always advantage elites at the expense of the poor. Meanwhile, Figure 3.10 gives a more practical illustration of the degree to which growth-inducing facilities may remain highly concentrated in core regions. The figure shows the location of both domestic and foreign research and development facilities in Europe in 1999–2000. The map also shows the location of major universities. The contrast between the north-west European core and the rest of Europe could hardly be greater.

The mercantile model and spatially uneven development

Writing just a few years after the appearance of Friedmann's much-cited model, an American geographer Jay E. Vance (1970) noted that it was with the development of mercantile societies from the fifteenth

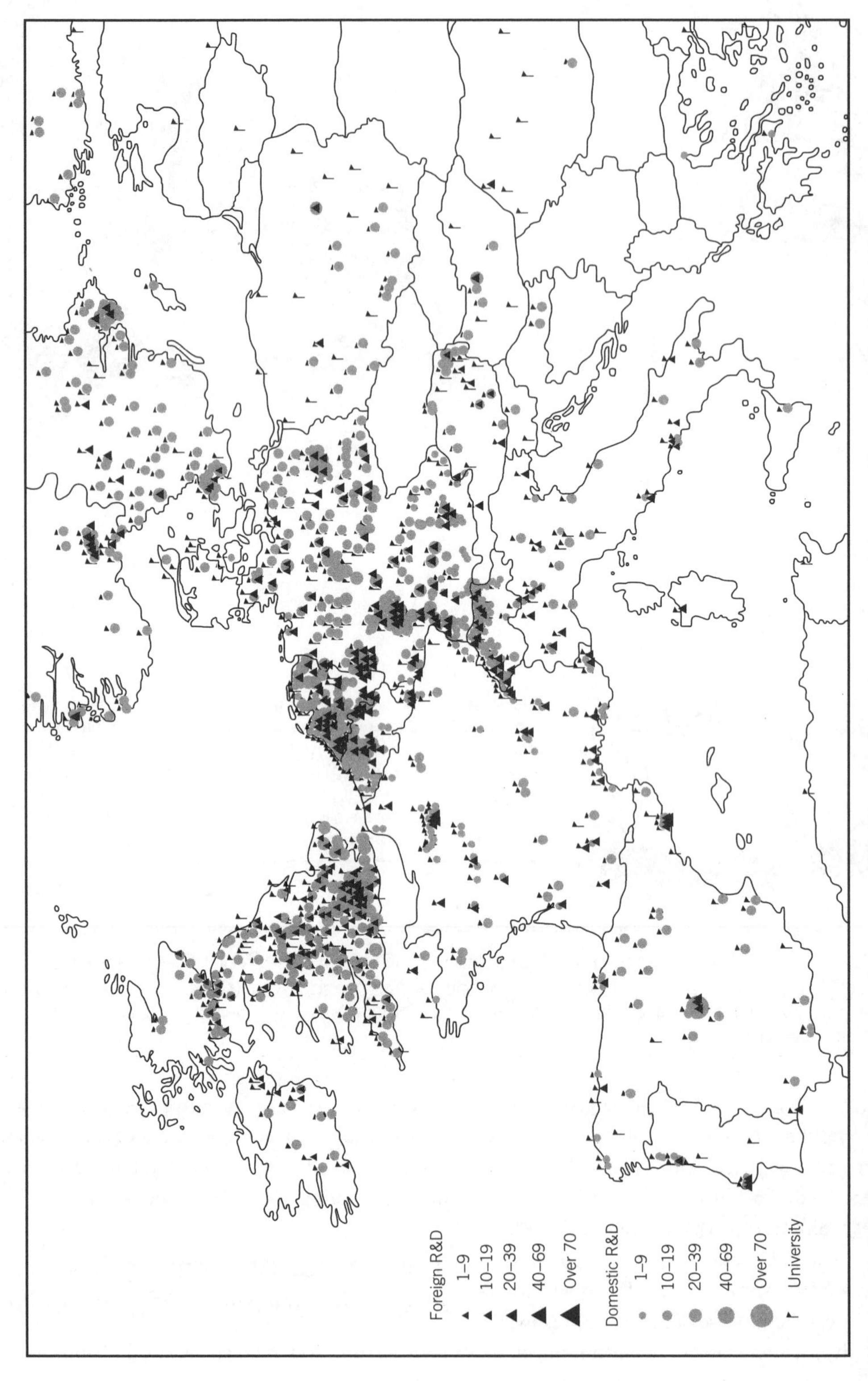

Figure 3.10 Research and development facilities and major universities in Europe, 1999–2002
Source: Adapted from UNCTD (2001)

century onwards that settlement systems started to evolve along more complex lines.

The main development came with colonialism, for continued economic growth required greater land resources. Frequently, this requirement was initially met by local colonial expansion via trading expeditions. By the seventeenth and eighteenth centuries, however, this need was increasingly fulfilled by distant colonialism, the transoceanic version of local colonialism. The implications of these historical developments have been well summarised by Vance (1970: 148):

> **The vigorous mercantile entrepreneur of the seventeenth and eighteenth centuries had to turn outward from Europe because the long history of parochial trade and the confining honeycomb of Christaller cells that had grown up with feudalism left little scope there for his activity. With overseas development, for the first time the merchant faced an un-organised land wherein the designs he established furnished the geography of wholesale-trade location. By contrast, in a central-place situation (such as that affecting much of Europe and the Orient), to introduce wholesale trade meant to conform to a settlement pattern that was premercantile.**

During the period of mercantilism (see also Chapter 2), ports came to dominate the evolving urban systems of both the colony and the colonial power. In the colony, once established, ports acted as gateways to the interior lands.

Subsequently, evolutionary changes occurred that first saw increasing spatial concentration at certain nodes, then lateral interconnection of the coastal gateways and the establishment of new inland regions for expansion. The settlement pattern of the homeland also underwent considerable change, for social surplus product flowed into the capital city and the principal ports, thus serving to strengthen considerably their position in the urban system.

These historical facets of trade articulation led Vance (1970) to suggest what amounted to an entirely new model of colonial settlement evolution; one that was firmly based on history. This is referred to as the mercantile model, and its main features are summarised in Figure 3.11. The model is illustrated in five stages. In each of these, the colony is shown on the left of the figure, and the colonial power on the right:

1. The first stage represents the initial search phase of mercantilism, involving the quest for economic information on the part of the prospective colonising power.
2. The second stage sees the testing of productivity and the harvest of natural storage, with the periodic harvesting of staples such as fish, furs and timber. However, no permanent settlement is established in the colony.
3. At the third stage, the planting of settlers who produce staples and who consume the manufactures of the home country occurs. The settlement system of the colony is established via a point of attachment. The developing symbiotic relationship between the colony and the colonial power is witnessed by a sharp reduction in the effective distance separating them. The major port in the homeland becomes pre-eminent.
4. The fourth stage is characterised by the introduction of internal trade and manufacture in the colony. At this juncture, penetration occurs inland from the major gateways in the colony, based on staple production. There is rapid growth of manufacturing in the homeland to supply the overseas and home markets. Ports continue to increase in significance.
5. The fifth and final stage sees the establishment of a mercantile settlement pattern and central-place infilling occurs within the colony; there emerges a central-place settlement system with a mercantile overlay in the homeland.

The mercantile model stresses the historical–evolutionary viewpoint in examining the development of national patterns of development and change. The framework offers what Vance sees as an alternative and more realistic picture of settlement structure, based on the fact that in the seventeenth and eighteenth centuries mercantile entrepreneurs turned outwards from Europe. Hence the source of change is external to developing countries. In contrast, the development of settlement patterns and systems of central places in the developed world was based on endogenic principles of local demand, thereby rendering what is essentially a closed settlement system (Christaller, 1933; Lösch, 1940).

The hallmark of the mercantile model is the remarkable linearity of settlement patterns, first along coasts, especially in colonies, and then along the routes that developed between the coastal points of attachment and the staple-producing interiors.

These two alignments are also given direct expression in Taaffe *et al.*'s (1963) model of transport development

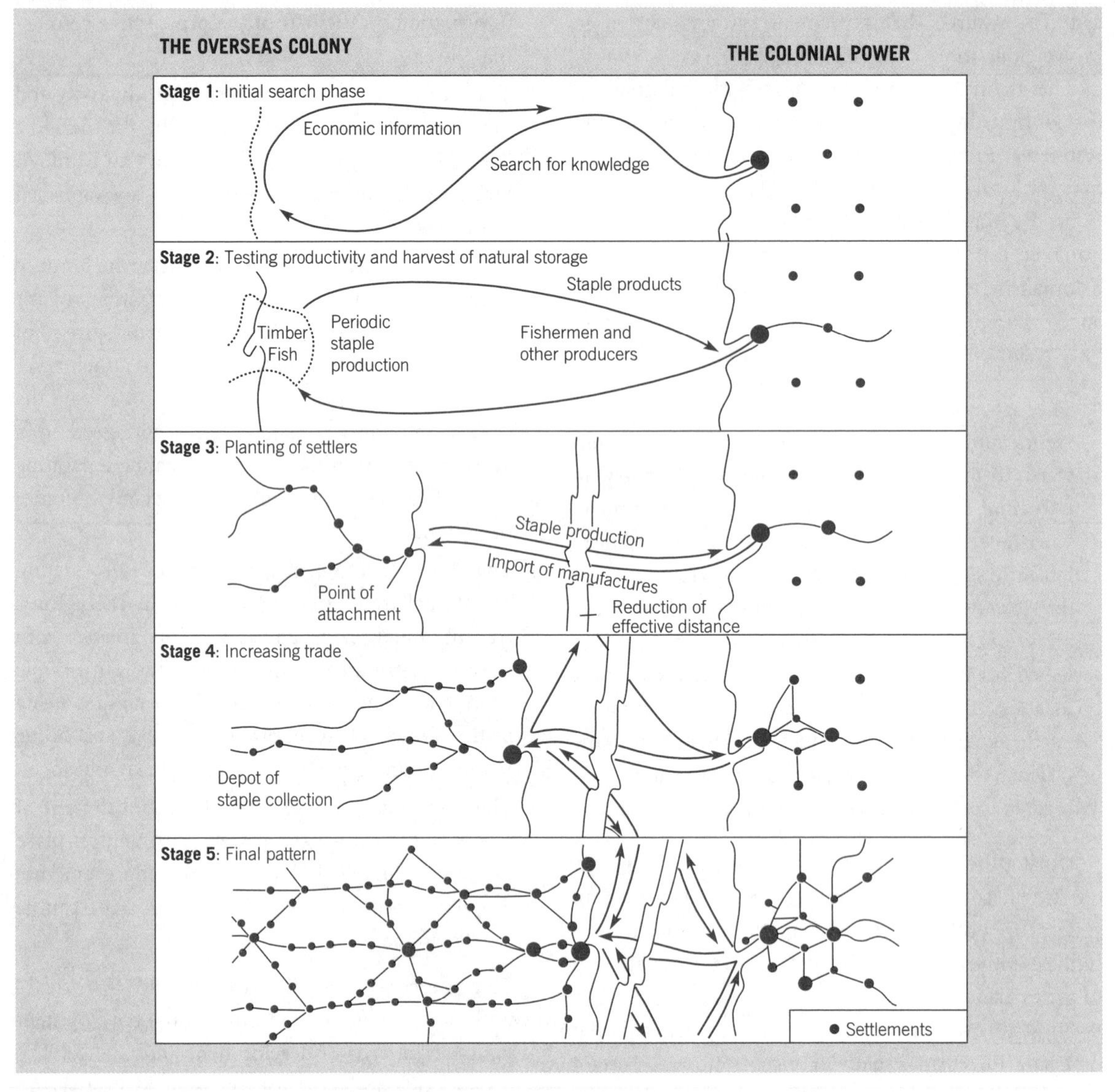

Figure 3.11 Vance's mercantile model: a simplified version
Source: Adapted from Potter (1992a)

in less developed countries, as shown in Figure 3.12a. The model was based on the transport histories of West African nations such as Nigeria and Ghana, plus Brazil, Malaya and East Africa. Hoyle's (1993) application of the framework to East Africa is shown in Figure 3.12b.

Figure 3.13 deals specifically with the example of Brazil, showing the foundation of the earliest towns. The early ports along the Atlantic coast were small and served thinly settled hinterlands. The evolving plantation economy later focused on sugar mills and gave rise to small inland settlements. The concentration of the first 12 towns of Brazil on the seaboard is very apparent from Figure 3.13.

Plantopolis and uneven development

In plantation economies such as those of the Caribbean, a local historical variant of the mercantile settlement system is provided by the plantopolis model. A simplified representation of this is shown in Figure 3.14. The first two stages are based on Rojas (1989), although the graphical depiction of the sequence and its extension to the modern era have been effected by Potter (1995a, 2000):

1 In the first stage, plantopolis, the plantations formed self-contained bases for the settlement pattern, such

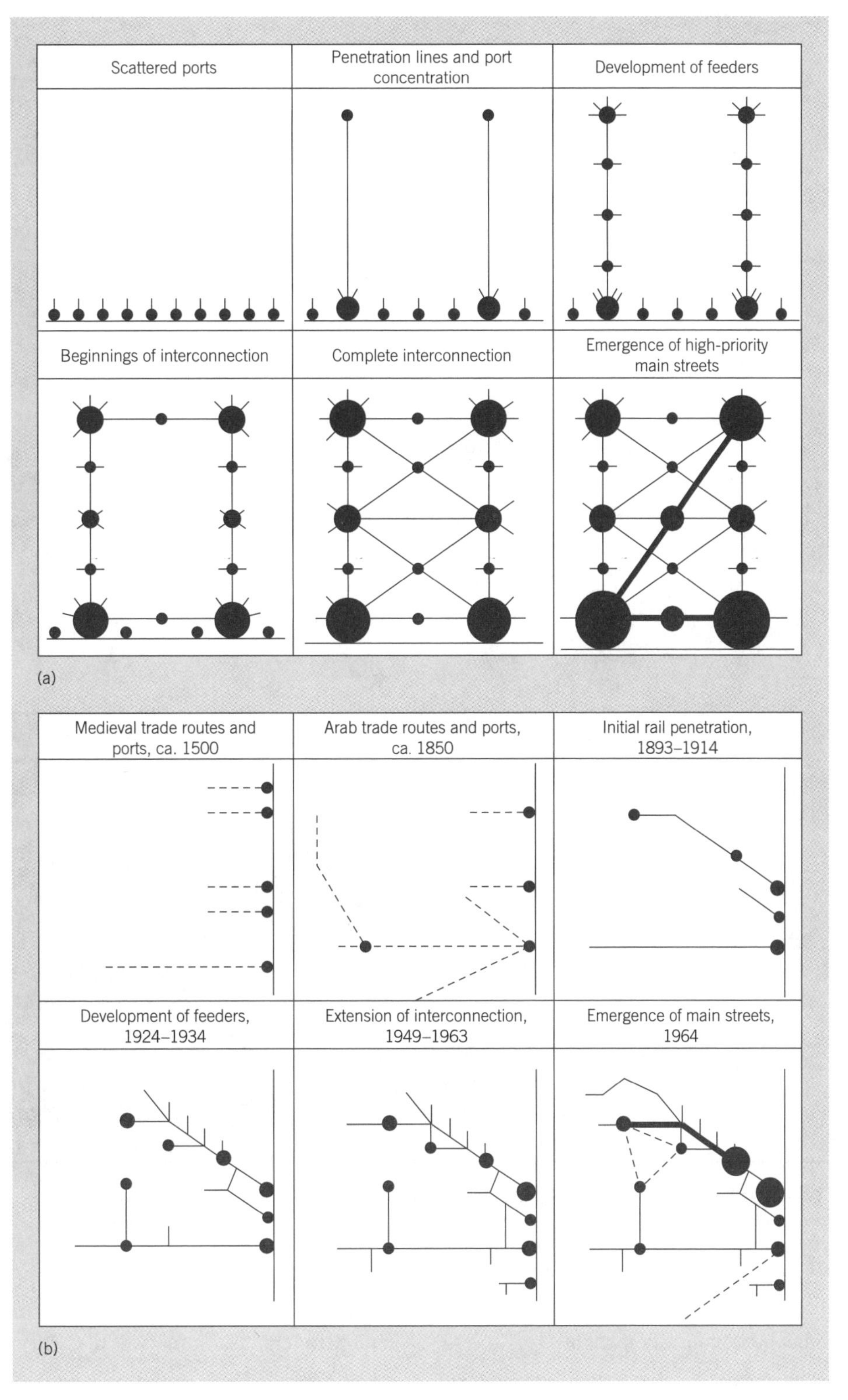

Figure 3.12 The Taaffe, Morrill and Gould model of transport development and its application to East Africa
Source: Adapted from Hoyle (1993)

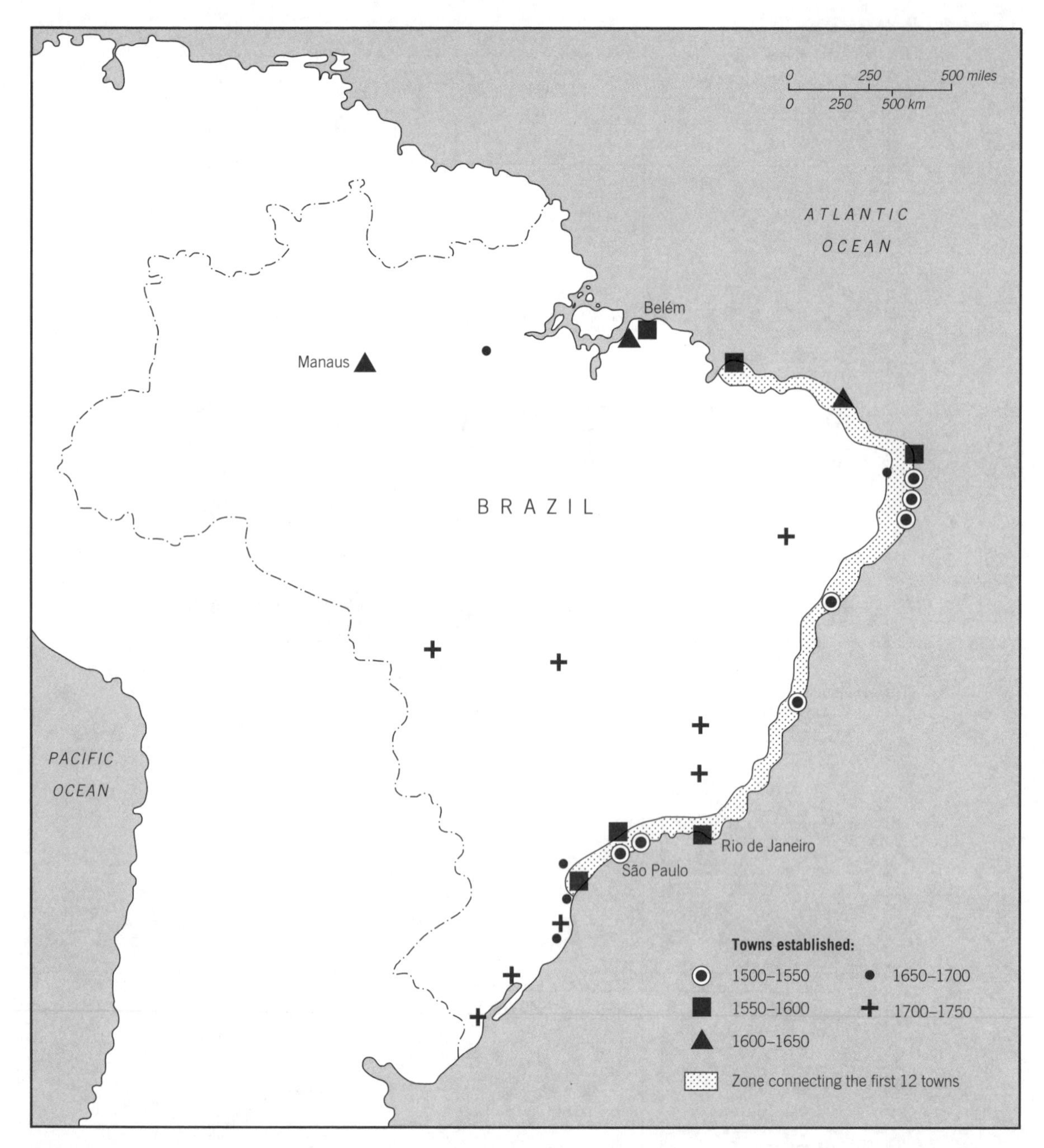

Figure 3.13 The foundations of the earliest towns in Brazil, 1500–1750

Source: From *Latin America and the Caribbean: A Systematic and Regional Survey*, (Blouet, B.W. and Blouet, O.M.), © 2002 John Wiley & Sons Inc. Reprinted with permission of John Wiley & Sons, Inc.

that only one main town was required for trade, service and political control functions.

2 Following emancipation, the second stage, small, marginal farming communities – clustered around the plantations, practising subsistence agriculture and supplying labour to the plantations – added a third layer to the settlement system. The distribution of these communities would vary according to physical and agricultural conditions.

3 Figure 3.14 suggests that, in the Caribbean, the modern era has witnessed the extension of this highly polarised pattern of development; this is the third stage. The emphasis is placed on extension, for this may not in all cases amount to intensification per se. This has come about largely as the result of industrialisation and tourism being taken as the twin paths to development. In 1976, Augelli and West (1976: 120) commented on what they

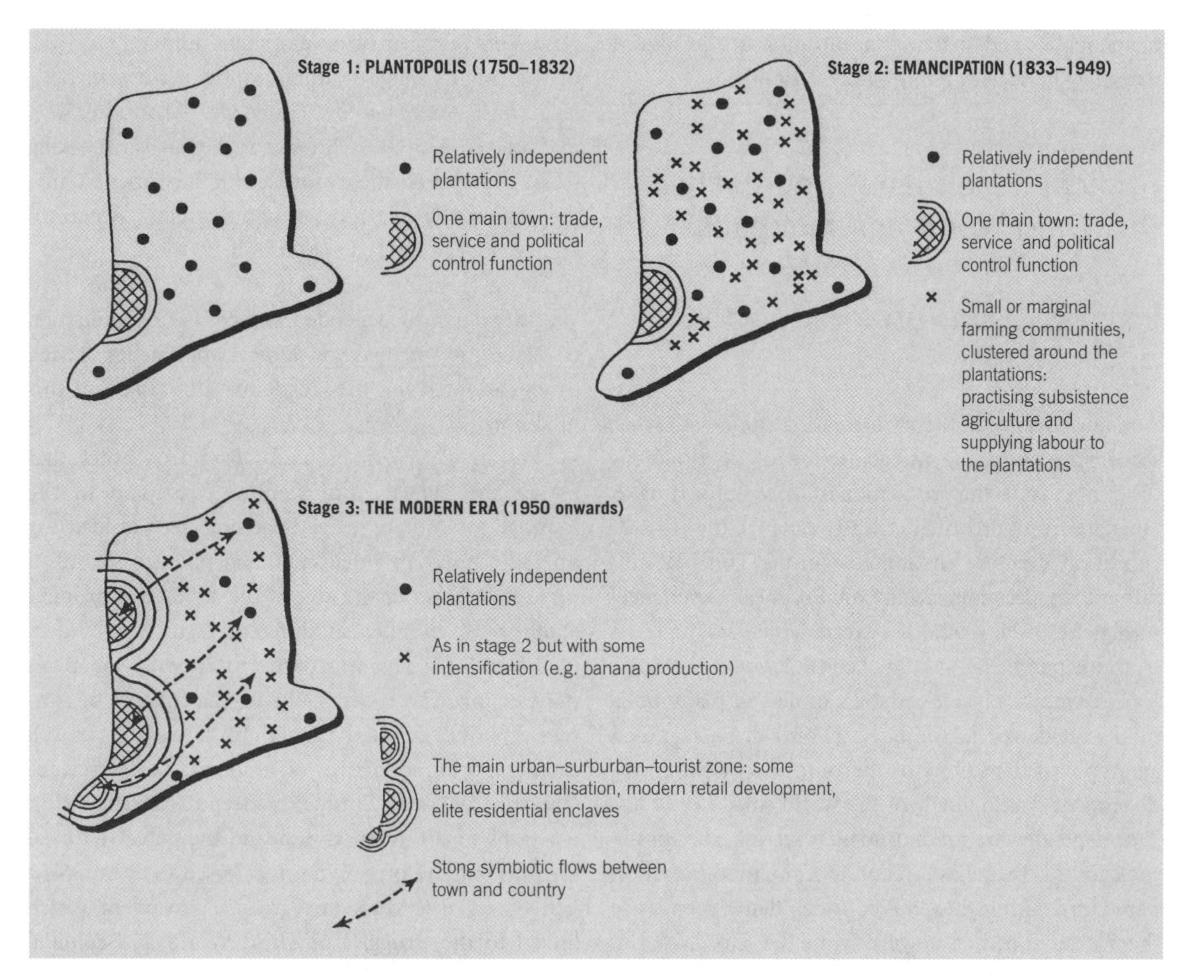

Figure 3.14 The plantopolis model and its extension to the modern era
Source: Adapted from Potter (1995a)

regarded as the disproportionate concentration of wealth, power and social status in the chief urban centres of the West Indies. As shown in Figure 3.14, such spatial inequality is sustained by strong symbiotic flows between town and country. This theme is picked up in detail in Chapter 9 in respect of the nature of rural–urban interrelations in developing countries.

An evaluation of the mercantile and plantopolis frameworks

The virtues of the mercantile and plantopolis models are many. Principally, they serve to stress that the evolution of most developing countries amounts to a highly dependent form of development. Certainly we are reminded that the high degree of urban primacy and the littoral orientation of settlement fabrics in Africa, Asia, South America and the Caribbean are all the direct product of colonialism, not accidental happenings or aberrant cases: hence the comment that modernisation surfaces essentially chart colonial and neo-colonial penetration.

According to these models, ports and other urban settlements became the focus of economic activity and of the social surplus that accrued. The concept of surplus product, defined as an excess of production over need or consumption, is developed more fully in the next major section, which deals with radical approaches to development. A similar, but somewhat less overriding spatial concentration also applies to the colonial power. Hence a pattern of spatially unequal or polarised growth emerged strongly several hundred years ago as the norm due to the strengthening of this symbiotic relationship between colony and colonial power.

The overall suggestion is that, due to the requirements of the international economy, far greater levels of inequality and spatial concentration are produced than may be socially and morally desirable.

Radical dependency approaches: the Third World answers back?

Introduction: an indigenous approach?

A major advance in theory formation came with what some refer to as the indigenisation of development thinking; that is, the production of ideas purporting to emanate from, or which at least relate to, the sorts of conditions that are encountered in the Third World, rather than ideas emanating from European experience. As Slater (1992a, 1992b) has averred, the core can learn from the periphery – and it needs to learn.

The empirically derived mercantile and plantopolis models reviewed in the last section can be regarded as graphical depictions of the outcome of the interdependent development of the world since the 1400s. The dependencia, or dependency school, specifically took up this theme as a rebuttal of the modernisation paradigm. Although some consider that dependency theory developed as a voice from the Third World, others have maintained that its most cogent formulation represents Eurocentric development thinking by virtue of the origins of its leading author, a German-born economist, Andre Gunder Frank. Or, put another way, as Clarke (2002) states, although in the English-speaking world dependency theory focused on the work of Frank, the main body of contributions was Latin American and Caribbean in origin (Marshall, 2008).

However, before we look in more detail at the Third World origins of dependency theory we should first consider the reasons why radical approaches started to be adopted in the academic literature at this juncture.

Preston (1996), in a useful overview, notes that before the 1960s little attention was paid to the Marxian tradition of social theorising (see Key idea box on Marxism). The intellectual and political revival of interest in Marx at the end of the 1960s was brought about by a number of different factors, as shown in Figure 3.15. This was partly to do with America's military involvement in Vietnam, the collapse of consensus politics due to the civil rights movement, as well as the moribund nature of academic social science (Preston, 1996). In Europe, the trend was reflected in reactions to the Vietnam war and the perceived need for university reform. What is referred to as the 'New Left' emerged as a broad progressive movement, which linked to the struggles of Third World anti-colonial movements (Figure 3.15).

Key idea

Marxism

The terms 'Marxism' or 'Marxist' can be applied to sets of ideas or practices that are based on the writings of Karl Marx (1818–1883) and Friedrich Engels (1820–1895). Marx was a German-born political scientist and revolutionary, who together with Engels published the *Communist Manifesto* in 1848, an early statement on their general principles. Their writings were strongly class-based, stressing the struggles of the working classes and the need to replace the capitalist system if the lot of ordinary people was to be improved. They regarded all history as the history of class struggles. An essential view was that control over the forces of production is critical, giving rise to a two-fold class division. On the one side are the owners of the means of production (land and capital), and on the other there are the workers, who only have their labour to sell. Workers have to toil long hours and the value of what they produce is far higher than what they are paid. Thus, an economic surplus is accrued by the owners of capital and the economy is exploitative of labour. The early forms of surplus were created by merchants' 'primative accumulation', acquired by 'raiding' non-capitalist societies for commodities such as gold and silver. Later the surplus represented the value expropriated from workers. In the views of Marx and Engels, due to the inhumane nature of capitalism and its inherent contradictions, in time the system would ultimately fail and be replaced by socialism.

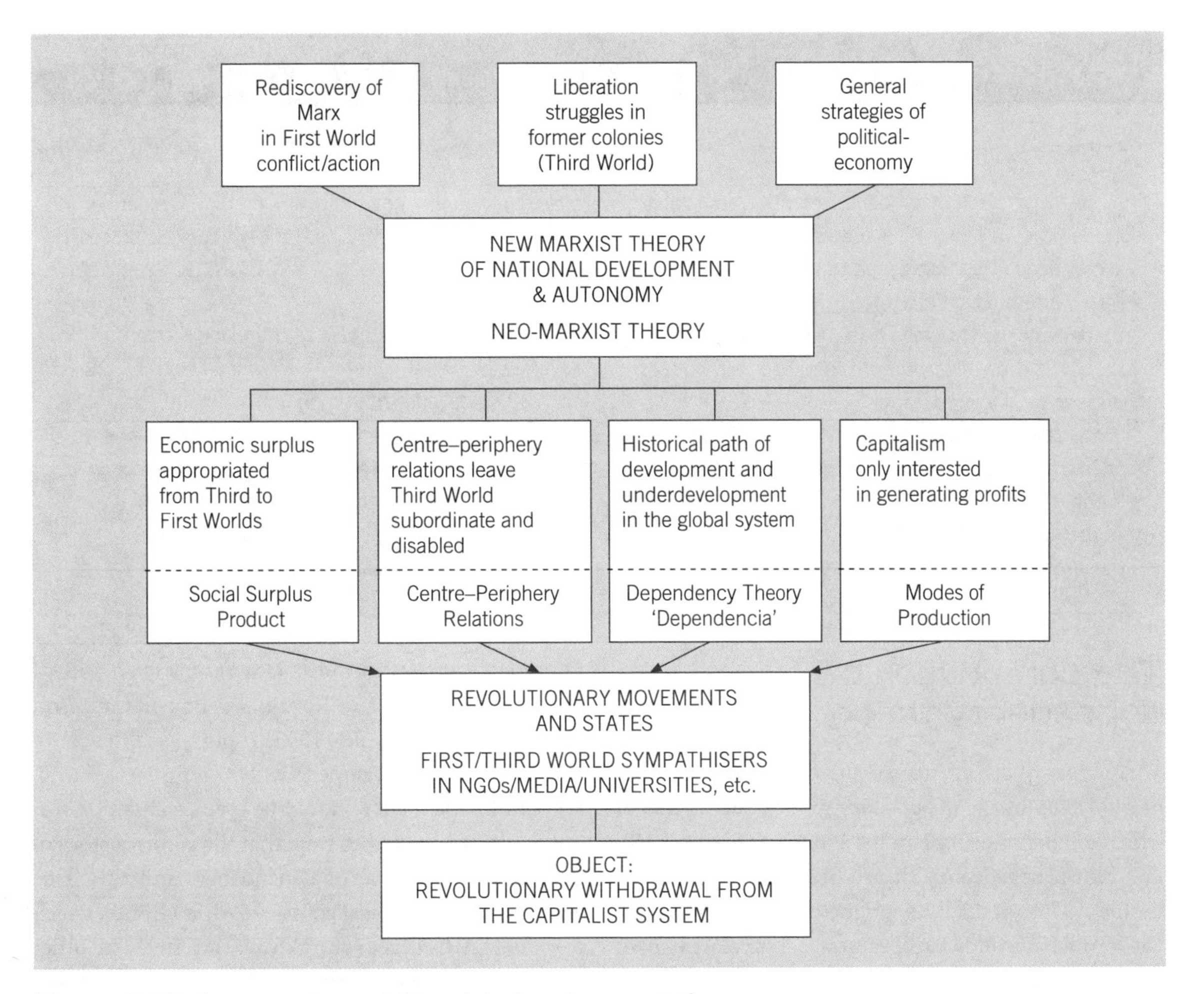

Figure 3.15 An overview of Marxist development theory
Source: Adapted from Preston (1996)

Paul Baran and the critical role of the economic surplus

Paul Baran, reflecting the general mood of the time, turned to a radical perspective in respect of development theory, promulgating a clear neo-Marxist approach (Baran, 1973; Baran and Sweezy, 1968). The principal idea was that of economic surplus, which is created by the inherent working of the capitalist system. It is clear that Baran was defining a surplus as an excess of production over the needs of consumption, and which therefore exists as a material quantity of goods.

The basic point is that once surplus production is redistributed within society, it effectively becomes a surplus of time and energy (Potter and Lloyd-Evans, 1998). Those who do not have to produce their own means of subsistence are freed for other activities. Part of the surplus can be redistributed to parasitic groups, and part can be used for conspicuous consumption or for state monumentalism – in other words it can be concentrated both in space and among different groups. It can be argued that this surplus redistribution has characterised the mercantile period onward, and has witnessed surplus value being expropriated from developing countries.

Baran's Marxist framework also envisaged that during the colonial period advanced nations entered into special partnerships with powerful elite groups in less developed and pre-capitalist countries. By such means, surplus was extracted and appropriated by elite groups. Thus, for Third World countries, Baran saw the key to development as disengagement from the deforming impact of the world capitalist economy, as capitalism creates and then diverts much of the economic surplus into wasteful and sometimes immoral consumption (Baran, 1973).

Key idea

Social surplus product

The idea of a surplus can be looked at as a social surplus product – that is, a surplus of production over societal need. This first occurred when human groups were able to produce more food than they needed on a daily basis. This meant that some members of society could be released from the need to produce their own food and could take on other roles in society, such as religious, military and political leadership. In other words, the food surplus became a surplus of time and energy within society, and it is this that can be referred to as social surplus product. This surplus can be concentrated in space and distributed highly unequally among the members of society. It is thus central to the development process.

Critical reflection

It is an interesting exercise to think about the different places and contexts in which social surpluses can be witnessed – from monumental buildings, regional variations in house prices, art collections, to personal displays of wealth. Make a list of regions, places, organisations, buildings where signs of the accumulation of social surplus value have been accrued and can be witnessed.

The contribution of dependency theory

These types of arguments were perhaps most cogently promulgated by the dependency school, the origins of which can be traced back to the 1960s (see Figure 3.15). Full-blown dependency theory became a global force in the 1970s. It had its origins in the writings of Latin American and Caribbean radical scholars known as structuralists, because they focused on the unseen structures that may be held to mould and shape society (Clarke, 2002; Conway and Heynen, 2002; Girvan, 1973; Marshall, 2008).

The approach was the outcome of the convergence of pure Marxist ideas on Latin American and Caribbean writings about underdevelopment. Essentially, by such a process, the Eurocentric ideas of Marxism became more relevant to the Latin American condition. Marx and Engels had seen capitalism as initially destructive of non-capitalist social forms, and thereafter progressive (Preston, 1996). Prebisch (1950) and Furtado (1964, 1965, 1969), who wrote about Argentina and Brazil respectively, are the best-known Latin American structuralists. Raul Prebisch stressed the importance of economic relations in linking industrialised and Latin American less developed countries, and observed that these were primarily prescribed in the form of centre–periphery relations (see Figure 3.15). Furtado (1969) argued that present-day Latin American socio-economic structures were the result of the manner of incorporation into the world capitalist system, an argument which is close to the full dependency line. Other later writers included Dos Santos (1970, 1977) and Cardoso (1976) (see Key thinker box on Cardoso).

As noted by Hettne (1995), opinions differ as to whether the well-developed Caribbean or 'New World' school of dependency should be seen as an autonomous form. It is salient to note that the whole history of the Caribbean is one of dependency, and this issue was central to the so-called 'New World Group', which first met in Georgetown, Guyana, in 1992, in order to discuss Caribbean development issues. Key names include George Beckford (1972), Norman Girvan (1973) and Clive Thomas (1989).

The development of the Caribbean-based New World Group has recently been charted in some detail by Marshall (2008). The founders of the group aspired to identify indigenous paths for the region's development. They were convinced that the twin strands of modernisation and industrialisation were not suited to the region. They were particularly resistant to Arthur Lewis' (1955) call for industrialisation by invitation. Beckford (1972) then added a strong case that the historic plantation slave economy had led to underdevelopment in the region and the evolution of what he termed 'persistent poverty'.

The contribution of A.G. Frank

However, the dependency approach proper is strongly associated with the work of Andre Gunder Frank (see Key thinker box). Frank's key ideas were outlined in an article published in 1966, 'The development of underdevelopment', as well as in the book *Capitalism*

Key thinker

From radical Latin American structuralist to President – Fernando Cardoso

Plate 3.6 Fernando Cardoso
Source: PA Photos

Fernando Henrique Cardoso (1931–) offers a fascinating case of an academic interested in development who later became a political figure on the world stage, thereby exemplifying afresh the close relation between politics and development thinking. Cardoso was born in Rio de Janeiro, Brazil, in 1931 and trained as a sociologist. He made a major contribution with his study of *Dependency and Development in Latin America* (1969). Cardoso regarded dependency as neither stable nor permanent and he rejected any simple link between dependency and underdevelopment (see Sanchez-Rodrigues, 2006). In the 1970s Cardoso became actively involved in the pro-democracy movement in Brazil and from there he moved into politics, becoming a member of the Brazilian Social Democrat Party in 1982. He was made Foreign Minister in 1992/93 and in 1995 he was elected as President of Brazil. He served two terms of office, with his presidency extending through to 2003. Some argue that in power he abandoned the Marxist roots he showed as a structuralist, and that he served the neo-liberal interests of multinational business elites (Sanchez-Rodrigues, 2006).

Key thinker

Andre Gunder Frank

Plate 3.7 Andre Gunder Frank
Source: with kind permission from Andre Gunder Frank

Andre Gunder Frank (1929–2005) was born in Germany, but his family fled to Switzerland during the ascendancy of Adolf Hitler, and then in 1941 moved to the United States. Frank undertook a PhD at the University of Chicago, where in 1957 he received his doctorate for a thesis on agriculture in the Ukraine. After working in a number of American universities, Frank moved to South America, and at one point was Professor of Sociology and Economics at the University of Chile. During this period he underwent a rapid and thorough radical conversion (Brookfield, 1975). The time he spent in Chile is seen as the foundation of his work on dependency theory. His seminal ideas were summarised in the phrase 'the development of underdevelopment'. The large body of books and papers written by Frank were directly shaped by the ideas of Marx, especially the concept of accumulation on a global scale (Watts, 2006). Throughout his career Frank moved from post to post and from country to country. Watts (2006: 90) has argued that Andre Frank was 'at once too radical, too ornery and too unconventional for most universities on both sides of the Atlantic'.

and Underdevelopment in Latin America, which appeared in 1967. Although a scholar of European origin working in the USA, Frank had researched in Mexico, Chile and Brazil.

Frank (1967) maintained that development and underdevelopment are opposite sides of the same coin, and that both are the necessary outcome and manifestation of the contradictions of the capitalist system of development.

The thesis presented by Frank was devastatingly simple. It was argued that the condition of developing countries is not the outcome of inertia, misfortune, chance, climatic conditions or whatever, but rather a reflection of the manner of their incorporation into the global capitalist system. Viewed in this manner, so-called underdevelopment, and associated dualism, are not a negative or void, but the direct outcome and reciprocal of development elsewhere (Figure 3.16). The only real alternative for such nations was to weaken the grip of the global system by means of trade barriers, controls on transnational corporations and the formation of regional trading areas, along with the encouragement of local or indigenous production and development (see Plate 3.8).

The phrase 'the development of underdevelopment' (Frank, 1966) has come to be employed as a shorthand description of the Frankian approach, which, as noted previously, has a strong graphical tie-in with the mercantile and plantopolis models. Quite simply, if the development of large tracts of the Earth's surface has depended upon metropolitan cores, or 'metropoles', then the development of cities has also depended principally upon the articulation of capital and the accumulation of surplus value (Figure 3.17).

The process has operated internationally and internally within countries. Viewed in this light, so-called

Figure 3.16 Dependent relations according to the *New Internationalist*
Source: Cartoon by Paul Fitzgerald

Plate 3.8 'Organisation for Rural Development' poster advocating more domestic production and fewer imports in St Vincent
(photo: Rob Potter)

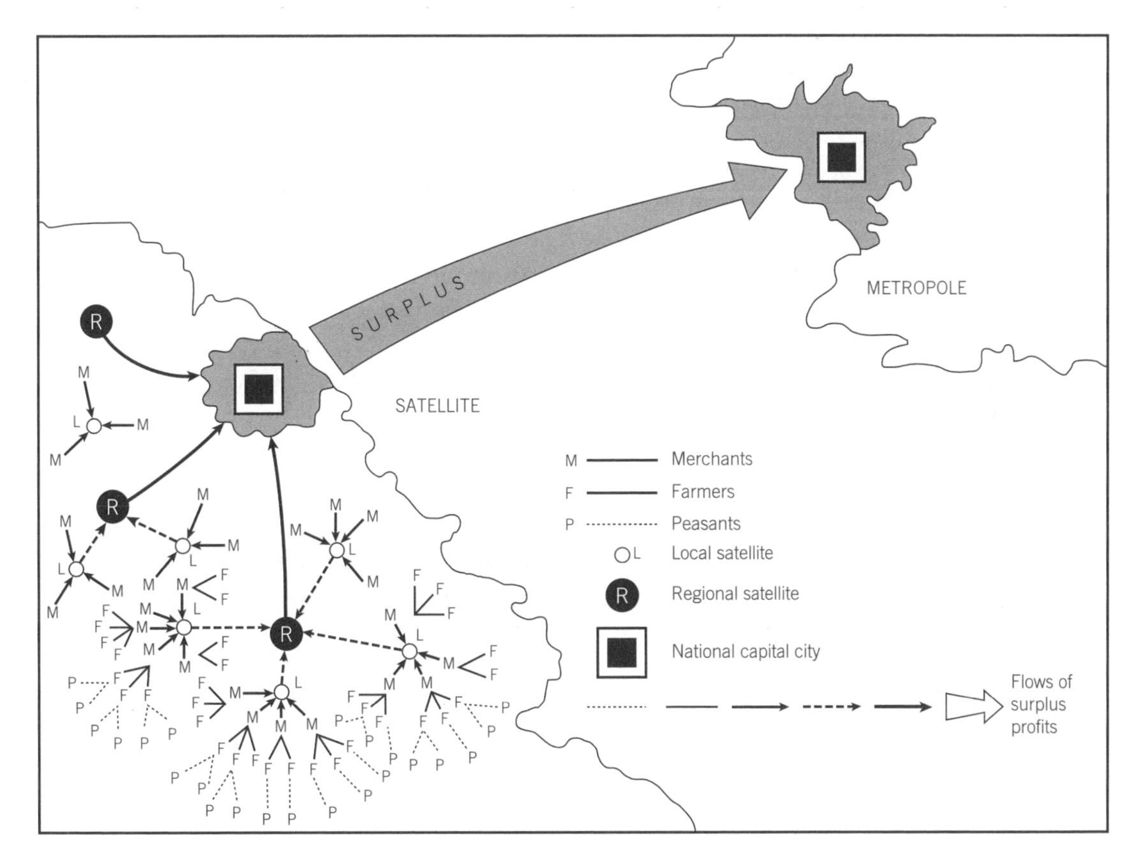

Figure 3.17 Dependency theory: a graphical depiction
Source: Adapted from Potter (1992a)

backwardness results from integration at the bottom of the hierarchy of dependence, not a failure to integrate within the global economy. Frank argued that the more 'satellites' are associated with the metropoles, the more they are held back, and not the other way around. In this connection, Frank specifically cited the instances of north-east Brazil and the West Indies as regions of close contact with the core, but where processes of internal transformation had been rendered impossible due to such close contact.

Conway and Heynen (2008) note how Frank presents Brazil as the clearest case of national and regional underdevelopment. He argued that the expansion of capitalism beginning in the sixteenth century sequentially incorporated urban cores and their extensive hinterlands into the global economy as exporting nodes. Cities such as Rio de Janeiro, São Paulo and Paraná figured most prominently in this expansion. Frank (1967) argued that this process witnessed the apparent development of these nodes, but in the long run led to the underdevelopment of the wider region.

As shown by the example of Brazil, dependency theory represents a holistic view because it describes a chain of dependent relations which has grown since the establishment of capitalism as the dominant world system, so its expansion is regarded as coterminous with colonialism and underdevelopment.

The chain of exploitative relations witnesses the extraction and transmission of surplus value via a process of unequal exchange, extending from the peasant, through the market town, regional centre, national capital, to the international metropole, as shown in Figure 3.17. The terms of trade have always worked in favour of the next higher level in the chain, so that social surplus value becomes progressively concentrated (Castells, 1977; Harvey, 1973).

By such means, dependency theorists argue that the dominant capitalist powers, such as England and then the USA, encouraged the transformation of political and economic structures in order to serve their own interests. According to this view, colonial territories were organised to produce primary products at minimal cost, while simultaneously becoming an increasing market for industrial products. Inexorably, social surplus value was siphoned off from poor to rich regions, and from the developing world to the developed.

The critique of dependency and the rise of world systems theory

The chief criticism of dependency theory is that it is economistic, seeing all as the outcome of a form of economic determinism, conforming with what Armstrong and McGee (1985: 38–9) have described as the 'impersonal, even mechanical analysis of structuralism'.

Furthermore, the theory only appears to deal with class structure and other factors internal to a given nation, insofar as they are the outcome of the fundamental economic processes described. Another point of contention is how dependency theory suggests that countries can only advance their lots by delinking from the global economy, whereas the capitalist world system is busily becoming more global and interdependent. For all these reasons, dependency theory has largely been out of fashion in the First World since the 1980s (Preston, 1996). But again we should stress that in radical quarters such ideas and ways of thinking have never gone out of fashion.

Wallerstein (1974, 1980) attempted to get around some of the criticisms of basic dependency theory, including the internal–external agency debate, by stressing the existence of a somewhat more complex and finely divided 'world system' (Taylor, 1986). The essential point is that Wallerstein distinguishes not only between the core nations, which became the leading industrial producers, and the peripheral states, which were maintained as agricultural providers, but also identifies the semi-peripheries (see also Klak, 2008).

It is argued that the semi-peripheries play a key role, for these intermediate states are strongly ambitious in competing for core status by increasing their importance as industrial producers relative to their standing as agricultural suppliers. For the semi-peripheral capitalist nations, read the NICs. Within the world system since the sixteenth century there have been cyclical periods of expansion, contraction, crisis and change. Hence it is envisaged that the fate of a particular nation is not entirely externally driven, but depends on the internal manner in which external forces have been responded to and accommodated.

Dependency theory: final comments

Frank's ideas are seen by many as being near to the orthodox Marxist view that the advanced capitalist world at once both exploited and kept the Third World underdeveloped. Although many would undoubtedly refute this view as extreme, if one can clear away the moral outrage, it may be argued that elements of the analysis, even if in a world-systems form, are likely to provide food for thought for those interpreting patterns of development.

As Hettne (1995) notes, dependency theory stressed that the biggest obstacles to development are not a lack of capital or entrepreneurial skill, but are to be found in the international division of labour. As already observed, certainly the graphical representation of pure dependency theory exhibits many parallels with the spatial outcomes of the core–periphery, mercantile and plantopolis models of settlement development and structure (see Figure 3.17).

Articulation theory

One final element of radical development theory stressing exploitation is the so-called theory of the articulation of the modes of production. The basic argument is that the capitalist mode of production exists alongside (and is articulated with) non-capitalist and pre-capitalist modes of production. The capitalist system replaces the non-capitalist system where profits are to be accrued in so doing. However, the pre-capitalist form is left intact wherever profits are unlikely to be made and it is therefore advantageous to do so.

A frequently cited example is low-income housing, where the poor are left to provide their own folk or vernacular homes whereas the formal sector provides for the middle and upper classes (Burgess, 1990, 1992; Drakakis-Smith, 1981; McGee, 1979; Potter, 1992b, 1994; Potter and Conway, 1997).

A macro-spatial but more specific example of the modes of production approach was provided by the operation of apartheid in South Africa. In this, the so-called 'homelands' preserved the traditional mode of production in order to conserve black labour, which was then allowed to commute to 'White South Africa'.

Hence, according to this radical perspective, dualism is a product of the contradictions of the capitalist system, not some form of aberration. Similarly, according to this view, underdevelopment can be described as a stalemate in the process of articulation. In other words, underdevelopment continues where there is no direct interest in the capitalist system in transforming the existing situation. Case study 3.2 discusses the position of aborigines in Australia, using the modes of production framework of analysis as an example.

Case study 3.2

Aborigines, development and modes of production

As noted by Drakakis-Smith (1983), Australia has experienced many different facets of colonialism. For much of the last 200 years it was a direct colony of Britain and was exploited as such for its mineral wealth and agricultural produce. However, throughout its history as a White nation state, whether colony or independent, Australia has harboured its own internal process of exploitation – that of its indigenous Aboriginal population by Whites. There can be little doubt that Aborigines do in fact comprise a subordinate, deprived and exploited group within Australian society.

Close investigation reveals how government measures have effectively institutionalised Aboriginal dependency in an unequal relationship which brings benefits principally to the White middle classes. Before the initial settlement by the British in 1788, Aborigines lived in what Meillassoux (1972, 1978) has termed a natural economy where land is the subject rather than the object of labour. The technology was simple but effective, with human energy alone being employed to tap the environmental resources through systems of hunting and collecting. Although life was primarily organised around the collection and consumption of subsistence food, a considerable amount of time was given over to ritual and ceremonial activities and to the production of consumer durables in the form of hunting weapons, tools and items with religious significance.

The mode of distribution was based on sharing the hunted and gathered food, and was therefore strongly related to the mode of production. One of the most important features of the pre-capitalist Aboriginal economy was an intense involvement with the land, both in physical and spiritual terms, with a strong emphasis on the spiritual. The small numbers, simple lifestyle and apparent lack of political organisation among the Aborigines convinced the British that there was no need for negotiation or treaties with such a 'primitive' people. Accordingly, all land was declared to be Crown land from the outset, so no compensation was paid to the Aborigines and no pre-existing rights were recognised. This legalistic appropriation of the land itself has been a fundamental

▶

Case study 3.2 (continued)

factor in the subsequent exploitation of the Aboriginal people in Australia. Simply on the grounds that the indigenous population appeared to be disorganised and primitive, the British removed at one administrative stroke both the economic and spiritual basis of Aboriginal society.

For several decades after these developments, Aborigines appeared to be unimportant within the Australian colonial economy. In the first instance, labour was provided by assigned convicts, and later by larger landowners buying out many of the initial smallholders. By the 1830s, however, the rapid development of sheep farming to supply wool for export had led to the encouragement of large-scale labour migration from Britain.

But this rapid expansion of pastoral farming in the second half of the nineteenth century brought Aborigines once more into direct contact with the vanguard of White settlement. Hitherto, they had been virtually ignored – despised, destitute and decimated by starvation, anomy and disease. By the mid nineteenth century pastoral settlement was beginning to push into the central and northern regions of the country, where conditions were harsher and Aborigines more numerous. Thus, Aboriginal labour was for the first time becoming a necessity on those properties where sheer size and harsh climate led to a sharp reduction in the enthusiasm of recruited White labour to move north.

Thus, Aboriginal labour, either as station hands or domestics, was extensively used from the 1880s onwards. Bands and families were encouraged to stay on the land after its appropriation, where they received payment in kind for work undertaken on the station. Labour relations were at best paternalistic, but more often than not the property owners or managers had little interest in the reproduction of Aboriginal labour, considering the supply to be limitless and the individuals unworthy of detailed attention. Living conditions and diet were often totally inadequate and populations were decimated. As the value of the Aboriginal labour began to be more appreciated, so the Australian government began to establish a series of expanded reserves and settlements.

Australia has changed markedly since 1945. Although it still makes a notable contribution to the production process, the Aboriginal community is now more important as a consumer group for the goods and services of an extensive tertiary system operated almost entirely by Whites. In effect, this comprises a third stage in the institutionalisation of Aborigines within a dependency framework, following the appropriation of their land and labour power.

In the contemporary situation in Aboriginal Australia, therefore, the dominant capitalist mode of production has conserved the Aboriginal pre-capitalist mode of production largely for its role as a consumer of goods and services. The class position and economic prosperity of the White population is largely dependent on this relationship.

Alternative, bottom-up and participatory approaches: perspectives on 'another' development

Introduction

The somewhat inelegant and uninformative expression *another development* has been used to denote the watershed in thinking which has characterised the period since the mid 1970s (Brohman, 1996; Hettne, 1995).

The concept was born at the Seventh Special Session of the United Nations General Assembly and the allied publication by the Dag Hammarskjöld Foundation of *What Now*? The session stressed the need for self-reliance to be seen as central to the development process, and for the emphasis to be placed on endogenous (internal) rather than exogenous (external) forces of change (see also Chapter 1).

It also came to be increasingly suggested that development should meet the basic needs of the people (see Chapter 1). At the same time, development needed to be ecologically sensitive and to stress more forcefully the principles of public participation (Potter, 1985).

Thus, from the mid 1970s a growing critique of top-down policies, especially growth-pole policy, argued that such approaches had merely replaced concentration at one point in space with concentrated-deconcentration at a limited number of new localities. In other words, the status quo had been maintained, even if the pattern had changed.

However, at long last, assertions that there is only one linear path to development and that development is the same thing as economic growth came to be seriously challenged, at least in some quarters. Liberal and radical commentators averred that top-down approaches to development were acting as the servants of transnational capital (Friedmann and Weaver, 1979). In a similar vein, other commentators argued that what had been achieved in the past was economic growth without development, but with increasing poverty (Hettne, 1995).

The territorial bases of development

In their book *Territory and Function*, John Friedmann and Clyde Weaver (1979) presented the important argument that development theory and practice to that point had been dominated by purely functional concerns relating to economic efficiency and modernity, with all too little consideration being accorded to the needs of particular territories, and to the territorial (indigenous) bases of development and change (Plate 3.8).

Since 1975 a major new paradigm has come to the fore, which involves stronger emphasis being placed on rural-based strategies of development. As a whole, this approach is described as 'development from below'. Other terms used to describe the paradigm include agropolitan development, grassroots development and urban-based rural development. In the context of wider societal change, such developments can be related to the rise of what is called neo-populism. Neo-populism involves attempts to recreate and re-establish the local community as a form of protection against the rise of the industrial system (Hettne, 1995: 117). The territorial manifestation of neo-populism has been the rise of the green ideology as a global concern, allied to green politics.

Basic needs and development

The provision of basic needs became a major focus during the early 1970s. The idea of basic needs originated with a group of Latin American theorists, and was officially launched at the International Labour Organization's World Employment Conference, which was held in 1976. Preston (1996) argues that the pessimistic view of the Club of Rome's Limits to Growth (Chapters 5 and 6), was the motivating force behind basic needs strategies.

The approach stressed the importance of creating employment over and above the creation of economic growth. This was because the economic growth that had occurred in developing countries seemed to have gone hand in hand with increases in relative poverty. Development, it appeared, was failing to improve conditions for the poorest and weakest sectors of society.

The argument ran that what was needed was redistribution of wealth to be effected alongside growth. During this period, the basic needs approach was accepted and adopted by a range of international agencies, not only the International Labour Organization (ILO), but also United Nations Environmental Programme (UNEP) and the World Bank. However, it has to be recognised that many basic needs approaches used the aegis of the poor in cheap basic needs programmes in place of greater state commitment to poverty alleviation. This is why the World Bank became such a rapid convert to the approach. In these circumstances it has to be acknowledged that the practical implementation of basic needs had little to do with socialist principles per se.

The principal idea is that basic needs, such as food, clothing and housing, must be met as a clear first priority within particular territories. In the purest form, it is argued that this can only be achieved by nations becoming more reliant on local resources, the communalisation of productive wealth, and closing up to outside forces of change. This aspect of development theory is known as selective regional and territorial closure.

Development from below or bottom-up development

In basic terms, therefore, it is argued that Third World countries should try to reduce their involvement in processes of unequal exchange. The only way round the problem is to increase self-sufficiency and self-reliance. It is envisaged that later the economy can be diversified and non-agricultural activities introduced. But it is argued that, in these circumstances, urban locations are no longer likely to be mandatory, and cities can in

Plate 3.9 Environmental costs of development; pollution near Witbank, Eastern Transvaal, South Africa
(photo: Eric Millex, Panos Pictures)

this sense be based on agriculture. Thus, Friedmann and Weaver (1979: 200) comment that 'large cities will lose their present overwhelming advantage'.

Clearly, such approaches are inspired by, if not entirely based on, socialist principles. Classic examples of the enactment of bottom-up paths to development have been China, Cuba, Grenada, Jamaica and Tanzania under the *ujaama* policies inspired by African socialism. As Hettne (1995) notes, self-sufficiency has frequently been perceived as a threat to the influence of superpowers, as in the case of tiny Grenada (Brierley, 1985a, 1985b, 1989; Potter, 1993a, 1993b; Potter and Welch, 1996). This example is discussed in some detail in Case study 3.3.

Case study 3.3

Paths to development: the case of Grenada

The experience of Grenada in the eastern Caribbean is useful in demonstrating that alternative paths to development do not have to be revolutionary in the Marxist political sense (Potter and Lloyd-Evans, 1998). In March 1979, Maurice Bishop, a UK-trained lawyer, overthrew what was regarded as the dictatorial and corrupt regime of Eric Gairy. Maurice Bishop led the New Jewel Movement (NJM), the principal theme of which was anti-Gairyism allied with anti-imperialism. The movement also expressed its strong commitment to genuine independence and self-reliance for the people of Grenada (Brierley, 1985a, 1985b; Ferguson, 1990; Hudson, 1989, 1991; Kirton, 1988; Potter, 1993a, 1993b).

On the eve of the revolution, Grenada suffered from a chronic trade deficit, strong reliance on aid and remittances from nationals based overseas, dependence on food imports and very substantial areas of idle agricultural land. After the overthrow of Gairy, the NJM formed the People's Revolutionary Government (PRG), the movement taking a basic human needs approach as the core of its development philosophy. The PRG stated its intention of preventing the prices of food, clothing and other

Case study 3.3 (continued)

basic items from rocketing, along with its wish to see Grenada depart from its traditional role as an exporter of cheap produce. The government also set up the National Cooperative Development Agency in 1980, the express aim of which was to engage unemployed groups in villages in the process of 'marrying idle hands with idle lands'.

Between 1981 and 1982, two agro-industry plants were completed, one producing coffee and spices, the other juices and jams. A strong emphasis was placed on encouraging the population to value locally grown produce together with local forms of cuisine, although the scale of this task was clearly appreciated by those concerned (Potter and Welch, 1996). The PRG also pledged itself to the provision of free medicines, dental care and education. Finally, it was an avowed intention of the People's Revolutionary Government to promote what Bishop referred to as the New Tourism, a term which is now widely employed in the literature. New Tourism meant the introduction of what the party regarded as sociologically relevant forms of holidaymaking, especially those which emphasised the culture and history of the nation, and which would be based on local foods, cuisine, handicrafts and furniture-making (Patullo, 1996). Such forms of tourist development, it was argued, should replace extant forms based on overseas interests and the exploitation of the local environment and socio-cultural history.

The salient point is that, throughout the period, 80 per cent of the economy of Grenada remained in the hands of the private sector, and a trisectoral strategy of development that encompassed private, public and cooperative parts of the economy was the declared aim of the PRG. In this sense, the so-called Grenadian Revolution was nothing of the sort. The economy of Grenada grew quite substantially from 1979 to 1983, at rates of between 2.1 and 5.5 per cent per annum. During the period, the value of Grenada's imported foodstuffs fell from 33 to 27.5 per cent. Even the World Bank commented favourably on the state of the Grenadian economy during the period from 1979 (Brierley, 1985a).

For many it was a matter of great regret that Maurice Bishop was assassinated in October 1983, and the island invaded by US military forces, because this saw the end of the four-year experiment in alternative development set up in this small Commonwealth nation (Brierley, 1985a). This deprived other small dependent Third World states of the fully worked-through lessons of grassroots development that Grenada seemed to be in the process of providing.

Critical reflection

Grenada provides a good example of how a small island nation can endeavour to localise and indigenise development. Most countries are, of course, far larger than Grenada and have more extensive resource bases. Think of the wider strategies that nations can employ in a number of different economic sectors in order to promote bottom-up development. How easy is it to implement such approaches in the global economy of the present day?

Walter Stöhr (1981) provides an informative overview of development from below. In particular, his account stresses that there is no single recipe for such strategies, as there is for development from above. Development from below needs to be closely related to specific socio-cultural, historical and institutional conditions. Simply stated, development should be based on territorial units and should endeavour to mobilise their indigenous natural and human resources. More particularly, the approach is based on the use of indigenous resources, self-reliance and appropriate technology, plus a range of other possible factors, many of which are shown in Table 3.1.

It is noticeable that bottom-up strategies are varied, with alternative paths to development being stressed. They share the characteristic of arguing that development and change should not be concentrated at each higher level of the social and settlement systems, but should focus on the needs of the lower echelons of these respective orders. It is this characteristic which gives rise to the term *bottom-up* development, for such strategies are in fact often enacted by strong state control and direction from the political 'centre'.

Table 3.1 Stöhr's criteria for the enactment of 'development from below'

Broad access to land
A territorially organised structure for equitable communal decision making
Granting greater self-determination to rural areas
Selecting regionally appropriate technology
Giving priority to projects which serve basic needs
Introduction to national price policies
External resources used only where peripheral ones are inadequate
The development of productive activities exceeding regional demands
Restructuring urban and transport systems to include all internal regions
Improvement of rural-to-urban and village communications
Egalitarian societal structures and collective consciousness

Source: Based on Stöhr (1981)

Environment and development

Perhaps the major development since the 1970s has been the emergence of environmental consciousness in the arena of development thinking (see Plate 3.9). Central to this evolving concern was the Brundtland Commission on Environment and Development which reported in 1987 (WCED, 1987). Even more important was the Earth Summit held in Rio in the summer of 1992. This United Nations Conference on Environment and Development (UNCED) brought together some 180 nations. It was at this stage that principles of environmental sustainability became a political issue in the development debate (Pelling, 2008). This interest continued in the Rio plus 10 (years) conference, or the Second Summit on Sustainable Development held in Johannesburg, South Africa in August–September 2002.

Ecodevelopment becomes sustainable development

Ecodevelopment, now known as sustainable development, has become one of the leading development paradigms since the 1990s, stressing the need to preserve the natural biological sytems that underpin the global economy (Elliott, 1994; Redclift, 1987); and such approaches are fully explored in Chapter 6.

Sustainability constitutes the ecological dimension of territorialism discussed previously (Hettne, 1995). Territory, it is argued, should be considered before function, and developing countries should not look to developed nations for the template on which to base their development. Rather, they should look towards their own ecology and culture (see Case study 3.3).

In this context, too, it is recognised that development does not have a universal meaning. Strongly allied to this, the need for emancipatory views on women and development, and ethnicity and development has started to receive the attention that it merits, not least in the guise of ecofeminism.

It is in this sense that sustainable development means more than preserving natural biological systems. There is the assumption of implicit fairness or justice within sustainable development, so the poor and disadvantaged are not forced to degrade or pollute their environments in order to be able to survive on a day-to-day basis (Drakakis-Smith, 1996; Eden and Parry, 1996; Lloyd-Evans and Potter, 1996).

The much-quoted definition of sustainable development was provided by the Brundtland Commission as development 'that meets the needs of the present without compromising the ability of future generations to meet their own needs' (WCED, 1987: 43). This is a far cry from the unilinear, Eurocentric functional perspectives advanced during the 1960s and 1970s, and demonstrates the wide-ranging changes that have occurred in development theory, development policies and geographies of development over the past 40 years.

However, as noted at the outset of this chapter, the promotion of sustainable development is occurring in a context where neo-liberal economic policies are dominant, and many would stress the incompatability of these two forces of change at their extremes – as is witnessed in the current debates about cheap air flights and global warming.

Alternative development: a summary

Brohman (1996) provides a useful summary of what he sees as the main elements of alternative development strategies, and this provides a useful summary of the discussion so far in this section:

- A move towards direct redistributive mechanisms specifically targeting the poor.

- A focus on local small-scale projects, often linked to urban or rural community-based development programmes.
- An emphasis on basic needs and human resource development.
- A refocusing away from growth-oriented definitions of development, towards more broadly based human-oriented frameworks.
- A concern for local and community participation in the design and implementation of projects.
- An emphasis on self-reliance, reducing outside dependency and promoting sustainability.

New forms of governance: civil society, social capital and participatory development

The account thus far has considered the provision of basic needs, and the promotion of redistribution and self-reliance, and these can certainly be seen as some of the basic characteristics of alternative development, both in terms of its origins and its early practice. But alternative development has now come to be associated with new and wider conceptualisations of planning and development.

The main distinguishing feature here is the fostering of participatory development, associated with more equitable principles of growth. Given the long hegemony of so-called 'top-down', Western, rational planning and development, increasing the involvement of people in their own development is seen by many as imperative. Chambers (1983) averred that it was time for the last to be put first (see also Mohan, 2008).

In this context, participation means much more than involvement or mere consultation (Conyers, 1982; Potter, 1985). While these calls seem eminently reasonable, how is this to be achieved, and who exactly is to participate? Clearly, not everybody can participate in all decisions all of the time. Indeed, it has to be recognised that it is a democratic right not to participate.

These changes have become involved with wider issues, suggesting the need for the evolution of new forms of governance. In turn, the account on governance is closely associated with buzz words and new concepts. Not that many years ago, the terms 'civil society' and 'social capital' had little or no currency. But as the state has progressively withdrawn from specific areas during the neo-liberal era, so the organisations existing between family groups and the state (leaving aside businesses), have become more and more important. This is referred to as the rise of 'civil society' (Edwards, 2001a; Fukuyama, 2001). Civil society may be regarded as forming a so-called 'third sector', in addition to the traditional two of the state and the marketplace.

Non-governmental organisations (NGOs) form a vital part of civil society, along with other types of civil associations. These have come to play an increasingly important role in local and community-based initiatives in developing societies (Desai, 2002; Mercer, 2002). Brohman (1996) cites 2200 developed world NGOs operating in 1984, with as many as 10,000 to 20,000 in the Global South working in partnership with northern counterparts. NGO projects tend to be small-scale and stress the employment of local resources and appropriate technologies, in tandem with grassroots participation.

One of the principal merits of NGOs is seen as their extreme sensitivity to local conditions and their dedication to the tasks at hand (Brohman, 1996). NGOs are also often linked to wider forms of global citizen action in the form of popular movements in the fields of health, education, welfare and employment. Further, much of the work of NGOs has involved poverty reduction measures. On the negative side, there is a strong argument that NGOs have been used to fill the vacuum left by the rolling back of the state as part of structural adjustment programmes and neo-liberalism more generally.

The rise of civil society is also based on the importance of social capital (Bebbington, 2008; Fukuyama, 2001). The expression 'social capital' first emerged in the early 1990s, and has quickly become a key term used by international agencies, governments and NGOs.

An American academic, Robert Putnam (1993), who studied community linkages in southern Italy and the USA, is often mentioned as the key figure in the field. However, many difficulties surround both the definition of, and attempts to measure, social capital. One commonly employed definition sees social capital as comprising the informal norms that promote cooperation between two or more individuals. These norms also lead to cooperation and the pursuit of mutual benefit in groups and organisations. The idea that promoting development is about increasing the stock of social capital within a given societal context is one that has come to be articulated by a number of authorities since the 1990s. Perhaps the most realistic perspective is to see an awareness of social capital as critical to understanding, fostering and guiding development (Fukuyama, 2001).

But it would be naive to see 'social capital' as a miracle cure for development problems wherever they are encountered. In any given territorial context, social capital is likely to be the product of a wide range of factors, including shared historical experiences, local cultural norms, traditions and religion. Thus, it is hard to suggest ways in which the ties which bind people together can easily be created or manufactured as a part of developing different areas.

Further, it has to be recognised that social capital can have as many negative connotations as positive ones. As a simple example, the ties which serve to bond together members of a criminal fraternity also, simultaneously, serve to exclude non-members, who are among those likely to be exploited. In this manner, we are left with a similar argument to that encountered in relation to wider civil society; namely, that the concept can be employed to paper over the cracks left by the withdrawal of the state under World Bank and IMF neo-liberal economic packages. Thus, while some may see social capital as a key concept in promoting participatory development, it must be recognised that it can also meet many of the needs of the neo-liberal right.

A changing paradigm is also discernible in the field of planning, where since the late 1990s there has been a clear move away from expert-based and top-down systems. The argument runs that all too often planning and development in the past have been directly associated with 'outside' experts being brought in to solve problems (see, for example, Potter and Pugh, 2001; Pugh and Potter, 2000).

It is increasingly being argued that all stakeholders relating to particular issues need to be brought into the framework as part of a 'good governance' agenda (see Chapter 7). Indeed, the World Bank itself has increasingly adopted this stance in its public pronouncements.

An early representation of such an argument in the field of development was Chambers (1983; see also 1997). Chambers stressed the Eurocentric and other biases which have customarily pervaded development and urged the use of participatory rural appraisal (PRA) to counter such tendencies. PRA rejects written means of investigation. Rather, it is argued that in exploring local community development issues, visual and oral techniques should be employed. Thus, PRA advocates the use of oral histories, mapping exercises, and the ranking of preferences to explore community-based issues. The aim is to articulate and listen to a wider set of local voices as part of community planning and development.

Fundamentally, it must be recognised that all meaningful participation in planning and development practice is about changing existing power relations in the arena of decision making. Thus, changing the ways in which planning and development are carried out involves the empowerment of new groups of stakeholders. More recently, this move towards 'people power' or 'citizen control' (Nelson and Wright, 1995) has given rise to what are referred to as collaborative approaches to planning.

In collaborative planning, the accent is placed on developing collaboration among the various stakeholders, in respect of both policy development and delivery. The approach has been explored in the European context by Healey (1997, 1998, 1999) and Tewdwr-Jones and Allmendinger (1998). Essentially, it is recognised that reactive institutional frameworks need to be fostered which will allow a wide range of stakeholders to be involved in decision making, and not just trained experts, professionals and elites. It is a prime requirement of the 'communicative turn' that the many different forms of local knowledge that exist are taken into account.

All such approaches basically involve consensus-building in decision making. The approach is sometimes also referred to as 'communicative planning', as it is in part based on Habermas' concept of communicative rationality. It also takes in Foucault's ideas concerning the centrality of power in all social spheres.

Debate continues concerning exactly how practical such ideas are in reality, and whether consensus can ever be established in areas where fragmentation and conflict seem to be built into the system. However, collaborative and communicative planning must be seen as new perspectives in development thinking, which are likely to have a direct bearing on development theories and strategies.

All such approaches, involving greater participation and empowerment, witness a movement toward considering development at the local scale. But such approaches are not without their potential problems. Thus, Mohan and Stokke (2000) argue that the focus on the local carries the danger that the power relations and inequalities that underlie development issues and problems may be lost sight of. The authors argue that a stronger emphasis must be placed on the politics of the local if such 'dangers of localism' are to be avoided.

More recently, Purcell and Brown (2005) have argued that there is no inherent reason why decision making at the local level should be more efficient or just. Scale, they argue, is a backdrop to good decision making.

Development theory, modernity and postmodernity

A postmodern age?

Abandoning the evolutionary and deterministic modernisation paradigm associated historically with the enlightenment opens up several postmodern options for future development, and some of them have been reviewed in the previous section on bottom-up, alternative and participatory conceptualisations of development.

These trends in development thinking can be linked with the idea that, globally speaking, we are entering a postmodern age, associated with the rise of a knowledge-based post-industrial economy. It is also associated with greater plurality and hybridity. This theme is briefly addressed here, and receives, further, more explicit attention in relation to globalisation trends and development in the next chapter. Some aspects of postmodernity in relation to development ideologies have already been outlined in Chapter 1.

Postmodernism as twenty-first-century development

In simple terms, postmodernity involves moving away from an era dominated by notions of modernisation and modernity (Plate 3.10a,b and Figure 3.18). It is therefore intimately associated with development theory and practice. It involves the rejection of modernism and a return to premodern and vernacular forms, as well as the creation of distinctly new post-modern forms (Harvey, 1989; Soja, 1989; Urry, 1990). It can be seen as a reaction against the functionalism and austerity of the modern period in favour of a heterogeneity of styles, drawing on the past and on contemporary mass culture (Plate 3.11).

As argued earlier in this chapter, there is much in the idea that the whole ethos of the modern period privileged the metropolitan over the provinces, the developed over the developing worlds, North America over the Pacific Rim, the professional expert over the general populace and men over women (see Figure 3.19).

In contrast, the postmodern potentially involves a diversity of approaches, which may serve to empower 'other' alternative voices and cultures (Figure 3.19). A strong emphasis on bottom-up, non-hierarchical growth strategies, which endeavour to get away from the international sameness, depthlessness and

Plate 3.10a Modern Hong Kong towers above traditional sampans
(photo: Rob Potter)

Plate 3.10b Modern high-rise apartments in Havana, Cuba
(photo: Rob Potter)

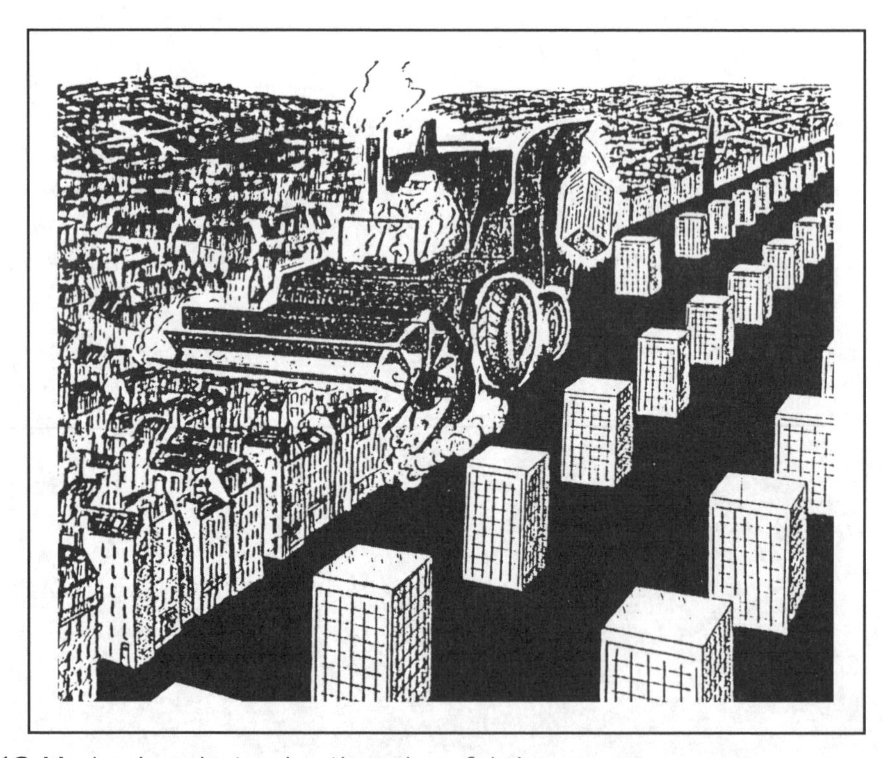

Figure 3.18 Modernism destroying the urban fabric
Source: Cartoon by J.F. Batellier

ahistoricism of the modern epoch, can be seen as part and parcel of the postmodern world (Figure 3.19 and Table 3.2). The accent can potentially be placed on growth in smaller places rather than bigger, and in the periphery not the core.

Postmodernism as late capitalism?

However, although postmodernism may in certain respects be seen in this optimistic manner, as a liberating

Plate 3.11 Resonances of postmodernity in a tourist setting: tourists and members of a traditional 'Tuk' band in Barbados
(photo: Rob Potter)

Table 3.2 Some polar differences between modernism and postmodernism that have relevance to development

Modernism	Postmodernism
Form	Antiform
Conjunctive, closed	Disjunctive, open
Purpose	Play
Design	Chance
Hierarchy	Anarchy
Finished work	Performance, happening
Synthesis	Antithesis
Centring	Dispersal
Root, depth	Rhizome, surface
Interpretation, reading	Against interpretation, misreading
Narrative	Antinarrative
Master code	Idiolect
Determinancy	Indeterminancy
Transcendence	Immanence

Source: Adapted from Harvey (1989)

force associated with small-scale non-hierarchical development and participatory change, there is another distinct facet to the trend of postmodernism.

This is very much like the arguments presented in the last section regarding the different ways in which civil society, social capital, NGOs and collaborative planning can all be regarded as serving the imperatives of both the political right and the left. As well as the rejection of the modern and a hankering for the premodern, there is the establishment of 'after the modern'. This is frequently interpreted as 'consumerist postmodernism', involving the celebration of commercialism, commercial vulgarity, the glorification of consumption and the related expression of the self (Cooke, 1990).

As will be explored in Chapter 4, these are trends which are of interest in relation to what is happening in parts of the developing world, an example being the promotion of international tourism as a major plank of development (Plate 3.11; see also Chapter 4). It is this aspect of postmodernism that gives rise to the suggestion that it maintains significant affinities with both right- and left-wing lines of thought and associated policy prescriptions (Jones *et al.* 1993).

Such a condition is related to a conflation of trends in which aspects of art and life, high and low culture are fused together, or 'pastiched'. Images, signs, hoarding and advertisements are potentially more important than 'reality'. Mass communications lead to mass image creation (Massey, 1991; Robins, 1989, 1995). Substance gives way to style, and policy to spin in the public and political spheres.

History and heritage may be rewritten and reinterpreted in order to meet the needs of international business and consumerism. This may all lead to further

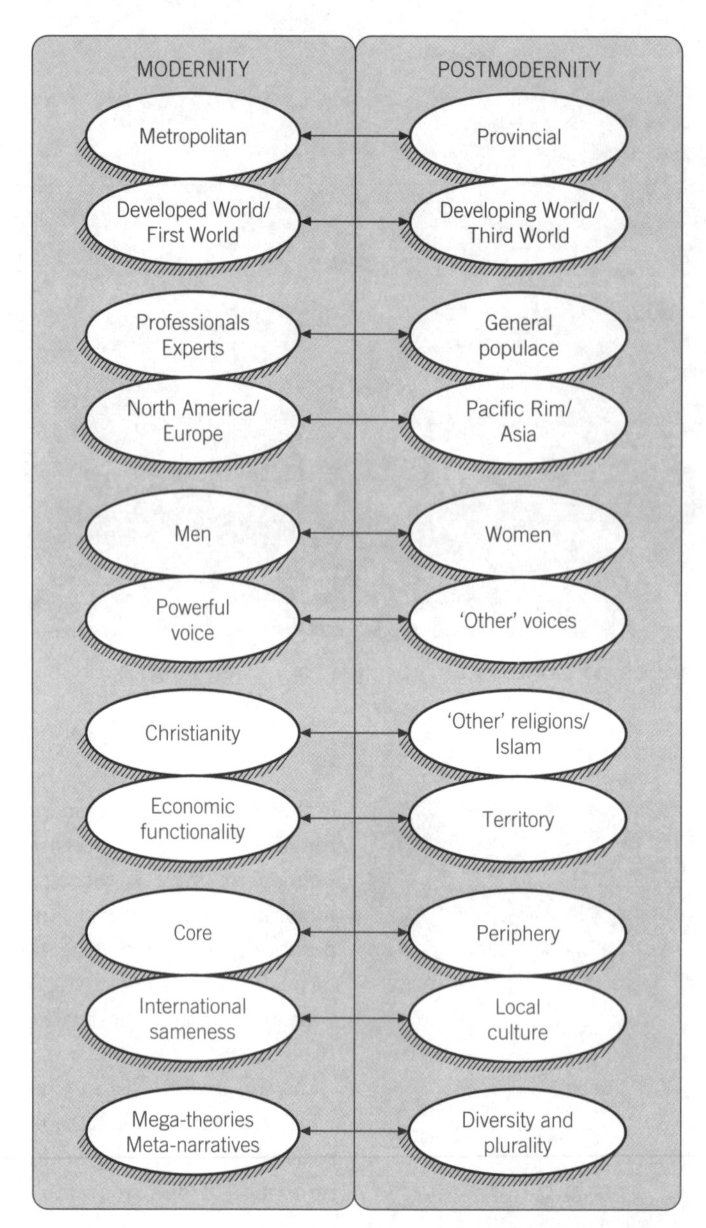

Figure 3.19 A graphical depiction of some common ideas about modernity and postmodernity

external control, exploitation and neo-colonialism. Many of these features can be interpreted in terms of the overconsumptive lifestyles which were offered to members of the upper white-collar strata of society in the Reagan–Thatcher era. Several of these themes will be further explored in the next chapter.

In this regard, rather than being seen as a freeing and enabling force, postmodernism may alternatively be interpreted as essentially the logical outcome of late capitalism (Cuthbert, 1995; Dann and Potter, 1994; Harvey, 1989; Jameson, 1984; Kaarsholm, 1995; Potter and Dann, 1994, 1996; Sidaway, 1990). Sardar (1998) argues that with its roots in colonialism and modernity, postmodernity, in fact, operates to marginalise the realities of the non-Western world further.

The role of TNCs in the promotion of tourism in the Caribbean may be seen as another instance where advertising and promotion campaigns may be interpreted as being aimed directly at increasing both the environmental and social carrying capacities of the nation. Such developments have interesting implications in contexts where nations themselves are still endeavouring to modernise. Indeed, the situation may give rise to all sorts of ambiguities (Austin-Broos, 1995;

Plate 3.12 Images of development ideology in Cuba: 'Revolution Yes!'
(photo: Rob Potter)

Masselos, 1995; Potter and Dann, 1994; Thomas, 1991). This theme is picked up and developed in Chapter 4 using the example of the prime Caribbean tourism destination of Barbados.

The development of postmodern trends that influence developing societies, in particular via the activities of TNCs and tourists, is of further interest in the new world order following the collapse of the communist world in 1989. We are certainly entering a noticeably less certain, less monolithic and unidirectional world. Hence the already wide diversity of development, with many varied geographies of development, seems likely to get more complicated, rather than less, over the coming years.

For example, some authorities are now talking of the tripolarity of development, with the Americas, Europe and Pacific Asia each presenting a particular version of industrial capitalism (Preston, 1996). The recent problems being faced by Asian economies are another sign of increasing global volatility and dynamism. The events of 11 September 2001 confirm only too cogently the validity of such a view.

This perspective is reflected in what has been called the impasse in development studies (Booth, 1985; Corbridge, 1986; Preston, 1985; Schuurman, 1993, 2002; Slater, 1992a). The world has become a more complex place since the collapse of communism in 1989, making the division into the First, Second and Third Worlds much more questionable, or perhaps even meaningless, as discussed in Chapter 1 (Plate 3.12).

Hettne (1995) argues that the rise of neo-conservatism in the global political realm, and monodisciplinary trends in the academic world have both presented development thinking with fundamental challenges. Some have pointed to what they regard as an impasse in theorising development itself, although this seems unduly pessimistic given the range of ideas considered in this chapter. Hettne (1995) also refers to the failure of development in practice as contributing to self-criticism, pessimism and 'development fatigue', especially in relation to the ultimate relevance of Western-developed research and ideas.

Notwithstanding these justifiable concerns, it is axiomatic that 'development', defined as change for the better or for the worse (Brookfield, 1975), will proceed in each and every corner of the globe. This being the case, there continues to be the need for the generation and discussion of realistic, although challenging and often conflicting, sets of ideas concerning the process of development, as well as the conditions that are to be encountered on the ground in developing countries themselves.

Key points

- Development theories and strategies have been many and varied, with new approaches generally being added alongside existing ones.
- Classical and neo-classical economic approaches generally stress the need for unrestrained, polarised growth and of letting the market decide for itself.
- Neo-liberalism as a generic development paradigm stems from the New Right and emphasises what is seen as the continuing need for market liberalisation and for the economy to be market- and performance-driven.
- Historical models give a normative impression of the degree to which in the past, since mercantilism and colonialism, development has been highly uneven and spatially polarised.
- Both dependency (radical) approaches and alternative/another development can be seen as direct critiques of modernisation theory. Thus the economic growth paradigm of the 1950s was challenged by socialist and environmentally oriented paradigms in the 1960s and 1970s respectively.
- Development thinking reflects political views – views on how economies should work and how societies should be structured. Postmodernity is leading to less monumental approaches to development.

Further reading

Cowan, M.P. and Shenton, R.W. (1996) *Doctrines of Development.* London: Routledge.
A comprehensive text dealing with principles of development.

Desai, V. and Potter, R.B. (eds) (2008) *The Companion to Development Studies*, 2nd edn. Part 2: Theories and strategies of development, London: Hodder-Arnold.
Contains a range of short essays covering the most important aspects of development theories and strategies.

Greig, A., Hulme, D. and Turner, M. (2007) *Challenging Global Inequality: Development Theory and Practice in the 21st Century*. Basingstoke: Palgrave Macmillan.
Stresses global poverty and inequality in reviewing contemporary development theory and practice.

Hettne, B. (1995) *Development Theory and the Three Worlds*, 2nd edn. London: Longman.
Although produced in the mid-1990s, this book still affords a very clear introduction to the principal paradigms of development.

Preston, P.W. (1996) *Development Theory: An Introduction*. Oxford: Blackwell.
Another book dealing with the theory and practice of development.

Simon, D. (ed.) (2006) *Fifty Key Thinkers on Development.* London: Routledge.
A useful source which brings together short essays on those who are deemed to have had a noticeable impact on studies of development.

Websites

www.iedconline.org
The website of the International Economic Development Council based in Washington DC. IEDC is a non-profit membership organisation dedicated to assisting 'economic developers' to do their job more effectively. With over 4500 members worldwide, IEDC offers support for professional development, advisory and legal services and regular conferences on development topics.

www.ids.ac.uk
The Institute of Development Studies (IDS) at the University of Sussex offers an extensive website giving access to the online catalogue of the IDS Library, with good holdings on development strategies and ideologies.

www.id21.org
Also out of IDS, and known as 'Communicating Development Research', this is a reporting scheme, bringing a selection of the 'latest and best' UK-based development research.

Discussion topics

- Assess the extent to which 'modern' development theories have been discredited by the rise of so-called 'alternative' and 'postmodern' approaches.
- Examine the extent to which Vance's mercantile model can be seen as a graphical representation of classical dependency theory.
- 'Old development theories never die. In fact, they don't even seem to fade away!' Discuss.
- For one developing nation, outline and assess the national development strategies employed since 1947.

Chapter 4

Globalisation, development and underdevelopment

It is tempting to suggest that the word 'globalisation' has increasingly come to characterise the times in which we live. This chapter seeks to explore the implications of globalisation for the process of development. On the one hand there is the essentially neo-liberal argument that for development to occur there has be globalisation – to allow countries to buy and sell goods and to buy into new forms of technology, cultural change and the like. At the other extreme, those who are far less happy with the outcomes of contemporary processes of globalisation argue that the process is 'distorting' patterns of development and creating ever-increasing global inequalities. At the extreme, globalisation has been branded as neo-modernisation, a renewed twenty-first-century faith in the notion that modernisation will progressively develop the world in a benign, efficient and positive manner. At their extremes, these two arguments are associated with pro-globalisers/ultra neo-liberals on the one hand, and anti-globalisers on the other. In detail, this chapter:

- Explores the links between globalisation and development looking at arguments about both the negative and positive aspects of the relations between the two;
- Examines the realities of a shrinking world as an increasingly unequal world;
- Presents an updated picture of the global digital divide showing just how unequal the world is in terms of communications;
- Reviews economic aspects of globalisation, especially what is referred to as 'global shift' in the manufacturing sector;
- Defines and explores the realities of the joint processes of global convergence and global divergence as key processes leading to homogeneity in patterns of consumption for those who can afford it, and heterogeneity in respect of production and ownership;
- Reviews aspects of cultural globalisation;
- Considers political aspects of globalisation;
- Overviews protests against the current forms that globalisation is assuming in the form of the anti-globalisation and anti-capitalism movements;
- Uses the linkages existing between tourism, development and globalisation to provide an extended example of current globalisation trends and prospects.

Defining globalisation

Just as the first pictures of the whole earth from outer space made us aware of the interdependence and ecological fragility of the planet, so the events of 11 September 2001 and subsequently have served to show just how globalised and socio-politically fragile the world is today, certainly in respect of major events and strategic upheavals.

Over the last 20 years or so, one of the major trends has been that the world in which we live is being seen as ever more global in character and orientation. This trend has been witnessed in increasing actual and potential interactions between different parts of the globe. People are increasingly thinking in terms of an 'era of global change' and a 'globalising world'.

Such global change is intimately connected with a battery of developments which define our age, including e-mail, the internet, and the digital world in general, along with older technologies which are still acting as agents of diffusion and change, including the telephone, fax, television, video recorder and the wide-bodied jumbo jet.

Schech and Haggis (2000: 58) define globalisation as the intensification of global interconnectedness, a process that they see as associated with the spread of capitalism as a production and market system. In an essentially similar manner, Kiely (1999a: 3) avers that 'globalisation refers to a world in which societies, cultures, politics and economies have, in some sense, come closer together'. The same author goes on to note, however, that globalisation involves substantially more than interconnectedness. The process also involves the intensification of worldwide social relations, serving to link events in widely geographically separated places (see Schuurman, 2001; Murray, 2006; Conway and Heynen, 2006).

Contemporary globalisation is also associated with changing experiences of time and place and with the development of new communications technologies and the rise of what has been referred to as the 'information society' (Castells, 1996). Governments in their turn frequently stress that globalisation involves enhancing the free movement of goods, services, capital, information and, in some instances, people, across national boundaries. The process of globalisation is, therefore, inescapably plugged into the neo-liberal world order that was discussed in the previous chapter.

In addition, on occasions, globalisation is directly equated with a distinctly post-modern world within which states and boundaries are becoming less-powerful agents of change.

Yet, there remains much controversy about the likely developmental consequences of the diverse strands which make up globalisation (Murray, 2006; Conway and Heynen, 2006). Thus, the major theme of this chapter is that these global tendencies are highly uneven, both spatially (between places and regions), and socially (between peoples and groups). The impacts of globalisation vary from region to region, and group to group, in ways that are clearly contributing to the further development of diverse and plural geographies of development at the beginning of the twenty-first century. This overall diversity and plurality which is contingent upon globalising tendencies is the major theme of this chapter.

A theme that we touch on several times in the present chapter is that, however we define it, globalisation has, in fact, been around in different forms for many centuries. But it is a frequently heard assertion that over the last 20 years or so the rate of globalisation and its intensity have both increased dramatically – and it is the developmental connotations of this claim that form the principal focus of this chapter.

Globalisation and development: 'for and against'/'solution or problem'?

Introduction

There are at least three distinct aspects to the current processes of global change that we are witnessing.

First, the world is effectively 'shrinking' in terms of the distances that can be covered in a given period of time, due to faster and more efficient transport, although this process works more for the affluent 'jet-set' than for everyone else.

Second, better communications, such as cable and satellite television, mean that many – though by no means all of us – hear about what is happening elsewhere in the world more swiftly than we ever did before. The global web now has far more connections than it had in the past (Allen, 1995) and those who have ready access to the technology have much more up-to-date information at their disposal.

Third, the ascendancy of global corporations and global marketing activities is resulting in the availability of many standardised and globalised products (Robins, 1995) and television programmes throughout the world, again for those who can afford them. In addition, not only do we live in a world of near ubiquitous Big Macs, Coca-Cola and Levi jeans, we are also witnessing the emergence of global financial markets. This is associated with a dramatic acceleration in the speed of financial flows and transactions, with money now moving in purely electronic form, so that it takes only fractions of a second to send sums from one part of the world to another (Leyshon, 1995: 35).

Strands of globalisation: economic, cultural and political

Following this argument, Allen (1995) has recognised three broad strands to globalisation: the economic, the cultural and the political.

In respect of economic globalisation, distance has become less important to economic activities, and large corporations frequently subcontract to branch plants in far distant regions, effectively operating within a 'borderless' world.

Secondly, the stereotype of cultural globalisation suggests that as Western forms of consumption and lifestyles spread across the globe there is an increasing convergence of cultural styles on a global norm, with that norm being codified and defined by the global capitalist system (Plate 4.1). Figure 4.1a takes a humorous look at this apparent tendency towards global homogenisation.

Thirdly, as already noted, in the arena of political globalisation, internationalisation is regarded as leading to the erosion of the former role and powers of the nation state.

Critical reflection

Globalisation in the Third World?

The photograph shows a 'modern' advertising poster in what appear to be strongly contrasting surroundings in a street in the capital city of Nigeria, Lagos. The product is boldly advertised in terms of its 'power and goodness'. Beaming, prosperous female and male consumers are depicted. On the other hand, the advertisement appears in what is clearly a traditional sector of the city, and in which one can only infer that the women and children photographed share a general lack of power in the marketplace. How do you find yourself reacting to this contrasting street scene? What does it say about the role of advertising in such circumstances and what outcomes do you think the advert might have? What influence might 'first world' advertising have on 'third world' inhabitants in a more general sense?

Plate 4.1 Globalisation in the Third World – modern advertising poster provides a contrast to the surrounding streets in Lagos, Nigeria
(photo: Mark Edwards, Still Pictures)

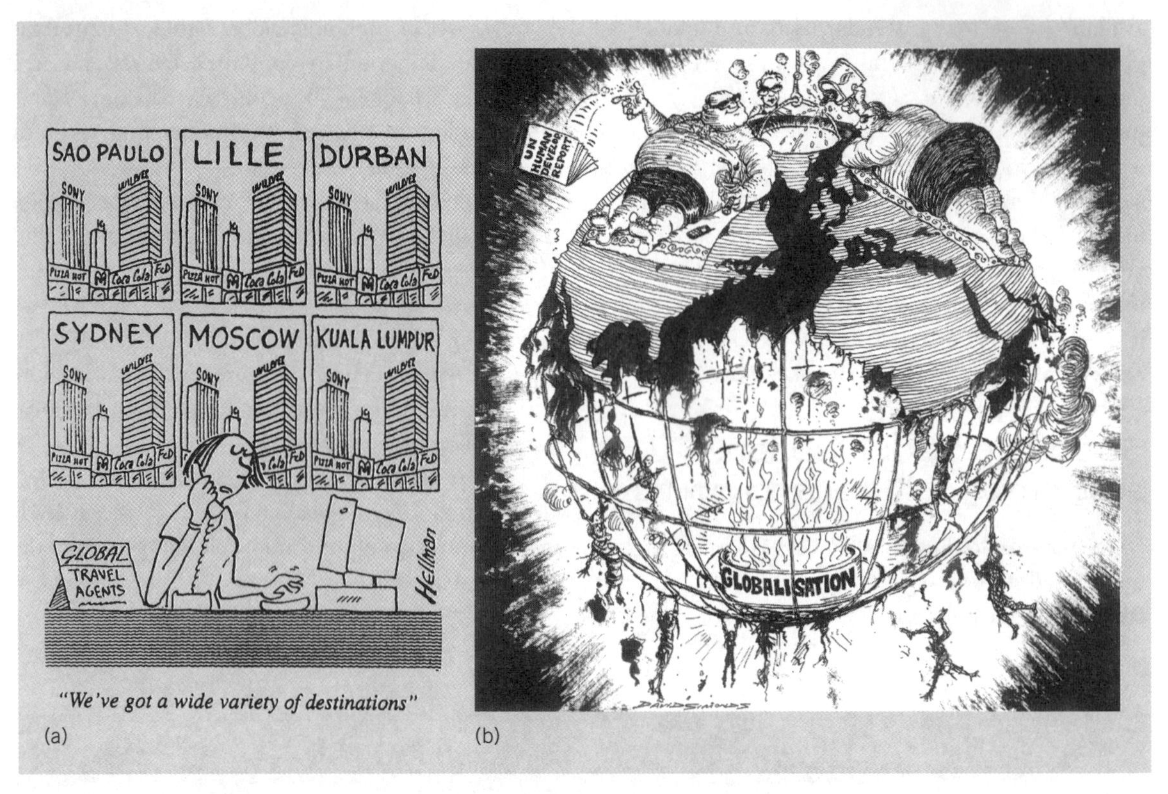

Figure 4.1 Contrasting views of globalisation as:

(a) the homogenisation of the whole world – or at least its cities
Source: Private Eye (2000)

(b) the conflagration of the South by the North
Source: David Simonds, Copyright Guardian News & Media Ltd 1999

Globalisation and development/ underdevelopment: a contentious issue

Following the overview of development theories and strategies presented in Chapters 1–3, the present account focuses on the question of what development means in a contemporary context that is dominated by processes of globalisation and global change.

One of the important questions to be addressed is whether contemporary globalisation is in fact a new process in any sense of the term. Is there any sense in which globalisation does mean that the entire world is becoming more uniform? Is there any chance that it means that the world will become progressively more equal over time? If not, is it the case that such a process of accelerated homogenisation will come about in the forseeable future?

Or does the available evidence point to increasing inequalities between the Global North and the Global South as the result of current global change? This less than positive view of globalisation is depicted in the cartoon reproduced in Figure 4.1b. In short, does globalisation mean that change and development will trickle down, and that this will occur with more speed than in the past, or will it be associated with increasing polarisation? These are just a few of the basic but highly contentious issues that will be addressed in this fourth and concluding chapter of Part I.

In other words, globalisation has to be seen as a highly contentious issue involving the current operation of the neo-liberal world order. This has clearly been witnessed in the '*anti-globalisation protests*' that have occurred outside major international financial gatherings since 1999. These anti-capitalist and anti-globalisation protests are more fully considered later in this chapter. On the other hand, many governments and world financial organisations such as the World Bank, International Monetary Fund and World Trade Organization have repeatedly stated the positive case they see for the role of globalisation in enhancing the overall process of global development.

Thus, as already intimated, two generalised views concerning the relationships between globalisation and development have emerged over the last decade, and these are outlined below.

Globalisation as development

The first view is the familiar claim that, to all intents and purposes, places around the world are fast becoming if not exactly the same, then certainly very similar (see Figure 4.1a). This view essentially dates from the 1960s' belief in the process of modernisation (see Chapters 1 and 3). Such a perspective tacitly accepts that the world will become progressively more 'Westernised', or, more accurately, 'Americanised' (Massey and Jess, 1995).

The approach stresses the likelihood of social and cultural homogenisation, with key American traits of consumption being exemplified by the 'coca-colonisation', or 'coca-colaisation', and the Hollywoodisation, or Miamisation of the Third World, replete with McDonald's golden arches. Thus Westernisation is seen as a natural and desirable reflection of the globalised spread of development. Globalisation is seen by those subscribing to modernisation theory as the outward flow of Western know-how, capital and culture to the rest of the world. It is in this connection that 'globalization studies can be seen as one step on from modernization' (Schech and Haggis, 2000; 57).

This is very much the view of globalisation that was presented by United Kingdom Government's White Paper on International Development presented at the start of the twenty-first century under the title *Eliminating World Poverty: Making Globalisation Work for the Poor* (Department for International Development, 2000a). In the words of Clare Short, the then Secretary of State for International Development, contained in the foreword:

> **This second White Paper analyses the nature of globalisation. It sets out an agenda for managing the process in a way that could ensure that the new wealth, technology and knowledge being generated brings sustainable benefits to the one in five of humanity who live in extreme poverty.**
>
> **(Department for International Development, 2000a: 7)**

It is also noted in the White Paper how 'encouragingly, in recent years we have seen the beginnings of a serious political debate about the equitable management of globalisation' and that 'making globalisation work more effectively for the world's poor is a moral imperative' (Department for International Development, 2000a: 14).

Box 4.1 summarises some of the arguments presented in the White Paper concerning the relationships between the degree of openness of economies and the generation of inequalities and economic growth, and the associations between globalisation and environmental conditions.

In overall terms, the White Paper is clear in its claim that, managed carefully, globalisation will bring specific benefits to the world's poor: 'The UK Government believes that, if well managed, the benefits of globalisation for poor countries can substantially outweigh the costs, especially in the long term' (Department for International Development, 2000a: 19).

Globalisation and marginalisation

However, even the White Paper is aware of the counterargument, noting of globalisation in respect of developing societies that, 'managed badly . . . it could lead to their further marginalisation and impoverishment' (Department for International Development, 2000a: 15). And the UK Government goes on within the same account to set out the beginnings of the anti-globalisation argument that the process is little more than one of neo-modernisation:

> **For some, globalisation is inextricably linked with the neo-liberal economic policies of the 1980s and the early 1990s. For them, globalisation is synonymous with unleashing market forces, minimising the role of the state and letting inequality rip. They denounce the increasingly open and integrated global economy as an additional more potent source of global exploitation, poverty and inequality.**
>
> **(Department for International Development, 2000a: 15)**

This cautious perspective regards globalisation as akin to the spread of advanced capitalism. Rather than suggesting that the net outcome is a more equal and more homogeneous world, the stance emphasises the reverse view, that globalisation is resulting in greater flexibility, permeability, openness, hybridity, plurality and difference, both between places and between cultures (Massey, 1991; Massey and Jess, 1995; Potter,

BOX 4.1

Making globalisation work for the poor: the 2000 White Paper on international development

The White Paper started by adopting a definition of globalisation which lets the reader know right away that it is coming from the pro-globalisation camp. Thus, globalisation is defined as the growing interdependence and interconnectedness of the modern world, but the central role is ascribed to the increased movement of goods, services, capital, people and information across national borders in rapidly creating a single global economy.

The White Paper goes on to associate globalisation with the diffusion of global norms and values and the spread of democracy. The proliferation of global agreements and treaties, including environmental and human rights agreements, is also emphasised in defining the character of current globalisation processes. Potter and Lloyd-Evans (1998), in stressing that not all the effects of globalisation are negative, cite the case of the Americanisation of Puerto Rico in relation to gay rights, where in much of the Caribbean homosexuality is still regarded as taboo.

The White Paper notes that some people argue that globalisation is creating increasing inequality and poverty. Although the document does not mention them directly, this is most likely through the imposition of structural adjustment packages (as described in Chapter 3) and the drive towards export-oriented economies. However, the White Paper presents a strong rebuttal, maintaining that there is no systematic relationship between the openness of economies and the generation of inequalities. The document argues that after increasing between 1960 and 1990, inequality has started to fall. Thus, it is argued that in 1960 the average real income in the countries containing the richest one-fifth of the world's population was some 12 times greater than that of the nations containing the poorest one-fifth. By 1990, this ratio had risen to 18 times greater. The White Paper reports that by the late 1990s this had fallen to 15 times. But it should be noted that even by these statistics, inequality was still higher in the late 1990s than in the 1960s.

The White Paper continues by making a strong argument for open economies and the virtues of free trade: 'everywhere it is clear that openness is a necessary – though not sufficient – condition for national prosperity' (Department for International Development, 2000a: 17). It is noted that no developed country is closed, while those poor nations that have successfully caught up in recent years, the NICs of East Asia and China, seized the opportunities offered by more open, world markets to build strong export sectors and to attract inward investment. The paper specifically stresses that in most East Asian countries the proportion of the population living in poverty is now under 15 per cent, down from around 40 per cent some 40 years ago.

The White Paper also addresses the suggestion that globalisation may be damaging the global environment. It notes that environmental conditions continue to deteriorate, but argues that this was occurring before global integration speeded up. Notwithstanding, the White Paper maintains that openness and integration into the global capitalist economy can assist countries to meet the new environmental challenges, stating that 'poverty and environmental degradation are often linked' (p. 16). In support of this, it is emphasised that over the past 50 years it has been the more closed economies that have had the poorest records of industrial pollution and urban environmental degradation. It is argued that stronger international institutions are needed to encourage more sustainable development. While many environmentalists would concur with this suggestion, it is likely that they would argue that the White Paper fails to address the issue of the environmental costs which are contingent upon moving goods around by air, sea, rail and road from one world region to another.

1993b, 1997; Robins, 1995). Following on from this perspective, far from leading to a uniform world, globalisation is viewed as being closely connected with the process of uneven development, and the perpetuation and exacerbation of spatial inequalities.

This view of globalisation argues that, by such processes, localities are being renewed afresh. This is particularly so in respect of economic change, where production, ownership and economic processes are highly place- and space-specific.

Even in regard to cultural change, it may be argued that although the hallmarks of Western tastes, consumption and lifestyles, such as Coca-Cola, Disney, McDonald's and Hollywood are available to all, such worldwide cultural icons are reinterpreted locally, and take on different meanings in different places (Cochrane, 1995). Further, it is obviously the case that access to them varies sharply by virtue of income and social standing. This view sees fragmentation and localisation as key correlates of globalisation and postmodernity.

A further major point substantiates this view. Evidence shows that globalisation is anything but a new process – it has been operating for hundreds of years. The process of globalisation can be seen as having started with the age of discovery (Allen, 1995; Hall, 1995). This argument has been clearly summarised by Stuart Hall (1995: 189):

> **Symbolically, the voyage of Columbus to the New World, which inaugurated the great process of European expansion, occurred in the same year as the expulsion of Islam from the Spanish shores and the forced conversion of Spanish Jews in 1492. This . . . [is] as convenient a date as any with which to mark the beginnings of modernity, the birth of merchant capitalism as a global force, and the decisive events in the early stages of globalization.**

This perspective usefully highlights how globalisation has always been intimately connected with power differentials and changes in culture. Early globalisation was associated with the conquest of indigenous populations, great rivalries between the major European powers in carving up colonial territories, and the eventual establishment of the slave trade, as detailed in Chapter 2.

Thus, globalisation has always been associated with increasing differences between peoples and places, rather than with evenness and uniformity. And globalisation has been a gradual, as well as a partial and uneven process, which has spread heterogeneously across the globe.

This overarching theme is addressed in the contemporary context in this chapter, first in relation to economic aspects of globalisation, and then in relation to cultural change. But before that, we turn to examine the overarching conceptualisation of a shrinking world as one of the fundamental cornerstones of contemporary globalisation.

Global transformations: a shrinking world or a more unequal world?

A shrinking world?

Whatever the respective arguments for and against globalisation as an agent of development, both critics and proponents point to the major changes that have occurred in the fields of transport and communications.

Thus, over the past 40 years there has been much talk about the world becoming a global village, and the associated 'compression' or 'annihilation' of space by time, in the context of what is referred to as the 'shrinking world'.

The phrase 'annihilation of space by time' is commonly attributed to Karl Marx (Leyshon, 1995: 23). Leyshon (1995) credits Marshall McLuhan (1962) with the first use of the expression 'global village', noting that the world was becoming compressed and electronically contracted, so that 'the global is no more than a village'. McLuhan went on to observe that due to the evolving electronic media, humans were beginning to participate in village-like encounters, but at a global scale, thereby cogently anticipating the development of electronic mail and the internet (see Chapter 8).

The main aspects of this change were outlined at the start of the present chapter. First, the world has effectively become a smaller place than it was 50 years ago, in terms of the time it takes to travel around it. At the global level, this is illustrated by the much-reproduced representations of the world shown in Figure 4.2. In the period between 1500 and 1840 the best average speed of horse-drawn coaches and sailing ships was about 10 miles per hour. In 1830 the first railway was opened between Liverpool and Manchester, and the first telegraph system was patented.

By 1900 a global telegraph system was in place, based on submarine cables, giving rise to the world's first global communications system. By the end of the period 1850–1940 steam trains averaged 65 miles per hour and steamships around 36 miles per hour. But, as shown by Figure 4.2, the real change came after 1950, with propeller-driven aircraft travelling at 300–400 miles per hour. After the 1960s commercial jet aircraft took speeds into the 500–700 miles per hour range. As a result of these progressive changes, the Earth has effectively been shrunk to a fraction of its effective size of some 500 years ago (Figure 4.2 and Plate 4.2).

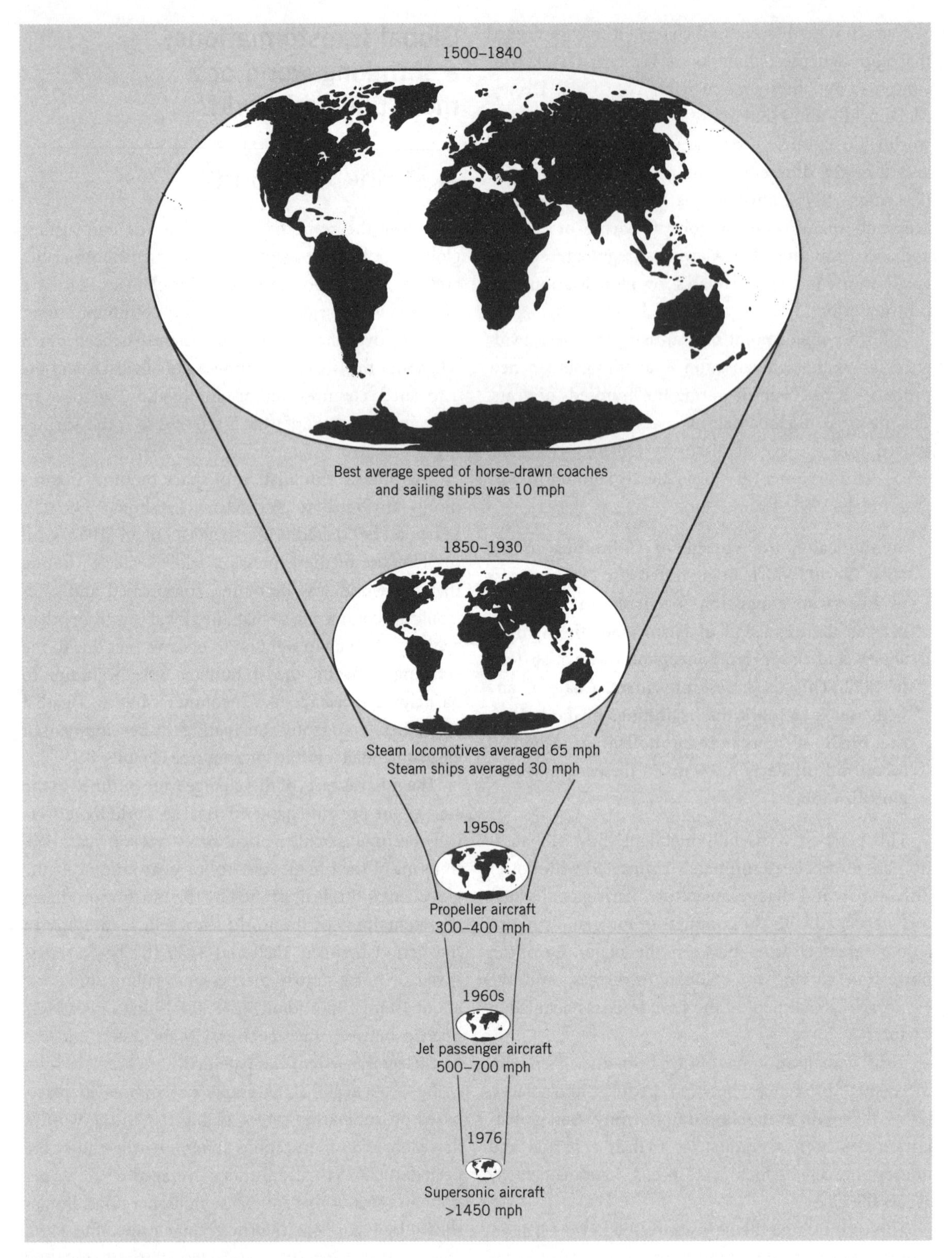

Figure 4.2 The shrinking world
Source: Adapted from Dicken (1998)

Plate 4.2 A380 Airbus 'double-decker' aircraft
(photo: Alamy Images/Andrew Hasson)

In the 1960s there was an exponential increase in the number of scheduled international flights globally. At the national level, large-scale highway construction proceeded in North America and Europe in association with rapidly increasing levels of car ownership. Between 1950 and 1960, domestic television was disseminated, followed by the exploration of space and the launch of communication satellites (Leyshon, 1995).

At the beginning of the 1970s Janelle (1973) referred to the '30-minute world', this being the time it would take for an intercontinental missile to travel from its launch site to its target on the other side of the world.

The process by means of which improvements in transport technologies have effectively moved places vis-à-vis one another within the settlement system was described by Janelle (1969) as time–space convergence. To give a national example, in 1779 it took four days or 5760 minutes to travel the 330 miles that separate Edinburgh from London. By the 1960s the time taken to travel between them had effectively been reduced to less than 180 minutes by plane, so the two places had been 'converging' at the rate of approximately 30 miles per year.

Of course, the world is also shrinking in another sense, in that many of us are potentially increasingly aware of what is happening in other far-distant places, without the need to move from our home localities. This is now achieved via the mass media. As Leyshon (1995: 14) notes, it is 'in the area of news and current affairs that television's ability to shrink space is best illustrated', as was all too clearly demonstrated by the coverage of the Gulf War by cable news early in 1991.

But cable news depends on relatively sophisticated and expensive technologies, so the relatively rich have also tended to become the information-rich (Leyshon, 1995), a point to which we shall return at several junctures in this chapter. There are other development-related implications to this set of changing circumstances in that they invite a redefinition of our ethical and moral responsibilities in relation to people who live far away from us.

Such 'responsibility to distant others' (Corbridge, 1993b; Potter, 1993a; Smith, 1994, 2002) is not unrelated to the observation that the global mass media frequently tend only to refer to developing countries when reporting natural disasters, social disturbances, poverty, mass starvation and other crises and mishaps. Some writers, especially those concerned with Africa, have observed how this is leading to the notion that

Africa is literally 'bad news', gradually desensitising the relatively wealthy from the real daily plight of Africans (Harrison and Palmer, 1986; Milner-Smith and Potter, 1995). Such an opinion leads to the implication that the Third World is literally viewed as a disaster zone, and serves to emphasise its status as something quite separate, representing the global 'Other'.

A shrinking world, but an increasingly differentiated one

It is all too easy to conclude that, as the world shrinks, all parts of the 'global village' share in the benefits of global development. However, this leads to a vital argument, for the places that share in development are

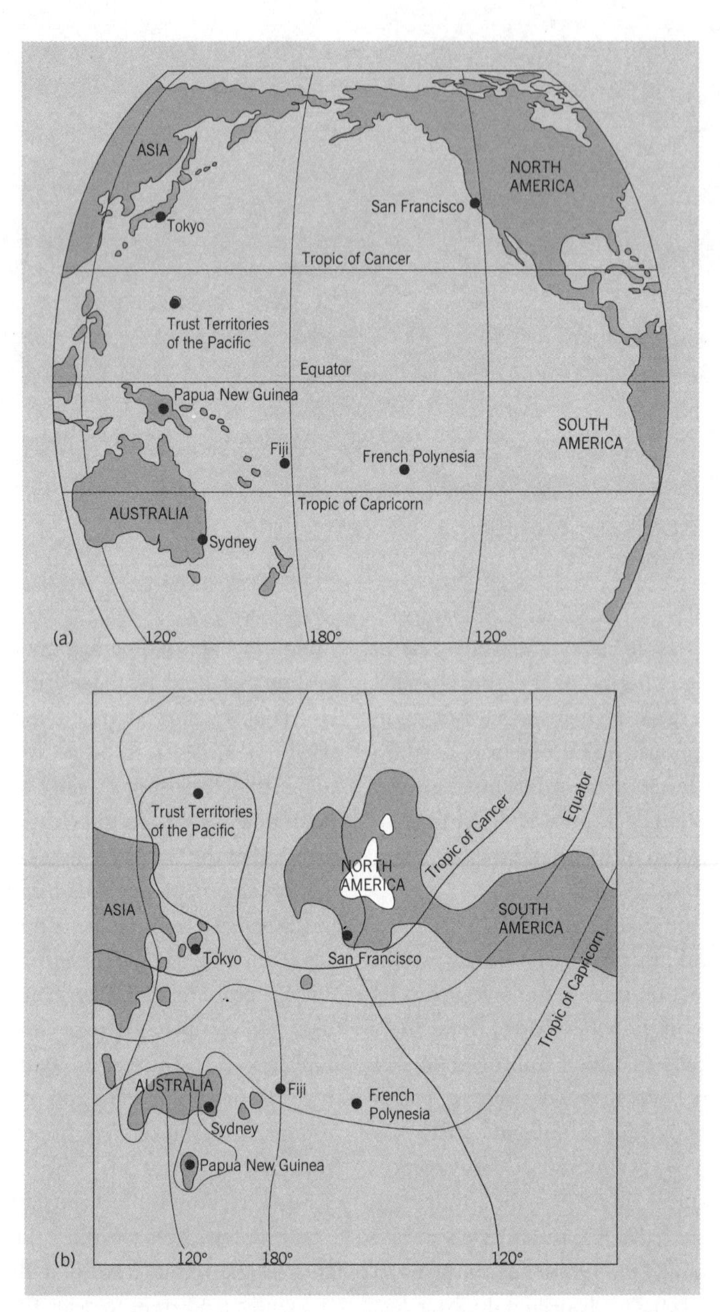

Figure 4.3 Time-space convergence and divergence: (a) the conventional projection of the Pacific; (b) time-space map of the Pacific based on travel times by scheduled airline in 1975

Source: Adapted from Haggett (1990 and 1990b) and Leyshon (1995) from 'Annihilating space? The speed-up of communications' in Allen, J. and Hamnett, C. (eds) *A Shrinking World?* Oxford University Press and the Open University. By permission of Oxford University Press, Inc.

those that are already well connected on the network. Places which are eccentric to it, or which are off the network altogether, are by definition massively disadvantaged. This is a fundamental point, and pursuing it at the sub-global level makes a very telling point about the developmental impact of globalisation. As well as relative distances being reduced by the process of global development, distances to other places can *increase* in *relative terms* within the overall context of a shrinking world.

Figure 4.3 gives a specific and very telling example of this. The maps show the Pacific Basin. Figure 4.3a shows the conventional cartographic projection, whereas Figure 4.3b has been redrawn according to travel times between places by scheduled airline in the mid 1970s. The figure is adapted from Haggett (1990) and Leyshon (1995). At first sight, North America has 'moved' closer to Asia, and Australia has 'drifted' north towards Asia. And if we look in a little more detail, we find that places like Tokyo, San Francisco and Sydney have indeed 'moved' closer to one another.

But if we look at Figure 4.3b more carefully, it is evident that some places have in fact become more 'distant' from each another. Thus, South America has 'trailed behind' North America in its 'convergence on' Asia. But if we look at specific places in more detail, they seem to have 'moved' quite substantially relative to one another. In particular, it is noticeable that poorer and less frequent air transport links mean that Papua New Guinea appears to have moved to the south of Australia, away from Asia, and the Trust Territories of the Pacific appear to have moved north, apparently now existing outside the Pacific Basin altogether. This clear example of overall time–space convergence shows that the process is far from homogeneous. In fact, it is sufficiently heterogeneous to produce instances of what may be called relative time–space divergence.

The idea that globalised improvements in transport and communications are leading to the intensification of the functional importance of certain places or nodes is confirmed if we look at world airline networks even as early as the 1990s.

Keeling (1995) produced a map showing the magnitude of the international air connections between major cities. The map, reproduced here as Figure 4.4 shows the number of outward and return non-stop flights per week from various global airport nodes. The data on which the diagram is based are reproduced in Table 4.1.

Table 4.1 Total number of non-stop flights per week to the major world cities in 1992

City	Notation on map	Global	Regional	Domestic
London	L	775	3239	1063
New York	N	644	634	8837
Paris	P	565	2264	1436
Tokyo	T	538	401	1814
Frankfurt	FRA	482	1376	771
Miami	MIA	311	1389	2146
Cairo	CAI	277	34	114
Los Angeles	LAX	245	419	7150
Bangkok	B	231	483	307
Singapore	SIN	221	831	0
Hong Kong	HKK	154	713	0
Sydney	SYD	144	89	1541
Rio de Janeiro	RIO	93	44	933
Moscow	MOW	87	400	1430
Bombay	BOM	64	111	313
São Paulo	SAO	64	97	1418
Buenos Aires	BUE	52	336	414
Johannesburg	JNB	40	108	450

Source: Adapted from Keeling (1995)

The outcome illustrates all too clearly the predominance of three global cities, London, New York and Tokyo, and the role these cities play as dominant global hubs. Together these three cities receive a staggering 36.5 per cent of total global non-stop flights to the world airline network's 20 dominant cities.

Beyond these three principal cities, Paris, Cairo, Singapore, Los Angeles and Miami are shown to be secondary global hubs, and Johannesburg, Moscow, Bombay, Bangkok, Hong Kong, Sydney, São Paulo, Rio de Janeiro and Buenos Aires as secondary hubs. Looking at the flows in Figure 4.4 shows just how marked is the concentration. Globalisation is leading to strong local concentration within continents and can again be seen as increasing 'differences'.

The almost complete marginalisation of Africa from the global airline network is clearly apparent, with only Johannesburg and Cairo appearing as salient nodes. In fact, much the same lesson is learned if we look at the global distribution of airports with more than 10 million passengers, as shown in Figure 4.5. Three global clusters emerge once more, centred on North America, Europe and Southeast Asia. The 'peripheral' status of both Africa and South America stands out noticeably.

A final but very important example of the concentrated nature of global transport is provided if we examine the global air cargo network in the 1990s. The

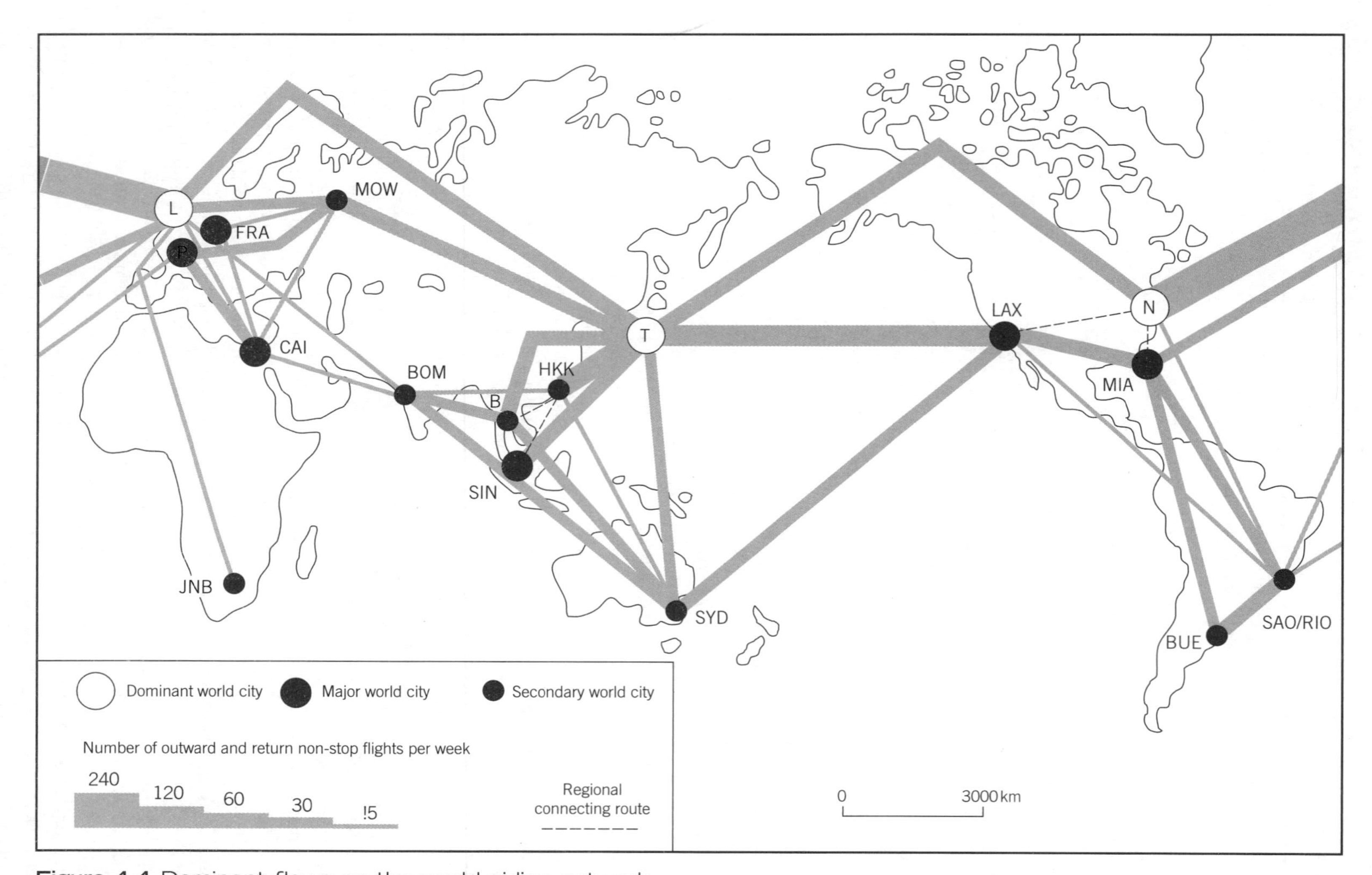

Figure 4.4 Dominant flows on the world airline network
Source: Adapted from Keeling (1995)

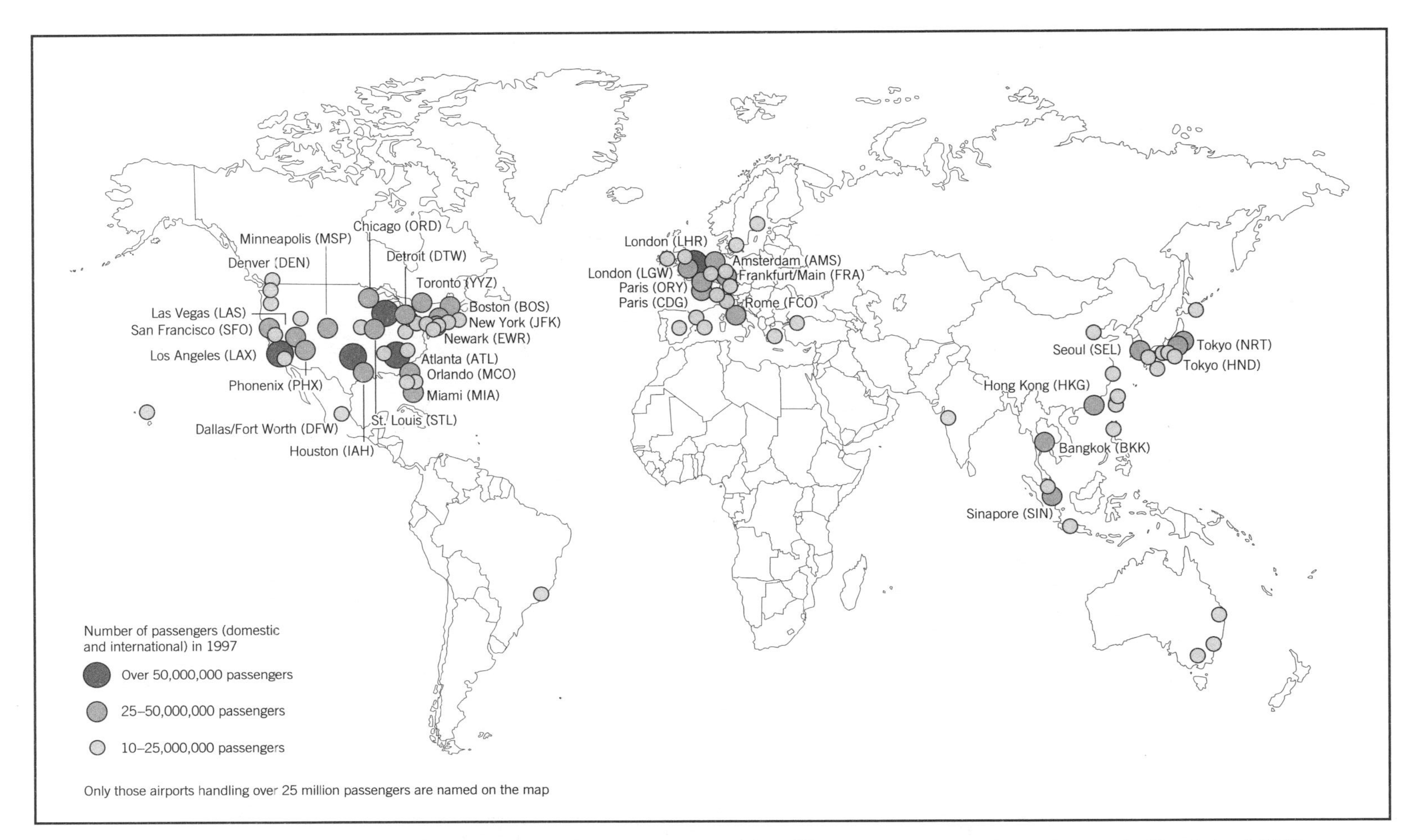

Figure 4.5 The distribution of world airports with more than 10 million passengers per annum in the late 1990s
Source: AND Cartographic Publishers Ltd, 1997, © RM Education plc (2008) All rights reserved. Helicon Publishing is a division of RM Education plc

pattern, shown in Figure 4.6, reveals once again the three-centred structure of the world economy, with the highest volume flows occurring between Western Europe, North America and Japan. This sharp polarisation reflects the evolution of the network of specialised parcel services following the introduction of the wide-bodied Boeing 747 in the 1970s, which saw the establishment of regular routes handling large volumes of high-value freight (Knox and Marston, 2001).

Finally, a commonly experienced event is the phenomenon of time–space compression as noted by Harvey (1989). The capitalist system demands efficiency and leads to the economic logic of reducing barriers to movement and communications over space, as time costs money. This leads to the progressive acceleration in the pace of life that seems to be universally experienced in the 'modern' world, and which seems to affect countries whether developed or developing, although perhaps in contrasting and locally specific manners. For those who are part of the network, communication with far-distant others can be a daily reality for hours at a time.

The next section looks at current examples of this in the field of information flows, and essentially comparable conclusions are reached concerning global tendencies and uneven processes and patterns of development.

Globalisation and the information society: the digital divide and an unequal world

Another major trend over the past 15 or so years has been the increased potential communications interconnectivity between people and organisations located in different parts of the world. Where once letters, telegraphs and the occasional telephone call were the principal means of communication, now faxes, mobile telephone calls, text messages and e-mail are dominant in the business world, and increasingly outside it, in education, commerce, entertainment, leisure and social life. The ability to move information and data quickly and cheaply has been greatly facilitated by the internet and the mobile telephone. We have already noted how some refer to the coming of the 'information society'.

The internet is seen by some as an optimistic possibility for global change and development and in 2007 1.13 billion people had access to it (World Internet Usage Statistics, 2007). Similarly, since its inception in the early 1970s, the growth of mobile telephone adoption means that there are now 1.5 billion subscribers (Dicken, 2007). Thus, many argue that new techniques in information processing offer poor nations new opportunities (Department for International Development, 2000a).

Some commentators have pointed to what they see as the internet's potential for democratising development. Others have even referred to it as allowing nations to 'leapfrog a stage of development', in what sounds like a direct reference to the Rostowian framework (Chapter 3). But caution needs to be exercised in respect of the uptake and spread of such new technologies. As suggested in Chapter 3, this is best understood if we look at the spread of previous technologies.

For example, even the UK White Paper noted that more than one-half the population of Africa had never used a telephone, and that fewer than one in 1000 Africans had access to the internet at that juncture (Department for International Development, 2000a). Indeed, by 2007 this figure had only increased to around 3.6 per cent. Others have warned that without care, where it is available, the internet may well serve to Westernise the Global South, so that 'e-imperialism' will be the outcome rather than any simplistic and comfortable notion of 'e-democracy'.

Kiely (1999a) noted at the time he was writing that 80 per cent of the world's population still lacked access to the most basic communications technologies, and nearly 50 countries had fewer than one telephone line per 100 people. Thus, Kiely (1999a) gave a slightly different gloss on the situation faced by Africa, emphasising that there are more telephone lines in Manhattan than there are in the whole of sub-Saharan Africa. The latest statistics show that such is the digital divide that while Developing Countries account for 75 per cent of total world population, they account for only 12 per cent of the world's telephone lines (Dicken, 2007).

The massive global inequality in access to telephones, for example, is exemplified in Figure 4.7, based on Clarke (2001) and UNDP (2001). The graphs show mainline telephone and cellular mobile telephone subscribers per head of the population in 1990 and 1999 for a selection of countries. The countries are classified according to their assessed levels of development using the United Nations Human Development Index (HDI) (see Chapter 1 for a full explanation of this measure).

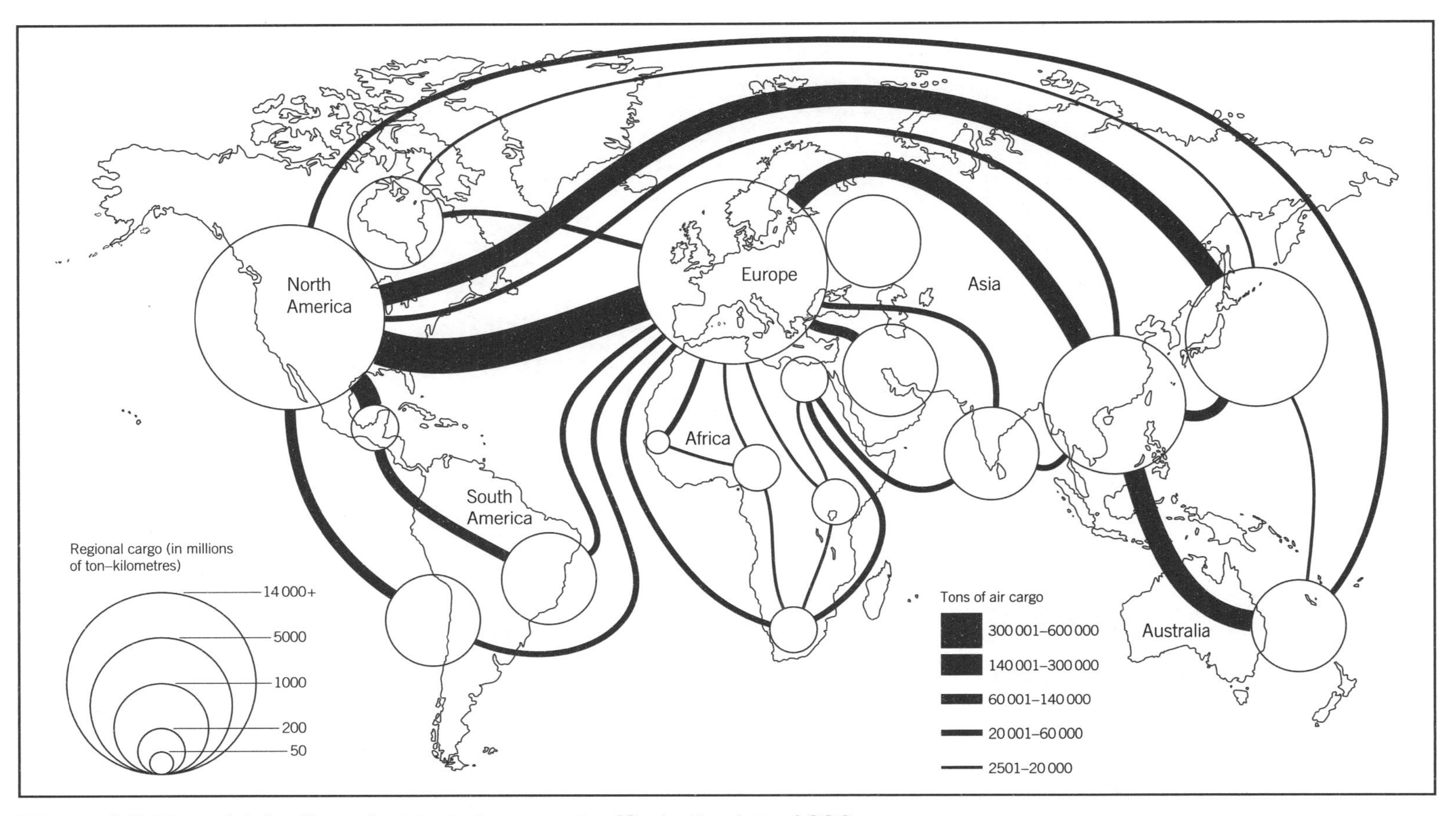

Figure 4.6 The distribution of global air cargo traffic in the late 1990s
Source: Adapted from Knox and Marston (2001)

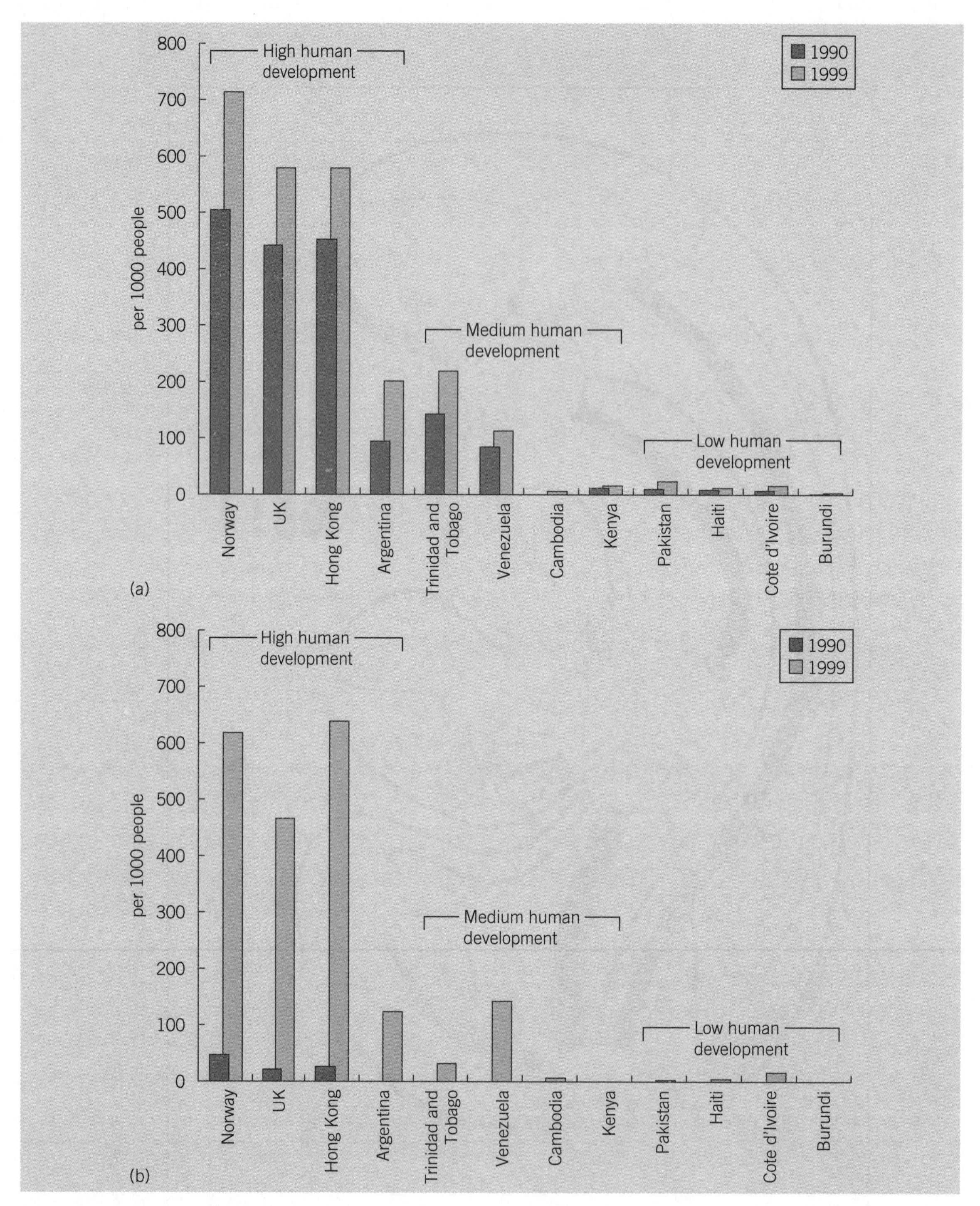

Figure 4.7 Mainline (a) and cellular mobile (b) telephone connections per 1000 of the population for a sample of nations with high, medium and low United Nations Human Development Index scores

Source: Adapted from Clarke (2001) and UNDP (2001) (*Human Development Report, 2001*, UNDP, Oxford University Press. By permission of Oxford University Press, Inc.)

It is, of course, nations like Norway, the United Kingdom, Hong Kong and Argentina which are assessed as being characterised by high levels of development that have by far and away the highest mainline and cellular connectivity, and, since 1990, especially the latter. Countries characterised as having low levels of development with respect to the HDI are typified by noticeably few telephone connections. A few medium-level

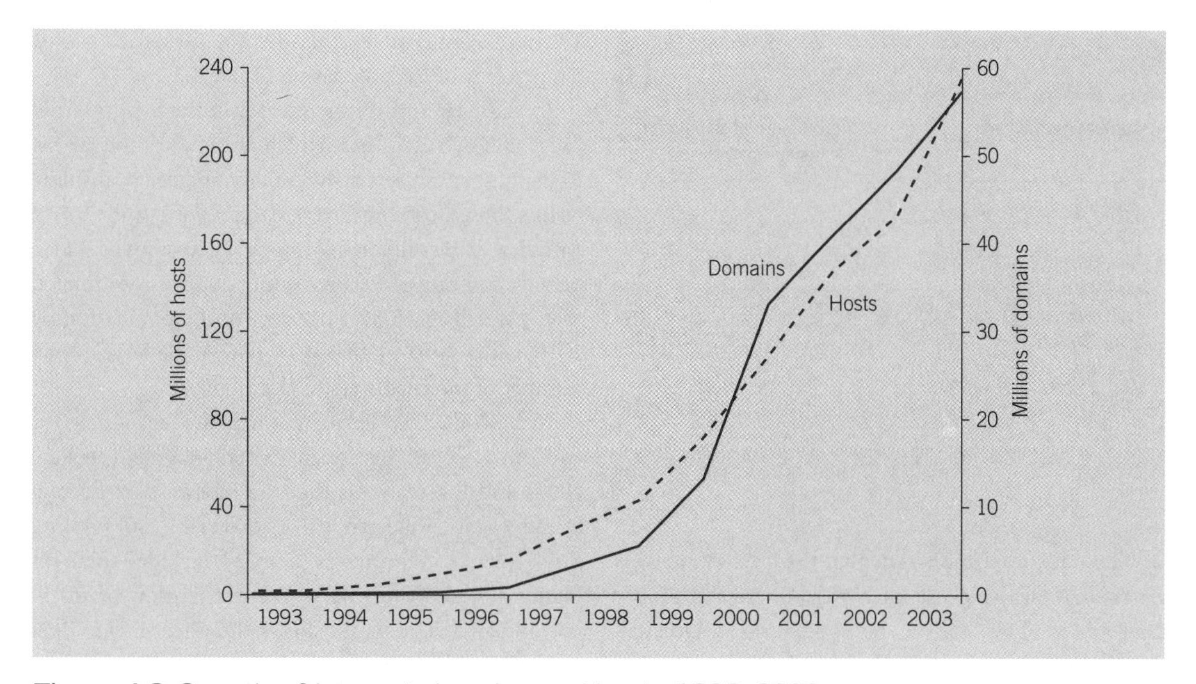

Figure 4.8 Growth of internet domains and hosts 1993–2003
Source: Dicken (2007), based on Zook (2005)

HDI nations, such as Venezuela and Trinidad and Tobago, are starting to show enhanced levels of subscribers, especially with respect to mobile phones (Figure 4.7).

At the turn of the twenty-first century, Kiely (1999a) gave further insights into how unequal access is to computers. Thus, while at that stage the USA was reckoned to have 35 computers per 100 people, South Korea had 9 per 100 people, and Ghana only a minuscule 0.11 per 100 people.

Thus, despite the relative ease with which computers can be linked through the internet, undue optimism regarding the role of the internet in developing nations would be misplaced. Even the pro-globalisation White Paper of the UK Government acknowledges that 'there is a real risk that poor countries and poor people will be marginalised, and that the existing educational divide will be compounded by a growing digital divide' (Department for International Development, 2000a: 40).

Others have referred to the massively unequal global distribution of communications. Quite simply, without telephones and computers, areas of the developing world cannot race ahead as an outcome of the existence of technologies such as those associated with the internet and e-mail.

As Zook (2005) and Dicken (2007) have recently observed, within a decade the internet has revolutionised the way people with access to it communicate, replacing part of the use formerly made of telephones and faxes. Hundreds of millions of networked computers link people around the world. What started in the early 1970s in the US Department of Defense as a specialist technology, and spread via academic computer networks, has by the mid to late 2000s rapidly become a household basic for those with the requiste resources.

From the early 1990s, when the number of networked computers (hosts) was virtually zero, the number grew exponentially to reach 120 millions by 2001 and 230 million by 2003 (Figure 4.8). At the same time formal sites for interaction and commerce, or domains, have similarly increased exponentially since 1995 (Figure 4.8).

As noted previously, a large proportion of business communications is now made through the internet and information is available on a huge array of topics via the World Wide Web. For those with access to the technology, personal communications via e-mail has become the preferred form of daily communication.

Statistics show that, of a world population of 6.58 billion people, 1.13 billion are internet users, representing 17.2 per cent (www.internetworldstats.com, 2007). The percentage of the population that has access to the internet is referred to as the internet penetration rate, and such rates are shown by broad continental divide in Table 4.2.

Table 4.2 Internet usage by major world region in 2007

Continental division	Percentage of population using the internet
Africa	3.6
Asia	11.0
Europe	39.4
Middle East	10.0
North America	69.0
Latin America	18.4
Oceania/Australia	54.4
World	17.2

Source: www.internetworldstats.com, 2007

The statistics demonstrate that the level of access is by far and away highest for North America at 69 per cent and is above half of the population in Oceania, and just under 40 per cent for Europe (Table 4.2).

The general correlation between levels of income and internet penetration in 2007 is confirmed if we look at rates for the remaining continental divisions. The level of penetration is close to the overall world average for Latin America and the Caribbean, and then falls progressively for Asia, the Middle East and Africa. We have already noted that the rate for Africa is only 3.6 per cent of the population (Table 4.2).

In the near future, the greatest growth in internet usage will occur in Asia, which already has the largest total number of users at 409 million, higher in absolute terms than both Europe (319 million) and North America (231 million). Although China only had a penetration rate of 7.9 per cent in 2002, it is estimated that this will reach 56.4 per cent of the population by 2010, with China by that stage having the largest total number of internet users (235 million).

In 2007, the US, Germany and the UK accounted for two-thirds of all registered domain names (Dicken, 2007) and the reality is that the internet is reflecting already well-connected places. Dicken (2007), using Zook (2005), reproduces a map, included here as Figure 4.9, showing the world distribution of internet domains. The figure shows the phenomenal dual clustering of internet domains on Western Europe and North America.

Finally, although the worldwide growth of the internet is likely to slow down in the future, the main trend now is in internet users upgrading from dial-up to fast broadband connections. In 2007, about one-third of all 1.1 billion world users of the internet are accessing it by

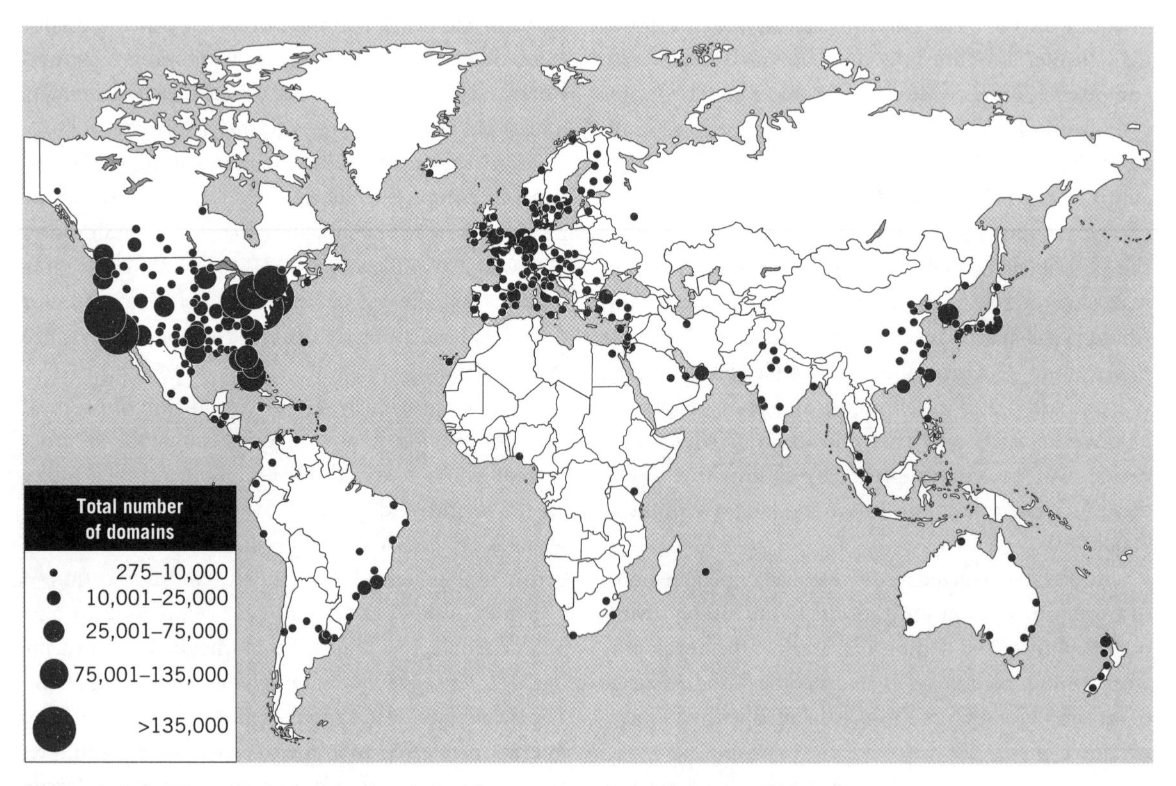

Figure 4.9 The global distribution of internet domains, c. 2005
Source: Dicken (2007), based on Zook (2005)

means of high-speed broadband connections (Dicken, 2007; Internet World Stats, 2007; Zook, 2005). And again, as would be expected, the statistics further paint a picture of a sharply divided digital world. Thus, of the 300 million globally who are accessing the internet using broadband, the USA accounts for the highest number with 60 million subscribers, representing 53 per cent of the total subscribers. At the other extreme most sub-Saharan African states hardly even register, with South Africa being the highest at only 1.79 per cent.

Critical reflection

Global interconnectivity and the internet

If you will be using the internet, e-mail, chat-rooms or social networking sites during the coming week, keep a list of the people you are in communication with, and the sites you have visited. Where in the world are these various contacts and sites based? You could classify and count them by continent or broadly by the 'developed world' and 'developing world' categorisation. What does the list say about your own global interconnectedness? Does the outcome of this reflection accord with the patterns outlined in this section of the book?

Thus, just like improvements in transport technologies, improvements in communications are tending to emphasise global and regional differences in the first instance. Notwithstanding the claims of governments, including those of the UK, much has to be done if these aspects of globalisation are to improve conditions for the poor and moderately poor in developing countries.

However, the situation is changing quickly and mobile phones especially can act as powerful agents of change, but much remains to be done (Kiely, 1999a). It is undoubtedly the case that the mobile phones and the internet can help those who are in marginal locations to do their shopping, communicate with family, establish and run businesses or, indeed, gain education by distance learning programmes (Unwin and de Bastion, 2008).

For instance, Urbach (2007), in the context of Botswana, notes how mobile phones can reduce communication costs, increase labour mobility and afford enhanced access to banking facilities and information on market prices. Although only 20 per cent of the population have access to mains electricity, enterprising village entrepreneurs offer recharging services using car batteries.

In respect of learning and education, the UK Government's Imfundio Programme sought to use information technology to improve primary education in Africa (Department for International Development, 2000a). However, we have reviewed enough evidence suggesting that without stronger and more all-embrassing intervention, the digital divide is likely to exacerbate the differences existing between the world's haves and have nots in the twenty-first century, although there is huge potential where the infrastructure can be provided.

Economic aspects of globalisation: industrialisation, TNCs, world cities and global shifts

Industrialisation

As explored in Chapter 3, the pursuit of unequal development as a matter of policy came to affect the newly independent, formerly colonial territories in the 1960s. It was almost inevitable that in seeking to progress during the post-colonial era, developing countries should associate development with industrialisation. This was hardly surprising given that the conventional wisdoms of development economics stressed so cogently this very connection (Potter and Lloyd-Evans, 1998; Chapter 3).

For many Third World countries, decolonisation afforded political independence and promoted the desire for the economic autonomy to go with it. In the words of Friedmann and Weaver (1979: 91), such nations:

> took it for granted that western industrialised countries were already developed, and that the cure for 'underdevelopment' was, accordingly, to become as much as possible like them. This seemed to suggest that the royal road to 'catching up' was through an accelerated process of urbanisation.

Import substitution industrialisation (ISI)

In the early phase, the trend towards industrialisation in developing countries was closely associated with the policy of import substitution industrialisation (ISI).

Plate 4.3 Import substitution industrialisation – a brewery in Ouagadougou, Burkina Faso
(photo: Jorgen Schytte, Still Pictures)

This represented an obvious means of increasing self-sufficiency, as such nations had traditionally imported most of their manufactured goods requirements in return for their exports of primary products such as sugar, bananas, coffee, tea and cotton.

During the era of import substitution industrialisation, key industrial sectors for development were those which were relatively simple and where a substantial home market already existed; for example, food, drink, tobacco, clothing and textile production (Plate 4.3).

While most developing countries have followed this path towards import substitution industrialisation, as Dickenson *et al.* (1996) observe, few have managed to progress much beyond it. An exception, perhaps, is Taiwan, where between 1953 and 1960 the ISI policy was put into practice, focusing on textiles, toys, footwear, agricultural goods and the like. During this era, manufacturing output increased by 11.7 per cent per annum. Only after the 1960s did Taiwan develop export-oriented manufacturing, and after 1980 the focus was on technologically advanced, high-value-added manufacturing, so as to stay ahead in the industrialisation stakes.

However, for most developing countries, the expansion of heavy industries such as steel, chemicals and petrochemicals, along the lines of the former Soviet model, has not been possible. Such a policy – which might seem attractive when following Rostow's (1960) linear model of development (Chapter 3) – requires a level of population and effective demand not normally present in most developing countries.

Furthermore, the competition from developed nations, along with capital and infrastructural shortages and problems of lumpy investment, technological transfer and capital rather than labour intensity, also militate against such heavy industrial development. An exception, however, is provided by India, which has achieved a high level of industrial self-sufficiency since 1945 (Johnson, 1983) and is now around the thirteenth largest industrial producer in the world (Dicken, 2007).

Light manufacturing and industrialisation by invitation (I-by-I)

However, from the 1960s onwards a number of developing countries embarked upon policies of light industrialisation by means of making available fiscal incentives to foreign companies (Potter and Lloyd-Evans, 1998; Chapter 3). This policy of so-called 'industrialisation-by-invitation' was strongly recommended by the Caribbean-born economist Sir Arthur Lewis (1950, 1955). Reviewed at length in Chapter 3, industrialisation-by-invitation involved the establishment of branch plants by overseas firms, with the products being exported back to industrialised countries.

The approach became closely associated with the setting up of free-trade zones (FTZs) and export-processing zones (EPZs). The FTZ is an area, usually located in or near to a major port, in which trade is unrestricted and free of all duties (Plate 4.4). The EPZ is normally associated with the provision of buildings and services, and amounts to a specialised industrial estate. Firms locating on EPZs frequently pay no duties or taxes whatsoever, and may well be exempt from labour and other aspects of government legislation. The approach is often known as enclave industrialisation.

According to Hewitt *et al.* (1992), the first EPZ established in a developing country was at Kandla in India in 1965, and this was quickly followed by further such developments in Taiwan, the Philippines, the Dominican Republic and on the United States–Mexico border.

In the case of Mexico during the 1960s, legislation was enacted permitting foreign, especially American, companies to establish 'sister plants', called *maquiladoras*, within 19 kilometres of the US borders, for the duty-free assembly of products destined for re-export (Figure 4.10). By the early 1990s, more than 2000 such assembly and manufacturing plants had been established, producing electronic products, textiles, furniture, leather goods, toys and automotive parts. In aggregate, the plants generated direct employment for over half a million Mexican workers (Dicken, 1998; Getis *et al.*, 1994).

Returning to the global context, by 1971 nine countries had established EPZs, and this increased to 25 by 1975 and 52 by 1985. In 1985, it was estimated that there were a total of 173 such zones around the world, which together employed 1.8 million workers.

Plate 4.4 Ice cream salesman outside the Koggala free-trade zone, Sri Lanka
(photo: Ron Giling, Still Pictures)

In 1998, the International Labour Organization estimated there were 850 EPZs employing around 27 million people, 90 per cent of whom were thought to be females. The global distribution of EPZs, those

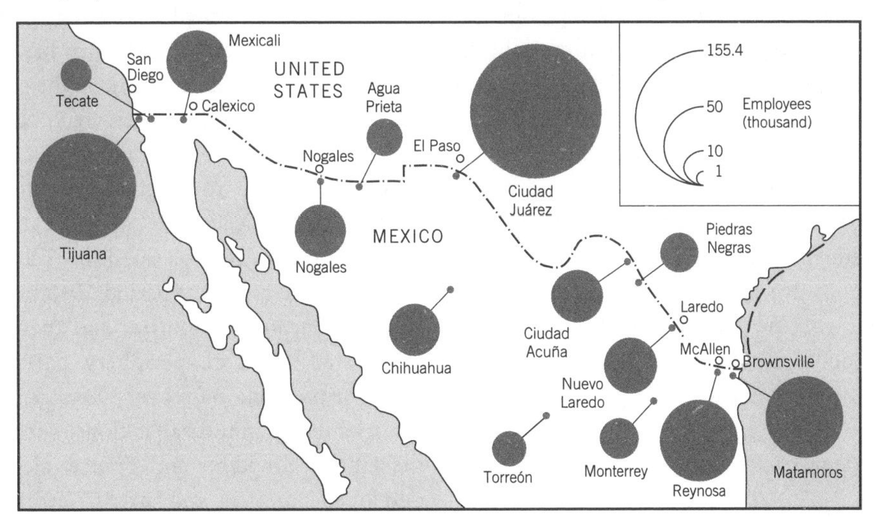

Figure 4.10 The principal *maquiladora* centres on the United States–Mexico border
Source: Adapted from Dicken (1998)

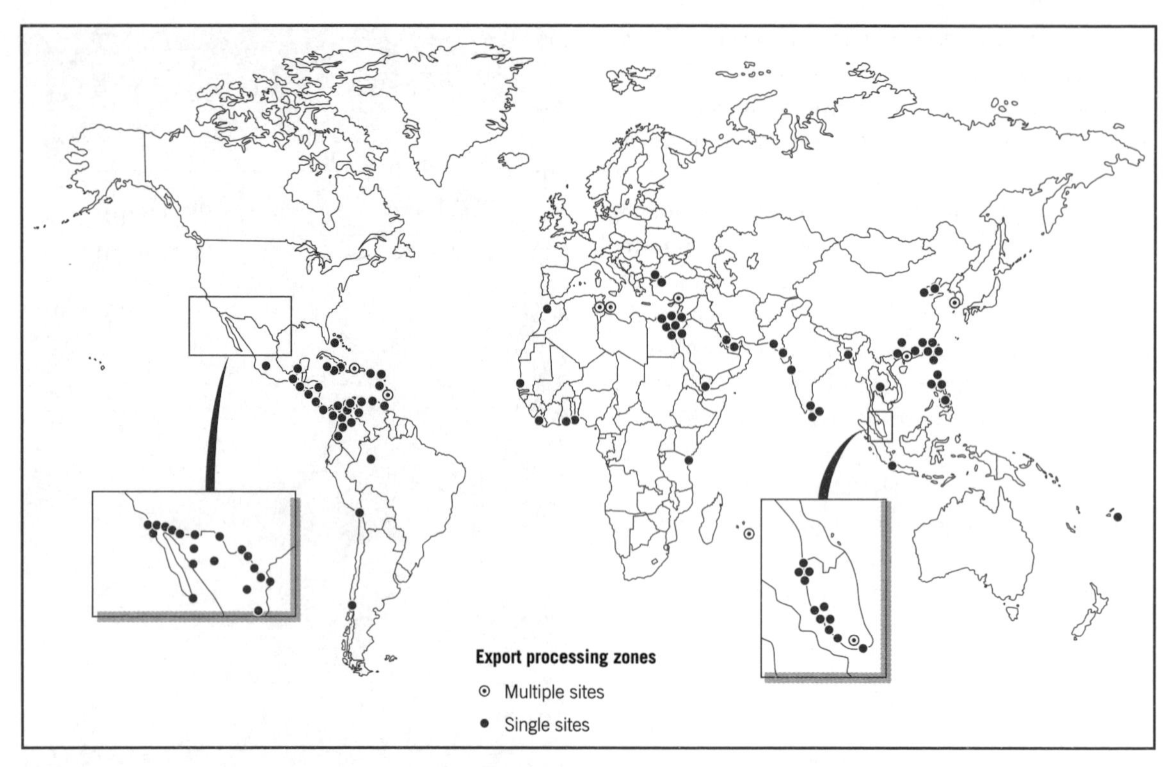

Figure 4.11 The distribution of export-processing zones in developing countries in the mid 1990s
Source: Adapted from Dicken (1998)

with both single and multiple sites, is summarised in Figure 4.11. China alone had 124 EPZs at the end of the 1990s, and together these employed a total of 18 million workers.

Frequently, programmes of industrial development have been strongly urban-based, as in the case of Barbados from the 1960s, where ten industrial estates were established, all within the existing urban envelope (Potter, 1981). Recently, data processing and the informatics industry have become very important on one of the central Bridgetown industrial parks (Clayton and Potter, 1996).

However, such schemes are not without their critics, and even the International Labour Organization has branded EPZs 'vehicles of globalisation', arguing that few have meaningful links with domestic economies and that most involve large numbers of low-waged, low-skilled workers. The accusation is also often made that firms run out of the country as soon as the financial incentives come to an end.

Global shift

By such means, developing countries have increased their overall level of industrialisation. From 1938 to 1950, developing nations experienced a 3.5 per cent growth rate of manufacturing per annum, and from 1950 to 1970 this annual rate increased to 6.6 per cent (Dickenson *et al.*, 1996).

But this growth has been minuscule compared with the growth of the urban population (see Chapter 9), and in many instances it has been based on a non-existent prevailing level of industrial activity. Furthermore, industrial growth has been characterised by several additional features. The first has been its highly unequal global distribution, and in the post-war period this has been associated with major changes in the global distribution of industrial production. Together, these are referred to as giving rise to 'global shift' (Dicken, 2007; Kiely, 2002).

This process is illustrated in Table 4.3, which deals with the historical period 1948 to 1984. Britain, Western European countries and then America had dominated the core–periphery pattern of manufacturing production for over 300 years. But from 1948 the traditional industrial nations, such as the USA, the United Kingdom and France, along with other developed countries, and latterly Germany, all showed reductions in their percentage share of world industrial production.

This went hand in hand with rising industrial production in Japan, which by 1985 had increased its share of the world total to 8.2 per cent. From 1948, industrial production also rose sharply in what were then referred to as centrally planned economies. A key feature was the increasing importance of the newly industrialised countries (NICs), which by 1984 accounted for 8.5 per cent of global production.

However, the remaining less-developed nations actually showed a declining proportion of total manufacturing production, from 9.1 per cent in 1948 to 5.4 per cent in 1984 (Table 4.3).

The emergence of the NICs, such as China, Brazil, India, South Korea, Mexico and Taiwan, is shown by their inclusion among the top 25 industrial nations in the mid 1980s (Table 4.4; see also Courtenay, 1994; Dickenson, 1994).

The global distribution of manufacturing production in the early 2000s is shown in Figure 4.12, and, although the USA, Western Europe and Japan between them account for three-quarters of total production, the importance of the Asian tigers (Hong Kong, Korea, Singapore, Taiwan, Indonesia and Malaysia, plus Japan), together with Brazil and Mexico, is clear from the figure. And the major change in the last 10 years or so has been the rapid growth of manufacturing in China and to a lesser extent in India.

Table 4.3 Changes in the global distribution of industrial production, 1948–1984

Country/region	Percentage of world industrial production		
	1948	1966	1984
USA	44.6	35.1	28.2
United Kingdom	6.7	4.8	3.0
West Germany	4.6	8.1	5.8
France	5.4	5.3	4.4
Japan	1.6	5.3	8.2
Other developed countries	14.7	12.5	11.1
Centrally planned economies	8.4	16.7	25.4
Newly industrial countries	4.9	5.7	8.5
Other less-developed countries	9.1	6.5	5.4
Total	100.0	100.0	100.0

Source: Chandra (1992)

It is this process of change in the world distribution outlined above that has been referred to by Dicken (1998, 2007) and others as 'global shift', whereby economic activity is becoming increasingly internationalised or globalised. This is sometimes regarded as the

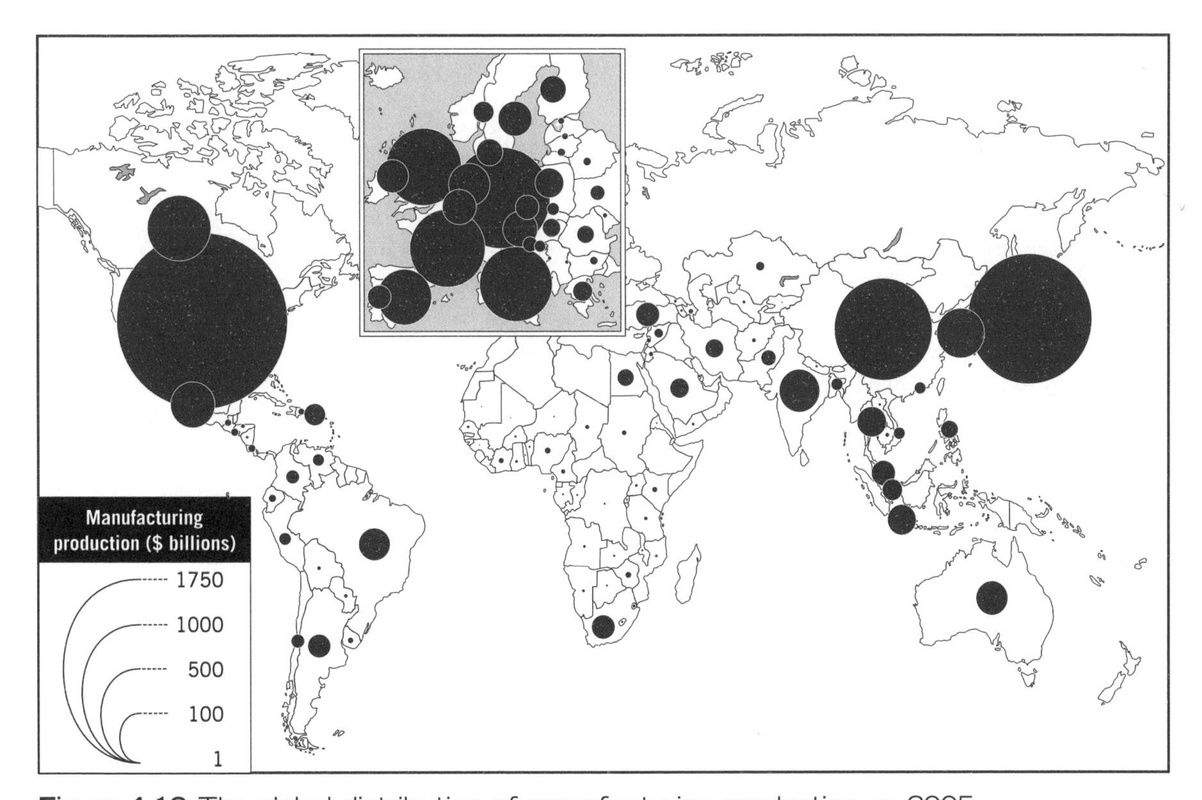

Figure 4.12 The global distribution of manufacturing production, c. 2005
Source: Dicken (2007), based on World Bank (2005)

Table 4.4 The world's leading manufacturing nations in 1986

Rank	Country	Percentage of world total manufacturing value added
1	USA	24.0
2	Japan	13.7
3	Former USSR	12.2
4	China	10.5
5	West Germany	6.5
6	France	4.0
7	United Kingdom	3.5
8	East Germany	2.2
9	Italy	2.2
10	Canada	1.8
11	Brazil	1.7
12	Spain	1.1
13	India	0.9
14	South Korea	0.9
15	Mexico	0.8
16	Taiwan	0.8
17	Switzerland	0.8
18	Sweden	0.7
19	Netherlands	0.7
20	Romania	0.6
21	Poland	0.6
22	Czechoslovakia	0.6
23	Yugoslavia	0.6
24	Belgium	0.6
25	Argentina	0.6

Source: Dicken (1992)

Table 4.5 The world's top 10 nations by manufacturing value added, 2005

Rank	Nation	Manufacturing value added (US$ million)	Percentage of world total
1	United States	1,422,999.9	26.33
2	Japan	865,809.7	16.02
3	China	407,513.6	7.54
4	Germany	385,923.9	7.14
5	United Kingdom	220,429.0	4.08
6	France	217,534.7	4.03
7	Italy	203,247.7	3.76
8	Canada	130,612.8	2.42
9	Korea	117,575.8	2.18
10	Mexico	110,381.6	2.04
	World	*5,404,373*	*100*

Source: World Bank (2005: Table 4.3) © International Bank for Reconstruction and Development/The World Bank

latest phase in the New International Division of Labour (Gilbert, 2002, 2008). An example of the complex changes that are occurring is given in Box 4.2. However, once again, this changing global pattern has been highly uneven in terms of its geography.

The present-day culmination of these global trends can be seen in Table 4.5. The table shows the world's top ten industrial nations in the first half of the 2000s, ranked by manufacturing value added (World Bank, 2005). While the United States remains dominant, accounting for just over one-quarter of world manufacturing production, its dominance has continued to decline in relative terms, as other competitors have emerged.

The resurgence of Asia, and in particular East Asia, is witnessed by the inclusion of Japan, China and Korea in the table in second, third and ninth places respectively. China's manufacturing value added increased from US$116,573 million in 1990 to US$407,514 million in 2001, roughly a 3.5-fold increase in ten years. Korea's manufacturing value added doubled from US$64,604 million in 1990 to US$117,576 million in the same time period. India is also now showing a notable increase in its manufacturing production, with the value added of the sector increasing from US$48,808 million to US$67144 million between 1990 and 2001. As Dicken (2007) notes, some are suggesting that in the not too distant future the world economy will be dominated by what they style 'Chindia'. Outside of Asia, Mexico, ranked in tenth place globally, saw its manufacturing value added double from US$49,992 million in 1990 to US$110,382 million in 2001.

Although the data exemplify the emergence of a growing interconnectedness within the world economy, as we have seen for the immediate post-war period, it remains the case that manufacturing production is highly uneven and clustered. Thus, 80 per cent of global manufacturing and service production is presently concentrated in just 15 nations and 90 per cent of outward FDI stock originated in 15 countries. Just seven East Asian countries – Hong Kong, China, Singapore, Korea, Thailand, Malaysia and Taiwan – account for 50 per cent of all FDI located in developing countries, this level of concentration having increased from 33 per cent in 1990 (Dicken, 2007).

In comparison, other less developed countries remain more poorly integrated into the global economic system. The story is once again of 'concentrated deconcentration' within the globalised economic system.

Transnational corporations (TNCs)

The second characteristic feature of post-1948 industrial change has been the rise of transnational corporations (TNCs), which now represent the most important

BOX 4.2

Globalisation and the production of athletic footwear

The footwear industry is labour-intensive but it is also highly dynamic. In a paper published in 1993, Barff and Austen show how sales tripled in the USA over the preceding ten-year period. And they show the industry's volatility, measured in spatial and geographical terms. In 1989, US market leader Nike Inc. had about 2 per cent of its shoes made by Chinese-based subcontractors. Just four years later, in 1993, almost 25 per cent of Nike athletic shoes came from Chinese factories (Barff and Austen, 1993).

Although characterised by such dynamism, the majority of athletic footwear production continues to occur in Southeast Asia. The three US companies which account between them for over 60 per cent of sales in the USA have the vast majority of their production based there. However, the details of this pattern are quite volatile, and many producers of athletic footwear have developed a complex set of long- and short-term subcontracting agreements with other firms that change from year to year as a result of factory improvements, market fluctuations and technological change (Donaghue and Barff, 1990).

On the other hand, several athletic footwear firms still manufacture in the USA. In particular, the cheapest sport shoes continue to be produced in the USA, whereas the more complex, expensive models tend to be manufactured in Asia. Barff and Austen (1993) show that, in order to understand this complex global geography, one must move beyond the basic consideration of international labour-cost differentials.

By means of case studies, the authors demonstrate that domestic production involves very different labour processes from those of production based in other countries. As in many sectors of the economy, domestic producers gain advantage by carrying smaller inventories via faster lead times. However, the best explanation for the globalised pattern of differential production centres is the nature of shoes themselves. The athletic shoes produced in the USA tend to have far fewer stitches in them than those manufactured elsewhere, and this minimises the most expensive component of the production process. Furthermore, the authors explain how tariffs on athletic shoes massively discriminate against imported shoes of a particular construction.

This example of global-scale production therefore demonstrates how processes of globalisation are based on subtle aspects of differentiation between world regions, and suggests that new forms of economic localisation may well be the outcome.

Critical reflection

Make a list of where in the world the goods you use regularly day to day are produced. Are there particular kinds of goods and services that are associated with particular world regions? What are the implications of transporting manufactured goods so far around the world to reach you? What are the benefits and what are the costs? Think who gets the benefits – and who gets the costs.

single force creating global shifts and changes in production (Dicken, 1998, 2007).

TNCs can be traced back to the late nineteenth century; to begin with they focused on agricultural, mining and extractive activities, but in the period since 1950 they have become increasingly associated with manufacturing (Department for International Development, 2000a; Jenkins, 1987, 1992; Dicken, 2007).

In 1985, the United Nations identified 600 TNCs operating in the fields of manufacturing and mining, each of which had annual sales in excess of US$1 billion. These corporations between them generated more than 20 per cent of the total production in the world's market economies. Over 40 per cent of total world trade takes place between the subsidiaries and parent companies of TNCs (Corbridge, 1986; Department for International Development, 2000a; Hettne, 1995).

During the 1960s, the foreign output of TNCs was growing twice as fast as world gross national product. By 1985, developing countries accounted for 25 per cent of total foreign direct investment (FDI). The largest share of this was in Latin America and the Caribbean (12.6 per cent of the world total), followed by Asia (7.8 per cent), Africa (3.5 per cent), with other areas accounting for 1 per cent (Dicken, 1998, 2007).

However, while data show that FDI flows to developing countries are still rising, their relative share has declined since the rapid expansion experienced in the 1980s and 1990s (UNCTAD, 2001). In 2000, FDI grew by 18 per cent, to US$1.3 trillion, although such flows were expected to decline in 2000.

By 2005, the global expansion of investment flows was driven by more than 70,000 parent company TNCs with over 700,000 foreign affiliates (Dicken, 2007), and TNCs account for approximately two-thirds of world exports of goods and services. Developed countries remained the prime destinations of FDI, accounting for more than three-quarters of global inflows.

While FDI inflows to developing countries also rose, reaching US$240 billion, their share of the world total fell in 2000 for the second year in a row, to 19 per cent. This compares with a peak of 41 per cent recorded in 1994. The countries of Central and Eastern Europe, with inflows of US$27 billion, retained their share of 2 per cent.

The 49 least developed countries (LDCs) remained marginal with only 0.3 per cent of FDI inflows in 2000. Most of these trends can be deciphered from Figure 4.13. However, it is also useful to stress that much FDI is still related to the resource production sector, and a considerable amount of it now emanates from Third World NICs such as South Korea and Singapore (Figure 4.14).

Figure 4.15 shows the global distribution of the largest 100 TNCs in the world, the largest 50 TNCs in developing countries, and the largest 25 TNCs based in central and Eastern Europe in 2000 (UNCTAD, 2001).

The big TNCs are headquartered in a handful of countries, namely the USA, Japan and those of Europe. The location of TNCs based in developing countries is very limited geographically, and focuses on Southeast Asia, South Africa, Mexico and parts of South America. Only South Africa stands out within Africa as a whole. The pattern is a very concentrated and uneven one.

The same point is revealed if we list the world's largest TNCs (Tables 4.6 and 4.7). Of the world's 20 largest TNCs ranked by foreign assets in 2000, six were American, five German, two were from the United Kingdom, two were Japanese and two French. One each was based in Spain, Switzerland and one jointly between the Netherlands and the UK.

The largest ten TNCs from developed countries are also shown to be geographically very concentrated, with half based in the Republic of Korea, and two hailing from China/Hong Kong (Table 4.7).

It is argued here that these contributory trends of globalisation are far too complex to be dealt with adequately by means of traditional models of development. In particular, the hierarchical model of change linked to modernisation reviewed in Chapter 3, put forward by Berry (1972), Hudson (1969), and Pederson (1970), may be seen as far too simplistic.

Given the omnipotence of TNCs, new production, innovations, capital and social surplus are not likely to

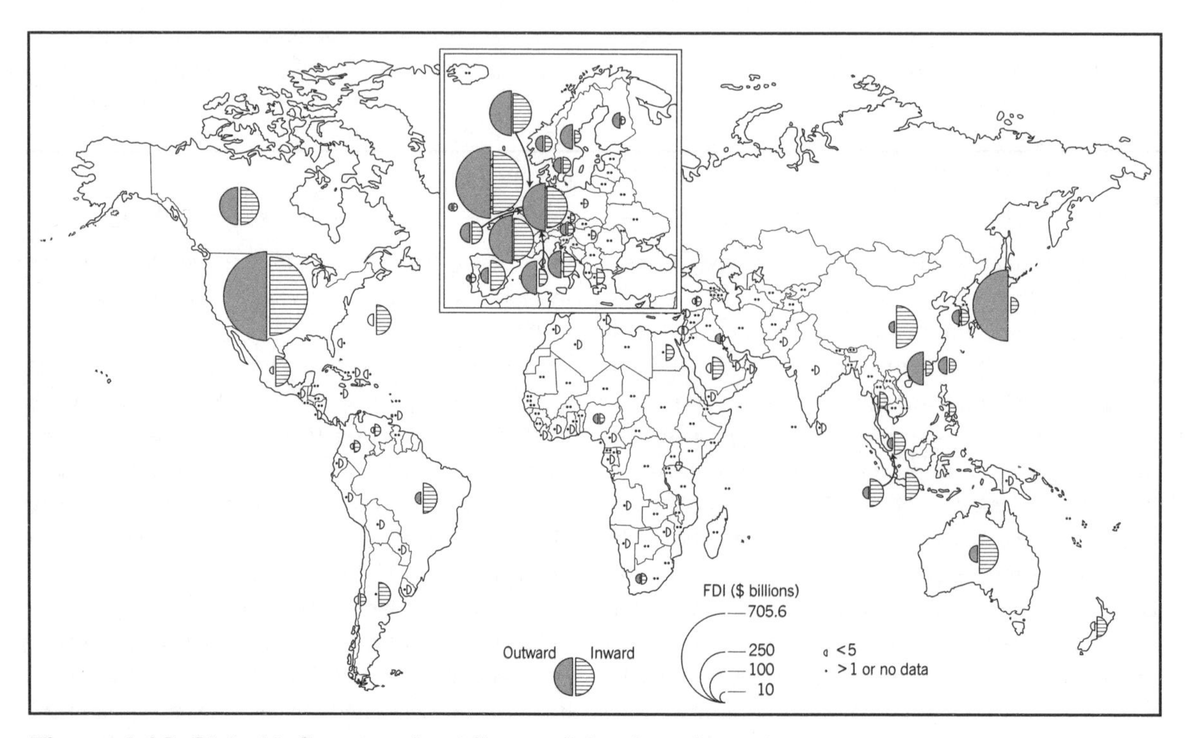

Figure 4.13 Global inflows and outflows of foreign direct investment

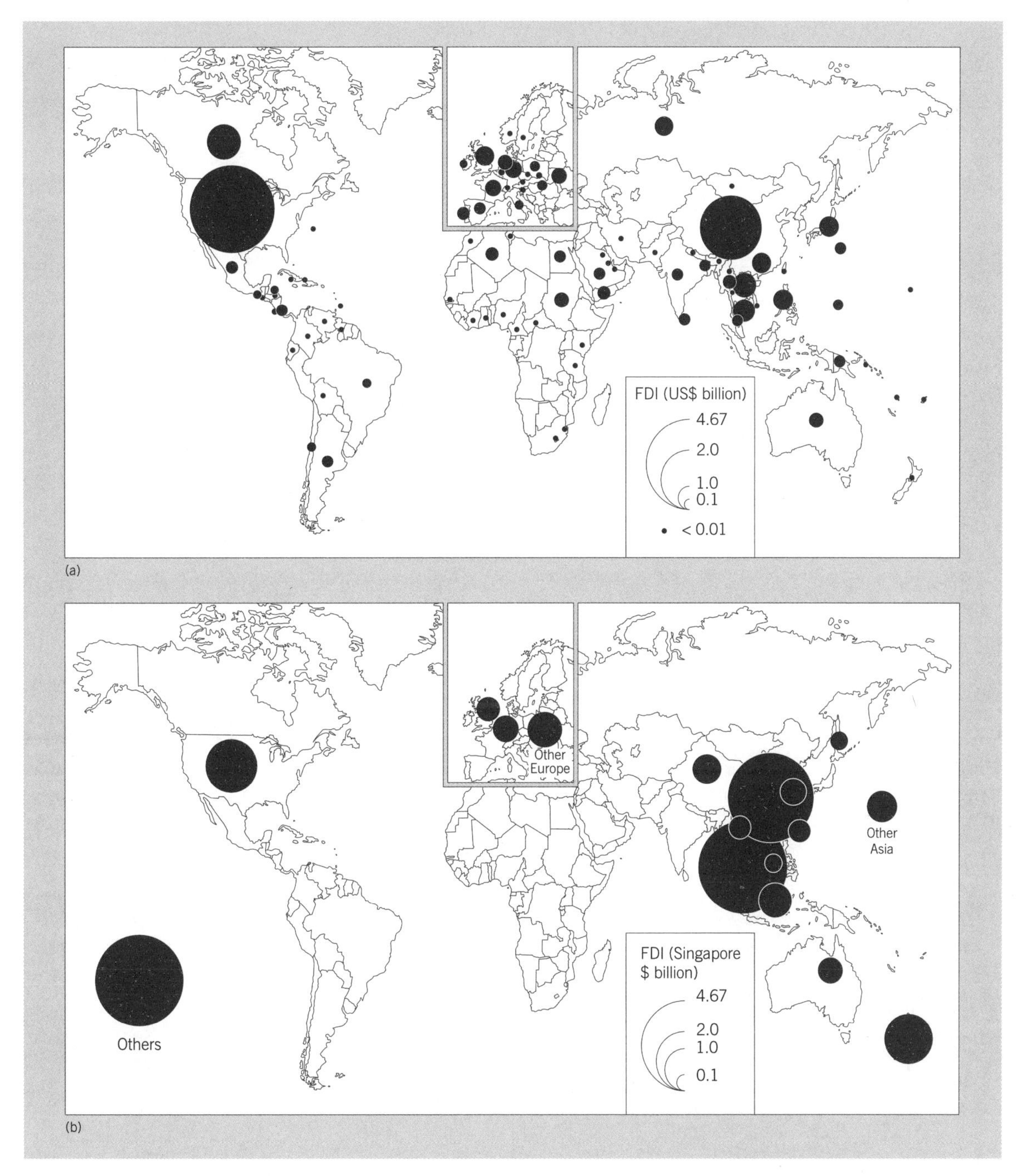

Figure 4.14 Foreign direct investment by (a) South Korea and (b) Singapore
Source: Adapted from Dicken (1998)

trickle down the national space economy in a smooth step-by-step manner, from the top to the bottom. Ownership and production are likely to be much more concentrated, an important theme which is picked up in the next major section of this chapter.

Furthermore, it follows from this that the decision to base production in one developing nation rather than another will have considerable impact on the geography of development and change, especially when it is remembered that many TNCs have annual turnovers that greatly exceed the gross national products of some small developing nations.

A broad appreciation of this is gained if we look at the global expansion of major TNCs. For example, the expansion of the Ford Motor Company between 1970 and 2000 is shown in Figure 4.16. Between these

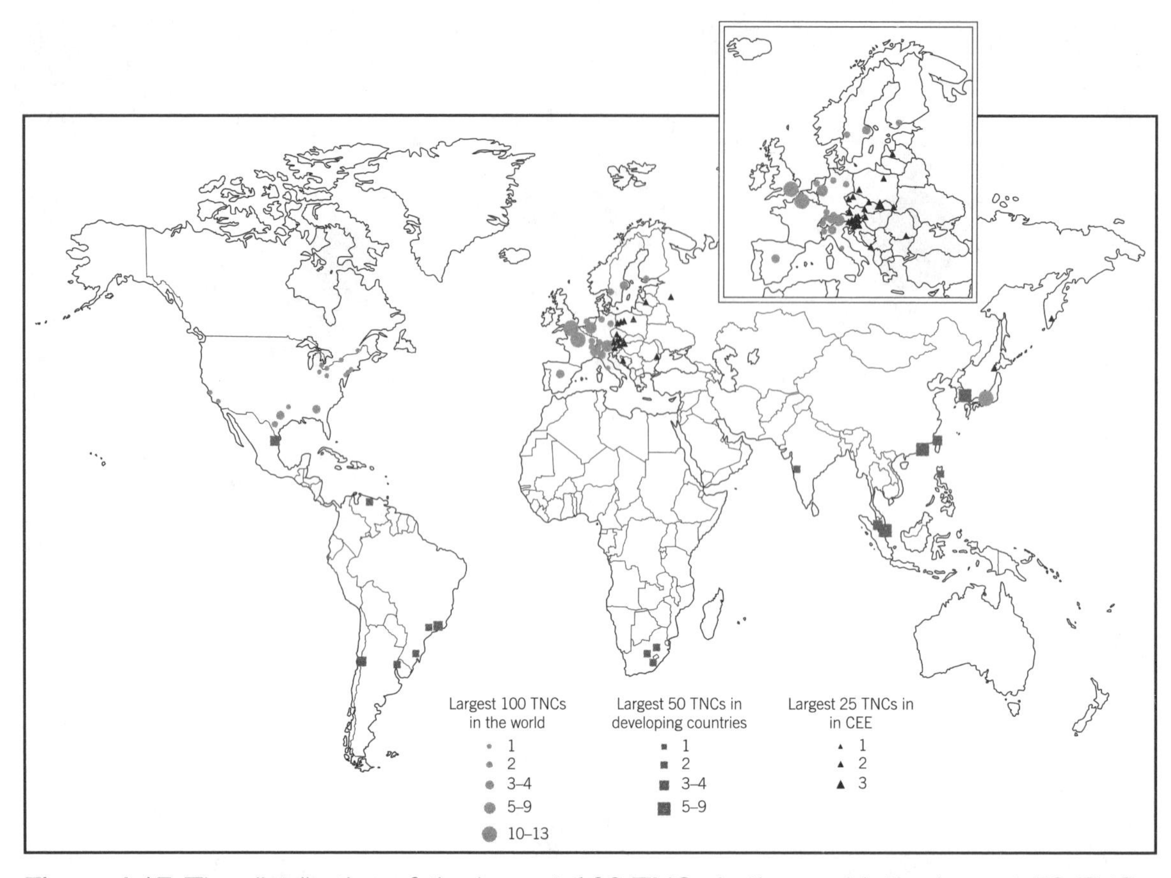

Figure 4.15 The distribution of the largest 100 TNCs in the world, the largest 50 TNCs in developing countries and the largest 25 TNCs based in central and Eastern Europe in 1999

Source: Adapted from UNCTAD (2001)

Table 4.6 The world's 20 largest TNCs ranked by foreign assets, 2000

Rank	Corporation	Country	Foreign assets (US$ billion)
1	General Electric	USA	141.1
2	ExxonMobil Corporation	USA	99.4
3	Royal Dutch/Sheil Group	Netherlands/UK	68.7
4	General Motors	USA	68.5
5	Ford Motor Company	USA	no data
6	Toyota Motor Company	Japan	56.3
7	Daimler Chrysler AG	Germany	55.7
8	TotalFina SA	France	no data
9	IBM	USA	44.7
10	BP	United Kingdom	39.3
11	Nestle' SA	Switzerland	33.1
12	Volkswagon Group	Germany	no data
13	Nippon Oil Co., Ltd	Japan	31.5
14	Siemens AG	Germany	no data
15	Wal-Mart Stores	USA	30.2
16	Repsol-YPF SA	Spain	29.6
17	Diageo Plc	United Kingdom	28.0
18	Marnesman AG	Germany	no data
19	Suez Lyornoise Eaux	France	no data
20	BMW AG	Germany	27.1

Source: UNCTAD (2001)

Table 4.7 The ten largest TNCs from developing countries ranked by foreign assets, 2000

Rank	Corporation	Country	Foreign assets (US$ million)
1	Hutchison Whampoa Ltd	Hong Kong, China	no data
2	Petróleas de Venezuela	Venezuela	8009
3	Cemex SA	Mexico	6973
4	Petronas-Petrolian Nasional Berhad	Malaysia	no data
5	Samsung Corporation	R. of Korea	5127
6	Daewoo Corporation	R. of Korea	no data
7	Lg Electronics Inc.	R. of Korea	4215
8	Sunkyong Group	R. of Korea	4214
9	New World Development Co., Ltd	Hong Kong, China	4097
10	Samsung Electronics Co., Ltd	R. of Korea	3907

Source: UNCTAD (2001)

dates, Ford went from 65 foreign affiliates to 270. This initially saw an intensification of affiliates in Canada, Western Europe and Australia. But from 1985 the number of affiliates increased in several of the big five economies of the south – India, Brazil, Mexico, plus China, Argentina and Venezuela. Again, however, the nations of Africa remain off the map.

Second, the global expansion of Unilever shown in Figure 4.17 shows a strong spread of affiliates to South America, China and Cambodia. Again, the African continent is largely unrepresented, save for Cameroon and the Ivory Coast.

Polarised and persistent patterns of change were identified by Pred (1973, 1977) in his early historical examination of the growth and development in the USA. Pred noted that the growth of the mercantile city was based on circular or cumulative causation, linked to multiplier effects. In addition, Pred argued that the growth of large cities was based on their interdependence, so that large city stability was often characteristic.

However, Pred maintained that key innovation adoption sequences were not always hierarchic, frequently flowing from a medium-sized city up the urban hierarchy, or from one large city to another large city.

Pred (1977) looked at the headquarters of TNCs in post-war America, and stressed the close correspondence with the uppermost levels of the world system. Thus, growth within the contemporary global system is increasingly linked to the locational decisions of multinational firms and government organisations, and with clustered forms of development and change.

The world city concept

These types of developments outlined in the last section have been given expression in the concept of the world city or global city. Although nebulous in terms of size definitions, the basic idea is that certain cities dominate world affairs.

At one level, this is a very straightforward and obvious proposition, but its contemporary relevance has been elaborated by Friedmann (1986), Friedmann and Wulff (1982) and Sassen (1991, 2002). Friedmann (1986) put forward six hypotheses about world cities, observing that they are used by global capitalism as 'basing points' in the spatial organisation and articulation of production and markets, and that they act as centres for capital accumulation (see also Potter, 2008).

Friedmann also suggested that the growth of world cities involves social costs which in fiscal terms the state finds it hard to meet. World cities have large populations, but more important, they have large and/or sophisticated manufacturing bases, sophisticated finance and service complexes, and they act as transport and communication hubs, the locations for corporate headquarters, involving TNCs and NGOs (Simon, 1992a, 1993; see also Friedmann, 1995; Knox and Taylor, 1995; Potter, 2008).

The principal world cities, such as New York, Paris, London, Amsterdam and Milan, are located in the developed world. But Singapore, Hong Kong, Bangkok, Taipei, Manila, Shanghai, Seoul, Osaka, Mexico City, Rio de Janeiro, Buenos Aires and Johannesburg (Plate 4.5) have all been recognised as part of an emerging network of world cities (Friedmann, 1995).

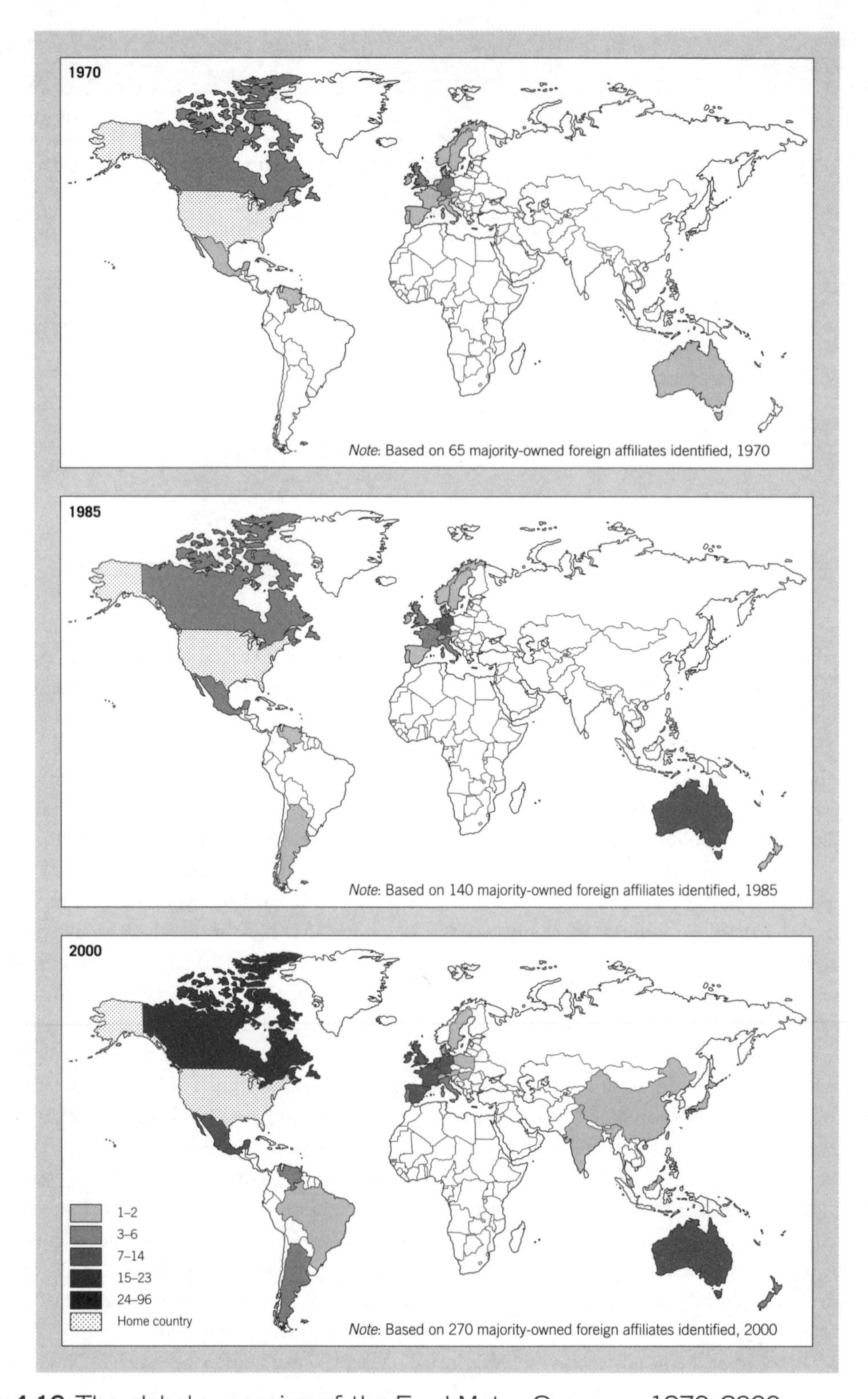

Figure 4.16 The global expansion of the Ford Motor Company 1970–2000
Source: Adapted from UNCTAD (2001)

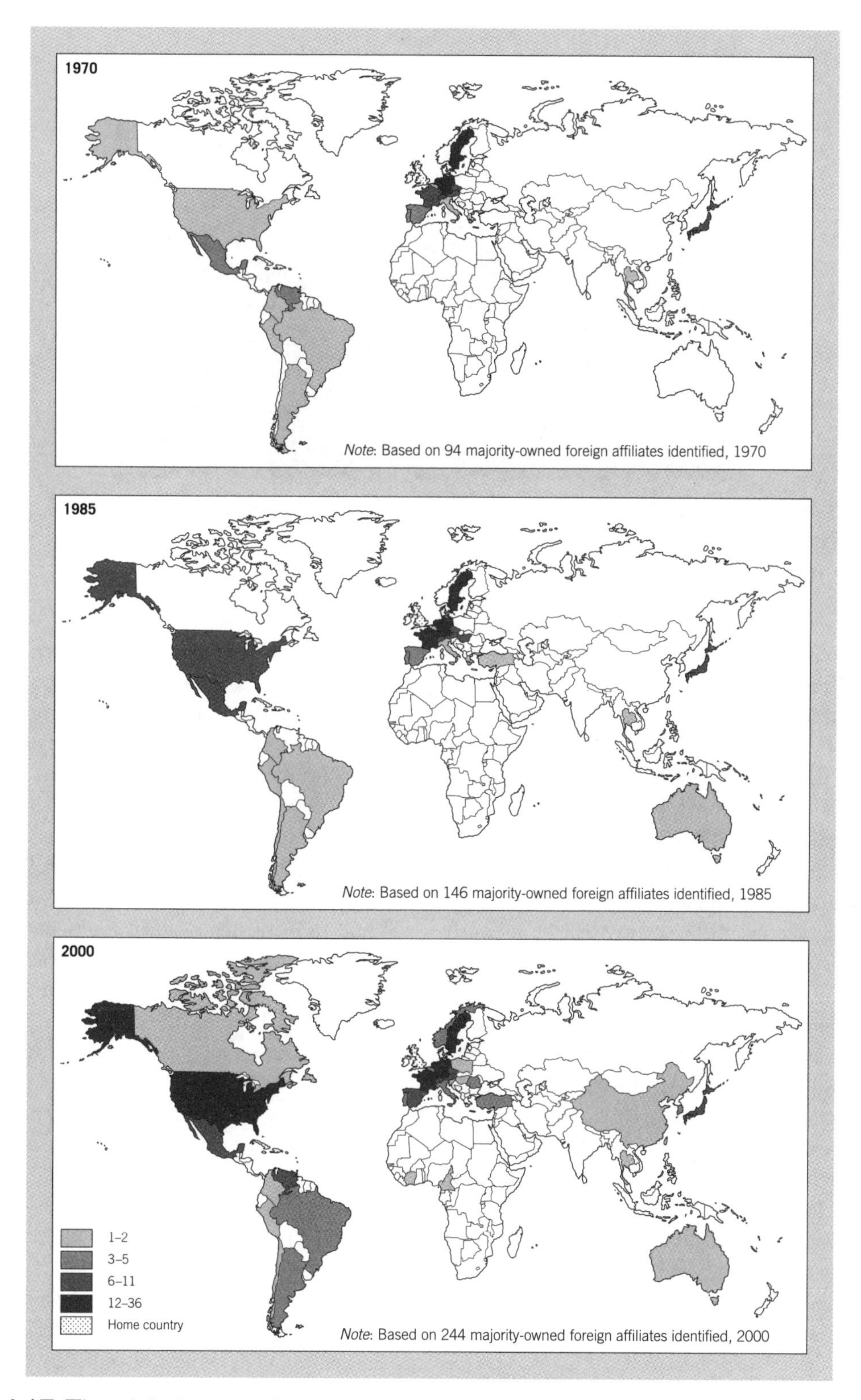

Figure 4.17 The global expansion of Unilever, 1970–2000
Source: Adapted from UNCTAD (2001)

Plate 4.5 A world city: the centre of Johannesburg from the air
(photo: Tony Binns)

This emergence is given spatial expression in Figure 4.18. In short, world cities may be seen as points of articulation in a TNC-dominated capitalist global system. But data show that world cities exhibit a very centric (or centred) structure, with major world cities such as London at the core (Taylor, 1985).

The implication is, therefore, that uneven development is particularly likely to be associated with developing countries, and that their paths to development in the late twentieth century will be infinitely more difficult than those which faced developed countries.

Quite simply, the world is already highly centred. This argument has been reviewed in the case of poor countries by Lasuen (1973). He started from the premise that, in the modern world, large cities are the principal adopters of innovations, so that natural or spontaneous growth poles become ever more associated with the upper levels of the urban system. Lasuen also observed that the spatial spread of innovation is generally likely to be slower in developing countries, due to the frequent existence of single plant industries, the generally poorer levels of infrastructural provision, and sometimes the lack of political will.

Thus, developing countries facing spatial inequalities have two policy alternatives. The first is to allow the major urban centres to adopt innovations before the previous ones have spread through the national system. The second option is to attempt to hold and delay the adoption of further innovations at the top of the national urban system, until the filtering down of previous growth-inducing changes has run its course.

This may sound somewhat theoretical, but these options represent the two major practical strategies that can be pursued by states. The first option will result in increasing economic dualism but, classical and neo-classical economists and neo-liberalisers would argue, the chance of a higher overall rate of economic growth. On the other hand, the second option will lead to increasing regional equity but potentially lower rates of national growth. Most developing countries have adopted policies close to the first option of unrestrained innovation adoption, seeking to maximise growth rather than equity. This theme is re-examined in Chapter 9.

The account presented in this section has shown that industrialisation in developing countries has been far from characterised by uniformity and homogeneity. In fact, it has been associated with global shifts, non-hierarchic adoption sequences and the growth of global or world cities.

In short, globalisation is leading to increasing differences between regions and places; for example, giving rise to centres, peripheries and semi-peripheries at the broadest scale, as noted in Chapter 3 from a theoretical viewpoint.

But, in reality, global patterns of differentiation and localisation are much more complex than this in the contemporary context. This chapter now turns to consider this argument in further detail.

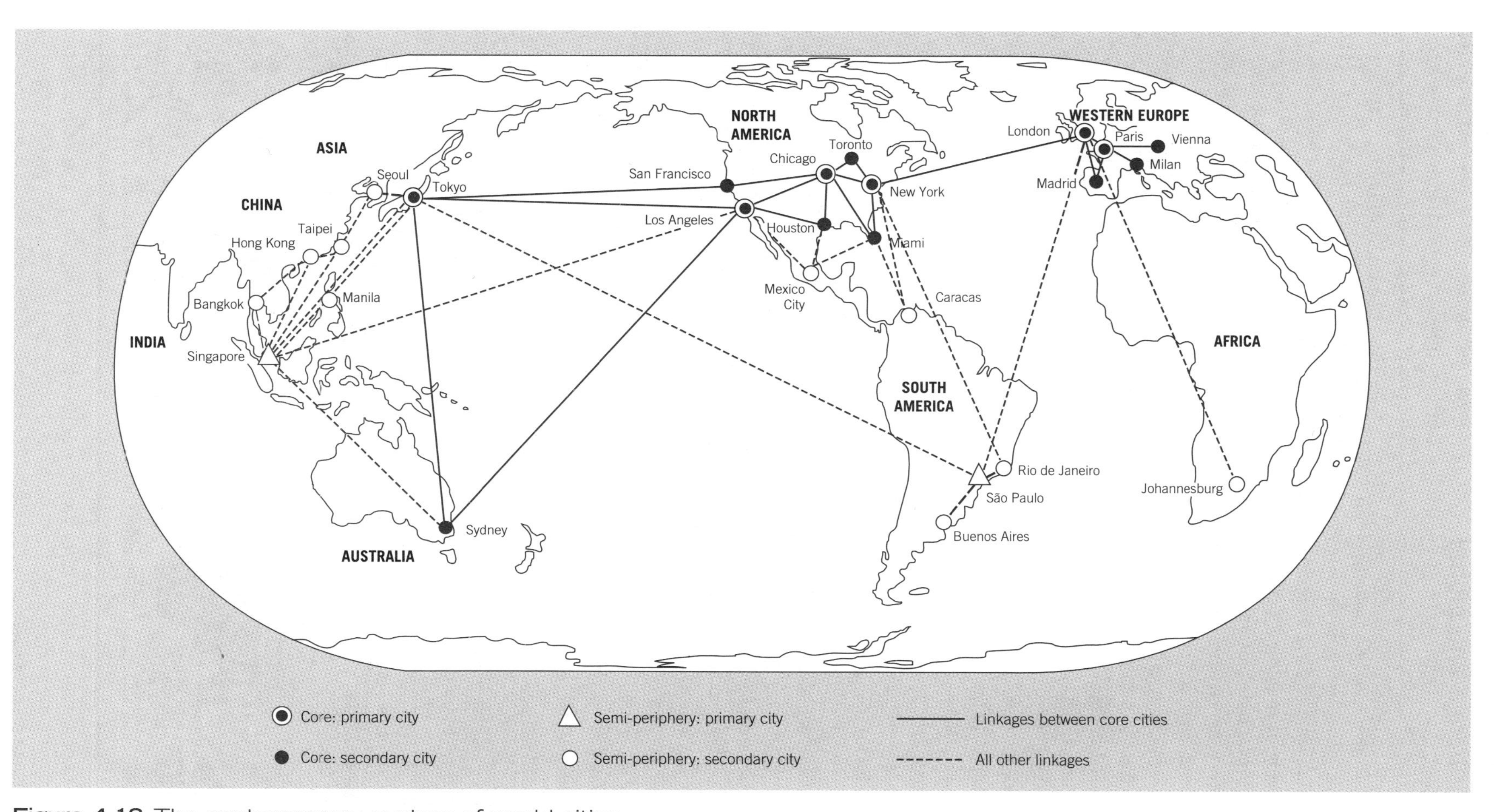

Figure 4.18 The contemporary system of world cities
Source: Adapted from Friedmann, J. 'The world city hypothesis' *Development and Change*, 17, Wiley-Blackwell Publishing Ltd, 1986

Economic change and global divergence

This leads to a major conceptualisation of what is happening to the global system in the contemporary world, and what this means for growth and change in present-day developing countries.

The basic argument is that the uneven development that has characterised much of the Third World during the mercantile and early capitalist periods has been intensified post-1945 as a result of the operation of what may be called the dual processes of global convergence and global divergence, terms which originate in the work of Armstrong and McGee (1985; see also Potter, 1990, 2000, 2008).

Together these processes may be seen as characterising globalisation. In this section, divergence is the focus of attention. We turn to convergence in the next major section.

Divergence relates to the sphere of production and the observation that the places which make up the world system are increasingly becoming differentiated, i.e. diverse and heterogeneous, as discussed in the first part of this chapter. Starting from the observation that the 1970s witnessed a number of fundamental shifts in the global economic system, not least the slowdown of the major capitalist economies and rapidly escalating oil prices, Armstrong and McGee stressed that such changes have had a notable effect on developing nations.

Foremost among these changes has been the dispersion of manufacturing industries to low-labour-cost locations, and the increasing control of trade and investment by TNCs. It is this trend which has witnessed the establishment of Fordist production line systems in the NICs, whereas smaller scale, more specialised and responsive, or so-called flexible systems of both production and accumulation have become more typical of advanced industrial nations.

In this fashion, productive capacity is being channelled into a limited number of countries and metropolitan centres. The increasing global division of labour and the enhanced salience of TNCs are leading to greater heterogeneity or divergence between nations with respect to their patterns of production, capital accumulation and ownership.

Thus, the industrialising export economies of Taiwan, Hong Kong and South Korea are to be recognised, along with the larger internally directed industrialised countries, such as Mexico, raw material exporting nations, like Nigeria, and low-income agricultural exporters, such as Bangladesh, and so on.

As argued in the previous section, in the contemporary world such changes are highly likely to be non-hierarchic in the sense that they are focusing development on specific localities and settlements. Armstrong and McGee (1985: 41) state that, 'Cities are . . . the crucial elements in accumulation at all levels, . . . and the *locus operandi* for transnationals, local oligopoly capital and the modernising state'. It is these features that gave rise to the title of their book, which characterised cities as 'theatres of accumulation' or global palladiums.

Global convergence: perspectives on cultural globalisation

Other commentators point to what initially appears to be the reverse trend: the increasing similarity which seems to characterise world patterns of change and development (Figure 4.19).

Figure 4.19 Globalisation and Third World societies: Rip Kirby airs the stereotypical argument
Source: Yaffa Advertising and King Features

There is at least one major respect in which a predominant pattern of what may be called global convergence is occurring. This is in the sphere of consumer preferences and habits. Of particular importance is the so-called demonstration effect, involving the rapid assimilation of North American and European tastes and consumption patterns (McElroy and Albuquerque, 1986; see Plate 4.1).

The influence of the mass media, in particular television, videos, newspapers, magazines and various forms of associated advertising, is likely to be especially critical in this respect. The televising of North American soap operas may well lead to a mismatch between extant lifestyles and aspirations (Miller, 1992, 1994; Potter, 2000, 2008; Potter and Dann, 1996), although there is equally the chance that such events will be reinterpreted and reconstituted from a local perspective. This argument is developed in Case study 4.1.

Case study 4.1

Global mass media, metropolitanisation and cultural change

British anthropologist Daniel Miller examined the popularity of soap operas produced in metropolitan regions of developed countries in a study published in 1992. This phenomenon can be seen as part of the evolution of 'global forms'. Such global forms have received a good deal of attention in relation to shifts in global production, but less has been written concerning the parallel process in global mass consumption.

Miller was researching on households and culture in Trinidad in the Caribbean, but he observed that 'for an hour a day, fieldwork proved impossible since no-one would speak to me, and I was reduced to watching people watching a soap opera'. The author goes on to note that much of the relevant research has been carried out on the pioneer coloniser of this type of television programme, *Dallas*. However, the programme that was receiving so much attention in Trindad was *The Young and the Restless*. This has been produced since 1973, and has always had a strong emphasis on sexual relations and associated social breakdown.

It is noted that many people went to extreme lengths to watch the programme. Those with low income, e.g. a large squatter community, were found to be the most resourceful in gaining access to the programme. Although most householders had neither domestic electricity nor water, many homes had televisions connected to car batteries so they could watch the show. The car batteries were recharged for a small fee per week by those residents who had electricity.

Although the programme has little to do with the environmental context of Trinidad, Miller notes that it was regarded as realistic in portraying key structural problems of Trinidadian society and culture. In particular, in fashion- and style-conscious Trinidad, local audiences identified with the clothing worn by the characters. Thus, a retailer observed: 'What is fashion in Trinidad today? The Young and the Restless is fashion in Trinidad today.' The programme was also seen to match with the local sense of truth as revealed by exposure and scandal.

The author concludes that 'Trinidad was never, and will never be, the primary producer of the images and goods from which it constructs its own culture', and 'Trinidad is largely the recipient of global discourses for which the concept of spatial origin is becoming increasingly inappropriate', however different they may be in terms of the physical environment. But Miller also stresses it would be wrong to assume that such developments mean an end to Trinidadian culture, which has always been derived from here, there and everywhere – Africa, India, France, Jamaica, USA and UK among others.

Source: Miller (1992)

Critical reflection

How do you respond to this example? To what extent do you feel that television programmes can promote new realities? For example, it has been common for people to blame rising violence in society on the incidence of violence on TV. How much can TV promote the demand for new goods and services, new lifestyles and, indeed, new forms of development? What about the sorts of houses people live in and their lifestyles as depicted on TV – do these have implications for development, do you feel? Or is it more the case that TV reflects reality in one place and is then re-interpreted to suit local circumstances? This is worth thinking about – and perhaps discussing as a group if you are in a classroom situation.

Table 4.8 Availability of household appliances by house type in Barbados in 1990

Number of occupied dwellings having household appliances in use (percentage of total occupied dwellings)

Material of outer walls of dwelling	Radio	Television	Video recorder	Telephone	Refrigerator	Washing machine	Solar water heater	Other water heating	None of these	All of these	Not stated	Total occupied dwellings
Wood	25,566 (85.21)	21,797 (72.65)	8,347 (27.82)	13,719 (45.72)	21,769 (72.55)	1,424 (4.75)	286 (0.95)	845 (2.82)	1,610 (5.37)	273 (0.91)	854 (2.85)	30,004
Wood and concrete block	13,971 (92.58)	13,968 (92.56)	7,282 (48.26)	12,026 (79.70)	14,210 (94.17)	3,098 (20.47)	1,256 (8.32)	2,204 (14.61)	72 (0.48)	1,120 (7.42)	201 (1.33)	15,090
Wood and concrete	829 (91.40)	843 (92.94)	443 (48.84)	714 (78.72)	843 (92.94)	197 (21.72)	83 (9.15)	148 (16.32)	4 (0.44)	76 (8.38)	14 (1.54)	907
Concrete block	23,615 (92.50)	23,765 (93.09)	14,548 (56.99)	21,569 (84.49)	24,291 (95.15)	11,963 (46.86)	9,400 (36.82)	4,503 (17.64)	96 (0.38)	6,961 (27.27)	394 (1.54)	25,529
Stone	2,130 (88.71)	2,149 (89.50)	1,191 (49.60)	2,085 (86.84)	2,208 (91.96)	1,367 (56.93)	840 (34.99)	702 (29.24)	30 (1.25)	775 (32.28)	71 (2.96)	2,401
Concrete	1,082 (94.09)	1,075 (93.48)	663 (57.65)	994 (86.43)	1,100 (96.65)	555 (48.26)	496 (43.13)	217 (18.87)	6 (0.52)	340 (29.57)	13 (1.13)	1,150
Other	96 (83.48)	84 (73.04)	42 (36.52)	80 (69.57)	91 (79.13)	48 (41.74)	23 (20.00)	25 (21.74)	6 (5.22)	21 (18.26)	5 (4.35)	115
Not stated	7 (46.67)	6 (40.00)	4 (26.67)	7 (46.67)	7 (46.67)	1 (6.67)	1 (6.67)	0 (0.00)	0 (0.00)	1 (6.67)	8 (53.33)	15
Barbados	67,296 (89.48)	63,687 (84.68)	51,194 (43.24)	64,519 (68.07)	18,634 (85.78)	12,384 (24.78)	12,385 (16.47)	8,644 (11.49)	1,824 (2.43)	9,567 (12.72)	1,560 (2.07)	75,211

Source: Potter and Dann (1996)

A major reality is that such media systems became truly global in character from the 1990s (Robins, 1995). Even in the 1990s, Potter and Dann (1996) and Potter (2000) showed that the ownership of televisions and radio receivers is almost universal, even among low-income households in Barbados in the eastern Caribbean.

The data on which this observation is made are reproduced in Table 4.8. It is clear that a surprisingly high proportion of households have a video recorder, some 43.24 per cent in 1990. Video ownership was as high as 27.82 per cent for the occupants of all-wood houses, and 48.26 per cent for the occupants of combined wood and concrete houses, those which are generally in the process of being upgraded.

Other aspects of the wider trend of convergence involve changes in dietary preferences, and the rise of the 'industrial palate', whereby an increasing proportion of food is consumed by non-producers (Drakakis-Smith, 1990; MacLeod and McGee, 1990).

Developing cities may be seen as the prime channels for the introduction of such emulatory and imitative lifestyles, which are sustained by imports from overseas along with the internal activities of transnational corporations and their branch plants. These in turn are frequently related to collective consumption, indebtedness and increasing social inequalities. These changes towards homogenisation are ones that are particularly true of very large cities.

Such a view sees globalisation as a profoundly unsettling process both for cultures and the identity of individuals, and it suggests that established traditions are dislocated by the invasion of foreign influences and images from global cultural industries.

The implication is that such influences are pernicious and are extremely difficult to reject or contain (Hall, 1995). Following this line of argument, Hall (1995: 176) has observed that the view is expressed that 'global consumerism, though limited by its uneven geography of power (Massey, 1991), spreads the same thin cultural film over everything – Big Macs, Coca-Cola and Nike trainers everywhere' (Plates 4.1 and 4.6).

However, once again the suggestion of homogeneity looks fragile when subjected to closer scrutiny. The impact of standardised merchandising is likely to be highly uneven, especially when viewed in terms of social class.

It stands to reason that it is the urban elite and the urban upper income groups who are most able to adopt and sustain the 'goods' provided by standardised merchandising – health care facilities, mass media and communications technologies, improvements in transport and the like.

It may be conjectured that the lower income groups within society disproportionately receive the 'bads' – for example, formula baby milk and tobacco products. Thus, once again, forces of globalisation may be seen to etch out wider differences on the ground. This heterogenising effect is true within urban areas too, with the residential subdivisions of the rich contrasting with those extensive areas that are inhabited predominantly by squatters and the low-income residents of the city.

But the capitalist system must inevitably be recognised as having a vested interest in globalising

Plate 4.6 McDonald's in Nanjing Road, Shanghai
(photo: Harmut Schwarzbach, Still Pictures)

the expectations of consumption and tastes. This, of course, can be related directly to the theme of the articulation of the modes of production under capitalism, as outlined in Chapter 3.

Global convergence and divergence: patterns of hierarchic and non-hierarchic change – a summary

Introduction

A direct and important outcome of the above suggestion is a strong argument that the form of contemporary development which is to be found in particular areas of the developing world is the local manifestation and juxtaposition of the two seemingly contradictory processes of convergence and divergence at the global scale.

In terms of examples, Armstrong and McGee (1985) look at the ways in which these trends are played out in Ecuador, Hong Kong and Malaysia. Potter (1993c, 1995a, 2000) has examined how well the framework fits the Caribbean, where it has been argued that tourism has a direct effect on the trends of convergence and divergence. This is another way of saying, via changes in production and consumption, that globalisation is not leading to uniformity, but to heterogeneity and differences between places.

This is also reflected in contexts where the Third World is represented in the First World as in connection with music, fashions and the like. Thus, it is necessary to acknowledge that the 'flow' is not one way, and although the North to South flow is dominant, it can be argued that the nature of the South to North flow is of increasing salience. Examples are found in Asian influences in high-street fashion, music and food, and in many other arenas.

Consumption and convergence

We are now ready to reconcile a number of closely linked arguments. It can be argued that it is the key traits of Western consumption and demand that are potentially being spread in a hierarchical manner within the global system, from the metropolitan centres of the core world cities to the regional primate cities of the peripheries and semi-peripheries, and subsequently down and through the global capitalist system (Plate 4.6). But the actual impact of these trends will be highly specific to locality, and by age, class, gender, religion etc. It is interesting to observe that the innovations cited by Berry and others in the 1960s and 1970s as having spread sequentially from the top to the bottom of the urban system of America were all consumption-oriented – for example, the diffusion of television receivers and stations. But the spread is one of potential, and many real differences are evolving.

Production and divergence

In contrast, aspects of production and ownership are becoming more unevenly spread; they are becoming concentrated into specific locations or what may be referred to as spatial nodes. This process involves strong cumulative feedback loops. Hence considerable stability is likely to be maintained at selected points within the global system, frequently the largest world cities. In other words, key entrepreneurial innovations are likely to be strongly concentrated in space, and are not likely to be spread through the urban system. This argument has parallels with the view that sees dependency theory as the diffusion of underdevelopment rather than development, as outlined in Chapter 3.

A graphical summary

The key elements of the argument presented above are summarised in Figure 4.20. On the one hand, the culture and values of the West are potentially being diffused on a global scale. By such means, patterns of consumption are spread through time (T1, T2, T3, etc.), and there is an evolving tendency for convergence on what may be described as the global norms of consumption. The figure recognises that such consumption aspects of global change are primarily expressed hierarchically, and are essentially top-down in nature.

In contrast, cities appear to be accumulating and centralising the ownership of capital, and this process is closely associated with differences in productive capabilities.

The tendency towards divergence is expressed in a punctiform, sporadic manner, which stresses activities in area and by region (A1, A2, A3). TNCs and associated industrialisation are the most important agents

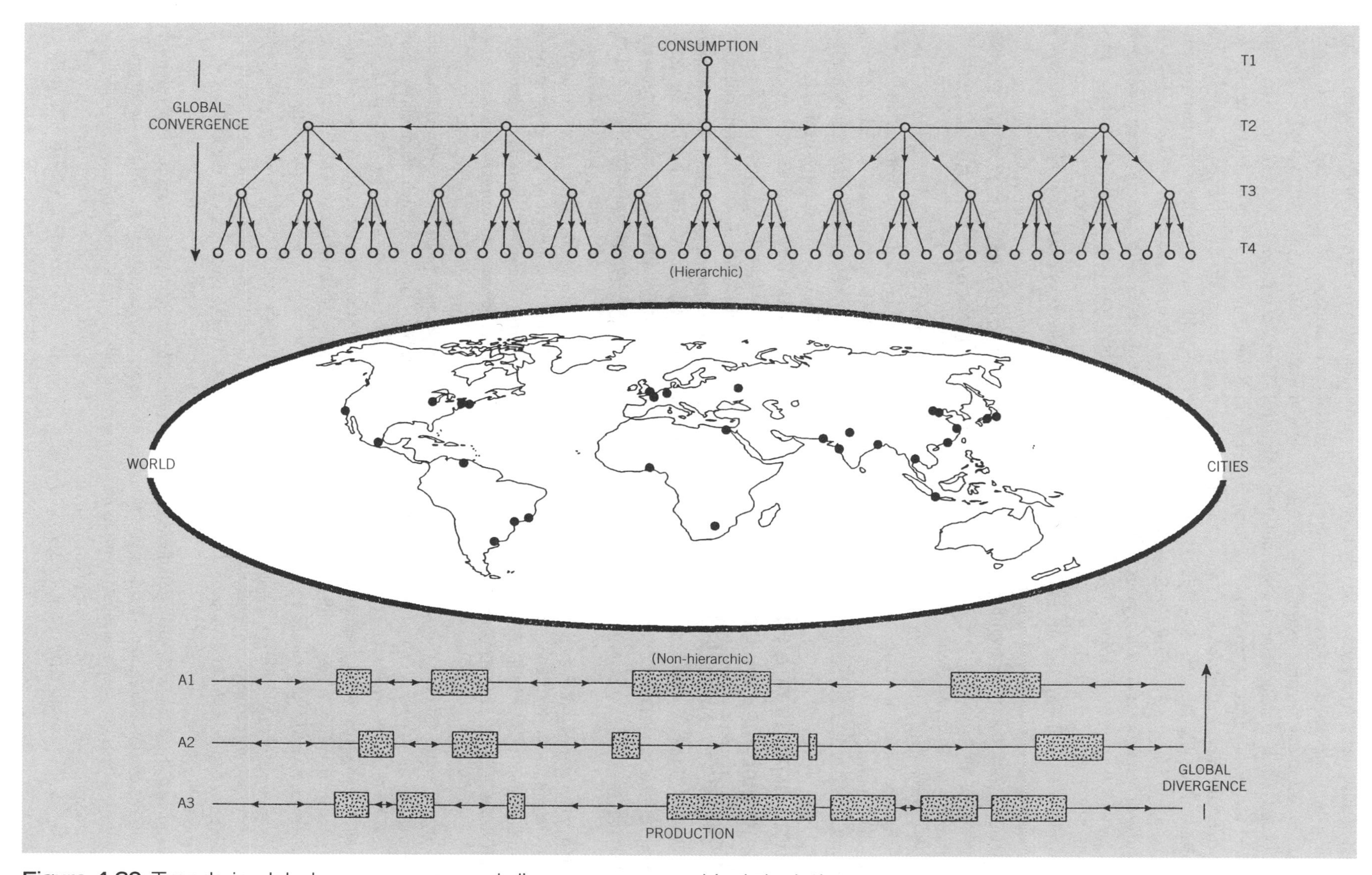

Figure 4.20 Trends in global convergence and divergence: a graphical depiction
Source: Adapted from Potter (1997)

involved in this process. This goes a long way towards explaining the plural and sometimes contradictory nature of the postmodern world system.

Cities and urban systems have to be studied as important functioning parts of the world economy. In such a role, cities act as agents of both concentration and spread, at one and the same time. Viewed in this light, the age-old argument as to whether cities are 'generative' or 'parasitic' of growth appears as naïve in the extreme (Potter and Lloyd-Evans, 1998).

Similarly, it is far too simplistic to ask whether cities spread change in a hierarchical or non-hierarchical manner, for in fact they are doing both simultaneously (Chapter 9). In this regard, it is tempting to argue that the breaking down of rigid hierarchical systems at a global level is very much part of the postmodern world. What we can certainly conclude is that globalisation has much to do with new and perpetuated forms of uneven development.

Political aspects of globalisation: the anti-globalisation and anti-capitalist movements

Introduction: the role of the state

The principal theme of this chapter has been that it is a gross oversimplification to think in terms of enhanced globalisation and the unfolding of the neo-liberal world order as giving rise to a more equal and uniform world.

In virtually every instance, spread at one scale or in one arena seems to have been matched by polarisation at another. Many grassroots development-oriented organisations, especially NGOs, seem to be increasingly aware of this. In addition, as noted in Chapter 3, both governments and international agencies are de-emphasising structural adjustment programmes and talking about poverty reduction programmes, which they argue need to be enhanced under globalisation (Desai and Potter, 2008; Simon, 2008).

A frequently cited view is that globalisation is witnessing the erosion of the former role and power of the nation state. This view emanates from the fact that certain large TNCs have annual turnovers that are substantially larger than the GDPs of small developing nations.

It also relates to the transnational movement of capital and the ascendancy of uncensored forms of global communication, such as the World Wide Web and the internet. Thus, in reviewing the changing role of the state in the field of development, Batley (2002) has noted that while for the first quarter of the twentieth century states held clear authority within their borders, the period since 1975 has seen the evolution of a more porous nation state, with fewer directly performed functions, and more partnerships with other actors. Indeed, public–private partnerships have become one of the buzz concepts of our time.

The anti-globalisation and anti-capitalist movements

The foregoing observations are reflected in what may be described as the 'anti-globalisation movement', which since the 1990s has mounted resistance to what is seen as the negative consequences of globalisation. This 'resistance' has been cogently depicted in an advertisement for the *New Internationalist Magazine*, reproduced here as Figure 4.21.

In fact, the anti-globalisation movement refers to a very broad array of interest, lobby and protest groups, many of which have taken direct action at major capitalist summits and meetings throughout the world (Murray, 2006). The anti-globalisation movement can be seen as part of a wider anti-capitalist movement against what are regarded as the excesses of neo-liberalism. Again a wide group of environmentalists, anarchists, feminists, consumers, unionists, workers, peasant farmers are brought together by the movement.

Saliently, the most visible of the anti-globalisation protests have been mounted outside international and supra-national finance meetings, particularly those of the World Trade Organization (WTO) (see Chapters 7 and 8 for further details of the WTO).

The WTO is responsible for formulating the rules that govern the conduct of world trade. Its predecessor was known as the General Agreement on Tariffs and Trade (GATT). The philosophy of the WTO is almost wholly based on free trade.

Indeed, the WTO has been staunch in arguing that wider issues, such as those pertaining to labour conditions and labour rights, health and safety issues at work, and environmental pollution and degradation, cannot be allowed to stand in the way of large TNCs trading freely and efficiently.

Figure 4.21 The *New Internationalist Magazine*'s take on globalisation and the battle for the twenty-first century

Such a position contrasts with the public concern which is openly being expressed concerning child labour under globalisation, as with the attention recently given to the children who work long hours in mines and factories – for example, stitching footballs in Pakistan, which retail so cheaply in high-street sports shops.

From 29 November to 2 December 1999 the WTO held its major conference in Seattle, and this was the focus for some of the largest protests seen in the USA since the Vietnam War (Madeley, 2000; Murray, 2006; Panayiotopoulos and Capps, 2001; Smith, 2000;). It is thought that in the region of 60,000 protesters took to the streets in what has been referred to as the 'Battle of Seattle'. Linking arms, they effectively blocked the exits from hotels and the entrances to the theatre where the opening ceremonial was scheduled to be conducted. It is thought that over 700 different groups took part in the 'Battle'.

Madeley (2000) cited Walden Bello, Co-Director of the NGO 'Focus on Global South' in explaining why the WTO was arousing so much protest:

I think it's because it's seen as standing for the subordination of so many aspects of human existence to trade, as an organization that represents primarily the interests of transnational corporations, and, from the South, as an organization with a very anti-development philosophy.

The role of the WTO in 'defending capitalism' was cogently depicted in a lead political cartoon published in December 1999 in *The Independent* newspaper at the time of the protests. The targeting of symbols of capitalism and globalism, in the form of major international stores, restaurants and coffee houses is also highlighted in this graphic (Figure 4.22). Seattle was home to the Starbucks' and Microsoft companies. At the same time, protests also occurred in London, Paris and Bombay.

Following Seattle, organised protests occurred at other locations, including Washington DC and Prague in 2000, although smaller numbers were involved (Smith, 2000; Wills, 2002). In July 2001, anti-globalisation protestors converged on Genoa in Italy, where the

Figure 4.22 One view of the Seattle anti-globalisation protests in December 1999
Source: Peter Schrank cartoon from *The Independent on Sunday*, 5 December 1999, © 1999 *The Independent*

leaders of the eight largest economies were meeting (the so-called 'G8 nations'). Estimates suggest that between 200,000 and 300,000 protestors assembled in Genoa – four times the number at Seattle – in order to voice their anger at the economic policies being followed by the largest nations, chanting 'a different world is possible'. Regrettably, however, a 23-year-old protestor was shot dead during these events (*The Guardian*, 2001).

Some commentaries, particularly those in newspapers, have suggested that as a result of these types of protests, views on globalisation have started to show a change since the turn of the millennium. Thus, Elliott (2000: 27) avers that

> **globalisation is no longer viewed as a force of nature . . . but something that can and must be shaped by human endeavour and integrity. It has been recognised that there are inherent problems in a system where capital calls all the shots . . . while labour is voiceless.**

The clarion call is that globalisation must be more people friendly, and must build core labour standards, with the developed countries being as responsible for this as developing nations. L. Elliott (2001) emphasises the then British Chancellor Gordon Brown's statement in November 2001, that managed badly globalisation will lead to 'wider inequality, deeper division and a dangerous era of distrust and rising tension'.

The linking of globalisation, inequalities, division and danger in the post-11 September 2001 world is highly salient. On the one hand, a few writers have pronounced the anti-capitalist/anti-globalisation movements to be in tatters in the aftermath of the terrorist attacks of the twin towers of the World Trade Center. On the other hand, one of the movement's leaders Naomi Klein (2001: 31) has warned that the 'debate about what kind of globalisation we want is not "so yesterday"; it has never been more urgent'. Such a stance argues the need for new forms of multilateralism/internationalism, occupying the space between what Klein (2001) refers to as 'McWorld and Jihad'.

As we saw in Chapter 3, in the scholarly world the same kind of argument is being presented in calls for the increasing localisation and territorialisation of development. This is often linked to the concept of neo-populism and the operation of selected regional closure (Chapter 3).

In the political domain, it is increasingly allied to the call for the relocalisation of production and reductions in 'food miles' (see also Chapter 7 on resources and development). This has been strongly advocated in respect of food production by the British Green Party (see Lucas, 2001a, 2001b). Lucas argues that the ever more international nature of the food trade is serving to increase greenhouse gases and leading to global warming. She also argues that it forces down food and animal welfare standards, and contributes to disasters such as foot and mouth disease and BSE.

Looking at the rise in the trade in meat, live animals and other agricultural products in Europe, Lucus points to what she sees as the folly of this huge 'food swap'. For instance, in 1999 Britain imported 61,400 tonnes of poultry meat from the Netherlands, while in the same year exporting 33,100 tonnes of poultry meat to the Netherlands. Essentially the same interchange occurred with respect to a range of other agricultural commodities. The British Green Party referred to this as 'Meals on Wheels: the mad dash around Europe', as shown in Figure 4.23 (see also Chapter 7 on resources).

A global solution to global problems? Tobin-type taxes

Critics of globalisation argue that it is characteristic of the extreme pro-globalists that while they argue for global free trade, they are often noticeably less global in their outlook on other important issues. For example, such commentators are frequently far less keen on the unrestricted movement of labour across borders. There is a major area where this argument seems to fit: and this is in relation to the proposal for globally based taxes in order to fight world poverty and underdevelopment.

What some people argue is urgently needed is strong forms of global redistribution. Way back in 1972, James Tobin (see Key thinker box), Professor of Economics at Yale University, suggested what he saw as the need for the global taxation of financial speculation. Tobin once commented that the initial idea 'sank like a rock'. But in 1978 he formalised his proposal, and in 1981 he was awarded the Nobel Laureate for Economics for his work on global taxation.

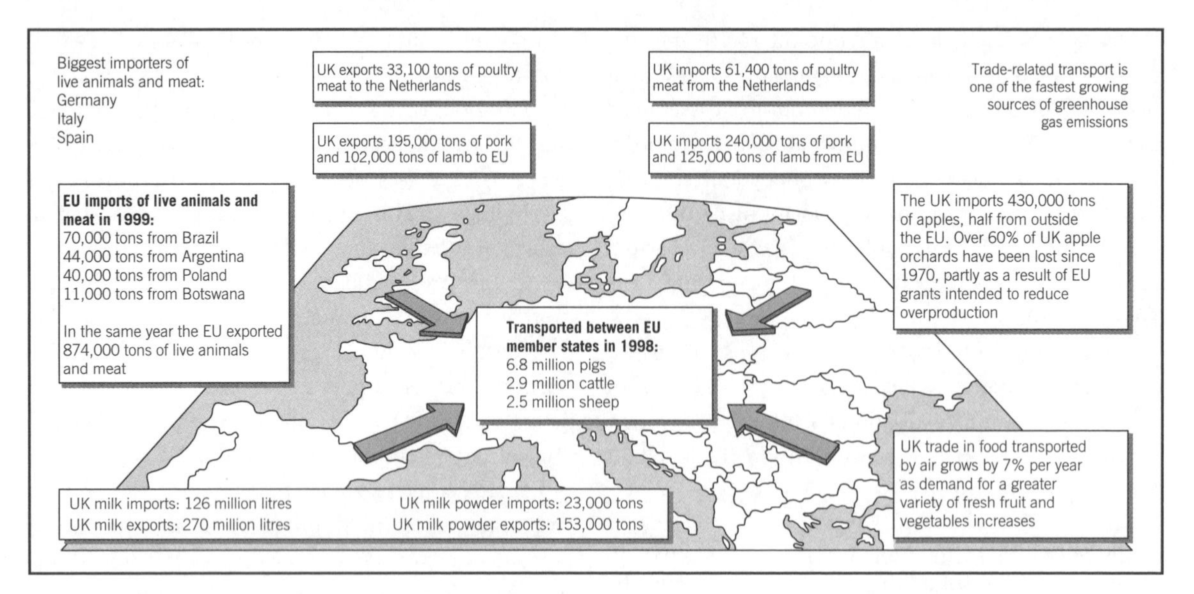

Figure 4.23 The European Food Swap and its implications for relocalisation of production

Source: Lucas (2001b) from 'The crazy logic of the continental food swap' in *The Independent*, 25 March (Lucas, C. 2001), © 2001 *The Independent*

Key thinker

James Tobin and Tobin-type taxes on speculation

Plate 4.7 Professor James Tobin, Economist
(photo: Getty Images/AFP)

Born in 1918, and educated in economics at Harvard University from 1935, James Tobin had a long interest in financial markets and investment decision-making extending back to the 1960s (Simon, 2006). It was in this context that Tobin suggested the need for global currency speculation to be taxed. Tobin referred to this as potential 'sand in the wheels' of international financial markets, which would serve to reduce their overall volatility. At first this suggestion was ignored both by professional economists and policy makers, who were generally against any market interference (Simon, 2006).

Initially at least, Tobin himself seemed to regard the funds raised by such taxation as a mere by-product. But even at 0.1 per cent, around half the rate initially suggested by Tobin, between US$50 and US$300 billion would be raised annually, a sum broadly equal to existing levels of development assistance.

In the late 1990s, the NGOs War on Want and Oxfam, as well as the governments of Canada, France and Belgium, moved to support the introduction of Tobin-type taxes. Anti-globalists also find it relatively easy to align with Tobin taxes as there is a strong argument that they would serve to dampen down aspects of financial globalisation.

Tobin formally retired in 1988 and died in 2004 aged 84 years. As Simon (2006) points out, as a key thinker he was also associated with the suggestion that development needs to be defined and assessed in terms of human welfare rather than by measures of GNP and income per head alone. Thus, in the 1970s, James Tobin proposed what he referred to as a Measure of Economic Welfare (MEW). This can be regarded as a forerunner to the United Nations' Human Development Index (HDI), which we reviewed in Chapter 1, and again serves to denote James Tobin as a key thinker in the field of global development.

Some US$1.5 to 2.0 trillion are traded daily on foreign exchange markets. It is believed that only 5 per cent of this sum is necessary to finance global trade. The remainder effectively amounts to speculative trading, that is making profits from changes in currency rates. Tobin-type taxes involve a levy of around 0.20 to 0.25 per cent on such global financial activities. This would presently yield US$250 billion per annum. This is over five times the total amount that is given in aid around the world. Although formidable issues would have to be faced in collecting and allocating such monies, it is generally argued that revenues should be collected by national central banks and then deposited with a United Nations body such as United Nations Development Programme (UNDP) or United Nations Educational, Scientific and Cultural Organization (UNESCO) (see also Chapter 7).

The US$250 billion that could be raised each year only makes sense when we consider it alongside what might actually be achieved with such tranches of money. For example, it has long been estimated that as little as US$8 billion per year would be enough to establish universal primary education on a global basis. Meanwhile, UNDP has calculated that US$80 billion is needed to eliminate the worst forms of global poverty. Further, the Jubilee 2000 campaign argued that US$160 billion per annum would be the cost of wiping out the Global South's unpayable debts.

But there is another reason for the introduction of global Tobin taxes, over and above poverty alleviation. A major consideration is that it would also serve to promote greater financial stability by damping down financial markets, rather than the extreme volatility which seems to have characterised them over the years, and which is regarded as having contributed so forcefully to the Asian crisis, for example (Rigg, 2002).

At the start of the 2000s, the British NGO War on Want ran a very strong and extensive campaign supporting the introduction of a Tobin-style tax. At the same time, the Canadian Parliament voted two-to-one in favour of introducing such a tax, which was staunchly advocated in the mid 1990s by the French President Francois Mitterand, shortly before his death.

However, it appears that, in general, those very politicians who espouse globalisation are those who dismiss out of hand a globalised tax to tackle world poverty. Thus, in 1995, the then Managing Director of the International Monetary Fund, Michel Casessus, is reported as having commented that 'financing an attack on poverty should be left to governments'.

Notably, the possibility of global taxation is not even mentioned in the UK Government's White Paper in dealing with globalisation and poverty reduction. In the words of L. Elliott (2001: 15): 'It is politics that is the killer . . . the political will for a Tobin tax is absent in the places which matter: Washington, London, Tokyo, Frankfurt.'

Globalisation and development: the example of tourism

In order to conclude this account, it is worth turning to a concrete example – one that exemplifies many of the themes addressed thus far.

Tourism is now regarded as the world's leading industry. Even by 1987, it recorded sales of US$2 trillion, and employed an estimated 6.3 per cent of the global workforce, making it the global premier industry (Gale and Goodrich, 1993).

Tourism is also quintessentially linked to globalisation and the phenomenon of time–space compression. In the socio-cultural realm, tourism is emblematic of globalisation, hyper-reality, fantasy and postmodernity, in that it is involved with the creation of vacation landscapes, places, icons and experiences. Tourism is a productive enterprise that is actively etching out differences between places. In addition, tourism is closely connected with conspicuous consumption and the adoption of outside norms, via the operation of the demonstration effect (see also Chapter 6).

Many developing nations have adopted specific programmes to promote the growth of tourism as part of their development strategies. One of the first examples, however, was afforded by the First World European nation of Spain, as discussed in Case study 4.2.

Barbados in the Caribbean is a good example from the developing world; the growth of tourism in Barbados dates from the late 1950s. In 1955, there were only 15,000 visitors to the island annually (Potter, 1983). With the advent of modern jet aircraft, there was dramatic growth in the tourist sector from 1966 to 1972; there was an overall increase in visitors of 165.9 per cent over this seven-year period. Visitors totalled 210, 349 in 1972, having increased from 79,104 in 1966. It was the heyday of the expansion in tourism, with yearly increases never falling below 14.2 per cent. By 1980, total tourist numbers had reached the dazzling heights of 369,915, compared with a national population of 248,983 (Potter, 1983). By 1992, tourist arrivals in Barbados amounted to 385,472 (Dann and Potter, 1997; Potter and Phillips, 2004).

Case study 4.2

Tourism and development: the example of Spain

Since the 1960s, tourism has been the mainstay of Spain's economic miracle. Spain now has a GDP per capita which stands at US$27,400, significantly above Portugal (at around US$19,800 per capita). Indeed, in 1995 Spain was the twelfth most powerful industrialised nation in the world (Dicken, 1998). Today, Spain is almost synonymous with the words *holiday* and *tourism*, and perhaps even the expression *package holiday*. In 1984, Spain recorded 43 million visitors, and by 1995 this had increased to a staggering 55 million. This should be put alongside the indigenous population figure of 39.1 million. Spain now accounts for some 9 per cent of total world tourists. The number of tourist arrivals rose from 6 million in 1960.

Tourism in Andalucia, which covers 17 per cent of Spain in the south of the country, accounted for nearly 13 million visitors in 1996 and yielded Pta 1 billion, amounting to some 10.5 per cent of Andalucia's GDP. The area has an exceptionally mild climate, with 3000 sunshine hours reported per year. Its 812 km of coastline is comprised of more than 300 beaches, of which 63 meet the European Blue Flag criteria for excellence.

The development of tourism in Spain started in the late 1950s. Years of ultra-right-wing rule under

Plate 4.8 High-rise tourism development in Benidorm, Spain
Source: Corbis/Guenter Rossenbach/Zefa

Case study 4.2 (continued)

Franco had seen Spain fall further behind the rest of Europe in economic terms. As a result, General Franco consciously decided to open the Spanish economy up to the outside world. Thus, the peseta was devalued in 1959, internationally making Spain a very cheap place to visit, in terms of accommodation, food and drink. These developments occurred at the same time as advances in cheap jet travel for the masses. The development of tourism in Spain can therefore be seen as an important step in the progress of its industrialisation and internationalisation. It also marked the first direct effort to use tourism as a central plank of development policy.

The speed with which tourism was developed meant there was little or no state or local planning. Effectively, the free market was left to sort things out. This was demonstrated most cogently by the fact there was little or no control over developments. Almost literally overnight, former small fishing villages such as Torremolinos, Fuengirola and Lloret de Mar became large tourist-led cities. Some would see this as early global neo-liberalism in practice.

Planning regulations were blatantly flouted during the Franco era, with virtually all land projects being left entirely to the private sector. 'Anarchic growth' is how some have described the type of development that occurred. It has been argued that many early developments in the tourist industry were at the expense of both the environment and the local population (Plate 4.8).

Critical reflection

The argument that tourism helps promote cultural contact and understanding between people from different places and cultures is sharply contradicted by those who maintain that it can just as easily serve to reinforce stereotypical views of 'Others'. It is also argued that tourism from relatively wealthy nations to relatively poor ones can even promote new patterns of dependency and servitude. This is particularly salient where the tourist/host relationship cuts across lines of race, ethnicity and class. Consider this argument in the light of your own experiences as a tourist.

Barbados has, through time, pulled in most of its visitors from Canada, the USA, the United Kingdom and other parts of Europe. This aspect of internationalisation has its downside, as it entails dependency on economic conditions and inevitable fluctuations elsewhere. This openness of the economy was witnessed in the vicissitudes of the mid 1970s, when recessionary tendencies and increasing oil prices had a marked effect on tourist arrivals. The number of visitors actually fell during 1975, and between 1973 and 1976 the increases at no time exceeded 5.6 per cent per annum.

A major aspect of globalisation in relation to tourism in Barbados is the foreign ownership of hotels. In Barbados, as elsewhere in the developing world, foreign ownership and participation in the tourist sector is particularly conspicuous in the larger, up-market tourist sector. Approximately 74 per cent of all class I hotel bed spaces are owned by non-nationals, and foreign ownership accounts for just over half of all hotel bed spaces (52.6 per cent). Such influence does not end there, however, with foreign ownership being directly responsible for 47 per cent of class I apartments, and a grand total of 44.2 per cent of all bed spaces.

There is another area of increasing globalisation and that involves the suggestion that Third World locations such as Barbados are playing host to First World postmodern tourists who are seeking to handle the pace of life and 'future shock' (Toffler, 1970) in their own countries. Such tourists are provided with selective glimpses of the simple life of yesteryear in a Third World or developing context (Dann and Potter, 2001; Potter, 2000). It is a short step to argue that this may represent one of the most lucrative and exploitable trends in the contemporary tourism industry (Dann and Potter, 1994, 1997; Potter and Dann, 1996).

Another frequently debated series of issues concerns the relations between tourism, globalisation, culture and learning about 'other' areas of the world. One side of the argument has it that tourism is a major positive force in promoting knowledge and employment about far distant places and ways of life.

Although this is undoubtedly true at a relatively simple and trivial level, there is another side to the argument, which focuses on the production of false information about places and the propagation of enduring stereotypes. In the case of Barbados, it can be argued that by putting plantation great houses and sugar factories on the tourist 'consumption' map, the past ignominious history of slavery is being openly discussed in a culturally sensitive, educationally sensitive and progressive manner. However, Potter and Dann (1996) and Dann and Potter (1997, 2001) demonstrate that it is undoubtedly part of the postmodern turn to conveniently disregard anything which may be deemed unpleasant and which might reduce enjoyment of the tourist 'product'.

In Barbados there is considerable evidence that a growing number of concerns have begun to capitalise on the postmodern ethos of their guests. Mock villages consisting of traditional houses have sprung up as commercial outlets at the very same time as the state refuses to see such houses as representing a cogent force in the future housing equation of the nation (Watson and Potter, 2001). 'Pirate cruises' are offered to holidaymakers and traditional chattel houses appear on hotel premises as the locales for serving buffet dinners.

Meanwhile, plantation floorshows and spectaculars even invite historically ignorant, or at best ahistorical, vacationers to see the cultures of 'Spanish, French and *African settlers*', completely ignoring the condition of slavery and grossly misrepresenting the inhumanity of the system of slavery, presumably in the interests of increasing the enjoyment of the tourist product (see Box 4.3). Recognition of the potential for this fragmentation of history in the so-called developing world by inhabitants of the core can have enormous financial rewards for the latter.

Although it is possible to put forward the argument that such 'staged authenticity' (MacCannell, 1976) means the private lives of Barbadians are relatively shielded from the tourist gaze (Urry, 1990), and that such developments may be making a real contribution

BOX 4.3

Whither the real Barbados?

This is an article written by Rob Potter and which first appeared in the newspaper Caribbean Week, *25 November–8 December 1995. It is reproduced here in an effort to afford further insights into the interplay between tourism, modernity, postmodernity and development in the Caribbean context*

It is pleasing indeed to see the traditional Barbadian housing display which is now mounted as a permanent exhibit at the Barbados Museum. This explains clearly the origins and architectural features of the chattel house. Just as I was enjoying the display, and reflecting on the way in which the efforts of local conservationists such as Henry Fraser are at long last beginning to pay off, I was intrigued by the comments of a young local teacher or tour guide who was escorting a party of Barbadian school children around the display. On reaching the housing display, he asked – rather surprisingly I thought – 'How many of you have seen a chattel house before?' At the end of a brief but interesting account of factors such as the transportability of the basic house form, the guide concluded by remarking 'You can still see *a few* of these houses around today!' (emphasis added). I stood for a short while and witnessed another such party of school children visit the stand, which was again dealt with as a relict feature of the landscape.

At first, it struck me that I was witnessing premature nostalgia. My current work on the 1990 Census indicates that almost exactly 40 per cent of all houses in Barbados remain constructed entirely of wood. It is true, of course, that this proportion is now falling very fast indeed, having stood at 57.31 per cent in 1980, and 75.25 per cent in 1970. It is also the case that there is a difference between chattel houses and modern wooden houses. However, in 1990, the proportion of houses built exclusively of wood was 45 per cent of the total housing stock in respect of six out of the eleven parishes – the predominantly rural ones of – St John, St Thomas, St Joseph, St Andrew, St Peter, and St Lucy. Change may be occurring rapidly, but to describe chattel houses as things of the past does seem somewhat premature. More importantly, it seems to suggest the

BOX 4.3 (continued)

continued operation of a sharply divided nation: the modern and the not so modern. One might assume that the school parties were from the urbanised–suburbanised–touristised coastal belt of St Michael, Christ Church and St James. But the 1990 census data show that even in these parishes, wooden houses account for 41.69 per cent, 29.91 per cent and 28.54 per cent of the total housing stocks respectively. More to the point perhaps, in 1991 some 559 house move permits were issued to enable chattel-type houses to be moved from one part of the country to another.

But on reflection, I found it more surprising that the children had not answered the teacher's original question with the reply, 'Sure, we've seen chattel houses before – why, I visited a tourist area only last week!' First there was the Chattel House shopping village in St Lawrence Gap and the chattel house and rum shop used to serve buffet meals at Sam Lords (a leading hotel). Then there was the Chattel Plaza, then Sandy Bank, and then the St James Chattel Village, opened over the past few months. These are all laudable signs of a revived interest in the local house form, but it is notable that this veritable explosion of replica traditional houses seems to be linked to commercial retail/tourist initiatives.

The situation seems to reflect a downgrading of Barbados' premodern past, in favour of its continuing and very successful efforts to modernise. Thus, some members of the public are quick to see the traditional house form as a relict feature, somehow far less worthy of attention than its modern counterpart.

In fact, this basic lesson had been brought home to me in another incident a few years ago, when I wrote an article about Barbadian housing for the first issue of the relaunched *Bajan* magazine. Some readers wrote asking why such an article stressing the past of Barbados had been featured. The new editor, John Wickham, wrote an editorial rejecting this view, and explaining that such houses continued to be an important part of the local cultural landscape. So much so, in fact, that he drew attention to the fact that the front cover logo of the magazine actually featured a drawing of a chattel house!

In a manner which parallels such downgrading, housing policy in Barbados has largely rejected the chattel house as part of a policy for providing houses for those on low incomes. The chattel house is an excellent example of self-help on the part of Barbadian people. Such houses can, of course, be put on a plot of land with basic services, together with a toilet (wet core), as part of a 'site-and-services' scheme. Slowly, the inhabitants can upgrade, converting to a walled structure when finances and time permit. In a sense, this idea formed a component of the tenantries programme launched in 1980. But the idea of using the chattel house as a plank of housing policy has never been fully pursued by government. Instead, concrete starter homes costing thousands of dollars were piloted, just as barracks-style and terraced row concrete houses were a feature of earlier government housing schemes (Watson and Potter, 2001).

Once again, it seems that the need to modernise is placing the premodern local system in a subordinate position. Even if bungalow-type houses are now the predominant feature of today's Barbadian housing scene, wouldn't it be nice for some of the architectural features of the chattel house to be retained, such as box-pelmets, fretwork and finial? Doesn't it seem odd that large swathes of the public and much of government policy rejects outright what the commercial–mercantile–tourist sector now suddenly seems to be embracing with open arms? Why should the chattel house be a thing of the past and a failure in its intended role, but a cogent contemporary emblem of commercial success?

Perhaps it is near to the truth to see all of this as being one of the manifestations of Barbados modernising as quickly as possible, but in a context where much of the rich, developed world is now moving into what is referred to as the postmodern condition. The postmodern world is one of globalisation and rapid telecommunications. The potential to advertise – by means of the mass media of television, newspapers and international magazines – means that advertising and image creation are central to our daily lives. The scenes and images created in this way are in many ways more important than 'reality'. But in such circumstances what exactly is reality? Reality becomes multifaceted, and there are different realities serving different purposes. So there is one reality which rejects the chattel house as a thing of the past; and another which embraces it

▶

BOX 4.3 (continued)

wholeheartedly as a symbol – an icon and an image – of traditional Barbados. But this image adopts the chattel house for purely commercial purposes. Similarly, there is a real Barbados of the rural zones, and of plantation tenantries in particular, which for years have been tucked away from view, off the main roads and highways. And then there is the 'real' Barbados of the tourist hotels, where images of a tropical paradise and rest and relaxation, replete with the chattel houses of a bygone era, are made 'authentic' for the visitor.

A final instance of the multiple realities which make up our modern society is also drawn from the realm of tourism: but this time in respect of the ways in which the realities of the history of Barbados are handled. Take the example of the floorshow that is offered to visitors several nights a week. Adverts appear here, there and everywhere, inviting the potential audience to come and 'see the cultures of the Caribbean as influenced by the Spanish, French and *African settlers*' (emphasis added). You perhaps wouldn't even notice if the advertising copy had been written as 'African, Spanish and French settlers'. But should a reality so central to the people, history and culture of Barbados be represented in this fashion? Should the greatest enforced movement of slaves be transmuted by verbal sorcery into an apparently voluntary migration? What reality is being served here, that one of the most salient aspects of the history of the island should be rewritten so blatantly for tourist consumption. A less cogent but interesting aspect of the same event is the exclusion of any reference to British history in the blurb. Presumably, this aspect of reality might be construed as too near to reality, and distract from the Latin connotations of Spain and France. Reinterpretations of history – which are presumably intended to ease collective guilt about the realities of slavery – abound in the Barbadian mass media aimed at the tourist. Hence, plantation house tours and dinners are advertised, stressing elegance, antiques, pine furniture and opulence, with no historical reference being made to the conditions for slaves that formed their counterpart. Elsewhere we are invited to trek across the original Barbados, as unspoilt now as it was 350 years ago, without any acknowledgement that the island would have been entirely wooded! It is hard not to reach the conclusion that all of these 'realities' are intended to increase the ability of the social and physical environments to handle tourists: to increase what the tourism expert would refer to as the carrying capacity of Barbados.

Whither the real Barbados? The truth is that it is a very successful semi-peripheral country which is modernising very rapidly. In a number of situations modernising Barbados is clearly overthrowing premodern Barbados. But in other arenas, particularly those relating to tourism and cultural change, postmodern influences in Barbados are busily reviving aspects of the premodern Barbados. The net result is highly interesting and vibrant, but is in a number of respects highly contradictory, ambivalent and uneven. But that's the postmodern world for you and there seems to be no doubting the fact that Barbados is a part of it.

Critical reflection

Some of what it is argued has been happening in this article can be related to the theory of the modes of production covered in Chapter 3. Where there is an advantage or a profit in doing so, the products of the past are retained or are converted. Where there is no advantage, they are left as they are. So, old and new, premodern, modern and postmodern are intimately linked, rather than being separate and contradictory. Are there other regions or localities you know where you might hypothesise that the same type of things are happening? Or are there sectors – like housing, clothing, music, dance – where to the best of your knowledge the same applies?

to promoting long-term sustainability, it is again possible to follow an entirely different line of argument. Thus, there is the real fear that by rewriting the history, and indeed in some instances reinventing the physical environment, of the Caribbean region (Potter and Dann, 1996), the carrying capacity of tourist destinations may be exceeded, both in socio-cultural and environmental terms. In such circumstances, the only ones to profit are likely to be the developers.

We clearly live in a globalised world, but the technological changes that have allowed the development of mass tourism are also the means by which simplified,

Plate 4.9 Barbados: 'just beyond your imagination', 1996 advert
(*Source*: © 1998 Barbados Tourism Authority. *Note*: This advert is included as a historical time-piece only; it is no longer used by the Barbados Tourism Authority.)

inaccurate or downright misleading images and representations can be propagated. It seems hard to refute the suggestion that the popular misconception of the entire Caribbean as a 'beach replete with swaying coconut palms' is the direct outcome of tourist advertising and promotion campaigns (Potter and Lloyd-Evans, 1998). At the same time, the daily realities of urban concentration, poverty and poor housing in the Caribbean are selectively weeded out from the stereotype. Thus, in a 1996 advertisement the Barbados Tourism Authority comments on the island as one 'with 340 days of sunshine a year, imagine how that reflects on you', and describes Barbados as 'just beyond your imagination', complete with the time-honoured view of sand, sea and coconut palm (Plate 4.9).

In a similar vein, the local population is also represented as a group of 'smiling, servile natives' ready to respond to the bidding of predominantly white, affluent tourists. Such images and perceptions have unfortunate connotations in a context where pride in African origins and negritude (Lowenthal, 1960) are increasing in the post-colonial era. Images can now be spread widely by means of colour brochures, television programmes, promotional video recordings and CDs, and almost instantaneously via websites. All of this can be said to be giving rise to new forms of dependency, or neo-dependency.

The complexities of the development implications of tourism promotion in developing countries are thus considerable in a globalising postmodern world. The case of Barbados illustrates the veracity of the argument that neither modernity nor postmodernity can exist without the other, and that rather than being different conditions, they are closely related. In an evolving world order characterised by globalisation, fluidity and change, it is not altogether surprising that the notion of postmodernity represents a 'handy category to employ in the struggle for emancipation and a virtual synonym for postcolonialism' (Jones *et al.*, 1993: 18).

Concluding comments: globalisation and unequal development

Throughout this chapter it has been argued that the notion of a basic sameness in respect of global culture is clearly a distortion and a gross oversimplification. Clearly, we live in a more globalised world, in which multinationals, increasingly, are coming to dominate world patterns of consumption and production. But there are many reasons why it is wrong to regard the outcome as increasing uniformity.

First, strong resistance is sometimes shown by local and national cultures, especially in response to the influences of North America. The idea of a single global culture is clearly misplaced. As an example, the opening of McDonald's was fiercely contested for some time in Barbados, despite its status as a leading tourist destination for North Americans and Europeans. When those in power relented, the fast-food chain only lasted six months, largely because Bajans prefer eating chicken to red meat. This is a simple and direct example illustrating that local customs and tastes can run directly across, and indeed against, apparently hegemonic global trends. The regionally based fast food outlet Cheffette/Barbeque Barn is strongly based on chicken meals, and is very popular – but does also serve beefburgers.

Second, rather than serving to erode local differences, global culture often works alongside them; and sometimes it even works via them. Particular groups within society may be targeted for the sale of certain products. In this manner, local differences may be explored and exploited wherever possible (Robins, 1995).

Furthermore, increasingly within the global economy cultural products are being assembled from all over the world and are being turned into commodities for an emerging cosmopolitan marketplace. This is particularly true of fashion, music and tourism (Crang, 2000; Robins, 1995). Thus, from reggae to soca to African indigenous music, Asian politics and menus, and in respect of Rastafarianism, the flow is not a one-way movement, and Third World products are being promoted and sold in First World marketplaces. Thus, globalisation has brought the possibility of the colony 'invading' the colonial power, and the periphery taking on and 'winning against' the centre (Robins, 1995). Hence, in the new globalising system we are encountering many incidences of what, viewed historically, is a reverse or counter flow. Again, such conflations can be interpreted as typical of the postmodern condition.

There is also the important argument that increasing globalisation and time–space compression in the end make us value more strongly than hitherto the notion of place as secure and stable. Thus, it can be posited that globalisation may well serve to engender localisation. This is sometimes described rather inelegantly as 'glocalisation', where there are multiple global–local relations through which locality becomes more salient than hitherto within the world system.

Furthermore, we have witnessed all too clearly that culture has always been characterised by hybridisation, difference, rupture and clashes, so it is possible to argue that nothing very new, strange or different is currently happening. Western European nation states may be seen as masters of modernity, whereas hybridised forms of culture are characterising the postmodern world. This, of course, reflects the fact that culture can never be seen as settled, finalised, complete and internally coherent (Hall, 1995).

It has to be appreciated that cultures (systems of shared meanings), products and lifestyles will inevitably spread, and sometimes contract, in a highly heterogeneous manner. But aspects of production and ownership are far from evenly spread, due to the process of divergence and differentiation reviewed in this chapter.

Finally, the economic competition between places is now intense, given that major corporations can select

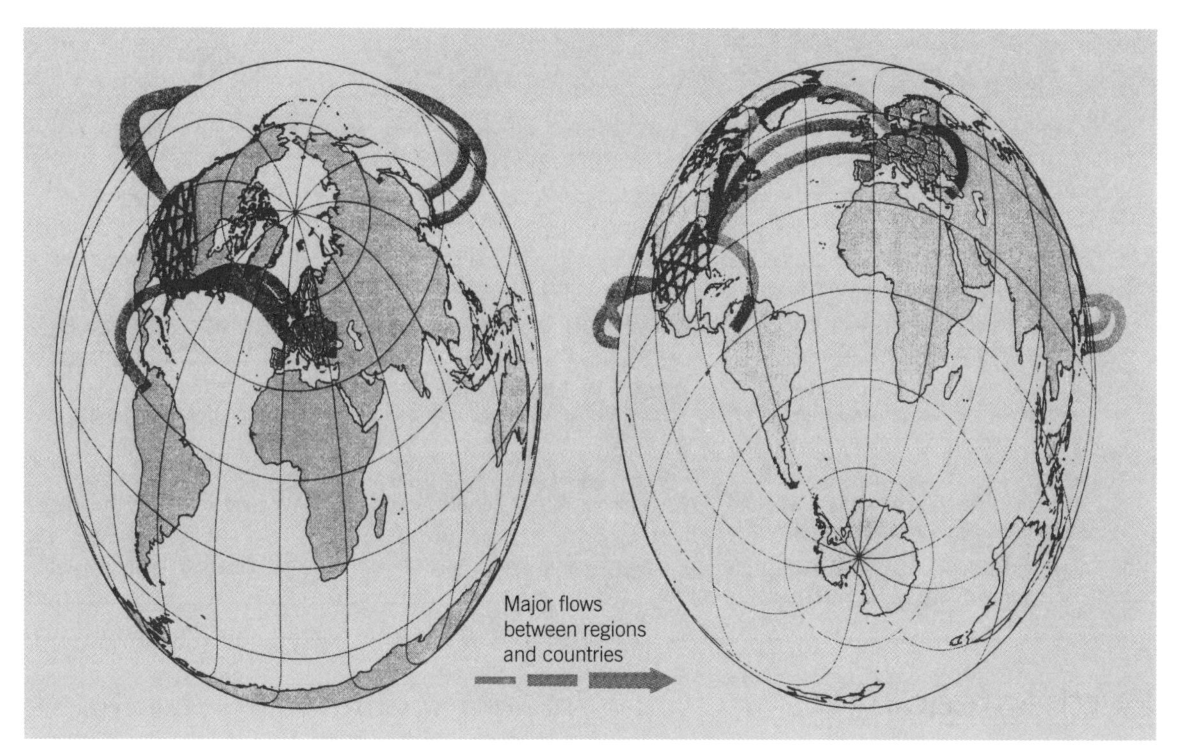

Figure 4.24 On and off the map: contemporary globalisation and the heterogeneity of global change
Source: Adapted from Allen and Hamnett (1995)

between them, and this is leading to the possibility of ever-sharper differences between areas, regions and places. Thus, trade liberalisation has brought a wave of anti-globalisation protests.

For all these reasons, we can conclude that uneven and unequal development are still characteristic of the global capitalist system (Figure 4.24). Globalisation is not all-encompassing and there is much that remains uneven about global relationships and global processes. All of these aspects of dynamic change are strongly skewed towards the developed North (Allen, 1995). The world may effectively be getting smaller, but the majority of its population does not yet have access to a telephone. For example, de Albuquerque (1996) stated that over half the world's population have never made a telephone call, and resonates a comment made earlier that there are more telephones in the New York–New Jersey metropolitan area alone than in all of Africa combined.

In discussing further 'net fever' in Haiti, the same author stresses that in a country where the average wage at the time of writing was US$3 per day and the per capita gross national product was US$220, very few could afford the US$2000–3000 required in 1996 for a personal computer. Thus, access to cyberspace in Haiti, as elsewhere, is restricted to a well-heeled elite minority living in a state-of-the-art, twenty-first century world, far removed from the impoverished majority (de Albuquerque, 1996). The same author summarises that internet access is paralleling class systems of stratification, and threatens to increasingly splinter the globe into haves and have-nots based on access to information/communications technologies.

In considering development, it has to be recognised that places in the globalising world system are not linked together in a uniform way. They are interrelated in very unequal ways, and such basic inequality would seem to be poised to increase rather than decrease in the near future. Competition between places for global capital is making the world more uneven and differentiated, reflecting the trends of global divergence (Armstrong and McGee, 1985; Cochrane, 1995; Potter, 1993c, 2000). In the words of Cochrane (1995: 276), 'Globalisation and localisation are not the polar opposites which one might expect them to be', because 'globalisation is underpinned by the realities of uneven development' (Cochrane, 1995: 277). One might even go further and say that uneven development is actively being promoted by contemporary processes of globalisation.

Key points

- The single word 'globalisation' seems to summarise our contemporary age, even though, in reality, globalisation has been a feature of the world economy and development patterns since 1492.
- The precise relationships that are envisaged to exist between globalisation and processes of development are central to thinking about planning and change.
- Some analysts, especially neo-liberals, see globalisation as the mechanism that will spread growth in the twenty-first century, so that neo-liberalism can be regarded as akin to neo-modernisation.
- The world is effectively getting smaller for those who have the resources to travel and use the most up-to-date communications technologies.
- But when diverse aspects of globalisation are viewed – be it digital technology, transport, manufacturing, cultural globalisation – while certain spread effects are recognisable, at a higher resolution the outcome appears to be a sharper polarisation – between both places and people.
- The framework of global convergence and divergence suggests that while patterns of consumption are promoting greater homogenisation for those who can afford it, global patterns of production are creating greater diversity and differentiation among world regions.
- The anti-globalisation and anti-capitalism movements argue forcefully that the effects of globalisation urgently need to be controlled, and in some respects, curtailed.

Further reading

Conway, D. and Heynen, N. (2006) *Globalization's Contradictions: Geographies of discipline, destruction and transformation*. London and New York, Routledge.
A critical overview of globalisation and neo-liberalism.

Department for International Development (2000) *Eliminating World Poverty: Making Globalisation Work for the Poor*, Cmnd 5066. London: The Stationery Office.
Worth reading as a strong template for the argument that globalisation is the way forward in delivering countries from poverty.

Desai, V. and Potter, R.B. (2008) *The Companion to Development Studies*, 2nd edn. London: Arnold.
An accessible source book that brings together over 100 key essays dealing with all aspects of the field of development studies.

Dicken, P. (2007) *Global Shift: Mapping the Changing Contours of the World Economy*, 5th edn. London: Sage.
A must read for those concerned with the realities of economic globalisation.

Knox, P. and Marston, S. (2001) *Human Geography: Places and Regions in a Global Context*. Englewood Cliffs, NJ: Prentice Hall.
Contains some useful sections dealing with contemporary global change.

Murray, W.E. (2006) *Geographies of Globalization*. London and New York: Routledge.
An overview of globalisation processes for the undergraduate student market.

Schenk, S. and Haggis, J. (2000) *Culture and Development: A Critical Introduction*. Oxford: Blackwell.
Endeavours to fill a gap by focusing attention specifically on cultural aspects of the development process.

Schuurman, F. (2001) *Globalization and Development Studies: Challenges for the 21st Century*. London: Sage.
A good collection of readings dealing with globalisation.

Websites

www.dfid.gov.uk
As noted at the end of Chapter 1, the DFID site gives direct access to a range of development items, including those on globalisation and related topics.

www.eldis.org
The site aims to share the best in development policy, practice and research.

www.greenparty.org.uk
Issues relating directly to globalisation and development are often dealt with here.

www.waronwant.org
The site contains details concerning War on Want's '*It's time for Tobin*' campaign.

www.internetworldstats.com
Provides updated figures on internet usage levels.

Discussion topics

- Consider the statement that contemporary globalisation is bringing forth new forms of localisation.
- Assess the view that while the world may be shrinking, it is also becoming noticeably more unequal.
- 'Globalisation dates back over 500 years.' What do you see as the development implications of this statement?
- Explain the basis of Tobin-type taxes and consider their potential for changing the face of international development.

PART II

Development in practice: components of development

Chapter 5

People in the development process

This chapter focuses on various aspects concerning people in the development process. Following consideration of issues such as population growth and distribution, we move on to examine a number of important factors which affect the quality of life for individuals, households and communities – notably health, education and human rights. It is suggested that if meaningful development is to be achieved in the world's poorest countries, these issues need to receive high priority both nationally and internationally.

This chapter:

- Examines the relationship between population growth and key resources needed for human survival;
- Considers the rate of world population growth and distribution;
- Explains the key elements of the demographic transition model;
- Examines why some countries pursue anti-natalist population policies whilst others adopt pro-natalist policies;
- Considers the changing structure of national populations in different parts of the world and the particular roles of children and older people;
- Reviews some of the key health issues facing people in poor countries;
- Examines different policies towards the provision of education and human rights.

Introduction: putting people at the centre of development

People are, or certainly should be, central to the development process and an essential element in all development strategies. But all too often in the past the needs of people have been ignored and there has been a failure to consider the possible implications of development policies on individuals, households and communities. As we have seen in Part I, there are many different and often conflicting views as to the meanings of development, and the most appropriate strategies to be followed at different points in time and space. However, for one influential development economist, Dudley Seers, development was unequivocally about improving the quality of people's livelihoods, and he argued that the reduction of three key variables – poverty, unemployment and inequality – should be central to the development process. As Seers observed: 'The questions to ask about a country's development are therefore: What has been happening to poverty? What has been happening to unemployment? What has been happening to inequality? If all three of these have become less severe, then beyond doubt this has been a period of development for the country concerned' (Seers, 1969).

Seers also emphasised the need for the true fulfilment of human potential and improvements in the quality of life. In a later paper entitled 'The new meaning of development', written after the oil crisis of the 1970s, he suggested that 'self-reliance' should be another important goal of development plans (Seers, 1979; see also Chapter 3). In order to reduce poverty, unemployment and inequality, Seers and others have argued that development strategies must fulfil basic human needs such as nutrition, water and sanitation, health and education. It is also important to recognise that these basic needs are inextricably linked, and policies must adopt a holistic approach towards improving human welfare. Too often in the past development strategies have been driven by economic goals, whereas fulfilling basic needs has received less priority, commonly assuming that economic growth will somehow 'trickle down' spontaneously to the most marginal elements of society and space, as reviewed in Chapter 3. In fact, the Third World is littered with so-called development projects which, far from empowering people, supplying their basic needs and raising living standards, have instead produced greater inequality, poverty and unemployment.

A further problem with many development strategies is that 'people' and 'communities' have all too frequently been perceived by developers as being homogeneous and passive, rather than as diverse and dynamic entities. The peculiar needs, knowledge and skills of different individuals and groups within communities have often been ignored in favour of a broad and less sensitive approach. As a result, although development projects might have benefited certain sections of the population, other elements have lost out. For example, in The Gambia, West Africa, where rice is a woman's crop and women possess the detailed knowledge and understanding of its production and processing, a series of overseas-funded irrigated rice development projects in the 1960s and 1970s achieved poor results precisely because women's considerable expertise was ignored by the development teams. As Dey comments, 'By failing to take into account the complexities of the existing farming system and concentrating on men to the exclusion of women, the irrigated rice projects have lost in the technical sense that valuable available female expertise' (Dey, 1981: 122). It is essential, therefore, that future development strategies are built upon a detailed understanding of the individuals or communities which are the target of such policies, rather than being based upon the assumption that people and societies are homogeneous.

This chapter will investigate the diversity of people and their role as a key resource in the development process. First, a number of important demographic features will be considered, and this will be followed by an evaluation of some broad issues affecting the quality of life.

Population and resources: a demographic time bomb?

The question of the rate of population growth and its relationship to the availability of food and vital natural resources has exercised the minds of many scholars for centuries. Although some commentators see population growth as the 'big issue' in world development, painting a 'gloom and doom' scenario of population growth outstripping food supply, others are much less pessimistic and view population growth more as an 'engine of development' playing an important role in the development process.

A frequently cited starting point in the population and resources debate is Thomas Malthus' *An Essay on the Principle of Population*, published in 1798. Malthus described a highly pessimistic scenario of population growing more rapidly than food supply, and he advocated the need for 'preventative' and 'positive' checks on population growth. He further argued that the tension between population and resources was a fundamental cause of misery for much of humanity (Crook, 1997). More recently, Paul Ehrlich in his book *The Population Bomb* (1968) has commented:

> **Americans are beginning to realize that the undeveloped countries of the world face an inevitable population–food crisis. Each year food production in undeveloped countries falls a bit further behind burgeoning population growth, and people go to bed a little bit hungrier. While there are temporary or local reversals of this trend, it now seems inevitable that it will continue to its logical conclusion: mass starvation.**
>
> **(Ehrlich, 1968: 17)**

The Club of Rome's *Project on the Predicament of Mankind* in the early 1970s further echoed Malthus' and Ehrlich's warnings, suggesting that

> **demographic pressure in the world has already attained such a high level, and is moreover so**

Key thinker

Thomas Robert Malthus (1766–1834)

Thomas Malthus was born in 1766 into a prosperous family near Dorking, Surrey, England. He was educated at home until 1784, when he went to Jesus College, Cambridge. Nine years later, after gaining Bachelor's and Master's degrees, he was elected a fellow of Jesus College. He took holy orders in 1797 and took charge of a small parish in Surrey. In 1798, he published the first edition of his most famous work, *An Essay on the Principle of Population as It Affects the Future Improvement of Society, with Remarks on the Speculations of Mr. Godwin, M. Condorcet, and Other Writers.* The *Essay* received much attention and was subsequently enlarged upon in a total of six editions, the last being published in 1816. The 1803 edition introduced the concept of a preventive check on population growth through what he called 'moral restraint'.

In 1805, Malthus became a professor of modern history and political economy at the East India Company's college at Haileybury (Hertfordshire), a position he kept until he died of heart disease in 1834. Malthus argued that the main aim of his research was to promote the happiness of mankind by identifying the real possibilities of progress. In 1819, he was elected a Fellow of the Royal Society, and in 1821 joined the Political Economy Club, whose members included influential writers such as political theorists David Ricardo and James Mill.

Despite generating a long-standing interest in the relationship between population and resources, Malthus and his followers greatly exaggerated the 'population question', and failed to anticipate the effects of the agricultural revolution (c.1750–1850), which led to a significant increase in food production in Western Europe, and the later influence of contraceptives, which led to a decline in the fertility rate.

Plate 5.1 Thomas Malthus
source: Getty Images/Hulton Archive

unequally distributed, that this alone must compel mankind to seek a state of equilibrium on our planet. Underpopulated areas still exist, but, considering the world as a whole, the critical point in population growth is approaching, if it has not already been reached.

(Meadows *et al.*, 1972: 191)

An alternative and much more positive perspective on the relationship between people, environment and resources was provided by economist Ester Boserup in her important book *The Conditions of Agricultural Growth* (Boserup, 1965, 1993). Boserup presented a convincing argument to show that population growth and increasing population density can in fact be key factors in generating innovation and intensification in traditional food production systems. She suggested that, provided the rate of population growth is not too rapid, populations will over time adapt their environment and cultivation strategies such that increased yields can be obtained without any significant degradation of the resource base. This viewpoint has gained greater popularity in recent years, as detailed empirical research has revealed the considerable capacity of indigenous peoples to raise the productivity of their farming systems in the face of increasing population numbers (Tiffen *et al.*, 1994). The continuing credibility of Boserup's thesis is indicated by the re-publication of her book, with a foreword by Robert Chambers, himself a key figure in development research. Chambers comments:

The Boserupian thesis will continue to stimulate argument and inspire research on the links between population change, agricultural technology, and now sustainability. It has stood the test of time: it is repeatedly referred to by those who have read this book and by many who have not.

(Boserup, 1993: 8)

Key idea

Who is right, Malthus or Boserup?

The debate concerning the relationships between people and environmental resources goes on and on. There are many instances in reports during the colonial period where European colonial officials made scathing remarks about indigenous farming practices in Latin America, Africa and Asia. It was often assumed that environmental management strategies developed in Europe and North America were far superior to those in tropical developing countries. But in the last 20 years or so, research in many rural developing areas has revealed that farmers have a remarkably good understanding of environmental resources and are capable of managing the environment so that it becomes more productive and can satisfy the needs of growing populations. Flights of terraced fields and intricate irrigation systems in places as diverse as south-west China, the Peruvian Andes and the High Atlas mountains of Morocco, testify to the ability and endeavours of people living in poor communities who have spent many hours adapting the environment to suit their needs in a perfectly sustainable way. On the contrary, many former colonies are littered with failed development schemes, where the technology which was introduced by European 'experts' proved to be totally inappropriate and machinery now lies idle and decaying. The ill-fated East African Groundnut Scheme in the 1950s, in what is now Tanzania, is a case in point, where, following the collapse of this ambitious project, the local Wagogo pastoralists referred to the scheme as 'The white man's madness'. Had the British 'developers' undertaken detailed climate and soil analysis, and interacted with the Wagogo to discover more about local environmental characteristics, it is likely that many of the pitfalls might have been avoided (Binns, 1994a). Ester Boserup's emphasis on understanding how communities manage their local environmental resources seems a sensible way forward in formulating rural development strategies.

The relationship between population and resources also featured strongly in the Brundtland Report of 1987 (WCED, 1987) and at the United Nations Conference on Environment and Development (the so-called Earth Summit), held in Rio de Janeiro, Brazil, in June 1992. This much-publicised event, attended by many world leaders and non-governmental organisations, was concerned with key global resources and a number of major environmental issues, such as global warming and climatic change. It was at Rio, and in subsequent publications, that the concept of sustainable development was popularised. The essence of sustainable development is the need to achieve an equilibrium between the world's basic resources and their continuing exploitation by a growing world population, so as not to jeopardise these resources for future generations. Agenda 21, the comprehensive programme of action adopted by governments at the Earth Summit, continually emphasises the links between environment, population and development (UN, 1993).

Ten years after the Rio summit, the World Summit on Sustainable Development was held in Johannesburg, South Africa, in August/September 2002, reputedly the largest conference ever held. Despite the much-publicised non-attendance of President George W. Bush of the world's wealthiest and most powerful nation (the USA), the Summit provided an opportunity to strengthen global commitments on sustainable development, as well as taking stock of progress towards the 2015 International Development Targets (Earth Summit, 2002). The many different viewpoints expressed in the long-running debate on the dynamic relations between population and resources themselves constitute distinctive 'geographies of development'.

Where do the world's people live?

At three minutes after midnight on 12 October 1999 Fatima Nevic gave birth in Sarajevo to a boy who was symbolically heralded by the media as the six billionth person in the world. In 2007 the world's population was estimated to be 6671 million (UN, 2007a), but these people are by no means distributed evenly across the Earth's surface and population density varies widely (Figure 5.1). With the massive populations of

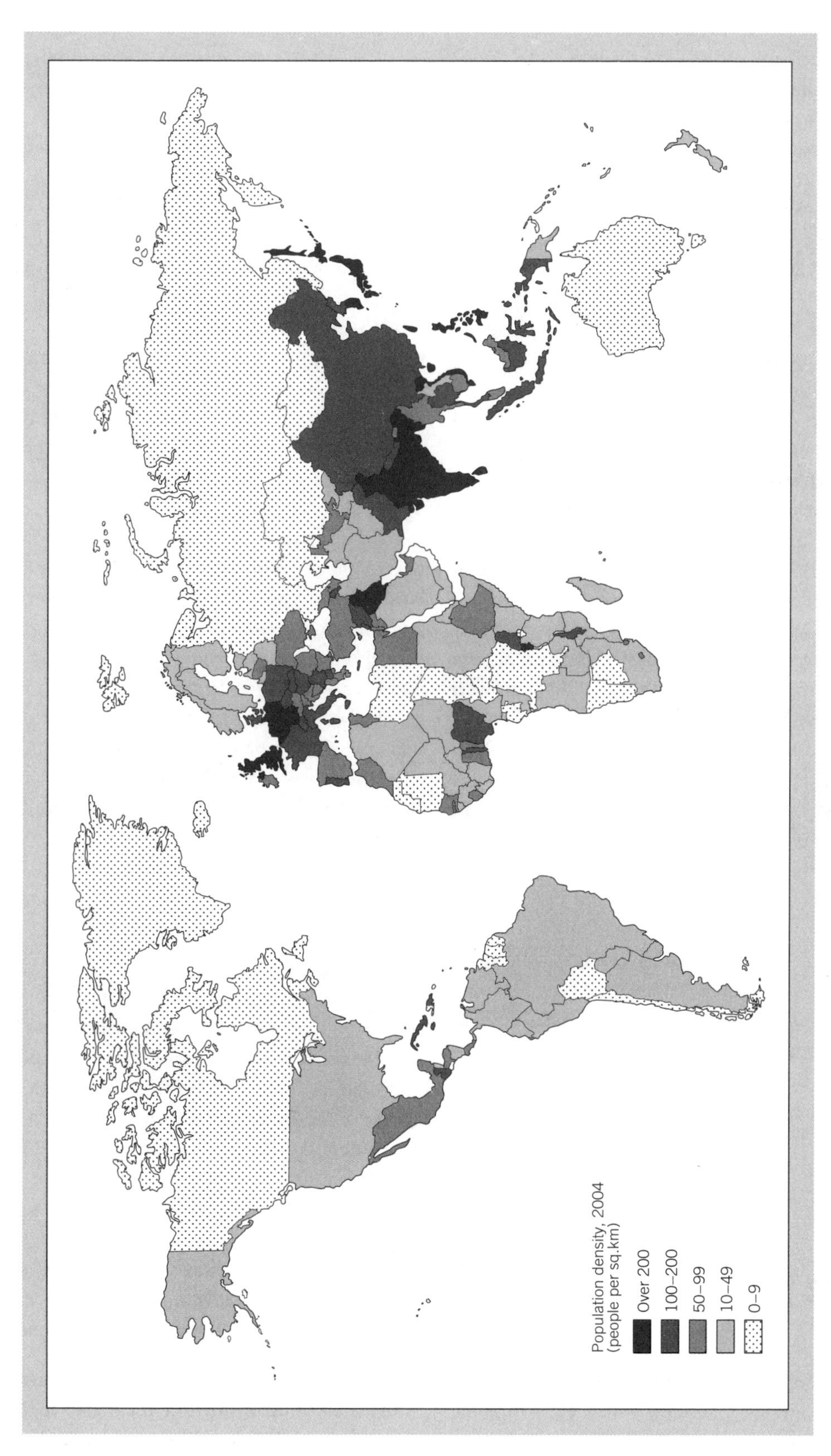

Figure 5.1 World population density

China and India, the dominance of the Asia-Pacific region in the world distribution of population is particularly striking, and well over half the total population (4064 million) live in this region, significantly more than the combined populations of the Middle East, Africa and Latin America. India joined China on 15 August 1999 as the only other nation on Earth with a billion people, and, on present projections, India's population could actually overtake China's by 2045 (World Bank, 2002a).

Apart from some small and densely settled island states such as Barbados (with 626 persons per square kilometre), the Maldives (1077) and Bermuda (1280), and small enclaves such as Singapore (6869), the most densely settled countries are to be found in Asia and Europe. India, with its massive population of 1087 million, has a density of 343 persons per square kilometre, whereas Japan has 339, South Korea 482 and Bangladesh a staggering 967. Although European population densities nowhere reach the magnitude of Bangladesh, discounting densely settled small enclaves such as Monaco (17,949), Europe does have some relatively high population densities, such as the United Kingdom with 244 persons per square kilometre, Belgium with 341 and the Netherlands with 388.

With the exception of the polar wastes of Greenland and Antarctica, the world's most sparsely settled areas include Australia and Canada, with only three persons per square kilometre, and large parts of Africa. In fact, Africa is still the world's least densely settled continent, with the desert states of Mauritania and Libya having only three persons per square kilometre. In southern Africa, Botswana and Namibia have, respectively, three and two persons per square kilometre. There are, however, certain countries in Africa with unusually high densities of population, notably the tiny countries of Rwanda and Burundi in central Africa, with densities of 337 and 262 persons per square kilometre, whereas Nigeria, the continent's most populous country with over 127 million people, has a density of 139 per square kilometre (World Bank, 2002a).

Counting the people

A word of caution is warranted concerning the reliability of national, and therefore regional and global, population statistics. Censuses, which even in some richer countries are not always reliable, are also costly, time-consuming and require considerable expertise to administer and analyse. Consequently, some poor countries, and/or those which are politically unstable, have often been unable to conduct regular censuses, so the available data are frequently old and unreliable. Furthermore, national planning has been hampered by the lack of reliable and up-to-date figures.

Nigeria, for example, undisputedly Africa's most populous country, had its first census for almost 30 years in November 1991. The previous census with any reliability (though this was questioned) was held in 1963, after which there were three further attempts to hold censuses, on each occasion failing because regional leaders submitted exaggerated head counts in an effort to show that their people were more numerous and thus entitled to 'a bigger share of the national cake'.

The Nigerian federal government made great efforts to conduct a successful census in 1991, with three years of careful preparation, an investment of some £75 million and enlisting the help of half a million enumerators, including women who worked in Muslim areas. Estimates before 1991 had put Nigeria's population at well over 100 million and so many Nigerians seemed genuinely shocked when the 1991 census revealed a total population of only 88.5 million! Fifteen years later, the provisional results of the March 2006 census suggest that Nigeria's total population has increased to 140 million.

Population change

For most of human history population growth, averaged over long periods, has remained near zero. In fact, the modern expansion of the world's population only started in the eighteenth century with the slow decline of the death rate in Europe and North America. Population growth then accelerated steadily in the twentieth century and has been particularly rapid in the developing countries since 1950 (Table 5.1).

Merrick (1986) points out that more people have been added to the world since 1950 than in all of human history before the middle of the twentieth century – a sobering thought. Bongaarts (1995: 8) comments:

> **The acceleration in growth is well demonstrated by the shortening of the time intervals needed to add successive billions to the world population. The**

Table 5.1 World population growth, 1900–2100

	Population (billions)			% increase	Estimated population (billions)		% increase
	1900	1950	1990	1950–1990	2025	2100	1990–2100
Developing countries	1.07	1.68	4.08	143	7.07	10.20	150
Developed countries	0.56	0.84	1.21	44	1.40	1.50	24
World	1.63	2.52	5.30	110	8.47	11.70	121

Source: From Bongaarts (1995). Adapted and reproduced with permission from the International Food Policy Research Institute.

first billion was reached early in the nineteenth century, the second billion took 120 years, the third 33 years, the fourth 14 years and the fifth (between 1974 and 1987) just 13 years.

If projections prove to be accurate, the next three billion will be added at an even faster pace, each taking just over a decade, to reach eight billion before the year 2020 (Bongaarts, 1995: 8). Such projections suggest that it is likely the world's population could reach around 11.7 billion in 2100.

Where and how these people will live in the future is of much interest and concern for development, since they will be by no means evenly distributed across the globe. The World Conference on Population and Development, held in Cairo in September 1994, had the aim of drawing up a 20-year programme (up to 2015) to combat overpopulation in the world, but reached deadlock on a number of issues, most notably the question of decriminalising abortion, a move which is rejected by the religious authorities of both the Catholic Church and Islam. The United Nations Population Division estimate that the total world population in 2015 will be 7295 million, of which no less than 6050 million (83 per cent) would be in the developing countries (UN, 2007b).

The phenomenal growth of population in the developing countries is revealed in Table 5.1, which shows that, whereas population in the developed countries increased by 44 per cent between 1950 and 1990, the developing countries experienced a massive growth rate of 143 per cent in the same period. Projected growth rates for the period 1990–2100 indicate a significant decline in the rate of population growth for the developed countries (24 per cent) compared with the earlier period, but there is a marked acceleration in growth in the developing countries to 150 per cent. These countries have growth rates more than four times higher than developed countries, and the sizes of their populations are also generally much larger than those of the developed countries. Such growth is going to place even greater pressure on resources, which in many poor countries are already stretched. According to the World Bank, in 2005 Bangladesh was ranked 175 out of 208 in the world in terms of per capita Gross National Income (GNI) (US$470), and in 2004 it had a total population of 139.2 million with an average population density of no less than 967 people per square kilometre (World Bank, 2007, 2005).

Projecting the size of national and global populations is fraught with difficulty, since it is affected by such unpredictable events as natural disasters, wars and medical advances. For example, the production and wide availability of an effective and cheap vaccine to combat malaria, which affects 40 per cent of the world's population and kills an estimated 2.7 million people every year, could have a massive impact on reducing death rates, which in turn would affect population growth rates nationally and globally. The introduction of government policies designed to increase or reduce population, will also affect growth rates, as in China, where its one-child policy has had a marked effect on national population growth rates (Plate 5.2; Box 5.1).

A further issue which makes the future prediction of the size and growth of population in specific countries particularly difficult is the question of population redistribution. The movement of people within countries, perhaps from rural to urban areas, and between countries as voluntary migrants, or maybe as refugees escaping from drought or civil war, has had a significant impact on population dynamics in certain countries and regions. Chapter 8 explores this in more detail within the context of movements and flows of both people and commodities.

Plate 5.2 'One-child' poster, Guangzhou, south-eastern China
(photo: Tony Binns)

BOX 5.1

China's one-child population policy

The population of the world's most populous country, China, passed the 1 billion mark in 1981 and by 2000 it had reached 1261 million. However, with an estimated average annual growth rate of 0.7 per cent between 2000 and 2015, this represents a considerable slowing down of population growth, from the 1.8 per cent per annum experienced between 1960 and 1993, and is well below the 2.2 per cent average growth rate for all developing countries over the same period (World Bank, 2002a). This significant decline in China's population growth rate is due to the implementation of a rigid birth control policy.

In 2004 the UN predicted that China's population would reach well over 1.4 billion by 2025 but, with steadily declining population growth rates, the 2050 population would be under 1.4 billion.

China: population size

1950	2005	2015	2025	2050
554,760	1,315,844	1,392,980	1,441,426	1,392,307

China: average annual rate of population change (%)

1995–2000	2000–2005	2010–2015	2020–2025	2045–2050
0.88	0.65	0.56	0.24	−0.35

Source: (UN, 2007a)

Following a marked decline in China's population during the late 1950s and 1960s due to loss of life from disasters such as typhoons and flooding and the severe famines which followed, the country then experienced a 'baby boom' from 1963. The government increasingly questioned the value of a rapidly growing population in relation to resources under increasing pressure, and during the 1970s various attempts were made to encourage both family planning and delaying marriage. For example, in Shifang County of Sichuan Province, the local authorities responded in 1971 by raising the age at which a person could get married; it went from 18 to 23 for women and from 20 to 25 for men. Additionally, the county government made great efforts to spread birth control information, and to

BOX 5.1 (continued)

provide free services for contraception. When contraceptive methods failed, induced abortions became more common (Endicott, 1988). Reducing the fertility rate became a key priority and the slogan 'one is not too few, two will do and three are too many for you' was publicised nationally. Communities were encouraged to recommend which women should be able to have a baby and the practice of 'giving birth in turn' became widespread.

In 1980, as those born during the baby boom of 1963 were approaching the age of marriage and childbearing, the policy of one child per family was officially adopted by the National People's Congress and was incorporated into the country's new 1982 Constitution (Jowett, 1990: 117). The State Council deemed it 'necessary to launch a crash programme over the coming twenty or thirty years calling on each couple, except those in minority nationality areas, to have a single child . . . Our aim is to strive to limit the population to a maximum of 1200 million by the end of the century' (quoted in Jowett, 1990: 117). A number of relaxations to the policy were permitted, particularly where the first child was a girl, also among minority peoples living mainly in western China, and in the rural areas of certain provinces. It is probably in the urban areas where the policy has been most strictly enforced (Leeming, 1993: 61). To enforce the one-child policy, a system of economic rewards and penalties was introduced, such as parents being offered a 5–10 per cent salary bonus for limiting their families to one child and a 10 per cent salary deduction for those who produced more than two children (Jowett, 1990: 119).

In 1988, Mrs Lui, Director of Number 2 Neighbourhood Committee, Hua Long Chao Sub-district in the Sichuan city of Chongqing, commented that the main job of the neighbourhood committee is to 'educate the people to realise the importance of the one-child family' (Binns, personal communication, 1988). Permits to have children were given annually to about 100 women in the neighbourhood, with priority being given to older women. The local factory, where most neighbourhood dwellers worked, evidently played a key role in awarding points to female workers, points that affected their relative position in the queue to receive a permit. Strong sanctions were imposed if a woman became pregnant without a permit, and the committee would notify the government authorities and the factory. An abortion was usually required, a fine imposed, wage increases frozen and a permit to become pregnant again would be further delayed.

China's total fertility rate fell from 6.66 in 1968, to 2.32 in 1987, and in urban areas decreased from 3.2 in 1970 to 1.3 in 1987. However, despite slowing down the population growth rate, the one-child policy has had widespread social implications. The policy fundamentally conflicts with traditional Chinese family values, in which children are seen as a source of happiness and fulfilment, as well as guaranteeing the continuation of the family line. The policy is undoubtedly unpopular, but much as they would like another child, many women accept official arguments that they must go without. The policy has also led to even greater prestige being given to the birth of a son, particularly in the rural areas where there are no state pensions. Much concern has been expressed in China about the long-term social effects of the creation of a generation of pampered 'little emperors' with no sisters and cousins. Furthermore, the under-registration of female births, abandonment or neglect of girl babies, prenatal sex testing followed by selective abortion, and instances of female infanticide were frequently reported in the Chinese press in the 1980s. In 2000, the sex ratio at birth was 117 boys for every 100 girls, with the imbalance even greater in rural areas. The Chinese government is trying to reduce the cultural discrimination against girls, with a 'care for girls' campaign in rural areas. Some have also criticised the one-child policy on the grounds that it represents an infringement of a woman's control of her reproduction process 'since many women, whilst perhaps not wanting to have to go on until they have one or more sons, do wish to have at least two children' (Endicott, 1988: 179).

A further concern is that the policy will completely transform the country's age structure, and in time will result in a declining workforce and a very aged population. The baby boom of the 1960s will eventually result in a large number of retired people in the 2030s. As Jowett comments:

> **By then, over-65s could constitute more than 25 per cent of the population and thus, within a lifetime, the number of retired will have increased from one in twenty to one in four. Such a high level of old-age dependency is unprecedented even in today's developed countries where the over-65s generally constitute 10–15 per cent of a country's population.**
> **(Jowett, 1990: 121)**

In spite of reducing China's population growth rate, the policy has received much criticism, not least because it represents a massive 'experiment' in social engineering.

Table 5.2 Population statistics for selected low-, middle- and high-income countries, 2005

Country	GNI per capita (US$)	Birth rate	Death rate	Infant mortality rate	Under-5 mortality rate	world rank	Life expectancy at birth	Mean annual population growth rate (%) 1990–2005
Bangladesh	470	26	8	54	73	57	64	2.1
Brazil	3,460	20	7	31	33	86	71	1.5
China	1,740	13	7	23	27	96	72	0.9
India	720	23	9	56	74	54	64	1.7
Jamaica	3,400	20	8	17	20	113	71	0.7
Japan	38,980	9	8	5	4	182	82	0.2
Mali	380	49	17	120	218	7	48	2.8
Sierra Leone	220	46	23	165	282	1	41	2.0
Sweden	41,060	11	10	3	4	182	80	0.4
United Kingdom	37,600	11	10	5	6	161	79	0.3
USA	43,740	14	8	6	7	156	78	1.0
Zimbabwe	340	29	23	81	132	31	37	1.4

Source: UNICEF (2007)

Understanding population statistics

Changes in population growth rates over time are affected by a wide range of factors, but are essentially controlled by the changing relationship between birth rates and death rates. Table 5.2 presents population statistics for a sample of low-, middle- and high-income countries from across the world.

The crude birth rate is the most common index of fertility and is a ratio of the number of live births to the total population, usually expressed as so many per 1000. Although the United Kingdom and the USA had birth rates of 11 and 14, respectively, in 2005, Mali and Sierra Leone in West Africa had rates of 49 and 46, respectively, indicating relatively little change from the figures of 51 and 49 for these African countries in 1970. The crude death rate is the number of deaths per 1000 of the population. Considering the same four countries, the United Kingdom and the USA had death rates of 10 and 8 in 2005, whereas Mali and Sierra Leone recorded death rates of 17 and 23 in 2005 (UNICEF, 2007).

Two other important indicators of the quality of life and levels of development are infant mortality and life expectancy. The infant mortality rate measures the number of deaths of infants under one year old per 1000 live births. This variable reflects general living conditions and also the health and nutritional status of pregnant and lactating mothers. Infant mortality rates in 2005 varied from 3 in Sweden and Switzerland (5 in the United Kingdom), to 165 in Sierra Leone, one of the world's poorest countries. In fact, in terms of both infant and under-five mortality, Sierra Leone had the dubious distinction of being ranked lowest in the world along with Afghanistan (UNICEF, 2006). However, there is not always a direct relationship between wealth and infant and child mortality. For example, Vietnam, ranked 166 by the World Bank in 2005 according to GNI per capita, with a per capita GNI of only US$620, had an infant mortality rate of 16, whereas Gabon in central Africa, with a per capita GNI over eight times greater (US$5010), had an infant mortality rate which was more than three times that of Vietnam (60) (World Bank, 2007; UNICEF, 2006).

Life expectancy at birth also reflects general living standards, nutrition and health care, and it reveals tremendous inequalities between developed and developing countries. For example, a child born in Japan in 2005 could expect to live for 82 years, but in Sierra Leone an average life span of only 41 years is all that a child could expect (UNICEF, 2006).

The demographic transition

The changing relationship over time between fertility and mortality rates is clearly demonstrated through the demographic transition model, which usually identifies four or five key stages (Figure 5.2).

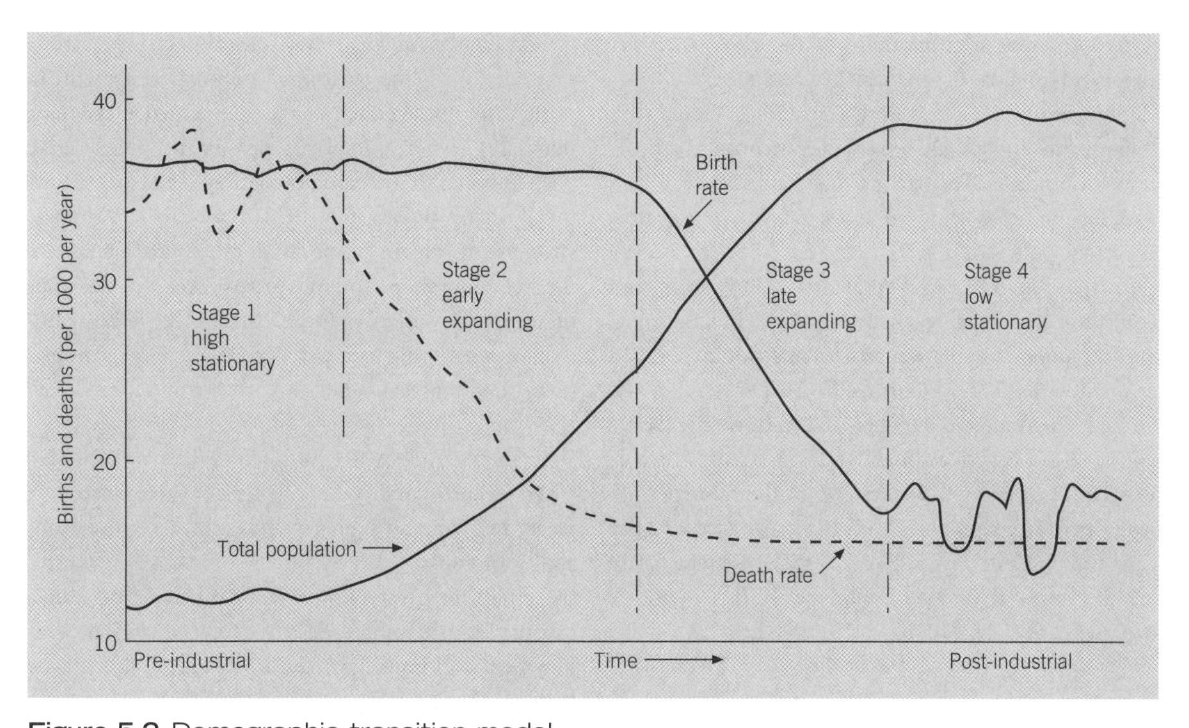

Figure 5.2 Demographic transition model

Source: Adapted from *A Dictionary of Geography*, Oxford University Press (Mayhew, S. 1997). By permission of Oxford University Press, Inc. and adapted from Jones and Hollier (1997)

Stage 1 (high stationary or pre-transition phase) is characterised by high birth rates and high death rates, such that population growth is static or negligible. This situation applied to pre-eighteenth-century Europe and North America, but in many developing countries this remained the position up to the Second World War.

In the next phase (Stage 2: early expanding or early transition) the death rate (mortality) begins to decline with better living standards, largely due to improvements in nutrition and public health. The incidence of famines and epidemics also falls. However, the birth rate (fertility) remains high, so population growth accelerates. In much of Africa, Asia and Latin America mortality did not begin to decline until the first half of the twentieth century. By the late 1960s the annual average death rate had dropped to 15 per 1000, whereas the birth rate remained high at 40 per 1000, resulting in an annual growth rate of 25 per 1000 people or 2.5 per cent. In 1979 Kenya recorded an annual population growth rate of 4.1 per cent, one of the highest in the world. Some African countries today still have the highest birth rates in the world. Kenya's crude birth rate in 2005 was 39, whereas Uganda and Niger had even higher rates of 51 and 54, respectively (UNICEF, 2006). These growth rates are well above those observed in European populations when they were at the same stage in the transition.

Stage 3 (late expanding or mid transition phase), occurs when improved technology in agriculture and industry, together with better education systems and legislation controlling child employment, lead to a decline in the economic and social value of children. Furthermore, the breakdown of the extended family system places greater physical, emotional and financial costs on the parents. Couples start to use contraception to limit family size and the birth rate begins to fall. The death rate continues to decline and population growth at this point reaches its maximum. This stage occurred in the developed countries in the late nineteenth and early twentieth centuries. Elsewhere, in the last two decades there have been marked reductions in birth rates in East and Southeast Asia, due to a combination of family planning programmes and socio-economic development. Thailand had an annual population growth rate of 0.9 per cent for the period 2000–2005 and South Korea 0.4 per cent (UN, 2005).

In stage 4 (low stationary or late transition phase), the death rate reaches its lowest level and as fertility steadily declines, the rate of population growth begins to fall. Most developing countries are currently in the mid and late transition stages. Hong Kong and Singapore, however, have nearly completed the transition and their respective projected population growth rates (2010–2015) of 0.92 per cent and 0.96 per cent are not too dissimilar from those of the USA (0.85 per cent) and Canada (0.8 per cent) (UN, 2005).

Some writers, such as Bongaarts (1994), add a fifth stage to the model, known as the 'declining or post-transition phase'. This stage is characterised by a new equilibrium being achieved between births and deaths, such that population growth is close to zero; in some cases there is even negative growth. Many European countries have now reached this stage. For example, average annual population growth rates for 2010–2015 are as low as 0.21 in Portugal, 0.30 per cent in the United Kingdom and 0.05 per cent in Germany. Some of the most negative growth rates are found in the Russian Federation, 0.50 per cent for the same period, whereas in Hungary and Latvia the figures are 0.32 per cent and 0.52 per cent, respectively. The average figure for the 'more developed regions' over this period is 0.20 per cent (UN, 2007b).

Population policies

The reasons underlying the various changes reflected in the demographic transition are highly complex. It is unwise to blindly advocate the strict control of population growth in poor countries. As O'Connor argues in the case of Africa:

> **there is no evidence to suggest that rapid population growth is the main cause of poverty . . . which was just as widespread when the growth rate was much slower. In so far as the rapid growth results from high fertility it can be seen alternatively as a consequence of poverty.**
>
> **(O'Connor, 1991: 54)**

Population growth is undoubtedly a problem in some countries, but it is invariably a symptom of other problems such as poverty and lack of security. Having more children among poor families is an insurance strategy to ensure household survival in places where infant and child mortality rates are high. Children frequently play a vital role in generating household income, as well as providing social security in the absence of any state provision, such as under structural adjustment programmes (see Chapter 3). The most successful initiatives to reduce family size start by tackling underlying causes and finding ways of ameliorating poverty and insecurity, helping to improve the health of mothers and children. Experience has shown that, with such strategies, both fertility and mortality rates should begin to fall.

Some governments have, however, taken a strong interventionist line to control population growth. In some cases, these policies have been aimed at reducing population; whereas in others, pro-natalist policies have been adopted to increase population. The Nazi regime in Germany during the 1930s and early 1940s was strongly pro-natalist, generating much propaganda on the need to create a master Aryan race. A wide range of measures was introduced, such as generous family allowances and tax concessions for large families, taxes on unmarried adults and prosecution for induced abortions.

Romania under the Ceausescu regime also adopted a strong pro-natalist line, with abortion becoming illegal in 1966 and tight controls placed on the availability of contraceptives. In the following year, 1967, the crude birth rate almost doubled to 27 per 1000, causing considerable pressures on education, employment and housing as the 1967–1970 bulge moved through the age groups.

Elsewhere in the world, Israel and Saudi Arabia have also encouraged population growth, primarily to strengthen their political power. Israel's average annual population growth rate in the period 1975–2004 was 2.3 per cent, whereas Saudi Arabia's rate was 4.1 per cent (UNDP, 2006), and reached as high as 5.2 per cent per annum during the period 1980–1990 (World Bank, 1996a). The upsurge of Islamic fundamentalism has resulted in some states becoming pro-natalist. By introducing its New Population Policy in 1984, Malaysia reversed a long-standing policy of promoting family planning to encourage women to 'go for five', to catch up with more populous neighbouring states and to prevent ethnic Malays from being dominated by the Chinese population. In response to a shortage of labour, Singapore from 1987 adopted a more selective pro-natalist stance, encouraging educated and professional couples to 'have three, or more if you can afford it' rather than 'stop at two' (Drakakis-Smith *et al.*, 1993). Singapore's average annual population growth rate was 2.2 per cent during the 1975–2004 period (UNDP, 2006a).

However, with growing concern about the population–resource balance, anti-natalist policies are now rather more common than pro-natalist measures. India launched a family planning programme as early as 1951 (see Box 5.2), and during the 1960s many other countries followed in response to current development thinking and the production of the contraceptive pill and intrauterine devices. In 1965, US

BOX 5.2

Planning the growth of India's population

The population of India was estimated at 1103 million in 2005 and is projected to increase at an average annual rate of 1.2 per cent to reach 1400 million by 2026. The crude birth rate is projected to decline from 23.2 in 2005 to 16.0 by 2025 due to declining fertility. Problems associated with rapid population growth were recognised by the Indian government soon after independence in 1947, and since then there has been a succession of policies designed to control the burgeoning population. In 2003, India spent only 1.2 per cent of GDP on public health, compared with 6.9 per cent in the UK and 6.8 per cent in the USA.

India's Family Welfare Programme was introduced four years after Independence in 1951 and initially focused on improving the health of mothers and children. However, from the Third Five-year Plan period (1961–1966) there was a marked shift in emphasis from the welfare of women and children towards the objective of achieving population stabilisation. The 1977 Population Policy emphasised the need for an educational and motivational approach to strengthen the voluntary acceptance of family planning.

The International Conference on Population and Development in 1994 and the Beijing Women's Conference in 1995 had a significant effect on India's family planning policy, such that in 1997 the Community Needs Assessment Approach was introduced with decentralised participatory planning that was concerned primarily with addressing clients' needs. The Reproductive and Child Health Programme, also launched in 1997, emphasised the principle of client satisfaction and the provision of high quality comprehensive and integrated health services. This Programme seeks to integrate services for the prevention and management of unwanted pregnancy, the promotion of safe motherhood and child survival, and the prevention and management of reproductive tract infections and sexually transmitted infections. Attention is directed particularly towards underserved and neglected population groups, including adolescents, and economically and socially disadvantaged groups, such as urban slum dwellers and tribal populations. For the first time in the history of the Family Welfare Programme attention is now focused firmly on gender concerns.

The National Population Policy (NPP), launched in February 2000, supports the provision of client-based services, and provides a framework for achieving the objectives of population stabilisation and promoting reproductive health within the wider context of sustainable development. In the medium term, the NPP aims to bring the total fertility rate down to replacement level by 2010, using strategies such as grassroots service delivery, empowering women, encouraging male involvement, meeting the need for family welfare services, addressing the needs of disadvantaged and underserved population groups, and establishing public–private partnerships.

Various promotional measures have been introduced, not only for sterilisation, but linked to poverty, delaying marriage, antenatal and delivery care, birth registration, birth of a girl child and immunisation. More specifically, cash incentives are provided for women who have their first child after 19 years of age, and couples below the poverty line are rewarded if they postpone their marriage and have their first child after the mother reaches the age of 21. Rewards are also offered if parents adopt a terminal method after the birth of their second child. In some states stricter policies have been introduced, such as in Madhya Pradesh, where individuals who marry before the legal age are prevented from seeking jobs, gaining admission to educational institutions and applying for loans. Furthermore, couples with more than two children are prevented from contesting local elections. It remains to be seen whether such strategies will be effective in controlling population growth.

President Lyndon Johnson, addressing a United Nations audience argued, 'Let us act on the fact that less than five dollars invested in population control is worth a hundred dollars invested in economic growth' (Stycos, 1971: 115). The United Nations began to provide advisory services to family planning programmes in 1965, and by 1976 some 63 developing countries had initiated such programmes. The 1974 World Population Conference in Bucharest brought population issues and family planning initiatives under the political spotlight. At this meeting, Western countries were strongly criticised by developing countries for placing too much emphasis on the population issue in development strategies, to the neglect of promoting social and economic progress.

Indonesia, the world's fourth most populous nation, adopted family planning in 1970 as a key element in the drive for economic growth. Considerable success was achieved between 1970 and 1999, with the average number of children per woman falling from 5.6 to 2.5 (UNICEF, 2001). China's one-child policy (see Box 5.1) is probably one of the most well-known and most rigid anti-natalist policies, affecting one-fifth of the world's population and enforced through tight community control and a series of rewards and sanctions. As a result, the 1970 crude birth rate of 33 per 1000 fell to 18 in 1979 and 13 in 2005 (UNICEF, 2001, 2006).

Other family planning schemes have commonly rewarded individuals with various tax concessions, with free or preferential medical treatment (South Korea), priority schooling (Singapore) or even a new sari (Bangladesh). It is in Asia that family planning programmes have been most widely implemented, whereas progress has been slower in Latin America, largely due to the proscription of 'artificial' methods of contraception by the Roman Catholic Church. Although evidence suggests that individual families frequently disregard the Church's views on birth control, government policy makers are likely to be more strongly influenced.

Another important factor affecting the size of families is the status of women in society. Muslim societies are generally strongly patriarchal; men can have up to four wives and female education is a low priority, since women are expected to stay in the home. The 1979 fundamentalist revolution in Iran swiftly overturned the Western-style policies introduced by the Shah, which had given a considerable degree of emancipation to women. Under the new regime a woman was required to gain her husband's consent to work; access to sterilisation and abortion was tightly controlled and the minimum age of marriage for women was reduced from 18 to 15 years. As a result, from 1980 to 1990 Iran had an average annual population growth rate of 3.5 per cent (World Bank, 1996a).

Since such population policies led to shrinking family size in the world's most populous countries, one British newspaper in January 1998 claimed that the 'population bomb' had been defused (Figure 5.3), although many of the world's poorest nations, and particularly those in Africa, continue to have rapidly growing populations. According to Figure 5.3, it is projected that in 2020, for the first time, Africa's most populous country, Nigeria, will be one of the world's six largest countries (*The Independent*, 1998: 11).

Population structure

Key elements in the demographic transition process, together with the impact of disasters such as war, famine and disease epidemics are manifested in changes to the population structure. Typically, developing countries with declining death rates, high birth rates and low life expectancy have youthful populations (Plate 5.3). Table 5.3 clearly shows how some of the poorer countries with high population growth rates have a large proportion of their populations under the age of 18. In Mali and Sierra Leone, respectively 55.0 per cent and 49.3 per cent of the populations are under 18, whereas in more developed countries these figures are much lower (Japan 17.0 per cent, Sweden 21.5 per cent). These statistics have implications for child welfare, especially educational provision.

The structure of a nation's population is clearly revealed in an age–sex diagram, commonly known as a 'population pyramid'. The population pyramids for Ghana and the United Kingdom in 2000 reveal quite different population structures (Figure 5.4). In fact, only the diagram for Ghana actually resembles a pyramid, with a wide base indicating a youthful population and a steeply tapering top; there are fewer people in the older age groups due to an average life expectancy at birth of only 57 years. In sharp contrast, the diagram for the United Kingdom is hardly a pyramid, and is typical of a more stable population situation, with a smaller proportion of the population in the lower age groups and a greater proportion in the upper age groups, reflecting an average life expectancy that is 20 years greater than for Ghana. Interestingly, the UK figure also shows a higher proportion of women over 70 years, indicating the differential in life expectancy between males and females, a common feature in a number of developed countries.

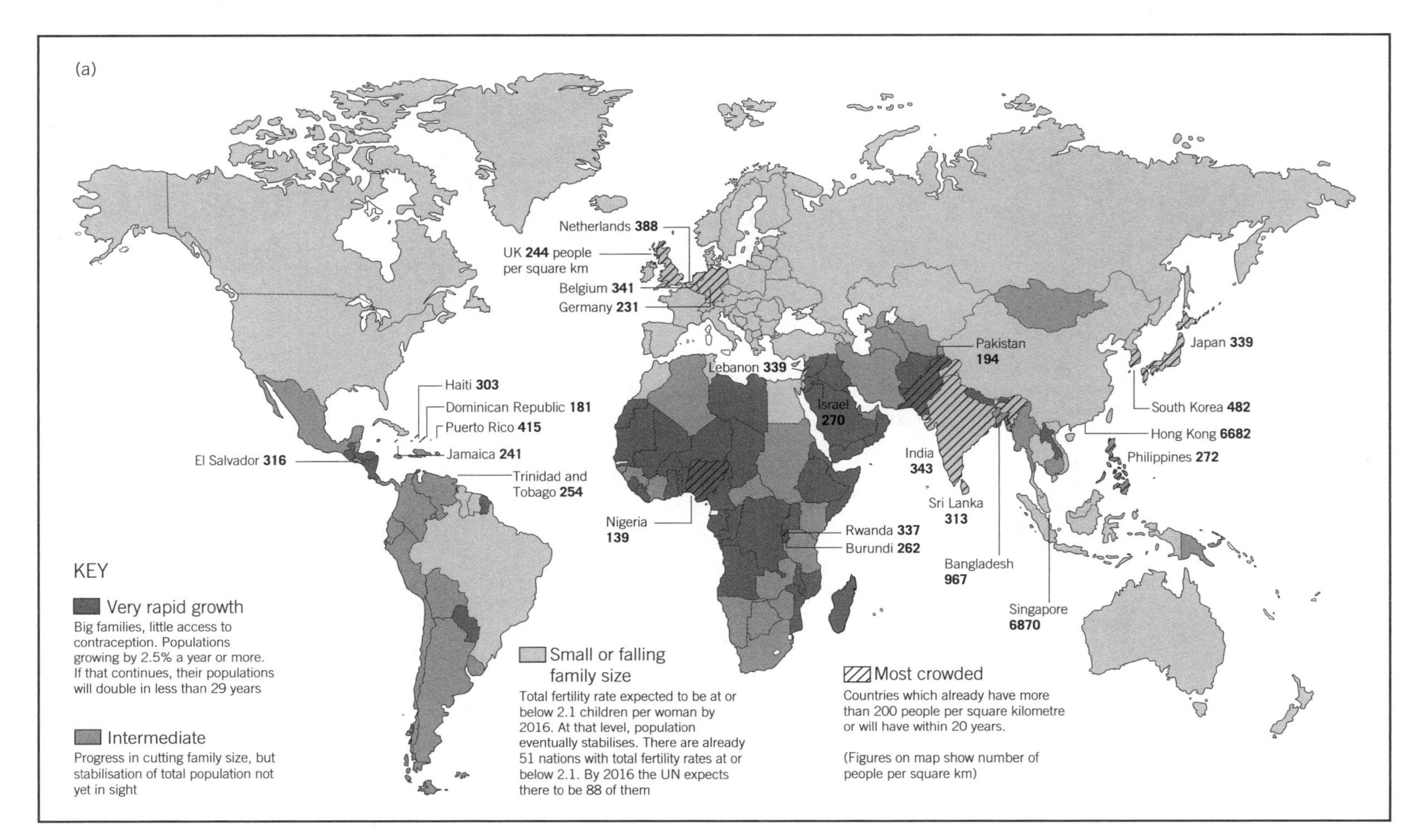

Figure 5.3 (a) Family sizes are shrinking rapidly, although human numbers are still soaring in many of the world's poorest nations. *The Independent*, 12 January 1998

Source: © reproduced by permission of *The Independent*

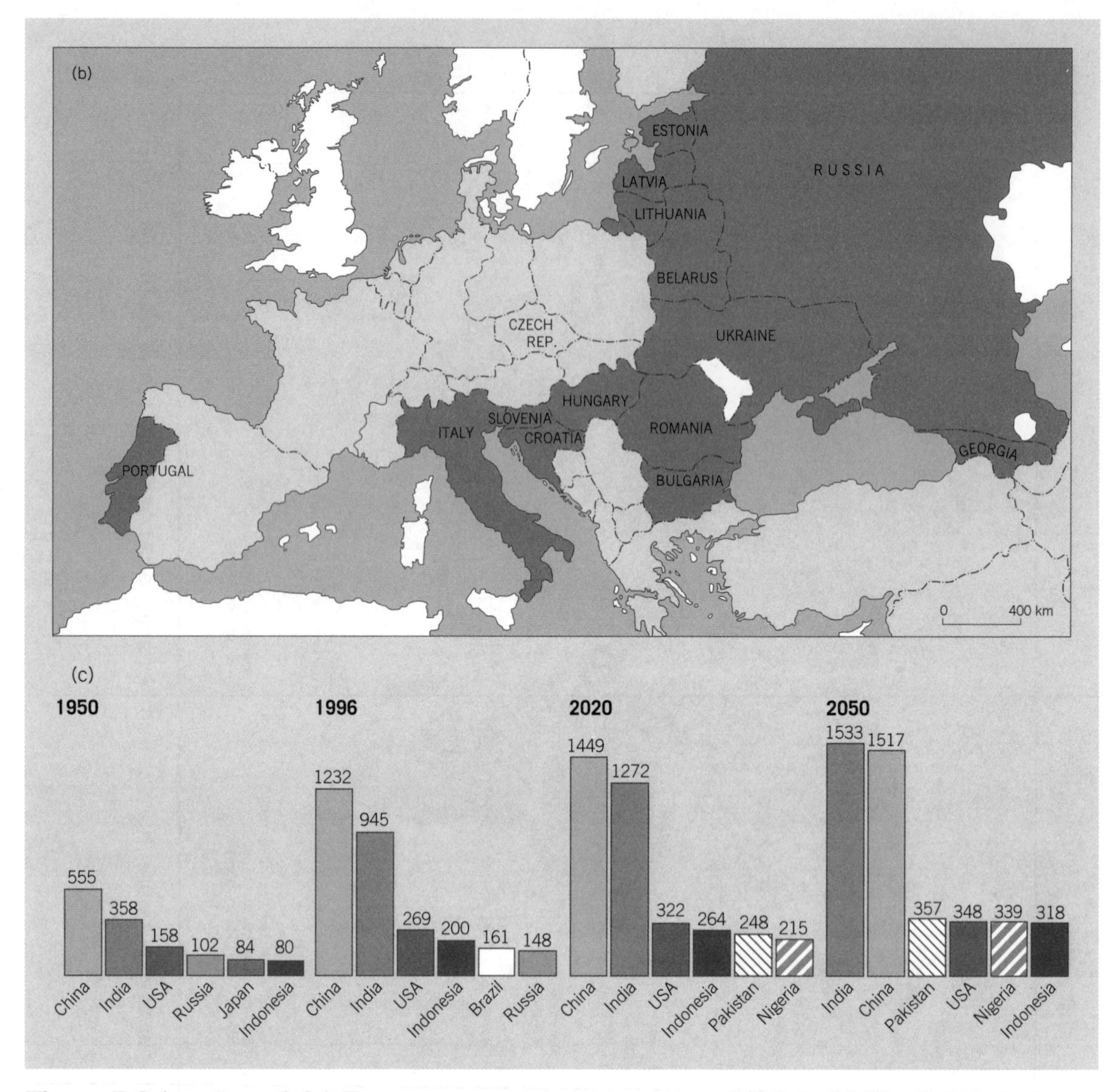

Figure 5.3 (continued) (b) European countries where population has already begun to fall. (c) The six biggest countries (in millions of people). The data are from the United Nations Population division and the projections to 2050 are from the 'medium variant'
Source: Adapted from *The Independent* (1998)

Ageing populations

Many of the richer countries in the world have been concerned for some time about the social and economic implications of their steadily ageing populations, particularly in terms of the impact on the size of the workforce and the increasing cost of providing health care and social security to older groups. In the United Kingdom, for example, the average age of the population is likely to rise from 38.8 years in 2000 to 42.6 years in 2025, and by 2040 the number of people aged 80 and over is expected to double to 4.9 million from 2.4 million in 2000. Some of the 'newly industrialising countries', such as Singapore and South Korea, are also experiencing a 'greying' of their populations, with an increasing proportion in the higher age groups. Globally, the number of people aged over 60 is the fastest growing section of the population, registering an increase of 63 per cent between 1960 and 1980, and with a predicted quadrupling of the size of this age

Plate 5.3 Children in The Gambia – a youthful population
(photo: Tony Binns)

Table 5.3 Proportion of the population under 18 years old, in selected countries, 2005

Country	Total population (millions)	Population under 18 years (millions)	Proportion of total population under 18 years (%)
Bangladesh	141.8	59.4	41.9
Brazil	186.4	62.2	33.4
China	1315.8	352.7	26.8
India	1103.4	420.7	38.1
Jamaica	2.7	1.0	37.4
Japan	128.1	21.8	17.0
Mali	13.5	7.4	55.0
Sierra Leone	5.5	2.7	49.3
Sweden	9.0	1.9	21.5
United Kingdom	59.7	13.1	22.0
USA	299.2	74.9	25.0
Zimbabwe	13.0	6.3	48.1

Source: UNICEF (2007)

group between 1955 and 2025. Interestingly, between 1980 and 2000 the biggest increases in the elderly population occurred in Africa and Asia, despite their relative poverty and poorer health and welfare systems compared with many richer countries (Morrissey, 1999). The United Nations declared 1999 the International Year of Older Persons, advocating the fostering of a 'culture of ageing', that considers older persons as both beneficiaries and agents of development. While the ageing of populations in richer countries gives cause for concern, it is likely that many poorer countries, with multiple calls on limited funding, will experience even greater difficulty in meeting the needs of their elderly populations (Ageing and Development, 2002).

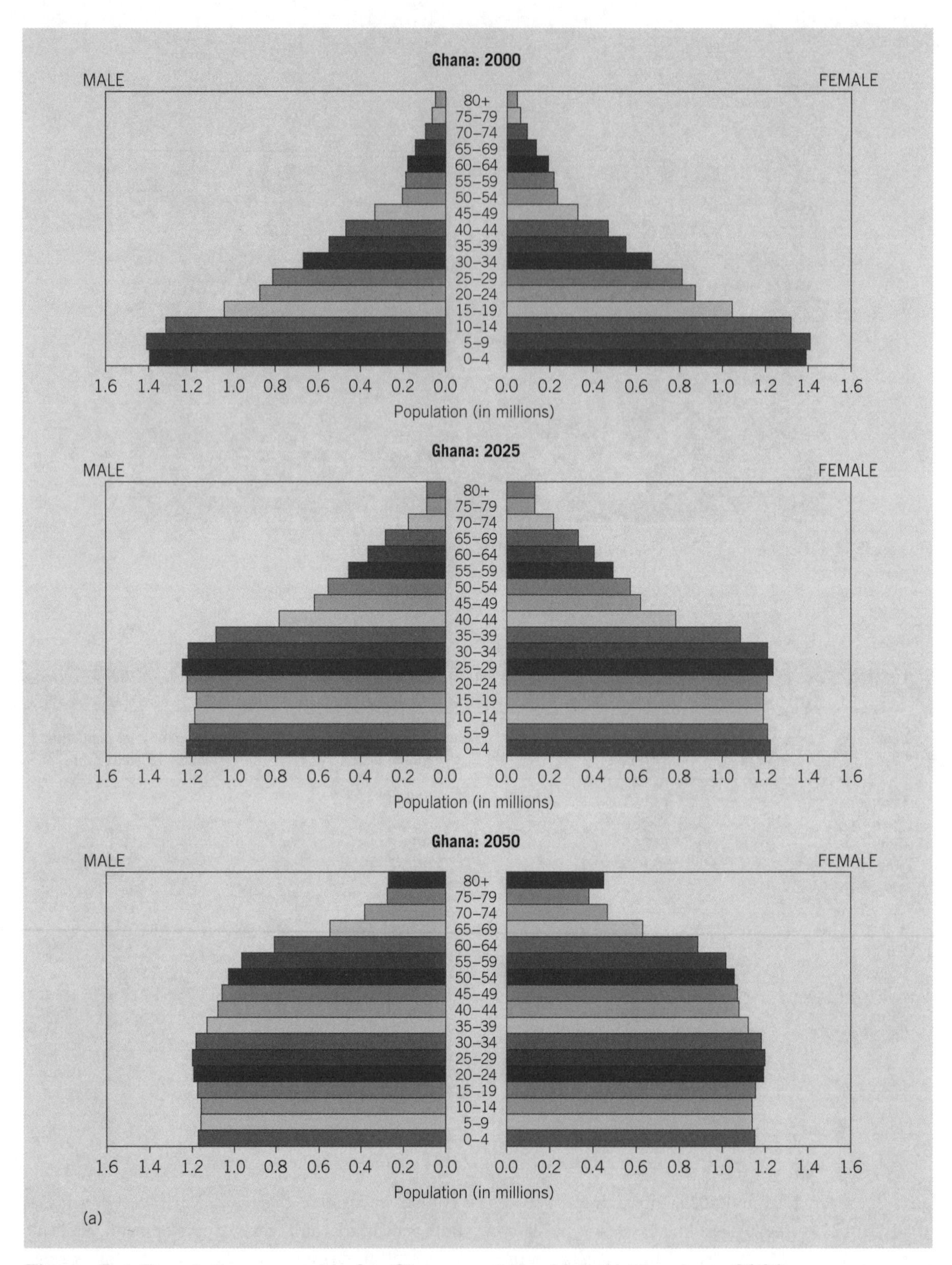

Figure 5.4 Population pyramids for Ghana and the United Kingdom, 2000
Source: Adapted from Binns (1994b)

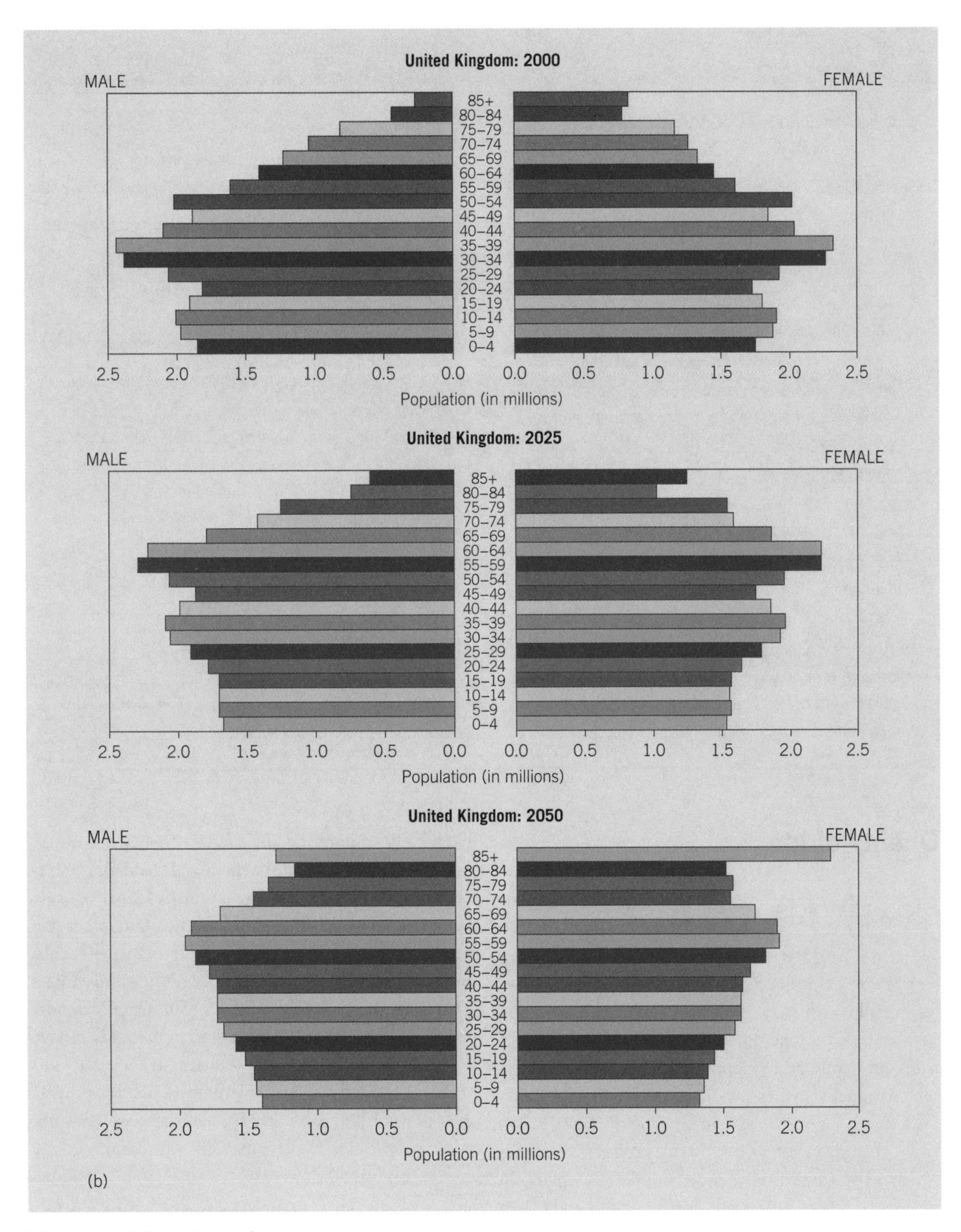

Figure 5.4 (continued)

Critical reflection

Grey Power New Zealand

Grey Power is a lobby organisation in New Zealand which works to 'promote the welfare and well-being of all those citizens in the 50 plus age group'.

Grey Power's aims and objectives

1. To advance, support and protect the welfare and well-being of older people.
2. To affirm and protect that statutory right of every New Zealand resident to a sufficient New Zealand Superannuation (retirement pension) entitlement.
3. To strive for a provision of a quality Health Care system to all New Zealand residents regardless of income and location.
4. To oppose all discriminatory and disadvantageous legislation affecting rights, security and dignity.
5. To be non-aligned with any political party, and to present a strong united lobby to all parliamentary and statutory bodies on matters affecting New Zealanders.
6. To promote and establish links with kindred organisations.
7. To promote recognition of the wide-ranging services provided by senior citizens of New Zealand.
8. To gain recognition as an appropriate voice for all older New Zealanders.

Source: http://www.greypower.co.nz/ (accessed 5 August 2007)

Grey Power New Zealand has successfully raised the profile of issues and concerns affecting the country's older people and achieves a good amount of media attention. There are organisations such as Grey Power New Zealand in other developed countries which have been successful to a greater or lesser extent in giving older people a voice in shaping government policy and improving their quality of life.

What do you believe is the potential role for such organisations in poorer countries? From literature and internet searches, can you find any examples of such organisations in poorer countries?

Quality of life

Households

Sadly, the world is a very unequal place and not all people enjoy equal access to basic needs and a satisfactory quality of life. As we have already seen, variables such as life expectancy and infant mortality vary greatly from one country to another. We should be aware that national statistics produced by the World Bank, the United Nations Development Programme (UNDP) and other agencies also conceal marked variations which exist within and between different regions in specific countries, within and between rural and urban areas and also between different communities and households. The household is still the key living unit in most developing countries, which also usually controls production, consumption and decision making. Households have become an increasingly important focus of study in recent years, since it has been recognised that 'the very success of development policy is likely to be undermined by a failure to view the household and family in a holistic manner' (Haddad, 1992: 1). Households in developing countries are typically larger than those in developed countries, and rather than merely comprising parents and children, they often also include grandparents, unmarried aunts and uncles, as well as more distant relatives. The term 'extended family' is often used to describe such households. Urban-based households frequently retain strong links with their rural relatives, particularly those who live in the ancestral village, to which regular visits are common and where urban dwellers may have farmland.

Household development cycle

All households pass through a developmental cycle, during which their size and composition may change and the all-important ratio between workers and dependants changes over time. At certain stages in this developmental cycle, households may be under considerable pressure and this may directly impact on the quality of life, for example in terms of disposable income and nutritional intake. Households with a high proportion of very young or old members may encounter difficulties, since relatively few active workers in the household may have to support a large number of dependants. From a study in rural Malaysia, Datta and Meerman (1980) have suggested that most households experience a four-stage per capita income cycle, which is linked closely with 'dietary stress' in the following way:

1 Immediately following marriage, both husband and wife tend to earn income. Per capita income is quite high, reducing the risk of dietary energy strees.
2 With the arrival of small children, there are more mouths to feed, but no additional income earners. Indeed, the mother usually devotes time to childcare, which frequently reduces her income-earning capabilities. The household suffers increased risk of energy stress.
3 As the children grow older, they are able to contribute to income earning and there are decreased demands on the mother's time for childcare. The risk of energy stress falls.
4 Finally, the income-generating children leave home and parents are at a greater risk of being ill, so there is increased risk of energy stress caused by a decline in command over food production.

As a result of this 'cycle', per capita household income in Malaysia typically fell as the age of the household head rose from 25.0 to 37.5 years and again from 50.0 to 62.5 years (Datta and Meerman, 1980).

In most developing countries there are also marked inequalities between households. More powerful households may be ethnically distinctive and better educated; they are key elements in the local, and possibly national, power structures, having good links with government officials, police, large landowners and traders; and they are frequently well endowed with assets and income. Poor households, however, are often less well educated, have poorer nutrition, are likely to be ignorant of the law and have few assets apart from their labour. Poor households are often vulnerable and susceptible to exploitation as they become locked into cycles of debt, and they may depend upon the assistance of one or more richer patrons. The concept of social capital (as discussed in Chapter 3) is useful in recognising the levels of loyalty and response that a household can tap during vulnerable or hard times.

Pressures and responsibilities within households

Different pressures and responsibilities also exist within households. In some rural African societies, for example, husbands and wives may live in separate houses in a village compound, have very different household responsibilities and quite separate incomes. Age and gender differentiation in household labour inputs and expenditure patterns has long been appreciated, with women often displaying greater concern for others, particularly children (Barrett and Browne, 1995).

The gender of the household member who directly benefits from development policy interventions may be relevant to the intra-household distribution of wealth and resources, and there is a suggestion that 'enhancing and securing female earnings appears to make sound policy sense' (Kabeer, 1992: 51). In fact, women's ability to maximise both their own welfare and the welfare of their dependants may be severely constrained by power relations within the household.

The contributions of women to household income and welfare need to be much better understood. It is estimated that 70 per cent of the food of tropical Africa is produced by women, and yet far too frequently it is assumed that all farmers are men: 'Many of the most demanding farm jobs, such as hoeing, weeding and harvesting are done by women, in addition to taking care of young children, collecting firewood and water and processing and cooking food' (Plate 5.4; Binns, 1994a: 88). Some of the many responsibilities of rural African women are depicted in Figure 5.5.

Many development schemes have failed and tensions within households and communities generated through misunderstanding the different roles and responsibilities of women and men. Household analysis should take note of relationships between households (inter-household), as well as the composition of, and relationships within, individual households (intra-household). For example, in eastern Sierra Leone among the Mende, Leach (1991) found that in addition to working on the household farm, women and men undertake separately a range of productive activities. Whereas men gained independent incomes from selling bushmeat, palm wine tapping and undertaking day labour, many women made individual rice swamps, cassava and groundnut farms and vegetable gardens. Some women also traded commodities, such as salt and dried fish, from their homes or were active

Plate 5.4 Women farmers harvesting rice in The Gambia
(photo: Tony Binns)

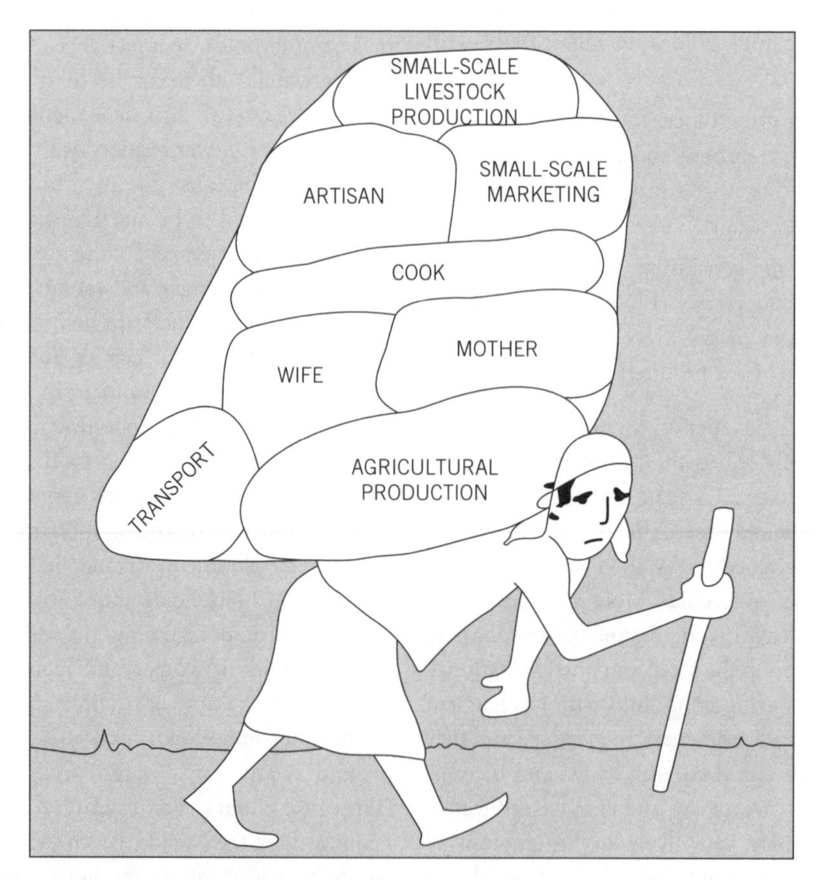

Figure 5.5 The multiple roles of Africa's rural women
Source: From www.coop.org/women/ica-ilo-manual/transparencies.htm, 2 May 2001

in local markets (Leach, 1991). In polygamous households, co-wives usually had separate individual enterprises and their consumption unit consisted of themselves and their own children.

Agricultural change has altered the division of labour between women and men, such that with the increasing production and sale of coffee and cocoa, 'wives are expected to assist with harvesting and processing, receiving only a discretionary gift of cloth and the right to glean fallen produce in return' (Leach, 1991: 47). Women have also increased their work on the family rice farm, whereas work on tree crops was regarded as a woman's duty for her husband. These responsibilities could take significant time away from a woman's independent vegetable gardening or trading. However, Leach found that the life cycle of individuals and households was significant in that middle-aged men, who had built up resources through farming or trade, might relieve their wives' workloads by recruiting labour to help with the tree crop harvest or to plant rice, or might possibly help with clearing the swamp and the provision of trading capital. As Leach comments, 'Such wives tend to complain less of overburden, although they are just as concerned to maintain their own income streams' (Leach, 1991: 48).

In the south Indian state of Kerala, state government policies have significantly improved the position of women relative to other parts of India. It is common for women to inherit and own land, giving them financial independence and power of their own. Keralese women are regarded as an asset rather than a drain on a family's finances, such that instead of paying out an expensive dowry when daughters marry, parents in Kerala receive money from the bridegroom's family.

Although Kerala is one of India's poorer states, state government policies have led to a female life expectancy of 75 years (compared with 64 years for India as a whole), infant mortality is only 14 per cent (India 56), and the female adult literacy rate is 85 per cent (India 48). The average age of marriage for women in Kerala is the highest in India and, together with the widespread use of contraceptives, this has led to smaller family sizes, with an average of only 1.8 children per family compared with 3.2 for India as a whole (Global Eye, 2002).

Children

Another important, but frequently ignored, factor in many household budgets and survival strategies concerns the cost and benefits of having children, especially the important work they do (Box 5.3). Pryer's study of some ultra-poor households in an urban slum in Khulna, Bangladesh, reveals how, faced with low, unreliable and seasonal incomes, households attempted to achieve a diverse employment profile with as many family members working as possible. The extreme poverty of the community is reflected by the fact that in 1984 no fewer than 67 per cent of children under five in the slum were second- or third-degree malnourished; that is, they were under 75 per cent of the expected weight for their age (Pryer, 1987: 133).

All seven severely undernourished households surveyed were deeply indebted, usually to landlords, employers, shops and neighbours. Since the only productive asset of these households is labour, all able-bodied adults and many of the children needed to do some form of paid work. Employment opportunities were invariably poorly paid in the informal sector, with men engaging in rickshaw pulling, petty trade, hawking and labouring on a daily basis. But during the monsoon months (June to September) less work was available. In two of the households women were major earners, typically engaged in domestic work or home-based piece-rate work, which involved long hours and was poorly paid both absolutely and in relation to male wages.

In the seven households, an average of 68 per cent of income was spent on food; the rest went towards fuel, rent and repayment of debt. Food intake was well below the recommended daily allowance.

One mother, whose husband died of tuberculosis, went out to work from 7 am to 4 pm and again from 6 pm to 11 pm every day, while her 10-year-old daughter assumed responsibility for childcare and domestic work. In another family, the wife, with a chronically sick husband who was unable to work, sold saris illegally on the black market, assisted by her 17-year-old niece, and her 12-year-old daughter was a servant in the main market. Given the family's circumstances, the wife was clearly the economic and social household head.

In both families, illness of the chief earners had led to sale of assets, indebtedness and deepening poverty, resulting in inadequate, unreliable and seasonal flows of food. One of the two women household heads was forced into heavy dependence on a patron who was her employer and landlord, while the other was engaged in illegal trading, both highly precarious strategies (Pryer, 1987). This study clearly illustrates both the different roles of members in attempting to ensure the household's survival, and the cycle of poverty and malnutrition from which it can be extremely difficult to escape.

BOX 5.3

Children: a neglected piece in the development jigsaw

Perceptions of children and childhood

Whereas children in developed countries can usually expect to be well fed, have good clean clothes, attend school until they are 16 or older, be well protected against ill-health and have a life span of well over 70 years, their counterparts in poor countries have few, if any, of these assurances. Furthermore, the welfare of children is a constant focus of attention in developed countries and a prime concern of many households, but children in developing countries are themselves expected to help in maintaining the household in so many ways. Fetching water and firewood, scaring birds from the fields, processing food and selling items in the urban informal sector, are just some of the many tasks which children commonly perform. Yet, all too often, the vital contribution which children make to the welfare of households in poor countries receives remarkably little attention. As Robson comments, 'By the age of 10–12 years, some children may contribute as much to household sustenance as adults' (Robson, 1996: 43).

Whereas in the last two decades there has been, quite appropriately, much debate and writing about the position and role of women, children still remain a somewhat neglected element in the development process. A report from Actionaid (1995) clearly shows that children and children's work have been neglected. Empirical research has also revealed the importance of this in different social and environmental contexts. The Actionaid report comments, 'The inclusion of children and their participation in the development process is as much their basic human right as is their right to health provision and protection from hazardous work and living conditions' (Actionaid, 1995: 5). With their numerous household responsibilities, so many children in poor countries are denied the stereotypical 'childhood' which young people experience in rich countries.

Children in conflict situations

When children are caught up in conflict situations such as in Afghanistan and Sierra Leone, normal daily routines of school, housework and play are completely transformed, and in some cases children (boys and girls) have been drawn into using weapons in combat scenarios. As Amnesty International reflects on the situation in Afghanistan, two decades of civil war from the 1980s meant that,

> **Families have been torn apart in the fighting, many children have lost parents or siblings. Others have been forced to flee from their homes, either abroad or to other parts of Afghanistan. All have suffered from disrupted schooling and economic hardship. The physical, emotional and mental development of generations of Afghanistan's children has been severely affected by the ongoing fighting.**
>
> **(Amnesty International 2002a: 1)**

A study undertaken in Afghanistan by UNICEF in 1997 indicated that most of Kabul's children were suffering from serious traumatic stress, with 72 per cent of those children interviewed having experienced the death of a relative between 1992 and 1996. Virtually all the children had witnessed acts of violence, and two-thirds of them had seen dead bodies or body parts. As many as 90 per cent of the children feared that they would actually die in the conflict (UNICEF, 1998).

Meanwhile, in Sierra Leone an estimated 8000 children under the age of 18 fought in the civil war during the 1990s. Many of these children were abducted from their homes and families and forced to fight. During the rebel invasion of the capital, Freetown, in January 1999, when some 2000 civilians were killed, over 500 people had limbs severed and the raping of girls and women was widespread. It was estimated that some 10 per cent of the rebel combatants were children. Child combatants lived in constant fear of being beaten and killed and their individual stories are often harrowing (Amnesty International, 2002b). In February 2002, the Optional Protocol to the UN Convention on the Rights of the Child came into force, representing a strong expression of international views against the use of children as soldiers. However, by July 2002 only a small number of states had agreed to the Protocol and, realistically, it is unlikely it will be respected in the chaos of the most serious conflict situations.

BOX 5.3 (continued)

The status and roles of children

Examining the status and role of children more broadly, research from a variety of different contexts has indicated that children need to be viewed as 'social actors', and the conventional model of childhood must be modified to support the notion of a 'plurality of pathways to maturity' (Save the Children Fund, 1995). In relation to this, gender is often important in determining children's roles and responsibilities. For example, a study in Jamalpur, a rural area in Bangladesh, revealed that boys are mainly involved in cultivating family plots of land and occasionally they go to neighbouring farms to work as wage labourers. They enjoyed their work and in earning cash from casual labour they felt that they were an asset to their families. Girls in the same area, however, were somewhat resentful that their work, which involved mainly household duties and looking after younger children, was invisible. The girls saw education as a way of earning greater respect from their families and communities. Both girls and boys agreed that they should work hard to help their parents and to contribute to family income and/or welfare (Actionaid, 1995: 49).

Children and education

For children in many poor countries, attending school is regarded as a privilege rather than a right. Commonly, far fewer girls than boys in developing countries go to school and this is reflected later in life in the form of a considerable differential in adult literacy rates between males and females. The West African state of Niger, with one of the poorest adult literacy records in the world, had a male adult literacy rate in 2000–2004 of 43 per cent, while the rate for females was only 15 per cent.

Although levels of educational attainment are better in India, the gender differentiation is significant, with a female adult literacy rate of 48 per cent compared with the figure for males of 73 per cent (UNICEF, 2006). This differential is also apparent in some richer countries where Islam is the dominant faith, and where girls are more likely to stay at home to assist their mothers with household chores rather than go to school. In Libya, whereas the male adult literacy rate is 87 per cent, the female rate is only 67 per cent. Saudi Arabia similarly has a large male–female literacy differential, with a male literacy rate of 91 per cent and a female rate of 70 per cent (UNICEF, 2001). In the mainly Moslem West African country of Chad, the female adult literacy rate in 2004 of 12.8 per cent was only 31 per cent of the male rate (UNDP, 2006a).

Whereas girls in poor countries may often attend the first few years of primary school, the proportion of girls in school classes frequently falls sharply among higher age groups. In Uganda, while 87 per cent of girls were enrolled in primary schools in 2000–2004, only 12 per cent were enrolled in secondary schools. In Saudi Arabia, the female enrolment falls from 57 per cent in primary schools to 51 per cent in secondary schools (UNICEF, 2001).

A study in the Kyuso area of Kitui district in eastern Kenya found that the major reason for school non-enrolment and drop-out was poverty and the high costs of school materials and uniforms. During drought conditions, parents frequently withdraw their children from school to assist with water collection and to look after younger children (Actionaid, 1994). In many developing countries, more and more children, particularly girls, are being withdrawn from school, as investment in their education is seen as being lost when they marry.

Gender and the roles of children of different ages need to be much better understood. Development agencies should seek to identify the effects of development projects on children and to reduce the need for children to be sent out to work as part of household survival strategies. Appropriate forms of education should be introduced, and, if necessary, this should be vocationally oriented and undertaken in the workplace.

Some progress has been made, notably through the International Year of the Child in 1979, the 1989 UN Convention on the Rights of the Child, and the 1990 UNICEF-sponsored World Summit on Children. These were important landmarks in attempting to raise the profile of children in all communities, but in some countries there is still a very long way to go in improving the lives of children.

Table 5.4 Commitment to health: access, services and resources

Country	Public health expenditure (% of GDP)	Physicians (per 100,000 people)	Population with sustainable access to improved sanitation (%)		Population with sustainable access to an improved water source (%)
	2003	1990–2004	1990	2004	2004
Bangladesh	1.1	26	20	39	74
Brazil	3.4	115	71	75	90
China	2.0	106	23	44	77
India	1.2	60	14	33	86
Jamaica	2.7	85	75	80	93
Japan	6.4	198	100	100	100
Mali	2.8	8	36	46	50
Sierra Leone	2.0	3	..	39	57
Sweden	8.0	328	100	100	100
United Kingdom	6.9	230	..	..	100
United States	6.8	256	100	100	100
Zimbabwe	2.8	16	50	53	81

a. Data refer to the most recent year available during the period specified.

Source: column 1: calculated on the basis of data on health expenditure from WHO (2006b); correspondence on health expenditure, May, Geneva.
column 2: WHO (2006c); *World Health Statistics 2006*, Geneva.
columns 3,4,5: UN (2006c); Millennium Indicators Database. Department of Economic and Social Affairs, Statistics Division, New York. (http://mdgs.un.org), accessed July 2006, based on a joint effort by the United Nations Children's Fund (UNICEF) and the World Health Organization (WHO).

Health and health care

The health status of a population, or elements of a population, can be crucial in the development process. It might be argued that a healthy population is more able to contribute to development efforts and will also be better placed to benefit from the fruits of these efforts. We have already seen that many developing countries have high rates of infant mortality and their life expectancy levels are well below those of richer, more developed countries.

Variables such as these are often a good reflection of the health status of a population and the quality of health care. In many poor developing countries health facilities are inaccessible to a large proportion of the population, especially those living in remoter rural areas. Where hospitals and clinics do exist, there are frequently shortages of trained health workers, drugs and basic equipment.

Providing good health care for everyone is an expensive undertaking for the governments of poor countries. In many parts of the world, but notably tropical Africa, religious missions continue to play a crucial role in providing health care in certain areas, together with a variety of non-governmental organisations (NGOs). Many developing countries still cling to a top-down style of health care inherited from the colonial period, with a considerable proportion of health expenditure being allocated to a few key hospitals, particularly in the main towns and capital city, whereas the rural areas remain relatively neglected. Table 5.4 shows some key variables relating to the status of health care in selected countries.

The proportion of populations with adequate sanitation varies widely from 100 per cent in the richer countries to only 33 per cent in India. In 1999 only 44 per cent of Sierra Leone's population had access to essential drugs, while in India and Brazil the figures are even lower (India 35 per cent and Brazil 40 per cent). It should be remembered that these are countrywide figures and they therefore conceal considerable spatial and social inequalities in service provision. In terms of national expenditure on health, all the richer countries spend more than 5 per cent of their GDP on public health care, whereas in India it is only 1.2 per cent of GDP and 1.1 per cent in Bangladesh.

In China under Mao Zedong various efforts to improve public health were introduced, particularly from 1968 onwards, when a central government document called the 'June 26th Directive on Public Health' demanded that the focus of medical and public health work should be transferred to the countryside. Mao suggested that although city hospitals should keep some doctors, a greater proportion should be sent to work in the villages (where 85 per cent of the population lived) to teach medical knowledge to the peasant youth (Endicott, 1988: 157).

Although these 'barefoot doctors', as they were popularly called, had a lot to learn, Mao believed they were better than 'fake doctors' and 'witch doctors'. Furthermore, he argued that villages could afford them as medical funding was redistributed such that the bulk was now directed to prevention and cure of the common, frequent and widespread diseases. Village commune hospitals received much more finance, which could be used to purchase equipment such as X-ray machines, as well as to develop both Chinese and Western medicine. There was also a campaign to extend the recruitment of barefoot doctors, midwives and medical orderlies.

The effects of this campaign are partly reflected in statistics relating to the number of doctors for every 100,000 of the population (Table 5.4). In the period 1990–2004, China had 106 doctors per 100,000 people, not far short of the figure for Brazil (115). In this regard China is well ahead of many other countries in the UNDP's 'medium human development' group, which have significantly higher GDP per capita figures than China.

The impact of these policies in China was clearly seen in rural areas such as Shifang County in Sichuan Province:

> **In three-month, sometimes six-month, courses, qualified doctors, sent down to the countryside by rotation during the Cultural Revolution, trained 658 barefoot doctors in basic first aid, Chinese medicine, acupuncture, the use of thermometers, the dispensing of vaccines by injection and drugs for influenza, stomach upsets and other common ailments. [As a result] the total number of medical personnel in the County rose from 592 in 1965 to 3420 a decade later . . . [The barefoot doctor initiative represented] . . . a good start on creating an accessible, experimental, non-elitist public health system biased in favour of prevention.**
>
> **(Endicott, 1988: 158)**

A notable achievement in Shifang County's health programme was the virtual eradication of schistosomiasis (bilharzia), such that the number of people affected was reduced from 5700 in 1959 to only four in 1982. By the end of the 1970s, 85 per cent of Chinese villages had a health station staffed by one or more barefoot doctors.

Health problems in developing countries are frequently the result of poverty, notably inadequate or poor-quality food and water and the lack of proper sanitation. Children are particularly susceptible to diarrhoea, and measles also claims many victims, whereas vitamin A deficiency may lead to blindness and infection, particularly after measles. Iron deficiency often leads to anaemia, causing weakness and particular risks for newborn children. Children are also very susceptible to a number of nutrition-related diseases such as pellagra, beriberi, rickets and kwashiorkor.

Diseases associated with water are a major problem in many developing countries, and there is some evidence that the expansion of irrigated agriculture has encouraged their spread with large areas of slow-moving water in dams and reservoirs. Schistosomiasis is transmitted by snails in slow-moving water, whereas onchocerciasis (river blindness), which is endemic to large parts of tropical Africa, is also associated with water and is transmitted by the black fly.

Possibly the most serious threat to health in tropical regions is malaria, transmitted by mosquitoes which breed close to stagnant or slow-moving water. It is estimated that in tropical Africa alone as many as 200 million people are affected by malaria, which weakens victims and lowers their resistance to a wide range of other possible infections. Although draining swamps and spraying pools with insecticide might help, mosquitoes are becoming resistant to certain chemicals and anti-malarial drugs. Mosquitoes are also responsible for transmitting dengue fever and yellow fever.

Key idea

Malaria: the scourge of Africa

Malaria is the main cause of childhood mortality in sub-Saharan Africa – a child dies from malaria every 20 seconds in Africa. Between 1 and 3 million child and adult deaths each year in Africa are due to malaria, representing 80 per cent of the global total. The economic and development costs of malaria are considerable, slowing economic growth by about 1.3 per cent annually and contributing to a decline in per capita GDP in many sub-Saharan African countries.

Malaria is a parasitic infection which is transmitted to humans through the bites of infected female *Anopheles* mosquitoes. The parasite spreads rapidly through the bloodstream to the liver and then settles in the red blood cells, where it multiplies and emerges in bursts of new organisms. The parasites can cause considerable damage to the kidney, liver and nervous system. Children and adults who have not recently been infected, and have therefore not developed natural immunity, can die in a short time from cerebral malaria, and others may later die from anaemia or liver and kidney failure. If left untreated, up to 20 per cent of infected persons will die.

A number of possible strategies exist for dealing with malaria:

- Controlling the breeding of mosquitoes by getting rid of stagnant water where they breed.
- Using chemical pesticides such as DDT, but mosquitoes have in some cases developed a resistance to the pesticides.
- Limiting human exposure to mosquito bites, by using nets and window screens treated with insecticide.
- Using anti-malarial prophylactic drugs, particularly for travellers visiting infected areas for short periods of time. But mosquito resistance to the drugs is a common problem.
- Developing a vaccine to combat the disease. Work on producing a vaccine for malaria has been going on for some time, but the complexity of the organism is making this difficult. A team of scientists is working on developing a vaccine through the European Malaria Vaccine Initiative (EMVI), which was started in 1998 with funding from the European Commission, Denmark, Ireland, The Netherlands, Norway and Sweden http://www.emvi.org/. In 1999, another group, the PATH Malaria Vaccine Initiative (MVI) began work, funded by the Bill and Melissa Gates Foundation http://www.malariavaccine.org/about-mvi.htm.

In the absence of complete protection against malaria, African communities have often developed their own strategies for dealing with the problem of malaria. For example, in the Tigray region of Ethiopia 'mother coordinators' have been trained to educate other mothers about the symptoms of fever and malaria. Mothers were provided with low-cost chloroquine and information on how to administer the drug. By educating mothers in this way under-five mortality has been reduced by 40 per cent, and the burden on hospitals in dealing with severe cases of malaria has been reduced (World Bank, 2004).

The health and nutritional status of households and household members may vary over time. For example, maternal health is an area of much concern. The 1994 International Conference on Population and Development (ICPD) held in Cairo transferred the focus on women's, and particularly maternal, health from a demographically driven approach to a human rights focus. Concern was expressed about the continuing high maternal mortality ratio in many poor countries, which leads to some 514,000 maternal deaths a year during pregnancy and after childbirth, the majority occurring in the first 24 hours after delivery. In 2005, globally, only an estimated 63 per cent of births were attended by skilled health personnel and in poorer countries, most notably in sub-Saharan Africa, this figure is considerably lower (UNICEF, 2006). In Malaysia, where the government signed the Convention on the Elimination of all Forms of Discrimination Against Women in 1995, considerable success has been achieved in reducing the maternal mortality ratio, largely through home deliveries being conducted by trained community midwives, and by establishing a nationwide system to detect early complications in pregnancies and closely reviewing all maternal deaths (World Bank, 1999).

We have already seen how a particular stage in a household's life cycle may affect income and nutrition. This longer-term variation may be compounded by marked seasonal pressures in tropical countries, where the period of hardest work commonly coincides with the rainy season, which is also the time when the occurrence of many diseases, such as malaria, increases. Furthermore, in many communities, towards the end of the rainy season, but before the harvest, food stocks are often getting low and the quantity and quality of food intake declines. This time of year, sometimes called the 'hungry season', is associated with greater susceptibility to infection, though people still have to work hard in the fields. Certain elements of communities and households, such as pregnant and breast-feeding women and young children, may suffer disproportionately at such times.

HIV/AIDS

Since the early 1980s the HIV/AIDS pandemic has introduced a sinister new dimension to the world health scene, and it is unfortunately the poorer countries and people who have suffered disproportionately. AIDS (acquired immune deficiency syndrome) is a disease in which the body's natural protection or immune system is damaged. Since the first AIDS case was reported in the USA in 1981, 'the world has been facing the deadliest epidemic in contemporary history' (UN, 2000a: 76). The extensive spread of human immuno-deficiency virus (HIV), the aetiologic agent that causes AIDS probably began as early as the 1960s, but spread rapidly during the mid to late 1970s and early 1980s. Two strains of HIV have been identified (HIV1 and HIV2) and infection occurs when blood from an infected person passes directly into another's bloodstream, and also through sexual intercourse. In almost all cases, those infected by HIV develop AIDS, which is inevitably fatal.

BOX 5.4

Entitlements, food security and nutrition

Much interest has been shown in recent years in the question of 'food security' – what it is, how it can be achieved, and the reasons why some individuals, communities and geographical areas are 'food secure' while others are not. This complex debate involves examining aspects of nutrition, vulnerability, coping strategies, and what Amartya Sen has called 'entitlements'. There have been many attempts to define 'food security'. The World Bank (1986) suggested that it means, 'Access by all people at all times to enough food for an active, healthy life'. Kennes (1990), meanwhile, believes it is 'The absence of hunger and malnutrition'. Maxwell (1988: 10) provides a more detailed definition:

> **A country and people are food secure when their food system operates efficiently in such a way as to remove the fear that there will not be enough to eat . . . In particular, food security will be achieved when the poor and vulnerable, particularly women and children and those living in marginal areas, have secure access to the food they want.**

Whether or not individuals or communities are 'food secure' or 'food insecure' requires very careful investigation beyond the issue of quantifying food supply and must take into account various 'entitlement' relationships. Sen's important work on entitlements starts with the provocative statement, 'Starvation is the characteristic of some people not having enough food to eat. It is not the characteristic of there being not enough food to eat. While the latter can be a cause of the former, it is but one of many possible causes' (Sen, 1981: 1). The entitlement approach provides a useful framework for analysing the relationship between rights, interpersonal obligations and individual entitlement to things. An individual's 'entitlement set' is a way of characterising his or her overall command over things, taking note of all relevant rights and obligations.

Some of these entitlement relationships might include:

- 'trade-based entitlement', where an individual (or group) is entitled to own what they obtain by trading something that they own with a willing party;
- 'production-based entitlement', where an individual (or group) is entitled to own what they get from arranging production using their own resources,

▶

BOX 5.4 (continued)

or resources hired from willing parties meeting the agreed conditions of trade;

- 'own-labour entitlement', where an individual (or group) is entitled to their own labour power, and thus to the trade-based and production-based entitlements related to that labour power;
- 'inheritance and transfer entitlement', where an individual (or group) is entitled to own what is willingly given to them by another who legitimately owns it, possibly to take effect after the latter's death (if so specified by them) (Sen, 1981).

These relationships may exist and develop between individuals, or groups such as households. In relating this to food security, Blaikie suggests, 'there have been famines where the total food availability has not declined at all, but instead there has been a failure of effective demand (not need!) for food through the failure of their entitlements. People simply became too poor to afford food which is physically available' (Blaikie, 2002: 302).

The entitlement approach has helped in understanding the differential impacts of famines and how famines might occur at times when food is plentiful. It has also drawn attention to the functioning of markets and how certain catastrophic events (such as droughts or floods) do not necessarily lead to famine and food insecurity.

However, Devereux and Maxwell (2001) indicate some shortcomings of the entitlements approach, suggesting that in reality entitlements are often much less clear than the model suggests, and they may occur outside the legal framework suggested by Sen. The model has little to say, for example, about 'informal' or 'illegal' entitlements, which are often most important at times of food security crisis. Research has shown that food-insecure households are often remarkably resilient, adopting many ingenious coping strategies for securing nutrition, for example, borrowing food, gathering wild foods or disposing of assets such as livestock to purchase food. It is suggested that social capital, intra-household distribution and well-adapted social networks are also not sufficiently articulated in the entitlement approach, which gives prominence to formal exchange mechanisms.

Much recent work on food security has focused on individuals and households, examining the diversity of causes within different situations and the nature and effectiveness of the coping strategies adopted. In an attempt to reduce household vulnerability to nutritional insufficiency, food security planning has been moving away from a broad 'blue-print' approach towards developing flexible, locally based responses which build upon indigenous knowledge and skills (Maxwell, 1996).

BOX 5.5

Globesity

In some parts of the world a new health problem is emerging – obesity. While this has long been a concern in the world's richer countries, recent evidence suggests that it is also becoming a major health issue in some poorer countries. Obesity is a disease with many causes that leads to an imbalance between energy intake and output, resulting in the accumulation of large amounts of body fat. It is most commonly measured as excessive weight for a given height, using the body mass index (BMI) – weight in kilograms (kg) over height squared (m^2). The World Health Organization (WHO) defines overweight as a BMI between 25.0 and 29.9 kg/m^2 and obesity as a BMI of 30.0 kg/m^2 or greater. The World Health Organization estimated that in 2005 approximately 1.6 billion adults (c.25 per cent of the world's population) were overweight and at least 400 million adults (c.6 per cent) were obese (WHO, 2006a).

Obesity is rare in South Asia, and although generally rare in sub-Saharan Africa, it is becoming an increasing problem among urban and educated women. Some 44 per cent of black women living in the Eastern and Western Cape provinces of South Africa were found to be clinically obese (Grummer-Strawn *et al.*, 2000a). In Latin America, obesity was once restricted to those with high socio-economic status, but in Brazil and Mexico it is emerging as a feature in poor households. Childhood obesity is a

BOX 5.5 (continued)

problem because it generally leads to adult obesity (Grummer-Strawn *et al.*, 2000b; Martorell, 2001). In 2005 there were globally an estimated 20 million overweight children under the age of five. In China, the number of overweight people increased from less than 10 per cent to 15 per cent in three years. There is much concern in China that the one-child policy is leading to overfeeding and obesity in children, particularly boys. Some of the Pacific islands have the greatest incidence of obesity, with an estimated 65 per cent of men and 77 per cent of women being obese in Samoa and Tonga.

Global rises in the numbers of overweight and obese people are due to a variety of factors, including a shift in diet towards a greater intake of foods which are high in fat and sugars, and a trend towards less physical activity due to the increasingly sedentary nature of lifestyles. Higher levels of obesity in

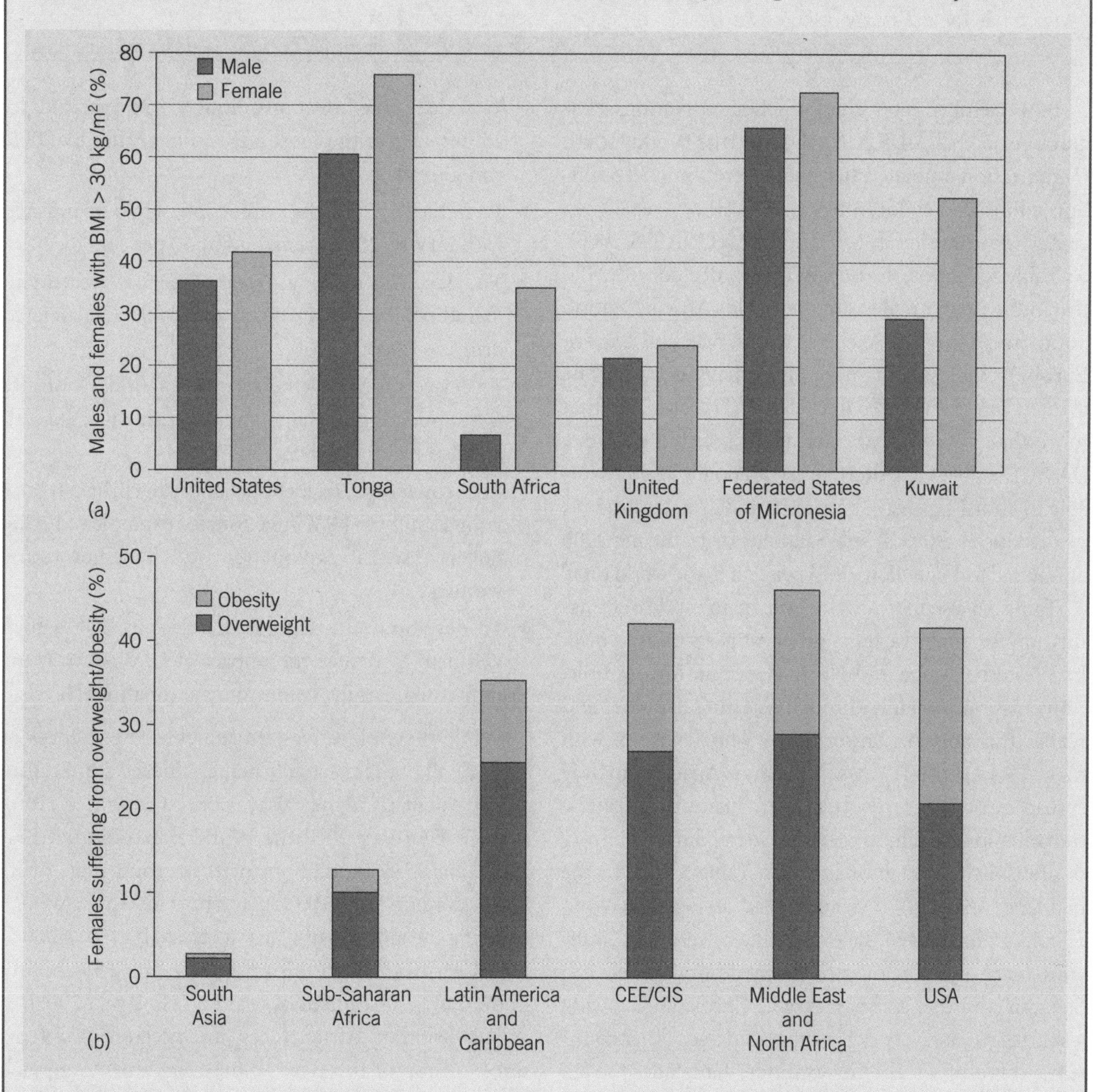

Figure 5.6 (a) Levels of obesity among men and women, according to selected countries

Source: From *The Independent* (2002)

(b) Overweight and obesity in women, according to world regions

Source: From Grummer-Strawn *et al.* (2000a) Adapted by permission from Macmillan Publishers Ltd: *European Journal of Clinical Nutrition* 54, (3), p.250 © 2000

BOX 5.5 (continued)

developing countries are leading to a greater incidence of diabetes, hypertension, stroke, cardiovascular disease and some cancers. Mortality rates from these diseases are increasing in poorer countries, whereas they were once regarded as diseases of the richer developed countries. However, since the main focus continues to be on undernutrition in poor countries, the incidence of obesity and its link with chronic diseases is receiving little attention. In Singapore, which has the necessary resources to deal with the problem, the government introduced a 'Trim and Fit Scheme' in 1992, a ten-year programme involving teacher education, a reduction of sugar content in children's beverages and an increase in physical activity during the school day. More attention needs to be given to the problem of obesity with public health awareness campaigns, more recreational facilities in urban areas, nutrition labelling and agricultural research programmes that will lead to lower-fat meat and other products.

By December 2006, the Joint United Nations Programme on HIV/AIDS (UNAIDS) and the World Health Organization estimated that there were some 39.5 million adults and children living with HIV, over 60 per cent of whom are in sub-Saharan Africa (UNAIDS, 2006). As Table 5.5 shows, the bulk of those infected with HIV live in the poorer regions of the world (Africa; Central, South and Southeast Asia; Latin America; and Eastern Europe), with sub-Saharan Africa having by far the greatest infection rate (5.0 per cent). It is estimated that more than 7500 Africans are newly infected each day (UNAIDS, 2006). Furthermore, with poor countries unable to afford expensive anti-retroviral drug treatments, deaths due to AIDS in sub-Saharan Africa during 2006 amounted to 1.6 million, or 76 per cent of the world total.

Table 5.5 also indicates that the main mode of transmission for adults varies in different parts of the world. In Western Europe and North America, for example, transmission is mainly through injecting drug use and sexual transmission among men who have sex with men. However, HIV in sub-Saharan Africa is mainly spread through heterosexual intercourse and perinatal transmission, which can occur *in utero*, during delivery or after birth through breast milk. Table 5.6 shows the incidence of HIV/AIDS among adults aged between 15 and 49 in selected developed and developing countries. Once again the overwhelming dominance of sub-Saharan countries is apparent, with Zimbabwe showing a staggering, and very depressing, lead over other countries, some of which (for example, Bangladesh, Mali and Sierra Leone), in relation to other development indicators, have both lower Human Development Indexes and levels of per capita GNP than Zimbabwe.

In June 2001 the UN General Assembly Special Session on HIV/AIDS (UNAIDS, 2001) agreed on six main targets:

- To reduce HIV infection among 15–24-year-olds by 25 per cent in the most affected countries by 2005, and globally by 2010.
- To reduce by 2005 the proportion of infants infected with HIV by 20 per cent, and by 50 per cent by 2010.
- To develop national strategies for treatment, including the provision of affordable HIV-related drugs by 2003.
- To develop and implement by 2005 national strategies for supporting orphans and children infected and affected by HIV/AIDS.
- To formulate strategies by 2003 which will help reduce vulnerability to HIV infection, for example reducing poverty, sexual exploitation and by empowering women.
- To develop multi-sectoral strategies by 2003 which will help to reduce the impact of HIV/AIDS at the individual, family, community and national levels.

While these objectives are laudable, the chances of some or all of them actually being achieved seem rather slim, particularly in the world's poorest countries. However, in February 2002 the World Bank approved an additional US$500 million for the second stage of its Multi-Country HIV/AIDS Programme for Africa (MAP), bringing the amount of its no-interest HIV/AIDS lending to Africa through this programme to US$1 billion in the 2001–2002 financial year (World Bank, 2002b).

Sub-Saharan Africa faces the greatest problems since, on top of the general impoverishment and lack of any meaningful 'development' in many countries over the last two decades, HIV/AIDS adds further to the challenges which cash-starved African governments face today. In Botswana, Malawi, Zambia and Zimbabwe, AIDS is now considered to be the leading cause of death between the ages of 15 and 39. It was estimated

Table 5.5 Regional HIV/AIDS statistics and features, December 2007

Region	Epidemic started	Adults and children living with HIV	Adults and children newly infected with HIV	Adult (>15) prevalence (%)	Adult and child deaths due to AIDS	Main mode(s) of transmission for adults living with HIV/AIDS
Sub-Saharan Africa	Late 1970s, early 1980s	22.5 million	1.7 million	5.0	1.6 million	Hetero
Middle East and North Africa	Late 1980s	380,000	35,000	0.3	25,000	Hetero, IDU
South and Southeast Asia	Late 1980s	4.0 million	340,000	0.3	270,000	Hetero, IDU
East Asia	Late 1980s	800,000	92,000	0.1	32,000	
Oceania	Late 1970s, early 1980s	75,000	14,000	0.4	1300	
Latin America	Late 1970s, early 1980s	1.6 million	100,000	0.5	58,000	MSM, IDU, Hetero
Caribbean	Late 1970s, early 1980s	230,000	17,000	1.0	11,000	Hetero, MSM
Eastern Europe and Central Asia	Early 1990s	1.6 million	150,000	0.9	55,000	IDU
Western and Central Europe	Late 1970s, early 1980s	760,000	31,000	0.3	12,000	MSM, IDU
North America	Late 1970s, early 1980s	1.3 million	46,000	0.6	21,000	MSM, IDU, Hetero
TOTAL		33.2 million	2.5 million	0.8	2.1 million	
Range		*(30.6–36.1 million)*	*(1.8–4.1 million)*	*(0.7–0.9)*	*(1.9–2.4 million)*	

Source: UNAIDS (2007) *Aids epidemic update*, UNAIDS/WHO (Table 1, p. 7)

Table 5.6 Incidence of HIV in selected countries and regions, 2003 and 2005

Country	Population ages 15–49 living with HIV (%)	
	2003	2005
Bangladesh	<0.1	<0.1
Brazil	0.5	0.5
China	0.1	0.1
India	0.9	0.9
Jamaica	1.5	1.5
Japan	<0.1	<0.1
Mali	1.8	1.7
Sierra Leone	1.6	1.6
Sweden	0.2	0.2
United Kingdom	0.2	0.2
USA	0.6	0.6
Zimbabwe	22.1	20.1

Source: WHO (2007)

in 1994 that in Nairobi (Kenya) and Abidjan (Côte d'Ivoire) the prevalence of HIV among prostitutes was well over 80 per cent (US Census Bureau, 1994). Although the problem is greatest in urban areas, it is not insignificant in rural areas. But given the fact that Africa is still mainly rural, in absolute numbers AIDS cases in rural areas predominate, though accurate data are difficult to obtain.

The HIV/AIDS epidemic has shifted south in Africa since the early 1980s, when the greatest incidence was in a band from West Africa across the continent to the Indian Ocean. But, while infection rates in West Africa stabilised at lower levels, by the late 1980s infection rates in southern African countries had increased, such that in 2002 they were the highest in the world (Daniel, 2000). According to the United Nations Development Programme (UNDP, 2006), no fewer than 24.1 per cent of adults in Botswana were infected with HIV in 2005, whilst Swaziland had a figure of 33.4 per cent, the highest level of infection in the world (Table 5.7).

Table 5.7 HIV/AIDS, prevalence and mortality in selected countries

Country	HIV prevalence (% ages 15–49); range of estimate given in brackets 2005	HDI rank 2007	Life expectancy at birth (with AIDS) 2000–2005	Years lost with AIDS	Deaths of adults and children due to AIDS (thousands); range in brackets 2005
Swaziland	33.4 (21.2–45.3)	146	32.9	30.7	16 (10–23)
Botswana	24.1 (23.0–32.0)	131	36.6	32.1	18 (17–25)
Lesotho	23.2 (21.9–24.7)	149	36.7	27.2	23 (20–27)
Zimbabwe	20.1 (13.3–27.6)	151	37.2	26.3	180 (120–250)
Namibia	19.6 (8.6–31.7)	125	48.6	19.9	17 (8–27)
South Africa	18.8 (16.8–20.7)	121	49.0	18.0	320 (270–380)
Zambia	17.0 (15.9–18.1)	165	37.4	16.9	98 (77–120)
Mozambique	16.1 (12.5–20.0)	168	41.9	10.7	140 (100–200)
Malawi	14.1 (6.9–21.4)	166	39.6	17.1	78 (38–120)
Central African Republic	10.7 (4.5–17.2)	172	39.4	14.0	24 (10–39)
Gabon	7.9 (5.1–11.5)	124	54.6	8.4	5 (3–7)
Côte d'Ivoire	7.1 (4.3–9.7)	164	46.0	8.0	65 (39–96)
Uganda	6.7 (5.7–7.6)	145	46.8	9.7	91 (54–130)
Tanzania, U. Rep. of	6.5 (5.8–7.2)	162	46.0	12.0	140 (110–180)
Kenya	6.1 (5.2–7.0)	152	47.0	13.5	140 (110–170)
Cameroon	5.4 (4.9–5.9)	144	45.8	7.8	49 (36–55)
Congo	5.3 (3.3–7.5)	140	51.9	8.3	11 (7–17)
Haiti	3.8 (2.2–5.4)	154	51.5	7.8	16 (10–24)
Cambodia	1.6 (0.9–2.6)	129	56.0	4.0	16 (9–26)
Ukraine	1.4 (0.8–4.3)	77	66.1	1.9	22 (13–33)
India	0.9 (0.5–1.5)	126	63.1	1.4	n/a (270–680)
USA	0.6 (0.4–1.0)	8	77.3	0.3	16 (10–24)
Brazil	0.5 (0.3–1.6)	69	70.3	0.7	14 (8–21)

Sources:

HIV prevalence	UNDP (2006)
HDI rank	UNDP (2006)
Life expectancy with AIDS	UN (2005)
Years lost to AIDS	UN (2005)
Deaths to AIDS	WHO (2007)

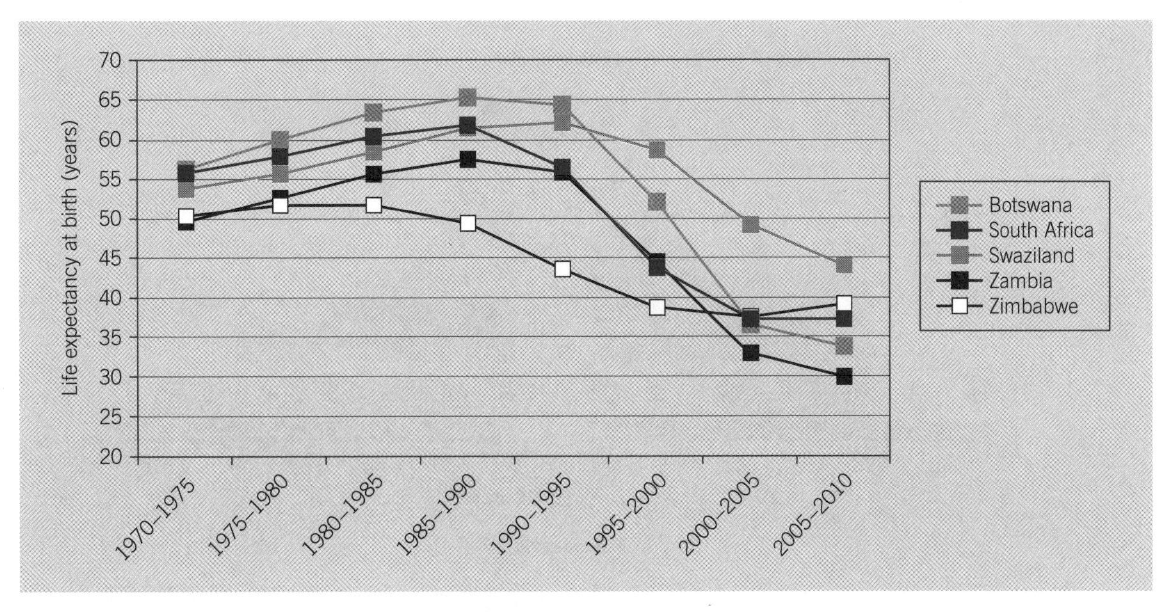

Figure 5.7 Impact of AIDS on life expectancy in five African countries, 1970–2010
Source: UNAIDS (2006) *Report on the global AIDS epidemic*, Figure 4.1

Although Botswana has a relatively small population of only 1.8 million, it was reckoned that between 1995 and 2015 there would be 385,000 additional deaths due to AIDS.

The significant effect of HIV/AIDS on life expectancy is also shown in Table 5.7 where, in the case of Swaziland, life expectancy with AIDS is only 32.9 years, representing 30.7 years lost to AIDS. In many African countries there was a steady increase in life expectancy before the 1980s, followed by a sharp decline due to the impact of HIV/AIDS (see Figure 5.7 above).

Uganda, one of the earliest countries affected by the epidemic, is unusual in having experienced an improvement in life expectancy since the mid 1990s, due to a decline in the overall prevalence of HIV/AIDS from 14 to 6.7 per cent between 1990 and 2005. This has resulted from a massive AIDS awareness programme, which was initiated in 1986, four years after AIDS was first recognised in the country. Political commitment has been at the highest level, from President Museveni downwards, such that every ministry has an AIDS Control Programme. The campaign has been remarkably open in attempting to demystify AIDS, involving religious and traditional leaders and being reinforced in schools, where young people were told of the merits of delaying sexual relations and engaging in safe sexual behaviour through the use of condoms. Although Uganda has achieved much success, the Ministry of Health estimated that in 2000 there were 1,438,000 people living with HIV/AIDS, and there were 838,000 deaths from AIDS, 83,000 being children (Evans, 2001).

The HIV/AIDS epidemic is having a significant effect on population structure in some countries, as can be seen by comparing two countries with similar development indices, but where one country has a low HIV/AIDS infection rate, and the other has a high rate of infection. Morocco and Botswana are ranked 123 and 131, respectively, by UNDP in terms of the Human Development Index (UNDP, 2006). In economic terms, Botswana has a much higher per capita GDP of US$9945 compared with Morocco's US$4309 (UNDP, 2006). However, in 1999, Botswana had an HIV/AIDS infection rate of 24.1 per cent among adults between the ages of 15 and 49, while the infection rate in Morocco was only 0.1 per cent (UNDP, 2006). Comparing population pyramids for the two countries, while there is a broad similarity in 2000, by 2025 population projections suggest that the Botswana pyramid has developed into what the US Census Bureau refers to as 'the population chimney' (US Census Bureau, 2002) (Figure 5.8).

Whereas the pyramid for Morocco has the 'normal' broad base, the Botswana pyramid has a much narrower base, since many sexually active HIV-infected young people die prematurely, while women may become

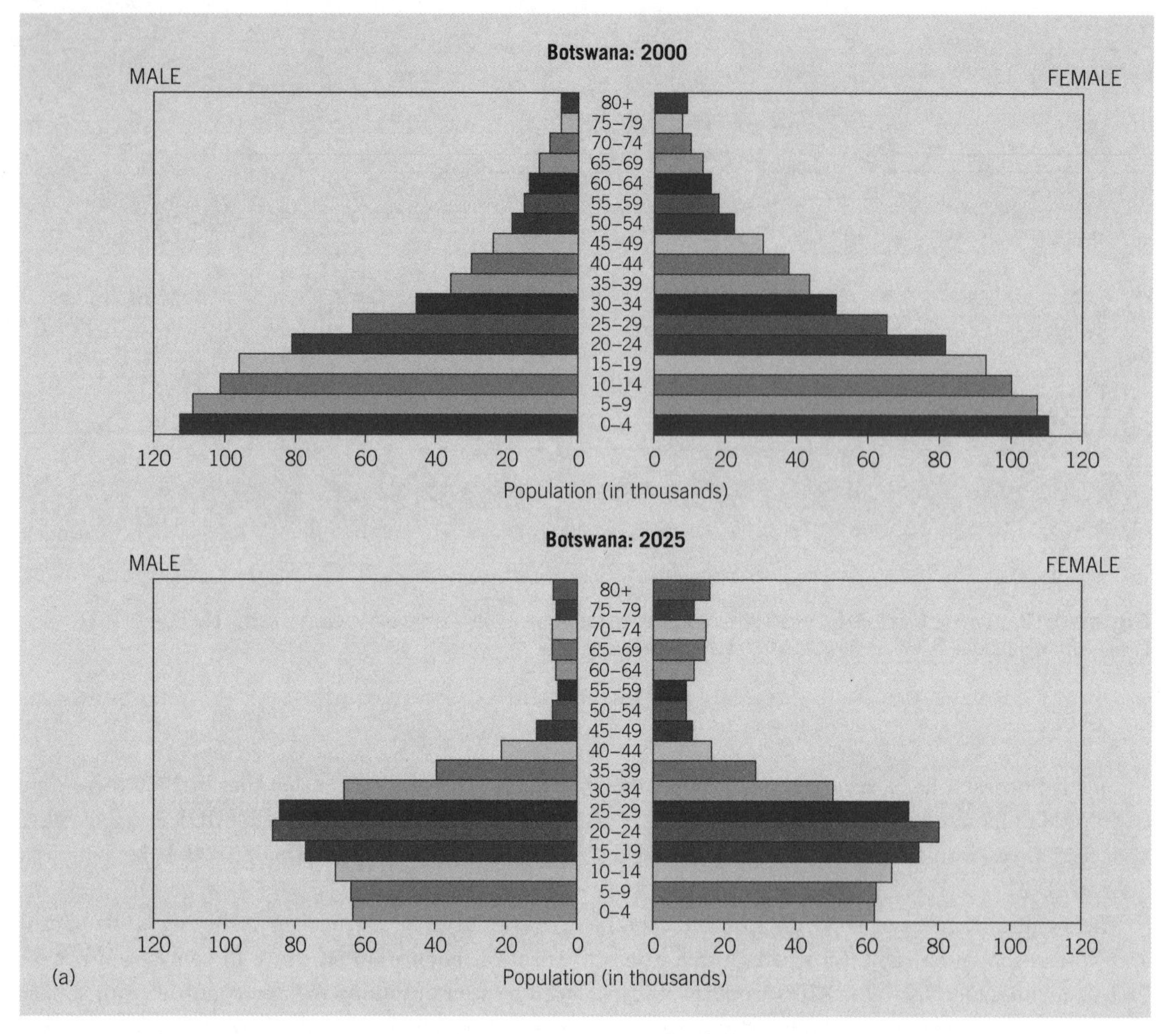

Figure 5.8 (a) Population pyramid Botswana
Source: From US Census Bureau (2002)

infertile well before the end of their childbearing years, leading to fewer babies being born. Furthermore, as Barnett suggests, 'Up to a third of the infants born to HIV-positive women become infected themselves before or during birth, or through breast milk. Hence fewer babies survive to childhood and adolescence' (Barnett, 2002b: 393). The smaller number of females under 50 years old in the 2025 Botswana pyramid indicates the higher infection rate among young women than men.

Prothero (1996) has considered the possible effects of population migration on the transmission and diffusion of AIDS in West Africa, concluding that there is a need for more research on the complex interactions of socio-economic, cultural and biomedical mechanisms. High rates of infection have been found among truck drivers, and the second stage of the World Bank's MAP programme, implemented from February 2002, is targeting the Abidjan–Lagos transport corridor in West Africa which passes through Ivory Coast, Ghana, Togo, Benin and Nigeria (World Bank, 2002b). HIV/AIDS is also a significant problem among military personnel. UNAIDS reported that in peace time, infection rates of sexually transmitted diseases, including HIV, among armed forces are generally two to five times higher than in civilian populations, and in times of conflict, when mobility is greater and troops are often away from their families, the difference can be 50 times higher or more (UNAIDS, 1998a).

AIDS has now been added to the list of other child-killers in sub-Saharan Africa: diarrhoea, malaria and measles. Browne and Barrett (1995) specifically examine the impact of the African AIDS epidemic on children

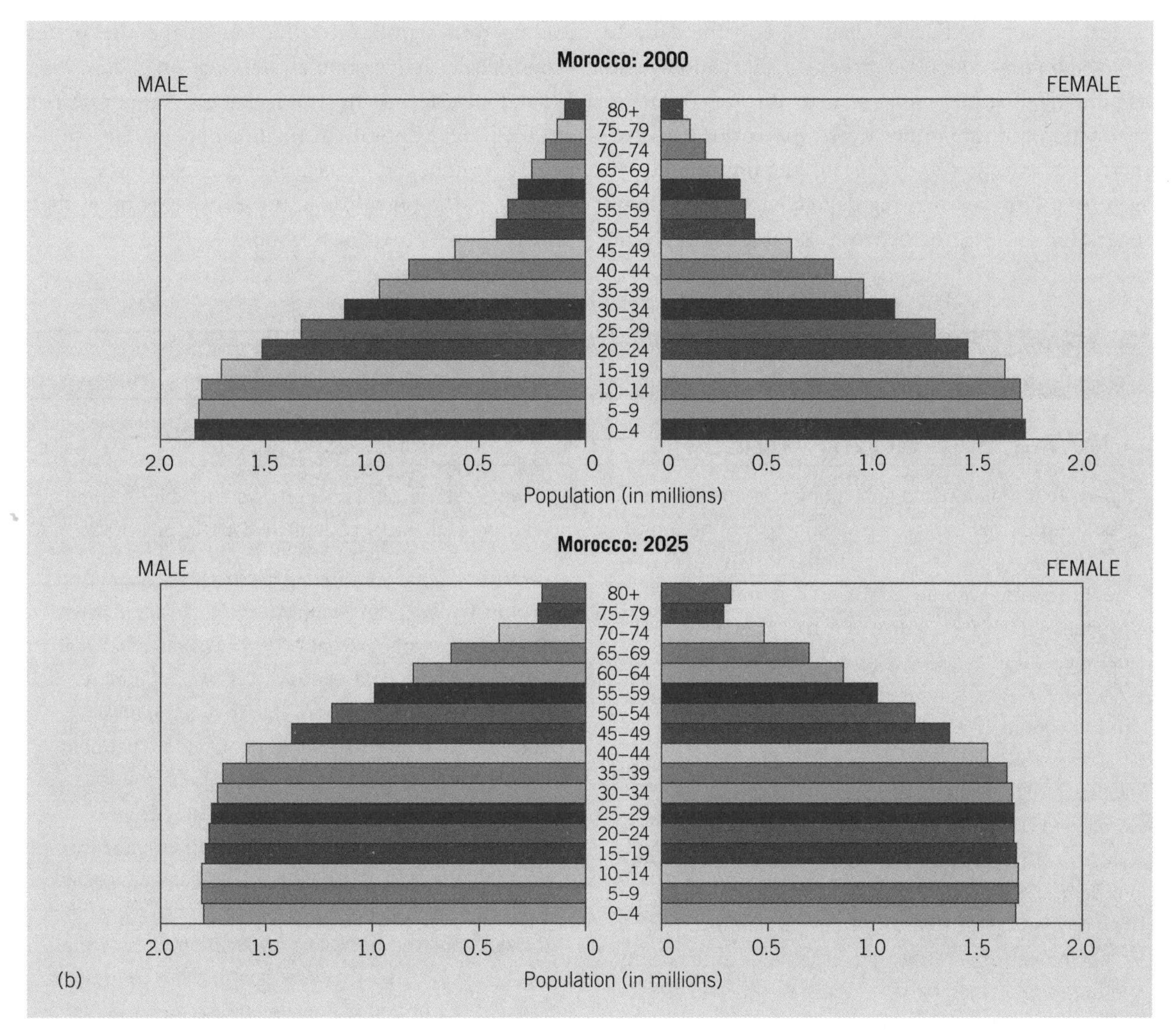

Figure 5.8 (continued) (b) Population pyramid for Morocco
Source: From US Census Bureau (2002)

and suggest that although more children still die from malaria, diarrhoea and acute respiratory infections, the long-term effect on economic and human development at national, community and household levels gives much concern. HIV-infected children have a short life expectancy, with 80 per cent dying from AIDS-related causes by the age of five. In Zambia the under-five mortality rate has increased from 125 per 1000 live births in 1989 to 182 in 2005, due at least in part to AIDS-related causes. A further problem concerns the number of children orphaned as a result of AIDS, such that the rate of orphanhood has doubled in some countries. It was estimated that in ten Central and East African countries there would be between five and six million orphans by 2000, representing about 11 per cent of the total 10–15-year-old child population (Browne and Barrett, 1995). A more recent estimate from UNAIDS suggests that in 2005 there were 12 million children under 17 in sub-Saharan Africa who had lost one or both of their parents (UNAIDS, 2006).

AIDS could also have a wider impact on population growth rates, which are likely to fall, having a Malthusian effect that could lead to an improvement in the ability of the world to sustain and feed itself. However, the spread of AIDS will increase the demand for curative health care, placing considerable pressure on poor African countries and perhaps meaning less health care for the rest of the population. For example, treating the estimated number of AIDS cases could represent 23 per cent of 1990 public health spending in Kenya and as much as 65 per cent in Rwanda. However, as Brown (1996a: 17) suggests, 'The most critical impact of AIDS is likely to be in the damage inflicted on the productive capacity of an economy and its potential to achieve food security through domestic production or economic access in world markets'. Increasing infection and mortality among the 20–40 age group has caused rising dependency ratios; some suggest an increase of between 18 and 29 per cent, leading to a significant reduction in the size of the economically active population (Gregson *et al.*, 1994). What, sadly, does seem clear is that sub-Saharan Africa already has the dubious distinction of being the world's poorest region, and AIDS is likely to further exacerbate poverty as the poor lose access to what is often their only resource – their own labour.

BOX 5.6

HIV/AIDS in South Africa

Apart from India, with a population 23 times larger, South Africa has the dubious distinction of having the largest number of people infected with HIV/AIDS in a single country, with an estimated 4.7 million South Africans being HIV positive in 2001. It is likely that between 1995 and 2015 South Africa will have 7.4 million excess deaths due to AIDS, the largest in the continent. Furthermore, life expectancy at the age of 20 with AIDS will be 15 years lower than without AIDS, while under-five mortality with AIDS is projected to be 98, whereas without AIDS it would be only 39 (UN, 2000a). The high incidence of HIV/AIDS in South Africa is due to factors such as the long-established migrant labour system, which involves predominantly men leaving their families and moving to mines and cities for work where they live in single-sex hostels. Other aspects of migration also play an important role in spreading infection. For example, KwaZulu-Natal province had South Africa's highest infection rate in 1995 at 18.2 per cent. One reason for this, it is suggested, is that Durban, its largest city, is situated on a major truck route from Malawi which has been called 'the highway of death', since 92 per cent of truck drivers visiting the city were infected with HIV (Webb, 1997).

The gravity of the situation in South Africa is reflected in a report published in May 2000, suggesting that:

- an estimated £720 million was spent on educating those of a productive age who died of AIDS in 1999–2000;
- by 2003, 12 per cent of highly skilled workers, 20 per cent of skilled workers and 27.2 per cent of low-skilled workers will be infected;
- medical aid claims are expected to rise rapidly and some schemes could face bankruptcy;
- it will cost the public health system £1690 a year to treat each AIDS patient (ING Barings, 2000).

Another report, considering the macro-economic impact of HIV/AIDS in South Africa, suggests that by 2008 the difference in real GDP growth rates between an 'AIDS scenario' and a 'no-AIDS scenario' could reach 2.6 per cent. However, due to a cumulative effect, real GDP by 2010 could be about 17 per cent below the level attained in the 'no-AIDS' scenario, as a larger proportion of the economically active workforce becomes infected (Arndt and Lewis, 2000).

In sharp contrast to Uganda, it has been suggested that, 'In the field of HIV/AIDS South Africa is a land of missed opportunities and prevarication. One of the highest rates of HIV/AIDS infection in the world makes the tragic lack of political leadership so much worse' (Haffajee, 2001: 46). There has indeed been much controversy in South Africa over the AIDS issue. President Thabo Mbeki has been strongly criticised for his stance, in which he was reluctant to accept the link between HIV and AIDS, and was less than enthusiastic about AIDS testing and the availability of anti-retroviral drugs. While the country's Anglican bishops agreed to be tested, Mbeki refused to have an AIDS test, saying it was 'irrelevant'. As the influential *Mail and Guardian* newspaper commented in May 2001,

> **Rather than appearing as Solomonic wisdom, Mbeki's equivocation on HIV/AIDS, AIDS tests and anti-retroviral drugs sounds like a dissident without the courage of his convictions . . . Government's schizoid attitude has already – and is now – taking a terrible toll, not only among the people smitten by HIV, but also among those in government trying to combat it . . . The underlying cause of AIDS is HIV. No one has provided a plausible opposing paradigm. Until**

BOX 5.6 (continued)

Mbeki can either admit or rebut it, our advice to the president is to shut up.

(Mail and Guardian, 2001a: 26)

A significant step forward was achieved in April 2001 after a protracted court case in which the Pharmaceutical Manufacturers' Association of South Africa (PMA) disputed the legality of the Medicines and Related Substances Control Amendment Act, which would allow the Minister of Health to take action to procure cheaper generic drugs for South Africa. Eventually, the PMA and 39 of its members dropped their court action, placing the ball firmly in the Government's court to implement major improvements in the health care system, especially for people with HIV/AIDS. As the *Mail and Guardian* commented, 'Now the world will see whether there is any truth in the pharmaceutical company arguments that high prices have simply been an excuse for lack of action by a government that in reality lacked the capacity or political will' (*Mail and Guardian*, 2001b: 2). In October 2001, Mbeki repeated in Parliament his view that anti-retroviral drugs are toxic, asserting that they 'are becoming as dangerous to health as the thing they are supposed to treat' (*Financial Times*, 2001: iv).

The key question is how to deal with the AIDS crisis as swiftly, effectively and as economically as possible. Government policies have been generally successful in promoting awareness of the disease and most people know their 'ABC': Abstain, Be faithful, Condomise (Plate 5.5), but the chairman of the activist group Treatment Action Campaign (TAC), Zachie Achmat, commented, 'A model that pushes prevention without providing treatment is fundamentally flawed' (*Financial Times*, 2001: iv). The Government announced in November 2001 that it would increase spending on HIV/AIDS in the period up to 2004–2005 from £11 million in 2000 to £37 million. Yet, unlike neighbouring Botswana, South Africa had no plans to make anti-retroviral drugs available because of (according to government) the drugs' perceived toxicity, their excessive cost and the poor infrastructure to support their administration. Meanwhile, Nobel Peace Prize winner Desmond Tutu complained about the Government's 'dithering' and suggested that AIDS is 'the new enemy, the new apartheid' (*Financial Times*, 2001: iv).

Plate 5.5 Roadside HIV/AIDS poster in South Africa
(photo: Bjorn Omar Evju, Panos Pictures)

BOX 5.6 (continued)

But despite government indecision there are a number of NGOs in South Africa which are actively involved in campaigns to promote HIV/AIDS awareness. For example, 'loveLife', supported by the President's wife, Zanele Mbeki and other leading South Africans, and funded by such bodies as the US-based Henry J. Kaiser Family Foundation, sees itself as 'a new lifestyle brand for young South Africans promoting healthy living and positive sexuality', and has been active, through high-powered media publicity and outreach support programmes, in targeting the 15–20 age group (loveLife/Henry J. Kaiser Family Foundation, 2001). Another organisation, GIPA (Greater Involvement of People Living with AIDS), which started in 1997, has implemented the 'Workplace Model', which involves placing articulate, open and often healthy HIV-positive people in workplaces to raise awareness and encourage debate (Haffajee, 2001).

The pressure group Treatment Action Campaign achieved a notable success in July 2002 when it won its court battle against the government to provide the anti-AIDS drug Nevirapine free of charge to all HIV-positive pregnant mothers in public hospitals. The Constitutional Court ruled the government's policy to be a violation of the constitution, a judgment that represents a landmark in the campaign for free anti-retroviral drugs in South Africa. As a result of pressure from the worldwide campaign for cheaper medicines in developing countries, pharmaceutical manufacturers have reduced significantly the price of many drugs, and it remains to be seen if and when the governments of other poor countries will be both willing and able to make anti-retroviral drugs freely available. Meanwhile, projected trends suggest there are likely to be 635,000 AIDS-related deaths in 2010 in South Africa, and as many as 2.5 million orphans under 15 years.

Education

If individuals and households are to fulfil their true potential, in addition to the provision of adequate nutrition and health care, it might be argued that an entitlement to education and freedom of expression should be key elements in the development process. Such an approach is embodied in the 'development as freedom' perspective (Chapter 1). Like health care, education is an expensive item for poor countries and its quality and availability show considerable variations between and within countries, as does student attainment. And also like health care, education systems are frequently a legacy of the colonial period, often totally inappropriate for the present-day needs of individuals, communities and nations. Indeed, there has been much debate on what is the most appropriate form and structure of educational provision in poorer countries. For example, what proportion of the budget should be allocated to the different sectors (primary, secondary and tertiary), and should more attention be given to non-formal education, such as farmer training and the acquisition of craft skills, rather than formal classroom tuition? Most commentators would probably agree, however, that providing everyone with basic primary education, especially literacy, should be the first priority of all countries. Table 5.8 shows how primary school

Table 5.8 Primary school enrolment and literacy for selected countries, 2000–2005

Country	Gross primary school enrolment ratio		Adult literacy rate	
	Male 2000–2005	Female 2000–2005	Male 2000–2005	Female 2000–2005
Bangladesh	107	111	*51.7*	*33.1*
Brazil	145	137	88	89
China	118	117	95	87
India	120	112	73	48
Jamaica	95	95	74	86
Japan	100	101	99	99
Mali	71	56	27	12
Sierra Leone	169	122	47	24
Sweden	99	99	99	99
United Kingdom	107	107	99	99
USA	100	98	99	99
Zimbabwe	97	95	*93.8*	*86.3*

Source: UNICEF (2007)
Primary school gross enrolment ratio – the number of children enrolled in primary school, regardless of age, expressed as a percentage of the total number of children of official primary school age.
Estimates for Zimbabwe and Bangladesh are based on outdated survey information and should be treated with caution.

Note: The values given are for the latest available year in the given time-frame, so care should be taken in any comparison between estimates.

enrolment and adult literacy rates in two of the world's poorest countries, Mali and Sierra Leone, are well below those of other countries. In many countries there is also a marked difference in the number of boys who attend school compared with the number of girls. Generally, fewer girls attend school, particularly in Muslim countries, and this is reflected in later years in male and female adult literacy rates (see Critical reflection box).

The 1990 World Conference on Education for All, held in Jomtien, Thailand, proclaimed the need for diverse, flexible approaches within a unified national system of education (UNICEF, 1997). The conference agreed a number of objectives for primary education:

- *Teach useful skills.* Courses should be relevant and linked to community life.
- *Be more flexible.* Use child-centred approaches; adjust school timetables to the daily routine and the seasonal farming calendar.
- *Get girls into school.* Be sensitive to social, economic and cultural barriers to ensure equal participation.
- *Raise the quality and status of teachers.* Improve pay and conditions and retrain teachers with negative and stereotypical ideas.
- *Cut the family's school bill.* School fees and equipment charges deter participation; basic education that deters child labour must be free of such costs for poor families.

The difference in enrolment ratios between developed and developing countries becomes significantly greater at secondary and tertiary levels. As Gould (1993) has shown for secondary enrolment, Africa lags very far behind other regions, whereas for South Asia the relatively higher proportion of the secondary age group enrolled in India is pulled down by much lower proportions in Bangladesh and Pakistan, both with populations of well over 100 million. Both Pakistan and Bangladesh are Muslim countries with low female participation in secondary education. At the tertiary level, although 40 per cent of the age group is enrolled in education in the high-income countries, an average figure of under 10 per cent is common in low- and middle-income countries, ranging from 5 per cent in sub-Saharan Africa to 11 per cent in Latin America and the Caribbean (Gould, 1993; UNESCO-UIS, 2006).

Some governments have introduced wide-ranging reforms to their education systems in an effort to make

Critical reflection

Educating girls

Millennium Development Goals 2 and 3 are concerned with achieving universal primary education, promoting gender equality (especially in education) and empowering women. In many developing countries far fewer girls attend school than boys, which is reflected in the difference between male and female adult literacy rates. Some North African, and predominantly Muslim countries have among the highest disparities between male and female adult literacy rates. In Algeria, for example, a 'medium human development' country according to UNDP, almost 80 per cent of males over the age of 15 are literate, but the proportion of females is only 60 per cent. In Tunisia, the corresponding figures are 83 per cent and 65 per cent respectively. In Niger, the country with the lowest Human Development Index in 2006, less than half (43 per cent) of adult males could read and write, but only 15 per cent of females had these important skills (UNDP, 2006a).

Numerous studies have clearly shown that educating girls has a significant impact on the health and welfare of households. Educated girls generally marry later and are more likely to engage in economic activity outside the home. Furthermore, they tend to have fewer children and seek medical attention sooner for themselves and their children. They typically provide better care and nutrition for themselves and their children, which leads to a reduction in disease and lower child mortality. A reduction in child mortality over time leads to smaller families, increased use of contraceptives and smaller households. Childcare improves with smaller households and with lower fertility the school-age population gradually declines.

Examine some of the different strategies for improving the education of girls. What do you believe is the best way forward in the world's poorest countries?

them more appropriate for national development needs. Whereas the small West African state of The Gambia in 1999 established its own university with Canadian and Cuban assistance, rather than sending students overseas for higher education, Africa's most populous state, Nigeria, has debated whether it should rationalise its university system to reduce expenditure. Meanwhile, Ghana has also undertaken major educational reforms. In the 1950s the country probably had the highest proportion of its children in school in Africa, and this expanded further after independence. However, as the national economy deteriorated so did the schools and the quality of education, such that education spending fell from 6.5 per cent of GDP in 1976 to only 1 per cent in 1983. Since 1985, under the Economic Recovery Programme of Jerry Rawlings' government, schools have been encouraged to make a more positive contribution to economy and society. A single Ministry of Education and Culture was created along with decentralised planning in 110 districts. The school system was reduced from 17 to 12 years, the cost of boarding was passed on to pupils and student loans were introduced in the tertiary sector. The curriculum was restructured to emphasise practical and life skills rather than academic subjects. A strong emphasis was placed on expanding school enrolments, especially for girls (Binns, 1994a). Investment in female education must receive top priority, not least because studies have revealed strong links between education and health, notably a strong correlation between high levels of infant and child mortality and low levels of maternal education, particularly basic literacy (see Critical Reflection box on p. 223).

Human rights

The Vienna Conference on Human Rights in 1993 revealed significant differences in opinion on the nature of human rights and related policies. For example, some Asian countries questioned external criticism of their human rights records; in particular, they showed their resentment at having imposed on them a set of values based upon Western traditions (Drakakis-Smith, 1997). However, many would agree that an important issue affecting the quality of life is the ability of all people to voice their opinions freely and without fear of retribution. In some countries it is apparent that certain elements of the population, such as women, are denied complete freedom of speech because of religious and/or cultural attitudes. Across the world, there are many examples of repressive regimes, both military and civilian, which have clamped down with varying degrees of severity on any opposition.

China's continuing 'occupation' of Tibet is a source of much controversy. Although the extent and nature of Chinese influence and control over Tibet through history is disputed, in 1950 the People's Liberation Army clashed with Tibetan troops as the new Chinese government sought to integrate Tibet into the Chinese state. This policy was given a significant boost when, in 1954, China and India reached an agreement, whereby India recognised Tibet to be an integral part of China in return for China undertaking to respect religious and cultural traditions. However, in 1959 there was a rebellion in Tibet against Chinese control and the Tibetan Buddhist spiritual leader, the Dalai Lama, and many of his followers, fled to India, where they still remain. Meanwhile, China has steadily strengthened its control over Tibet, increasing the number of troops, introducing inappropriate policies and suppressing religious and cultural activity. China has even installed a young boy as its own 'puppet' spiritual leader in place of the Dalai Lama. In 1987, Beijing reasserted that 'Tibet is an inalienable part of Chinese territory' (*Beijing Review*, 19 October: 14).

In other countries, freedom of speech has been denied to specific racial groups, and nowhere was this more entrenched than under the apartheid regime in South Africa (Lester *et al.*, 2000). Racial discrimination and separation in South Africa can be traced back long before 1948, when the National Party took power and formally introduced a policy of 'apartheid' or separateness. Apartheid was a policy based on fear, notably fear of the minority White population being dominated by the majority Black population. But the White regime was also well aware that economic survival was completely and unavoidably dependent on the plentiful supply of cheap non-White labour. The National Party government argued that different racial groups should be allowed to live and develop separately, each at its own pace and in accordance with its own cultural heritage, resources and abilities.

In reality, however, the regime was harsh and introduced a wide range of oppressive legislation to control the lives of non-White groups, which together comprised over 85 per cent of South Africa's population. Black political parties such as the African National Congress (ANC) were banned in 1960, police powers increased in 1962, Black newspapers suppressed in 1976 and censorship of political pamphlets introduced.

Meanwhile, the Group Areas Act of 1950 extended the principle of separate racial residential areas on a comprehensive and compulsory basis. Its application was felt most strongly in cities such as Pretoria, where Indian traders were moved out of the city centre, and Cape Town, where coloured inhabitants in the suburbs were relocated in segregated areas despite local council objections. Under the 1955 Natives (Urban Areas) Amendment Act, the rights of Blacks to live in a town were restricted to those who had either been born there or who had worked there for 15 years, or ten years with a single employer. All other Blacks required a permit to stay for longer than three days.

A catalogue of legislation enforced what was known as 'petty apartheid', which took such forms as segregated transport, public toilets and even beaches. The broader national development strategy known as 'grand apartheid' was manifested in the creation of homelands called Bantustans. Through the 1959 Promotion of Bantu Self-Government Act, eight (later extended to ten) distinct 'Bantu homelands' were created, each with a degree of self-government and based largely on the historic homelands of different Black tribal groups. Some of these homelands were geographically nonsensical, such as KwaZulu comprising 48 large and 157 small isolated tracts, and Bophuthatswana with one of its six segments located 320 kilometres from the others. All Black South Africans were given the citizenship of a particular homeland in 1970, but then some subsequently lost their South African citizenship when four homelands were given 'independence': Transkei (1976), Bophuthatswana (1977), Venda (1979) and Ciskei (1981). Although these homelands had all the trappings of independent states, their independence was not recognised by any country other than South Africa (Lester *et al.*, 2000).

Many of the homelands were located on poor-quality marginal land away from major urban areas, yet the South African economy, and particularly the mines and industries, were totally dependent on a plentiful supply of cheap Black labour. Consequently, male family members were frequently absent from the homelands, spending much of their time living in crowded hostels close to the mines and on the edge of the large cities. Meanwhile, the women and children were largely left to fend for themselves in the rural areas, but were heavily dependent financially upon remittances sent from their absent male relatives.

Growing internal and international pressure, however, gradually forced the minority government to consider dismantling certain elements of apartheid, and this process was accelerated after F.W. de Klerk took power in 1989. In February 1990 an important and historic signal of intent was given to the world community when Nelson Mandela and several other ANC leaders were released from prison. The country's first democratic elections were held in April 1994 and the charismatic Mandela was proclaimed as first President of the 'new' South Africa. Mandela and the new ANC government then set to work on implementing a range of policies through its Reconstruction and Development Programme, designed to dismantle the structures of apartheid and address the practical problems facing one of the world's most 'unequal' nations (Binns and Robinson, 2002).

Conclusion

Returning to a point made at the beginning of this chapter, people are central to the development process. Unfortunately, in recent years people have too often been a secondary consideration after the quest for wealth and profit. There is a need to reshape development strategies so they place people at the heart of development. There have been some brave calls to the world community to make this a reality, but sadly their impact has been small. For example, the Independent Commission on Population and Quality of Life (ICPQL, 1996) has argued that the world faces a linked crisis of environment, quality of life and population, and proposes a number of guiding principles in relation to population growth and improving the quality of life: equity, caring, sharing, sustainability and human security. The commission takes issue with the prevailing concept of development, describing it as 'exclusively economic and obsessed with deregulation [it] . . . inevitably produces massive exclusion, inside every society, among nations, on all continents. This requires a shift in the way policies and measures are shaped and in how political decisions are made' (ICPQL, 1996: 4). The commission argues that a number of issues must be tackled urgently, including

> **making life more liveable through improved individual and collective health and security; dealing with the scourges of poverty and exclusion; raising the levels of literacy, education and access to needed information; rationalising production and consumption in terms of what the planet's**

resources can continue to provide and bringing fairness and equity to all through better-balanced exploitation and use of these resources [such as keeping more profits from raw materials 'at home'; utilising them in a sustainable manner]; more effective policies of aid and assistance; and finding new funding mechanisms between North and South. And, last, but hardly least, caring for ourselves, our neighbours, and the environment by observing the rights pertaining to all of humankind.

(ICPQL, 1996: 286)

The commission stresses the importance of not only environmental sustainability but also social sustainability, and it emphasises the synergy between the two. Importantly, the commission places much emphasis on improving the quality of life, which 'should become the chief focus of governments north and south'. It suggests:

We urgently need a new synthesis, a new balance between market, society and environment, between efficiency and equity, between wealth and welfare. A new balance between economic growth on the one hand, and social harmony and sustainability on the other.

(ICPQL, 1996: 16)

These are admirable sentiments, but given the lamentable record of national government and international community action following earlier well-intentioned initiatives, such as the Brandt Commission (1980) and the Brundtland Commission (1987), there is inevitably some scepticism about possible future progress.

Key points

- Improvement in the quality of life for all people should be at the core of development processes at all levels – internationally, regionally, nationally and locally.
- Relations between people and resources are often complex. Before development strategies are implemented it is important to understand these relationships in detail.
- Reliable data on population are essential for planning development interventions at all levels.
- Households are often the key unit of production, consumption and decision making. Household structure, dynamics and needs must be understood if appropriate development strategies are to be implemented.
- Achieving universal primary education for both girls and boys is one of the Millennium Development Goals (MDG 2), and is a key priority for all countries.
- Access to good health care for everyone is an important goal, particularly in poor countries, and is reflected in several Millennium Development Goals.
- Equality and empowerment for women and basic human rights for all are fundamental entitlements, but some countries still have a long way to go to achieve these objectives.

Further reading

Ansell, N. (2005) *Children, Youth and Development.* London: Routledge.
An excellent book, which examines the position and role of young people in different societies, with particular reference to the process of development in poor countries.

Binns, T. (ed.) (1995) *People and Environment in Africa.* Chichester: John Wiley.
This collection of essays provides a number of very useful case studies of how people interact with and manage environment in Africa south of the Sahara.

Boserup, E. (1993) *The Conditions of Agricultural Growth.* London: Earthscan.
A classic text, which is frequently cited and counters the Malthusian view concerning the relationship between population and resources.

Crook, N. (1997) *Principles of Population and Development.* Oxford: Oxford University Press.
A helpful and straightforward text which examines demographic aspects in the context of development.

Devereux, S. and Maxwell, S. (2001) *Food Security in Sub-Saharan Africa.* London: ITDG Publishing.
An important text which defines the concept of food security and examines aspects of food security with reference to case studies from Africa.

Gould, W.T.S. (1993) *People and Education in the Third World.* Harlow: Longman.
One of relatively few texts that focus on the importance of education as a key element in the process of development.

Momsen, J.H. (2004) *Gender and Development.* London: Routledge.
A frequently quoted book examining the significance of gender in the development process.

Websites

www.amnesty.org
Amnesty International.

www.earthsummit2002.org
Earth Summit 2002.

http://esa.un.org/unpp/index.asp?panel=2
UN (2007b) *World Population Prospects: The 2006 Revision Database*, online database, United Nations Population Division.

http://www.malariavaccine.org/about-mvi.htm
Malaria vaccine initiative.

http://www.unaids.org/en/HIV_data/epi2006/
UNAIDS (2006) AIDS epidemic update, December 2006. Geneva: UNAIDS.

http://hdr.undp.org/hdr2006/
UNDP (2006) Human Development Report 2006. New York: UNDP.

http://hdr.undp.org/hdr2006/statistics/build_your_table/default.cfm#
Useful UNDP statistical database to search a variety of development indicators.

http://www.unicef.org/sowc07/docs/sowc07.pdf
UNICEF (2006) *State of the Worlds Children 2007.* New York: United Nations Children's Fund.

http://www.census.gov/ipc/www/idb/html
US Census Bureau, International Data Base, 2007.

Discussion topics

- With reference to specific case studies, examine the relationships between people and resources. Evaluate the merits and problems of Malthus' and Boserup's perspectives.
- Select two countries, one with a pro-natalist population policy and another with an anti-natalist policy. Consider the reasons for the adoption of these policies and the likely effects on population growth and structure.
- Summarise the key features of the 'entitlement approach', and evaluate the advantages and problems of this approach in ensuring food security in specific countries or regions.
- Investigate the demographic, social and economic effects of high rates of HIV/AIDS infection in selected countries.

Chapter 6

Resources and the environment

This chapter investigates the relationship between development and the environmental resources of the globe. Since we live in a closed system in which matter and energy cannot be created or destroyed, there is no doubt that human social development ultimately depends on the physical resource base. However, just as Part I has shown something of the diversity of opinion on normative questions of how wealth and well-being should be created and/or redistributed, this chapter reveals the substantial debate that persists as to the precise relationship between prospective development achievements and environmental resources. Questions of resource scarcity, of environmental degradation and global environmental change and the search for new patterns and processes of development that are more sustainable are major challenges currently in development studies and beyond.

This chapter:

- Details major theories on the links between resources and development at the international scale;
- Identifies the core features and debates underpinning the notion of sustainable development;
- Examines historical patterns of resource use in key sectors such as water and energy that have created economic and social opportunity in development;
- Considers how resource scarcity constrains development processes;
- Examines the environmental impacts of development through core global issues such as climate warming, deforestation and desertification;
- Illustrates the place-specific nature of the relationship between environment and development but also how the challenges of sustainable development are shaped by processes operating at wider scales.

Introduction: the search for sustainable development

Debates concerning the relationship between human social development and the environmental resources of the globe are of long standing. In 1945, for example, the geographer Huntington modelled resource inadequacies as the cause of underdevelopment in the tropical regions of the world (Huntington, 1945). In direct contrast, in 1968 Erhlich modelled development as having put 'mankind on the brink of extinction' (Ehrlich, 1968) through the ways in which it has led to environmental resource destruction. More recently, similarly deterministic understandings of the role of 'geography' (in terms of climate, land productivity and location) in explaining levels of economic development at an international scale have appeared (Sachs and Gallup, 1998 cited in Pearlstein, 1998). Similarly, as mounting research exposes the role of human action in global climate changes, for example, understandings of the Earth as seriously damaged by contemporary processes of development become prominent and predictions of quite disastrous impacts on economic and social prosperity follow (see, for example, Stern, 2007).

Amid these debates, the notion of sustainable development is based on a recognition that past patterns and processes of development cannot be sustained environmentally over time. However, it also forwards an understanding of the complex relationships that occur between a lack of development and environmental degradation, for example, and the unsustainable nature of contemporary processes and patterns of economic development in terms of social concerns such as justice, human rights and security (see Elliott, 2006).

What is sustainable development?

It has been suggested that the concept of sustainable development can mean anything or everything you want (O'Riordan, 1995). The most often quoted definition of sustainable development dates back to the World Commission on Environment and Development of 1987: 'Development that meets the needs of the present without compromising the ability of future generations to meet their own needs' (WCED, 1987: 43). Ecologists may stress sustainability in terms of the future productivity of biomass; economists in terms of capital and natural environmental asset stocks; and sociologists in terms of cultural diversity and social justice, for example (see Elliott, 2006).

Distinct views of sustainable development can certainly be identified according to different perspectives on the fundamental relationship between human societies and the environment. For example, ecocentrists view humankind as part of nature and of a global ecosystem in which nature must be respected regardless of its value to society; technocentrists see nature as separate from society and as an instrument for exploitation and material gain for human benefit (Pepper, 1996). It is therefore not surprising that there are also quite varied orientations to environmental management and action emerging from these fundamentally different philosophical entry points to the debate on sustainable development. For example, 'deep ecologists', the 'ultimate ecocentrists' (O'Riordan and Jordan, 2000: 496), favour radical changes in political and economic structures that are in contrast to the 'self regulation through enlightened conscience' (O'Riordan, 1995: 13) promoted by the 'dry greens' (who accept the present economic system but believe in more responsible and accountable institutions, for example). Such differences can be seen below in terms of the current debates concerning what is needed to address climate warming.

The need for change

Despite this evident diversity of views surrounding sustainable development in theory and in practice, there is evidence, including from recent international summitry on the environment, that sustainable development is considered to be inherently desirable and a policy objective which should be striven for (Elliott, 2006). In September 2002, the largest international conference in history, the World Summit on Sustainable Development, took place in Johannesburg, South Africa. A decade on from the first 'Earth Summit' held in Rio de Janeiro, the complex interdependencies of global environmental, economic and social development concerns are understood to constitute the fundamental challenge of global sustainability. It is acknowledged that overcoming poverty is central for effective strategies for sustainable development. It is also recognised that processes of globalisation bring threats as well as opportunities for the environment. In addition, there is a growing awareness into the twenty-first century of the close links between environmental resources and civil conflict, humanitarian disasters and human rights abuses, for example. Basic human security is understood to be a precondition for sustainable development, as discussed in Chapter 7.

As Redclift has stated in relation to sustainable development, 'Like motherhood, and God, it is difficult not to approve of it' (1997: 438). For Adams, sustainable development is a 'statement of intent, not a route-map' (2001: 383). Indeed, the attractiveness of the concept may lie in precisely the way in which it can be used to support varied political and social agendas.

Resources and development

The importance of resources in development is easily evidenced. As Emel *et al.* (2002: 377) suggest:

> **Everything we use to communicate, move, stay warm, stay cool, sit, sleep, cook, and refrigerate comes from the 15 billion tons of raw material that humans extract from the earth each year.**

All forms of productive activity make demands on the resource base, as raw materials and as energy sources in industrial and agricultural processes and in terms of the varied 'sink' functions that the environment provides in absorbing, dissipating and transporting the by-products of these activities. Resources are also consumed through human social activities associated with fulfilling the basic need for shelter, sustaining urban lifestyles, and so on. Intrinsically, the environment supports life itself through the regulation of the Earth's temperature and atmosphere.

It is true that sections of the world's population now live in a 'post-industrial' society that has seen a decline in the traditional dependence on local or regional resources and environments. However, the livelihoods of many more people around the globe remain very directly linked to immediate survival needs secured through their local environmental resources (see Chapter 10). Furthermore, the apparent delinking of development from the natural resource base in the more affluent nations has been referred to as a 'geographical illusion' (Emel *et al.*, 2002: 338). The suggestion is that the increased spatial separation of production and consumption in the post-industrial age, whereby such lifestyles depend on longer distance flows of resources, foodstuffs and material goods via such complex commodity chains, obscures the view of the political–economic, social and environmental patterns and processes through which production, exchange and consumption occur. In fact,

> **far from quaint anachronisms, practices of forestry, fishing, agriculture, and mining remain key components of the contemporary global economy . . . In developed countries, the emergence of the 'new economy' has decreased the relative significance of primary industries, yet the products of these industries remain central to the experience of post-industrialism and are increasingly sourced from the peripheries of the global economy.**
>
> **(Emel *et al.*, 2002: 389)**

Furthermore, experiences such as the 'foot and mouth' crisis in British farming through 2001 and evidence of 'bird flu' entering human food chains in 2006/7 are promoting substantial debate and increasing exposure of these processes underpinning more affluent lifestyles. The Critical reflection on 'food swaps' in Europe

Critical reflection

'Food swaps' in Europe

Sections of the world's population (largely, but not exclusively, within the more developed countries of the North), now live in a 'post-industrial' society that has seen a decline in the traditional dependence on local or regional resources and environments. This has been enabled by developments in transport and communications in particular, and, in the case of food supplies, by the increase in air-freight and refrigeration technologies. However, it has also produced patterns of trade such as shown in Table 6.1, where trade in animal products between Britain and European countries has increased dramatically in the last 30 years, but a reverse trade in precisely the same products has occurred back into the country.

Table 6.1 'Food swaps' around Europe

Product	Transfer 'out' of UK	Transfer 'in' to UK
Poultry	331,100 tons to Netherlands	61,400 tons from the Netherlands
Pork	195,000 tons to EU	24,000 tons from EU
Lamb	102,000 tons to EU	125,000 tons from EU
Milk	270 million litres to EU	126 million litres from EU
Milk powder	153,000 tons to EU	23,000 tons from EU

Source: Lucas (2001b) © 2001 *The Independent*

Critically consider what the benefits and costs of such patterns of trade are and for whom. Identify the processes (and interests) that sustain such trade.

considers what has been termed the 'crazy logic' (Lucas, 2001b: 15) of aspects of the trade in animal products between Britain and European countries.

Resource 'limits' in development?

The predominant view of resources is that they are given value by society in respect of the functions they can perform and according to the levels of development and aspirations of society. As Zimmerman (1951: 7) has stated, 'Resources are not, they become'. However, there is also the view of environmental resources as stocks of substances or materials found in nature. The substantial debate concerning the adequacy of resources to support the demands of modern society, particularly over the last 200 years, flows from these conceptual differences (Mather and Chapman, 1995: 3):

> **If environmental resources are simply stocks of substances found in nature, then they are inevitably fixed and limited in quantity. Limits to resource use must inevitably exist. If, on the other hand, resources reflect human appraisal, then the conclusion is quite different. In this case, their limits are not imposed by the non-human environment, but rather by human ingenuity in perceiving usefulness or value.**

In Chapter 5, the doomsday predictions of Thomas Malthus were noted in which he considered that there were definite environmental limits (a fixed amount of land) to human development. Subsequently, influential publications such as *The Limits to Growth* (Meadows *et al.*, 1972) promoted similar views of an ultimate limit on economic development presented by the availability of resources.

In contrast, authors such as Ester Boserup (1965) and Julian Simon (1981) were central in promoting the functional view of resources encapsulated in Zimmerman's definition, and in arguing against the inevitability of societal collapse. They point rather to the social, institutional and technological factors that serve to extend the boundaries of development. In contrast to Malthus's emphasis on the physical limits of resources, these authors focused on the stimulus to, and opportunities for, innovations and developments in resource use which population growth brings.

Human ingenuity is certainly displayed in the development over recent years of 'new forms of nature' – the new plants, animals and materials produced through the application of biotechnology. However, there is substantial debate (and scientific uncertainty) concerning the opportunities and risks for human society and development presented through biotechology, as discussed in Box 6.1 in relation to genetically modified crops.

Indeed, it will be seen in the course of this chapter that discussions of resources and development currently tend to focus less on issues of scarcity (and 'what resources can do for development') and more on how development is impacting (often detrimentally) on a range of resource functions of the Earth, particularly those related to the 'sink functions'. The contemporary concern is less that resources will run out, and more in relation to the environmental and social costs of current consumption and how these constitute significant barriers and obstacles to sustainable forms of development.

BOX 6.1

Creating nature: the opportunities and hazards of genetically modified (GM) crops

In short, genetically modified crop technology involves inserting alien genes (from different plants, animals, viruses or bacteria) into plants to give certain traits that would not occur naturally; genes are transferred 'horizontally' between species that do not interbreed.

Much research and development to date has focused on genetic modification of seeds to provide resistance to herbicides (enabling crops to be 'protected' while unwanted pests and weeds, for example, can be controlled through blanket applications of such chemicals). Other traits that can be developed artificially within plants include drought resistance and vitamin enrichment.

In recent years there has been a substantial case made for GM technology as a route to solve world hunger, to produce more food more efficiently and

BOX 6.1 (continued)

to conserve resources. The basic premises are that world hunger is caused by a shortage of food; that larger, technology-intensive farms are more efficient for food production; that 'low-input' agricultural systems require more land and therefore threaten remaining forests, valued ecosystems and wetlands; and biotechnology will deliver more food with less pollution via decreased chemical use (Kimbrell, 1998).

Cultivation of GM crops has expanded rapidly in recent years, from an estimated 1.7 million hectares in 1996 to 52.6 million hectares in 2001 (Branford, 2002). The majority of GM crops are grown in the USA, where approximately 68 per cent of global GM production takes place, the next largest producer being Argentina with 22 per cent (Maddox, 2002).

The opportunities provided by and the risks associated with GM crops and foodstuffs are an arena for substantial debate (as with wider biotechnology and genetic engineering interventions in areas such as health and human reproduction). Substantial scientific uncertainty is starting to be acknowledged concerning the ecological repercussions of horizontal breeding and the human health implications of GM foods. In addition, recent research concerning the experiences of GM crops in America concluded that GM crops had delivered few, if any, of the economic benefits promised to farmers (Soil Association, 2002). Most widely perhaps, critics of the expansion of GM crops and the further industrialisation of world agriculture are fearful of the rising corporate control over the food chain as a whole encompassed in these developments.

The majority of research and development into GM crops has been financed by private companies (rather than public research institutions) and has therefore focused on the needs of large-scale agriculture in the industrialised world, not on the basic food crops or needs of small farmers in the developing world (Madeley, 2002). Furthermore, corporate control over the production of GM crops is heavily concentrated in the hands of a few companies; the world supply of GM cotton seeds, for example, is controlled by just four companies (Vidal, 2002).

Of the total land area under GM cultivation worldwide, about two-thirds in 2001, according to Branford (2002), was under soya beans genetically engineered by a single company, Monsanto. Early in 2002, India opened its doors to genetically modified cotton varieties after many years of resistance. By that time Monsanto (the world's leading GM cotton seed producer) had already bought up several of India's largest seed companies in order that it would be in a position to promote GM varieties across the continent once government approval was forthcoming (Vidal, 2002). Shiva (2000: 81) notes that Monsanto spent over US$8 billion buying seed companies between 1995 and 1998.

There is substantial concern among opponents to the further expansion of GM cropping as to the implications for small farmers in the developing world of this rising corporate control over agriculture. Fundamentally, small farmers would experience an increased dependence on external inputs within production (sourced from multinational companies). It is suggested that this will create greater economic challenges among farmers who already struggle to afford certified seeds, for example. Such fears do seem to be supported by aspects of the experiences of the 'green revolution' in agriculture (considered in some detail in Chapter 10), that included the increased marginalisation of smaller farmers and a concentration of benefits among already more wealthy farmers (Brac de la Perriere and Seuret, 2000).

One of the latest GM developments, dubbed 'terminator technology', perhaps most clearly illustrates the fears of critics. It is now possible to create plant varieties that are biologically sterile, ensuring that farmers will have to buy new seeds each year. Vandana Shiva has been a vociferous opponent of such developments, pointing out, for example, that historically 'the seed, for the farmer, is not merely the source of future plants and food; it is the storage place of culture and history' (Shiva, 2000: 8). For centuries, farmers have evolved crops, experimented, innovated and exchanged knowledge as well as seeds, and these processes have been an essential part of local culture and heritage. Shiva refers to a 'growth illusion' and a 'corporate myth' concerning the extension of industrial agriculture that 'hides theft from nature and the poor, masking the creation of scarcity as growth' (2000: 1), a view that is now shared by most mainstream environmentalists and development organisations.

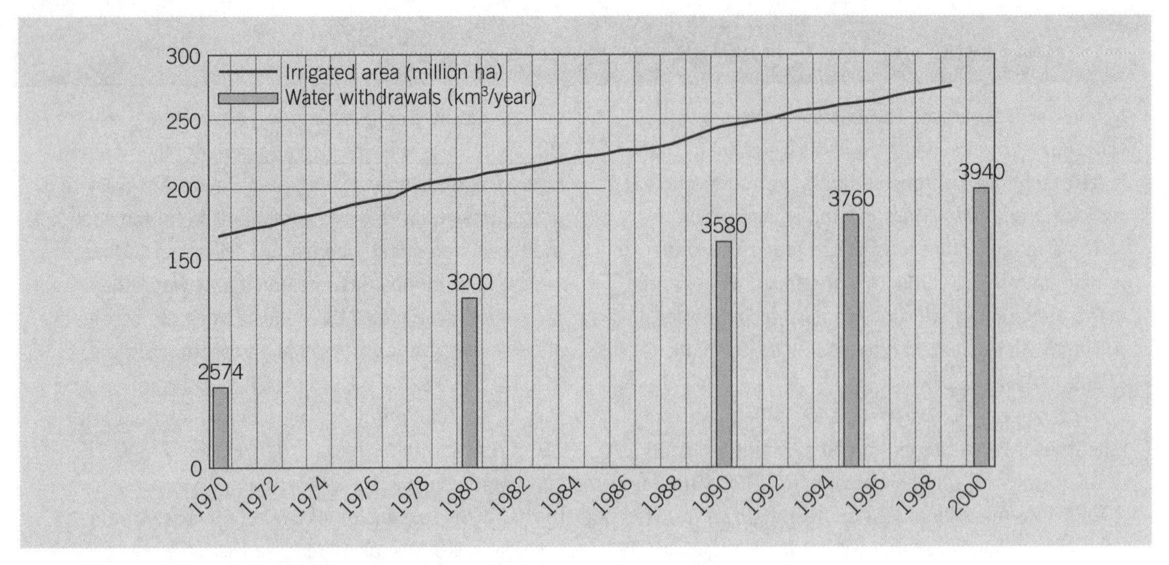

Figure 6.1 Global irrigated area and water withdrawals
Source: From UNEP (2002), UNEP/GRID-Arendal

Water resources in development: where will new sources come from?

Water resource management is fundamental to human existence and economic development. Global water withdrawals accelerated sharply over the twentieth century and at rates in excess of population growth (Mather and Chapman, 1995). Since 1970, total global water withdrawals have increased from a little over 2500 km^3 to almost 4000 km^3, as shown in Figure 6.1. In just five years between 1990 and 1995, global water consumption rose six-fold and at more than double the rate of population growth (Emel *et al.*, 2002: 378).

Total and per capita water consumption are generally higher in the developed world and with economic development relatively more water is consumed in the industrial and domestic sectors (see Figure 6.2a and b).

At a global level, agriculture is currently the largest user of water, accounting for approximately 67 per cent of water withdrawals at the turn of the century and reflecting the importance of agricultural production in developing countries. In comparison, industry accounted globally for 19 per cent of total water withdrawals (much of it in chemical manufacturing industries) and municipal and domestic usage for 9 per cent (World Commission on Dams, 2000: 5). Evaporation losses from large reservoirs in dry climates are also estimated to be close to 5 per cent of total water withdrawals and therefore represent a significant consumptive use of water worldwide.

It has been suggested that increases in agricultural output since the 1950s and the era of the green revolution in regions of the developing world (Chapter 10) have been 'roughly proportional to the amount of water supplied' (Ohlsson, 1995: 7). Certainly, a close association between patterns of increased global water withdrawals and the rise in irrigated area is evident in Figure 6.1.

Furthermore, the investment in the construction of dams worldwide was most rapid during the 1970s, as shown in Figure 6.3. Globally, it is estimated that there were more than 45,000 large dams (generally taken to be at least 15 metres high) at the end of the twentieth century in over 140 countries (World Commission on Dams, 2000: 8; Plate 6.1). Approximately two-thirds of these are in the developing countries, largely in India and China. It should be noted that Figure 6.3 excludes data for China where it is estimated that more than 25,000 large dams have been built since the late 1940s (International Rivers Network, 2007).

Whilst most donors have moved away from financing major dam projects (including because of the extensive social and environmental impacts with which they are associated), governments within developing nations such as India and China remain committed to them as a means for delivering both water supply to agriculture and power for rural and urban households.

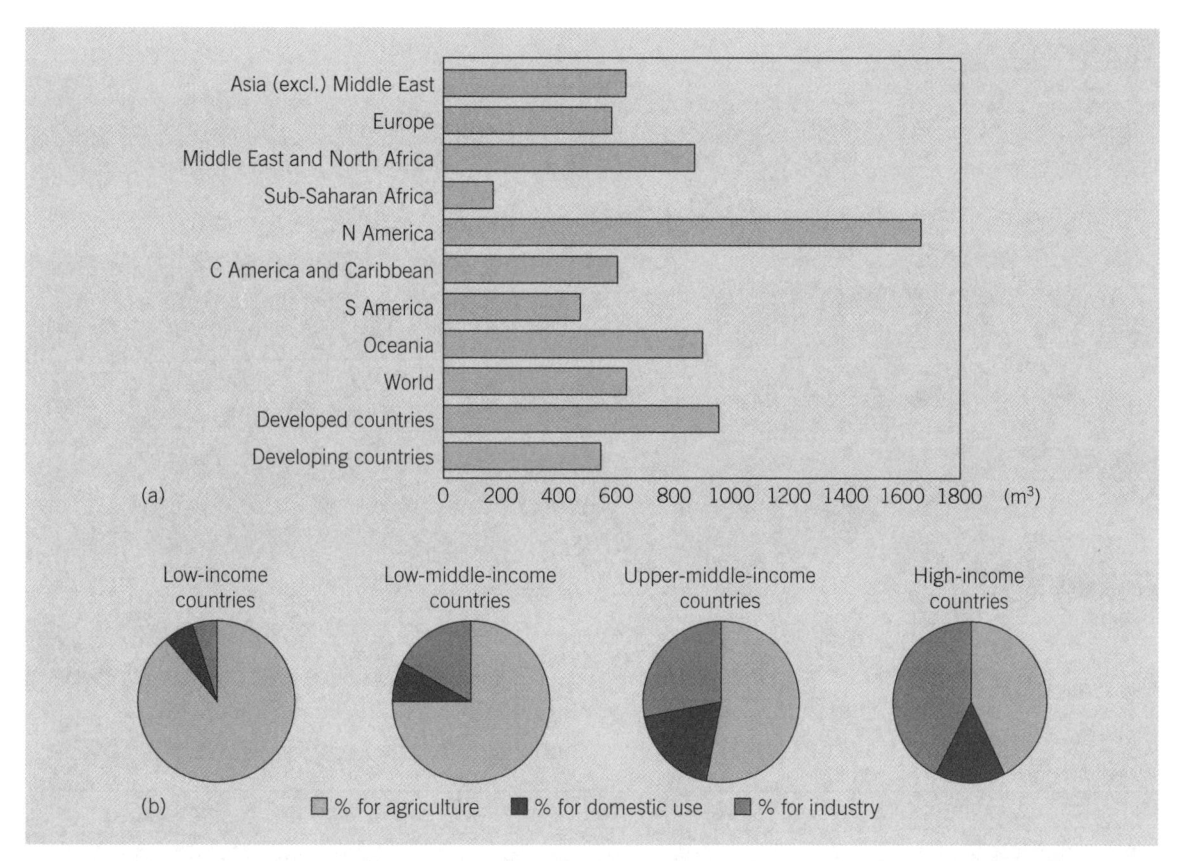

Figure 6.2 Annual freshwater withdrawals: (a) per capita average, by world region (2000); (b) as a percentage of total resources withdrawn (2002)
Source: Compiled from (a) World Resources Institute (2005b), (b) World Bank (2006)

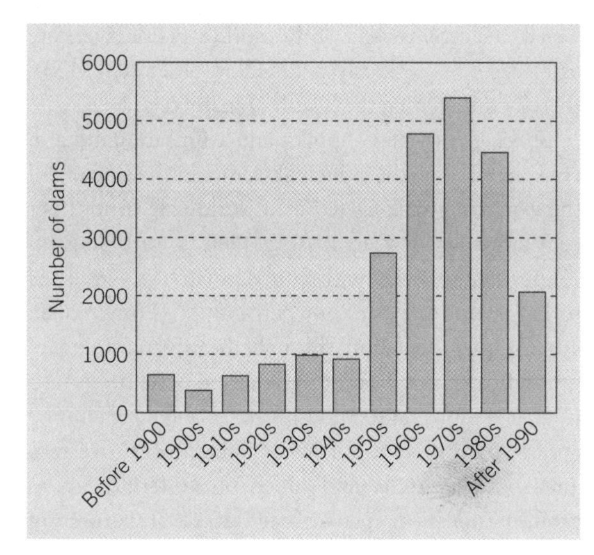

Figure 6.3 Construction of dams by decade, 1900–2000
Source: From World Commission on Dams (2000)

In contrast, new dam construction in Europe and North America has been falling since the 1970s. In the USA, the rate of dam decommissioning now exceeds that of dam construction (McCully, 2001). Many are being decommissioned for safety reasons – 25 per cent of US dams are now over 50 years old – but increasingly it is also for environmental reasons such as to allow for more natural flow regimes and for migratory fish passage. On the Snake River in Washington state (the main tributary of the Columbia), for example, environmentalists, fishing interests and Native Americans are campaigning to breach four large hydro dams. For a review of the impacts of river basin control and dam building on development, see Adams (2001) or McCully (2001).

Water resources, however, are fundamentally limited; the amount of water which can be made available to various groups of people is determined by the precipitation which falls. Augmenting water supply in

Plate 6.1 Three Gorges Dam, China
source: Corbis/Xiaoyang Liu

one place over a particular period therefore necessarily impacts on supplies elsewhere.

Although there is some indication that historical increases in per capita water consumption may be slowing with the more conservation-oriented approach in some developed nations, access to quality water supplies remains a primary factor in realising future development aspirations in many regions. This is highlighted in the example provided in Box 6.2.

Indeed, water was called 'the oil of the 1990s', referring to the centrality of the resource for future world development, for regional political and economic stability and in reference to the likelihood of many more people having their lives adversely affected by issues of water in the future than was the case with respect to oil in the 1970s and 1980s (Biswas, 1993). In 2000 UNEP estimated that by 2025 two-thirds of the global population will be living in water stressed countries (UNEP, 2000).

Furthermore, at the turn of the century, 1.1 billion people still lacked access to safe drinking water (UNEP, 2002). In addition, the issues are of water quality as well as availability; half of the world's major rivers are assessed to be 'seriously polluted' (UNEP, 2002: 153), thereby threatening the health of many people.

Shortage of water supplies and salinisation are also considered to be the principal factors in the slowing of the expansion of irrigated area worldwide from 2 per cent annually between 1970 to 1980 to an average of 1.3 per cent in 1990s, with future growth rates expected to be around 0.6 per cent per annum (Postel, 2000: 40). Aquifer depletion through the overpumping of groundwater supplies for agricultural activities is also now a serious and widespread problem including across central and northern China, north-western and southern India and much of western USA, as groundwater is pumped at faster rates than nature can replenish (Postel, 2000). In northern China (where approximately 40 per cent of China's grain is produced), the water table is dropping by between 1 and 1.5 metres a year while water demands go on increasing (Postel, 2000: 42).

BOX 6.2

The challenge of transboundary water resources for international institutions in development

Although there is much progress to be made globally in the more efficient management of existing water sources to enhance supply, increasingly (and particularly in the developing world), the prospects of securing the water resources required to facilitate future development depend on the use of transboundary river and lake basins. In 2002, UNEP identified a total of 261 rivers covering 45.3 per cent of the total land area of the globe (excluding Antarctica) being shared by two or more countries, 'making transboundary water resources management one of the most important water issues today' (UNEP, 2002: 154). The significance of the challenge was encapsulated by Biswas (1993: 167):

> **There is no question that it is going to be an increasingly complex task to provide an adequate quantity and quality of water for various human needs. Difficult though it is going to be to institute more rational and efficient management policies and practices for water sources that are contained wholly within the geographical boundaries of individual sovereign states for a variety of interrelated technical, economic, social, institutional and political reasons, the problem is likely to be intensified by several orders of magnitude when the management and development processes for water sources that are shared by two or more countries are considered.**

Table 6.2 highlights the predominance of transboundary river and lake systems within the developing world. Although the majority of international water bodies are shared by only two countries, there are nine river and lake basins which cut across more than six countries. Except for the Danube (12 countries) and the Rhine (eight countries), all of these systems – the Niger, Nile, Zaire, Zambezi, Amazon, Lake Chad and the Mekong – are in the developing world.

The particular and challenging standing of Africa as host to many of the largest river systems of the globe is clearly evident. However, part of the explanation of the large number of international basins in Africa is due to the characteristics of the political boundaries within the continent, which were 'drawn by the European powers with scant regard even for the physical geography of Africa, let alone the Africans' (Griffiths, 1993: 66). Despite the potential for resource disputes presented by the geography of the African continent, conflicts over water have surfaced more predominantly in the more developed and faster developing regions of the world, as shown in Table 6.3.

Table 6.2 Distribution of transboundary river and lake basins by region

Region	Number of rivers and lakes extending into two or more countries	Percentage of area within international basins
Asia	53	39
Europe	71	54
North America	39	35
South America	38	60
Africa	60	62
Total	261	45.3 (excluding Antarctica)

Source: Wolf *et al*., (1999)

International initiatives towards the improved management and governance of freshwater sources have proliferated in recent decades, with over 2000 treaties relating to common rivers and water basins (World Resources Institute, 1994). However, in many cases these regulations have been either disregarded or inadequate (Biswas, 1992). For example, in the Middle East, the Nile Water Agreements signed in 1959 between Sudan and Egypt are regularly ignored, with Egypt exceeding its quota for water extractions on an annual basis (Lee and Bulloch, 1990). The agreements are also insufficient in that they do not accommodate the upstream needs of Ethiopia (Biswas, 1992).

The resultant struggles over limited water resources could threaten already fragile ties between states in this region. Considering the case of the Euphrates alone, Syria and Iraq nearly went to war in 1975 after Syria and Turkey filled reservoirs behind two new dams, causing a sharp drop in the level of the river (Vesiland, 1993). Iraq depends on the Euphrates

▶

BOX 6.2 (continued)

Table 6.3 The location of major international water disputes

River	Countries in dispute	Issues
Nile	Egypt, Ethiopia, Sudan	Sitation, flooding, water flow and diversion
Euphrates, Tigris	Iraq, Syria, Turkey	Reduced water flow, salinisation
Jordan, Yarmuk, Litani, West Bank Aquifer	Israel, Jordan, Syria, Lebanon	Water flow diversion
Indus, Sutlei	India, Pakistan	Irrigation
Ganges–Brahmaputra	Bangladesh, India	Sitation, flooding, water flow
Salween	Burma, China	Sitation, flooding
Mekong	Cambodia, Laos, Thailand, Vietnam	Water flow, flooding
Parana	Argentina, Brazil	Dam, salinisation
Lauca	Bolivia, Chile	Dam, salinisation
Rio Grande, Colorado	Mexico, USA	Salinisation, water flow, agrochemical pollution
Rhine	France, Netherlands, Switzerland, Germany	Industrial pollution
Maas, Scheide	Belgium, Netherlands	Salinisation, industrial pollution
Elbe	Czechoslovakia, Germany	Industrial pollution
Szamos	Hungary, Romania	Industrial pollution

Source: Middleton *et al.* (1993)

and the Tigris for its water, most of which originates in Turkey and Syria.

In recent years, Turkey has once again begun to harness the waters of these rivers within the Greater Anatolia Project (GAP). The project includes the great Ataturk Dam, a further 22 dams, 19 hydraulic power plants and a whole series of very large irrigation tunnels to divert water to the Harran Plain (the land between the two great rivers which hosted the ancient agricultural civilisations of Mesopotamia). In 1990, when Turkey started to fill the Ataturk Dam, it stopped the flow of the Euphrates and 'suddenly achieved what years of diplomacy had failed to do – the bringing together of Iraq and Syria' (Lee and Bulloch, 1990: 13).

In the longer term, the Anatolia project is expected to reduce Iraq's receipt of water from the Euphrates by 90 per cent and Syria's to 60 per cent of normal flow (World Resources Institute, 1994: 183). The water resources development component of the GAP is estimated to be completed by 2010 at a cost in excess of US$32 billion (Greater Anatolia Project, 2007).

The problems associated with individual transboundary water systems are very country-specific. They accommodate factors including fears over national sovereignty, political sensitivities, historical grievances and national self-interest. Thus, developing international principles for the management and control of such resources remains problematic:

> **To a great extent, international organisations such as the UN system have deliberately stayed away from the issue of international rivers and lakes primarily because they have considered such issues to be politically sensitive.**
>
> **(Biswas, 1992: 7)**

Furthermore, it is also recognised that realising the outcomes of international agreements will depend on close consideration of local interests and appropriate understanding, tools and capacity at that level if measures to achieve those goals are to be implemented (Milich and Varady, 1998).

Moves towards a single unified framework for the environmental governance of the world's transboundary water resources continue to evade the international community (Uitto, 2004). Indeed, Biswas (2004: 81) suggests that subsequent to the United Nations Conference on Water held in 1977 in Mar del Plata, 'water disappeared from the international political agenda in the 1980s and 1990s' and was 'confined to the wings' during the Earth Summit at Rio de Janeiro (see Chapter 7) as issues of climate change and biodiversity, for example, took centre stage.

In 1997 the UN General Assembly adopted the convention on non-navigational uses of international watercourses after several decades of negotiations, but few countries to date have ratified that convention. In

BOX 6.2 (continued)

1998 the International Conference on Water and Sustainable Development initiated a priority action programme for the management and protection of transborder water resources, emphasising the need to facilitate the exchange of accurate and harmonised information among riverine countries; promote consultation at all levels, especially within pertinent international institutions and mechanisms; and define medium-range priority action programmes of common interest to improve water management and decrease pollution (UNEP, 2002).

Since 1996, 125 member organisations from 49 countries have been part of the International Network of Basin Organisations whose objectives include: facilitating the exchange of experiences and expertise among network members; promoting the principles and means of sound water management in sustainable development cooperation programmes; promoting the exchange of information and training programmes for the different actors involved in water management.

In 2000, at the request of governments within the Commission on Sustainable Development (see Chapter 7), the United Nations founded the World Water Assessment Programme (WWAP) to 'take on the task of systematically marshalling global water knowledge and expertise to develop over time the necessary assessment of the global water situation, as the basis for action to resolve water crises' (UNESCO, undated: 2). Whilst not focused on the issues of transboundary water resource issues specifically, this is an initiative towards assisting national governments in the development and implementation of national water management programmes, to initiate a global monitoring system and to pool the perspectives of the various UN agencies involved, and includes the triennial publication of a World Water Development Report.

Despite the relative lack of direct international governance of water resources, international institutions and bilateral organisations do play a role. Few developing countries have adequate funds for capital-intensive water development projects, for example, such that donor conditions can be influential in promoting the adoption of international treaties and the management of needs across the whole system. The Global Environment Facility (see Chapter 7) is considered to be a major facilitator of the implementation of international water agreements and of action plans around transboundary freshwater resources (Uitto, 2004).

Energy resource developments: the search for equity and efficiency

On the global scale, there is a strong relationship between GNP per capita and commercial energy use, as seen in Figure 6.4. In the mid 1990s, over 60 per cent of all commercial energy sources were consumed in the high-income economies as defined by the World Bank (World Bank, 1997a: 228). It is predicted that by 2030 world consumption of energy will be two-thirds more than it is currently and that the developing world will overtake the more industrialised world as the largest consumers (IEA, 2004). Stark regional variations in the use of energy sources persist, however, and are highlighted by the much-cited reference to the annual commute by car into New York alone using up more oil than is consumed by the whole of Africa excluding South Africa (Edge and Tovey, 1995).

Energy consumption is influenced by many factors including climate and the size of a country (see Canada in Figure 6.4). Past transformations of society from a predominantly agricultural and rural base to an urban and industrial base have been associated with a rise in total energy consumed and an increased reliance on commercial fuels rather than traditional fuels such as wood, charcoal and other biomass sources. This pattern led to the classification of 'high- and low-energy societies' on the basis of the level and type of energy consumed (Mather and Chapman, 1995). However, more recent discussions have centred on the nature and challenges of the 'third energy transition' (Seitz, 2002) that the world now faces (the first being the transition from wood to coal, the second from coal to oil) as industry, governments and consumers face the transition to a non-carbon-based global future (Dorian *et al.*, 2006).

'Energy intensity' is an indicator of how much energy is required to yield set increases in GDP. The

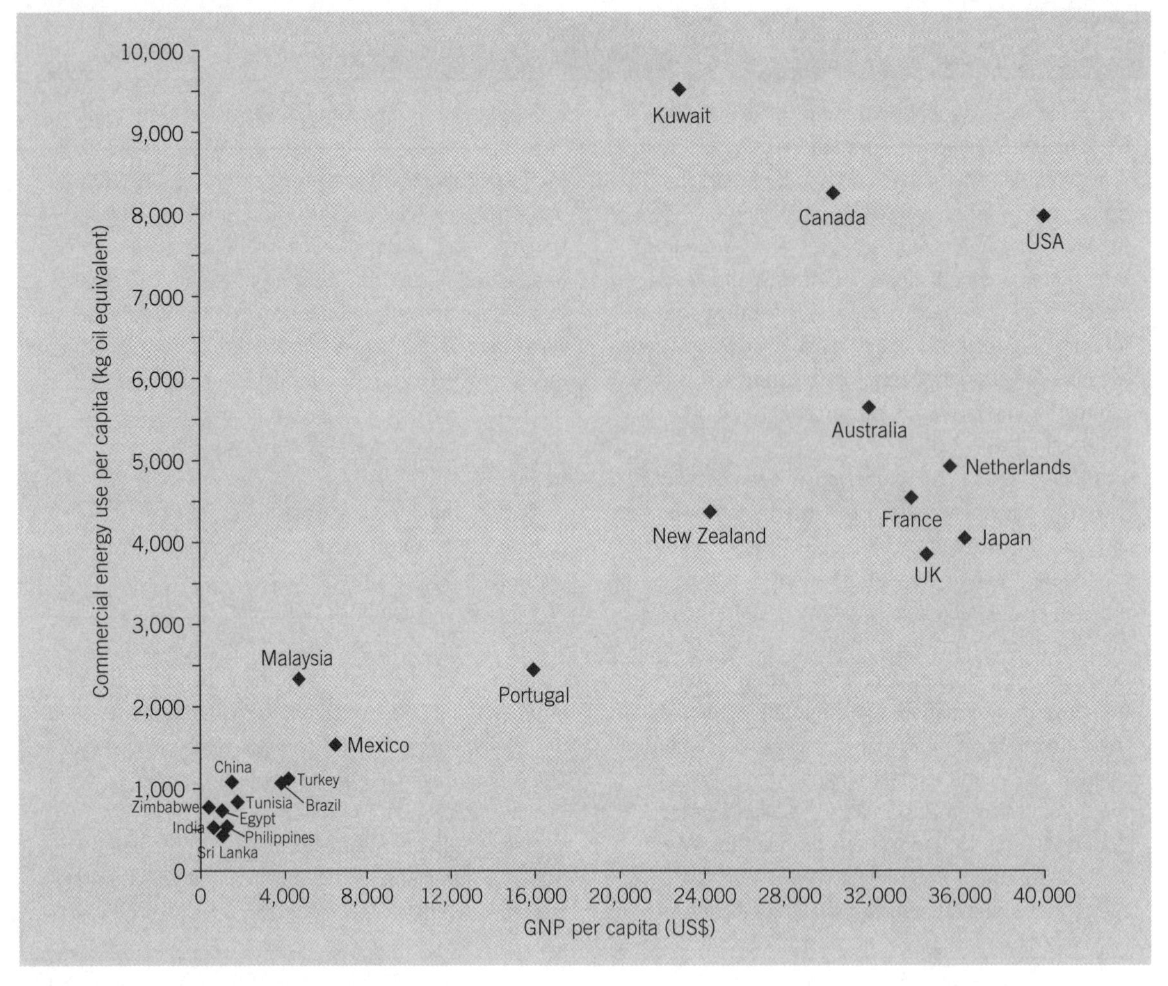

Figure 6.4 Energy consumption and economic development for selected countries
Source: Based on UNDP (2006) and World Bank (2005c)

experience of the industrialised countries has been that this ratio of energy consumption to Gross Domestic Product rose and then fell over time as shown in Figure 6.5 with improvements in materials science and energy efficiency (Edge and Tovey, 1995). Furthermore, the maxima reached during these processes has progressively declined. In part, this has been due to the decoupling of energy growth from economic expansion. In countries such as Japan, for example, in the era of oil price rises, GDP rose much more quickly than energy use. But while improvements in energy efficiency were important in explaining this pattern, there was also a shift in Japan towards development based on less energy-intensive service industries (and away from 'heavy' industry).

> **The energy-intensive industries did not disappear, however, they went abroad, to wherever the energy could be obtained cheaply. The energy is still 'used' by the Japanese economy, in the form of embodied energy imported as products.**
>
> **(Edge and Tovey, 1995: 325)**

Therefore the global energy demand remains unchanged (or in fact increased, through the energy expended in transportation). The Critical reflection box, which flowers should I buy?, illustrates how the extent of energy use in the production and delivery of products is often not only difficult to quantify but difficult to 'see' for consumers who may wish to make energy informed consumption choices.

In 2005, China surpassed Japan as the second largest consumer and importer of oil after the United States (Flavin and Gardner, 2006). Whilst coal continues to deliver more than two-thirds of China's energy demands (supplied domestically in the main), China currently looks to over 20 countries for its oil imports, with over 30 per cent coming from sub-Saharan Africa (Furniss, 2006).

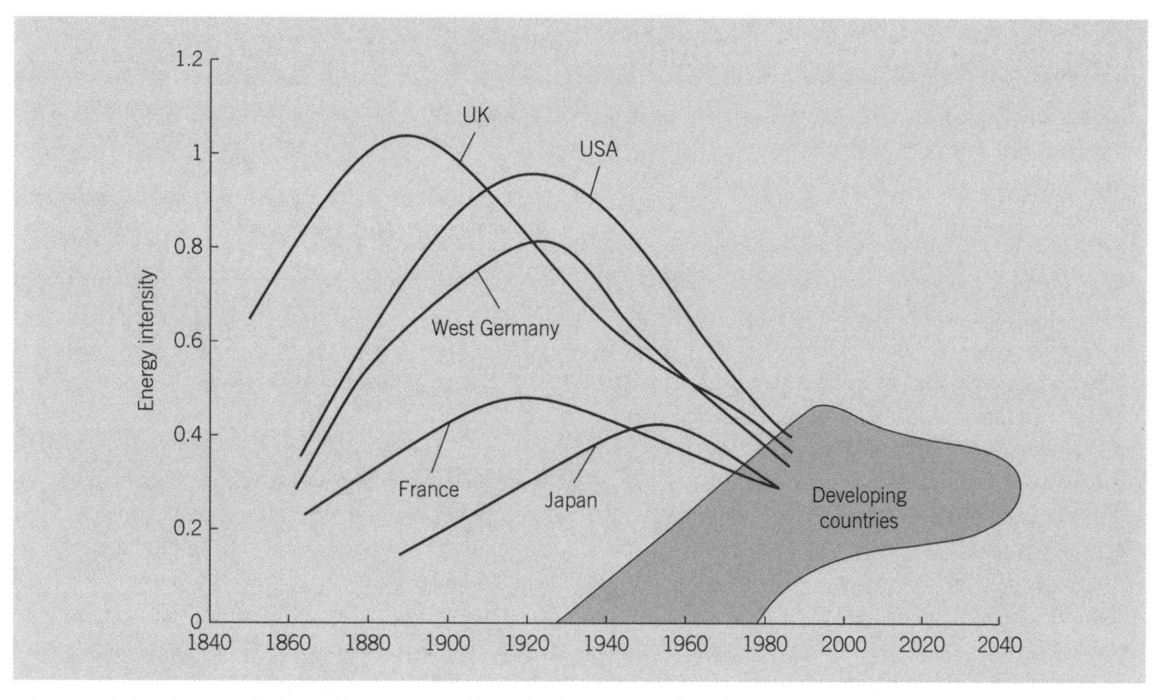

Figure 6.5 Energy intensity versus time in industrialised and developing countries
Source: Holdern and Pachauri (1992)

Critical reflection

Which flowers should I buy?

Consumers in the UK are increasingly interested in making consumption choices that are less harmful to the environment (in terms of their production and transportation) and make a positive contribution to economic and social development in the areas of production. This is revealed, for example, in their purchases of organic products (those where the production methods largely exclude the use of synthetic fertilisers, pesticides, feed additives and growth regulators) and Fair Trade products (a certification system for products sourced from companies who pay a premium price to producers to invest in social projects or business development and cover the costs of sustainable production and living). Sales of organic grocery products grew by £2.3 million per week in the UK during 2005 to reach over £1.2 billion for that year and an 11 per cent increase on the previous year (Soil Association, 2005). There are now 548 certified producer organisations representing an estimated 5 million people within over 58 countries in Africa, Latin America and Asia certified under the Fairtrade Labelling Organisations International (an umbrella body for independent national initiatives including the UK's Fairtrade Foundation) (Fairtrade, 2007).

A general understanding has emerged that buying goods closer to home is less harmful to the environment as a result of the reduced carbon dioxide expelled in transportation. However, the wisdom of such choices can be challenged with increased understanding of the nature of the production process. For example, the UK is the joint largest importer of flowers, with Valentine's Day (14 February) and Mother's Day (a Sunday in March) being dates when demand is at a peak – an estimated 7 million bunches being bought for Mother's Day in 2007 (War on Want, 2007). Three-quarters of flowers imported to the UK come from the Netherlands, although many originate from Columbia, Israel and Kenya. A recent report has suggested that if the environmental costs of artificial heating and

Critical reflection (continued)

lighting in the production of Dutch flowers is included, it may be less damaging to purchase flowers flown in from further afield, from Kenya, for example.

Williams (2007) conducted a Life Cycle Assessment (LCA) for the production and delivery of roses from two production centres, one at Oserian, Kenya and one near the Hook of Holland. LCA methods quantify all the resources used and emissions produced through production (including packing and cooling) and transport to Hampshire, UK (a regional distribution centre for World Flowers, which commissioned the research). In all, over 500 different data inputs were part of this assessment, encompassing all the energy consumed directly in the production, the energy use associated with the manufacture, use and delivery of fertilisers and pesticides, for example, and energy use and emissions associated with vehicle use and maintenance and the materials used within buildings.

Evidently, the Kenyan operation was found to use substantially less primary and fossil energy (the main Kenyan energy source being geothermal) and to emit smaller quantities of the principal greenhouse gas than the Dutch operation. Furthermore, the Kenyan operation was also able to achieve annual yields of marketable stems that were 70 per cent above those of the Dutch operation due to the more optimal year round growing conditions.

Table 6.4 The environmental costs of roses

Production and transport of 12,000 cut rose stems	Consumption of energy (MJ)	C02 emissions (kg)
Kenyan	53,000	2,200
Dutch	550,000	35,000

Source: Compiled from Williams (2007)

What further information would you wish to have before you decided which roses to buy?

Fears concerning the finite nature of fossil sources of energy were paramount in the energy debates of the 1970s and 1980s (and indeed underpinned writings concerning global future scenarios more widely, as seen above). Whilst there remains concern currently over increasing oil scarcity and the timing of 'Peak Oil' at a global level (after which the global economy will face inevitably declining production), in the immediate term it is increasingly questions regarding the security of energy supplies (including in an era of terrorism) and the environmental impacts of the continued dependence on fossil fuel sources that are shaping world energy production and trade and the search for alternative energy sources and measures for energy conservation.

Mineral resources: curse or cure for development?

The presence of diverse mineral supplies within a nation is a potential source of comparative advantage in economic development, as shown by the historical experience of Europe and North America. Currently, mineral production is extremely important in the economies of many developing nations, as shown in Table 6.5. As well as contributing to the export earnings of a country, mineral development can assist in attracting foreign capital, in creating jobs and stimulating demand for local goods and services, in raising taxes, in prompting infrastructure developments and in providing options in a country's route to industrialisation (Plate 6.2). Yet evidence suggests that, since the 1960s, the 'mineral economies' of the developing world, defined as having over 40 per cent of their export earnings from the mineral sector, did not perform as well on conventional economic development indicators as was predicted or in relation to the less well-endowed countries. Whilst the conventional view had been that favourable natural resource endowment was particularly important in the early low-income stages of the development process, this has been increasingly challenged through the development of the idea of the 'resource curse' as considered below.

Table 6.5 The importance of mineral production in the economies of developing nations. Percentage of total merchandise exports via fuels, ores and metals for selected countries, 2004

Country	Percentage of merchandise exports via fuels, ores and metals	Country	Percentage of merchandise exports via fuels, ores and metals
Algeria	97	Nigeria	98
Armenia	24	Norway	51
Australia	35	Oman	84
Azerbaijan	83	Papua New Guinea	71
Bolivia	57	Peru	53
Chile	57	Russian Federation	58
China	4	Saudi Arabia	86
Colombia	39	Senegal	23
Ecuador	54	South Africa	31
Guinea	72	Syria	69
Indonesia	25	Togo	13
Jordan	13	Trinidad and Tobago	60
Kazakhstan	79	Tunisia	11
Kuwait	79	Venezuela	88
Libya	95	Yemen	92
Niger	57		

Source: World Bank (2006e)

Key idea

The 'resource curse thesis'

In 1993, Auty, in his book *Sustaining Development in Mineral Economies*, proposed the 'resource curse' to explain how and why high levels of resource endowment at a country level were not translating into higher levels of economic development (and indeed could explain the underperformance of such countries relative to those 'less well endowed').

The suggestion of a 'resource curse' was based on detailed research of the economic performance of countries from the 1960s through to the early 1980s where exports of hard minerals such as copper, bauxite and tin accounted for more than 40 per cent of export earnings (and at least 8 per cent of GDP). Whilst these countries sustained significantly higher levels of investment than the non-mineral economies, their GDP growth per capita was found to be lower (and could not be explained by declining terms of trade for those products).

The conclusion was that 'the cause of under-performance of the hard mineral economies lies not so much in a lack of investment resources as in the inefficiency with which those investment resources were deployed' (Auty, 1993: 5). In particular, the roots of the underperformance were found to be in a number of characteristics of the mining sector and its production function.

For example, mineral production is highly capital-intensive and is regularly controlled by a few multinational companies that are able to raise the necessary capital. In turn, there may be little impact on national employment, production is often heavily mechanised with expatriate labour being used to provide specialist skills. Mineral exploitation also shows marked enclave tendencies and may yield only modest local production linkages due to factors including the often remote location of mineral resources, the importation of specialist technologies and few local factories being established to supply imports or process ores prior to export.

In 2004 Christian Aid undertook further comparative research of the economic, poverty and human development performance over four decades between six oil-producing nations (Angola, Iraq, Kazakstan, Nigeria, Sudan and Venezuela) and six non-oil-producers (Bangladesh, Bolivia, Cambodia, Ethiopia, Peru and Tanzania) that further confirmed the operation of a 'resource curse'. The oil economies were found to have achieved slower growth, an average of 1.7 per cent per annum compared to an average annual growth of 4 per cent in non-producing countries. The military expenditures (and the size of armies maintained) within oil economies averaged 6.8 per cent per annum (against 2.9 per cent in non-oil economies), supporting suggestions that oil-producing countries are highly prone to conflict.

Whilst the findings in relation to human development were less strong, both life expectancy at birth and literacy rates were found to have improved slightly more in non-oil economies than in the oil economies.

> **At the very least, they show that oil has done nothing to significantly improve the lives of ordinary people in oil-rich countries.**
>
> **(Christian Aid, 2004: 5)**

Ainger (2004: 10) summarises a 'paradox of plenty' as 'the more a country is enriched by the extraction of primary resources and the more its dependency on them deepens, the lower its ratings for human development drop'.

Authors such as Watts (2004) have included factors of the geostrategic significance of oil to capitalism and to US hegemony in particular as part of the explanation for how high resource endowment can undermine economic growth and democracy and be a source of civil unrest. In 2001 approximately one-quarter of ongoing wars and armed conflicts were considered to be linked with resource exploitation (Azinger, 2004). It should be considered also that in addition to these 'resource wars' there are many more local struggles and conflicts that occur over natural resources and there is now a sizeable literature that puts environmental concerns – from resource mining through to urban pollution as central issues on the human security agenda (Najam, 2003). These issues are further explored in Chapter 10 (see Critical reflection box on China in Africa).

Resource constraints and the development process

The assertion that resources are fundamental to economic development has the corollary that limitations in the quantity or quality of resources will hamper these processes. This section investigates the concept of resource scarcity and, through the detailed example of water, it shows that resource constraints do operate in development. However, the nature of the 'constraint' is highly dynamic; it needs to be considered in relation to particular groups of people and specific locations; and it may owe less to the characteristics of the resource base than to issues of control, use and management of resources.

Different notions and sources of resource scarcity

Whilst the works of Malthus, Ehrlich and Meadows referred to above continue to attract support and to be applied in contrasting socio-economic and historical contexts (see also Chapter 5), their envisaged limits to human development due to insufficient quantities of resources to meet demands have not materialised. The work of Malthus typifies 'physical resource scarcity' ideas, whereby resources and resource functions focus on those that are physically finite (non-renewable) and/or the rates of consumption of renewable resources are in excess of the rates of such renewal. In such circumstances, resources are predicted to become increasingly scarce and costly with conflicts over resources likely to increase.

It has been recognised for some time that 'threats to the sustainable use of resources comes as much from inequalities in peoples' access to resources and from the ways in which they use them as from the sheer numbers of people' (WCED, 1987: 95). Rather than absolute resource scarcity, it is evident that particular groups of people and locations have suffered (and continue to experience) resource constraints in development relative to others. The concept of resource scarcity becomes much more complex when applied beyond the 'stock resources' such as non-renewable minerals and fossil fuels. Indeed, the physical existence itself of a resource does not ensure its availability for development; access may be difficult or there may be a lack of sufficient capital or appropriate skills and technology to bring the location into production.

Plate 6.2 Diamond mining, Namibia
(photo: J. Elliott)

In contrast, a resource which exists only in small quantities in a few locations may not be scarce if demand is low (perhaps due to issues of income levels and distribution). Furthermore, 'geopolitical' resource scarcity (Rees, 1990) may also be produced in circumstances where a resource is heavily localised and producing countries are able to restrict output and/or exports, such as prompted by the Organization of Petroleum Exporting Countries (OPEC) during the Arab–Israeli conflict in 1973.

Issues of human influence on resource 'scarcity' are also to the fore in 'Cornucopian' models that conceive of resource scarcity in economic terms. Within such thinking, price increases with rising scarcity stimulate human responses such as resource substitution, the exploration of new areas and innovations in technology to raise efficiency.

Rees (1990) forwards a further category of resource scarcity referring to scarce 'qualities' of resources such as with respect to attractive landscapes, wildlife or clean air. Here quality may refer to aesthetics (in the case of valued landscapes) or physical characteristics (such as the ability of the atmosphere to absorb pollutants).

Historically, market forces have not operated for these resource qualities as they have with resources such as oil to influence demand or to generate alternatives (Mather and Chapman, 1995). However, this is changing, such as through government payments to landowners to maintain certain kinds of uses and features and charging for public entry to protected areas and wildlife reserves.

Furthermore, as seen later in this chapter, scarcity in terms of quality is in many instances increasing under the impacts of development. In short, questions of resource scarcity may be shaped by factors not only of supply, but also of demand, the economics of the market and political decision making. These ideas are now developed in relation to water resources.

How water resources may 'constrain' development

Despite the evident complexity of the factors shaping the relationship between resources and development, it is possible to identify resource constraints that are common across a number of locations and groups of people, particularly in the developing world. Within agricultural production in the tropics, for example, it is precipitation not temperature that controls plant growth. This is in direct contrast to the 'boundary conditions' (Biswas, 1992) within temperate agriculture that are set by the large changes in temperatures between summer and winter months, for example. For the majority of farmers of the tropics dependent on rainfed production (as discussed in Chapter 10), precipitation is highly variable (particularly where rainfall totals are low), often localised in distribution and having a high intensity. Challenges for the development of tropical agriculture therefore include managing the variable extent and timing of rainfall and the potential implications for soil erosion and fertility presented by the high kinetic energy of rainfall.

In the arid zones of the developing world at least two distinct strategies for survival evolved in response to the varied pattern of water resources over space. One is based on the low-intensity use of dispersed resources (water being of fundamental importance), such as within pastoral societies; the other is based on the intensive use of more favoured locations, such as where plentiful water resources enable permanent cultivation and higher densities of population. Although distinct, they may be operated in conjunction (see Chapter 10). Box 6.3 looks at the characteristics of the irrigation technologies and agro-ecosystems which the Marakwet peoples in Kenya developed in order to sustain occupation and year-round agriculture within the arid Kerio Valley. It is an example of how the constraints of the physical environment are being overcome through technological

BOX 6.3

Extending the ecological margins for development: indigenous irrigation in the Kerio Valley, Kenya

The system of stream diversion and canal networks along the east wall of the Rift Valley in Kenya, the 'Kerio cluster', is referred to as the most complex and extensive indigenous water management system in Africa south of the Sahara (Adams and Anderson, 1988). Since the nineteenth century at least, the Marakwet peoples have applied considerable engineering skills to exploit the challenging ecologies of the Rift Valley, to enable permanent settlement and the cultivation of cereals and vegetables across the valley floor.

The Kerio Valley itself is an arid area lying around 1000 m above sea level, receiving less than 600 mm annual rainfall and with scrub-like natural vegetation. However, less than ten miles away, but separated by the precipitous Rift Valley escarpment, is the substantially different natural ecology of the Cherangani Plateau at 3000 m above sea level. This is a well-watered zone, receiving around 1500 mm of rainfall annually, supporting evergreen forests, and crossed by two major and several minor rivers and streams.

Over the years the Marakwet people have constructed dams, furrows, channels and terraces to divert and carry the water from the minor streams of the plateau to the fields of the valley floor. Earth and stone channels with simple brushwood and stone dams are used to modify the variable flow of the natural streams and to carry water to the majority of fields between the foot of the escarpment and the east bank of the Kerio River. Irrigation furrows are also used to irrigate kitchen gardens and fields among the villages of the hillsides (Adams and Anderson, 1988). Complex systems of water allocation for the maintenance of structures have developed within the Marakwet communities (Adams, 1996).

By the 1930s, the system of irrigation and livelihood was the focus of much commentary by outsiders, including colonial officials such as the district officer in the area at that time (Henning, 1941: 270):

> **The plan of using streams at the top of the escarpment to water parched fields 3000 feet [1000 m] below shows a practical imagination which has sometimes been supposed not to exist in the African.**

Subsequently, there have been many different views concerning the environmental sustainability of the system and projects to modify and extend it. It is considered, however, that the current extent of cultivation and settlement along approximately 50 kilometres of the Rift Valley above the Kerio River is at least as extensive as it was in the early nineteenth century (Adams, 1996).

innovation but also through new social institutions for management of the water and land resources.

Constraints beyond physical factors

In 1987 the Brundtland Commission identified a large group of the world's agricultural producers who could be considered 'resource-poor'. It was estimated that one-quarter of the global population at this time lived in very poor rural environments in terms of the inherent characteristics of water resources, among other things, and the acquired features of those environments (the outcomes of development interventions).

These farmers, because they are poor in income terms, lack the financial resources to invest in the capital equipment and necessary inputs to raise production or manage the resources in these areas appropriately, yet they live in areas of the world which require precisely such levels of investment if they are not to be degraded further. In addition to what could be considered their economic and ecological marginality, these farmers are often also marginal in a political sense. For example, they may have little political power in the sense of their participation in, and control over, institutions which structure their lives.

At the turn of the century, over 1 billion people around the world were suffering serious water resource scarcity in terms of their lack of access to safe drinking water (Department for International Development, 2001a: 13). 2.4 billion people also lacked access to adequate sanitation (ibid.). Many of those people living without such basic services are in Asia, as shown in Table 6.6.

The centrality of these issues in human health and development is confirmed in Table 6.7, which shows the potential impact of resource improvements on mortality from the most common causes of death. Such diseases are a function not only of the presence of disease vectors (i.e. water quality), but also the quantity of water a household can command (through public piped supply, purchase or collection) and the provisions for removal of water once used.

Table 6.7 The potential health benefits of water supply and sanitation improvements

(a) Effects of improved water and sanitation on sickness

Disease	Millions of people affected by illness	Median reduction attributable to improvement (%)
Diarrhoea	900a	22
Roundworm	900	28
Guinea worm	4	76
Schistosomiasis	200	73

(b) Effects of water supply and sanitation improvements on morbidity from diarrhoea

Type of improvement	Median reduction in morbidity (%)
Quality of water	16
Availability of water	25
Quality and availability of water	37
Disposal of excreta	22

[a] Refers to number of cases per year

Source: World Bank (1992) © International Bank for Reconstruction and Development/The World Bank

Table 6.6 Water supply and sanitation coverage by region, 2004

Region	Population millions	Percentage with access to		Number unserved (millions)	
		water	sanitation	water	sanitation
Northern Africa	152,085	91	77	13,423	34,617
Sub-Saharan Africa	734,641	56	37	322,351	463,208
Eastern Asia	1,388,052	78	45	302,336	760,750
South Asia	1,528,108	85	38	225,881	955,041
South Eastern Asia	548,525	82	67	97,659	183,460
West Asia	194,170	91	84	17,983	31,187
Latin America and Caribbean	553,725	91	77	50,374	124,870
Oceania	8,712	50	53	4,346	4,133
CIS (Russia)	278,264	92	83	21,485	48,060
Europe, North America and Australia	1,002,984	99	99	12,951	6,521

Source: WHO and UNICEF Joint Monitoring Programme – water and sanitation data. www.wssinfo.org/en/233_wat_africaS.html © WHO and UNICEF Joint Monitoring Programme. All right reserved.

There is plenty of evidence that those people who have to depend on water vendors use less water and suffer greater ill health than those that have piped supply (Hardoy *et al.*, 1990). In addition, the costs of piped water supplies are often higher for poor urban residents than for richer (as well as accounting for a higher proportion of their more limited budgets). A poor resident of Jakarta typically pays ten times more than a rich resident for a litre of clean water and suffers two to four times more gastroenteritis, typhoid and malaria (World Bank, 2002a: 141).

There is also an increasing consensus that privatisation of public services in many countries (and most notably in the water sector) over the last decade has led to deepening poverty through higher tariffs, inability to afford sufficient quantities of water and households cutting back on other essential expenditures including food and school fees for children (War on Want, 2004).

Furthermore, water resource constraints operate in conjunction with other features of the urban environment – such as the cramped housing conditions which make for the rapid transmission of disease – to aggravate ill-health. In these ways, many urban households remain in poverty and individuals continue to lack the basic rights of good health and the ability to participate in their own development. A target to halve the proportion of people without sustainable access to safe drinking water and basic sanitation is now encompassed within the Millennium Development Goal number 7.

No simple resource–development–environment models

Change is an inherent characteristic of all systems for securing livelihood, be they within more or less developed economies. The nature of resource constraints in development also varies over time and, indeed, processes of change do not impact on all individuals or groups of people equally within any society.

Continuing the example of water, research into drought-coping mechanisms and the consequent impacts on food security conducted in Zimbabwe during a period of exceptionally low rainfall in the early 1990s revealed significant differences in the nature and sequencing of household responses according to factors including socio-economic status and length of residence in the areas under study (Campbell *et al.*, 1991).

The poorest groups in all areas under study liquidated their assets earlier in response to food shortages. As the food deficit persisted, these groups also relied increasingly on the sale of their labour. In contrast, the wealthier groups took on alternative options to raise cash, such as the brewing of beer for sale.

Irrespective of socio-economic status, those households resident in the long-settled communal villages were able to draw upon established social networks to facilitate the redistribution of food in the early stage of a deficit situation, whereas those newly arrived in resettlement areas were more reliant on the sale of their labour and crafts.

It is evident in this research that water shortage impacts most detrimentally on the poorer groups and the newly settled households who become more vulnerable to food insecurity. Further research is needed into the capacity of those households that have responded via the diversion of labour away from subsistence production to cope and, critically, to become more secure in the future.

The Zimbabwean example confirms that whether or not resource conditions constitute a 'constraint' on development is defined at the level of the individual and is place-specific. This is in response to varied and dynamic political, economic and social forces in addition to environmental factors such as meteorological drought. All shape the nature of the resource 'constraint' and the ability of certain groups in society to respond and take action to overcome such limitations.

On a wider scale, and in relation to resource sectors other than water, national governments have been able to import food to overcome internal limitations of land and agricultural production. Similarly, innovations related to the mobility of energy supplies have enabled the transformation from a low-energy to high-energy society despite the relative paucity of indigenous energy resources as seen in the case above of Japan.

In summary, it is evident that resource constraints on development do exist; many people in the developing world experience the harsh reality of hunger, disease, pollution and hazard, and the persistence of poverty is testimony to the unresolved challenge of finding the means to overcome continued resource constraints in development at the local level. However, it is increasingly acknowledged that the persistence of these development challenges owes less to the geographical characteristics of the natural resource base than to issues of control, use and management of resources.

These factors also underpin much of the discussion of resource degradation in the following sections. However, the fundamental challenge for sustainable development in the future is in overcoming the resource constraints of the poorest groups in society, where options for development are most restricted. The examples given so far show how these groups have a very close relationship with the physical resource base, live in some of the most impoverished environments of the world, lack precisely the means required to prosper in those areas, and through their poverty may contribute to the further degradation of those resources.

Environmental impacts of development

Just as the characteristics of the resource base shape the challenges and opportunities for development, so development processes themselves impact on the environment and the varied functions it performs. Indeed, it has been understandings of the detrimental impacts of development (as well as resource scarcities) that have shaped environmentalism in the modern era. In the 1960s/1970s, for example, environmentalists argued strongly that development was incompatible with conservation. Particularly in America, the undesirable side effects of industrial development such as air and water pollution were being experienced and an emerging environmental movement campaigning on these issues was fuelled by the generally anti-establishment middle-class sentiments which prevailed at that time (see Elliott, 2006).

The dissonance between development and the environment was further reinforced in development thinking during the 1960s by the primacy given to economic growth (Chapters 1 and 3), within which the limiting factors were considered to be finance and technology, not natural resources. Although it is now widely acknowledged that a lack of development can also be a prime factor in resource degradation, this section emphasises the ways in which the characteristics of past development patterns and processes have contributed to some of the major contemporary global environmental issues.

What is degradation?

Whilst it is easy to find patterns in the environment that suggest change for the worse over time, it is important to retain a consideration for how what is valued as a resource and therefore what is considered degradation are socially constructed. For example, previous sections have highlighted how the environment provides a number of interrelated resource functions for the development of society: as inputs into the economic system; as a sink for the waste products of human activities, including economic production; and in terms of 'services' such as the maintenance of the gaseous composition of the atmosphere or for aesthetic pleasure and recreation. Subsequent sections confirmed that such functions are neither discrete nor time-bound; resources may serve multiple ends and the significance of particular functions may change with use, political interest or economic developments.

Resources rarely cease to exist in absolute terms as a result of development. Instead, they become 'degraded' in relation to the actual or possible future functions they can perform. For example, as land becomes 'reduced to a lower rank' (Blaikie and Brookfield, 1987: 1), agricultural productivity declines, requiring capital and labour inputs to rectify the situation and prevent further losses. Although absolute exhaustion of the resource may be avoided, in this case through the application of chemical fertilisers, the costs of resource degradation in terms of required remedial actions may be substantial.

Furthermore, environmental impacts are not necessarily felt solely by a particular land user at that point in time. In the case of chemical fertilisers, for example, these may contribute to the nitrification of water courses and impact on downstream users over time. Indeed, Blaikie and Brookfield (1987) refer to land degradation as the 'quiet crisis'; over longer periods of time, processes of degradation make land users more vulnerable to adverse conditions such as drought. Critically, any discussion of environmental degradation needs to be not only in relation to particular resource functions at specific times and places, but also in relation to identified interest groups:

> **To a hunter or herder, the replacement of forest by savanna with a greater capacity to carry ruminants would not be perceived as degradation. Nor would forest replacement by agricultural land be seen as degradation by a colonising farmer. Usually there are a number of perceptions of physical changes of the biome on the part of actual or potential land-users. Usually too, there is conflict over the use of land.**
>
> **(Blaikie and Brookfield, 1987: 4)**

Plate 6.3 Deforestation in Surinam – rainforest along the Tapanahony river is cut down to make room for agriculture
(photo: Ron Giling, Panos Pictures)

Deforestation

The removal of forests and woodlands through cutting or deliberate fire at rates in excess of natural regeneration processes is perhaps the most obvious global pattern of resource degradation (Plate 6.3). In 1982 the Food and Agriculture Organization (FAO) estimated the rate of tropical forest loss to be around 114,000 square kilometres per year. In 1990 the World Resources Institute reported a substantially higher figure of 204,000 square kilometres lost annually throughout the 1980s. In 2001 the FAO calculated the global loss of forest area during the 1990s to be 94 million hectares, equivalent to 2.4 per cent of total forests (UNEP, 2002: 92).

However, understanding of these figures and trends requires caution. As Grainger (1993) suggested, all estimates of tropical deforestation rates are unsatisfactory in some sense, although some can be considered more unsatisfactory than others. Table 6.8 identifies the principal issues involved and the Critical reflection on deforestation data investigates these challenges of data and their interpretation in more detail. But too often such problems of data remain unexplored.

In considering the particular case of Nepal, Thompson and Warburton (1985) argued that the 'uncertainty' surrounding forest degradation must be made explicit and part of future planning and development:

> **If the most pessimistic estimates are correct, the Himalayas will become as bald as a coot overnight . . . if the most optimistic estimates are correct, they will shortly sink beneath the greatest accumulation of biomass the world has ever seen.**
>
> **(Thompson and Warburton, 1985: 116)**

Table 6.8 Some problems with data on 'deforestation'

1. Lack of uniformity in the definitions of *forest* and *forest land* between different countries
2. Variations in the definition of *forest destruction* (it may include selective felling or it may be complete clearance only)
3. Differing techniques in undertaking forest inventories
4. A common concentration on commercial timbers to the exclusion of non-commercial trees and other woody plants
5. Data are often for individual countries rather than subdivided by forest type
6. Data may be withheld by governments or companies for strategic reasons

Source: Adapted from Reading *et al.* (1995)

Critical reflection

Getting beneath the deforestation data

The Food and Agriculture Organization is acknowledged to provide the most comprehensive information on global forest cover. Since 1946 it has been compiling data that is used by ecologists, climate change scientists, policy makers, educators and environmental activists (Matthews, 2001). In 2000 and 2005, the FAO published a Global Forest Resources Assessment report (see Table 6.9). But caution is required when considering this data. For example, the FAO continues to depend on national inventories supplemented by satellite information and expert opinion. Yet in over half of the developing countries' inventories used within the 2000 assessment, for example, the data used were over ten years old (Matthews, 2007) and of 137 countries included, only 22 had systems for continuous modelling and monitoring. Very often, national data (including within more developed countries (MDCs)) is sourced and reported over different time scales and in different ways, such that even national data sets are not completely comparable.

Importantly, the definition of 'forest land' used by the FAO has changed; in 1990 'forests' were defined as areas of more than 20 per cent cover for MDCs and more than 10 per cent canopy within less developed countries (LDCs). In 2000 the definition was standardised as more than 10 per cent, but whilst this has given greater comparability, it has had the effect of 'revising upward' the extent of forest lands within MDCs.

> **The results of the FRA 2000 are not what they seem to be. Changes in assessment methodologies explain much of what appears, at first, to be real change.**
>
> **(Matthews, 2001: 7)**

Interpreting trends over time in relation to forest status is also difficult. Changes in the extent of forests are often reported as aggregates of both natural and plantation forestry, yet these are very different in terms of the biodiversity hosted, their productivity, management requirements and amenity value, for example. Even where they are differentiated, it is now understood that there are very substantial problems in attempting to 'go back' to attest the extent of 'original' vegetation, including for the way in which such modelling depends on estimates of historical vegetation, population levels and activities, climate interrelationships and so on.

The work of Melissa Leach, Robin Mearns and James Fairhead is well known for their detailed research into environmental change (particularly in relation to forestry) within Guinea and the West African region more widely. Drawing on data from oral histories, participant observation, archival sources and sequential aerial photography, they have revealed quite different understandings of forest change to that generated and sustained within official thinking (including colonial policies but also as portrayed within global assessments). They found that extensive 'forest islands' were created by human occupation

Table 6.9 Change in forested land by region, 1990–2005

Region	Forest area (1000 ha)			Percentage change in forest area per year	
	1990	2000	2005	1990–2000	2000–2005
Africa	699,361	655,613	635,412	–0.64	–0.60
Asia	574,487	566,562	571,577	–0.14	0.18
Europe	989,320	998,091	1,001,394	0.09	0.07
Caribbean	5,350	5,706	5,974	0.60	0.90
North and Central America	710,790	707,514	705,849	–0.05	–0.05
Oceania	212,514	208,034	206,254	–0.21	–0.17
South America	890,818	852,796	831,540	–0.44	–0.50
World	4,077,291	3,988,610	3,952,025	–0.22	–0.18

Source: Food and Agriculture Organization (2005 and 2005–06) and UNEP (2002) UNEP/GRID-Arendal

▶

Critical reflection (continued)

and were being sustained by active agro-forestry management activities rather than being the last remnants of ancient forests 'assumed' over long periods of time to be the natural vegetation of the region between the Equatorial belt to the south and the Sahara desert to the north.

Their work has also promoted investigations into how and why certain narratives of environmental change (even when substantially erroneous) become so entrenched and continue to shape development interventions via the activities of government forestry departments and donor projects and programmes, for example.

Consider why we hear and see so much more in the academic literature and popular media about global patterns and processes of deforestation than local experiences of forestry and woodland change?

Deforestation itself is not a new phenomenon. Much of Europe was largely cleared of forest in the Middle Ages, and the forest cover of North America has been reduced from an estimated 170 million hectares to 10 million hectares (1 hectare = 10,000 square metres = 0.01 square kilometres) over the period since colonisation (Goudie, 1990: 38). However, it is the contemporary rates and extent of forest removal across the humid tropics which are unprecedented and the focus of much global environmental concern at the present time.

In part, the degree of popular and scientific interest in deforestation reflects the varied functions woodland resources currently fulfil, and are expected to perform in the future, for diverse interest groups across all spatial scales. Indeed, forests and woodlands could be considered the archetypal multiple resource in that they provide a range of raw materials, including fuelwood, fruits and medicines; they also play a vital role in the maintenance of plant and animal life support systems and global bio-diversity; they modulate climate; and they perform a host of less obvious resource functions which are now acknowledged to be essential for societal well-being, including opportunities for recreational enjoyment.

Complex drivers of deforestation

A close association at the global level between shrinking forests and expanding areas of cropland has been identified. Some 70 per cent of world forest loss in the 1990s was attributed to agricultural conversion (FAO, 2001), predominantly under permanent rather than shifting systems although the patterns and scale of these differed regionally (UNEP, 2002).

The driving factors behind both deforestation and reforestation, however, are quite poorly understood (Grainger, 1993). In particular, there is a need to distinguish between the proximate and fundamental causes of deforestation. For example, although 'farmers are probably the most significant anthropogenic influence' (Adams, 1990: 124) particularly in Africa and Latin America, the ultimate cause of deforestation may lie in a complex mix of factors that operate in any particular locality (Chapter 10 looks in more detail at the role of woodland resources in rural livelihoods and the complexity of forces in environmental transformations).

In the Rondonia region of Brazil, where impoverished farmers colonised areas bordering newly constructed highways and were accepted to be the principal agents in the exponential increase of deforestation rates from the mid 1970s, Colchester and Lohmann (1993: 15) stressed the underlying failure of economic development planning as fundamental in the explanation of these patterns. They argue that agrarian reforms

> **have failed to achieve their targets, they have failed to alleviate rural poverty, they have failed to secure peasant tenure, they have failed to effect adequate redistribution of land, they have failed to stem the rising tide of landlessness and, above all, they have failed to respond to the needs and demands of the peasants themselves.**

In summary, the inadequacies of past development are prevalent throughout the most regularly proffered explanations of tropical deforestation, whether they are stated as the inappropriateness of government policies, the lack of agricultural skills on behalf of farmers or the penetration of capitalism and the modern debt crisis. The very direct impacts of deforestation, ranging from local livelihood options to global resource functions including climate change and biodiversity, are seen in subsequent sections and chapters. Less direct impacts have also been revealed recently in the increased vulnerability of coastal residents to cylcones and tsunamis in regions of mangrove destruction.

Soil erosion and desertification: local environmental issues experienced globally?

Soil erosion is a further and related global environmental issue that arises fundamentally from resource use at rates in excess of regeneration through natural processes. Soil-forming processes occur at slow rates (a few millimetres per century) and are controlled at the local scale by highly specific factors of geology, climate and topography. Processes of soil erosion may in contrast be very rapid, perhaps several centimetres a year.

Soil degradation may occur naturally as a result of the action of wind and water over time, but accelerated rates of erosion occur through the interaction of human activities with such biotic agents. The removal of tree cover exposes soil surfaces to the direct impacts of rainfall and also removes the effect of root binding on soil stability. Soil may subsequently be degraded 'quantitatively' through its physical removal from one location to another or 'qualitatively' referring to losses in fertility, in moisture and nutrient content or changes in chemical composition, in soil flora and fauna (Barrow, 1995).

Stocking (2000) argues that land degradation is the single most pressing current global problem. It is estimated that since 1945 an area roughly the size of China and India combined has been eroded 'at least to the point where the original biotic functions are impaired' (Stocking, 2000: 287). On the basis of assessments for the purposes of the World Map of Soil Degradation, Stocking reports 22 per cent of the Earth's vegetated and potentially cultivatable surface as now degraded to a measurable extent.

In continuity with processes of deforestation, however, assessments of soil erosion also suffer from problems of data quality and comparability of measurements. It has been suggested that estimates of land damaged or lost for agricultural use through soil degradation can range from 'moderate to apocalyptic' (World Bank, 1992: 55). As Stocking (2000: 292) warned:

> **certainly, soil erosion and land degradation are serious in some places; equally, their degree and extent have been overstated, often for political ends. Rhetoric on soil erosion can be a useful device to displace people, point fingers of blame, promote the cause of environmental agencies and the professionals who work in them, and mobilise international aid.**

As with deforestation, it is also extremely difficult to attribute the complex processes of soil erosion to a single cause. In 1994 the German Advisory Council on Global Change (cited in UNEP, 2002: 64) suggested that soil degradation worldwide could be attributed to factors of overgrazing (accommodating 56 per cent of soil degradation), deforestation (30 per cent), agricultural activities (27 per cent), overexploitation of vegetation (7 per cent) and industrial activities (1 per cent). Clearly, these processes themselves may be interlinked. However, there is evidence to suggest that soil degradation is getting worse on the global scale (Stocking, 1987) and human impacts on soil degradation could be expected to widen in the future through processes of acid deposition, radioactivity and pesticide pollution (Barrow, 1995).

Soil erosion is one component of the wider processes of desertification; desertification occurs wherever

> **land is periodically deprived of adequate moisture, where soils are infertile, or poor drainage leads to saline conditions, crusts or pans, if vegetation is present it will probably be easy to disturb and slow to re-establish. Because the end product often resembles desert the process has been termed desertification.**
>
> **(Barrow, 1995: 105)**

In continuity with soil erosion, desertification may be triggered through natural processes, such as drought or the actions of wild animals in destroying vegetation, e.g. rabbits, termites or locusts. However, human actions most regularly introduce or speed up these processes, fundamentally through resource use in drylands at rates in excess of natural processes of soil, moisture and vegetation regeneration. Population pressure, overgrazing and deforestation are the most widely proposed explanations of desertification, but, as already seen, these processes themselves may reflect deeper underlying forces of change.

Conserving soils

In large measure, the physical processes of soil erosion are well understood and technical packages for conservation are available. However, soil erosion clearly persists. Blaikie in 1985 suggested that the most promising

Plate 6.4 Gully erosion, Zimbabwe
(photo: J. Elliott)

direction in terms of the explanation of continued soil degradation was that 'conservation is as much about social processes as physical ones and the major constraints are not technical, but social' (Blaikie, 1985: 50).

This work did much to raise understanding of the individual land user and the wider political economic context in which resource management decisions have to be taken. For example, many colonial interventions in African agriculture were justified on the basis of the conservation of soil. Measures such as compulsory construction of contour banks and limitations on the number of stock owned, however, may have done more to protect settler interests in agriculture in southern Africa than they did to prevent soil erosion (Elliott, 1990). Many soil conservation programmes in developing countries are now delivered through foreign aid institutions within which the implicit assumptions of this 'colonial model' may persist (Blaikie, 1985).

In 1977 the United Nations Environment Programme (UNEP) organised a World Conference on Desertification (UNCOD) in Nairobi, which was instrumental in forwarding desertification as a major contemporary environmental issue. It was suggested that 35 per cent of the world's land surface (Thomas, 1993) and one-sixth of the world's population (Barrow, 1995) were at risk from desertification. Since this landmark conference, there have been substantial efforts to assess and address the problem of desertification (Plate 6.5). For example, improvements

Plate 6.5 Anti-desertification scheme – building a digue (barrage) to hold water in the river bed during the rainy season to irrigate nearby crops, near Timbuktu, Mali
(photo: Jeremy Hartley, Panos Pictures)

since that date include a consensus on the definition of desertification, an agreed assessment database and 191 signatories to the 1995 UN Convention to Combat Desertification.

Between UNCOD and UNCED (the United Nations Conference on Environment and Development held in Rio de Janeiro in 1992), images of advancing deserts have been refined, particularly as improved understanding of dryland ecologies has forced a distinction between natural fluctuations and long-term degradation. It has also been recognised that many actions to combat desertification had been costly, focused on technical interventions, were largely initiated by the international aid community, and were rarely sustained beyond the initial donor input stage (Dowdeswell, undated).

At UNCED, often known as the Earth Summit, and within the subsequent UN Convention to Combat Desertification (UNCCD), the social dimensions of desertification (which had been almost totally lacking in the plan of action which emerged from UNCOD), were considered to be of paramount importance. However, it has been suggested that, in fact, there has been little progress over the last decade towards developing structures for the promotion of the required participatory, decentralised policies and projects to assist the most marginalised groups in drylands to cope with desertification (Mortimore, 1998; Toulmin, 2001; UNEP, 2002), and developing country efforts to fulfil commitments under the convention continue to be hampered by inadequate resource mobilisation among the international community (UNEP, 2002). There are also concerns over the potential increase in aridity and the expansion of deserts within the predicted climate warming scenarios discussed in the following section.

Global warming

Past development processes and patterns have also impacted negatively on the ability of the environment to absorb the waste products of those activities. A further function of the environment – maintaining key life support systems – is also often impaired. Up to a certain threshold, natural sinks, such as the atmosphere, oceans, vegetation and soils of the Earth may be able to absorb gases, particulate matter and chemicals which are created as waste in production processes. Beyond that threshold, however, pollution effects may occur, whereby the presence of those substances may compromise that sink function in the future and concurrently impact negatively on human health and the operation of plant and animal systems.

The issue

One of the most fundamental sink functions of the natural environment is in respect to the carbon cycle, within which oxygen is emitted through processes of photosynthesis and respiration as carbon is circulated between the atmosphere, oceans, soils and land vegetation. The amount of carbon dioxide in the atmosphere is a critical determinant of global surface temperature. The reradiation of solar radiation back into space (i.e. the amount of heat lost from the Earth's surface) is controlled by the insulating effect of 'greenhouse gases' in the atmosphere, in the presence of water vapour. These gases include carbon dioxide and also methane, nitrous oxide and a range of chlorofluorocarbons (CFCs).

If there were no such insulation or 'greenhouse effect' from the atmosphere then the average temperature of the Earth's surface would be at least 30 °C cooler than at present (Kelly and Granich, 1995: 77). However, the average temperature of the Earth's surface rose by approximately 0.5 °C over the twentieth century (Reading *et al.*, 1995; Seitz, 2002) as concentrations of these warming agents in the atmosphere have increased. For example, the latest report of the Intergovernmental Panel on Climate Change (the IPPC, created in 1988 by the World Meteorological Organization and the United Nations Environment Programme), reported that whilst natural ranging carbon dioxide concentrations 650,000 years ago (as established within ice cores) were between 180 and 300 ppm[3], the global atmospheric concentration of CO_2 in 2005 was 379 ppm[3]. Table 6.10 highlights a range of evidence for climate change from this report. This 'global warming' has been referred to as 'the major environmental challenge facing humanity over the next 100 years' (Hill, O'Keefe and Snape, 1995: 83).

Figure 6.6 points to the debate regarding the directions of climate change. Whilst a current period of warming is now widely accepted, it is not uncontested and there is certainly debate concerning the speed of change (and its causes, as seen below). For example, the 'warming consensus' from the scientific community in the mid 1990s was for global mean temperature to rise by 3 °C in the following century (Reading *et al.*, 1995: 353). In 2001 the IPCC suggested that its previous report had underestimated the likely average global

Table 6.10 Direct observations of recent climate change

- Eleven of the 12 years between 1995 and 2006 were the warmest years in the instrumental record of global surface temperature since 1850
- The linear warming trend for global surface temperatures over the last 50 years (0.13 degrees Celsius per decade) is nearly twice that for the last 100 years
- Average temperatures of global oceans have increased at depths of more than 3000 metres since 1961
- Global average sea levels have risen at an average rate of 1.8 mm per year between 1961 and 2003 and rose between 1993 and 2003 at a rate of 3.1 mm per year
- The extent of Arctic sea ice has shrunk by 2.7 per cent per decade since 1978
- The maximum area covered by seasonally frozen ground has decreased by about 7 per cent in the Northern Hemisphere since 1900
- Long-term trends in precipitation from 1900 to 2005 have included significant increases in eastern parts of North and South America, northern Europe and northern and central Asia. Decreases have been observed in the Sahel, the Mediterranean, southern Africa and parts of southern Asia. The frequency of heavy precipitation events has increased over most land areas
- More intense and longer droughts have been observed over wider areas since the 1970s, particularly in the tropics and subtropics
- Widespread changes in extreme temperatures have been observed over the last 50 years; frosts have become less frequent and heatwaves more frequent

Source: IPCC (2007)

Figure 6.6 Debate continues over the direction of global climate change
Source: Cartoon by David Hughes

temperature rise and predicted a 5.8 °C increase by the end of the next century. In 2007 the IPCC suggested a 'probable' temperature rise by the end of the century of between 1.8 and 4 degrees Celsius and a 'possible' rise of between 1.1 and 6.4 degrees Celsius (IPCC, 2007). In 2007 the 'Stern Review', undertaken by a former World Bank economist for the UK Government, reported that 'the scientific evidence is now overwhelming: climate change is a serious global threat, and it demands an urgent global response' (Stern, 2007: xv).

Global warming is a 'supranational' environmental issue in that the causes and impacts extend across all national boundaries to include all peoples and environments: 'A molecule of greenhouse gas emitted anywhere becomes everyone's business' (Clayton, 1995: 110). The explanation of global warming illustrates clearly how the sink function of the environment has been overloaded via the destruction of vegetation, the production of pollutants at rates in excess of those which can be rendered harmless by natural processes and through the creation of artificial substances which cannot be so absorbed. Furthermore, an increase in the Earth's surface temperature threatens many of the life-support functions which the environment serves through processes such as the disruption of sea levels and ocean currents and alterations of the world's major biomes and agricultural potentials, as discussed below.

Causes

As well as the debate concerning the direction of climate change noted above, there has been substantial uncertainty concerning the role of particular drivers of such change and the role of human versus natural factors and processes more widely. In 2007, the fourth assessment report of the IPCC, drawing on 'new and more comprehensive data, more sophisticated analyses of data, improvements in understandings of processes and their simulation in models, and more extensive exploration of uncertainty ranges' (IPCC, 2007: 2), was able to assert 'very high confidence' (defined as a 90 per cent chance of being correct) that the globally averaged net effect of human activities has been one of warming. 'Discernible human influences' were also found to extend to other aspects of climate, including ocean warming, continental average temperatures, temperature extremes and wind patterns (IPCC, 2007: 8) i.e. the patterns identified in Table 6.10.

For some time it has generally been understood that carbon dioxide is the principal anthropogenic greenhouse gas (the other principal forcing agents being methane and nitrous oxide). In the early 1990s it accommodated approximately 55 per cent of the warming effect (Barrow, 1995) and it is expected that the contribution of CO_2 in global warming will rise to 75 per cent by 2100 (Worldwatch Institute, 2002). Natural 'sinks', such as the oceans and green plants, are thought to absorb approximately one-half of the carbon dioxide produced by human activities (Foley, 1991). Increasingly, however, additional carbon is transferred into the atmosphere more quickly than these natural processes of diffusion and photosynthesis can remove it, and in consequence the proportion of carbon dioxide in the atmosphere rises.

Table 6.11 Percentage of total energy use supplied via fuelwood sources in developing countries

Country	1980*	1996*	2003**
Bangladesh	81.3	43.3	51.5
Brazil	35.5	29.2	29.1
China	8.4	5.6	4.6
Ecuador	26.7	14.3	18.7
Egypt	4.7	3.5	9.4
Ethiopia	89.6	93	96.5
Fiji	45.0	50	36
Ghana	43.7	78.1	84.7
Guyana	24.1	33.3	43.6
India	31.5	21.2	19.8
Kenya	76.8	78.9	83.1
Nepal	94.2	90.9	93.2
Nigeria	66.8	69	82.9
Papua New Guinea	65.4	62.5	62.2
Philippines	37.0	31.7	33.2
Venezuela	0.9	0.8	2.5
Zimbabwe	27.6	23.4	67.2

Source: *UNDP (2000), **UNDP (2006). Permission from Oxford University Press and UNDP

The major source of carbon dioxide production since the pre-industrial era has been the burning of fossil fuels. Development processes globally have required a progressive increase in per capita consumption of coal, oil and natural gas ('commercial fuels'), as noted earlier in this chapter. In many developing countries, however, in excess of 90 per cent of total energy needs may be supplied via non-commercial sources, as shown in Table 6.11. Furthermore, in many of these countries (including oil exporters) dependence on biomass sources of energy increased over the period 1980–2003, as also shown in the table.

Although the energy-related carbon dioxide emissions from the developed world have been declining and their relative share of world energy consumption within these regions has fallen in recent decades (in part due to the rising consumption in the newly industrialising countries in particular, but also due to the increased efficiency in energy use in many developed countries), the USA continues to have the highest carbon emissions from all sources (in total and per capita) as seen in Table 6.12. Although China is now the second largest global emitter of carbon, in per capita terms, emissions remain less than one-seventh of the US level.

In the first IPCC report of 1990, deforestation was estimated to account for approximately 10–15 per cent

Table 6.12 The major carbon emitters of the world 2004

Country or region	Carbon emissions (million tons)	Carbon emissions per person	Carbon emissions per unit of GDP (tons per million dollars)	Increase in carbon emissions, 1990–2004 (percent)
China	1021	0.8	158	+67
India	301	0.3	99	+88
Europe	955	2.5	94	+6
Japan	338	2.7	95	+23
US	1616	5.5	147	+19

Source: Worldwatch Institute (2006)

of the contemporary increase in atmospheric carbon dioxide (IPCC, 1990). In 2007 the Stern Review suggested that changes in land use account for 18 per cent of global greenhouse gas emissions 'driven almost entirely by emissions from deforestation' (Stern, 2007: 196).

The loss of tree cover itself reduces the uptake of carbon dioxide from the atmosphere (i.e. compromises the sink function of the environment), and the associated burning of logs and biomass sources contributes further to the production of carbon through oxidation processes. The continued reliance in many developing countries on fuelwood and biomass sources of energy for national development and for daily survival of the majority of the people in these regions is evidently in itself a major factor in deforestation and in global warming.

Other aspects of past development patterns in the developing world further compound these processes. The felling and export of timber to secure foreign exchange to finance national development and service debts; regional development strategies which encourage large-scale woodland clearance such as to support ranching, often owned and operated by multinational companies; and development policies which have failed to provide more sustainable livelihood options for small-scale cultivators – all three are factors in the rising contribution of the developing nations to global carbon emissions via deforestation. Currently approximately 30 per cent of all land use related greenhouse gas emissions are from Indonesia and 20 per cent from Brazil (Stern, 2007: 196).

In the longer term, the forcing contribution of carbon dioxide could decline. For example, CFCs may be up to 100,000 times more effective than carbon dioxide in forcing global warming. CFCs are used for purposes including refrigeration, air-conditioning, aerosol propellants, insulation foams, the cleaning of electronic components and fire-fighting gases. Although global production of CFCs declined through the early 1990s, production within the developing world is increasing. Furthermore, CFCs remain active in the atmosphere for periods up to 75 years in the case of CFC 11 and 110 years for CFC 12 (Thompson, 1992: 70) before they are eventually destroyed by ultraviolet radiation in the stratosphere.

CFCs are also the primary cause of ozone depletion in the upper atmosphere, which leads to the exposure of plant and animal systems to the damaging effects of increased ultraviolet radiation. Not only have global industrial production processes been a large consumer of fossil fuels and a major producer of carbon dioxide, they have also impacted on the environment through the development of synthetic chemicals such as CFCs. But there is the prospect that industrial developments will produce less environmentally damaging substitutes for these chemicals, such as has occurred with aerosol propellants.

Impacts

> **The likely effects of human-generated climate change will be experienced almost in inverse proportion to innocence and blame. Those people and countries who contribute most to the emissions of radiative forcing gases are, for the most part, least likely to be most inconvenienced, impoverished or physically vulnerable to the consequences of their behaviour.**
>
> **(O'Riordan, 2000a: 171)**

Much has been learnt in the last decade concerning the damage from climate change, including shrinking glaciers, regional changes in climate that have led to shifts in plant and animal ranges and the increasing experience of previously unprecedented climate events such as floods and hurricanes. Further extreme climatic phenomena and widespread impacts on human development are predicted to occur by the latest report of the IPCC (2007) and the Stern Review (2007), for example. These impacts are summarised in Box 6.4.

It is evident that the impacts of climate change are going to be regionally differentiated. They will also have varied impacts socially. As Adger *et al.* (2003: 179) state, 'the world's climate is changing and will continue to change into the coming century . . . but some sectors are more sensitive and some groups in society more

BOX 6.4

The impacts of global warming: the predictions of two major recent reviews

If no action is taken . . .

- 100 million people displaced by floods
- 1 in 6 of world population experiencing water shortage
- 40 per cent of species could become extinct
- Tens or even hundreds of millions of refugees created by drought
- The global economy could shrink by 20 per cent (Stern, 2007)

If greenhouse gas emissions continue at or above current rates, projected warming will lead to . . .

- Warming will be greatest over land and at most high northern latitudes and least over the southern oceans and parts of the North Atlantic ocean
- Contraction of snow cover and widespread increases in thaw depth over most permafrost regions
- Shrinking of sea ice in both Arctic and Antarctic
- Heat waves and heavy precipitation events will become more frequent
- Future tropical cyclones (hurricanes and typhoons) will become more intense, with high peak wind speeds and more heavy precipitation
- Extra-tropical storm tracks will move poleward with consequent changes in wind, precipitation and temperature patterns
- Increases in precipitation in high latitudes and decreases in most subtropical land regions

(IPCC, 2007)

vulnerable to the risks posed by climate change than others'.

The outcomes of climate change for particular places and people will also be very much linked to the nature of future development – i.e. what happens in terms of greenhouse gas emissions into the future, future population growth, economic development, non-fossil fuel energy consumption and multilateral climate initiatives such as the Kyoto Protocol, for example. As Stern (2007: 74) urges, 'clarity over adaptation is critical for work on the impacts of climate change'. Whilst some studies/models assume no adaptation, many assume individual or 'autonomous' adaptation and some assume economy-wide adaptations. The IPCC, for example, has six 'Emission Scenarios' that encompass different scenarios in terms of population, economics and political structure over coming decades as seen in Box 6.5. Interestingly, no scenario explicitly assumes implementation of the UN Framework Convention on Climate Change or meeting the emissions targets of the Kyoto Protocol, which are discussed in the following section.

BOX 6.5

The emission scenarios of the IPCC

A1 The A1 storyline and scenario family describes a future world of very rapid economic growth, global population that peaks in mid century and declines thereafter, and the rapid introduction of new and more efficient technologies. Major underlying themes are convergence among regions, capacity building and increased cultural and social interactions, with a substantial reduction in regional differences in per capita income. The A1 scenario family develops into three groups that describe alternative directions of technological change in the energy system. The three groups are distinguished by their technological emphasis: fossil intensive (A1FI), non-fossil energy sources (A1T), or a balance across all sources (A1B) (where balance is defined as not relying too heavily on one particular energy source, on the assumption that similar improvement rates apply to all energy supply and end use technologies).

▶

BOX 6.5 (continued)

A2 The A2 storyline and scenario family describes a very heterogeneous world. The underlying theme is self-reliance and preservation of local identities. Fertility patterns across regions converge very slowly, which results in continuously increasing population. Economic development is primarily regionally oriented and per capita economic growth and technological change more fragmented and slower than other storylines.

B1 The B1 storyline and scenario family describes a convergent world with the same global population, that peaks in mid century and declines thereafter, as in the A1 storyline, but with rapid change in economic structures towards a service and information economy, with reductions in material intensity and the introduction of clean and resource efficient technologies. The emphasis is on global solutions to economic, social and environmental sustainability, including improved equity, but without additional climate initiatives.

B2 The B2 storyline and scenario family describes a world in which the emphasis is on local solutions to economic, social and environmental sustainability. It is a world with continuously increasing global population, at a rate lower than A2, intermediate levels of economic development, and less rapid and more diverse technological change than in the B1 and A1 storylines. While the scenario is also oriented towards environmental protection and social equity, it focuses on local and regional levels.

An illustrative scenario was chosen for each of the six scenario groups A1B, A1F1, A1T, A2, B1 and B2. All should be considered equally sound.

The SRES scenarios do not include additional climate initiatives, which means that no scenarios are included that explicitly assume implementation of the United Nations Framework Convention on Climate Change or the emissions targets of the Kyoto Protocol.

Source: IPCC (2007)

It is evident that the major predicted impacts of global warming relate to water resources in some way. Global mean sea level under the most recent IPCC assessment is predicted to rise between 0.18 and 0.59 metres at 2090–2099 relative to 1980–1999 through the thermal expansion of the oceans and the melting of mountain glaciers and ice sheets. Low-lying countries such as Bangladesh and the Netherlands therefore will be particularly vulnerable to the inundation of lands and the physical loss of property resulting from global warming. The Dutch have already initiated a huge and costly programme of coastal defence construction to protect themselves from a possible 50-centimetre rise in sea level (Barrow, 1995). A 50-centimetre rise is also the 'best-case' scenario currently predicted by the IPCC for Bangladesh. A sea level rise of this magnitude would lead to substantial inundation, displacement of people and the loss of farmland in that country. Figure 6.7 illustrates the spatial extent of such impacts under an even more serious scenario of a 250-centimetre rise.

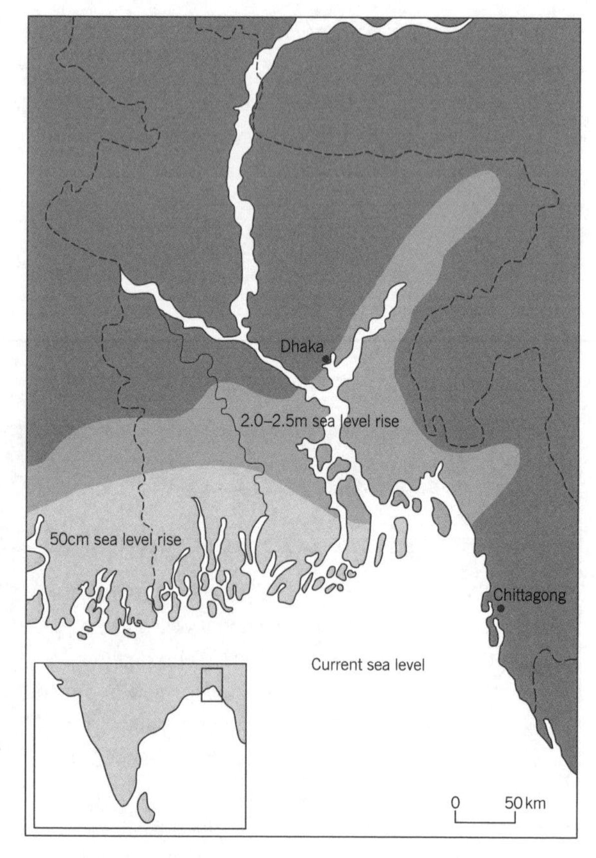

Figure 6.7 The inundation of Bangladesh under proposed sea-level changes
Source: Adapted from Reading *et al.* (1995)

The impacts of global warming on human health are also extensive, as considered in Figure 6.8. Not only does climate affect illness and premature death directly, as in the case of disasters through extreme weather events, for example, but it influences many of the key determinants of human health indirectly through altering the geographical range of diseases and

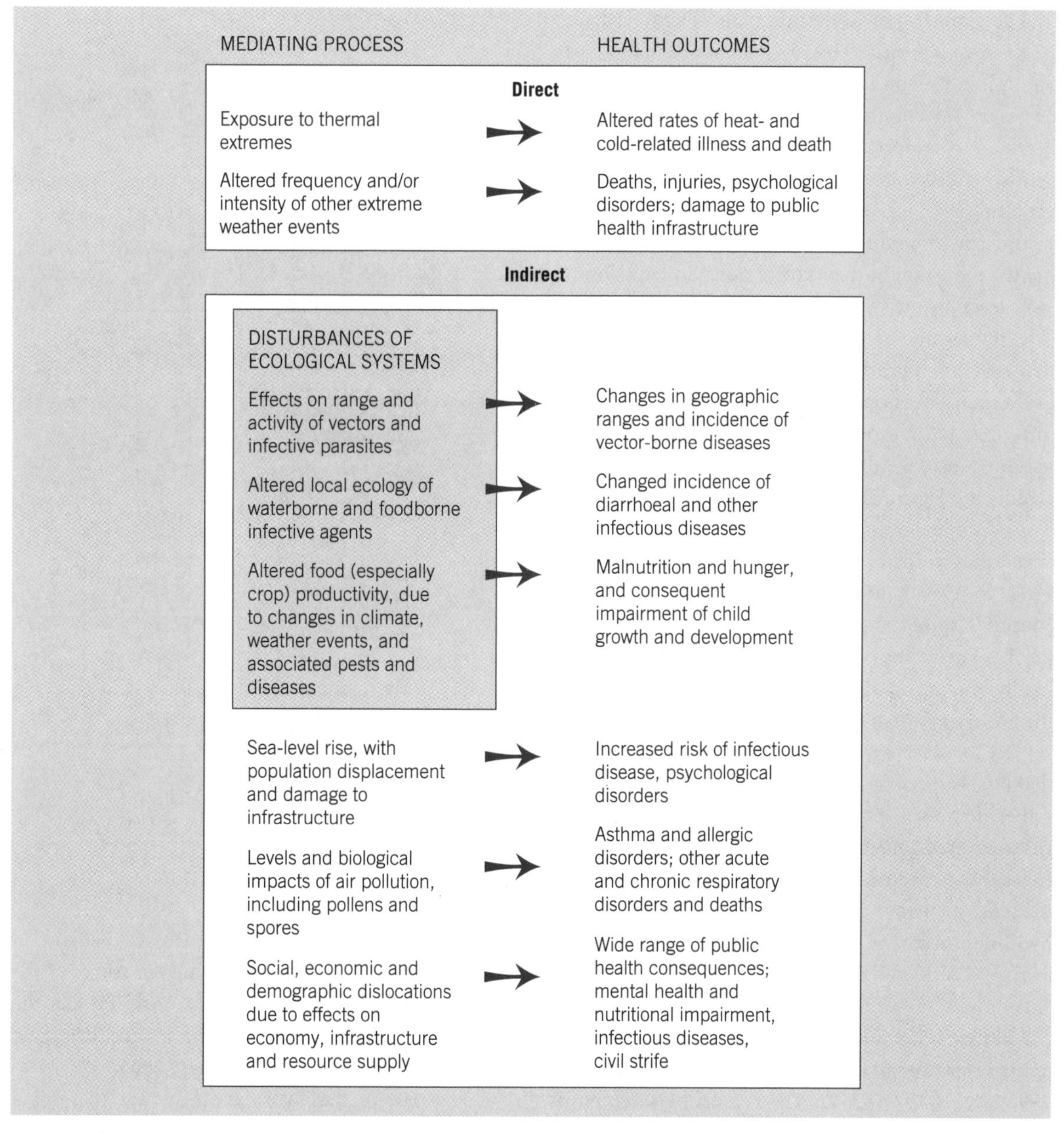

Figure 6.8 The direct and indirect health impacts of climate change
Source: World Resources Institute (1998)

the productivity of lands. Global warming must also be considered a factor in the displacement of increasing numbers of people from their lands and homes; environmental refugees are currently assessed to outstrip political refugees by two to one (Simms, 2002).

The response

It is evident that climate warming presents huge challenges for the global community into the future and suggests an urgent need for strengthening climate policy worldwide. As O'Riordan (2000a: 171) stated:

Climate change will test science, science–policy relationships, global environmental agreements, the economics of response, the politics of coalition building across interests and generations, the morality of individual 'lifestyle choices' in the light of the innocence–blame contradiction, and the collective ethics of responding, coping, adapting and sharing in a world that is not yet one. The coming decade will bring all these connections into a common strand of analysis and moral judgement.

The Framework Convention on Climate Change (UNFCCC) was signed by 167 nations at the Rio Earth Summit in 1992 and came into force in 1994. There are currently 189 countries signed to the convention and it remains the principal international agreement concerning efforts to respond to climate change. The convention was based on 'soft law', calling on all parties to voluntarily commit to a number of obligations, including the reduction of their greenhouse gas emissions to 1990 levels by the year 2000 (few in fact achieved this). The framework recognised several basic principles: that scientific uncertainty must not be used to avoid precautionary action; that nations have 'common but differentiated' responsibilities; and that industrial nations must take the lead in addressing the problem (Dunn and Flavin, 2002).

A 'Conference of Parties' (CoP) comprising the signatories to the convention has met each year since 1994. In 1997 it met at Kyoto in Japan, where 113 countries agreed the 'Kyoto Protocol' to supplement the UNFCCC and which, if fully ratified, would be legally binding on all parties. The principal features of the protocol are that they allow different countries to be subject to variable targets for emissions reduction and that targets should be met within a period (2008–2012) rather than by a set year. The number of greenhouse gases covered under the protocol was also increased (over those covered in the framework) to six (carbon dioxide, methane, nitrous oxide, perfluorocarbons, hydrofluorocarbons and sulphur hexafluoride). On average, the Kyoto Protocol requires so-called Annex 1 countries (OECD members, plus the countries of the former USSR and Eastern Europe) to reduce their annual emissions of these gases by 5.2 per cent from the 1990 level by 2008–2012. The protocol also commits developing countries to monitor further and address their greenhouse gas emissions.

In order for the protocol to be ratified and fully implemented (i.e. to become an instrument of international law), it required the signatures of 55 per cent of these 'Annex 1' nations and also the signatures of those nations whose 1990 carbon dioxide emissions accounted for 55 per cent of all such emissions in that year. In 1990, the USA accounted for 36 per cent of all carbon dioxide emissions from the industrialised countries (Lean, 2002). The Kyoto Protocol commits the USA to a 7 per cent reduction in greenhouse gas emissions between 1990 and 2008–2012, but the Bush government has refused to sign the protocol and has removed itself from the process (as has Australia, as seen in Figure 6.9). In late 2004, the Russian Federation (accommodating 18 per cent of total emissions of the Annex 1 countries) agreed to ratify, ensuring that the 'numbers test' was passed and the Protocol became legally binding on Parties in February 2005.

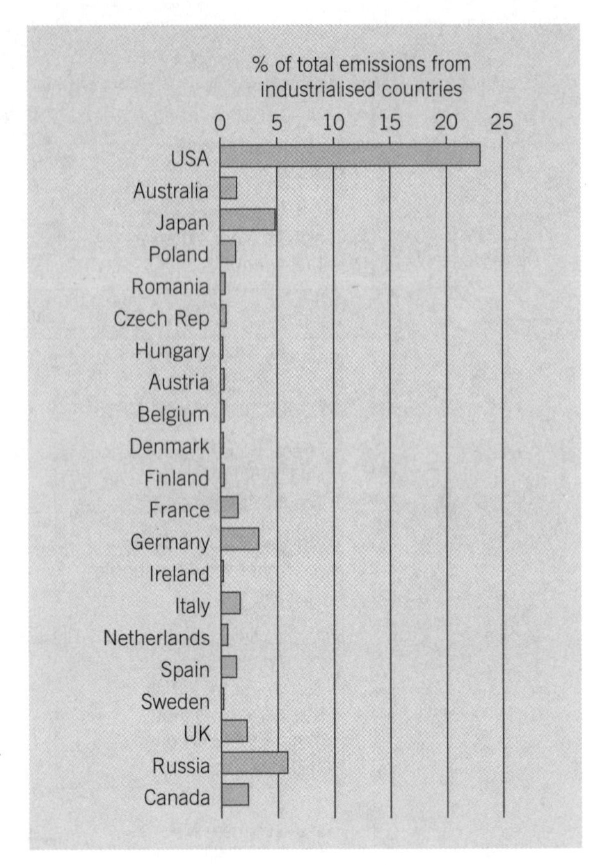

Figure 6.9 Share of carbon dioxide emissions and ratification of the Kyoto Protocol
Source: UNDP Human Development Report (2006)

A key part of the Kyoto Protocol was that some emission reductions required of developed countries could be met by 'transferring' to other countries under what are called 'flexibility mechanisms'. In short, these mechanisms as outlined in Table 6.13 introduce an official market in carbon through the trading and offsetting emissions. Within this market, nations and firms (or a combination of the two) can negotiate the trading of carbon credits or carbon emissions reduction units in order to achieve their legally binding targets for compliance under the protocol. Under emissions trading, for example, developed countries are allocated their emissions allowances based on the reduction target negotiated. Each allowance is made up of 'Assigned Amount Units' (each unit is equivalent to one metric tonne of CO_2). At the end of the compliance

Table 6.13 Flexibility mechanisms of the Kyoto protocol

1 Emissions trading: a developed country that has exceeded its Kyoto targets to reduce carbon dioxide emissions can sell its surplus reduction to another developed country that has failed to meet its target (Article 17).

2 Joint implementation: a developed country can fund a project in another developed country that will either reduce carbon dioxide emissions or enhance carbon sinks. The reduction of carbon dioxide in the atmosphere is allocated to the country financing the project (Article 6).

3 Clean development mechanism: a developed country can fund a project in a developing country that reduces emissions or enhances sinks. Such projects must produce reductions that are additional to any that would occur in the absence of certified project activity (Article 12).

Source: Adapted from UN (1998) Kyoto Protocol to the UNFCCC

period, a country has to hold an amount of AAUs equivalent to how much greenhouse gas it has emitted. Any reductions below the allocated allowance can be traded with countries that have exceeded their capped allowance. 'Offsetting' involves investment in projects that are designed to 'absorb' carbon from the atmosphere or that generate savings in emissions that would not otherwise have been made. The briefing box below investigates the outcomes of these mechanisms to date.

Critical reflection

The outcomes of emission trading and offsetting under the Kyoto protocol

In early 2007, the Clean Development Mechanism of the Kyoto Protocol registered its 500th project – a wind farm in Gujarat, India, that is estimated to reduce carbon dioxide emissions by more than 15,300 tonnes annually (UNFCCC, 2007 – cdm.unfccc.int/statistics). To date, CDM projects have been established in more than 20 countries, although there is a heavy concentration of projects in the larger, most rapidly developing economies as seen in Figure 6.10a.

Figure 6.10b identifies the partners within these CDM projects to date. They are displayed by country although a number of country partners may be identified for each project. It is evident that the UK and the Netherlands are the principal investors in such voluntary arrangements to date. In February 2007, for example, a Dutch bank with operations in 53 countries including the UK applied through the Department for Environment and Rural Affairs (the UK's Designated National Authority) to become the project partner with a private pulp and paper making facility in Andhra Pradesh, India as part of the CDM. The project focused on making energy savings in the production processes through the replacement of equipment with more energy efficient alternatives that required additional investment and technologies that were not prevalent within that industry and that region at that time.

Figure 6.10 Clean Development Mechanism (CDM) projects: (a) principal country locations for existing registered CDM projects

Country	Number of projects
India	178
Brazil	94
Mexico	78
China	42
Chile	15
Malaysia	14
Honduras	10
Republic of Korea	10
Ecuador	9
Indonesia	8
Philippines	8

Source: Compiled from cdm.unfcc.int/statistics, accessed 12 March 2007

Underpinning the dominance of European partners within CDM projects shown in Figure 6.10b is the factor of the European Union Emissions Trading Scheme. This was established on 1 January 2005 and is the largest emissions trading scheme currently in operation globally. It has been referred to as 'one of the most ambitious multinational policy programmes in history' (Haar and Haar, 2006: 2615). To date, it covers approximately 12,000 installations, within

Critical reflection (continued)

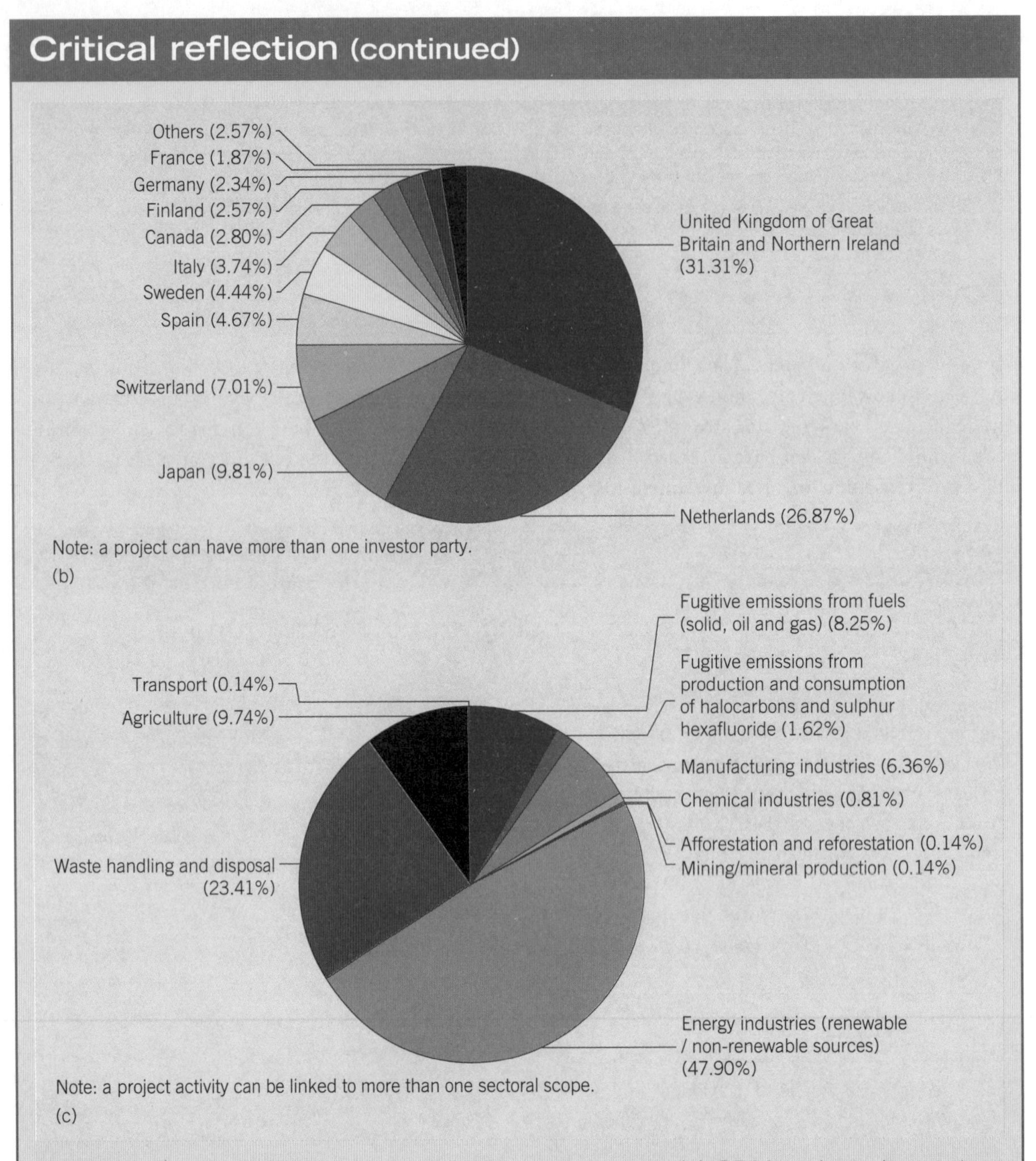

Figure 6.10 (continued) (b) projects by investor parties, (c) CDM projects by sector
Source: Compiled from cdm.unfcc.int/statistics, accessed 12 March 2007

six major sectors of industry and across 25 countries (Carbon Trust, 2006). The scheme involves the allocation and trading of CO_2 allowances across the European Union towards assisting Member States and the Community to fulfil their commitments to reduce greenhouse gas emissions under the Kyoto protocol.

In the same way as the flexibility mechanisms of the Kyoto protocol as a whole, the scheme is predicated on the understanding that the costs of abatement of greenhouse gas emissions are not uniform across the region, such that trading (i.e. making cutbacks where the costs of abatement are lowest) enables the overall costs for the EU of reducing emissions to be minimised. Through the scheme, companies can reduce their own emissions, buy EU allowances in the market and/or purchase credits through CDM or Joint Initiative products (within set limits).

Figure 6.10c shows the sectors within which the CDM projects to date have been developed and reveals

Critical reflection (continued)

that almost half are projects relating to the energy industry. The 500th project in Gujarat, for example, is on behalf of a private Indian company that has existing business interests (in component manufacturing) within the areas of the project and looks to establish a number of wind farms to generate electricity that is then delivered to their manufacturing plants by the State Electricity Board grid.

Whilst both the Gujarat and the Andhra Pradesh examples can readily be considered to be delivering environmental improvements in the short term and within those project areas, there are a number of emerging problems in the way that the flexibility mechanisms of the Kyoto Protocol are being delivered. For example, the CDM mechanism of the Kyoto Protocol was envisaged to deliver financial support to projects towards sustainable development within some of poorest countries of the world. But it is evident from Figure 6.10a that most of the money to date is going to the largest and most industrialised emerging economies. Indeed, only Bangladesh, Bhutan and Nepal from the UN's list of 50 least developed countries have registered CDM projects (New Internationalist, 2006). Furthermore, over two-thirds of the CDM projects are initiated from just three countries: the UK, the Netherlands and Japan.

It was also proposed that the Clean Development Mechanism would provide a stimulus for the development of renewable technologies such as wind, solar and geo-thermal energies. To date, however, only a very small percentage of the capital generated by CDM is going to these technologies. For example, the majority of energy sector projects shown in Figure 6.10c are credits generated by gas capture projects at major chemical and manufacturing plants. Critics suggest that such companies should not be being paid to clean up a mess of their own making (see New Internationalist, 2006) and also point to the substantial wider and ongoing pollution that such installations generate.

Consider the suggestion that the carbon market enabled by CDM mechanisms may detract from the wider changes in economy and society globally that are needed to tackle climate warming.

Many subsequent meetings of the UNFCCC Conference of Parties have been characterised by intensive negotiations, dealings and substantial confusion regarding the application of these mechanisms. In 2005 a working group was established to initiate new talks on the future of the climate change process as a whole and for negotiating the new commitments under the Kyoto Protocol for the second period, 2013–2017. Under the original Framework Convention, aggregate emissions from Annex 1 countries are scheduled to be reduced by at least 18 per cent through this period, for example (Greenpeace, 2006). Thirty per cent reductions are required for the third period, 2018–2022. It is anticipated that future meetings of the Conference of Parties will need to review the list of Annex 1 countries (those nations legally bound to reduction targets) in the light of new circumstances (including what has been revealed in the last decade concerning reductions and emerging polluters). There is a significant argument also to include emissions from international aviation (currently excluded from country records of carbon emissions), for example, and to revise how the 'sink activities' of land use change and reforestation activities are calculated and integrated into country accounts (Greenpeace, 2006).

While emissions limitations are an important part of the response to global warming, the original UNFCCC had much broader objectives. A key principle that Parties to the conventions agree, for example, is to cooperate to promote a supportive and open international economic system that would lead to sustainable economic growth and development. Under Article 4, all parties commit to implement measures to mitigate and adapt to climate change; to promote and to co-operate in the development, application and diffusion of mitigation and adaptation technologies; and to cooperate in preparing for adaptation to the impacts of climate change. It can be argued that all such intentions under the UN framework now need reinvigorating.

Urban atmospheric pollution

Climate warming is a major atmospheric pollution problem which stems from the production of substances in quantities in excess of their absorption by natural processes with implications for the functioning of

Plate 6.6 Urban air pollution
Source: Corbis/Kevin R. Morris/Bohemian Nomad Picturemakers

Plate 6.7 Smog enveloping Mexico City: air pollution has been estimated at 100 times above acceptable levels
(photo: Mark Edwards, Still Pictures)

the Earth as a whole. Additional atmospheric pollution problems, such as urban air pollution, may have quite local (largely industrial and transport related) sources and localised impacts, but are global environmental issues in the sense that all cities tend to experience similar problems.

For example, in 1993 every 'megacity' in the world surveyed as part of a UNEP and WHO programme had at least one major pollutant which exceeded the threshold determined by the World Health Organization, above which deleterious effects on human health could be expected (UNEP/WHO, 1993; Plate 6.6). In the late 1990s between 70 and 80 per cent of 105 European cities surveyed by the European Environment Agency were found to exceed the same guidelines for at least one pollutant (World Resources Institute, 1998).

It is evident that contemporary patterns and processes of development are compromising future development prospects very directly through urban air pollution particularly in the developing world. Figure 6.11 reveals the levels of deaths from urban air pollution around the world. The implications for health services are revealed in research within Jakarta, Indonesia, which shows that 1400 deaths, 49,000 Emergency Room visits and 600,000 asthma attacks could be avoided in Jakarta (Indonesia) every year if particulate levels were brought down to WHO standards (World Resources Institute, 1998).

Furthermore, for many people in the developing world, it is indoor air pollution that is a major cause of premature death and substantial ill-health. In 2000, indoor air pollution from solid fuel use was deemed to be responsible for more than 1.6 million deaths annually and 2.7 per cent of the global burden of disease in terms of Disability Adjusted Life Years, DALYs (WHO, 2002). This suggests that solid fuel use within the home is the second biggest environmental contributor (behind unsafe water and sanitation) to ill health.

Sulphur dioxide is probably the best-known urban pollutant, with most industrial cities having been subject to sulphurous smogs at some point (Elsom, 1996). The key health effects of sulphur dioxide are the impairment of respiratory function (including the aggravation and/or causation of bronchitis, asthma and emphysema) and the subsequent strain on the heart, leading to premature death. The primary sources of sulphur dioxide production are the smelting of metallic ores, the burning of coal and oil in power production and heating, and in transport. The World Health Organization recommends that sulphur dioxide exposures should not exceed an average of 40–60 micrograms per cubic metre over the course of a year. Whilst sulphur dioxide emissions have been falling over recent decades in cities in the more developed world, they

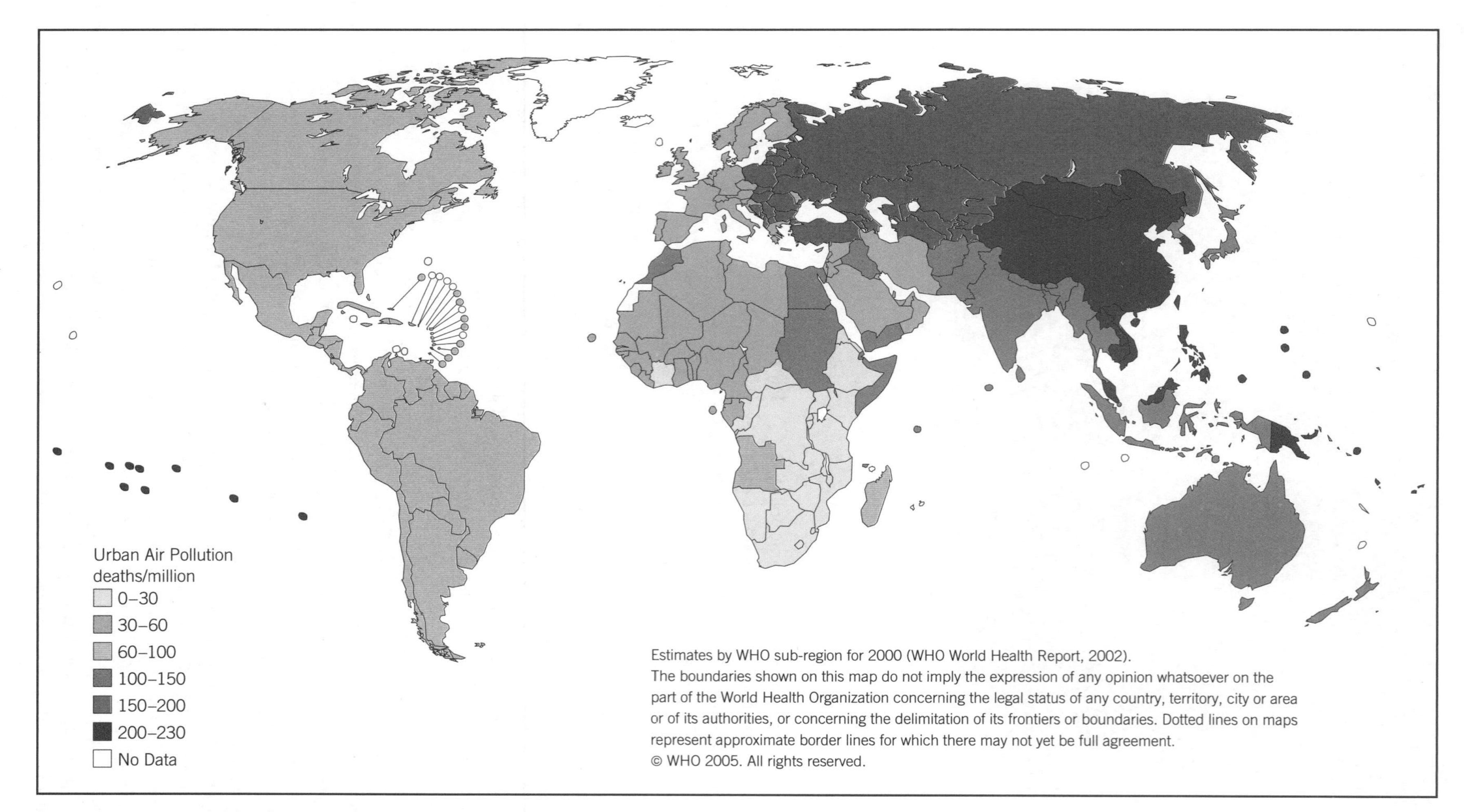

Figure 6.11 Deaths from urban air pollution
Source: www.who.int/heli/risks/urban/urbanew/en, accessed 9 May 2007

Table 6.14 Sulphur dioxide pollution in selected cities, 1995–2001

City	Micrograms per cubic metre
Cairo	69
Sao Paulo	43
Caracas	33
Beijing	90
Guangzho	57
Shanghai	53
Bangkok	11

Source: www.worldbank.org/data/wdi2006 © International Bank for Reconstruction and Development/The World Bank

remain high in many cities of the less developed nations as illustrated in Table 6.14.

In an attempt to reduce the localised health impacts of sulphur dioxide pollution at the ground level, some of the measures taken have in fact contributed to a wider pollution problem known as acid rain. This phenomenon includes the dry deposition of sulphur and nitrogen oxides onto damp surfaces, thereby converting to sulphuric and nitric acids, and the precipitation of acids formed in the atmosphere through similar processes of oxidation. As emissions of sulphur dioxide increased and strategies such as the building of taller stacks were implemented, pollutants have been placed into the larger atmospheric circulation system for longer periods, causing major problems in the natural environment often at substantial distances from the source of the pollutant. For example, drifting pollution from Europe was linked to acidification of Scandinavian water bodies in the 1960s. The average annual precipitation in Britain in the early 1980s was estimated to be between pH 4.5 and 4.2 in comparison to the 'normal level' for 'naturally acid' rainfall of pH 5.0. The impacts of acid rain extend from damage to fish species, through the disruption of soil nutrient cycles, and, as a result of the liberation of heavy metals from soils and bedrock, to possible human health impacts.

In addition to sulphur dioxide, particulates and ozone are further aspects of urban air pollution that constitute significant risks to human health and development opportunity, particular in less developed regions. Particulate air pollution is a complex mixture of small and larger particles of varying origin and chemical composition including smoke and dust, but also soot from vehicle exhausts (often coated with chemical contaminants or metals). According to the WHO and other organisations, 'no evidence so far shows there is a threshold below which particle pollution does not include some adverse health effects' (World Resources Institute, 1998: 64). Ozone is not emitted directly but is formed when nitrogen oxides from fuel combustion react with other 'volatile organic compounds' (VOLs) in the atmosphere. Ozone is the major component of photochemical smogs in urban areas. As a powerful oxidant, ozone reacts with most biological tissues and as such has widespread potential health impacts.

Biodiversity loss

It is now appreciated widely that the environment performs 'new' resource functions, including protecting the global commons and providing amenity services. However, there is evidence that development processes are also degrading these functions. For example, the first comprehensive global assessment of biodiversity undertaken by UNEP concluded that between 5 and 20 per cent of some groups of animal and plant species were threatened with extinction in the near future (World Resources Institute, 1996: 247). However, ascertaining the resultant loss of quality or value of these 'resources' is very problematic.

For example, diversity within the biosphere can be considered in terms of ecosystems, species and genetic material (Mather and Chapman, 1995). In addition, biodiversity can be valued for many different reasons, as highlighted in Box 6.6. Development processes have certainly depended on the manipulation of the gene pool of species considered useful to societies. The particular case of high-yielding varieties in agricultural development is discussed in Chapter 10. Earlier sections of this chapter have also detailed the substantial modifications of habitat which have occurred with 'development'.

The capacity of human society to impact on biodiversity has increased with economic development. Over the last few hundred years humans have increased species extinction rates by as much as 1000 times the background rates that were typical over Earth's history and projected future extinction rates are ten times higher than current (Millennium Ecosystem Assessment, 2005). However, there are substantial problems with such predictions. For example, only a small proportion of the Earth's species have been identified to date; species concepts have also been developed with respect to studies of invertebrates and insects that may not be relevant when considering micro-organisms such as fungi; and there is substantial debate even concerning what constitutes a 'species' (Dolman, 2000).

BOX 6.6

Values of biodiversity

Anthropocentric material values (empiricist/rationalist)

Consumptive use (life support): Local use and consumption of natural resources, such as food, fish, meat, building materials, timber, fuel, medicines

- Cultural value system: Socially regulated commons
- Resource use paradigm: Sustainable management, resource conservation

Productive use: Material resources (food, timber, etc. *Option value* opportunities, e.g. potential medical drugs, genetic resources

- Cultural value system: Market economics
- Resource use paradigm: Exploitation, genetic engineering
- Cultural value system: 'Utilitarian' societal ethic
- Resource use paradigm: Interventionist resource management regulated for the common good; emphasis on 'progress' and 'improvement'; increasing use of ecological management

Non-productive use, indirect values: Ecosystem services, e.g. pollination, regulation of pest species through predator complexes, nursery habitats for fisheries (e.g. mangrove swamps), soil fertility, nutrient cycling, aquifers, flood control, waste assimilation capacity, climate regulation

- Cultural value system: Societal utility, stewardship, intergenerational equity
- Resource use paradigm: Ecologically sensitive management, sustainable resource use, bioremediation, ecological engineering

Anthropocentric cultural values (subjective)

Value of biodiversity: Societal and individual well-being through contract with nature; human-oriented spirituality

- Cultural value system: Aesthetics, philanthropic ethics, spiritual beliefs
- Resource use paradigm: Nature and landscape conservation, interventionist management, ecological education

Ecocentric values

- Cultural value system: Belief that the natural world has unquantifiable intrinsic value; respect for non-human life
- Resource use paradigm: Fully inclusive biodiversity conservation; emphasis on natural dynamics; active restoration of ecological processes, or alternatively, non-intervention 'wilderness' areas

Source: From Dolman (2000)

The impacts of such biodiversity losses are also hard to determine since relatively little is known about ecosystems themselves, and because the potential value of diversity in ecosystems, species and genetic materials, even solely within future agricultural production, is impossible to quantify.

Sources of biodiversity loss

Despite the problems of data above, it is widely accepted that the primary cause of loss of biodiversity has been the intensification of agriculture. This includes the manipulation of the gene pool of species, the increased use of pesticides, the overcropping of animal species such as through hunting, and the modification and loss of habitat with the encroachment of agriculture. The food supply for over 85 per cent of the global population, for example, is based on only 20 plant species (Mather and Chapman, 1995: 119). Furthermore, the development of high-yielding cultivated varieties has led to a loss in genetic diversity and a vulnerability to predation by bacteria or aphids, which, 'unlike their genetically uniform prey, are constantly evolving' (Murray, 1995: 22). As an illustration, 70 per cent of all rice grown in Indonesia is descended from a single maternal plant (Mather and Chapman, 1995: 119).

Although pesticide use to combat vulnerability to predation has increased rapidly in agriculture, crop losses have remained constant over the same period as an ever-increasing number of pest species develop resistance to known pesticides. Further threats to

Plate 6.8 Photo of Wetlands at Black Moshannon State Park, a natural bog area in Pennsylvania, for biodiversity section
(photo: Alamy Images/H. Mark Weidman Photography)

genetic and species diversity are caused by development processes which have commoditised species for purposes other than food production; for example, rhino horn is commoditised as an aphrodisiac and exotic species are commoditised as pets. Fears for the further loss of plant genetic diversity is also a principal concern of opponents to the widespread adoption of genetically modified crops as highlighted in Box 6.1.

It is the modification and loss of habitat, especially forests and wetlands, resulting from the conversion of lands for agricultural production that has been the major factor in the loss of species diversity. Tropical forests are particularly 'species diverse'. They occupy less than 7 per cent of the Earth's surface but account for between 50 and 90 per cent of all known plant and animal species (Mather and Chapman, 1995: 122). Yet development, as earlier identified, accelerated the processes of tropical deforestation during the twentieth century; and wetland habitats worldwide are increasingly under threat from agricultural intensification, but also through urban developments. 'Notwithstanding the wide gaps in knowledge, there are many indications that marine biodiversity is in real trouble' (World Resources Institute, 1996: 249).

Wetlands include areas of marsh, fen, peatlands, swamp forests, estuaries and coastal zones. They provide many resource functions, including direct use value in terms of supporting agriculture, fishing, tourism and transport. Wetlands also have indirect hydrological uses, including groundwater recharge and discharge, sediment trapping and flood protection (Hughes, 1992). Their biodiversity function is in the diverse habitats supported for aquatic species, waterfowl and other wildlife. It is estimated that there are 60 species of trees and shrubs and over 2000 species of fish, invertebrates and epiphytic plants that depend for their survival on the wetland environments of mangrove swamps alone (Maltby, 1986). However, over half of the world's mangrove forests may already have been destroyed (Millennium Ecosystem Assessment, 2005).

Riparian and coastal zones have long been favoured areas for human population and settlement, particularly in arid zones, as illustrated in Box 6.3. Approximately 40 per cent of the world's population lives within 100 km of a coastline, an area that accounts for only 22 per cent of the Earth's landmass (World Resources Institute, 2000) and of which 19 per cent is considered to have been 'highly altered' in the sense that it has been converted to agricultural or urban uses. Distinct agricultural systems have evolved worldwide to exploit the productive habitats of wetlands such as in aquaculture, recessional cultivation and wet rice cropping.

Many modern agricultural systems involve substantial modification of natural ecologies and investment in drainage and irrigation works, as illustrated in the case of the fenlands of eastern England. Forces of environmental disturbance to wetlands in developing countries, including the intensification of wet rice cultivation in Asia and wider water and irrigation engineering works, are discussed in Chapter 10.

One of the most powerful contemporary forces of wetland degradation globally is the encroachment of urban, industrial and infrastructural developments. Riverside and coastal locations are highly valued sites for housing and tourism developments throughout the world and have exhibited some of the highest levels of urban growth to date (World Resources Institute, 1996). The impacts of such development on biodiversity may be very direct, such as in the case in eastern Calcutta, where 4000 hectares of inland lagoons have been filled to provide homes for middle-class families, but will also include less direct impacts via altering water flows, enhancing siltation and aggravating pollution (World Resources Institute, 1996).

Conclusion: towards sustainable resource management

This chapter has confirmed that development patterns and processes are intrinsically related to the environment and it is this interrelationship of environment and development that is central to the notion of sustainable development. Development challenges are environmental challenges and one end will not be achieved without the other. The chapter has also shown that the nature of the challenge to reconcile environment and development in the future is highly specific in time and space ensuring that there are no 'blueprints' for sustainable development. For example, it was seen that conceptions of the environment, of resources and of their value to society are constantly being redefined as development occurs. It has also been clear that the significance and implications of resource constraints, scarcity and degradation are similarly highly conjunctural at the local level.

Although the detail and shape of the challenge of sustainable development may be highly place-specific, both the challenges and opportunities of sustainable development depend on actions throughout the hierarchy of societal organisation. For example, the consideration of a number of environmental issues of global concern has shown how ideas about resource use and development may be shaped to a greater extent by forces and events at substantial distances from a particular society and its local environment. Many resources are now increasingly mobile through technological developments in exploitation, transport and communications; and through similar processes of development, the potential of human actions to impact on environments currently extends far beyond the site of production or use.

The analyses presented in this chapter have confirmed a number of core challenges for resource management and have alluded to some of the ways in which new processes and patterns of development need to be found. Addressing the welfare needs of the poor is essential; the persistence of poverty restricts the power of national governments and individuals to engage in development activities, to avoid degradation of resources and ill-health and to rectify existing environmental damage. Inequalities in access to resources, at all scales and across sectors such as land and energy, simultaneously inhibit the incentive to re-use, recycle or conserve on behalf of those who control such resources; and they restrict the options of marginalised people in avoiding actions which further degrade the resource base on which they depend for immediate survival. In such ways, these processes lead to highly contrasting and plural geographies of development.

In short, this chapter has illustrated areas of research and understanding that have been important in overcoming ideas of any simple, deterministic link between resources and development. Part III of the text provides insight as to how these challenges are being variously taken up, rejected, moulded or compromised by the activities of different organisations involved in influencing development policy and action within particular places and spaces.

Key points

- Past patterns of human development have been closely linked to the exploitation of environmental resources but the presence of such resources does not ensure high levels of development at a country scale.
- The notion of sustainable development centres on the interdependence of environment and development into the future but also highlights how past patterns and processes of 'development' cannot be sustained in economic and social terms.
- Whilst modern society as a whole has not experienced the absolute resource scarcities that have been predicted in the last century, there are many dimensions of resource scarcity that are experienced currently that are important factors in shaping the opportunities for human development locally and in many places around the globe.
- Questions of the future availability of water and commercial energy sources are particular environmental resource challenges worldwide that require new ways of working for national governments and international institutions as well as changed actions on behalf of business and individuals in resource use, consumption and resource management, for example.
- There are a number of global environmental changes such as climate warming and the decline of biodiversity that are widely understood to be underpinned by human actions and need to be addressed by multilateral actions.
- The persistence of poverty and rising inequality are major constraints on the prospects for sustainable development in the future, including in the ways that they restrict options in environmental resource management and inhibit incentives for conservation.

Further reading

Adams, W.M. (2001) *Green Development: Environment and Sustainability in the Third World*, 2nd edn. London: Routledge.
An important text that presents a thorough analysis of the evolution of environmental thinking, of both mainstream ideas of sustainable development and its opponents, and how development in practice has been influenced by such 'Greening'.

Elliott, J.A. (2006) *An Introduction to Sustainable Development*, 3rd edn. London: Routledge.
An accessible introduction to the challenges and opportunities of sustainable development with particular reference to the developing world but with close consideration of wider processes and institutions underpinning globalisation, debt, aid and trade, for example.

Leach, M. and Mearns, R (eds.) (1996) *The Lie of the Land: Challenging Received Wisdom on the African Environment*. London: James Curry.

A landmark text that brings together authors and long-term studies of particular aspects of environmental and social change within the African continent and considers critically how environmental problems (and 'crises') are constructed and acted upon.

Madeley, J. (2002) *Food for All: The Need for a New Agriculture*. London: Zed Books.
An accessible text that considers the requirements for science and policy to end hunger and famine.

O'Riordan, T. (2000) Climate change, in O'Riordan, T. (ed.) *Environmental Science for Environmental Management*. Harlow: Pearson, 171–213.
A clear introduction to the science of climate change and the politics of the responses by an author of long-standing authority on the widest issues of environmental resource management. The text as a whole also has dedicated chapters on issues of water, soils, biodiversity and air quality management, for example.

Stern, N. (2007) *The Economics of Climate Change: the Stern Review*. Cambridge: CUP.
A report commissioned by the British Government that emphasises the economic implications of climate change, but also draws together in a relatively clear manner, many of the major previous studies on the projected impacts of climate change.

Websites

www.climatecare.org and **www.carbonneutral.com**
Two companies through which individuals and companies can make payments to offset their contribution to carbon emissions such as when they travel by plane. Leaders in the 'voluntary offsetting market'.

www.ipcc.ch
Site of the Intergovernmental Panel on Climate Change. Hosts a wealth of technical, scientific and socio-economic information regarding understanding and responding to climate change.

www.maweb.org
Site of Millennium Ecosystem Assessment project that reports on five years of work by over 1000 scientists to report conditions and trends of global ecosystems and consequences for human well-being. Five synthesis reports on biodiversity, wetlands and water, desertification, health, business and industry can be accessed in full.

www.iied.org
Site for the International Institute for Environment and Development, which is an independent non-profit organisation promoting sustainable patterns of development. Gateway to substantial research and policy studies in areas of climate change, forestry, agriculture, biodiversity and mining, for example.

Discussion topics

- This chapter has focused on global patterns of the complex interrelationship between resources, the environment and development. Compile a list of evidence that you can find locally, regionally and nationally for this relationship.
- What is more damaging to the environment, poverty or affluence?
- Research in more depth the problems of data accuracy in relation to deforestation, biodiversity or soil erosion.
- Investigate the notion of 'common but differentiated responsibilities' within the ongoing development of the Kyoto Protocol. Should countries of the developing world have to meet targets for the reduction of greenhouse gases?
- Debate the strengths and weaknesses of carbon offsetting for addressing the challenge of climate warming.
- Consider evidence within recent films and novels, for example, that extreme weather events make good fiction.

Chapter 7

Institutions of development

Decision making in development is undertaken by various individuals, agents, bodies and organisations. Historically, this role most frequently fell to governments in defining and implementing policy and planning at a national level, for example. Currently, however, the role of the state in development has altered substantially for reasons including economic, political and cultural processes of globalisation reshaping the capacity of governments to operate in these ways. In addition, there are now new 'global' institutions such as the World Trade Organization, transnational business interests that have more economic power than whole nations, as well as allegiances of civil society groups across the globe that shape development internationally and within countries. Furthermore, there is an emerging understanding within the development literatures of the 'less formal' institutions that operate often at the local and community levels to shape livelihoods and outcomes. There has been substantial debate over the history of development thinking and practice as to which institutions are best able to take on the task of acting on behalf of others to promote development.

This chapter:

- Identifies the major institutions and organisations involved in development planning and practice;
- Describes the responsibilities and operations of the core institutions of global governance including the United Nations, the World Bank and the World Trade Organization;
- Considers the impacts of economic restructuring and the search for sustainable development as principal forces of change within institutions of development;
- Critically examines the practice of multilateral institutions, the state, business and civil society and the way in which these agencies increasingly 'work together' in development;
- Examines the moves towards good governance and democracy and the impacts on the key institutions in development;
- Explores the factors underpinning the rise of civil society in development and some of the challenges for new social movements, NGOs and CBOs.

Introduction

Whilst 'doing development' was for a long time the preserve of governments, the capacity of governments to shape the trajectories of human and resource development within the boundaries of the state changed substantially over recent decades, as considered in Part 1. Significant forces of change included those associated with globalisation, the end of the Cold

War, and the mounting influence of the international financial institutions (IFIs), such as the World Bank and the International Monetary Fund. In addition, civil society has a much more prominent role currently in international policy debates and global problem solving than a decade ago (Edwards, 2001a). What are now considered oversimplistic dualistic discussions of 'state' or 'market' were also replaced through the 1990s, for example by those concerning strengthening state capacities and building institutions for markets (World Bank, 2002a). Attaching liberal democratic conditionalities to aid packages came to dominate official development thinking, with a 'flow of pronouncements' being issued from international organisations and donor governments on 'governance, democracy and the relationship of either or both to development' (Leftwich, 1993: 611).

While poorly functioning institutions (particularly public sector institutions like health or education departments) have long been held as a barrier to development, as the critique of market-led strategies for development (and macro-economic policy changes in particular) gathered momentum, institutional strengthening as a complement to macro-economic policy changes became part of the development agenda:

> **Reforming countries were discovering that economic growth did not matter much to people if hospitals did not have medicines . . . A competitive exchange rate could not do much to bolster exports if inefficiency and corruption paralysed the ports, and fiscal reform did not matter much if taxes could not be collected.**
>
> (Naim, 2000: 514)

As the millennium loomed, the fundamental legitimacy of some of the major international institutions of development, particularly financial institutions such as the International Monetary Fund and the World Bank and the World Trade Organization, were increasingly challenged by broad based social reaction (the 'anti-globalisation' movement discussed in Chapter 4). In short, neo-liberalism (and the policy and project outcomes that flowed from that thinking) had led to widespread disruption felt in both developing and more developed economies and societies; 'Development's lead institutions had to be recast as more "inclusive", more responsive and "participatory", and thereby, somehow more legitimate' (Craig and Porter, 2006: 4).

All these forces of change have contributed to a rethinking not just in theory but in the practice of how these various institutions, organisations and agencies in development operate (and, importantly, work with each other). Whilst many of the new institutions created in recent decades may not set policy, they influence substantially the directions and outcomes of policy. A number of such examples are detailed in this chapter, including the increase in international organisations concerned with global environmental issues, the expansion of non-governmental organisations (NGOs) generally and the emergence of influential 'new social movements'.

Discussions of institutions in development have also been expanded recently to include less 'tangible' institutions that may lack a specific organisational form but are now understood as often critical in mediating the processes through which people construct their livelihoods, for example. These institutions include the regularised patterns of behaviour between individuals and groups, which are structured by 'unwritten codes' and norms that have widespread use in society. These informal institutions may have as much power over human action, decision making and what happens in terms of development as formal institutions governing markets or the activities of the state. Many insights into the operation and impacts of these kinds of institutions have come through work concerning natural resource management (see Mehta *et al.*, 1999; also see Chapter 10). It is also through this kind of thinking that the importance of gender relations as institutions structuring the lifestyles and livelihoods of individuals has been exposed (see Joekes *et al.*, 1995) and this is considered below in terms of understanding the impacts of macro-economic reforms.

In examining the role of institutions in shaping development outcomes it is very important to understand that institutions across all these scales are not neutral factors in the development process; 'they represent values, which in turn represent the interests of some political or social group' (Sharp, 1992: 55). As such there is a need for ongoing critical analysis of institutions, not just for the ways in which they 'do development' but also for the way in which they may constrain development opportunities and present barriers for particular groups and social actors.

Institutions under change

It is evident that institutions are dynamic. Indeed, this chapter is structured around the role of three interrelated forces and the ways in which they are directing change within and between the core institutions in development. First we look at the global search for sustainable patterns and processes of development from the late 1970s onwards. In short, this brought environmental issues into international relations and to the understanding of, and response to, the 'grass roots realities' (Chambers, 1983, 1993) of the most impoverished groups in developing societies. The concept of sustainability (considered in more detail in Chapter 6) is built on the fundamental interdependence of development and environmental conservation at all scales of analysis and the requirement for new norms of behaviour within institutions across all spheres of human activity (Elliott, 2006).

Second, recession within the global economy, emerging through the 1980s, demanded a worldwide reassessment of the role of particular institutions and the relationships between various organisations in the search for ways to ease debt, to stimulate economic growth and to make the most effective use of financial and human resources in the delivery of products and services. These economic concerns remain fundamental priorities for institutions at all scales into the twenty-first century.

Finally, this chapter also considers the drive towards 'good governance' that is increasingly held by the IFIs and donor agencies in particular as a crucial determinant of development (World Bank, 2006a: Craig and Porter, 2006). Back in 1993, Leftwich referred to good governance and democracy as a new orthodoxy dominating development thinking and aid policy. What he considered was 'new' was the way in which good governance and democracy was forwarded as a necessary condition for development, rather than an outcome of it (as understood within earlier modernisation theses).

The Key idea box examines a number of ways in which the notion of 'good governance' is and has been used to direct development thinking and practice. The various sections of this chapter consider ideas around good governance as a factor in institutional change that encompass: concerns for sound development administration and management; the responsiveness and accountability of institutions to public needs; the access of citizens to decision making and power; the relative autonomy of associations of civil society; and the presence of competitive democratic politics.

This chapter assesses the nature and activities of a number of institutions prominent in the development process in terms of these three forces of international change. Although they are considered separately, the interdependence of these forces for institutional development is highlighted throughout and it is perhaps at these junctures that the key challenges and uncertainties for future development lie. Further illustration of the impacts of particular initiatives on the practice of development can be followed within Part III.

Key idea

The multidimensional notion of 'good governance' in practice

- Protecting and building confidence around markets and capital security such as through anti-corruption measures and strengthening rules of law.
- Putting a 'human face' to macro-economic reforms such as through working with civil society to ensure better delivery of social services and increasing investments in health and education.
- Decentralisation of administration, resources and decision making to subnational authorities and communities as a key to less corrupt environments for business, better access to service delivery and more responsive and accountable agencies.
- New fiscal arrangements for decentralised governance as a means for lower transaction costs and greater allocative efficiencies that would lead to better outcomes for the poor.
- Good governance as a means to overcome the breakdown of civic order and address the phenomenon of 'failed states'.

Source: Compiled from Craig and Porter (2006)

The rise of global governance

International institutions are created by states, usually as a 'means of achieving collective objectives that could not be accomplished by acting individually' (Werksman, 1996: xii). At the end of the Second World War the challenges of the resurrection of the global economy (of avoiding a return to the national protectionism which created the Great Depression of the 1930s) and of reconstruction in Europe in particular, prompted the creation of new forums such as the United Nations and the International Bank for Reconstruction and Development (IBRD or, as commonly termed, the 'World Bank'). The intention was the development and coordination of international efforts towards preserving peace, resolving conflicts and promoting social and economic development.

Over 60 years on, the reconstruction of Europe is evident and peace has been retained in the region. However, poverty and armed conflict persist in many areas of the developing world, and there are new challenges for such international institutions, including the need to resolve development aspirations across the globe within the limits of natural resources and the environment (see also Chapter 6). Furthermore, these institutions are under change. Membership of the United Nations, for example, expanded from 51 countries in 1945 to 192 in 2001. In addition, many more intergovernmental fora have been established in recent decades, amounting to over 1000 institutions (Dodds, 2002: 291).

International institutions are highly varied in terms of their shape and function, the rules and practices concerning their activities and their power to influence the behaviour of other organisations. Fundamentally, the pursuit of collective goods, including conservation of the global environment, demands some devolution of sovereign power, but the capacity of international institutions to take on characteristics and powers distinct from the states which created them depends on the willingness of those states to make such investments (Werksman, 1996). This provides the context for any critical analysis of the operation and development outcomes of these institutions.

The United Nations system

The Charter of the United Nations (UN) was signed in San Francisco in June 1945 by 51 countries as the successor organisation to the League of Nations created in the inter-war years. The UN was a more complex and ambitious undertaking than the League of Nations, intended to go beyond merely ensuring stability in international relations in order to 'systematise the promotion of change' (Righter, 1995: 25; Plate 7.1). The UN today is a complex web of institutions, many

Plate 7.1 Repatriation of Cambodians from refugee camps in Thailand under United Nations protection
(photo: Teit Hornbak, Still Pictures)

with overlapping responsibilities, but retains the original broad purposes, including a commitment to equal rights for people of all nations, to free succeeding generations from the scourge of war and to promote social and economic progress. In 2003 the credibility of the UN system suffered substantially as the US (supported by the UK) declared war on Iraq without the support of the UN. There is currently substantial debate and uncertainty regarding its role, operations and future as encapsulated in the statements in Table 7.1.

The UN system, as illustrated in Figure 7.1, includes the United Nations organisation itself (comprising six main organs as shown), various programmes and funds (such as the United Nations Environment Programme), nine functional commissions (such as in human rights), five regional economic and social commissions, numerous standing and ad hoc committees, plus a number of specialised agencies (such as the International Labour Organization and the World Bank Group). It also includes varied associated but autonomous bodies and institutions such as the World Trade Organization, which liaises closely with the UN.

Almost every country in the world (a total of 192) is now a member of the UN. Its headquarters are in New York and it is funded through a mixture of member state assessments (ranging from a minimum of 0.001 per cent of the total for the lowest income countries to a ceiling, 25 per cent for the USA) and voluntary contributions (Department for International Development, 1999: 9). Although the US is the largest financial contributor to the United Nations ($3.8 billion to be paid for 2006/07), many smaller countries tend to contribute more per citizen to the UN budget. For example, Luxembourg pays $2.44 per capita, while Germany contributes $1.51 in comparison to $1.23 per capita in the United States (UN, 2006).

Table 7.1 Debating the future of the United Nations

- Disillusion with the UN has become one of the clichés of our time. The reality is much more complex (Ignatieff, 1995)
- Everyone expects everything of it (Evans, 1993: 24)
- Few would question the need for reform of the UN, crafted in another era, haunted by recent failures in Somalia and former Yugoslavia and facing grave financial woes (Littlejohns and Silber, 1997)
- The UN was set up to recognise the supremacy of the nation state; it now needs to factor in the impact of globalisation on the system (Dodds, 2002: 293)
- As the gap between rich and poor continues to widen and as a spiral of poverty, environmental degradation and violent conflict become all but a way of life in some regions, the difference between what the UN enshrines and what it is able to accomplish threatens its legitimacy (Whitman, 2002: 467)
- In Iraq a tipping point has been reached. After going to war in 1991 and overseeing the massacre of untold thousands of Iraqis, the UN became complicit in the slow death of thousands (Ransom, 2005a: 11)

Typically, many members have been, and continue to be, in arrears to the UN. For example, in August 2001, 103 member states had paid their regular budget contributions in full. However,

> **some major contributors have paid none or only part of their dues, forcing the UN to cross-borrow from the peacekeeping account to offset the earlier and larger than usual deficit currently experienced. It is clear that the UN cannot function effectively unless all member states pay their dues, in full, on time and without conditions.**
>
> **(UN, 2001: 48)**

The top ten debtors to the UN peacekeeping budget in 2006 amounted to $1.34 billion with the United States, Japan, Ukraine, China and the Republic of Korea heading the list (UN, 2006).

Each member state of the UN is represented on the General Assembly (GA) with equal voting power irrespective of the size, population or power of that country. The GA meets annually but serves largely as a 'forum for forging consensus and influencing state behaviour' (Werksman, 1995: 11) since decisions taken by the GA (unlike those of the councils) have no legally binding force for governments. Such decisions do, however, 'carry the weight of world opinion' (Buckley, 1995: 4), with many resolutions being subsequently incorporated into international treaties such as the Climate Change Convention of 1994. The GA controls the budget and staffing levels of the UN and deputes its work to six committees dealing with disarmament, economic and financial matters, social humanitarian and cultural matters, decolonisation, administrative and budgetary affairs and legal matters.

The UN and sustainable development

In 1992 the United Nations facilitated the largest gathering of heads of government in history, and it was a conference on the environment that brought

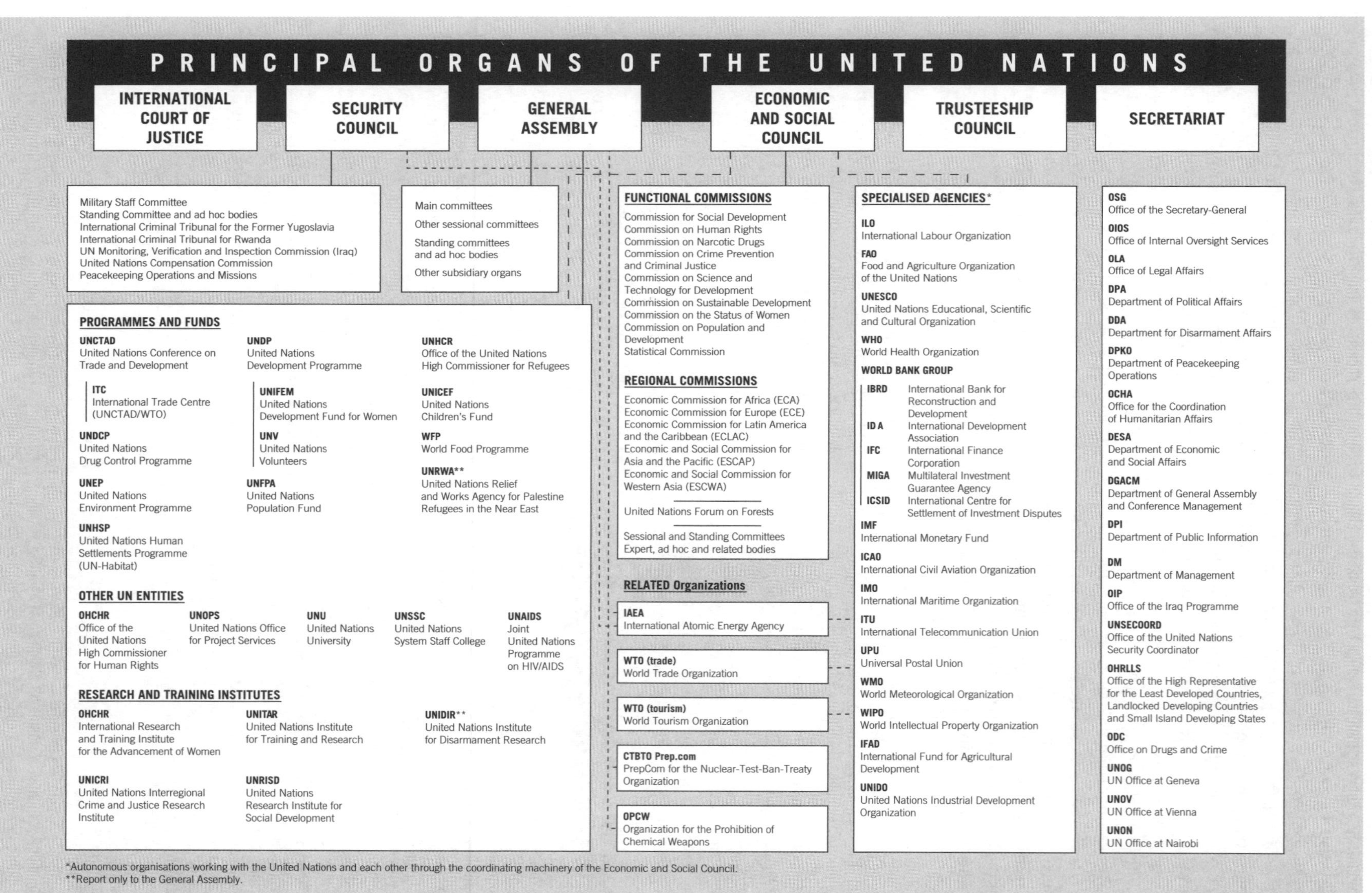

Figure 7.1 The UN System

Source: From UN Department of Public Information (2003)

them together in Rio de Janeiro at the United Nations Conference on Environment and Development (UNCED). Ten years on, an even bigger gathering of not just heads of state but members of business and global civil society was held in Johannesburg aimed at reinvigorating North–South partnerships for sustainable development. This was the World Summit on Sustainable Development (see Elliott, 2006). Through the overall processes of these summits (including the preparatory meetings as well as the direct outcomes of the conferences), further institutions within the UN family have been created and new activities of the UN facilitated towards achieving more sustainable patterns and processes of development. In addition, the UN itself has undertaken a process of internal reform, as considered in the Critical reflection box.

New institutions: the Commission on Sustainable Development

Potentially, the most notable achievement of the Earth Summit of 1992 was the adoption of Agenda 21, a 40-chapter action plan for the global community. However, it was also the point at which preparations for the creation of a new international institution within the UN system, the Commission on Sustainable Development (CSD), was initiated. It was recommended that such a commission was necessary as a means through which governments could review progress towards the goals of the Earth Summit (encapsulated in Agenda 21) and to formulate an integrated approach to economic and environmental policy making.

The recommendation was for a UN Board of Sustainable Development at the 'very apex of the UN' (Jordan

Critical reflection

The UN under reform

On 8 September 2000, all 189 member states of the UN adopted the Millennium Declaration. The Declaration was designed to reaffirm faith in the organisation and its Charter and identified a number of key objectives for the UN community considered to be of 'special significance' in the early years of the twenty-first century. These objectives related to

II. Peace, security and disarmament
III. Development and poverty eradication
IV. Protecting our common environment
V. Human rights, democracy and good governance
VI. Protecting the vulnerable
VII. Meeting the special needs of Africa
VIII. Strengthening the UN

(UNGA, 2000)

In the following year the UN detailed its 'road map' towards the implementation of the Millennium Declaration. In relation to objective IV, protecting our common environment, the Secretary General stated that 'one of our greatest challenges in the coming years is to ensure that our children and all future generations are able to sustain their lives on the planet' (UN, 2001: 32). Commitments were made to,

- make every effort to ensure the entry into force of the Kyoto Protocol, preferably by the tenth anniversary of the UNCED in 2002, and to embark on the required reduction in emissions of greenhouse gases;
- intensify our collective efforts for the management, conservation and sustainable development of all types of forest;
- press for the full implementation of the Convention on Biological Diversity and the Convention to Combat Desertification in those countries experiencing serious drought and/or desertification, particularly in Africa;
- stop the unsustainable exploitation of water resources by developing water management strategies at the regional, national and local levels which promote both equitable access and adequate supplies;
- intensify our collective efforts to reduce the number and effects of natural and man-made disasters;
- ensure free access to information on the human genome sequence.

Select one of these commitments. What are the principal barriers and opportunities facing the UN in meeting this commitment? Consider what changes are needed within the institutions of the UN itself and what it needs to do in influencing the activities of other institutions.

and Brown, 1997: 274), chaired by the Secretary General, with the capacity to assess, advise, assist and report on progress of sustainable development and with supreme powers to create interagency commitment to, and coordination for, sustainable development. However, 'in the event, the Rio conference fudged this issue' (Jordan and Brown, 1997: 274), creating a lower ranking body placed under the Economic and Social Council (rather than the GA itself).

The CSD met for the first time in 1993, comprising 53 elected members: 13 from Western Europe and America, 13 from Africa, 11 from Asia, ten from Latin America and the Caribbean, and six from Eastern Europe (www.un.org/esa). This 'overarching international environmental organisation' has been called the 'UN Committee of the Whole' (Righter, 1995: 305), relating to its principal function in monitoring progress in the implementation of Agenda 21 through evaluation of all reports from all relevant organisations, programmes and institutions of the UN system. The CSD was also specifically charged with monitoring commitments by UN member nations to provide financial resources at the interface of environment and development and to the transfer of technologies as outlined in Agenda 21.

However, the CSD in fact has no legal or budgetary authority, and it is essentially a forum for review, exchanging information, building political consensus and for forging partnerships. It also has a mandate to foster the participation of NGOs, industry, scientific and business communities through encouragement of such groups to make written submissions or to address meetings of the commission.

Dodds (2002) considers the CSD an important arena in which stakeholder involvement in the UN has been developed in recent years. However, the CSD is also a 'creature of state control, unable to inspect activities at the sub-national level directly for itself or do more than accept what states themselves report' (Jordan and Brown, 1997: 274). There continue to be problems of integrating and coordinating environment into development activities worldwide, not least because government representatives on the CSD tend to be environment ministers, 'with little influence over their more powerful colleagues in trade and industry departments' (Jordan and Brown, 1997: 275). Bruno (2005) considers that the CSD operates only as a 'talk shop'. However, the CSD has had successes including in convening 'hundreds of government- and major group-hosted intercessional meetings' (Dodds, 2002: 296) and bringing tourism, for example, into the Rio process through discussion for the first time in 1999.

Changing UN activities towards sustainable development

Since the Rio conference, the challenges of integrating environment and development have continued to prompt change within the institutions and activities of the United Nations. Indeed, Chapter 38 of Agenda 21 committed the UN itself to strengthening cooperation and coordination on environmental and developmental issues across the UN system. Towards these ends, Table 7.2 identifies a host of summits that were held to follow up aspects of Agenda 21 in more depth. Within the preparation meetings for the 2002 Johannesburg summit, the UN urged the need for varied institutions to work together in furthering sustainable development:

A focused agenda (for Earth Summit 2) should foster discussion of findings in particular environmental sectors as well as in cross-sector areas, such as economic instruments, new technologies and

Table 7.2 Following up the commitments made at Rio

1994	UN Conference on Population and Development (Cairo)
1994	Conference on Small Island Developing States (Barbados)
1995	World Summit on Social Development (Copenhagen)
1995	Fourth Conference on Women and Development (Beijing)
1996	Habitat II – Conference on Human Settlements (Istanbul)
1997	UN General Assembly to Review Implementation of Agenda 21
1999	UN General Assembly Review of Cairo
1999	UN General Assembly Review of Barbados Plan of Action
2000	UN Millennium Summit (New York)
2000	UN World Summit for Social Development and Beyond (Geneva)
2001	UN Conference on Least Developed Countries (Brussels)
2002	World Summit on Sustainable Development (Johannesburg)
2002	International Conference on Financing for Development (Monterrey)
2005	The 2005 World Summit (New York)

globalisation . . . Private citizens as well as institutions are urged to take part in the process. Broad participation is critical. If further action is to be effective in achieving the ultimate goal of sustainability, Governments cannot work alone.

(UN, 2001: 32/3)

The United Nations Environment Programme (UNEP) is the UN's primary environmental policy coordinating body, with its headquarters in Nairobi. It was created in 1973 following the United Nations Conference on the Human Environment in Stockholm of the previous year. UNEP's constituent act stressed the need to assist developing countries to implement environmental policies and programmes that are compatible with their development plans. Its coordinating role in environmental matters was given impetus by the report of the Brundtland Commission (WCED, 1987) and was reaffirmed at UNCED.

However, many consider that UNEP has in fact become weaker rather than stronger since the Rio conference (Dodds, 2002; Sandbrook, 1999). In particular, the number of multilateral environmental agreements (conventions, treaties and agreements) has soared over the last few decades and UNEP has been instrumental in many of these (e.g. the Convention on International Trade in Endangered Species (CITES) and the Basel Convention on the Control of Transboundary Movements of Hazardous Wastes and their Disposal). But 'each environmental treaty creates its own mini-institutional machinery' (French, 2002: 177) with their own offices (secretariats), located in various capitals around the world, and treaty members (Conference of the Parties, CoP) to convene regularly and oversee implementation.

As well as the challenges of coordination across continents presented by the geographical locations of these various offices, the relationship between these mini-institutions and UNEP remains unclear (Sandbrook, 1999). There are also problems of scarce funding, for these treaties as well as for UNEP itself. In 2004/05, UNEP's funding from the regular UN budget was only US$10.5 million and although it also receives voluntary contributions, particularly via the Environment Fund, finances fall short of what UNEP's council itself considers necessary to deliver its remit (www.unep.org/rmulen/financing_of_UNEP/index.asp).

The United Nations Development Programme (UNDP) is the UN's central funding agency for technical assistance activities, and is the principal in-country representative of the UN. In 1992, at UNCED, it launched its Capacity 21 strategy to assist developing countries to incorporate principles of sustainable development into national programmes and processes aiming to strengthen both national and local capacity to meet the goals of Agenda 21. The notion of capacity is used to refer to the abilities of individuals, institutions and societies to perform functions, solve problems, and set and achieve objectives in a sustainable manner (www.capacity.undp.org).

UNDP seeks to assist countries through the process of developing such abilities as are obtained, strengthened, adapted and maintained over time. However, although launched in Rio with a target of US$100 million (Williams, 1995), Capacity 21 has also suffered from decreasing and inadequate funding. UNDP is a programme of the UN, rather than a principal body or organ. It therefore depends largely on voluntary funding from member states, largely Official Development Assistance, and has been under considerable financial pressure in recent years as ODA has increasingly been diverted to areas of humanitarian relief rather than longer term development assistance (Thomas and Allen, 2000). For example, by the mid 1990s the World Bank had become a larger funder of technical assistance to developing countries than the UNDP (Gwin, 1995). A decade on, Bruno (2005: 27) suggests that UNDP has been impoverished to such an extent that it now sees itself as a broker between countries and its private sector partners.

In 2003, UNDP launched its Capacity 2015 strategy that has superseded Capacity 21. Its focus is now on working through a series of partnerships to build capacities at the local level in order to achieve the Millennium Development Goals (MDGs) as well as to continue the work of Capacity 21 in realising the goals of Agenda 21. The strategy aims to develop active partnerships through supporting networking and the exchange of ideas (within countries and across countries with similar economic, social and environmental conditions) and through supporting projects and programmes that explicitly support both capacity development and/or linkages between local communities and the global economy (www.capacity.undp.org).

UNDP shares with UNEP (as well as the World Bank), responsibility for managing and implementing the Global Environment Facility (GEF). The GEF was pioneered in 1991 and aimed to create new funds for the governments of low-income countries (defined as those with per capita GNP of less than US$4000) to undertake environmental actions which have clear global benefits in areas such as greenhouse gas emissions, biodiversity, ozone and international waters. GEF monies are therefore a core means for promoting national-level implementation of the major multilateral environmental treaties within developing countries (French, 2002). The donations are voluntary and are meant to be in addition to existing flows of aid. Whilst it has been referred to as one of the main 'winners' at Rio in that many world leaders made large pledges of new and additional money at the conference, raising monies for the GEF has proven a continued challenge (French, 2002). Some further issues concerning the future of the GEF are considered in the section on the World Bank below.

The future role of the UN in sustainable development

While the UN clearly provides many opportunities in terms of global governance towards sustainable development, there are evident limitations to date in its efforts to integrate social, environmental and economic management within the system as a whole that remain contrary to the fundamental principles of the concept; 'at the same time, [the UN] is the best and the worst place to address the challenges thrown up by the concept of sustainability' (Jordan and Brown, 1997: 273).

Further implications of these continued challenges for and within the UN to the prospects for sustainable development are considered in subsequent sections. In particular, the suggested 'move away' from the UN (towards the World Bank and IMF) in recent years on behalf of donor countries is discussed in terms of the implications for the capacity of developing countries to influence processes and outcomes. In short, while these countries have an influence and voice through the 'one member–one vote' system of the UN, it will be seen that within the international financial institutions there is 'an unequal decision-making structure where the richer countries who "pay the bills" have more influence' (Dodds, 2002: 300).

The UN and economic change

Through most of the UN's life, the bulk of its finance and effort has gone towards economic and social development, coordinated by the Economic and Social Council (ECOSOC) and effected through nine functional commissions and the five regional commissions shown in Figure 7.1. As much as 75 per cent of the UN effort (measured in terms of money spent and the people involved) went towards these activities through the 1980s and 1990s (Buckley, 1995).

Unlike the Bretton Woods institutions, the UN offers assistance in the form of grants, rather than loans subject to interest. There is substantial debate, however, as to the development impact of UN activities. For example, the high proportion of UN spending on economic and social development needs to be viewed in terms of the very limited overall budget of the UN system. Despite calls at UNCED to raise the level of donor funds for development administered centrally, only 14 per cent of ODA globally in the 1990s went through the UN (World Resources Institute, 1994: 226).

The development impacts of UN activities are also considered to have been limited by the number of agencies competing for the same monies and what has been termed the UN's 'butterfly' or 'laundry list' approach: 'thousands of mini projects working on mini objectives' (Righter, 1995: 59). Inevitably, problems of duplication and a lack of coherence in UN programmes (and often supported by excessive bureaucracies) have been identified.

However, such criticisms are not to deny the significant role played in directing development research and action by many individuals working within the UN system or for those as guided by the annual publications of UN organs such as the *Human Development Report* (initiated by UNDP in 1990). Thomas and Allen (2000: 200), for example, consider the Reports to have played an important role in critiquing the effects of neo-liberalism during the early 1990s and specifically for 'drawing attention to issues of poverty and deprivation', thereby representing an important contribution to more sustainable patterns of development.

In terms of responding to global economic recession and the international debt crises, central UN organs have exhibited perhaps less change than associated institutions such as the World Bank. These latter developments are discussed more fully in the next section.

UN agencies have traditionally been concerned with projects and technical cooperation, while the World Bank has historically provided the investment support. By the mid 1990s it was suggested (ul-Haq *et al.*, 1995: vii) that 'most of the finance and much of the action' in economic and social development are now concentrated in the international financial institutions, rather than the UN.

Indeed, Righter (1995: 268) suggests that even in terms of policy analysis in their own sectoral specialisations, UN agencies were being outperformed by the development banking system: 'The World Bank's expertise in education now exceeds that of UNESCO for example'. Similarly, Dodds (2002: 300, emphasis added) has referred to the 'move away from the UN towards the Bretton Woods institutions for discussion on finance, trade and *even poverty*'.

As seen above, programmes of UN organs such as the UNDP are now increasingly taking a multifaceted strategy in their activities, collaborating as much with private enterprise as with governments, other international and bilateral donor agencies and in supporting measures to promote capacity building within the business sector. In these ways, the distinction between the roles of agencies of the UN system and the Bretton Woods institutions, including the World Bank, has become less clear in recent years and is recognised by many as a problem (Department for International Development, 1999; UN, 2000b).

The UN and political change

Ensuring a level of democracy within member nations through the maintenance of international peace and security was the principal purpose of the UN as originally stated within its founding charter. Although it is generally considered that recent decades have been an era of unprecedented world peace, many of the inhabitants of the developing countries still do not enjoy basic freedom from conflict. Since 1945 the UN has undertaken over 60 peacekeeping missions and 17 of those were still active in 2007. World military expenditure in 2005 reached an estimated $1.1 trillion per year corresponding to approximately 2.5 per cent of the world Gross Domestic Product (UN, 2006). In the mid 1990s, fewer than half of the UN member states had elected governments (Buckley, 1995: 4).

Responsibility within the UN system for effecting peace and security lies with the Security Council, of which there are five 'permanent members': the USA, the UK, China, France and Russia (the leading powers at the time of the establishment of the UN). Any UN operation, including peacekeeping, can only proceed with the agreement of the member states. Each of the members of the Security Council has the 'power of veto', in that a no-vote by any one of them can stop a resolution being passed even if the other 14 (co-opted) members vote yes.

During the Cold War, the power of veto served to limit substantially the UN's peacekeeping role, as the USA or the former Soviet Union would veto any substantive proposal in the Security Council. Since the late 1980s, however, the power of veto has been used only rarely and, concurrently, the number of peacekeeping operations undertaken and the costs of UN military responses increased dramatically in the mid 1980s, as shown in Figure 7.2. Despite this increase in activity, many countries (including some of the world's richest countries and members of the Security Council) do not meet their peacekeeping dues each year, as identified above.

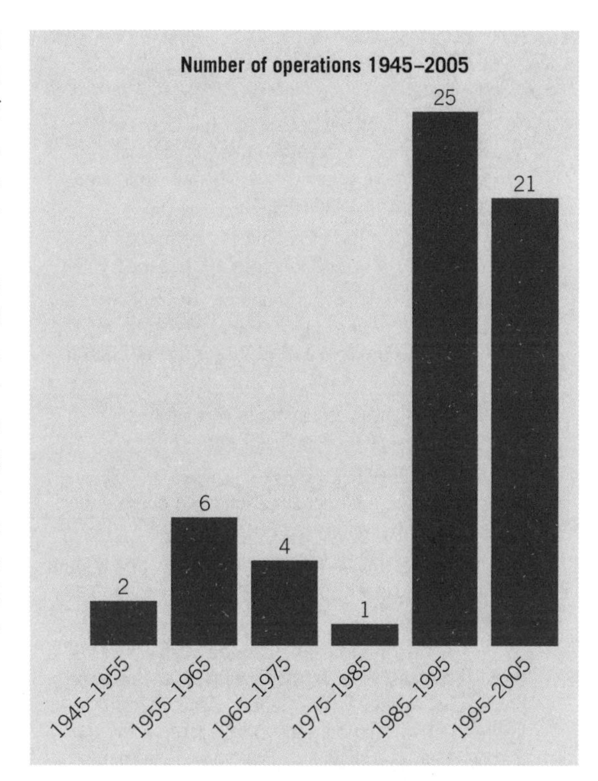

Figure 7.2 The number of UN peacekeeping operations, 1945–2005
Source: www.un.org/depts/dpko/list/list.pdf

Under the Millennium Declaration of 2000 the UN stated that 'we will spare no effort to promote democracy and strengthen the rule of law, as well as respect for all internationally recognised human rights and fundamental freedoms, including the right to development' (UN, 2000c: 6). Within the Road Map Towards the Implementation of the UN Millennium Declaration (the document within which the list of Millennium Development Goals was agreed), the continued violation of human rights was identified as an attack on the very founding principles of the UN that were committed to supporting human dignity (UN, 2001). The Road Map applauded the efforts of governments towards respecting the Universal Declaration

Table 7.3 Strengthening the United Nations

- To reaffirm the central position of the General Assembly as the chief deliberative, policy-making and representative organ of the UN, and to enable it to play that role effectively
- To intensify our efforts to achieve a comprehensive reform of the Security Council in all its aspects
- To strengthen further the Economic and Social Council, building on its recent achievements, to help it to fulfil the role ascribed to it in the Charter of the UN
- To strengthen the International Court of Justice in order to ensure justice and the rule of law in international affairs
- To encourage regular consultations and coordination among the principal organs of the UN
- To ensure that the UN is provided on a timely and predictable basis with the resources it needs to carry out its mandates
- To urge the Secretariat to make the best use of those resources, in accordance with clear rules and procedures agreed by the General Assembly, in the interests of all member states, by adopting the best management practices and technologies available
- To promote adherence to the Convention on the Safety of UN and Associated Personnel
- To ensure greater policy coherence and better cooperation between the UN, its agencies, the Bretton Woods institutions and the WTO, as well as other multilateral bodies
- To strengthen further cooperation between the UN and the Inter-Parliamentary Union
- To give greater opportunities to the private sector, NGOs and civil society in general to contribute to the realisation of UN goals and programmes

Source: UN (2001)

on Human Rights but noted the gap that regularly occurred between rhetoric and action. Significantly, the Millennium Declaration highlighted that human rights were also to be a central tenet of UN reform, as detailed in Table 7.3;

> **the cross-cutting nature of human rights demands that whether we are working for peace and security, for humanitarian relief or for a common development approach and common development operations, the activities and programmes of the system must be conducted with the principles of equality at their core.**
> (UN, 2001: 37)

It is evident throughout the sections above, that the constituent institutions of the UN system are currently undergoing substantial change. Bogert (1995) questioned whether the UN was the 'global emergency number', a 'development agency' or a 'floating cloakroom', in recognition of an uncertain future for the UN. Bogert was confident that there was a future for the UN, in addressing global issues beyond the reach of any one country, dealing with issues such as fighting terrorism, curbing nuclear proliferation and combating disease. However, the logic of the UN is predicated on the idea of members that are nation states, but, as discussed in subsequent sections and in Chapter 4, states are becoming increasingly fractured and multinational corporations are bigger than some countries:

> **It [the UN] is trying to re-invent itself for a world its founders never envisaged. It is no longer policing disputes between states, but within states. It no longer arbitrates between sovereignties, but struggles to keep sovereignties from disintegrating under the strain of civil war. It was intended as an organisation of states and yet it is now called upon, time after time, to protect people against their states.**
> **(Ignatieff, 1995)**

In 2001 the Secretary General of the United Nations, Kofi Annan, was awarded the Nobel Peace Prize (shared with the UN itself) in recognition of his 'pre-eminent role' in 'bringing new life to the UN' (Usbourne, 2001) and in fostering the kinds of institutional changes identified in Table 7.2. Whilst it should be acknowledged that these are ongoing processes of reform, there is a notable absence of any targets or indicators within the Millennium Development Goals referring to the activities of or relationships between major agencies in development (or for governance).

In 2005, the UN Millennium Development Project (an independent advisory body to the UN) reported on progress towards achieving the MDGs and identified a 'Practical Plan' to achieve those goals (Sachs, 2005). Whilst the report stressed that there was no 'one-size-fits-all' explanation for why the goals were succeeding or failing within particular countries, governance failures were considered to be one of four overarching obstacles to reaching the MDGs by 2015:

> **sometimes the problem is poor governance, marked by corruption, poor economic policy choices, and denial of human rights. Sometimes the problem is a poverty trap, with local and national economies too poor to make the needed investments. Sometimes progress is made in one part of the country but not in others, so that pockets of poverty persist. Even when overall governance is adequate, there are often**

Key idea

The World Bank Group

Figure 7.3 identifies the five institutions that comprise the World Bank Group. The first institution to be established was the International Bank for Reconstruction and Development in 1945. It was founded on the principle that many countries in Europe after the Second World War would be short of foreign exchange for reconstruction and development activities, but would be insufficiently creditworthy to borrow all the necessary funds commercially. In contrast to individual countries, IBRD as a multilateral institution, with share capital owned by its member countries, could borrow on world markets and lend more cheaply than commercial banks. The IBRD raises monies through selling bonds and other securities to individuals, other banks, corporations and pension funds around the world. It lends money over 15- to 20-year periods, and loans are subject to interest.

The International Development Association (IDA) is the Bank's concessional lending window, providing no-interest loans to the poorest countries, defined in 2007 as those with per capita incomes of less than US$1025 (www.worldbank.org). IDA loans typically constitute around 25 per cent of total World Bank lending in any year. The IDA cannot raise funds on capital markets as does the IBRD, so it depends entirely on wealthier nations for finance (supplemented by a portion of general World Bank profits). Loans are to be repaid within 35 to 40 years with a 10-year grace period. While the majority of financing from IDA is to investment projects, IDA also provides adjustment credits assisting governments to finance teachers' salaries and agricultural extension services, for example, during the processes of adjustment. Figure 7.4 illustrates where IBRD/IDA finances went geographically and by sector in 2006.

The International Finance Corporation of the World Bank Group lends directly to the private sector without the requirement of government guarantees. The Multilateral Investment Guarantee Agency also promotes private investment in developing countries through providing guarantees on investments against non-commercial risks such as war or nationalisation. Both these institutions of the World Bank Group are legally and financially independent and have different owners, clients, mandates and operational procedures such that this chapter uses the term World Bank to refer solely to the IBRD and IDA.

The World Bank Group				
The World Bank				
IBRD	IDA	IFC	MIGA	ICSID
International Bank for Reconstruction and Development	**International Development Association**	**International Finance Corporation**	**Multilateral Investment Guarantee Agency**	**International Centre for Settlement of Investment Disputes**
Established 1945 184 countries own, subscribe to its capital	Established 1960 165 members	Established 1956 178 countries	Established 1988 165 members	Established 1966 142 members
Lends to creditworthy borrowing countries, based on high real rates of economic return	Lends at a favourable rate to poorer countries with a per capita GNP of less than $885	Assists economic development by promoting growth in the private sector	Assists economic development through loan guarantees to foreign investors	Provides facilities for the conciliation and arbitration of disputes between member countries and investors who qualify as nationals of other member countries

Figure 7.3 The World Bank Group

Source: World Bank (2005d) © International Bank for Reconstruction and Development/The World Bank

▶

Key idea (continued)

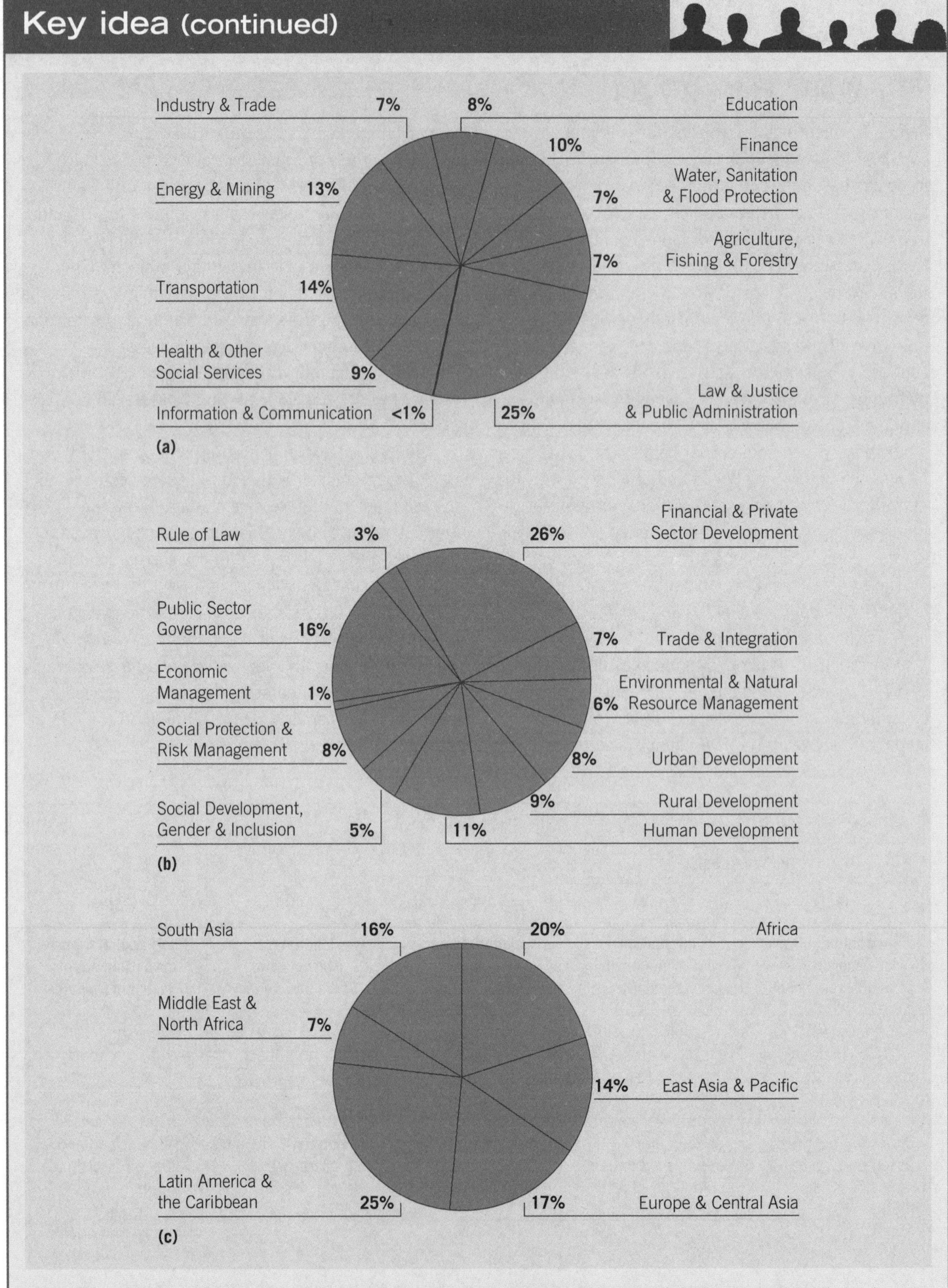

Figure 7.4 World Bank lending by (a) sector, (b) theme and (c) geographical region, fiscal year 2006 (share of total lending of $23.6 bn)

Source: World Bank (2006b) © International Bank for Reconstruction and Development/The World Bank

areas of specific policy neglect that can have a monumental effect on their citizens' well-being. Sometimes these factors occur together, making individual problems all the more challenging to resolve.

(Sachs, 2005: 15)

The World Bank Group and the International Monetary Fund

The World Bank Group consists of five closely associated institutions, as identified in the Key idea box. It was formed at the Bretton Woods Conference in 1944, convened to establish a new framework for world economic stability after the crisis of the 1930s and the Second World War. The World Bank is owned by its member countries that are represented by a Board of Governors and a Washington-based Board of Directors. Its President, by tradition, is a national of the largest shareholder, a US citizen, and operates on a five-year renewable term of office. The headquarters of the World Bank are in Washington and it has over 10,000 employees spread between headquarters and its 100 country offices (www.wb.org).

The International Monetary Fund (IMF) was also created at the Bretton Woods conference and is closely associated with the World Bank Group. The World Bank and the IMF are part of the UN system as seen in Figure 7.1, being specialised agencies reporting to ECOSOC. In practice, they function as institutions very separate from and with little accountability to the UN system (Thomas and Allen, 2000), fundamentally through their very different management and voting arrangements.

Core characteristics and functions

Although only 'peripherally conceived as a development agency' (Righter, 1995: 187), since 1950 the World Bank has increasingly lent monies to governments of the developing nations, loaning a total of US$23.6 billion in fiscal 2006 (World Bank, 2006b). However, the bulk of finances in recent years to developing countries have come from private rather than official sources, as shown in Figure 7.5. The Critical reflection box considers the transfer of finances both to and from the developing countries (see also Chapter 8).

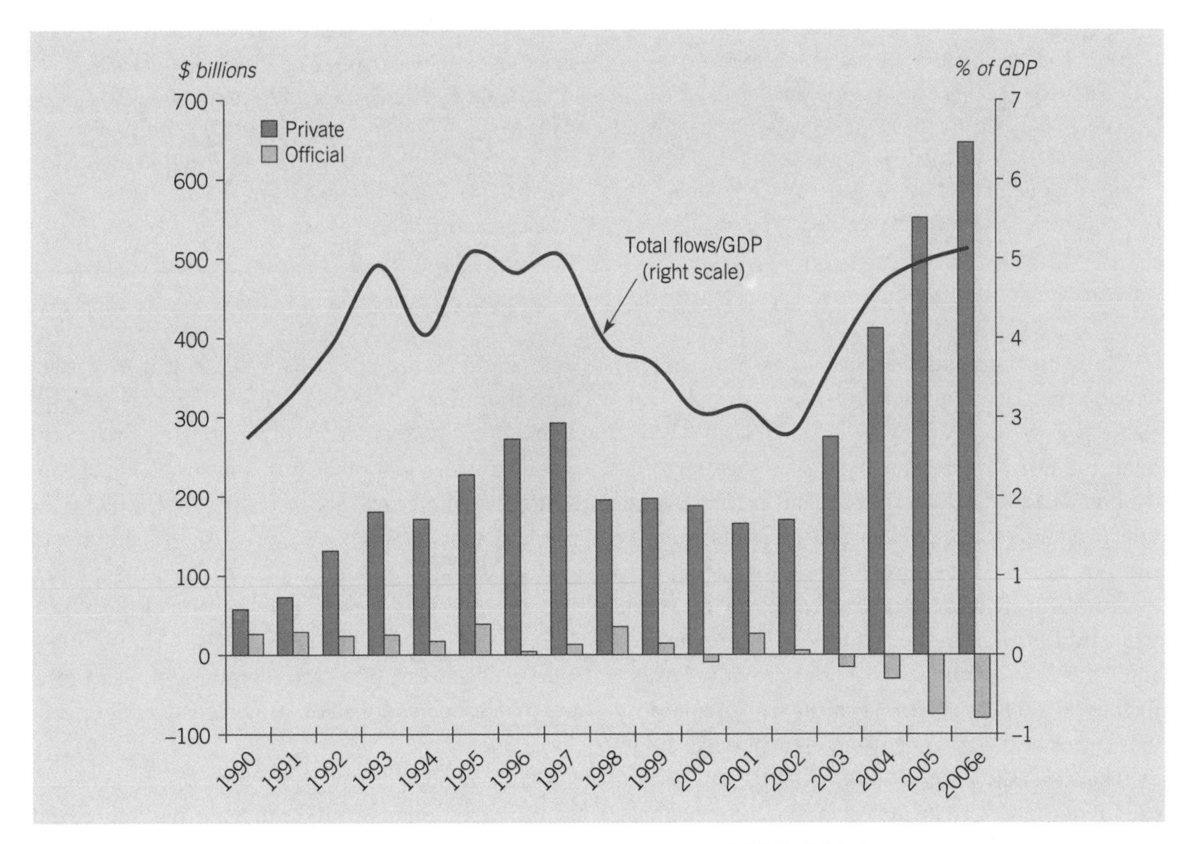

Figure 7.5 Net capital flows to developing countries, 1990–2006
Source: World Bank (2007b) © International Bank for Reconstruction and Development/The World Bank

Critical reflection

The changing nature of finance for development

The early years of the twenty-first century have seen a rapid growth in financial flows to developing countries, averaging 40 per cent in the first three years. In 2005/06 the rate of increase slowed to 19 per cent, although total capital inflows (from private and official sources) to developing countries in that year equated to $571 billion. Seventy per cent of total capital inflows, 2004–2006 came from private equity flows, including from private banks (particularly Chinese) and through foreign direct investments from business.

In 2005/06, repayments from the developing nations on loans to donor governments and the international financial institutions (public institutions) exceeded receipts of loans by $145 billion – this was largely because high oil prices enabled a number of middle-income oil-exporting counties such as Algeria, Nigeria and Russia to make repayments to the Paris Club of creditors (an informal group of 19 creditor governments from the major industrialised countries) and to multilateral institutions. Whilst loans to developing nations through Official Development Assistance increased by a record $27 billion in 2005, such 'aid' fell by $3 billion in 2006.

The year 2005 is considered a watershed for scaling up aid commitments and deepening debt relief to low-income countries. But these commitments risk remaining unfulfilled, as recognised by the World Bank itself: 'Debt relief is intended to be additional but may be counted toward fulfilling aid targets' (World Bank, 2006a: xviii). For example, the fall in ODA in 2006 over the record high levels of 2005 reflects the substantial debt relief arrangements that were made in that year to Iraq and to Nigeria by their Paris Club creditors. Whilst donors at the UN Conference on Financing Development held in Monterrey in 2002 (and again at the G8 conference in Gleneagles in 2005) had pledged that such debt relief arrangements would not displace other components of ODA and made commitments to enhance aid particularly to the low-income countries of SSA, the evidence above suggests that this is not being achieved: excluding debt relief, ODA disbursements remain static.

Source: World Bank (2007b)

Research the commitments that were made by the G8 group of countries under Tony Blair's Chairmanship at Gleneagles in 2005 toward increasing aid to Africa and progressing debt relief for particular countries. What progress has been made in terms of meeting these commitments? Websites of civil society organisations like Oxfam which took a prominent role in the Make Poverty History and Live8 campaigns (considered in Chapter 4) will be useful. You could also consider what the British Government has done in particular using the Department for International Development (DFID) website.

Whilst the international financial institutions may have become less significant in terms of overall financial flows to the developing countries, the World Bank and IMF have become increasingly central to the lending decisions of other public and private institutions. Many of these lenders and investors over the last two decades have linked their own spending to the presence in those recipient countries of World Bank/IMF programmes of reform (such as structural adjustment programmes and, more recently, Poverty Reduction Strategy Papers, which are discussed in more detail below). In order to access debt rescheduling or relief from the Paris Club of industrialised nations (discussed in the Critical reflection box), for example, countries have to have a current programme of economic and financial reform with the IMF.

In addition, the World Bank is a hugely important institution due to the way that it influences national development policy in terms of directing research, technology transfer and other forms of institutional support. The country and sector reports drawn up by the World Bank, for example, are used by commercial lenders and aid agencies to plan their own activities. The annual *World Development Reports* of the World Bank and associated publications are also an important source of information and opinion for academics and practitioners (not all of whom take the full institutional line).

Certainly, and particularly through structural adjustment programmes, the World Bank has moved far beyond its original function as a straightforward credit institution to become involved in policy determination and planning in developing countries to an unprecedented extent. As such, the World Bank is a major player in shaping development outcomes globally.

Financial resources and decision making

Like any other bank, the World Bank Group pursues profit and returns on its investments and has been successful in this respect every year since its inception. World Bank securities are considered to be among the world's safest, enabling it to borrow on very favourable terms. Unlike commercial banks, however, the World Bank takes no financial risks since its loans are to governments with whom it has 'preferred creditor' status, in that debts are paid out of the borrowing country's general revenue or from new loans. Werksman estimated in the mid 1990s that for every US dollar the US government has invested in the World Bank, US companies received back US$1.10:

> The other major shareholders are not faring badly either. Income accruing from procurement contracts when compared to every dollar contributed to the Bank leads to benefits amounting to US$1.90 for the United Kingdom, US$1.78 for France, US$1.47 for Germany. Only Japan with US$0.97 appears to receive a little less than its contributions.
>
> (Werksman, 1996: 131)

There are currently 184 member countries of the World Bank, each with a representative on the Board of Governors. A smaller group, the Board of Executive Directors, is responsible for general operations and policy making within the group, the Chairman being the World Bank President. Votes are made by member countries on the allocation of World Bank funds according to the financial contribution of that nation to the Bank ('one dollar one vote'). On this basis, the G8 group of industrialised countries have just over 45 per cent of voting rights in the IBRD (US 16.38 per cent), while the remaining 175 countries share the rest. Typically most of the poorest countries have less than 0.1 per cent of the votes. Table 7.4 shows the voting power of selected countries in the World Bank.

To gain membership to the World Bank Group, a country must first be a member of the IMF. Member countries of the IMF pay a subscription and agree to abide by a 'mutually advantageous code of economic conduct' (Crook, 1991: 3). Traditionally, the primary distinction between these two institutions is that the IMF is concerned with the health of the international monetary system and may lend (from subscriptions) to member countries briefly to overcome short-term financial instability. Each member of the IMF pays in a certain amount to the fund depending on the size of its economy, and can borrow from the fund on a short-term basis.

Table 7.4 Voting power at the World Bank

Country	Percentage of total votes in IBRD
G8 countries	
USA	16.38
Canada	2.78
United Kingdom	4.3
France	4.3
Germany	4.49
Japan	7.86
Italy	2.78
Russian Federation	2.78
Selected countries	
Argentina	1.12
Bangladesh	0.32
Brazil	2.07
Chad	0.07
China	2.78
Croatia	0.16
Ghana	0.11
India	2.79
Kenya	0.17
Malaysia	0.53
Nigeria	0.80
Pakistan	0.59
Philippines	0.44
South Africa	0.85
Uganda	0.05

Source: www.worldbank.org © International Bank for Reconstruction and Development/The World Bank

Whilst historically the World Bank has been more concerned with the financing of longer term development in the poorer countries of the world, through the 1980s insolvency of the middle-income debtors threatened the international financial system (the 'debt crisis' considered in Chapter 8), such that the IMF moved increasingly into the area of lending to members and to 'a lead role in deliberations over rescheduling' (Thomas and Allen, 2000: 206). In consequence, the distinctions between the roles of the World Bank and the IMF became less clear:

> The Fund had begun to worry about longer-term development, and the Bank was taking a new

> interest in short-term macro-economic policy . . . in one developing country after another, the two institutions devised overlapping programmes of economic reform and backed them with cash.
>
> (Crook, 1991: 4)

The IFIs and sustainable development

The search for sustainable patterns and processes of development has prompted many changes in the structure, function and activities of the World Bank Group in recent decades. In the 1970s, the World Bank itself recognised that in many instances environmental degradation was compromising the impact of its project lending on economic development. In addition, the role of World Bank lending in causing environmental destruction was taken up at this time by environmentalists, particularly in the USA.

Contributors to *The Ecologist* magazine were particularly prominent in their condemnation of World Bank practices (see, for example, vols 14–17, 1984–87). The impacts of the Carajas iron ore project in eastern Amazonia, which involved the clearance of tropical forest the size of England and France combined, were among those documented. Similarly, the Polonoroeste Highway project in the Brazilian north-west was exposed in the same publication as requiring the resettlement of 30,000 families and the large-scale destruction of tropical forests.

Such media exposure and emerging public pressure, particularly from US environmentalists, were important factors in prompting changes in the World Bank towards greater environmental sustainability within its operations. The USA contributes approximately one-fifth of all capital funds to the World Bank each year and US environmentalists have extensively lobbied congressional subcommittees to press for reform of the World Bank and to withhold funds if changes are not forthcoming.

In 1973 the World Bank established an Office of Environmental Affairs to review the prospective environmental impacts of its project lending. By 1985, that office had only five staff, which, as Rich (1994) suggested, raises questions in relation to its capacity to 'review all projects' as mandated. In 1987, under President Conable, a central Environment Department was created with four Regional Environment Divisions to oversee and promote environmental activities. In 1993 the World Bank articulated a 'four-fold environmental agenda' as shown in Table 7.5 that continues to underpin much of its activities towards sustainability in development.

Table 7.5 The four-fold environmental agenda of the World Bank

1. Assisting member countries in setting priorities, building institutions and implementing programmes for sound environmental stewardship
2. Ensuring that potential adverse environmental impacts from bank-financed projects are addressed
3. Assisting member countries in building on the synergies among poverty reduction, economic efficiency and environmental protection
4. Addressing global environmental challenges through participation in the Global Environment Facility

Source: World Bank (1994a) © International Bank for Reconstruction and Development/The World Bank

Currently (as discussed in more detail below), responsibility for 'mainstreaming' sustainable development into the bank's activities lies with the Environmentally and Socially Sustainable Development Network (ESSD). The goal of the ESSD is to:

> contribute to the Bank's mission of fighting poverty by improving poor people's livelihoods, health, and security today and in the future. ESSD does this by helping to: enhance environmental quality and natural resource management; maintain the global ecosystems; improve access to natural resources; and generally increase poor people's capacity to improve their lives and influence the decisions that affect them.
>
> (www.worldbank.org/essd)

New monies for the environment

Assisting member countries to establish and build programmes for the environment (the first dimension of the environmental agenda shown in Table 7.5) has encompassed a variety of measures, including support for national and regional environmental planning exercises. National Environmental Action Plans, for example, aim to integrate environmental considerations into a nation's overall economic and social development strategy over the longer term. This element of the agenda also includes financial resources for targeted, primarily environmental, projects such as in pollution control, land conservation and natural habitat protection.

At the end of 2006, the total active portfolio of projects with environment and natural resources components was US$9.7 billion (World Bank, 2006c)

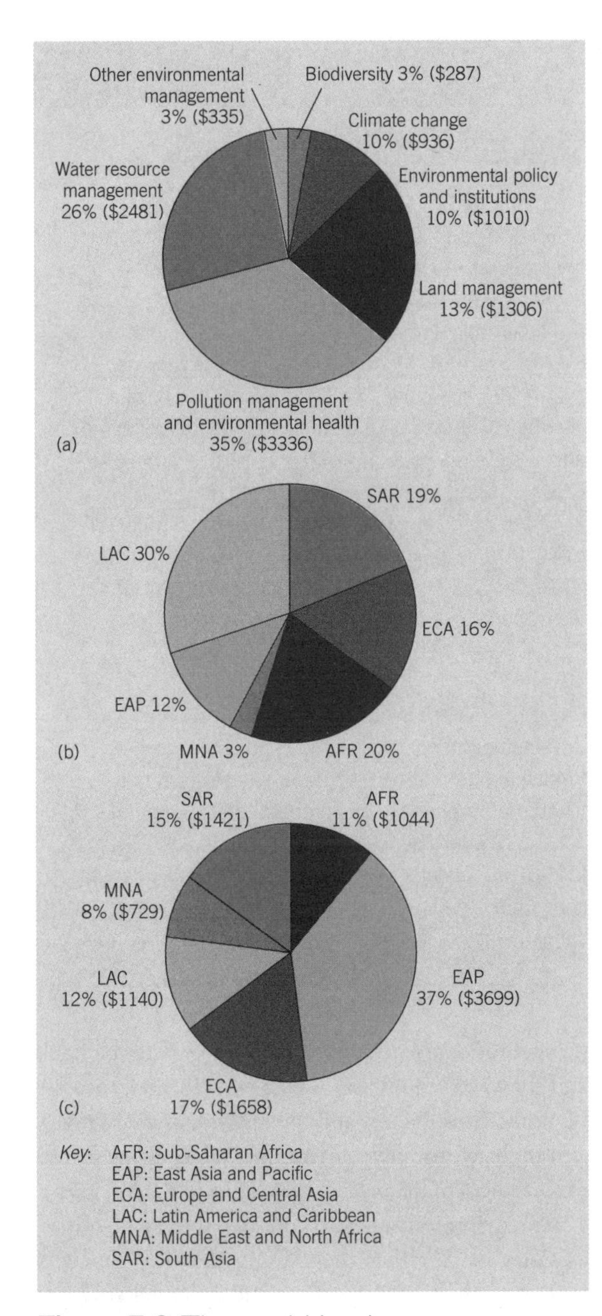

Figure 7.6 The world bank core environmental portfolio: (a) active projects in 2006, (b) regional distribution in 2001 and (c) regional distribution in 2006

Source: (a) and (c) from World Bank (2006c); (b) from World Bank (2001c) © International Bank for Reconstruction and Development/The World Bank

up from US$6 million in 2001 (World Bank, 2001c). Figure 7.6 shows the nature of the projects supported and their changing regional distribution. Some caution is required in that these figures on environmental lending include both targeted environmental projects and those that have environmental objectives within them. Some authors, such as Werksman (1996), point out that many projects involving natural resource management (and therefore incorporated into such lending figures) may not be actually environmentally beneficial. The changing regional composition of the 'environmental portfolio' reveals that in recent years Europe and Central Asia's share has increased whilst Latin America and the Caribbean's has fallen.

Some caution is also required in considering the World Bank's involvement with the Global Environment Facility (the fourth element of the agenda as seen in Table 7.5). As seen above, GEF is a programme in conjunction with UNEP and UNDP that is aimed at creating new finances for the least developed nations to tackle environmental problems explicitly in areas of the global commons. Since its foundation in 1991 there have been problems certainly in terms of its continued financing, but also in its relation to the role of the World Bank within this global initiative. These are reviewed in the Critical reflection box below.

Minimising environmental impacts of lending

Since 1989, the World Bank has had mandatory procedures for assessing and mitigating the potential adverse environmental impacts of its project lending. In short, 'Operational Directive 4' required that prior to the approval stage of a proposed investment, projects had to be 'screened' for prospective environmental impacts. Proposed projects were then assigned to one of four categories on the basis of the nature, magnitude and sensitivity of environmental issues which then required varying levels of subsequent environmental analysis before approval. Staff of the Regional Environment Divisions then assist the prospective borrower in carrying out the necessary environmental analysis.

In 1993, 14 per cent of projects were screened as Category A (defined as likely to have significant adverse environmental impacts that are sensitive, diverse or unprecedented) and required a full environmental impact analysis (World Bank, 1994a). In addition, 50 per cent of all projects at that pre-approval stage were identified as Category B (having potentially adverse environmental impacts, but that are less widespread, unlikely to be irreversible and in most cases mitigatory measures could be established). As such, these projects were subject to a lesser degree of environmental analysis.

In the late 1990s the World Bank started to convert Operational Directives (that combined elements of policy, procedure and guidance) into separate statements of mandatory policy, mandatory instructions for carrying out the policy and advice on good practice.

Critical reflection

Ongoing challenges within the Global Environment Facility

In 1994 GEF was extended from its initial pilot phase, with the World Bank acting as trustee, the UNDP providing technical expertise and preparing projects, and UNEP in an overseeing role. In 1994 total spending on GEF projects amounted to US$4.5 billion (World Bank, 1994b) rising to US$7.3 billion in 2001 (World Bank, 2001d). In that year, 72 per cent of GEF financing went to projects centred on climate change issues. By 2003, climate change projects accounted for 24 per cent of total finance with biodiversity projects accommodating 59 per cent of monies (World Bank, 2003b).

Critics of GEF (particularly in its early guise) wonder if its 'noble aspirations' (such as to provide a focus for multilateral cooperation on environmental issues) can be achieved (Werksman, 1995), particularly due to the influence of the World Bank in decision-making procedures.

Membership of GEF is dominated by those developed countries wealthy enough to contribute the minimum US$4 million required. The chair of GEF is appointed and employed by the World Bank. During the pilot phase of GEF, 80 per cent of projects were linked in some way to larger World Bank projects.

Jordan and Brown (1997: 278) referred to GEF having adopted many of the failings of its 'administrative parent' and to 'profound discrepancies' between the ways in which North and South perceive the need, role and scope of additional financial transfers to GEF. Many consider that the monies within GEF are fundamentally too small to have a significant impact on global environmental concerns and there remains a lack of consensus regarding how NGOs can and should participate in GEF projects.

These problems are not unrecognised within GEF itself. On the appointment of the new Chief Executive Officer of GEF in 2006, it was identified that 'an institutional sea change' was required to revitalise GEF (www.gefweb.org).

Using the materials of Chapter 6 on Resources and Development, consider whether the GEF may be fundamentally limited in the way that it fosters a 'Northern' rather than a 'Southern' environmental agenda. Sources such as Adams (2001) and Elliott (2006) will also be useful in considering this debate concerning the core environment and development concerns of the developing world.

An Operational Manual on Environmental Assessment is produced and continuously updated 'so that staff, clients, shareholders, and external stakeholders can understand what is *required* under Bank policies, as distinguished from what may be *encouraged* or *desired*' (World Bank, 2000a: 3).

In 1999 Operational Policy 4.01 replaced OD 4.01. A new category was added to include the screening of projects that involved Bank funds going through a financial intermediary or 'sub-borrower'.

Environmental assessment procedures for structural adjustment loans (i.e. policy reforms rather than projects) were also introduced although they remain recommended rather than required. In 2002, 8 per cent of all new projects in terms of lending amount were screened as 'Category A' and a further 34 per cent of all projects at that pre-approval stage required a simpler environmental analysis (www.worldbank.org).

However, a number of reviews of the impact of environmental assessment undertaken by the Bank itself has revealed ongoing challenges, particularly in terms of the institutional capacity for EA (within both the Bank and the client countries), weaknesses in the processes of public consultation and the analysis of alternatives, and the need to move environmental assessment into earlier stages of the project cycle (World Bank, 1997b). Too many projects are also considered to be identified as Category B rather than A, 'so that key elements such as analysis of alternatives and potential environmental impact on a wider area than the project site, public consultations, and supervision do not receive adequate attention' (World Bank, 2000a: xiv). Efforts towards further strengthening of compliance with these mandatory requirements on environmental assessment and related 'safeguard policies' is now part of the Bank's Environment Strategy, considered below.

Recognising the centrality of poverty in environmental outcomes

As shown in Table 7.5 above, assisting member countries in building on the synergies of poverty reduction, economic efficiency and environmental protection has been on the environmental agenda of the World Bank since 1996. As identified in Chapter 6, an enhanced appreciation of the interdependencies of poverty alleviation and environmental conservation was a central contribution of the Brundtland Report published almost a decade previously (WCED, 1987).

In 1990 poverty had been the focus of the Bank's *World Development Report* and in 1992 it was development and the environment. However, the persistence of poverty and the emerging understanding of the multidimensional nature of the challenges of poverty led to poverty being the focus once more for the *World Development Report* in 2000/01. Within that report, it was suggested that poverty was the principal global challenge for sustainable development.

It is evident that over the decade between the two 'poverty reports' sustainable development thinking has become more entrenched in the rhetoric and actions of the World Bank. The comprehensive and interrelated political, economic and social dimensions and implications of poverty have been identified. Similarly, the importance of change at the local and intra-household level as well as within international institutions themselves has been recognised.

It is considered that the 2000/01 WDR was instrumental in defining what is acknowledged to be the current international consensus on a strategy to reduce poverty into the twenty-first century (Maxwell, 2004) and as illustrated in Table 7.6. The 2000/01 report was informed substantially by participatory poverty assessments that were carried out with various groups of people in 60 countries, the 'voices of the poor' project (Narayan *et al.*, 2000). In addition, the WDR emphasised the multidimensional nature of poverty and outlined a three-pronged assault on poverty via multidimensional strategies that promote opportunity, facilitate empowerment and enhance security (World Bank, 2001a).

It was identified, for example, that increasing numbers of people worldwide continued to lack basic opportunities for development, including material assets, jobs and access to health care. Thereby, it was asserted that strategies for alleviating poverty must focus on creating opportunities through road building, widening credit facilities, enhancing markets for produce, and providing access to water and sanitation, for example.

The report also recognised the non-income sources of poverty, particularly the lack of political power among, and the failure of institutions to work well for, the poorer groups in society. But it was acknowledged that institutions also operate at a local level and empowerment therefore, in part, involves making institutions at all levels more responsible, egalitarian and democratic. The report highlighted that poorer groups are very vulnerable to all kinds of shocks and external events, such that strategies for sustainable development must enhance their security through reducing these risks and by increasing the ability of such groups to cope with those shocks.

Table 7.6 An international consensus on poverty

- The Millennium Development Goals that encompass what international development comprises and have poverty reduction at the centre
- A strategy on how to reduce poverty summarised in the *World Development Report*, 2000/01
- Mechanisms for operationalising poverty reduction at a country level – through Poverty Reduction Strategy Papers
- Technologies for delivering aid in support of PRSPs including frameworks for governments to identify expenditure and for donors to work together in sector-wide approaches
- A commitment to results-based management that shifts focus from inputs and activities to outputs and outcomes

Source: Compiled from Maxwell (2004)

Poverty planning is now considered to be 'on a roll' as the Millennium Development Goals, amongst other factors, are increasingly directing donor and national efforts (see Elliott, 2008). A poverty focus has also become central, for example to the way in which the World Bank and other IFIs allocate funds for debt reduction as considered below.

Change within the Bank

In July 2001, following a comprehensive assessment of the World Bank's past environmental performance undertaken by the independent Operations Evaluation Department, the World Bank's management and Board of Directors endorsed a new environment strategy entitled *Making Sustainable Commitments* (summary available through www.worldbank.org/essd). The report had concluded that despite significant progress within the World Bank's environmental work, there was a need to instil the environment as integral, rather than as an 'add on' to, the concerns for sustainable development and poverty reduction. Environmental objectives needed to be incorporated into the core strategy and operations of the World Bank itself.

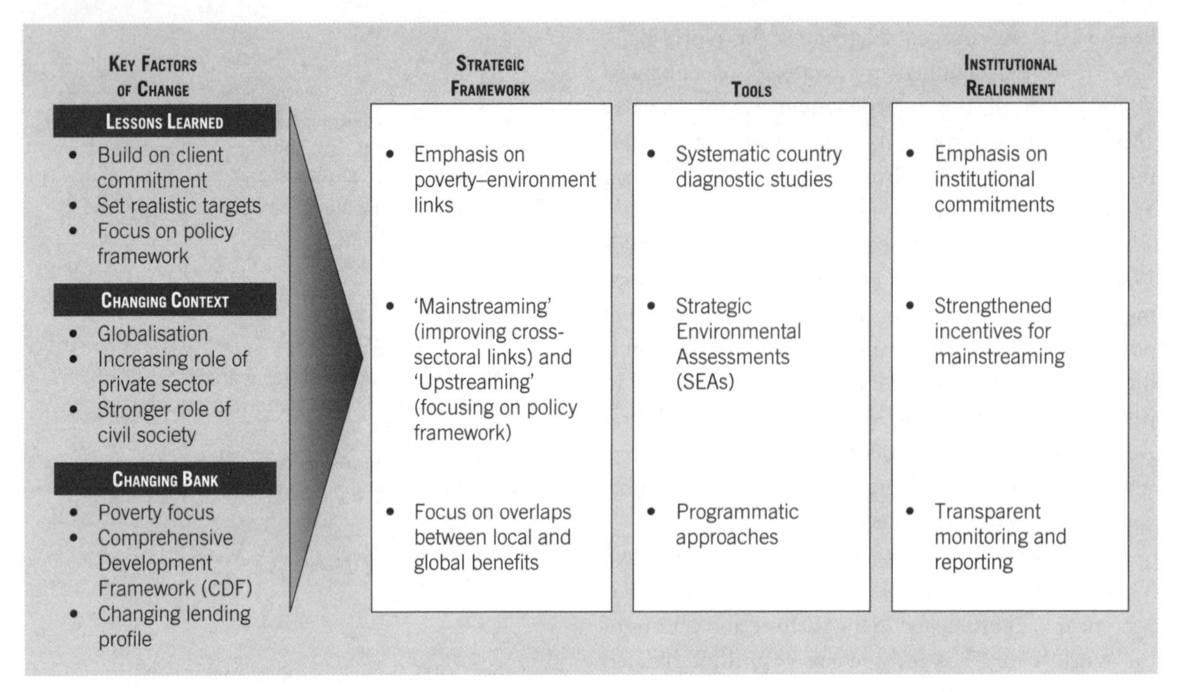

Figure 7.7 The changing context of the World Bank environment strategy
Source: World Bank (2001e) © International Bank for Reconstruction and Development/The World Bank

The report also identified that further work needed to be done in terms of improving the World Bank's environmental safeguard policies and their implementation. Figure 7.7 details both the changing context within which the new strategy is presented and the essential new components of that strategy. Critically, the strategy recognises that realising the goals requires institutional change within the World Bank itself:

> We need to align our incentives, resource allocation, and skills mix to accelerate the shift from viewing the environment as a separate, freestanding concern to considering it an integral part of our development assistance. We need to put this understanding into practice in our analytical work, policy dialogue, and project design.
>
> (World Bank, 2001e: 28)

It is apparent that in terms of its own accountability and incentives, the World Bank recognises that new staff skills are required as the emphasis has shifted from project-level safeguards to integrated portfolio-level risk assessment, for example. In addition, partnerships with civil society and the private sector are seen as essential to increase development effectiveness and for reducing transaction costs. New systems for monitoring and reporting to ensure accountability and the capacity to learn from experience are also identified as necessary to deliver on the World Bank's commitments to sustainable development. Box 7.1 provides further detail of the commitments and emphases within the environment strategy.

International financial institutions (IFIs) and the debt crisis

The debt crisis of the early 1980s prompted profound changes within the institutions of the World Bank. Since 1979 an increasing proportion of World Bank lending was not to projects, but to programmes of broad-based policy reforms in recipient countries. By 1994, 30 per cent of World Bank lending fell into this category (World Bank, 1994a) rising to over 50 per cent by the end of the decade (Brown and Fox, 2001).

As developing world governments became increasingly cash-strapped in the 1980s at a time of world recession and declining terms of trade (particularly for non-oil commodities), the World Bank noted that any developmental progress from its traditional portfolio of project-based lending was 'being swamped by macroeconomic imbalances in most of its client countries' (Reed, 1996: 9). At the same time, the IMF was looking beyond 'crisis management' in the monetary and financial sectors of developing countries, towards assisting countries in building up productive capacity.

BOX 7.1

The World Bank environment strategy

The environment strategy published in 2001 was intended as a single comprehensive strategy outlining how the World Bank will work with borrowing governments to address their environmental challenges. It is also designed to ensure that World Bank projects and programmes integrate the principles of environmental sustainability. The strategy is outlined in the following terms:

> **This Environment Strategy outlines the priority actions the World Bank plans to take to help its clients address the environmental challenges of development. In keeping with the World Bank's mission of reducing poverty within a framework of economic development, the environment strategy gives priority to issues where the links between poverty and the environment are particularly strong. Therefore, the strategy puts the environmental challenge into a local perspective, focusing on people in client countries and on the way environmental conditions and resources affect them.**
>
> **(World Bank, 2001e: 15)**

The environment strategy sets a course of action for the longer term as well as five-year targets.

Three main objectives are identified as representing the holistic approach that the World Bank is to pursue in order to link environment and development into the future. These are:

- Improving the quality of life (focusing on three broad areas where environment, quality of life and poverty reduction are strongly interlinked – in enhancing livelihoods, preventing and reducing environmental health risks and reducing people's vulnerability to hazards);
- Improving the quality of growth ('It is not enough to improve the quality of people's lives today; we have to ensure that short-term gains do not come at the expense of constrained opportunities for future development' (World Bank, 2001e: 18));
- Protecting the quality of the regional and global commons ('the search for solutions to sustainability needs to go beyond individual countries' (World Bank, 2001e: 19)).

In short, the work envisaged towards these objectives comprises:

- Assisting clients with analysis, technical assistance and training;
- Addressing environmental priorities through projects;
- Improving the safeguards systems.

Country-level environmental analysis will build on National Environmental Action Plans and other environmental work undertaken by the World Bank and its partners, to provide guidance in policy dialogue and to set priorities within countries. Such work will also be used to inform poverty reduction strategies and country assistance strategies.

Both the World Bank and IMF are encouraging low-income countries to consider environmental factors within PRSPs, 'because of the links between environment and poverty, and because a poverty reduction strategy must be environmentally sustainable over the long term' (World Bank, 2001e: 16). Technical analysis and training are central to the intended World Bank assistance towards helping integrate environmental issues into the PRSP process in individual countries.

While the World Bank recognises that many environmental problems are best addressed by dedicated projects, World Bank environmental lending has gradually shifted towards lending as a component of sectoral projects. The environment strategy commits the World Bank to work on both these fronts in the future 'depending on client demand and circumstances' (World Bank, 2001e: 23). Under this remit, the World Bank aims to ensure that best practice from environmental projects are disseminated and applied to new projects; that policy reform and specific investment projects are carefully sequenced in recognition that 'some investment projects are unlikely to bring lasting results in a distorted policy environment' (World Bank, 2001e: 24); and will 'pay special attention to reinforcing positive and minimising potentially negative outcomes (World Bank, 2001e: 25) of adjustment lending.

Improving the safeguard system is central to integrating environmental and social concerns into development policies, programmes and projects. This includes strengthening compliance with existing safeguard policies that are set as mandatory requirements to be followed by World Bank operations (i.e. the Operational Policies on Environmental Assessment, Forestry, Involuntary Resettlement and Indigenous Peoples, for example) as well as paying increased attention to the results of these on the ground.

▶

BOX 7.1 (continued)

The environment strategy also commits the World Bank to reforming the safeguard system in recognition of 'new challenges posed by a greater variety of lending instruments, including programmatic lending and projects implemented at the grassroots levels' (World Bank, 2001e: 26–7) and moving safeguard considerations into decision-making processes earlier. Figure 7.8 maps the environmental and safeguard input into the project cycle (World Bank, 2001e: 26).

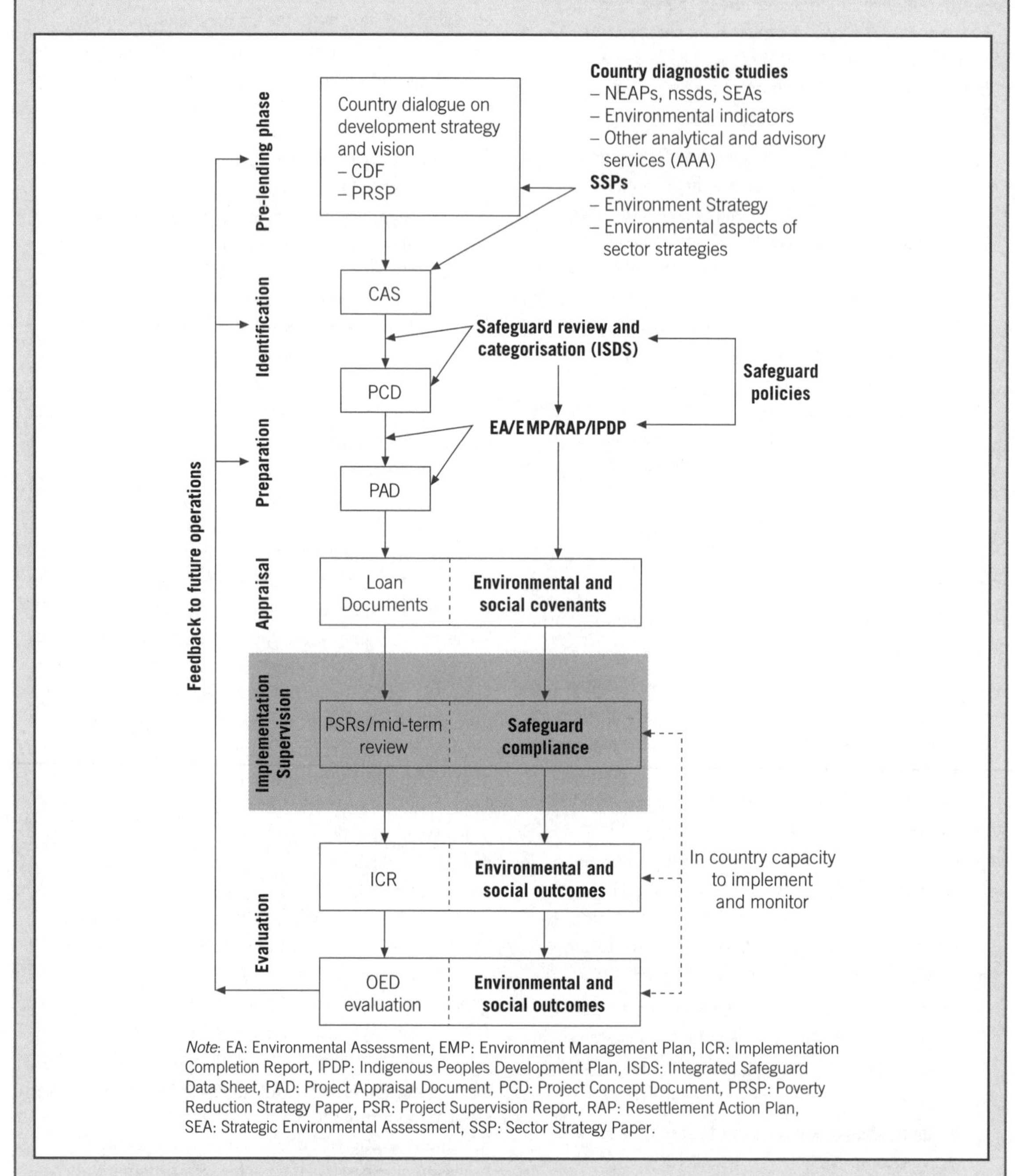

Note: EA: Environmental Assessment, EMP: Environment Management Plan, ICR: Implementation Completion Report, IPDP: Indigenous Peoples Development Plan, ISDS: Integrated Safeguard Data Sheet, PAD: Project Appraisal Document, PCD: Project Concept Document, PRSP: Poverty Reduction Strategy Paper, PSR: Project Supervision Report, RAP: Resettlement Action Plan, SEA: Strategic Environmental Assessment, SSP: Sector Strategy Paper.

Figure 7.8 The World Bank environment strategy: environmental and safeguard input to the World Bank project cycle

Source: World Bank (2001e) © International Bank for Reconstruction and Development/The World Bank

Comprehensive solutions to the debt crisis were considered to be essential. From within both institutions in the 1980s, the economic crisis in developing countries was conceived as more than a temporary liquidity issue (as it had been viewed in the 1970s). Multilateral cooperation and significant changes were required in the mandate and practices of the institutions of the World Bank and the IMF themselves.

The term structural adjustment programmes (SAPs) is used generically to describe the activities of the World Bank and the IMF in the design and support of packages of broad-based policy reform within a country. The central aim of these programmes was debt reduction via:

> **programs of policy and institutional change necessary to modify the structure of an economy so that it can maintain both its growth rate and the viability of its balance of payments in the medium term.**
>
> **(Reed, 1996: 41)**

The first SAP was initiated in Turkey in 1980 and by the end of the decade 187 SAPs had been negotiated for 64 developing countries (Dickenson *et al.*, 1996: 265); 'the two organisations used the programme-lending approach to promote structural adjustment vigorously throughout the 1980s' (Thomas and Allen, 2000: 206). The specific instruments of structural reform are varied but typically include those listed in Table 7.7 and confirm a close association with the principles of neoliberal development thinking discussed in Chapter 3.

The impacts of macro-economic reform

Identifying the impacts of SAPs on economic development is difficult, not least due to the far-reaching implications of the different elements of the 'package' shown in Table 7.7. It has been suggested that neither the World Bank nor the IMF themselves have been able to demonstrate a 'convincing connection in either direction' between SAPs and economic growth (Killick, 1995, cited in Mohan *et al.*, 2000: 58).

Table 7.7 The principal instruments of structural adjustment

- Currency devaluation
- Monetary discipline
- Reduction of public spending
- Price reforms
- Trade liberalisation
- Reduction and/or removal of subsidies
- Privatisation of public enterprises
- Wage restraints
- Institutional reforms

As suggested previously, debt burdens have increased rather than decreased since the 1980s, despite the varied and extensive efforts towards debt rescheduling as discussed below. Mohan *et al.* (2000: xiv) have also asserted that critical reflection on and monitoring of the impacts of SAPs were not a priority until relatively recently. They point to how the first 15 years of structural adjustment was:

> **relatively insidious, taking place in distant and impoverished places that were always in trouble and usually too complicated to comprehend. Once these distant places began to melt down (Russia, Indonesia, Brazil) in the mid 1990s and threaten 'our' development we began to take note.**

An important publication in exposing the negative impacts of SAPs came from UNICEF in 1987 (Cornia *et al.*, 1987). It was entitled *Adjustment with a Human Face* and pointed to the increased poverty and social polarisation created under adjustment experiences and the nature of local coping strategies to adjustment that often meant overexploitation of natural resources. In particular, the report pointed to the worsening fate of women and children under five years and the reversals in human development that were currently occurring after a period of previous advancements in welfare (see also Chapter 3). However, the UNICEF report did not question the need for adjustment, rather it urged that more needed to be done to protect the interests of the poor during the process.

Similarly, while the IFIs themselves started to consider poverty much more explicitly in the early 1990s (as discussed above), and adjustment was recognised to be a longer process than assumed and could not wait (morally or politically) for growth alone to reduce poverty, the rationale or basic design of SAPs not questioned. Rather measures to address the social costs of adjustment and to target groups particularly challenged in the course of adjustment (broadly termed social safety nets) were devised as 'add-ons' to the same basic design of SAPS.

Understanding gendered impacts of economic reform

In contrast to what can be considered a 'gender neutral' view of SAPs taken by the IFIs (and UNICEF, for example), it is widely considered elsewhere that SAPs had specific and 'probably negative' impacts on women 'given the complexity of gender relations, the invisibility of women's work both in the rural and urban sectors, and the multiple and complex roles women actually perform' (Pearson, 1992: 309).

Some of the negative impacts of SAPs on women can be understood in terms of accepted gender roles, whereby women generally assume greater responsibility for household reproductive tasks. Women as a result are intensive users of health services (as are children), and as government spending in health has declined under adjustment, women have tended to suffer more.

In Tanzania, cutbacks in public health services led to an increase in under-five mortality from 193 per thousand in 1980 to 309 per thousand in 1987 (Pearson, 1992: 310). Not only had the services on which women relied, such as clinics and maternity care, been reduced (necessitating greater travel and more expense for women), but women themselves had to supplement those public services through their own labour (Pearson, 1992).

Many research findings also revealed that women were eating less in times of food insecurity; research in Dar es Salaam found that 58 per cent of households had reduced the number of cooked meals from three to two under structural adjustment (Pearson, 1992) with consequent declines in women's health (and impacts on children if women were pregnant or lactating).

Similarly, gender role analysis has exposed how women's unpaid work often increased under adjustment (Sparr, 1994), whilst SAPs assume implicitly that women's time is infinitely elastic. The greater unemployment, decreased purchasing power and cutbacks in social services resulted in women adopting strategies to make funds go further, such as time spent in shopping for cheaper alternatives, purchase of foods that have had less processing (leading to increased food preparation time) and supplementing income through the growing of vegetables for household and/or sale. In agricultural production, women's unpaid labour may increase as households try to cut costs and raise production.

An understanding of gender relations, however, is required to understand how many more women than men became unemployed under structural adjustment policies due to the nature of their employment and women often being on temporary contracts and locked into apprenticeship grades. Similarly, the shift to export crops encouraged by SAPs often did not benefit women, the explanation for which requires an analysis of gender relations; women may produce and trade more in restricted or local markets and potentially non-monetary markets, for example.

The 'gender-neutral' stance of SAPs was encapsulated by the way in which they did not require any significant reform in terms of enhancing women's access to land, credit or other inputs, yet these are fundamental gender inequalities that impede the economic efficiency of SAPs. Indeed, it has increasingly been understood that there is an inherent male bias to the neo-liberal economic model and as imposed by the IFIs:

> **Neo-liberal economics is based on the assumption that individuals are free and rational, able to respond to market signals without being constrained by social relations. From a gender perspective, there are two crucial weaknesses in the thinking. First, this school of economics overlooks the way gender inequalities in a range of social institutions, including households, the market and the state, limit individual choices, especially women's. Gender segregation in labour markets is one example of this. Second, although women's unpaid and largely 'invisible' care work, e.g. looking after children, has no explicit place in this model, it is in fact present in the form of hidden assumptions about gender roles and relations.**
>
> **(Womankind World Wide, no date: 14)**

Negative environmental impacts

Evidence to suggest that improvements in the macroeconomy under adjustment may be at the cost of environmental sustainability mounted through the 1990s. Ghana, for example, was widely heralded by the World Bank as one of its success stories in terms of progress with economic adjustment, with average annual growth in GDP of 3.8 per cent throughout the 1980s (Rich, 1994). However, timber exports increased from US$16 million in 1983 to US$99 million in 1988 under the trade liberalisation required by structural adjustment. As a result, the forest area within Ghana was reduced to 25 per cent of its original size by the late 1980s (Rich, 1994).

In Java a dramatic increase in the terms of trade for horticultural goods through liberalisation of markets encouraged upland farmers to move from less profitable basic starch staples to more profitable fruits and vegetables (Conway and Barbier, 1995). Although this created an incentive to invest in soil conservation measures, the increased profitability of these crops also encouraged farmers to extend the cultivation into steeply sloped volcanic soils where run-off and soil erosion were enhanced.

In Cameroon the IMF recommended export tax cuts accompanied by devaluation of the currency under structural adjustment in 1995. This led to a substantial economic incentive to export timber resources from that country; the number of logging enterprises increased dramatically, from 194 in 1994 to 351 in 1995, and lumber exports grew by 49.6 per cent between 1995/96 and 1996/97 (www.foe.org/camps/intl/imf/cons/, accessed 11 February 2008; and IMF 1998).

Bryant and Bailey (1997: 79) concluded that 'multilateral institutions have rarely, if ever, been on the side of poor and marginal grassroots actors – rhetoric notwithstanding', pointing to the widespread enclosure of land and other environmental resources used by these local actors which has occurred under policies pursued by these institutions. Mohan *et al.* (2000) asserted that it took ten years before proponents and architects of SAPs 'gradually realised' their environmental impacts. Table 7.8 reproduces what they consider to be the 'environmentally blind' assumptions on which SAPs were built.

Table 7.8 Explaining environmentally blind SAPs

- The World Bank and other lenders did not consider the environment to be a priority at the time
- Those borrowing from the major lenders were also more concerned with pressing fiscal issues and did not specifically request funding for environmental protection
- Sustainable development in its broadest sense was not high in the public consciousness so that environmental protection was not an obvious policy element
- Environmental protection would incur further state expenditure which was the antithesis of adjustment programmes

Source: Mohan *et al.* (2000)

Poverty Reduction Strategies

In 1999, Poverty Reduction Strategy Papers (PRSPs) became the 'successor' to structural adjustment programmes (Simon, 2002: 90). PRSPs are now the strategic documents around which the World Bank and IMF and other donors coordinate their assistance to low-income countries. The poorest countries now have to produce PRSPs as a prerequisite both for WB/IMF concessional lending and for debt relief, such as within the Heavily Indebted Poor Countries Initiative (HIPC). PRSPs are also seen as the key tool for operationalising the World Bank's approach to poverty as identified in the 2000/01 *World Development Report* highlighted above and for meeting the Millennium Development Goals. At the end of 2004, 42 countries had PRSPs in place and 23 had supplied one or more annual reports on those strategies (World Bank/IMF, 2004).

PRSPs are written by national governments and should be formulated through broad participatory processes involving civil society and key donors as well as the World Bank and IMF. They are required to set out coherent macro-economic, structural and social sector reform focused on reducing poverty and identify the financing needs and sources. They should have clear links with the principles embedded in the CDF and with agreed International Development ('Millennium') Goals (detailed in Chapter 1). The core principles of the PRSP approach are identified in Box 7.2.

BOX 7.2

Core principles of the PRSP approach

Country driven and owned: PRSPs should involve broad-based participation by civil society and the private sector at all stages, including formulation, implementation and outcome-based monitoring

Results-oriented: PRSPs should focus on outcomes that will benefit the poor

Comprehensive: PRSPs should recognise the multidimensional nature of poverty and the scope of actions needed to effectively reduce poverty

Partnership-oriented: PRSPs should involve the coordinated participation of development partners, including bilateral and multilateral agencies and non-governmental organisations

Based on medium- and long-term perspectives: PRSPs should recognise that sustained poverty reduction will require action over the medium and long term as well as in the short run

Source: Compiled from World Resources Institute (2005)

As the experience of the PRSP process mounts, a number of institutions have undertaken reviews of progress to date. The World Bank itself, for example, suggests that whilst there has been some good progress in terms of some of the more straightforward problems of the approach, there remain some complex challenges (World Bank IMF, 2004). In particular, they raise the importance of enhancing the analysis on which PRSPs are based, strengthening of institutional capacity for implementation and enhancing aid effectiveness.

The World Resources Institute also reviews the evidence concerning how PRSP processes are actually engaging with the MDGs – in particular MDG7 that refers to environmental sustainability – and suggests that an 'environmental overhaul' of PRSPs is needed if the central role of ecosystems in the lives of the poor and their potential to reduce rural poverty is to be secured (WRI, 2005).

As suggested at the outset of this chapter, there was a strong imperative on the IFIS at the turn of the century to be seen to be more participatory and responsive to individual country needs. Whilst participation is certainly there as prominent thread in the PRSP process, it is suggested that some caution is required in the desire to build a national consensus on poverty reduction for the way in which it may obscure important trade-offs and conflicts of interest – both between the poor and non-poor and among the poor (Maxwell, 2004).

Most fundamentally, critics of the PRSP process and the new poverty emphasis within the conditions for accessing lending and debt relief question whether these programmes actually represent any real shift beyond neo-liberalism of the previous decades. They point, for example, to the continued emphasis on rolling back the state and engaging with the globalised economy as the solution to poverty. As such, external factors such as trade and international finance are seen as fundamental to overcoming poverty (rather than part of the dynamic of poverty creation). In turn, this continues to place responsibility on government and society of the countries in the developing world for both the cause and outcomes of poverty (Fairhead, 2004).

Progressing democracy and institutional change

This section considers explicitly how debates concerning good governance have influenced the activities and operation of the World Bank and the IMF. Clearly, as considered above, the IFIs have been important proponents of and contributors to these debates. The Key idea box looks more closely at the origins and

Key idea

The governance agenda at the World Bank

The World Bank formally launched its 'governance agenda' in 1991 when the Board accepted a paper on *Governance and Development* that stressed 'sound development management' across four themes (of public sector management, accountability, legal reform and transparency) as essential conditions for development. In short, according to Potter, D. (2000: 375), the basic position in the early 1990s was that a 'combination of liberal market capitalism in an international context and liberal democracy and "good governance" domestically were mutually reinforcing and provided core elements of a comprehensive strategy for development success'. It was considered that the governance agenda had 'universal developmental relevance for all cultures and societies in the modern world' (Leftwich, 1993: 605).

In 1997 the World Bank began monitoring (and reporting on) the quality of governance worldwide using six aggregate indicators (based on 33 individual data sources and hundreds of variables) as follows:

Voice and accountability

Political stability

Absence of major violence and terror

Government effectiveness

Regulatory quality of rule of law

Control of corruption

In 2007 it reported its findings of *A Decade of Measuring the Quality of Governance* (World Bank, 2007c). The headline of the findings was that some of the poorest countries (especially in Africa) had made significant progress in improving governance over a short period. However, it also found no evidence that, on average, governance worldwide had improved markedly.

development of this agenda within the World Bank. This section looks particularly at how the World Bank is working with non-governmental organisations as a means to promote institutional changes towards good governance. It also considers the role of NGOs in promoting change in the agenda, activities and impacts of the World Bank itself, particularly in terms of its own transparency and accountability. A later section of this chapter considers how the World Bank's governance agenda has influenced the role of the state in various spheres and the final section gives a fuller analysis of civil society and NGOs in wider development debates.

Working with non-governmental organisations

In continuity with organs within the UN system, the World Bank has stated a commitment to work with NGOs, including as a means of fostering local empowerment and the accountability of official institutions:

> The aims should be to empower ordinary people to take charge of their lives, to make communities more responsible for their development, and to make governments listen to their people. Fostering a more pluralistic institutional structure – including non-governmental organisations – is a means to these ends.
>
> (World Bank, 1989: 54–5)

The World Bank defines NGOs as private organisations that pursue activities to relieve suffering, promote the interests of the poor, protect the environment, provide basic social services, or undertake community development' (Gibbs *et al.*, 1999: ix). It distinguishes between national or international NGOs that act as intermediaries to support work at the grassroots, and (local) organisations within a particular community. All NGOs work in some way to serve others, and are identified separately by the World Bank from community-based organisations (CBOs) that exist to serve their members.

Since 1983 the World Bank has reported annually on its collaboration with NGOs. By 1993, 30 per cent of all projects were reported to involve collaboration with NGOs, rising to 50 per cent in 1998 (Malena, 2000) and 71 per cent in 2000 (World Bank, undated). However, it is suggested that despite this annual reporting of collaboration by the World Bank and lots of documents (including non-World Bank) and case studies, 'it is still difficult to determine the real extent and nature of NGO involvement in Bank-financed projects and the purpose and impact of that involvement' (Malena, 2000: 21).

Strictly, the World Bank's Social Development Department annually reports *intended* civic involvement figures as they are compiled from project appraisal reports that are prepared before project activities begin. Such figures also give little sense of the nature or 'quality' of World Bank–NGO collaboration, which, according to Malena (2000: 22):

> could indicate anything from the genuine and sustained participation of a large number of NGOs through all stages of the project cycle, to the contracting of an NGO for purposes of service delivery, to an informal lunch meeting with the local Oxfam representative.

Understanding the nature of the relationship between the World Bank and NGOs is important for considering the impacts of such collaboration on local empowerment and democracy, for example. Table 7.9 identifies three principal but different types of contact between the World Bank and NGOs.

Project collaboration between the World Bank and NGOs grew rapidly in the 1980s and 1990s principally because many governments' service delivery capacity was shrinking. In this sense, NGOs were seen as project implementers and as recipients or users of World Bank services or resources. An example would be NGOs receiving small grants through Emergency Social Funds (ESFs) that were created to provide 'social safety nets' in countries undergoing structural adjustment considered in the Key idea box. Nelson (2002: 500) notes that the World Bank invested more than US$3.7 billion to finance 108 such funds in 57 countries between 1987 and 1995, and 'NGOs had a role in most'.

Typically, this type of project collaboration between the World Bank and NGOs is with community-based organisations. Indeed, most World Bank projects with activities at the community level, within rural development projects, community health education, water

Table 7.9 Types of WB/NGO contacts

1 NGO collaboration in World Bank financed projects
2 Invited consultation in policy discussions
3 Confrontation over projects and controversial policies.

Source: Nelson (2002). Compiled from World Resources Institute, 2003

Key idea

Social safety nets

The term social safety net is used to describe various mechanisms that were implemented in conjunction with structural adjustment programmes 'designed to address either structural or transitional poverty and unemployment, to reduce the impact of adjustment measures on certain groups, or to create or improve both social and physical infrastructure' (Vivian, 1995: 5). Typically SSNs involved emergency, compensatory, employment and social investment funds designed to 'fill gaps', to supplement the activities of existing ministries and agencies that were unable to address social costs of adjustment and to target the poorest groups.

Initially, SSNs were seen as transitional measures in that they were not meant to replace standard social policies. The idea was that they would supplement existing policy and provision through the difficult adjustment period, therefore preventing further declines in poverty and unemployment until adjustment-led growth would make such support unnecessary.

Experience throughout the 1990s indicated that SSNs were seen increasingly as medium- to long-term policy shifts and were expected to tackle both transitional and structural poverty and unemployment.

and sanitation projects, as well as for social funds, now have some form of collaboration with NGOs. Although targeting CBOs in this way can bring important development benefits, it is suggested that examples of meaningful primary stakeholder participation and empowerment may be rare.

Participation, for example, is often limited to the implementation stages of projects and 'aimed at "mobilising" beneficiaries to fulfil predetermined roles rather than "empowering" them to influence and share control over development activities' (Malena, 2000: 25). Project design also remains firmly in the hands of the recipient government and the World Bank, rather than with the NGO. This kind of detail of the nature of NGO involvement suggests that the contribution to progressing democratic processes in recipient countries may be low.

Furthermore, there has been the expansion in recent years of NGOs whose primary purpose is to sell their services as project implementers, executing agents or managers in this same arena of service delivery. They are often created in direct response to the availability of World Bank funds and demand for such services. Typically, they do not have their own development agenda or, indeed, any popular support base. Again, this raises the question of the contribution of this kind of involvement between 'market-driven' NGOs and the World Bank towards forwarding democratic processes in host countries. However, this is often the type of collaboration with civil society organisations that (recipient) governments themselves tend to embrace.

The second form of collaboration between the World Bank and NGOs may also take place at the sector level through the World Bank sponsorship of formal meetings, conferences and electronic consultations concerning new policy directions. The World Bank itself reports that civil society involvement at the sector level tends to be highest where participatory approaches have been traditionally adopted, such as in education (95 per cent in fiscal 2001) and lowest where they have historically been low, such as in the economic policy sector (13 per cent in fiscal 2001; World Bank, undated: 5). In general, 'participation becomes more and more restricted as one moves from operations to decision making' (Tussie and Tuozzo, 2001: 108).

Indeed, perhaps of greatest concern for critics of the nature of World Bank–NGO collaboration has been the lack of capacity for civic involvement in the programmes of policy reform. Structural adjustment programmes, for example, were seen to accommodate significant proportions of lending yet often compounded rather than improved social and economic conditions within society.

As Brown and Fox (2001: 44) suggested, macroeconomic adjustment loans are 'inherently far removed from both civil society levers and the Bank's own social and environmental reform policies'. Similarly, Nelson (2000) has suggested that first generation SAPs had the general effect of making decision making in the World Bank less open to participation, including for the way that they privileged finance ministries over other ministries and economic reform packages decreased the influence of other interest groups. Whilst some optimism can be drawn from recent reviews of the implementation of PRSPs, where for example some of the best cases of environmental mainstreaming were found where civil society engagement had been most extensive (WRI, 2005), concerns remain as to how far such examples were driven by the donor community.

However, it is more generally accepted that the activities of NGOs have established significant concessions through their 'confrontational' (Nelson, 2002) or 'revolutionary' (Malena, 2000) involvement with the World Bank (the third type of contact identified in Table 7.9 on p. 303), particularly in the fields of environmental and social policy. Historically, the World Bank has been a target for policy advocacy of NGOs:

> **The World Bank has been a lightning rod for criticisms of the international economic system and of development aid. NGOs have pressed the World Bank to become more generous, egalitarian, and responsive to gender and minority concerns; more transparent and open to effective civil society participation; and more sensitive to natural resource use and human rights standards.**
>
> **(Nelson, 2002: 500)**

Although the 'least frequent' form of NGO involvement in World Bank projects (Malena, 2000), the most publicised roles are where NGOs intervene in opposition to World Bank/government activities to block or modify a planned project. Such activities often involve coalitions of affected groups and national or international NGOs who share their concerns. It is these kinds of activities that many advocacy-based international NGOs are keen on in that they combine policy dialogue (with the World Bank) with collaborative or confrontational actions at the field level through CBOs. NGOs may also seek involvement in World Bank projects as a way to influence World Bank and government actions. This may entail operational cooperation in carrying out contracted environmental or social assessments for the World Bank, facilitating public consultations, or monitoring a project's compliance with defined environmental or social safeguards.

If NGO collaboration with the international financial institutions is indeed to enhance the principles of good governance and democracy in recipient countries, it seems evident that the democratic credentials of the NGO will also need to be proven. Yet if formal standards are set on memberships or accounting then 'there is a real risk that only well-funded, highly organised and powerful NGOs will be recognised' (Woods, 2000: 835).

To date, it has tended to be the larger scale, service-orientated NGOs, and those that enjoy collaborative relations with governments, that are more likely to become involved in World Bank-financed projects. Smaller scale empowerment-oriented NGOs that 'see development as a process of social transformation, or openly oppose mainstream policies or actions' (Malena, 2000: 22) are less likely to be involved in collaboration.

Fundamentally, only a 'sub-set' of NGOs have the capacities (including a clearly defined mission, the relevant operational skills, adequate knowledge of the World Bank, negotiating prowess and private sources of income), for the kinds of interaction required to influence the World Bank (and government) agendas and practices.

There is also the real risk of rapid and excessive scaling up when NGOs become involved with the World Bank, 'targeting tens or even hundreds of thousands of beneficiaries' (Malena, 2000: 21). This can lead to an overstretching of management capabilities and 'compromising quality for quantity' within the NGO.

Furthermore, for Woods (2000: 836), a factor for consideration if World Bank–NGO collaboration is to enhance democratic processes in recipient countries is where IFIs may work through NGOs as a way of avoiding problems of local and central government:

> **there is a risk that rather than 'strengthening institutions', international agencies are engaging in 'governance-avoidance'. There may certainly be a case for doing so in undemocratic states, but in fragile, emerging democracies this may well have negative consequences on the process of democratisation.**

A fundamental consideration throughout this assessment of the capacities of NGOs to influence World Bank agendas and activities is that the World Bank lends to governments not NGOs. As a result, 'working with the Bank' usually means 'working with government' (Malena, 2000: 21). Yet NGO–state relations vary enormously from country to country, regime to regime and NGO to NGO. Working with the World Bank also means operating within the confines of a predetermined and time-bound project cycle during which it is the borrowing government and not the World Bank that owns and manages the project and maintains working relations with NGOs. As such,

> Even when the Bank and/or government do genuinely seek to benefit from NGO 'innovation and flexibility', predetermined project frameworks and rigid procedures often hamper NGOs' ability to innovate, and in practice, reduce their role to that of project implementor.
>
> (Malena, 2000: 29)

It has been suggested, however, that the increased engagement between civil society and the international financial institutions has been an important component of the reforms these institutions are undergoing in order to become more transparent, participatory and accountable themselves (Nelson, 2002; Tussie and Tuozzo, 2001; Woods, 2000). Both World Bank and independent reviewers have found clear evidence of changes in World Bank practice as a result of the 'high-volume encounters' between NGOs and the World Bank in the area of involuntary resettlement, for example (Woods, 2000). Similarly, proposals from environmental NGOs were the prompt for the development of the World Bank's information disclosure policies from 1993 (Nelson, 2000). This policy change has now made most project-related documents available to the public and given citizens of developing countries the fora to make grievances against the World Bank independently of their governments. Box 7.3 outlines the role of the World Bank Inspection Panel in forwarding this agenda.

The World Trade Organization

International trade is the most important aspect of the global economy and a determining factor in the social and economic development of nations. International trade has grown faster than world economic output in every decade since 1950 and flows of foreign direct investment (FDI) have grown faster than international trade since the mid 1980s (Neumayer, 2001; see also Chapter 4).

The majority of world trade takes place according to a set of rules administered by the World Trade Organization (WTO). The WTO was formed in 1995, although its origins go back to General Agreement on Tariffs and Trade (GATT) of 1948.

At the time of the formation of GATT, the 50 founding members of the United Nations had envisaged the creation of an institution, the International Trade Organization that would regulate world trade and international investment towards economic recovery and international stability. It would also serve to limit the restrictive and protectionist business practices that had brought about the Great Depression of the 1930s.

However, the charter drawn up proved unacceptable to many countries, particularly the USA, and led to 23 of the richest and most powerful countries withdrawing and forming the GATT. GATT was a 'weaker' institution than the proposed ITO in the sense that it was a contract between countries, rather than an organisation requiring formal ratification by member parliaments, for example.

All contracting parties (as they were known) to GATT committed themselves to upholding two 'liberal and unexceptional' principles (Hutton, 1993) designed explicitly to encourage free trade and to prevent the trade wars which had plagued the international economy in that era. These principles refer to 'national treatment' under which all countries must treat

BOX 7.3

The World Bank Inspection Panel

In 1993, the World Bank created the Inspection Panel in direct response to international environmental and human rights campaigns and criticisms. In particular, the broad-based North–South campaign against India's Narmada dam had exposed systematic violations of the World Bank's own social and environmental policies.

The creation of the Inspection Panel is illustrative of the power of civil society organisations to effect change in the operation of an international financial institution. It is also an indication of the broader processes of social and environmental reforms at the World Bank in the move towards more sustainable development. Furthermore, the Inspection Panel is a mechanism through which the World Bank itself is becoming more accountable through striving for greater compliance with its own social and environmental policies.

The Inspection Panel is composed of non-World Bank development experts (three persons at any time who are not Bank employees and reject any possible future World Bank employment). Through the

BOX 7.3 (continued)

Inspection Panel, citizens of developing countries can air grievances regarding the environmental and social costs of World Bank projects without going via the national government. To be eligible, they must be people directly affected or a local representative acting explicitly on behalf of such persons.

The Panel's mandate is to investigate charges that World Bank policies were not followed in the design and implementation of projects. Claimants must therefore be able to show that they have been or are likely to be adversely affected by a World Bank-financed project. In addition, they must be able to show that this harm or threat is related to a World Bank failure to follow its own policies, rather than, for example, being due to the borrowing government's failures. Claimants must also demonstrate that the problem had been previously brought to the attention of the World Bank but it had responded inadequately.

When the Inspection Panel receives a claim, a copy is sent to the World Bank and it is given 21 days to respond. The panel then weighs the evidence from both sides and determines whether an investigation into the alleged policy violations should be recommended. Such recommendations have to go back to the board of executive directors of the World Bank, comprising both donor and borrowing governments' representatives, which has authority over whether the Panel can investigate a case. If an investigation proceeds, the Panel sends its final report to the board and management of the World Bank, which has six weeks to prepare recommendations on what actions it will take in response to the findings. The Panel's report and the World Bank's recommendations are made public, although the board makes the decision on what action to take.

Over the first five years of experience of the Inspection Panel, only 11 claims were filed from civil society organisations. Nine of these involved infrastructure projects (five hydroelectric dams), and almost half involved projects in either Brazil or India. Most of the claims to date have focused on resettlement, environmental impact assessment and indigenous people's policy violations. These patterns can be considered consistent with the characteristics of what are known to be the most controversial World Bank projects, large-scale infrastructure projects, as well as the long history and prominence of locally based international environmental and human rights projects in those countries.

Only in one case, the Arun III dam in Nepal, did the claim lead to a clear-cut victory for the claimants and cancellation before construction began. In most cases, the tangible impacts of Panel claims have been limited and uneven, but not necessarily insignificant. For example, the Panel's intervention in the case of the Yacyreta hydroelectric dam in Argentina and Paraguay is considered to have opened a dialogue that was lacking between the binational dam-building agency, the affected population and the government institutions from both countries. Grassroots opposition in Paraguay has also increased significantly as local critics coalesced around the Panel claim and gained national allies through the process. Although the construction project continues, the international claim process has increased the local legitimacy and leverage of dam-affected people in those countries.

A number of factors constrain the potential impact of the Inspection Panel. First, the panel is relatively autonomous, but remains a World Bank institution. The World Bank's Board of Executives has the power to reject the Panel's recommendations, and in most cases has done so.

Second, many civil society actors affected by the World Bank's projects remain unaware of the Panel's existence or of the World Bank's social and environmental safeguard policies. Most widely, World Bank-funded investments appear to those affected to be nation-state projects, such that representation is not made to the Panel.

Third, the Panel's role remains restricted by a remit that focuses only on mandatory World Bank policies rather than recommended good practice. As such, many possible problems with World Bank projects remain beyond the Panel's activities. Critically, structural and sectoral adjustment loans are not part of the Panel's mandate and therefore elude many of the World Bank's social and environmental safeguard policies.

Fourth, there are substantial costs and risks involved for civil society groups in filing a claim, including the human resources required to prepare highly technical documentation and the threat of reprisals to potential claimants.

Source: Fox (2000) from *Global Governance: A Review of Multilateralism and International Organizations*, vol. 6, 11.

participants in their economies the same as domestic firms, and to the 'most favoured nation', which aims to ensure that any concession granted to one trading partner is extended to all.

These same rules continue to exist under the WTO. However, the WTO is a much stronger institution in that it now has legal status similar to the UN. Its rules are binding on members and the WTO has a dispute settlement body to enforce these rules and principles. The remit of the WTO, however, increasingly (and controversially) now extends beyond a concern for cross-border 'physical' trade issues into areas related to trade, such as services and patents, that impact substantially on domestic policy and economies, throughout health, environment and agriculture, for example, as will be highlighted below.

Membership of the WTO now includes the majority of the 'developing world' as well as the 'developed' countries (currently 148 members). The activities of the WTO have major implications for the working of markets worldwide, but also the prospects for sustainable development and the alleviation of debt and for progress towards democracy in the developing world. As Williams and Ford (1999: 273) noted,

compared with GATT, the increased scope, permanence and rule making authority of the WTO has alarmed environmentalists and other civil society actors who fear that the organisation and control of vital national decisions have been gradually and irretrievably displaced from national control to a supra-national organisation shrouded in secrecy.

The WTO hosts a series of multilateral trade negotiations, known as 'rounds', within which arrangements are negotiated and disputes resolved. Voting within the WTO is unweighted in that each member receives one vote. Decision making is generally based on consensus, although there are provisions for majority voting in certain instances. The length of each round gives insight into the challenges of achieving international consensus on trade matters. For example, the Uruguay round took eight years to complete (1986–1994) and it then took a further seven years to agree to the shape for the subsequent and current round of deliberations (finally agreed in Doha, Qatar, in November 2001). The structure of the WTO is shown in Figure 7.9.

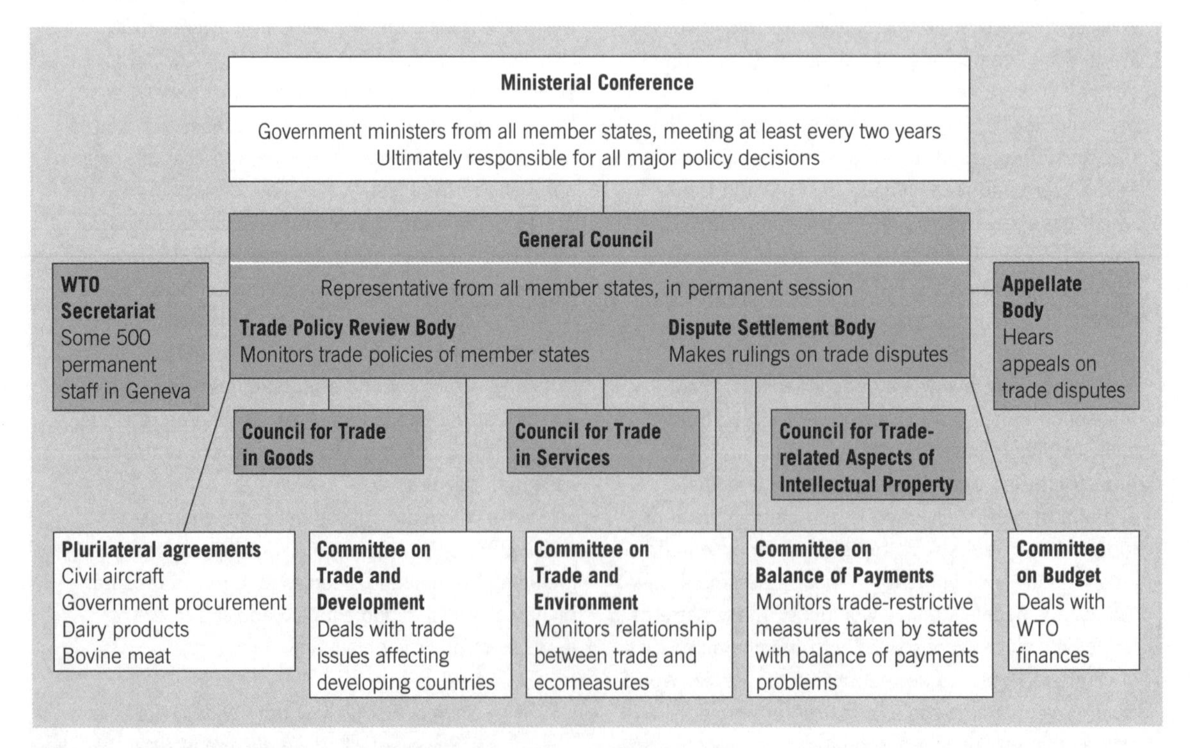

Figure 7.9 The structure of the World Trade Organization. WTO agreements are normally multilateral and apply to all member countries; plurilateral agreements apply only to certain members

Source: Buckley (1996) © Understanding Global Issues Ltd

World trade and the environment

> **Environmental collapse and deepening inequity are a visible, tangible, audible calamity everywhere on earth. The liberalisation of world trade is the most active agent in this calamity.**
>
> **(Ransom, 2001: 28)**

The World Trade Organization is increasingly a focus for environmentalists who fear that progress towards sustainable development is being compromised through the further liberalisation of trade generally and through particular WTO rulings. There is no explicit attention within the WTO mandate to the relationship between trade policy and other major policy areas such as the environment. However, there are articles in the agreement granting exceptions to the general free-trade requirements where the protection of human, animal or plant life or health and the conservation of exhaustible natural resources are concerned. WTO rules do not prevent a country from regulating trade in certain products in order to protect the environment or health, but any such measure must be applied to domestic and foreign firms and may not be used as a protectionist device: 'Thus tuna fish caught with nets which also take dolphin have to be treated in the same way as those caught with dolphin-friendly methods' (Buckley, 1994: 6).

In 1995, the WTO established a Committee on Trade and the Environment (CTE). Indeed, there had been a 'Group on Environmental Measures and International Trade' under GATT since 1971, although the group was supposed to convene upon request and there was no such request until 1991 (Neumayer, 2001). The terms of reference of the CTE are: to identify the relationship between trade measures and sustainable development; to make appropriate recommendations on whether the multilateral trading systems should be modified; and to assess the need for rules to enhance the interaction between trade and the environment, including avoidance of protectionist measures and surveillance of trade measures for environmental purposes (Williams and Ford, 1999).

However, the impact of the CTE and the outcomes of their meetings are generally considered to have been disappointing (Williams and Ford, 1999). Neumayer (2001: 11) suggests that 'instead of being a front-runner in triggering reform of the multilateral trade regime to make it more environmentally friendly . . . the CTE has proved to be a paper tiger'. The CTE is not a policy-making body and its terms of reference, as seen, focus generally on the negative impact of environmental measures on trade, rather than the effects of trade on the environment.

As Ransom (2001) noted, the WTO, as a 'key player in the neo-liberal project', is fundamentally concerned with environmental measures as distortions to free trade and the possibility of countries using such interventions to protect domestic firms. Table 7.10 highlights some of the key areas in which free trade and environmentalist views are divergent concerning the relationship between market economics and environmental protection. Environmental NGOs remain strongly critical of the CTE's slow progress (Williams and Ford, 1999), particularly through its failure to consider the extensive ways in which issues of sustainable development interact with WTO policy.

Transnational corporations are responsible for over two-thirds of global exports of goods and services (Dicken, 2007; see Chapter 4). Yet these companies remain largely untouched by any form of international regulation, including through the WTO, which 'makes no distinction between enterprises in terms of power, impact or scale of operation' (UNRISD, 1995: 153). In reference to the failure of the WTO to raise the issue of TNCs in international trade at all at its ministerial meeting in Doha in 2001, a representative of Christian

Table 7.10 Competing ideologies on trade and the environment

Environmentalists	Free-trade policy analysts
➤ Trade liberalisation creates unchecked growth and pollution	➤ Protectionism is inefficient and leads to even more wasteful use of resources
➤ Environmental costs must be internalised into all WTO activities and decisions	➤ Trade distortions and protectionist intentions of environmental measures
➤ Consideration of production processes must be part of WTO decisions	➤ Trade can create finances for design and enforcement of environmental measures
➤ Environmental solutions need to reduce consumption	➤ Well-targeted environmental policy can control environmental impacts of free trade

Aid suggested that 'it is like a conference on malaria that does not discuss the mosquito' (cited in Madeley, 2001: 27).

Although many TNCs now have well-developed corporate strategies on the environment, there remains much concern as to the prospects of sustainable development and poverty reduction through further trade liberalisation. As Bryant and Bailey (1997: 127) concluded, 'transnational businesses have generally had an adverse environmental effect, because these firms privilege profit maximisation over social justice and environmental conservation in their day to day operations'.

Whilst the Doha mandate now includes a commitment that negotiations within this 'development round' will aim to reaffirm the importance of trade and environmental policies working together and commits the Trade and Environment Committee to closer working with MEA Secretariats, for example, fundamentally, the WTO continues to assume that trade takes place between countries, when in fact an increasing proportion evidently occurs through and within transnational corporations.

The WTO and economic development

In continuity with other international institutions such as the World Bank and the IMF, the WTO prescriptions throughout recent decades have been centrally concerned with debt alleviation and the reinvigoration of the global economy. It is proposed that only through processes such as liberalisation, privatisation and deregulation will competition force the required changes in state and market institutions and overcome the generalised economic crisis. However, there is substantial scepticism regarding the potential of further trade liberalisation to overcome poverty in the developing world:

> By expanding markets, facilitating competition and disseminating knowledge, international trade can create opportunities for growth and promote human development. Trade can also increase aggregate productivity and exposure to new technologies, which can spur growth . . . but liberalizing trade does not automatically ensure human development, and increasing trade does not always have a positive impact on human development.
>
> (UNDP, 2003: 1)

A number of concerns regarding the rulings and operation of the WTO in this respect are detailed in Chapter 8. The raised costs for developing nations of technology transfers in future under TRIPS are examined, for example. Rulings on the protection of patents will also raise costs and decrease profits among small farmers, as discussed in Chapter 6 in relation to genetically modified crops. The work of international NGOs such as Oxfam and Christian Aid are seen in Chapter 8 to be particularly important in exposing how trade rulings of the WTO are working against the interests of developing nations and the poorest groups within these (see Box 8.5 in particular). Actions such as those of the Bush Administration in early 2002 in imposing steep tariffs on a range of steel imports to protect US domestic industry do much to support Oxfam's suggestion of 'rigged rules and double standards' (Oxfam, 2002) in the current international trade system.

Fundamentally, for most developing countries, engaging in world trade and increasing their foreign exchange earnings means expanding the export of primary products and raw materials. Some of the negative impacts of such resource dependence for development prospects were considered in Chapter 6 in terms of the 'resource curse'. 'Commodity dependency' is also discussed in Chapter 8. In recent years, there has been a boom in both fuel and non-fuel commodity prices, increasing by 157 per cent and 180 per cent respectively over the period 2002–2006 (IMF, 2006).

Whilst demand from China and other emerging markets are significant drivers of this pattern (see Chapter 6) and some suggest that the world has now entered a period of sustained high prices, particularly in metals, others suggest that 'prices will inevitably fall back and continue to decline gradually in real terms, *as during most of the past century* (IMF, 2006: 1, emphasis added).

Up until the last few years, the general pattern was of falling world commodity prices as developing countries strove to raise production and export of the same primary products (and crops) leading to gluts on the world market. Significantly, food and agricultural raw materials prices have increased by much smaller proportions than, for example, metal prices in recent years (IMF, 2006).

Furthermore, the terms of trade for developing countries (the costs of importing manufactured goods relative to the price received through their primary commodity exports) continue to deteriorate. Commodity prices remain outside the WTO remit or rulings as already identified. Yet historically low and certainly unstable prices for commodities are widely considered to be among the most powerful influences that prevent trade from working for the poor (Oxfam, 2002).

The WTO and democratic development

The dominance of TNCs within international trade, the economic strength of these companies in comparison to whole nations within the developing world and the limited opportunity for public participation within transnational corporations are all concerns in terms of the prospects for the transition to democracy in these countries (see also Chapter 4).

Whilst TNCs are not members of the WTO, they make very large contributions to the economic prosperity of particular (more developed) countries such that their concerns are 'practically guaranteed' to be listened to at a national level and taken forward to negotiations within the WTO (Taylor, 2003). Transnational corporations by definition are 'non-place-based actors' and therefore it could be suggested they have no loyalty to any community, government or people. Furthermore, initiatives towards enhancing the accountability of TNCs to people and places have been limited to date.

There is still no internationally binding agreement on corporate responsibility that is the focus for an ongoing campaign on behalf of international NGOs including Friends of the Earth International, for example. Trade-related investment measures (TRIMs) which had served to enable national governments to place conditions on foreign investors, including measures to ensure a degree of social responsibility such as in environmental protection or local participation, have been curtailed under WTO rulings (UNRISD, 1995) and there is a current concern that Multilateral Agreements on the Environment (MEAs) may also become subordinate to such rulings (Bigg, 2004).

Figure 7.10 identifies transnational companies (that are listed in the FTSE index of the top 100 companies and that are based in or have a large presence in the United Kingdom) that are investing in countries

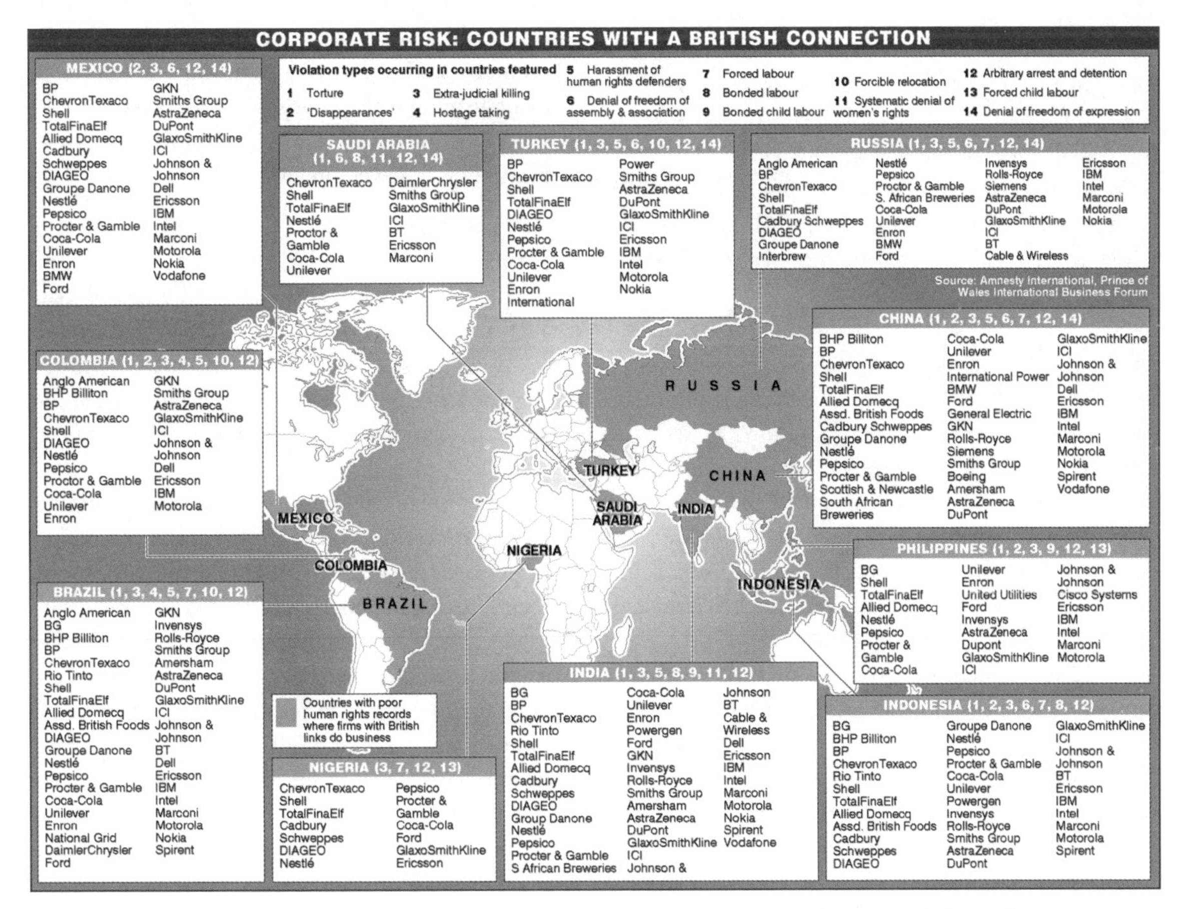

Figure 7.10 Multinational investment in countries with known human rights abuses
Source: © Amnesty International/International Business Leaders Forum (2002)

of known human rights abuses ranging from torture to child labour. The map originates from a text that examines the risks of damage to corporate reputation (in terms of litigation, lost production, problems of recruiting staff, higher security costs) that may be incurred to companies investing in the absence of transparent and properly enforced human rights policies (IBLF, 2002).

While the WTO, as principally a forum for intergovernmental negotiations, is not formally open to NGOs or other representatives of civil society, recent experience suggests that NGOs and less institutionalised movements have been successful in influencing negotiations at the WTO. Most obviously, the mass public demonstrations in Seattle in 1999 to coincide with the WTO ministerial conference were principal factors in the failure of that meeting to secure agreement on a new round of trade negotiations.

Environmental groups are also acknowledged to have been very important in shaping the debates that have resulted in some incorporation of environmental issues within the world trading system. This has been demonstrated through their lobbying, research and support for reform around, for example, the GATT ruling in the dolphin–tuna dispute between the USA and Mexico and the Multilateral Agreement on Investment (Williams and Ford, 1999).

Increasingly, social movements are also influencing international trade negotiations, through lobbying the WTO directly and through the more confrontational approaches that were seen in Seattle. In 1996, the WTO's General Council took two decisions that can be considered to have raised the capacity of NGOs to influence the WTO 'informally', as well as being significant in terms of raising the transparency of the WTO. Within *The Guidelines for Arrangements on Relations with NGOs*, the roles that NGOs can play in wider debates on trade issues are acknowledged, and under the *Procedures for the Circulation and De-restriction of WTO Documents*, most WTO documents should in future be circulated as unrestricted (Williams and Ford, 1999).

However, there remains concern that under certain conditions the 'blanket prescriptions' of the WTO may threaten democratic processes within countries, in particular through exacerbating social–economic inequality and vulnerability at the local level. The UNDP, for example, urges a new vision for trade wherein the multilateral trade regime recognises differences between developing and industrialised countries more effectively, allows diverse development strategies built on local priorities and permits rules that could favour the weaker countries where required (UNDP, 2003).

Fundamentally, markets are diverse and complex institutions which rarely operate to provide equal benefits to all participants. WTO rulings, however, do not distinguish between corporations, companies, individual producers, petty traders or consumers, nor do they recognise their differential power in accessing markets (which themselves are not uniform even within a region). Under certain conditions, it is increasingly evident that the trade liberalisation intentions of the WTO may be inappropriate and contrary to the pursuit of more equitable development patterns.

The role of the state

In most developing countries throughout the 1960s the state took a primary role in the design and implementation of development (Chapters 1 and 3). Indeed, as Mackintosh (1992: 61) has stated:

> **The very subject of 'development' was built on the idea of the state as the main lever for changing the economy and society. The ideology of 'developmentalism' and the concept of the interventionist state were inseparable in the optimistic post-war beginnings of development theory.**

States became involved in 'virtually every aspect of the economy, administering prices and increasingly regulating labor, foreign exchange, and financial markets' (World Bank, 1997b). Many developing nations built up large state enterprises in public utilities, nationalised mining and agricultural enterprises, for example, to lead industrial development in their post-independence period.

It was a stage of 'great optimism' (Mackintosh, 1992: 68) concerning both the benevolence and the competence of the state to work in the 'public interest' that only became seriously challenged in early 1980s (Thomas and Allen, 2000). In short, government expenditures within the developing world into and through the 1970s grew faster than GDP. Further state investment was therefore only enabled by borrowing from commercial and multilateral sources (that were easily available in the 1970s). However, the economic performance of many developing countries remained poor at the end of that decade and the state was beginning to be seen as part of the problem rather than the solution: 'The oil price shocks were a last gasp for

Plate 7.2 Botswana says no to corruption
(photo: J. Elliott)

state expansion . . . As long as resources were flowing in, the institutional weaknesses stayed hidden' (World Bank, 1997b: 23).

By the 1980s, the capacity of national governments in the developing world to continue to make investments in development were severely limited by rising oil prices, mounting debt burdens and recession within the global economy. Through the 1990s, substantial reforms of the state were held as central to global economic recovery and as conditions for the receipt of further development assistance from multilateral and bilateral sources. Reforms of the state also became acknowledged to be necessary for democratic and sustainable patterns of development in the future (Plate 7.2).

Defining terms and identifying roles

The state is a universal feature of the contemporary world (Johnston, 1996). The state is a network of government, quasi-government and non-government institutions that coordinate, regulate and monitor economic and social activities in society (UNDP, 1997). It is therefore a wider category than government, including the civil service and the legal system, for example. Exhibit 7.2 details and contrasts the concepts of state and government. The state is also one of many spheres of activity within society, others being arenas of civil society, as discussed below.

The state takes many different forms across the globe. As Johnston (1996: 146–7) argues, 'individual states have developed through conflict and accommodation, as various interest groups have contested for power within society'. States do not operate in any particular or predetermined way, it is people who interpret and define the roles and tasks of the state. As Thomas and Allen (2000: 191) highlight, the 'ensemble of political institutions – coercive, administrative, legal' comprising the state, 'may not always act as one or in concert'. In operation, therefore, the role taken by the state in any arena of activity is influenced by a host of forces including inter- and intrastate conflicts, access to finances and the decisions of international finance institutions as already considered above. Further

BOX 7.4

Concepts of state and government

State, in its wider sense, refers to a set of institutions that possess the means of legitimate coercion, exercised over a defined territory and its population, referred to as society. The state monopolises rule making within its territory through the medium of an organised government

Government has different meanings in different contexts:

- the process of governing, the exercise of power
- the existence of that process, a condition of ordered rule
- the people who fill the positions of authority within a state
- the manner, method or system of governing in a society, i.e. the structure and arrangement of offices and how they relate to the governed

Source: World Bank (1997a)

factors shaping the activities of the state include the mounting pressure from civil society organisations and grassroots movements within the developing countries (often in response to perceived shortcomings of the state) including for environmental reforms, which is the focus of the final section of this chapter.

The state and sustainable development

The Brundtland Report identified a key role for the state in fostering more sustainable patterns and processes of development worldwide in the future (WCED, 1987) and in ensuring that individuals within society behave in a responsible manner such that collective goods, including resource conservation, are achieved. Table 7.11 identifies the range of ways in which governments specifically can influence environmental outcomes that encompass both its key stewardship and development roles. Governments are also responsible for negotiating internationally to establish multilateral environmental agreements (as well as trade outcomes as considered above). Evidently, governments through these roles in establishing the policy, regulatory and institutional frameworks within a country have substantial impacts on the prospects for sustainable development within their own boundaries (Elliott, 2006).

Table 7.11 Government's role in environmental outcomes

- Establishing and enforcing laws that determine who has the right to use the environment and the duty to protect it
- Managing natural resources including state-owned and collective environmental goods
- Determining which environmental uses are to be taxed or subsidised
- Restricting environmental threats posed by individual or corporate behaviour
- Defining and enforcing the rules of formal markets
- Allocating funds for conservation and development
- Redistributing resources between groups in society

Source: Compiled from World Resources Institute (2003)

However, it has been asserted that 'rather than being an actor with possible solutions to environmental problems, the state has typically contributed to exacerbating those problems' (Bryant and Bailey, 1997: 55), particularly through intrastate conflicts such as bureaucratic resistance and corruption that have limited its stewardship role. For example, it may be that the most powerful groups have accessed that strength through their control over environmentally damaging activities, such as in mining and energy generation which they are reluctant to give up. The stewardship role of governments has also been compromised by often close associations between political leaders and business interests in many developing countries. These failings of the state and government were considered in some detail in Chapter 6.

In terms of the development role of the state in recent decades, it has also already been noted how international institutions are increasingly influencing national policy development in recipient countries. Yet that capacity of the state to respond to environmental problems was regularly undermined through the processes of structural adjustment via the pressure to cut the budgets and staffing of environment departments in particular (Bryant and Bailey, 1997).

Many governments of the developing world lack the fundamental finances to provide the kinds of investments, services and environmental controls to ensure healthy environments. In the urban sector, it is increasingly private (and often overseas) companies and corporations that shape access and opportunity in key resource sectors such as water supply and sanitation, yet there is substantial concern that such utility privatisation is not sustainable economically, socially or environmentally (Actionaid international, 2004; Budds and McGranahan, 2003; World Development Movement, 2006).

The state and economic reform

It was in response to the economic crises of the 1980s that potentially the most far-reaching changes in state institutions were promoted. By the late 1980s a new economic orthodoxy, termed 'market triumphalism' by Peet and Watts (1996), was gaining the sympathy of leaders in the industrialised nations.

Within this thinking, it is the market rather than the state that is seen as the prime instrument of economic development (see also Chapter 3). The state was conceived as a block to development rather than a protector of the public interest or a force for development. This agenda also included rediscovering civil society; 'the expressed aim of the neo-liberal project is to free the entrepreneurial potential of civil society from the omnipotent and omnipresent hostile state' (Zack-Williams, 2001: 218). This thinking soon spread into the policies and practices of the multilateral institutions, as identified earlier in relation to the World Bank.

Prescriptions for institutional change towards the regeneration of the global economy and debt reduction included reducing government spending, internal reforms of the state and the devolution of state activities and decision making. Conditions within structural adjustment programmes for internal reforms of the state included deregulation (e.g. giving up public monopolies in education), cost recovery (the use of fees for services such as health, aimed at cutting demand and promoting efficiency) and the targeting of remaining tax-financed spending on the most needy (such as subsidies concentrating on primary education and higher-level services being charged at nearer cost).

A further promoted means of saving on tax revenues, and thereby assisting with balance of payment difficulties in the developing world, was to devolve the activities of the state and its decision making to other institutions such as in the private or voluntary sectors. The state function was thereby reduced to one of co-ordination, 'or more vaguely, an "enabling role"' (Mackintosh, 1992: 83).

However, it has been suggested that the issues that dominated in the 1980s concerning states that were too strong were replaced by concerns that states may be too weak by the millennium:

> **Early in the 1990s, the fashionable code words commonly used by politicians, experts and journalists commenting on economic reforms were 'macro-economic stabilisation' and 'structural reforms'. *'Governance', 'transparency' and 'institutions'* have now replaced these terms . . . The obsession with crushing inflation, common in the late 1980s and early 1990s, has been substituted by the obsession with the need to curb corruption. Leaders of multilateral institutions today spend as much time highlighting the importance of strengthening the rule of law in some of the problem countries as their predecessors did 10 years ago about the need for their client governments to 'get the prices right'.**
>
> **(Naim, 2000: 521, emphasis added)**

For the World Bank, the interest in promoting good governance and democracy was certainly in part influenced by the experience of structural adjustment. It became apparent that the ability to implement the reforms was closely related to political commitment and capacity as well as bureaucratic competence and independence within recipient countries. Mohan *et al.* (2000: xiv) have referred to the irony of a theory that 'posits the "freedom" of the markets and the limited use of state power' that in fact 'required massive amounts of political interference to do so'.

Interest in governance issues within recipient nations also rose at the International Monetary Fund during the 1990s. For example, as financial crises spread throughout many countries of Asia in the late 1990s what had been considered a linchpin of Asia's success, the strong informal networks operating in the economy, were then 'relabeled and condemned as "crony capitalism"' (James, 1998: 46). According to James (1998: 47), such patterns within the World Bank and the IMF reflected a realisation increasingly shared that:

the world economy and world institutions, can be a better guarantee of rights and of prosperity than some governments, which may be corrupt, rent-seeking, and militaristic. Economic reform and the removal of corrupt governments are preconditions both for the effective operation of markets and for greater social justice.

Governance and the state

Whilst interest has risen in the role of the state in ensuring good governance within their boundaries, the relationship between democracy and economic development is far from straightforward. As Leftwich suggests (1993: 613), there is plenty of evidence worldwide that economic success has been achieved under conditions 'not remotely approximating continuous and stable democracy' such as in Brazil, South Korea, Taiwan, Indonesia or Thailand. Similarly, Potter (2000: 378) concluded through an analysis of data for 135 countries, that '"economic miracles" occurred in both democracies and dictatorships'.

The debate continues as to whether competitive elections, for example, will ensure democratic developments. India has been a liberal democracy for 60 years now but it also has the largest number of desperately poor people in the world. The validity of multiparty political systems for all societies equally remains contested, particularly by political leaders in the African continent who have asserted democracy as a foreign ideology and also as divisive by rousing ethnic identities (Zack-Williams, 2001).

While the perceived African solution to the problem of political governance in the past was the one-party state (sustained by the Cold War and superpower rivalry), there is now substantial pressure for political pluralism in the continent. The forces for change certainly come in part from 'outside' but are also increasingly coming from within the continent, from civil society challenges to the state as well as through the influence of the international financial institutions. Indeed, more autonomous and numerous associations in civil society are a further characteristic of democracy that is increasingly being recognised and receives further focus in the final section of this chapter.

However, governments are still identified as having a core role in the prospects for development within their boundaries. For example, in assessing the prospects for achieving the Millennium Development Goals, the report of the Millennium Project considers that

economic development stalls when governments do not uphold the rule of law, pursue sound economic policy, make appropriate public investments, protect basic human rights, and support civil society organizations – including those representing poor people – in national decision making

(Sachs, 2005: 16)

The final section of this chapter considers how local communities may be enabled to control their own development, including through new relationships with state as well as with voluntary and private institutions.

Civil society, NGOs and development

It has been suggested that the last decade of the twentieth century was witness to the 'rise and rise of civil society' (Edwards, 2001b: 2). Certainly there has been a rapid growth in the volume and depth of writing on civil society, from which are to come 'agents of change to cure a range of social and economic ills left by failures in government or the marketplace' (Van Rooy, 2002: 489).

It has already been seen that the role of non-governmental organisations in service delivery increased rapidly in the last 25 years, principally as many governments' capacity to deliver services declined. Autonomous and numerous associations in civil society have also been identified as a characteristic of, and a route to, more democratic societies. In previous chapters, the role of international advocacy groups, such as the 'anti-globalisation' movement, in prompting institutional change has been noted.

These brief examples hint at the diversity of actors encompassed within this arena of 'civil society'. Indeed part of the explanation for the proliferation of writings on civil society may be the various ways in which the term is used. But the examples also point to the prospect that the 'interest' in, or of, civil society to promote development changes can also be very contrasting. This is a further factor in the increased attention and debate. As Edwards (2001b: 1) suggests, 'Civil society is an arena, not a thing; although it is often seen as the key to future progressive politics, this arena contains difficult and conflicting interests and agendas'. Table 7.12 summarises some of the major forces behind the rise of civil society in development. This section critically examines the ways in which civil society actors are being brought into, and are shaping, interventions around the three themes for change: sustainability, economic restructuring and progress in governance and democracy.

What is civil society?

The concept of civil society is not easily defined, as discussed in Chapter 3. Most commonly, it is identified as an arena for association and action that is distinct and independent from both the state and the market, a voluntary, self-regulating, 'third' sector in which citizens come together to advance their common interests (excluding business). It includes both formal organisations, such as religious bodies, cultural societies, professional associations, trade unions and NGOs, as well as more informal types of association, including a host of informal networks and mutual support groups.

However, while this interpretation stresses the 'form' of a certain part of society, other definitions stress the 'norms' or particular characteristics of a society deemed 'civil'. Such interpretations focus on the social values and attributes, including trust, tolerance and cooperation, that characterise 'civil' ways of being and living in the world that are 'different from the rationality of either state or market' (Edwards, 2001b: 5). It is

Table 7.12 Explaining the rise of civil society

1. Changing ideas of international development and the move away from the Washington consensus of Western style democracy and market liberalisation as a blueprint for all: a new understanding has emerged that a strong social and institutional infrastructure is crucial, within which social capital as a rich weave of social networks, norms and civil institutions is just as important as other forms of capital towards these ends
2. More pluralistic forms of governance and decision making are now seen as key to sustainability: it is accepted that states no longer have a monopoly over governance as they are increasingly challenged by the influence of private, both business and not-for-profit, actors. Global governance in the twenty-first century also no longer means a 'single framework of international law applied through a unified global authority'. As such, a shared ownership of the development agenda is understood to be the way forward
3. Working with civil society is seen to make commercial sense: the roles of public, private and civic sectors is being reconceptualised throughout economics and social policy. It is seen that partnerships and a broader dialogue between all groups will deliver better rates of success at less political cost

Source: Edwards and Gaventa (2001)

Plate 7.3 World Bank headquarters in Washington
(photo: Getty Images/News)

in this sense that discussions of civil society are closely aligned to debates concerning social capital, as defined in Chapter 3.

The terms 'civil society' and 'non-governmental organisation' are often used interchangeably, most notably within the writings of the international financial institutions (Woods, 2000). However, NGOs are only part of civil society, as already seen, and 'any particular selection of non-governmental organisations may very unequally represent a society' (Woods, 2000: 835). Similarly, if civil society is taken to describe the society over which government governs it needs to be recognised that some parts of society may be highly organised, while other parts may have little access to government or indeed international organisations, 'even though they may be equally, if not more affected, by government policy and World Bank or IMF decisions' (Woods, 2000: 835). Box 7.5 looks at the evolving characteristics of NGOs in more detail.

Social movements are also part of civil society but are not equivalent to NGOs. The term tends to be used to refer to coalitions and networks that are less institutionalised than NGOs, although NGOs may provide an organisational focus for such movements. Social movements such as environmentalism and feminism are often referred to as 'new' in that they have mobilised people around issues of natural resources and gender relations that are distinct from the 'old struggles' concerning class and labour (Routledge, 2002). At the turn of the century, it was estimated that there may be over 20,000 transnational civic networks active on the global stage (Edwards, 2001b: 4).

A common feature of new social movements (NSMs) is also that they seek to challenge existing systems and structures and tend to operate outside the organised political sphere. In short, as expressed by McMichael (2000: 247), 'the new social movements are distinguished by their expressive politics and their

BOX 7.5

NGOs introduced

Conventionally, the principal defining feature of NGOs has been that they are 'not government' but private organisations operating on a 'not-for-profit basis. Further primary distinctions are often made between NGOs according to the scale of operation or between those that exist for members' or mutual benefits as distinct from those that exist for public benefit.

Some of the most 'well-known' NGOs, such as Greenpeace or Oxfam, are examples of NGOs that operate for public benefit at the international scale, that are based in the North, but have branches in the Southern countries in which they work. Such NGOs are evidently relatively permanent and institutionalised and are staffed by paid professionals. Oxfam, for example, was founded during the Second World War and had an annual income in 2005/06 in excess of £310 million (Oxfam, 2006 Annual Report).

Typically, public benefit NGOs work in two main areas; in campaigning and policy advocacy work or in service delivery (Greenpeace and Oxfam are illustrative of these kinds of work, respectively). At times, NGOs originally set up for advocacy work may move into service provision where this complements their promotional or lobbying work.

At a national level, within the South, there has been an expansion in recent years of NGOs working in both campaigning and service delivery for public benefit. Typically they centre on particular issues, such as human rights or conservation, or assisting particular groups, such as landless labourers or women. Many international NGOs increasingly work in partnerships with such national and provincial, particularly service providing, organisations. Southern NGOs benefit through such interaction with Northern NGOs in terms of access to resources, expertise and information, and opportunities for joint action and advocacy.

A further category of NGOs are those that form around and exist for members' interests and are typically local and issue-based, reflecting the goals and immediate concerns of members. For these reasons they are often termed 'grassroots' or 'community-based' organisations. Examples would include workteams, burial societies, village cereal banks and squatter associations. They may have varied degrees of formality, their formation may or may not have been stimulated from within or outside those communities and they tend to be highly dynamic, and 'may dissipate once their immediate concerns have been addressed' (Desai, 2002: 495).

challenge to the economism and instrumental politics of the "developed society" model'. For Ford (1999: 73), the critical distinction between NSMs and NGOs is that the former tend to reject engagement with institutions like the WTO to advocate a deeper, more radical challenge; 'it is not just about influencing the agenda but fundamentally altering the system and building a new agenda that puts people and the planet first'.

Typically, NSMs operate transnationally, as people across global regions are acting increasingly not on behalf of others as in the past, but in solidarity with each other as their common interests are recognised. While new social movements have been accused of being 'anti-globalisation' and lacking a coherent vision of social change, George (2002: 7) portrays a broader picture:

> Movement forces are anti-inequity, anti-poverty, anti-injustice as well as pro-solidarity, pro-environment and pro-democracy. While they may not agree on every detail of every issue, they share the basics. They refuse the 'Washington Consensus' vision of how the world should work. Often unjustly accused of 'having nothing to propose', they are, on the contrary, constantly refining their arguments and their counter-proposals.

NGOs do not set development policy or pass legislation, but they increasingly subscribe to development ideologies, as highlighted in Chapter 3, and their role in influencing policy formulation and implementation has risen rapidly in the past two decades, as discussed in previous sections.

> Vast areas of humanitarian or environmental endeavour are now identified in the public mind more with NGOs than with political parties or government programmes.
>
> (Ransom, 2005b: 2)

One indication of their expanding role can be suggested by the recent growth in the NGO sector. The number of international NGOs having consultative status with the Economic and Social Council of the United Nations, for example, rose from 200 in 1950 to 1500 in 1995 (Thomas and Allen, 2000), with current membership standing at over 2,800 (www.un.org/esa/coordination/ngo/). It has been estimated that there may be over 200,000 'grassroots' or community benefit organisations in Asia, Africa and Latin America (Thomas, 2001). However, problems of defining NGOs, the lack of formal registration and the often fluid nature of their existence present problems with such data.

The role of civil society in sustainable development

The search for sustainable processes and patterns of development has been a primary force in the mounting attention given to promoting the empowerment of local communities, whether through new or existing local institutions. As Pye-Smith and Feyerarbend (1995: 304) noted, 'an essential element of sound resource management is a local body that discusses, organises, plans, takes action and responds at a human scale'. It is also accepted that poverty, as a symptom of the disempowerment of local people, is a major constraint on sustainable development in the future (Elliott, 2006; Friedmann, 1992; Vivian, 1992).

Furthermore, many activities of social movements worldwide have been prompted by the loss of local control over environments, natural resources and indigenous land rights in recent years (see New Internationalist, 2001b). It has become evident that much presumed 'lack of care' regarding the environment at the local level arises because 'people do not feel in charge of or, indeed, do not have the power to act' (Pye-Smith and Feyerarbend, 1995: 303).

NGOs and community benefit organisations have a number of characteristics that could make these institutions particularly suited to effecting sustainable development interventions. These include their ability to innovate and adapt and their 'relative smallness' (Chambers, 1993). In addition, their social proximity (Malena, 2000), the tradition of working with the poorest groups, with women as well as men and from the grassroots (Craig and Mayo, 1995), and the calibre, commitment and continuity of staff (Conroy and Litvinoff, 1988) have proven to be essential characteristics in developing the relationships with local peoples required for sustainability. The World Bank's (2000) good practice guidance for involving NGOs in Bank-supported projects states that

> these organisations can make important contributions towards ensuring that the views of local people are taken into account, promoting community participation, extending project reach to the poorest, and introducing flexible and innovative approaches'.

Box 7.6 provides a number of examples of initiatives working to empower local organisations that are delivering environmental and livelihood improvements.

BOX 7.6

Civil society actions towards sustainable development in India and China

The Aga Khan Rural Support Programme

In Gujarat Province of India, the Aga Khan Rural Support Programme (AKRSP) is an NGO working to catalyse community participation in watershed management. At the outset, through techniques of participatory rural appraisal, the external team and local people work to assess the natural resources of the village, the indigenous and adapted practices and existing management systems and institutions.

A village natural resources management plan is then prepared by the villages, and local institutions are reinvigorated or created anew to present to external agencies for funding (local communities also invest their own resources) and to implement the activities identified.

Experience is showing that individual investments by farmers have risen since the initiation of the watershed programme. Village institutions have developed group credit schemes and taken on a number of operations such as ploughing and pooled marketing activities. Out-migration from participating villages has declined.

The Self-Employed Women's Association of India

The Self-Employed Women's Association of India (SEWA) is an organisation established to represent women working in the 'informal' sector in India, in hawking, petty vending and as home-based workers in the textile industry, for example. Before the formation of SEWA in 1972, such women were ignored by the organised unions.

SEWA aims to improve women's working environments and to enhance their income-earning opportunities. Initially, the focus was on work-related issues of women in urban areas. However, the organisation has since expanded into rural areas and training and welfare concerns. SEWA activities currently encompass savings and credit cooperatives, producer cooperatives, training courses (from radio repairs to midwifery), and banking and legal services.

The Center for Legal Assistance to Pollution Victims (CLAPV)

This NGO was set up in 1998 by a law professor at Beijing University. It undertakes a range of activities aimed at encouraging public awareness of environmental law and people's rights, improving the capacity of public agencies responsible for presiding over environmental conflicts, and promoting the enforcement of Chinese pollution laws generally. It runs entirely through volunteers – often staff and students within the universities – to assist pollution victims, undertake research and advise law makers. It has successfully brought to court and won over 50 cases. CLAPV has accessed funding and support from international NGOs as well as from donors.

The Center for Biodiversity and Indigenous Knowledge (CBIK)

This is a non-profit organisation established in 1995 in Yunnan province, China. Its mission is centred on biodiversity conservation but it is also dedicated to community livelihood development and to the documentation of indigenous knowledge. It has both research and outreach elements to its work. A major area of activity is watershed governance whereby the NGO is promoting dialogue within and between communities, government and NGOs towards understanding land use practices and the drivers of change affecting watersheds within a holistic framework.

Sources: AKRSP from Shah (1994); SEWA from UNDP (1993). CLAPV and CBIK from Turner and Zhi (2006)

Civil society and economic restructuring

Issues of economic restructuring and debt alleviation have also been important in prompting changes in the ways in which NGOs operate and the interactions between civil society organisations and other institutions in development interventions. It has already been seen that debt burdens (and the policy measures to address these) within developing countries forced cutbacks in state investments in development interventions, and community-based organisations have increasingly been promoted as a route to alternative solutions, to reduce state expenditure and 'fill gaps' where the state cannot provide. Indeed, Thomas and Allen (2000), as seen above, considered the decline of the state in service provision as part of donor-driven reform packages to be the principal factor in the growth of the NGO sector in recent years.

But it is also suggested that donors themselves, both bi- and multilateral, have been less willing to transfer resources through Southern governments in recent years, on the grounds of economic efficiency. An estimated US$5 billion of official development assistance in 2000, for example, was administered by NGOs rather than governments (World Bank, 2001a) and is a further factor in the growth of international service-providing NGOs in particular.

For Tussie and Tuozzo (2001: 112) it is the characteristics of NGOs that explain the interest of the multilateral development banks in these organisations; 'the ways in which this particular manner of organising collective action replicates self-regulation, individualism, and voluntarism – all principles embedded in the free-market system'. However, it has also been identified that the move away from the Washington consensus of market liberalisation has been important in explaining the expansion of civil society within development debates (Edwards, 2001b). Indeed, NGOs have been important in prompting such changes, as illustrated by the success of the Jubilee 2000 campaigns for the rescheduling and cancellation of debts of the poorest nations, for example.

Furthering democracy through civil society organisations

The desire for democracy has also been a driving force in the expansion of civil society organisations in recent years and in changing the ways in which other institutions in development operate. For Craig and Mayo (1995), the uniting characteristic of NGOs is their 'almost universal slogan' of 'empowering the poor' and, as such, is the basis for the contention that NGOs constitute a potential source of alternative development. It has also been suggested that NGOs may have a comparative advantage over other institutions for promoting democratic processes through their defining principles and structures. As Edwards and Hulme (1992) pointed out, membership of NGOs is more often based on a commitment to a normative purpose than to a narrow self-interest, and their organisational frameworks tend to be more democratic than hierarchical; 'simple, human concern for other people as individuals and in very practical ways is one of the hallmarks of NGO work' (Edwards and Hulme, 1992: 14).

While fostering the 'participation' of local people in development through the activities of community institutions has been widely promoted by various agencies since the early 1970s (Potter, 1985), it is now recognised that community participation and empowerment for an alternative development requires more than the occasional incorporation of localised community actions into wider development processes (see Chapter 6).

Democracy and empowerment depend on the reordering of social relations and the challenging of local and state power, for example. As such, local communities must move beyond actions in their immediate practical or survival/coping spheres (which are the traditional arenas for many people's organisations), to act strategically and to engage with the state and political processes at various levels (Taylor and Mackenzie, 1992).

Democracy also depends on the condition of social equality as considered in earlier sections in relation to the gendered impacts of SAPs. There are many examples, however, of community-based initiatives serving to aggravate social if not economic inequality (for a review of community-based natural resource management in Africa, see Elliott, 2002). Although community empowerment necessarily involves the challenging of social structures, it cannot be assumed that all interests are served equitably through 'community development efforts'.

For Edwards (2001a: 5), 'It is civil society's role in "good governance" that excites the donor imagination most of all' and therefore is a principal factor in explaining the current interest in NGOs as institutions for development. For example, where formal citizenship rights within a country may be weak, civic institutions can provide channels for people to challenge government to be more transparent and accountable.

Similarly, Korten (1990b) identified a critical role for NGOs in promoting local empowerment through the monitoring of, and protesting against, abuses of power, particularly on behalf of governments and businesses, and opening them for public scrutiny and action. Box 7.7 identifies a number of legislative changes that are opening the political space for civil society groups in China. Evidently, such processes

BOX 7.7

Emerging opportunities for civil society in China

1994: Rules for Registering Social Organisations granted legal status to independent NGOs for the first time. Environmental groups form the largest sector of civil society groups registered under this law.

2003: Environmental Impact Assessment Law strengthened citizens' rights to influence pollution prevention and natural resource management, extended the range of projects covered and requires that EIA reports be published and available for public comment.

2004: the State Council issued Guidelines for Full Implementation of the Law. This requires that most government information should be disclosed and open to public review. As such, this 'right to know' legislation overturns the previous Law of State Secrets that mandated that any government information that was not specified as public was assumed to be a state secret.

2004: the State Council's Administration Permission Law requires that when administrative institutions identify that a proposed project has direct impacts for a third party, stakeholders must be told that they have a right to demand a hearing and to express their opinions. Administrative institutions are required to organise such a hearing within 20 days of receiving that application.

Source: Compiled from Turner and Zhi (2006)

towards democracy may be unpopular and may be resisted by those with local as well as state power.

Competing priorities for civil society?

It is evident through this section that there are many varied expectations in development being placed on NGOs and CBOs, and these are likely to continue into the future. However, there is already evidence to suggest that the forces of change reviewed may well be competing under certain conditions. For example, as increasing aid to developing countries is channelled through NGOs there is a danger that control of public policy will become concentrated in the hands of the most powerful NGOs based abroad, which may be 'more difficult to counter or even monitor than conditions on aid to governments' (Mackintosh, 1992: 83). In Cambodia, almost the entire health service has been contracted out to international NGOs (Ransom, 2005b).

There is also a risk that the NGO's mission may become diverted by the need to focus on that of the donor (Malena, 2000). It has been seen that as service providers, NGOs are increasingly being required to operate in more market-driven ways. In order to be commercially competitive, a risk is that the smaller, perhaps more community-based organisations, will be driven out of that market. Between 1998 and 2004, for example, the proportion of total donations going to the largest charities rose from 41 per cent to 45 per cent. In Australia, a single NGO, World Vision, accounted for 40 per cent of all funds raised from the public for international development in 2003 (Ransom, 2005b).

Furthermore, the 'not-for-profit' characteristic of NGOs becomes compromised as increasingly they 'have to survive as bureaucracies, something which requires considerably more finance than that which is spent on their supposed beneficiaries' (Thomas and Allen, 2000: 210). The Wildlife Conservation Society, for example, has a Chief Executive Officer who has an annual salary of over US$0.5 million (Ransom, 2005b).

The need to focus on service provision within contracts may also compromise the broader watchdog or advocacy functions more traditionally associated with NGOs. There are fears of a general depoliticisation of NGOs as direct funding by official donors increases. Fundamentally, the assumption within governments and multilateral development banks that NGOs should be vehicles for channelling increased volumes of aid may compromise the ability of NGOs to define their own agendas.

Conclusion

This chapter has identified in some detail the nature and activities of a range of institutions that variously set and shape development policies, deliver investments and influence very widely the directions and outcomes of development policies, programmes and projects. These have included new international institutions with very overt structures and roles, such as the Commission on Sustainable Development, national legislative frameworks, networks of alliances between civil society organisations and institutions that structure individual livelihoods and opportunities in development.

The chapter was structured around the role of three interrelated forces of contemporary change within and between a range of core institutions in development, namely, the search for more sustainable patterns and processes of development, the means for stimulating economic growth and the alleviation of debt, and the pressures and desires for good governance. Through the chapter, it has been shown how previous, rather sterile debates such as the 'state versus market' of the early 1990s have now been replaced by more nuanced assessments of the extent of state interventions, of their nature and role in relation to other institutions of development, including the international institutions and civic organisations. Similarly, it is accepted that there is no single model of 'good governance' but that it may have different properties 'depending on the particular institutional context concerned' (Potter, 2000: 381).

In a globalised world, conceptions of governance in international relations are also no longer seen to be the monopoly of the state:

> As economic and cultural globalisation proceed, the state's monopoly over governance is challenged by the increasing influence of private actors, both for-profit and not-for-profit.
>
> (Edwards, 2001b: 3)

Furthermore, the challenges of governance are not just the responsibility of developing countries; 'all countries must take responsibility for strengthening global checks and balances and implementing strong anti-corruption standards' (World Bank, 2006a: xix).

Concurrently, it has been revealed that whilst there are considerable opportunities that can be and are being realised through an increased role for civil society in development, civil society is no 'magic bullet' for the future. Rather, there is a need to consider what particular combinations of public, private and civic institutions are most effective in specific contexts.

This chapter has considered substantial 'institutional learning' particularly in the early years of the twenty-first century towards the persistent challenges in development planning and practice, but it has also highlighted new challenges. For example, there is great uncertainty concerning the role of the United Nations despite substantial internal reform in recent years. Although there are now signs of an international consensus on the persistent and pervasive challenge of global poverty, there is a need to continue to monitor the commitments made to enhance aid and deepen debt relief, for example.

Critical assessment of the potential of Poverty Reduction Strategy Papers as a mechanism that can encompass truly participatory processes and capture national and local priorities as well as mainstream the environment is also required. Further thinking is needed on the role of international finance and trade as part of the dynamics of poverty creation and unsustainable environmental outcomes rather than seeing them as solutions.

Key points

- Decision making in development is undertaken by a range of organisations, bodies, agents and individuals. All these institutions of development are under constant change, as is the relationship between them.
- Institutions are not neutral factors in development processes, rather they are engaged in inherently political activities, creating both opportunities and barriers for particular groups and individuals in development.
- Recent decades have seen an increasing role for international financial institutions, transnational business and civil society in shaping development outcomes.
- Finding responses to the issue of debt, the challenges of sustainable development and mounting pressure (from various sources) for better governance and democracy have led to significant changes in the roles of and the relationship between the spheres of the state, the market and civil society. Recent years have witnessed more nuanced understandings of the capacities of any particular sphere to influence development outcomes.
- The challenge to the dominance of the international financial institutions in the last ten years has opened up new political spaces particularly for the role of civil society organisations.
- Ongoing critical analysis of institutions is needed, not just for the ways in which they 'do development' but also for the way in which they may constrain development opportunities and present barriers for particular groups and social actors.

Further reading

Adams, W.M. (2001) *Green Development: Environment and Sustainability in the Third World*, 2nd edn. London: Routledge.
A very clear overview of how the notion of sustainable development has come into development debates and influenced the practices of key institutions in development. Presents both 'mainstream' understandings and the challenges to these, including from the developing world.

Edwards, M. and Gaventa, J. (eds) (2001) *Global Citizen Action*. London: Earthscan.
Provides a very good overview of the processes underpinning the rise of civil society in development and the range of ways in which citizen action is shaping the agenda and practice.

Mohan, G., Brown, E., Milward, B. and Zack-Williams, A.B. (2000) *Structural Adjustment: Theory, Practice and Impacts*. London: Routledge.
A readable text that draws together the economic, social, political and environmental impacts of the adjustment experience and raises a number of alternatives to adjustment.

Nelson, P. (2000) Whose civil society? Whose Governance? Decision making and practice in the new agenda at the Inter-American Development Bank and the World Bank. *Global Governance*, 6, 405–31.
Identifies the emergence of the governance agenda in these two multilateral development banks and raises core questions such as which groups in society are able to participate in this agenda.

Oxfam (2002) *Rigged Rules and Double Standards: Trade, Globalisation, and the Fight Against Poverty*. Oxford: Oxfam.
The oft-cited report (that has both supporters and critics) which highlights the importance of international trade rulings in the prospects for economic and social development in the developing nations.

Thomas, A. and Allen, T. (2000) Agencies of Development, in Allen, T. and Thomas, A. (eds) *Poverty and Development in the Twenty-First Century*. Oxford: Oxford University Press, 189–216.
A useful introduction to the range of institutions, organisations and actors in development.

World Bank (2001) *World Development Report*, 2000–2001. Oxford: Oxford University Press.
The publication considered to be a landmark in the development of an international consensus concerning the new poverty agenda in international development.

Websites

www.Inweb18.worldbank.org/essd/essd.nsf/NGOs
Website established by the World Bank providing information, links and materials for Civil Society Organisations working with the bank. Accessed via the Environmentally and Socially Sustainable Development (ESSD) Network that hosts links to projects, programmes, research and reports in areas of environment and poverty alleviation, for example.

www.un.org/esa/sustdev
United Nations home page on sustainable development with detail of mandate, meetings and progress of the Commission on Sustainable Development.

www.oxfam.org
Website of long-established international non-governmental organisation. Hosts substantial materials relating to issues of international trade and debt alleviation, for example, from a more broadly Southern perspective.

www.50years.org
Website for the 50 Years Is Enough network, a US-based coalition of over 200 organisations committed to the transformation of the IMF and the World Bank.

www.unrisd.org
Website of the autonomous United Nations agency, the UN Research Institute for Social Development. Hosts research, information and publications relating to social and equity dimensions of contemporary development concerns including around corporate responsibility, economic transition, gender, human rights and sustainable development.

Discussion topics

- Review the arguments that NGOs may be best placed to promote more sustainable development processes. Consider the major challenges against these institutions operating in this way.
- How important do you consider the 'mega-summitry' hosted by the United Nations, such as the Rio (1992) and Johannesburg (2002) conferences, to have been in changing patterns of development worldwide?
- In what ways do you consider good governance to be a means or an end in development?
- Select one of the Millennium Development Goals. Identify the principal ways in which the various institutions of development considered in this chapter will shape the prospects for achieving that goal? Consider explicitly where the relationship between different institutions may be most important.
- Can further trade liberalisation deliver environmental conservation?
- Identify a heavily indebted developing country. Where does its debt originate from and from which institutions, private or official? Has it been able to access any of the debt rescheduling or relief from particular creditors or groups such as the Paris Club or through initiatives of the international financial institutions? What conditions have been required?

PART III

Spaces of development: places and development

Chapter 8

Movements and flows

This chapter examines the movements and flows of people, commodities and finance across the world and particularly in relation to developing countries. These movements and flows are numerous and often complex, but the chapter attempts to unravel some of the complexities and draws upon case study material to illustrate particular concepts and situations.

The main topics covered in this chapter are:

- Population movements – different types of population movements are discussed, including the relationships between rural and urban areas and the impact of migrants' remittances on their home communities;
- Tourism and development – tourism is a form of voluntary migration which has accelerated rapidly in recent years and contributes significant amounts of revenue to some developing countries. The future role and nature of tourism are considered, including possibilities for pro-poor tourism and ecotourism;
- Forced migration – sadly the world has no shortage of forced migrants and refugees often living in desperate conditions. Some resettlement programmes are examined, and the effects of civil unrest on communities and livelihoods are considered;
- Transport – the world is shrinking as communications improve, yet for many people in developing countries the basic necessities of life are still not available;
- World trade – the concept of an 'interdependent world' is examined, yet poorer countries have relatively little influence on world trade. The World Trade Organization is critically examined, and issues surrounding fair and ethical trade;
- Transnational corporations – the large companies operating in the world trading system have considerable power. Other players are strengthening their positions, notably from the BRIC countries (Brazil, Russia, India and China). Meanwhile, many transnational companies are giving greater attention to corporate social responsibility in response to consumer criticism of their operations;
- The debt crisis – many developing countries are faced with crippling levels of debt. The nature of this debt and how it might be reduced are examined, including the Heavily Indebted Poor Countries (HIPC) Debt Relief Initiative;
- Aid to poor countries – the quantity and quality of overseas development assistance (aid) given by the rich countries to the poor countries is examined.

Introduction: unravelling complexities

In addition to examining the nature of specific development variables in particular locations, geographies of development need also to consider relationships between people, environment and places in different locations and at a variety of different scales, ranging from the micro-level, such as the individual or the household, through the local community level to the regional, national, international and, ultimately, the global level. These relationships are, however, by no means static. On the contrary, the nature and relative significance of these relationships are changing constantly, both through time and space, and are themselves determined to a large extent by complex movements and flows of people, commodities, finance, ideas and information.

This chapter aims to simplify and explain some of these complex and interrelated movements and flows. Starting with a real-world example, in order to illustrate some of the interlinkages and flows which take place over time and space, we will then consider movements of people, before examining flows of commodities and finance through trade, aid and debt.

Movements and flows in the 'real world': Growing coffee for export

Reference to a real-world example should help to demonstrate some of the many possible interconnections between people, commodities and finance across the world. Let us consider the case of a resource-poor farming household in Ivory Coast (Côte d'Ivoire), West Africa, whose main income comes from growing coffee for export. Although Ivory Coast is Africa's third largest producer of coffee, and in 2006/07 was the world's twelfth largest producer (1.9 per cent), world coffee production in 2006/07 continued to be dominated by Latin American countries such as Brazil and Colombia, which produced, respectively, 34.3 and 9.4 per cent of the world crop (ICO, 2007). The expansion of coffee cultivation in Ivory Coast dates from the French colonial period when, both after 1930 and again in the 1950s, France assured a guaranteed market at very high prices for large quantities of coffee.

The producer household in Ivory Coast is heavily dependent on receiving a good return from cash crop sales in order to buy food and clothes, pay school and medical bills and hopefully, over time, steadily improve its standard of living. But there is a wide range of factors which can affect household income in any given year. Some of these factors may be environmental, whereas others are social, political and economic.

Inadequate rainfall, declining soil fertility and pest attacks are common, and usually unpredictable, environmental factors. Small farmers would generally receive little, if any, advice and practical help from agricultural extension officers, and they often lack the necessary finance to buy pesticides and fertilisers to raise productivity.

The availability of labour is another key element, perhaps the most important factor in poor households with low levels of technology. Such farmers can rarely afford to hire wage labourers and are often totally dependent on family labour. Family members between the ages of 15 and 40 years are likely to be the fittest and therefore particularly valuable for working on the farm. But if at crucial times in the cropping cycle just one key member becomes unwell, or perhaps decides to leave the village to seek work in the city, then this withdrawal of labour can have a major impact on the productivity and therefore the general well-being of the entire household.

In addition, rural producers are affected by a range of economic and political decisions that are well beyond their control. For example, farmers and others in developing countries have been affected to a greater or lesser extent since the mid 1980s by structural adjustment programmes (SAPs) imposed by major international donors such as the World Bank and the International Monetary Fund (IMF) (see also Chapters 3 and 7). In fact, it might be suggested that 'not since the days of colonialism have external forces been so powerfully focused in shaping Africa's economic structure and the nature of its participation in the world system' (Binns, 1994a: 163). Currency devaluation, raising interest rates and the removal of subsidies and price controls are just a few of the measures commonly introduced by SAPs (Mohan, 1996: 364).

In the case of Ivory Coast, with primary products representing over 75 per cent of exports, the country had a total external debt of US$13.3 billion in 2005, and debt service represented 26.2 per cent of exports of goods and services (UNDP, 2006b; World Bank, 2007c). As far as the coffee farmer is concerned, currency devaluation can also have a significant effect on returns from coffee sales. In fact, shorter or longer term fluctuations in the prices which producers receive for commodities such as coffee can have a major impact on household economy and well-being.

With the dominance of Brazil and Colombia in world coffee production, one or more of a variety of factors affecting production in these two countries could have a significant effect on the world coffee price. In simple terms, overproduction could lead to lower world prices, whereas a fall in production, perhaps due to changes in the Brazilian climate, might result in higher world coffee prices. This happened in early 1997, when prices reached a 20-year high due to extremely cold weather in Brazil, forcing producers to relocate entire coffee plantations to warmer areas. But in reality the situation is often much more complex.

Among farmers in developing countries, the decision to invest in the production of 'cash crops' such as coffee is usually taken when prices are high. It may involve a major reorientation of household activities and a significant switch of labour inputs from food to cash crop production. There is evidence to show that in some cash crop producing areas family nutrition has actually suffered due to a relative neglect of food production (Kennedy and Bouis, 1993). Furthermore, by the time the first cash crops are harvested, which may be up to ten years after planting, prices may well have fallen below the levels which existed at the time of planting.

The situation is further compounded by the fact that Third World producers generally only receive a small fraction of the final selling price of commodities such as coffee. Within producing countries there are usually networks of buyers, agents and sub-agents, each taking their share of the price, to say nothing of the substantial element taken by large transnational companies who process the final product in Europe or North America. An Oxfam study revealed that in Uganda, where coffee accounts for 90 per cent of all exports, coffee growers in 1993 received the equivalent of £0.08 (just 5 per cent) of the final value of a jar of coffee which was sold in UK supermarkets for £1.60. In sharp contrast, the shippers and roasters, who are generally part of one transnational corporation, received 65 per cent of the final selling price (Oxfam, 1994).

To reduce the impact of fluctuating world coffee prices on small producers, Oxfam's Bridge programme aims to provide a market for Third World producers, paying fair prices and purchasing through organisations which ensure that the bulk of the price actually reaches the producers. Bridge is involved with three other trade organisations in marketing 'Cafedirect', where the coffee is purchased directly from small farmers, who receive a price linked to the minimum floor price set by the International Coffee Organisation. As Oxfam points out:

> **When Cafedirect was launched during the trough in world coffee prices, producers were paid $1.20 per lb. Had they been selling in the international market, they would have been paid around 65 cents per lb.**
>
> **(Watkins, 1995: 148)**

Cafedirect is now the UK's largest Fairtrade hot drinks company, and has links with 37 producer organisations in 12 countries, 'ensuring that over a quarter of a million growers receive a decent income from trade' (http://www.cafedirect.co.uk/, 2007).

The purpose of presenting this case study, which is typical of so many situations, is to demonstrate that the smallholder coffee producer in Ivory Coast is just one element in a complex system of relationships involving local, national and international movements and flows of people, commodities, finance, ideas and information. Too often in the past, researchers have considered just one element in the system without appreciating the interconnectedness at different scales. As we have seen in earlier chapters (especially Chapter 4), globalisation is one of the most significant features of the late twentieth and early twenty-first centuries. With the increasing speed and frequency of international air travel since the 1960s, and most especially with the 'great leap forward' during the 1980s and 1990s in the transmission of information through satellite communication and the internet, the world is indeed becoming a 'global village' in one sense. Unhappily, however, as in so many communities (and particularly those in poor countries), there is a wide and ever growing disparity between the wealth and living standards of the 'haves' and the 'have-nots' (Chapter 4).

Although it would be impossible to identify, let alone discuss, the many movements and flows across the globe, this chapter aims to shed light upon some of those which involve people, commodities and finance. We will first examine movements of people and then consider trade, aid and debt, emphasising wherever possible the changing nature of these movements and flows over time and space.

Critical reflection

Rooibos tea production in Wupperthal, South Africa

Rooibos (red bush), a form of herbal tea that is indigenous to the Cape Floral Region in South Africa's Western Cape Province, is believed to possess health-giving properties, notably it is caffeine free, has a low tannin content and contains compounds which act as antioxidants. As a result, *rooibos* enjoys a rapidly growing international market. *Rooibos* production only occurs in a mountainous region some 200–300 km north of Cape Town, where micro-climates and soil conditions are particularly favourable. A community that has benefited from *rooibos* tea is Wupperthal, located in a remote area among the high valleys of the Cedarberg mountains, where 400 families currently live. During the 1990s, the community was struggling economically, and an unemployment rate of 80 per cent caused a steady stream of young people to head for Cape Town, since the only local work available was in vegetable farming, local shoe and glove factories and seasonal farm labouring.

In the 1990s, an NGO called A-SNAPP (Agribusiness in Sustainable Natural African Plant Products), visited the area to conduct a 'needs assessment', and began to develop a highly successful relationship with the community. A-SNAPP operates in nine African countries and helps community-based agribusinesses to compete in international markets. Grants worth £50,000 were accessed from various South African government departments and agencies. The funds were used to resuscitate the old 'tea court' in Wupperthal, buy a tractor, lay a new sheet of concrete in the tea-drying area, enlarge the drying floor to 1000 m^2, extend the storage shed and build a new store. These moves are important, since they enabled the community to process their tea locally, rather than having to transport the raw material to the town of Clanwilliam.

The community's customers have included European and North American Fair Trade distributors, such as TopQualiTea and Equal Exchange. Prices are favourable, since the high altitude of Wupperthal ensures good quality tea, which is also produced organically. The community gained accreditation from the Fairtrade Labelling Organisation (FLO) in 2005, which sets minimum standards for economic dividends and opens up more market opportunities.

Who benefits?

The Wupperthal initiative has achieved a significant level of success, such that since 1998 the number of farmers has increased from 25 to 170, while annual *rooibos* production has risen from 16 tonnes to 100 tonnes in good years. Incomes for farmers have risen noticeably. Cooperative members are now able to add value to the product that they grow by utilising their own tea court for processing. The establishment in March 2006 of the Fair Packers packaging venture in Cape Town has further strengthened the community's control over downstream elements of the value chain.

This expansion of the *rooibos* enterprise has already generated employment for Wupperthal migrants, as well as providing producers with an increased share of the value of their product. Most farmers now employ up to four labourers, and an additional 14 people work at the tea court. Furthermore, labourers are now able to gain employment locally without having to circulate around other farming regions in search of seasonal work. In this way, Wupperthal-sourced *rooibos* operates through an increasingly separate supply network to that of the vast majority of Cedarberg *rooibos*, which is produced on plantation style farms owned by white farmers, where opportunities for black people rarely extend beyond low-paid farm labour.

Funding from *rooibos* sales, which is boosted by guaranteed Fairtrade prices and the Fairtrade premium, has also been invested in farming equipment and the local school. Furthermore, the community's sense of self-determination has increased substantially. Thus, widespread benefits have been derived from tapping into a niche consumer market for a product that is healthy, environmentally friendly and ethically produced.

What are the likely advantages and disadvantages of communities such as Wupperthal linking with global trading organisations?

Source: Go to Bek *et al.* (2006) for further reading

People on the move

Population movements, or migrations, have been taking place in various shapes and forms for centuries, and there are many detailed studies and publications on this topic. As was recognised in Chapter 5, the movement of people within and between countries has played an important role in determining population growth rates and in affecting other factors such as ethnic composition and the spread of disease, including HIV/AIDS.

In broad terms, population movements may be divided into 'forced' and 'voluntary', but then more detailed classifications commonly focus on such aspects as the distance covered and the frequency and time span over which the migration occurs. Some writers differentiate between 'migration' and 'circulation' (Drakakis-Smith, 1992; Gilbert and Gugler, 1982; Gould and Prothero, 1975). Migrations are usually more permanent or irregular and can involve a lengthy change of residence, whereas circulations are generally shorter, sometimes daily, periodic or seasonal. Much interest has also been shown in the decision making which is involved in the migration process, and in the consequences of migration for the well-being of the migrant and the migrant's family, as well as the problems and benefits which migration causes in both the source and reception areas.

Seasonal migration and circulation

Although some population movements have a long history, there seems little doubt that colonial policies played a key role in accelerating the process. In Africa, the introduction by the colonial powers of taxation in the form of cash payments, together with the creation of many new towns, mines and cash-cropping areas, led to large-scale migration of wage labourers. In West Africa, for example, the movement of Mossi men from Burkina Faso to help with the cocoa harvest in Ghana and Ivory Coast has been going on for many years and the money they earn provides a vital addition to their poor villages in Burkina.

In northern Nigeria, with a long dry season when little can be cultivated on non-irrigated farms, there is a tradition of seasonal migration which is known by the local Hausa people as *cin rani*. Since the great majority of migrants in this strongly Muslim region are men, they are more correctly known as *masu cin rani*, 'men who while away the dry season' (Prothero, 1959). Commonly, men leave their homes in the dry season to visit relatives and/or to engage in craft industries, trade or irrigated farming.

Before the 1930s, Prothero reports that there was little reference to seasonal migration in the Sokoto region of north-western Nigeria, but a traffic census in 1928 recorded 3500 migrants per month passing southwards through Yelwa on the River Niger and heading mainly for the large towns of Yorubaland in south-western Nigeria. Dry season migration seems to have accelerated during the 1930s, such that the *Annual Report for Sokoto Province* in 1936 noted that 'there has been a large seasonal migration to the Gold Coast [Ghana] and elsewhere by men in search of money to pay their tax and support their families' (Prothero, 1959: 22).

A later survey, undertaken in Sokoto Province in the dry season of 1952–53, enumerated some 259,000 predominantly male migrants, with south-western Nigeria and Gold Coast (Ghana) as their main destinations. Some 92 per cent of migrants were seeking to supplement their income in various ways, through such occupations as labouring, petty trading, fishing and craft work. It seems likely that, on their return home, migrants probably brought more money into Sokoto Province than they could have created without migrating (Prothero, 1959). But much time is spent in travelling, and Prothero argued that the Province would gain a great deal if this labour could be diverted to local productive work, for example in the expansion of cotton and groundnut production for export. A particularly significant point, which has relevance for other poor migrant source areas, is that 'the total number of migrants away from the Province for several months of the year must go a considerable way towards conserving [food] supplies in the home areas' (Prothero, 1959: 34).

Seasonal migration is well established in other parts of the developing world. For example, much of the labour force for cutting sugar cane in north-west Argentina comes from neighbouring areas of Bolivia, where unemployment and poor wages provide an incentive to migrate. Whereas this generally involves migration on a seasonal basis, migrants have sometimes settled in Argentina, moving to cities such as Buenos Aires, where there are better employment opportunities (D. Preston, 1987).

Rural–urban migration

One particularly significant form of migration in many Third World countries is from rural areas to towns and cities, for reasons such as advancing education or to obtain specialist health care (see Chapters 5 and 9). Young people are frequently attracted to towns by the 'bright lights' syndrome – the idea that the towns have modern facilities compared with what are often perceived as being backward and traditional rural areas. In many parts of the Third World, economic reasons play the most important role in drawing people to the cities. For many, and particularly for young males, a spell in the big city is seen as an opportunity to earn a good income and is also regarded as an initiation into adulthood and Western culture.

In Central America, Mexico City's population grew from 14 million to a massive 19 million during the 1980s, and much of this growth was due to migration. Meanwhile, on Mexico's once predominantly rural Yucatan Peninsula there has been rapid development of tourism since the 1970s, which led to the city of Merida tripling its population in 25 years, to reach 650,000 in 1992, representing almost half of the state's total population. Ninety per cent of migrants to the city have been drawn from Yucatan's rural areas, whereas villages surrounding Merida have become dormitories for commuters working in the city (MoBbrucker, 1997).

A fascinating Indian study of some 50 years of migration from the rural village of Sugao, in Maharashtra State, to the city of Bombay, over 150 miles away, revealed some interesting features about both the migrants and the process of migration. Significantly, over a long period of contact between village and city, a valuable city-based network of village relatives and friends had been established, and this played a key role in locating jobs and providing shelter for newcomers. Most urban employment continued to be in the textile industry, though its relative importance has declined in recent years. Few rural families remained untouched by the migration process, which has been overwhelmingly male-dominated. Remittances from city workers have had a major impact in improving conditions back in the village, such that almost half the households have piped water and more than two-thirds have electricity. However, very few men actually left Sugao permanently, and generally returned home when their productive working life was over and it became too expensive to remain in Bombay (Dandekar, 1997).

Remittances

The significance of remittances sent by migrants to their families back home should not be underestimated. Remittances may take the form of various goods and commodities, as well as money, and they can have a significant impact on household livelihoods in developing countries. Many small Pacific island countries depend heavily on imported goods and finance, which is all too evident on flights from Auckland to Apia, the capital of Samoa, with travellers carrying large quantities of electrical goods, clothes, food and other items. Whilst the population of Samoa is 165,000, there are a further 120,000 Samoans living in New Zealand. In fact, it is often claimed that Auckland is the largest Samoan city! New Zealand Samoans have a material standard of living which is twice as good as Samoans living in Samoa. In 2003 an estimated $US60 million was sent in remittances from overseas Samoans to their families, churches and schools in Samoa – almost three times what the country received in foreign aid. The biggest sources of remittances were the USA (39 per cent) and New Zealand (31 per cent). The flows of income to Samoa and other Pacific islands are in fact substantially higher than most official records suggest, since visiting relatives frequently carry large quantities of cash, especially during the Christmas period.

In West Africa, a survey was undertaken in 1996 in eight villages close to Kayes in the Senegal River valley of Mali, a region which supplies large numbers of migrants to France. The Soninke migrants from Mali and neighbouring Mauritania and Senegal have been settling in France since the 1920s. Among the 305 households surveyed, there was an average of 2.6 migrants from each household, remitting about $US1500 in 1996 to their households in Africa. As Azam and Gubert comment,

> The amounts at stake are considerable: using the World Bank's poverty line of $US1 a day, remittances received per household represent on average no less than the annual consumption expenditures required for keeping three individuals just above the poverty line . . . It is far from certain that these 2.6 migrants on average would have been able to produce an output worth that amount, had they stayed with their family. It is almost certain that they would not have been able to produce such a surplus over and above their own consumption.
>
> (Azam and Gubert, 2006: 446)

Azam and Gubert conclude that remittances play a crucial role in the economies of the households which they studied, and in those households where there was at least one member overseas remittances represented a very significant 50.8 per cent of total gross household income (Azam and Gubert, 2006).

Growing urban populations

Although Africa is still overwhelmingly a rural continent, with some 70 per cent of the population in sub-Saharan Africa living and working in rural areas, the average annual urban growth rate of 4.3 per cent between 1990 and 2005 was more rapid than for any other part of the world, and much of this growth was due to rural–urban migration (UNICEF, 2006). In the immediate post-independence period, rapid rural–urban migration was fuelled by significant increases in urban formal sector wages. It is estimated that real average urban wages increased by 40 per cent in Zambia between 1964 and 1968, compared with only a 3 per cent increase in farmers' incomes. In Tanzania, the real urban minimum wage increased nearly four-fold between 1957 and 1972 (Jamal and Weeks, 1994).

Concern has been expressed about feeding these growing urban populations and the availability of employment, and terms like *over-urbanisation* have been used (Gilbert and Gugler, 1982: 163; Gugler, 1997: 114–23). For example, in Ghana during the first decade after independence (1957–1967), the urban population grew by 6.6 per cent, but employment in the modern sector increased by only 3.3 per cent, whereas registered unemployment rose by as much as 9.3 per cent each year (Knight, 1972).

Another early study by Gutkind in Lagos estimated that in 1967 approximately 21 per cent of all males over 14 were actively seeking work (Gutkind, 1969). While searching for work, many migrants relied on the support system provided by ethnic friends and relatives, who could assure a level of survival not radically different from that experienced in the rural village. Work is often gained through such ethnic and kinship links and the higher wages earned by a few manage to 'trickle down' to support those marginally employed and even unemployed. Others will in time find employment in the highly diverse 'informal sector', which might involve working in a family tailoring or carpentry workshop, or perhaps more likely begging or hawking on street corners, or even illegal activities such as prostitution and theft.

Governments and academics have given much thought to strategies for controlling rural–urban migration, notably the reduction of urban–rural differentials through the improvement of living standards in the rural areas and the redistribution and generation of more urban employment opportunities (Becker and Morrison, 1997). However, Riddell concluded in the late 1970s that

> there is no easy remedy for spatial differentials: urban incomes cannot be lowered because of the severe economic and political implications; jobs in the modern sector cannot be created in sufficient numbers because of the limitations upon the national economy; and rural agriculture cannot be transformed because of the scale of the problem and the very limited results likely to accrue.
>
> (Riddell, 1978: 260)

In South Africa, under the oppressive apartheid regime, rural–urban migration was tightly controlled, Black migrants were forced to carry identification passes and their movements were closely regulated by police and security forces. However, during the 1980s and 1990s, rural–urban migration in South Africa accelerated considerably with the progressive relaxation of apartheid controls.

In the 1980s and 1990s the situation changed dramatically in many countries, with structural adjustment programmes, falling urban incomes, deteriorating urban services and public sector retrenchments leading to a massive decline in urban living standards, which has resulted in urban growth rates and migration processes adjusting to urban economic conditions, in some cases leading to people actually moving back to rural areas. The rural–urban income gap collapsed in the 1970s and 1980s, and by the 1980s it seems that a 'new urban poor' had developed in many African cities (Jamal and Weeks, 1994). In Ghana, indexes of real minimum wages show a rise from 100 in 1970 to a peak of 149 in 1974, then a massive decline to 18 in 1984. In Tanzania, there was a rise from 100 in 1957 to 206 in 1972, then a fall to 37 by 1989.

The gap between rural and urban incomes in many countries either vanished or actually shifted in favour of the rural sector. In Sierra Leone before the 1990s civil war the average non-agricultural wage in 1985–1986 was estimated to have been 72 per cent less than average rural household incomes (Jamal and Weeks, 1994). Riley suggests that by 1986 the urban poor in Sierra Leone were 'a deprived group with fewer income or equivalent earning opportunities than the rural poor', which had adverse effects on levels of infant mortality and malnutrition (Riley, 1988: 7). Households have often responded by engaging more in informal sector activity, with wage earners taking on additional cash-earning activities, and by growing food on any available pieces of land.

Jamal and Weeks (1994) suggest that in Uganda another coping strategy is for urban residents, particularly from non-local ethnic groups, to migrate to rural areas. The 1980 Zambian census revealed a slowing down of urban growth rates and indicated a significant increase of urban–rural migration. Surveying the evidence, Potts suggests that a form of counter-urbanisation has been taking place in a number of African countries, 'where the number of urban residents opting to leave the city and move to rural areas has exceeded the number of rural–urban migrants' (Potts, 1995: 259). These 'return' migrants are often the poor and unemployed who are moving back to their rural homes, and this movement seems to be over and above the patterns of circulation between rural and urban areas which are such a long-established feature in many African countries.

International tourism and developing countries

Another type of population movement is represented by the phenomenal growth in international tourism since the Second World War, from some 25 million tourist arrivals in 1950 to over 800 million in 2005 (World Tourism Organization, 2006). Although Europe remains the dominant source and destination of international tourists, the expansion of air travel and the quest for adventure and exotic places has led to a massive increase in long-haul travel, particularly to the Caribbean and Central America, to Asian countries such as China, Hong Kong, Malaysia, Thailand, and to African countries such as Morocco, The Gambia, Tunisia, Kenya and South Africa (Harrison, 1992; see Chapters 4 and 6).

Tourism makes a valuable contribution to the economies of some developing countries (Table 8.1). For example, tourist arrivals in Thailand have grown on average 5.2 per cent per annum between 2000 and 2004, with 11.7 million arrivals in 2004 generating receipts of over US$10.0 billion. In 1988, tourism actually earned more for Kenya than either of the two

Table 8.1 International tourism receipts and tourist arrivals in selected countries (with large or fast-growing tourist industries)

Country	International tourism receipts (US$ million) 2004	International tourist arrivals (1000's) 2004	Average annual growth in tourist arrivals (%) 2000–2004
Angola	66	194	39.7
Armenia	86	263	55.5
Australia	15,191	4,774	3.4
Cambodia	603	1,055	22.7
China	25,739	41,761	7.5
Costa Rica	1,358	1,453	7.5
Egypt	6,125	7,795	11.1
France	40,841	75,100	−0.3
Kenya	486	1,199	*
Mexico	10,796	20,600	0.0
Mongolia	185	301	21.7
Morocco	3,924	5,477	6.4
New Zealand	4,790	2,348	7.1
South Africa	6,282	5,998	3.2
Thailand	10,034	11,737	5.2
Uganda	266	512	27.6
Ukraine	2,560	15,629	24.9
United Kingdom	28,221	27,800	3.5
USA	74,547	46,100	−2.6
Yemen	214	274	39.2

Source: World Tourism Organization (2006) © UNWTO, 9284400508

traditional exports of coffee and tea, and the 676,900 tourists who arrived in the country contributed no less than US$404.7 million to the economy. By 1995, despite exchange rate fluctuations, tourism accounted for 16 per cent of GDP and brought gross receipts of US$486 million, compared with US$350 million from tea. Almost half of the international visitors to Kenya in 1995 came from Germany and the United Kingdom (Binns, 1994a: 146). The systemic effects of the 11 September 2001 terrorist attacks in New York led to international tourist arrivals in 2001 being 4 million down (−0.6 per cent) on the 2000 figure. The most affected regions were South Asia (−6.3 per cent), the Americas (−5.9 per cent) and the Middle East (−3.1 per cent). Europe registered a decrease of −0.6 per cent, while Africa and the Eastern Pacific region experienced an increase in tourists of 3.8 per cent and 5.5 per cent, respectively (World Tourism Organization, 2002).

Tourism also provides a considerable amount of employment, both directly within hotels and also indirectly through taxis and transport, craft industries, restaurants and entertainment. The spectacular tourism development on Mexico's Yucatan Peninsula since the 1970s has transformed a formerly sparsely settled agricultural region around Cancun and has resulted in considerable in-migration, which is linked to work opportunities both in the construction industry and servicing the growing tourist population (MoBbrucker, 1997).

In many developing countries there is considerable potential for strengthening the linkages between the tourist sector and local food and drink production and supply systems, thus reducing the need for expensive imports. However, in Fiji, the linkages have so far been limited due to high imports of goods, hotel furnishings and services, and the repatriation of substantial profits by airline companies and foreign-owned hotels, which account for almost 60 per cent of bed capacity (Lockhart, 1993). In some countries there is an increasing trend towards providing all-inclusive tourist packages, whereby visitors receive all their meals and refreshments in hotels, and they therefore have little need to use local bars and restaurants (see Critical reflection box).

Critical reflection

Are all-inclusive tourist packages a good idea?

An all-inclusive package is when tour operators provide all food and beverages inside the hotel, in addition to providing transport and accommodation. The price paid by tourists is usually significantly lower than if they have to purchase their own meals and beverages at restaurants and bars outside the hotel. With the convenience and affordability of such all-inclusive holidays, there has been a marked growth in demand for this type of package over the past two decades. However, in many developing countries there has been considerable debate about the merits and problems of all-inclusive packages.

Hoteliers and tour operators point to a range of benefits, such as:

- Increased hotel earnings which lead to an increase in government revenues
- More employment opportunities and greater job stability in hotels
- If tourists spend most of their time within the grounds of the hotel, there is less risk of them facing crime and harassment.

However, a number of disadvantages of such packages are also apparent:

- Tourists with all-inclusive packages are likely to arrive with less spending money when their meals and beverages are included in the overall package price
- They are less likely to venture outside the hotel to spend their money in local bars and restaurants and in purchasing locally made crafts and souvenirs
- This can have a negative impact on local business revenue and employment
- It might also be suggested that tourists on all-inclusive packages have much less direct contact with local people, so there is less chance of breaking down stereotypical views of the country and its people.

With reference to particular case studies, do you believe that all-inclusive packages are a good way forward in promoting tourism in developing countries?

There are several concerns relating to the growth of tourism in poor countries. Hotel owners are often expatriates and a large proportion of tourism-related jobs are low paid; the wages of Kenyan hotel staff are lower than in many other sectors of the economy, except possibly agriculture and domestic service. Tourism may also be a strongly seasonal activity, such that during slack seasons staff may be laid off without wages. The impact of tourism on environment and society is a further controversial issue. The loss of valuable farm and grazing land and the 'dilution' of local cultures, with the transference of Western values and patterns of behaviour, are among the major concerns (see also Chapter 4). In small countries such as those of the Caribbean and the islands of the Indian Ocean, along with states such as The Gambia in West Africa, there is a danger of large numbers of tourists overwhelming the country and its people (Plate 8.1). The small Caribbean island of St Lucia has a population of only 158,000, yet in 1998 it attracted 629,598 visitors, of whom 252,237 were stay-over tourists, while the majority of the rest (327,068) were cruise ship passengers (Potter, 2001c).

In Africa, The Gambia is the smallest mainland country and one of the poorest, with an average life expectancy in 2005 among its 1.5 million people of only 57 years. South of the capital, Banjul, the 30-kilometre Atlantic coastal strip is being steadily taken over for hotel building, to cater for the growing number of tourists wanting to escape the European winter. In 2004, some 90,000 tourists visited the country and numbers are growing, particularly with stability restored soon after the July 1994 coup. However, following the coup, 1995 proved to be a low point for the Gambian tourist industry, which in the late 1990s contributed 12 per cent of GDP and provided direct and indirect employment to about 10,000 Gambians.

The great majority of The Gambia's tourist infrastructure is still confined to the coast, with most tourists seeking a beach holiday with the occasional day excursion. There are a few 'up-country' camps catering for more adventurous travellers, but so far they have had a limited impact on the rural hinterland. Indeed, there is some concern within the country about the possible negative effects of encouraging the penetration of larger numbers of tourists into the poor, remoter rural areas. Theft, begging and prostitution are already commonplace in the coastal area, as relatively wealthy visitors, invariably with little local knowledge, are seen as easy prey to those who are desperate to make a living. Similar problems are evident on the Kenyan coast.

In Kenya, farmers and pastoralists have had their traditional lands incorporated into national parks, where wild animals have caused damage to crops and livestock. The traditional pastoral economy of the Masai has been severely constrained, whereas some critics have questioned the profitability of allocating large areas of Kenya for wildlife-based tourism. Like other countries which are experiencing increasing pressures from tourism development, Kenya is keen to promote ecotourism which, in theory at least, is designed to have a more sensitive and sustainable approach to people and the environment. It remains to be seen whether this happens (Box 8.1).

Plate 8.1 Globalisation and international tourism: a tourist hotel in The Gambia
(photo: Tony Binns)

BOX 8.1

Pro-poor tourism and ecotourism

The literature on the developmental impacts of tourism, mainly in the developing world, but to a certain degree also in the developed world, has in recent years sought to identify whether tourism can actually be regarded as, and encouraged to become, a 'pro-poor' development strategy (Binns and Nel, 2002). Poverty alleviation/elimination is the core focus of 'pro-poor tourism' (PPT). But there is often some confusion as to how PPT relates to other tourism concepts such as 'ecotourism', 'sustainable tourism' and 'community-based tourism'. In an attempt to clarify the situation, the Pro-Poor Tourism Project explains:

> **PPT also overlaps with both ecotourism and community-based tourism, but it is not synonymous with either. Ecotourism initiatives may provide benefits to people, but they are mainly concerned with the environment. Community-based tourism initiatives aim to increase local people's involvement in tourism. This is a useful component of PPT. But PPT involves more than a community focus – it requires mechanisms to unlock opportunities for the poor at all levels and scales of operation.**
>
> **(Pro-Poor Tourism, 2002: 1)**

As Ashley and Roe (2002: 61) argue, 'despite commercial constraints, much can be done to enhance the contribution of tourism to poverty reduction, and a pro-poor tourism perspective assists in this endeavour'. In support of this approach, Sharpley (2002: 112) argues that, 'tourism has long been considered an effective catalyst of rural socio-economic development and regeneration'.

Sharpley does, however, question whether tourism can in fact be regarded as a developmental panacea. Even though positive evidence of the impact of tourism-based development on communities can be found in localities such as Taquile island (Mitchell and Reid, 2001), the reality is that in many countries control often remains vested in the hands of outsiders, such that local communities are often only incorporated at a subservient level. This can easily lead to negative effects, such as resource depletion and the loss, or commodification, of culture. As Tourism Concern Director, Patricia Barnett comments on Belize,

> **It now has a highly competitive tourism industry, more interested in marketing a product than ensuring that it is environmentally sound, or that the people are benefiting from it. Local people are marginalized as outsiders buy up the land. Locals are angry that they can no longer access their own forests, which have been their natural home for generations and their islands are sold out to American ecotourism developers.**
>
> **(Tourism Concern, 2002: 2)**

As Weaver (1998: 91) has shown, also in Belize, although ecotourism has spawned development initiatives in local communities, 'the overall number of local residents affected is probably quite low, due to the limited number of parks that accept significant visitor numbers and the tendency of groups to visit on a day-only basis'.

The Gambia in West Africa is attempting to restructure its tourist industry so that both poor people and the environment achieve greater benefits. For example, the government is opposed to 'all-inclusive' deals in hotels, since they limit many opportunities for local service providers to earn a living from tourism (see Critical reflection box above on all-inclusive tourist packages). The country's rich cultural heritage and its wildlife, particularly birds, are an important feature of the tourism development programme, and in 1998 an Ecotourism Task Force was established. An Ecotourism Map of the country has been produced and various sites have been identified as having good potential.

In 2000 one of the first ventures, the Tumani Tenda Ecotourism Camp, received an award from the German Institute for Tourism and Development as being a good model of community-based ecotourism. The camp, accommodating some 40 visitors in huts built from local materials, is located about 500 metres from a small village alongside a tributary of the River Gambia and adjacent to Kachokorr Community Forest, protected by, and for, the community since the mid 1980s. Meals are produced entirely from local ingredients, while water is provided by a solar-powered pump. Target groups are ornithologists and study groups who can take guided forest walks, canoe trips on the River Gambia to accompany fishermen or help women with oyster collection, and there are a variety of workshops such as batik production, traditional healing practices, and the production of vegetables and medicinal plants. The project has enhanced environmental awareness, as well as alleviating poverty among the local community. In the two years following its establishment in 1999 the project raised over £14,500 for the community, which was used to finance the village kindergarten and school. The entire village has been involved in the project, from designing and building the camp to arranging tourist programmes, while useful links have also been developed with Tourism Concern Gambia and the UK organisation Voluntary Service Overseas.

Perhaps Kenya and other countries could learn from the experience of the Central American state of Costa Rica, where ecotourism has been successfully developed. The Monteverde Cloud Forest Reserve (MCFR) in western Costa Rica was designated in the early 1970s and is an area of great biological diversity. Visitor numbers increased rapidly from 471 in 1974 to 49,552 in 1992, representing an annual increase of 578 per cent. The MCFR is now one of the main ecotourism destinations in Costa Rica, and the tourism industry has led to the creation of over 80 different businesses, of which a large percentage are locally owned, including hotels, restaurants, craft shops and bookstores. Additionally, tourism has led to the improvement of local education and the conservation of natural resources, and successful community participation has resulted in Monteverde becoming one of the most prosperous and successful communities in Costa Rica.

Forced migration

In the case of forced migration, the decision to relocate is made by people other than the migrants themselves. The Atlantic slave trade was one of the most massive forced migrations in history when, from the late sixteenth to the early nineteenth centuries, over 10 million Africans were transported to work on plantations in North and South America and the Caribbean (see Chapter 2). More recently, in West Africa, Nigeria shocked its neighbours by expelling 2 million foreign workers in 1983 and a further 700,000 in 1985. The purpose of the expulsion was supposedly to reduce unemployment among its own people during the slump following the 1970s oil boom, but other members of the Economic Community of West African States (ECOWAS), notably Ghana, whose nationals constituted the majority of those expelled, were appalled at Nigeria's action and argued that it contravened the 'spirit' of earlier ECOWAS agreements.

A further example of migration, where there is some controversy about the level of force involved, concerns the movement of over 6 million Indonesians from densely settled Java, where more than 60 per cent of the national population lives on 7 per cent of the land area, to some of the other 13,000 less populated islands and national territories. Population resettlement began in 1905 under Dutch colonial rule, but accelerated after Indonesia gained independence in 1945, and became even more intensive in 1969 under the government's aggressive 'transmigration programme'. Hancock (1997: 234) has described this migration as 'the world's largest ever exercise in human resettlement'. Furthermore, it has been supported by funding from a number of national and international development agencies. Between 1976 and 1986, the World Bank committed US$600 million to the programme, and further assistance has been given by the governments of the Netherlands, France and Germany, as well as USAID, UNDP, EEC, FAO, the World Food Programme and Catholic Relief Services.

The programme has been highly controversial in many ways, including migration to Irian Jaya (West Papua), the name given by the Indonesian authorities to the western half of the island of New Guinea, which they formally incorporated in 1969 with the approval of the UN. Land was taken by force for settlers from Java, which has fuelled a growing conflict between Indonesian armed forces and nationalist Irianese (Papuans). Reports suggested that villages have been bombed, people tortured and shot dead, with over 20,000 Irianese fleeing their homes and seeking refuge in neighbouring Papua New Guinea. Additionally, the Indonesian government wants to settle and 'assimilate' all Indonesia's tribal peoples, including moving (with force if necessary) Irian Jaya's entire indigenous population of 800,000 tribal people into resettlement sites on the island (Hancock, 1997: 235).

Transmigration has also happened elsewhere in Indonesia, with East Timor being seized by the Indonesian army in 1975, to provide for further resettlement from Java. An estimated 150,000 indigenous inhabitants of East Timor were either killed in fighting or died of hunger. In October 1999 the United Nations Transitional Administration in East Timor (UNTAET) was established, leading to East Timor eventually gaining its independence on 20 May 2002 under the democratically elected government of President Xanana Gusmão.

Other examples of forced migration in Southeast Asia (Plate 8.2) are the massive outflow of Khmer and Lao fleeing genocide and invasions in the late 1970s and early 1980s, and the many refugees escaping from Vietnam during the 1980s, in some cases as 'boat people' journeying to places such as Hong Kong in search of asylum. Meanwhile, in 1990s Africa streams of refugees have fled into neighbouring countries to escape civil wars and state collapse in Liberia, Sierra Leone, Rwanda, Burundi and Mozambique (Allen, 1999) (see Box 8.2).

After the Soviet invasion in 1979, Afghanistan lost millions of its people, fleeing mainly to neighbouring

Plate 8.2 Refugee camp for Rohingya Muslim refugees from Burma in south-east Bangladesh
(photo: Howard J. Davies, Panos Pictures)

countries such as Iran and Pakistan. By 1990 there were an estimated 6.3 million Afghans in exile, over 3 million in each of Pakistan and Iran and some further afield. Following the overthrow of the Taliban regime in November 2001, the Afghan Transitional Authority and the United Nations High Commission for Refugees (UNHCR) successfully repatriated more than 1.3 million refugees in the five-month period from March to August 2002 (UNHCR, 2002). In 2005 there were an estimated 20.8 million displaced people around the world under the care of UNHCR, 9 million of these being refugees (UNHCR, 2006).

BOX 8.2

Civil war and forced migration in Sierra Leone

Over the last 15 years, the small West African state of Sierra Leone has become synonymous with political instability, economic devastation and a brutal civil war that lasted most of the 1990s. The protracted conflict ravaged much of the country, and brought immense suffering to its people, with an estimated 50,000 deaths and the displacement of over half Sierra Leone's 5 million population. Many towns and villages were completely evacuated, particularly in the diamond mining areas of the country's Eastern Province, where fighting between RUF (Revolutionary United Front) rebels and government troops was most intense. The population of Freetown, the capital city, mushroomed during the conflict as refugees sought a safe haven and support from aid agencies.

As massive dislocation occurred, economic activities were severely disrupted, much of the country's infrastructure was destroyed or badly damaged, and poverty became widespread and deeply entrenched. Consequently, Sierra Leone now has the dubious distinction of being ranked 176 out of 177 countries according to the UN's Human Development Index (UNDP, 2006).

Since the completion of the disarmament process in January 2002, significant progress has been made towards peace and recovery in Sierra Leone, including the extension of civil authority throughout the country, peaceful parliamentary and presidential elections, and the return home of thousands of internally displaced persons (IDPs). Although these are encouraging signs,

▶

BOX 8.2 (continued)

much remains to be done to improve livelihoods and to set the country on a path towards sustainable development. The government has joined with national and international partners to make a concerted push towards reconstruction and development, and a National Recovery Strategy launched in 2003, based on needs assessments conducted in all districts, aims to provide a framework for these recovery efforts.

The causes of the debilitating conflict were complex, but it is commonly agreed that the origins of instability extend back well beyond the last 15 years, and embody a mixture of factors including bad governance, the denial of fundamental rights, economic mismanagement and social exclusion. The Eastern Province towns of Kayima and Panguma are typical of many settlements that felt the brunt of the war and its associated atrocities. A post-conflict survey in 2004 found that the proportion of demolished buildings was 34 per cent in Kayima and 32 per cent in Panguma, but vandalism was widespread and indiscriminate during the RUF incursions, such that very few buildings were left totally unscathed. The RUF also destroyed bridges, schools, hospitals, markets, community halls, water pipes, and in the case of Panguma, the electricity supply and sawmills. Several times during the conflict both communities had to be completely evacuated.

Although residents have been steadily returning since 2002, housing shortages are still a major problem. In addition to the widespread physical destruction of infrastructure during the civil war, the impact on the social fabric of rural society was equally damaging. Considerable efforts are now urgently needed to rebuild community cohesion and institutions, and to rehabilitate rural livelihoods in remote rural areas.

Source: Go to Binns and Maconachie (2006) for further reading.

The impact of migration on environment and health

The Indonesian resettlement programme has destroyed vast areas of rainforest in one of the most biologically diverse areas of the world. Sumatra alone has lost 2.3 million hectares of rainforest and the cleared land has rapidly become severely degraded. More than 30 per cent of Sulawesi has been reduced to a similar state. Hancock (1997: 237) suggested that by 1996 some 300,000 people were estimated to be living in 'economically marginal and deteriorating transmigration settlements', and were recognised by the Indonesian government itself as 'a potential source of serious political and social unrest in the future'. Infrastructure in the shape of clinics, schools and roads was often inadequate and the settlers suffered from malaria and other diseases. The land became so impoverished that many people moved back to the towns and cities. In recent years, although transmigration has been reduced from its aim of settling 20 million people, the scheme still continues, with private companies running enterprises such as the Barito Pacific plywood factory on Mangole Island, and the creation of new settlements to supply cheap labour for agribusiness, the timber industry and mining.

The potential environmental impact of the large-scale displacement of populations can be considerable, as was the case during the Ethiopian crisis in the 1980s (Box 8.3). However, two studies undertaken in the mid 1990s among Mauritanian refugees in the Sahel region of the middle Senegal River valley in Senegal (Black, 1997), and in areas of settlement by Liberian refugees in the remote forest zone of eastern Guinea (Black, 1996), revealed some decrease in woodland areas, but remarkably little other negative impact on the environment. In the Senegal study, Black questions whether changes in the fragile environment can indeed be attributed entirely to the influx of 50,000 refugees, and concludes that because the refugee population became dispersed over a large area and good relations were maintained with local populations, there has been little conflict over natural resources between the two populations. In Guinea, a similar number of refugees were also spread over a wide area and, as in Senegal, they were largely of the same ethnic groups as the local populations. Black (1996: 37) could find 'little or no evidence that refugees are using natural resources in a more "wasteful" manner than local people'. In both studies it seems that the degree of dispersal of refugees, the generally good relations between refugee and host populations, and the existence of strong local institutions seem to

BOX 8.3

The 1980s refugee crisis in Ethiopia

Some of Africa's poorest people are the millions who have been forced to flee from war, terrorism, persecution or natural disasters such as drought. It was estimated that in 1991 there were about 4 million refugees in tropical Africa (O'Connor, 1991), and in the same year Oxfam warned:

> **In Africa as many as 27 million people now face starvation as a result of drought and war . . . The famine threatening millions of Africans is on a greater scale than the famine of 1984/85.**
>
> **(Oxfam, 1993: 20)**

Famine and refugees in 1991 were associated with Liberia, Angola, Mozambique, Ethiopia and Sudan, but since then people have been forced to leave their homes in Somalia, Sierra Leone and in the Great Lakes region of central Africa, most notably the small but densely settled states of Rwanda and Burundi.

Mohamed Amin's film and Michael Buerk's commentary from Korem, Ethiopia, in October 1984 produced what was probably one of the most powerful pieces of television documentary ever; to millions of comfortable homes around the world it brought images of starving refugees. Here is part of the commentary:

> **Dawn, and as the sun breaks through the piercing chill of night on the plain outside Korem, it lights up a biblical famine, now, in the 20th century. This place, say workers here, is the closest thing to hell on earth. Thousands of wasted people are coming here for help. Many find only death. They flood in every day from villages hundreds of miles away, felled by hunger, driven beyond the point of desperation. Death is all around. A child or an adult dies every 20 minutes. Korem, an insignificant town, has become a place of grief.**
>
> **(Harrison and Palmer, 1986: 122)**

This report had a crucial impact across the world and was instrumental in generating such popular fund-raising efforts as Band Aid and later Live Aid, which took place simultaneously in London and Philadelphia in July 1985 and broke new ground in worldwide satellite communications, raising over US$100 million by mid 1986.

The causes and effects of Africa's continuing refugee problem are highly complex and each case has its unique features. Nowhere is this more evident than in the case of Ethiopia in the early 1980s, where drought, rural impoverishment and armed conflict all played a role in generating a serious refugee problem (see Figure 8.1). Ethiopia has a long history of famines. The great majority of rural Ethiopians in the 1960s were living in conditions which were similar to those of European peasants in the Middle Ages. Feudal landlords exacted heavy taxation and other obligations from their subjects; quite apart from the sheer volume of produce leaving the peasants' hands, the amount of time spent working for the landlord cost dearly in lost production.

It was the repercussions from droughts in Wollo during 1965–1966 and 1972 that eventually brought to an end the 44-year rule of Emperor Haile Selassie in September 1974, with the takeover of the Provisional Military Government of Socialist Ethiopia, known as the Derg. A land reform programme was quickly introduced, but in the years immediately following the revolution, food and cash crop production was disrupted by the many uncertainties caused by the radical transformation in land tenure arrangements. A poor distribution of surplus grain supplies, combined with an inadequate transport infrastructure, severely hampered redistribution to food deficit areas.

But the origins of the refugee problem which developed in the 1980s must also be examined in the context of the Derg's heavy military expenditure and its involvement in a series of costly and protracted conflicts. The civil war in Eritrea was ongoing, having begun in 1962 when Haile Selassie dissolved the federation between the two countries and annexed Eritrea to Ethiopia. The war in Tigray started in 1975, and was concentrated in the densely settled central highlands and the important agricultural region in the west of the province. To compound the situation further, the Derg's forces were engaged in a war with Somalia in 1977–1978 over the disputed Ogaden region. American military aid was withdrawn and Ethiopia requested Russian and Cuban aid, which had the effect of alienating many Western governments. In addition, the Derg massively increased its defence spending to US$378 million in 1981, representing a

▶

BOX 8.3 (continued)

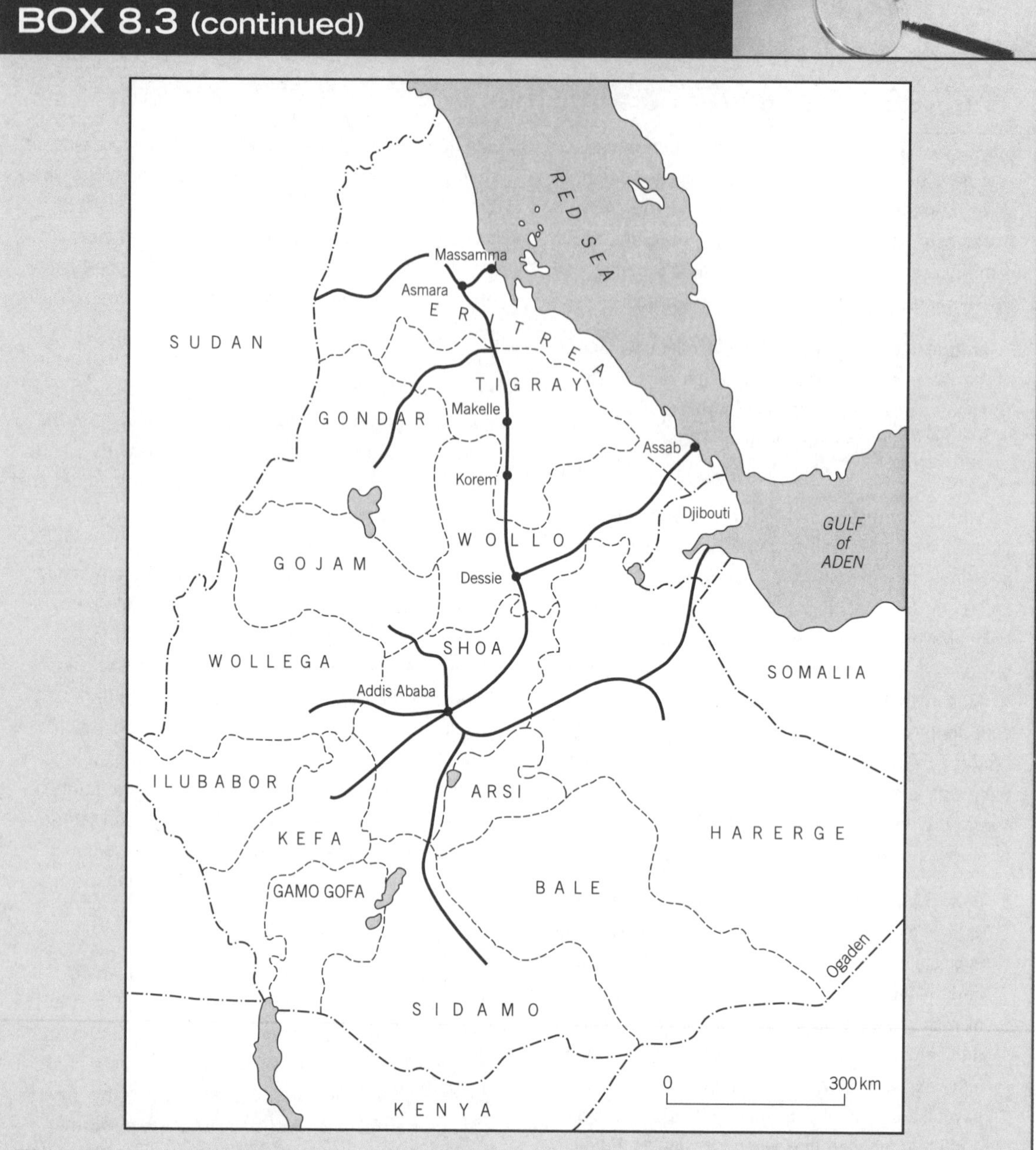

Figure 8.1 Map of Ethiopia in the 1980's
Source: Adapted from Oxfam (1984)

higher per capita military expenditure than any other Black African country and giving Ethiopia the second best equipped army on the continent, after South Africa (Harrison and Palmer, 1986: 94).

Against this background of conflict, poor rains in the early 1980s resulted in hardly any harvest in the northern regions of Wollo, Eastern Gondar and parts of Tigray in 1982 (see Figure 8.1). Refugees abandoned their homes and flooded across the Sudan border to await relief supplies. The main rains in July 1983 also failed, and it was estimated that more than 2 million people in Tigray and Eritrea were seriously affected by the drought and needed emergency assistance. The relief supply situation in both regions was further complicated by the lack of security, due to the separate armed conflicts with the government of Addis Ababa. The liberation fronts accused the government of withholding relief supplies and aid donors were also criticised for channelling all their relief through the government, when many of

BOX 8.3 (continued)

the people at risk were actually in areas not under government control.

The Relief and Rehabilitation Commission (RRC) of the Ethiopian government reported in May 1984 that the official population of 7800 in the Wollo town of Korem had been swelled by some 35,000 displaced people gathering at the feeding centre there, with a further 110,000 people registered to receive emergency food supplies from the RRC store in Korem. The RRC argued that it was too stretched to distribute seeds to ensure people could plant a crop during the July rains, and other seed distribution schemes were ineffective due to the security situation.

One long-term solution to the droughts in northern Ethiopia tried by the government was to move people out of the highlands to new RRC settlements in the south. However, this policy proved unpopular since highland farmers were reluctant to leave their lands and, when they did move away, they would require food and other subsidies for some years before they could regain self-sufficiency. The situation in Ethiopia deteriorated during 1984 as the central highlands and areas near the Kenyan border in the south were also affected by drought. In the eastern province of Harerge, close to the Somali border, the RRC estimated that some 350,000 people, mostly pastoralists, were at risk.

The Ethiopian refugee crisis, which reached its peak in 1984, was caused by an inability to break the constant cycle of drought–famine–emergency feeding, largely due to the continuing conflicts in the northern provinces and the Derg's heavy military expenditure. Both relief aid and long-term development projects were severely hampered by the conflict situation, but there was also a need for the improvement of basic infrastructure, particularly roads, health services and water supplies, and for the introduction of better agricultural techniques and conservation measures such as terracing and reafforestation to reduce people's vulnerability during future droughts.

In April 2000 Ethiopia was once again in a crisis situation, with 8 million people facing food shortages and over half the country's population living on less than US$1 a day. In May 1998 the governments of Ethiopia and Eritrea declared war over a border dispute. As a result, over £600,000 a day was spent by Ethiopia on funding the conflict, having serious effects on the country's economic and social progress and undermining the food security situation (Department for International Development, 2000c).

have been key factors in minimising environmental impact.

Migrants and disease

Population movements can have a significant impact on disease transmission and health. Prothero (1994) has demonstrated the significance of interactions between a variety of diseases and population mobility in tropical Africa, just as the spread of *falciparum* malaria, which is resistant to chloroquine, has been facilitated by movements of people, particularly of refugees, in South and Southeast Asia (Prothero, 1994). In African refugee camps, disease and high death rates have been associated with overcrowding, poor accommodation, inadequate water supply, sanitation and waste disposal, as well as the amount and quality of food available. More than half the deaths in the 'emergency phase' are due to measles, diarrhoeal diseases and acute respiratory infections.

Malaria is a major hazard when refugees from areas of low endemicity are forced to move into areas of high endemicity. Health problems were caused by forced resettlement in Ethiopia and Somalia, where people were moved from the relatively malaria-free Ethiopian plateau above 2000 metres to lower areas in the west and south-west where malaria was endemic. Irrigation projects at altitudes below 2000 metres also extended areas of endemic schistosomiasis (bilharzia). In addition, more contact between Ethiopian pastoralists and agricultural settlers increased pastoralists' risk of infection with schistosomiasis. Migrants also experienced nutritional problems, since the traditional cereal crops of the Ethiopian plateau could not be grown at lower altitudes (Prothero, 1994).

The spread of AIDS and sexually transmitted diseases can also be linked to population mobility (Chapter 5). In Burkina Faso, from where there is much migration to Ivory Coast, AIDS is known as *la maladie (ou la diarrhée) de la Côte d'Ivoire*. In Abidjan, Ivory Coast's largest city, some 25,000 deaths occurred from AIDS-related illnesses between 1986 and 1992. The incidence of AIDS along important national and international routes in Ivory Coast, Mali and The Gambia has been reported as being generally much higher than elsewhere in these countries, and lorry drivers, itinerant traders and prostitutes have higher than average levels of infection (Prothero, 1996). In Uganda, Cliff and Smallman-Raynor (1992) examined a number of possible factors underlying the spread of AIDS, including proximity to major roads and migrant labour. They found that the recruitment and movement of Ugandan soldiers in the 1970s and 1980s, and their contact with prostitutes, were key influences in the diffusion of AIDS.

Communications and transport

In the early twenty-first century we find ourselves in a world where people living and working in rich Northern countries can increasingly perform their business and everyday activities without even leaving their homes, through an ever changing array of technology, including telephone, fax, electronic mail (e-mail) and the World Wide Web (WWW). Virtually instant contact by e-mail is now possible across most of Europe and North America, and the WWW serves the needs of millions by providing vast quantities of information for business, education and entertainment. So-called 'telecommuting', where people work mainly from home, is being actively encouraged by governments and employers in crowded European countries in an attempt to reduce the number of car journeys and the associated pollution.

For many in the privileged North the world is shrinking day by day as innovations come on stream. As we have seen in Chapter 4, the 'globalisation' of a wide range of processes and transactions is a key feature of the last decades of the twentieth century, which is continuing and indeed accelerating in the new millennium. Meanwhile, in West Africa, one of the world's poorest regions, but only six hours' flying time from Europe, millions of people are still without electricity and fresh water supplies; their health and education services are inadequate; and they must work long hours in the fields using low-level technology to produce enough food to satisfy family needs. For many poor people in rural areas of developing countries, their main means of communication, in the absence of television and newspapers, is still by word of mouth; such is the 'digital divide' explored in Chapter 4. However, in some countries it is fair to say that the transistor radio has had a considerable impact in the transmission of knowledge and information.

With the high cost and lack of motorised transport, poor people are forced to cover thousands of miles each year on foot, often carrying heavy loads. A 1980s study in Ghana estimated that rural households typically spent some 4830 hours per year in transport activities, particularly collecting water and fuelwood, with most of this work being done by women (Porter, 1996). In such communities there has generally been little tangible improvement in living standards, and certainly no evidence of a shrinking world (Plates 8.3 and 8.4). It is astoundingly difficult at times to appreciate that people with such contrasting lifestyles actually inhabit the same planet!

It was during the colonial era that the first railways and surfaced roads were constructed in what are now the developing countries (see Chapter 2). In Africa, railways were built to ensure strategic and military control, but more especially to extract raw materials, whether cash crops or mineral resources, and typically linked major source areas with coastal ports. English, French and Portugese colonial powers, far from collaborating in the development of transport infrastructures to 'open up' Africa, were actively competing with each other, such that rail links between neighbouring anglophone and francophone countries never materialised. Despite ambitious plans, single unconnected lines were common within countries, and only states such as Morocco, Nigeria and South Africa can be said to have anything resembling a rail 'network'.

Today, in many African countries, rail transport suffers from a lack of investment and maintenance, and in relative terms is much less important than in the colonial period. For example, the 1146-kilometre line which links Abidjan in Ivory Coast with Ouagadougou in Burkina Faso carried 3 million passengers in 1988, but only 760,000 in 1993, due largely to the poor condition of the rolling stock. Freight tonnage also fell from 800,000 tons in 1980 to 260,000 tons in 1993 (Economist Intelligence Unit, 1996a). In 1995–1996 the Nigerian rail system was effectively closed down, awaiting rehabilitation by a team of Chinese rail engineers. However, the situation was little better in 2002, with virtually no operating train services.

In Latin America, by 1940 the railway systems of Argentina, Brazil and Mexico accounted for 75 per cent of the region's network, and in these three countries there was significantly more interlinkage of lines than in African countries. The railway system in São Paulo,

Plate 8.3 Cyclists in Kunming, southern China
(photo: Tony Binns)

Plate 8.4 Crowded water transport in Georgetown, Guyana
(photo: Rob Potter)

Brazil, was a key element in the state's industrialisation, facilitating the advance of the coffee frontier, increasing exports and the development of engineering enterprises linked to the railways.

Since the Second World War, road transport has increased in importance, and with growing populations and less traffic carried by rail, roads now have to shoulder a much greater burden (Plate 8.5). In many developing countries the road networks are similar to those constructed during the colonial period, although new capital cities such as Abuja (Nigeria), Brasilia (Brazil) and Islamabad (Pakistan) have necessitated further highway construction. In Nigeria, funds generated by the oil boom in the 1970s had a significant effect on the upgrading of the country's road network; but in the 1990s, like other African countries, and

Plate 8.5 Traffic congestion in Bangkok, Thailand
(photo: Mark Edwards, Still Pictures)

despite an injection of monies from the Petroleum Trust Fund for the rehabilitation of some major trunk roads, general road quality has steadily deteriorated. The stringencies imposed by SAPs in countries such as Ghana and Nigeria have had a major impact on road transport, with less road maintenance, fewer vehicles and a severe shortage of spare parts (Porter, 1996).

In tropical regions, road surfaces quickly become pot-holed in the rainy season, and unpaved roads can become completely impassable. Many Third World governments face the difficult dilemma of whether to invest limited funds in building a few all-weather highways to connect the main towns or, alternatively, constructing and upgrading many more kilometres of unpaved rural feeder roads, which could actually improve the lives of a greater proportion of the population in connecting them with clinics, schools and markets. In reality, it is probably most effective to try to achieve a balance between the two strategies.

For many rural producers, their main concern is how to transport their often perishable produce to markets as easily and quickly as possible. Women vegetable farmers in The Gambia, while praising the assistance from the NGO Action Aid in sinking wells and supplying tools and seeds, were concerned about the lack of reliable and refrigerated transport to carry their produce to large urban markets (Binns, personal interview, 1997).

North and South: an interdependent world

We live in an interdependent world where links and relationships have developed over time and space and where flows of commodities and finance reinforce these links. But it is also an unequal world, and many would argue that issues such as trade, aid and debt are crucial in perpetuating inequalities both between and within countries.

An important landmark in considering the ramifications of an 'interdependent' world was the Brandt Commission, established in 1978 under the chairmanship of Willy Brandt, former Chancellor of West Germany, and its influential report, entitled *North–South: A Programme for Survival* (Brandt, 1980). Possibly the most memorable thing about this report was the world map on its cover, across which a black line divided the rich North from the poor South. One of the main themes running throughout the report is the mutual interest of rich and poor countries in a better regulated world economy.

The Brandt Commission covered many issues, but its most important conclusions concerned the international monetary system, the transfer of resources from North to South, better trading opportunities for countries of the South and the nature of aid. A well-argued critique of the International Monetary Fund was presented, with the Commission proposing a system that would place less severe restrictions on borrowing countries and also establish greater stability in exchange rates. Brandt argued for a massive increase in resource transfers between rich and poor countries, with the aim of Northern governments first reaching, and then substantially surpassing, the target of allocating 0.7 per cent of their GNP to aid. A special initiative was proposed to cope with the world's 30 poorest countries, with a programme of long-term and flexible financial and technical assistance. Within poor countries, Brandt recognised the need for social and economic reforms to reduce inequality.

Reviewing the question of food aid, Brandt suggested that the best long-term solution to food shortages is for food production in poor countries to be increased to meet most of their own needs. Meanwhile, external aid should be devoted mainly to improving the capacity for local food production, rather than shipping in food supplies which, whether free or subsidised, compete with local production and may actually discourage it by depressing prices. The report concluded that the governments of poor countries must devote a large part of their development effort to increasing agricultural production.

Brandt was also concerned that rich countries, while reducing trading restrictions among each other, should consider dismantling trade barriers affecting the import of goods from poorer countries. A new set of trade rules and the negotiation of new commodity agreements were advocated, and both rich and poor countries were urged to liberalise their trading policies.

The powerful position of transnational corporations (TNCs) within the world economy was recognised; in 1980 TNCs controlled somewhere between one-quarter and one-third of world production, with just a few corporations controlling production, marketing and processing of important food and mineral commodities. Brandt advocated the establishment of a new mutually agreed 'investment regime' to ensure that host countries, as well as TNCs, benefited adequately from investment, through contractually agreed arrangements covering such aspects as foreign investment, transfer of technology and the repatriation of profits, royalties and dividends. In addition, the Commission favoured the introduction of legislation in each country to regulate the activities of TNCs in matters such as ethical behaviour, disclosure of information, restrictive practices and labour standards.

The Brandt Report received much attention at the time of its publication, and was regarded as visionary, though somewhat unrealistically idealistic, in the light of the strength and entrenched position of TNCs and other key actors on the world stage. In fact, its recommendations fell victim not only to apathy and intransigence, but also to international recession in the early 1980s. Just three years later, a sequel to the report commented:

> **Three years have passed since the publication of the Brandt Commission's Report: *North–South: A Programme for Survival* – years which have brought increasing economic hardship to the industrial countries, and little short of disaster to much of the developing world . . . The Commission offered hope. It expressed the belief that national problems could be solved, but only with a degree of collaboration and wider vision which is still lacking in international affairs . . . The Cancun Summit [October 1981], which brought world leaders together to consider North–South issues, was the first of its kind and was a direct result of the Report. The leaders present felt that their exchanges had been valuable, but while the Summit helped to keep alive the process of global negotiations within the United Nations, it did not make any immediate contribution to resolving the problems of developing countries; nor did it set up any continuing procedure to accelerate negotiations. Now, more than a year later, there is still little sign of action. The North–South dialogue remains much where it was when the Commission reported . . . Meanwhile the world economy continues its dangerous downward slide, and the desperate situation of many developing countries finds no new hope of relief.**
>
> **(Brandt, 1983: 11–12)**

Crisis and commodity dependency

Despite the good intentions of the Brandt Report, it certainly seems that little progress has been made in the world's poorest continent, Africa. An Oxfam report in 1993 presented an extremely depressing view, commenting:

> **Sub-Saharan Africa is on a knife edge. For more than a decade the region has been locked in a downward spiral of economic and social decline. That decline, unlike the tragedies of famine and drought,**

> which dominate news coverage of the region, has been largely invisible to the outside world. Yet it has spread human suffering and misery on an unprecedented scale. Hard-won gains in health and education have been reversed; living standards, already among the lowest in the world, have fallen; hunger is on the increase. And the tragedy is set to deepen. On current trends, the ranks of the 218 million Africans already living in poverty will increase to 300 million – equal to half the region's population – by the end of the decade (2000).
>
> (Oxfam, 1993: v)

So what has gone wrong? Earlier in this chapter the case of the Ivory Coast coffee producer was examined and the widespread implications of changing coffee prices for both national economies and poor rural households considered. In fact, it is the long depression in world commodity markets that has had such a profound impact on African economies, and therefore on the quality of life of Africa's people. The situation has been particularly serious where countries are heavily dependent for the generation of foreign exchange on a limited range of primary agricultural and mineral commodities such as coffee, cocoa, cotton and copper. Between 1992 and 1997 Uganda gained over 85 per cent of its export earnings from primary commodities, with coffee being the largest contributor at over 50 per cent (Table 8.2). In certain countries a single product dominates export earnings; tobacco provided 59.7 per cent of Malawi's total export earnings, whereas cotton generated 47.7 per cent of Mali's earnings.

The prices of such primary commodities fell dramatically during the 1980s, but import costs continued to rise, leading to a fall of about 50 per cent in the purchasing power of sub-Saharan Africa's exports in the decade from the early 1980s. The situation in some countries was much worse than the average picture:

> In 1986, coffee provided Uganda with US$365 million in foreign exchange earnings and financed about 70 per cent of its imports. By 1991 it yielded only US$115 million, and financed less than a quarter of imports . . . Overall, the slump in commodity prices cost Africa US$50 billion in lost earnings between 1986 and 1990 – more than twice the amount the region receives in aid.
>
> (Oxfam, 1993: 7)

The collapse in commodity prices and the deteriorating terms of trade, together with rising debt-service payments and a reduction in foreign investment, have made it even more difficult for many developing countries to purchase imports. Furthermore, these trends have seriously undermined structural adjustment programmes sponsored by the World Bank and the International Monetary Fund, which were so dependent on increasing exports. SAPs have probably exacerbated the situation by encouraging countries with a narrow range of exports to increase their production,

Table 8.2 African countries dependent on a single primary commodity for export earnings (annual average of exports, in US$, 1992–1997

	50% or more of export earnings	20–49% of export earnings	10–19% of export earnings
Crude petroleum	Angola, Rep. of Congo, Gabon, Nigeria	Cameroon, Equatorial Guinea	Algeria
Natural gas		Algeria	
Iron ore		Mauritania	
Copper	Zambia		Dem. Rep. of Congo
Gold		Ghana, South Africa	Mali, Zimbabwe
Timber (African hardwood)		Equatorial Guinea	Central African Rep., Gabon, Ghana, Swaziland
Cotton		Benin, Chad, Mali, Sudan	Burkina Faso
Tobacco	Malawi	Zimbabwe	
Arabica coffee	Burundi, Ethiopia	Rwanda	
Robusta coffee	Uganda		Cameroon
Cocoa	Sao Tome and Principe	Ivory Coast, Ghana	Cameroon
Tea			Kenya, Rwanda
Sugar		Mauritius	Swaziland

Source: Cashin *et al.* (1999)

depending on markets which are already saturated and have fixed levels of demand. This can be seen in the case of increased cocoa exports from Africa, the world's major producing region, which led to the collapse in world prices (Oxfam, 1993). Oxfam suggests that Africa's trading prospects can only be improved by establishing an African Diversification Fund to promote the increased processing of raw commodities in African countries, plus reducing protectionist barriers against Africa's exports as well as ending the subsidised disposal of agricultural surpluses on world and regional markets (Oxfam, 1993).

BOX 8.4

A stronger voice for Africa

Africa, the poorest and arguably the world's most marginalised continent, desperately needs a stronger voice which is listened to more seriously by the world community. In an attempt to achieve this, in July 2002, African leaders met in Durban, South Africa, for the Inaugural Summit of the African Union (AU) which replaced the frequently ineffective Organisation of African Unity (OAU). The OAU was formed in 1963 'essentially as a vehicle for pan-African unity and the coordination of the struggle against colonialism' (*Sunday Times,* 2002: 17). It has been suggested that, 'A major weakness . . . [of the OAU] is that member states have not found it easy to delegate their individual or collective powers to the OAU and the organization's role in the international system has therefore been limited' (Binns, 1994a: 165).

All countries on the continent are members of the new AU, with the exception of Morocco, which withdrew from the OAU in 1982 over the OAU's recognition of the Saharawi Arab Democratic Republic, following Morocco's invasion of Western Sahara in 1976. The aims and objectives of the African Union go well beyond those of the OAU and include:

> **acceleration of the political and socioeconomic integration of the continent; promoting democratic principles and institutions, popular participation and good governance; and, establishing the necessary conditions which enable the continent to play its rightful role in the global economy and in international negotiations.**
>
> **(African Union, 2002: 3)**

The AU aims to move away from the overly state-centric character of the OAU and the lack of civil participation, and is loosely based on the model offered by the European Union.

Much of the drive behind the establishment of the African Union and the New Partnership for Africa's Development (NEPAD) since 1999 has come from Thabo Mbeki, President of South Africa, supported by Nigeria's Olusegun Obasanjo, Senegal's Abdoulaye Wade and Algeria's Abdelaziz Bouteflika. Mbeki raised the issue of the motivation for an 'African Renaissance' when he was Vice-President in May 1996. A year later, in a landmark speech delivered to the US Corporate Council on Africa, Mbeki argued that an African Renaissance was a real possibility in which the current period of crisis might be seen as a time of opportunity, 'which the New Africa must seize for its own advantage' (Akosah-Sarpong, 1998, quoted in Lester *et al.*, 2000: 281). Mbeki envisaged the African Renaissance leading eventually to the emancipation of women and the mobilisation of youth, and he expressed 'hope for [achieving] sustainable development, together with the broadening, deepening and sustenance of democracy, with decision-making "trickling down" to the level of the actual people affected' (Lester *et al.*, 2000: 281).

The NEPAD initiative has developed from Mbeki's call for an African Renaissance, and is envisaged as a long-term vision of an African-owned and African-led development programme. NEPAD is, basically, an appeal to the West's conscience for help, although the initial presentation of NEPAD to the G8 Summit in June 2002 received a somewhat lukewarm response.

However, British Prime Minister Tony Blair publicly spoke about the need to move Africa higher up the world political and economic agenda. At the Labour Party Conference in September 2001 Blair said,

> **The state of Africa is a scar on the conscience of the world. But if the world as a community focused on it, we could heal it. And if we don't, it will become deeper and angrier.**
>
> **(The Independent, 2002: 10)**

▶

BOX 8.4 (continued)

The proposal from Africa's leaders asserts that NEPAD

> **is a pledge by African leaders, based on a common vision and a firm and shared conviction, that they have a pressing duty to eradicate poverty and to place their countries, both individually and collectively, on a path of sustainable growth and development and, at the same time, to participate actively in the world economy and body politic.**
>
> **(NEPAD, 2001: 1)**

The proposal suggests that the continued marginalisation of Africa, 'from the globalisation process and the social exclusion of the vast majority of its peoples constitute a serious threat to global stability' (NEPAD, 2001: 1).

There is much in the 70-page NEPAD document about Africa's historical legacies and the poor living standards of many Africans today. It is suggested that Africa's impoverishment is due to, 'the legacy of colonialism, the Cold War, the workings of the international economic system and the inadequacies of, and shortcomings in, the policies pursued by many countries in the post-independence era' (NEPAD, 2001: 1).

NEPAD argues that there is now greater democracy in Africa, and takes support from the UN Millennium Declaration of September 2000, which confirmed the global community's readiness to support Africa's efforts to address the continent's underdevelopment and marginalisation, and its commitment to enhancing resource flows to Africa by improving aid, trade, debt and private capital relationships between Africa and the rest of the world.

In essence, NEPAD calls for a 'new global partnership' and the importance of Africa negotiating a new relationship with its development partners. Such a relationship, it is argued, should lead to conflict prevention, debt reduction, increased development assistance to meet the 0.7 per cent of GDP target, progress in education and health with better access to inexpensive drugs, technical support and private sector investment. This is an ambitious agenda and the Heads of State present at the launch of the African Union in Durban reinforced a shared commitment to NEPAD, saying

> **We do not underestimate the challenges involved in achieving NEPAD's objectives, but we share a common resolution to work together even more closely in order to end poverty on the continent and to restore Africa to a place of dignity in the family of nations.**
>
> **(NEPAD, 2002: 10)**

Cynically speaking, both NEPAD and the African Union might be accused of generating yet more political rhetoric about a continent which is a lost cause. However, being more optimistic, there is also a real sense that a group of relatively new African leaders, notably Thabo Mbeki, is genuinely committed to achieving a fresh start for Africa and, rather than merely talking among themselves, they have engaged at the highest level with the leaders of the world's most powerful countries in a carefully planned dialogue.

One outcome of this dialogue was the launch In early 2004 by UK Prime Minister Tony Blair of the 'Commission for Africa', which brought together a task force of 17 people, including politicians, business people and pop star Bob Geldof, to define the challenges facing Africa, and to provide clear recommendations on how to support the changes needed to reduce poverty. The Commission reported in March 2005 and its key recommendations were:

- Building capacity in Africa, with better education systems and vocational training;
- Improving accountability in government and management, with greater transparency and less corruption;
- Building capacity to prevent and manage conflict;
- Investing in people, reducing poverty and rebuilding health and education systems;
- Encouraging entrepreneurship and investment in infrastructure, agriculture, small enterprises, women and young people;
- Improving Africa's capacity to trade by reducing tariffs and other non-tariff barriers to African products;
- Supporting an additional US$25 billion per year in aid to Africa, to be implemented by 2010.

(Commission for Africa, 2005)

Following the Commission's report, Africa was put at the top of the agenda at the G8 Summit in July 2005 held in Gleneagles, Scotland. The Summit agreed to increase aid to the developing world by $50 billion

BOX 8.4 (continued)

and to write off the debts of Africa's 18 poorest countries, although African nations had called for all African debts to be cancelled.

Since the Commission for Africa and the Gleneagles Summit, Africa has been changing rapidly, with economic growth fuelled by massive investment from China and India. However, growth is not yet translating into significant reductions in poverty. Furthermore, there is some doubt that the target of US$25 billion aid to Africa will be achieved by 2010, and aid needs to be delivered in such a way that it does not undermine African institutions. Perhaps most important is that Africa needs to be treated as an integral part of the global community. With increasing global investment in the continent, it should be recognised that if things go wrong in Africa there could be repercussions throughout the world.

World trade: the changing scene

There were major changes during the twentieth century in the geography of international trade. At the beginning of the century, Europe and the USA dominated the world scene. The European powers relied on their colonies in the developing world, but also places such as Australia, Canada and New Zealand, to produce raw materials to supply growing industries at home, industries that produced manufactured goods which could then be traded for more raw materials. Strong trading links still remain between many former colonies and their former European masters. Jamaica still exports most of its bananas to the United Kingdom; and in Africa there is still much trade between francophone countries and France and between the former British colonies and the United Kingdom. For example, some 30 per cent of francophone Mali's exports in 2000 comprised raw cotton, chiefly destined for France, its major trading partner (Economist Intelligence Unit, 2002a). Meanwhile, some 25 per cent of anglophone Kenya's exports to non-African countries were destined for the United Kingdom; and the United Kingdom supplied 15 per cent of the country's imports (Economist Intelligence Unit, 2001). The association between colonialism and export economies was emphasised by the concentration of the large-scale export trade in the hands of a few large, mostly European firms and by transnational corporations (see also Chapter 2).

Although there have been colonial links between Western Europe and Pacific-Asia – the British colony of Hong Kong was only returned to China in 1997 – international trade in the Pacific region is dominated by Japan and the USA. It has been suggested that 'the West Europeans withdrew from Pacific-Asia in the post-war decades not just politically and militarily, but also economically' (Shibusawa *et al.*, 1992: 30). The dominance of Japan and the USA in the Pacific-Asian region is reflected in the fact that the region's share of US imports was 37 per cent in 1988, whereas Japan's overall trade surplus with the eight major East Asian economies grew from US$1.2 billion in 1980 to US$20.4 billion in 1989 (Shibusawa *et al.*, 1992: 12).

GATT and the WTO

The General Agreement on Tariffs and Trade (GATT), established in 1947 at a time of world reconstruction, was designed to bring some order to world trade and prevent the instability of the inter-war years, at the same time advocating the pursuit of free-trade policies (see Chapter 7). The reduction of tariffs, prohibition of quantitative restrictions and other non-tariff barriers to trade, and the elimination of trade discrimination, were the main objectives of GATT. The Uruguay round of GATT negotiations began in September 1986 and only concluded in April 1994, after which GATT was replaced by the World Trade Organization (WTO). The talks focused more on debates between Europe and the USA on agricultural subsidies, whereas issues of greater relevance to poor countries, such as gaining better access to developed world markets, were sadly rather neglected.

The expansion of commerce through the deregulation of markets is the main aim of the WTO, and trade liberalisation, in the shape of measures such as removing tariff barriers, quotas, price supports and subsidies, is also increasingly central to economic policy in developing countries. The WTO has come under much attack in its relatively short history. As Watkins observes, 'issues of sustainable resource management, the regulation of commodity markets, and poverty reduction strategies, are conspicuous by their absence from the international trade agenda' (Watkins, 1995: 32). Others are critical of the way that at the WTO's ministerial meetings rich countries tend to lobby hard to shape the rules in their favour, and because poorer countries are less able to field delegates from powerful large companies they suffer as a result. Christian Aid is critical of the WTO agreement on investment where

> **developing countries are often prevented from favouring domestic investors over foreign ones, [and] . . . developing countries are also unable to provide short-term subsidies to help their agriculture and industry become competitive, again because of WTO rules that forbid favourable treatment to domestic over foreign firms.**
>
> (Curtis, 2001a: 12)

WTO rules have also changed the way that large multinational companies are able to tap into markets in poor countries, for example in agricultural trade and in services, and through foreign investment and securing patent rights to natural resources (Box 8.5). Curtis argues that new trade rules should be formulated which target the eradication of poverty and cover the activities of powerful multinational companies, as well as of governments. These rules, he suggests, should be decided democratically and carefully monitored (Curtis, 2001b).

In a 2002 report, Oxfam, a tireless campaigner for fair trade, stated that

> **World trade has the potential to act as a powerful motor for the reduction of poverty, as well as for economic growth, but that potential is being lost. The problem is not that international trade is inherently opposed to the needs and interests of the poor, but that the rules that govern it are rigged in favour of the rich.**
>
> (Oxfam, 2002: 3)

Like Christian Aid, Oxfam is keen for poor countries to have a stronger voice in the WTO and is particularly critical of such issues as the agreement on Trade-Related Aspects of Intellectual-Property Rights (TRIPs). It is suggested that

> **More stringent protection for patents will increase the costs of technology transfer. Developing countries will lose approximately US$40 billion a year in the form of increased licence payments to Northern-based TNCs, with the USA capturing around one-half of the total. Behind the complex**

BOX 8.5

Christian Aid's 'Seven deadly WTO rules'

Christian Aid argues that seven specific trade rules have a significant impact on poor countries:

1. *Rule one:* *WTO limits protection against cheap food imports*
   Developing countries are restricted from intervening in order to raise adequate barriers against cheap food imports, while export subsidies by rich countries are allowed to persist.
   The outcome: *a flood of food imports into developing countries which undermines or threatens to undermine the livelihood of many poor people.*
2. *Rule two:* *WTO limits government regulation of services*
   Countries which agree to sign up must open up their services sectors to foreign suppliers by abolishing restrictions on access to those markets.
   The outcome: *the renegotiation of the WTO's services agreement may result in health, education and water services being run and controlled by profit-driven foreign corporations.*
3. *Rule three:* *WTO limits regulation of foreign investment*
   The WTO's investment agreement bans policies and regulations favouring the use of domestic over foreign products.
   The outcome: *poor countries are denied some important ways of supporting the development of viable local industries over foreign producers. A new investment agreement would further strengthen this rule.*

BOX 8.5 (continued)

4. *Rule four:* *WTO limits use of agricultural subsidies*
Developing countries are limited in their freedom to increase subsidies to agriculture. Some (the non-least developed countries) are required to reduce them. Particular types of subsidy are banned altogether. Meanwhile the EU and US are still permitted to spend huge sums on agricultural subsidies themselves.
The outcome: *poor people's food security is being undermined by restrictions on subsidies which deny poor countries an important tool in development.*

5. *Rule five:* *WTO puts limits on industrial subsidies*
Governments are prevented from using industrial subsidies to promote the manufacture of domestic products over imported alternatives. Some subsidies of special use to rich countries are permitted.
The outcome: *poor countries' industrial development is being hampered by taking away a critical policy tool to help develop their own industrial sector.*

6. *Rule six:* *WTO blocks exports from developing countries*
Rich countries can retain high import barriers or other restrictions against key exports from developing countries.
The outcome: *poor countries lose much needed export revenues and the economic growth rates of affected poor economies will suffer potential losses.*

7. *Rule seven:* *WTO gives business rights over knowledge and natural resources*
This WTO rule requires countries to introduce effective patenting laws, including plant varieties and seeds, which can give TNCs rights over those products for 20 years. This in effect legalises biopiracy of natural resources and knowledge.
The outcome: *poor people's food and health security can be threatened if TNCs are successful in securing monopoly control over knowledge and natural resources.*

Source: Curtis (2001b)

> arguments about intellectual-property rights, the TRIPs agreement is an act of institutionalized fraud, sanctioned by WTO rules.
>
> (Oxfam, 2002: 14)

Oxfam is particularly concerned about the effects of reinforced patent protection on the costs of medicines in poor countries. The WTO, Oxfam argues, 'is old before its time, [and] . . . behind the façade of a "membership-driven" organisation is a governance system based on a dictatorship of wealth' (Oxfam, 2002: 15). More specifically, Oxfam advocates a number of reforms to the WTO, notably: an end to the universally applied intellectual-property blueprint, so that poor countries can maintain shorter and more flexible systems of intellectual-property protection; a commitment to put public health priorities before the claims of patent holders; a prohibition on patent protection for genetic resources for food and agriculture and the ability of poor countries to develop more appropriate forms of plant-variety protection and protect farmers' rights to save, sell and exchange seeds; to prioritise development objectives and strengthen national sovereignty; to strengthen WTO provisions for the 'special and different treatment' of poor countries; and to remove restrictions on governments to regulate foreign investment and protect their infant industries (Oxfam, 2002: 14–15). Ransom suggests that, 'the WTO must shrink, divest itself of "trade related" issues like services, patents and investment, open itself up to democratic control and close itself off from corporate manipulation' (Ransom, 2001: 28).

The WTO refutes the suggestion that it is undemocratic, saying that decisions are usually made by consensus with every country having a voice, and that WTO trade rules were in any case ratified in members' parliaments. Furthermore, it asserts that freer trade creates more jobs than are lost and that, 'while about 1.5 billion people are still in poverty, trade liberalization since World War II has contributed to lifting an estimated 3 billion people out of poverty' (World Trade Organization, 2000: 239). Adding to the debate,

Clare Short, then UK Secretary of State for International Development, in a speech to the WTO meeting in Seattle in November 1999 commented:

> **those who make blanket criticisms of the WTO are working against, not for, the interests of the poor and the powerless. International trade can be unfair and exploitative. The strong can deceive and defraud the weak. That is precisely why we need an institution like the WTO which is membership-based and rules-based – to prevent fraud, monopoly, predatory pricing and other abuses. Just as we need rules on these issues at the national level, so we need them at the international level.**
>
> **(Short, 2000: 11)**

However, the influence of countries such as the USA, the European Union and Japan within the WTO was clearly seen in the collapse of the WTO Doha Development Trade Round in July 2006 (Key idea box). The failure of over five years of trading negotiations was largely blamed on the USA, because it felt that developing countries would not open their markets in the same way that the USA was being asked to open its markets, so it saw no point in continuing the talks. It is a fact of life that in the case of one commodity, cotton, a kilo of cotton produced in the USA costs more than 3.5 times to produce as the same amount in the West African state of Burkina Faso. If EU and US cotton subsidies were removed, cotton exports from sub-Saharan Africa could increase by up to 75 per cent.

Key idea

The collapse of the 'Doha Round' trade talks

The so-called 'Doha Round' of WTO trade talks was launched in Doha in the Middle Eastern state of Qatar in November 2001 and was due to be concluded by December 2006. The talks aimed to prioritise the development of poor countries and were therefore referred to as the 'development' round. Subsequent meetings were held in Cancun, Mexico in 2003, Geneva (2004), Paris (2005), Hong Kong (2005) and Geneva (2006). It was in Geneva in July 2006 that the talks failed to reach an agreement about reducing farm subsidies and lowering import taxes. A successful outcome of the Doha Round seemed unlikely as the broad trade authority granted under the US Trade Act of 2002 to President George Bush expired in June 2007. A further meeting held in Potsdam in June 2007 failed to break through the impasse between the US, the EU, India and Brazil in relation to opening up agricultural and industrial markets and how to reduce rich nation farm subsidies.

As India's commerce minister, Kamil Nath, commented following the Geneva conference in 2006,

> **This is a Development Round, completing it is extremely important, but equally important is the content of the Round. The content has to demonstrate new opportunities for developing countries, primarily market access of developing countries into markets of developed countries. This Round is not for perpetuating the flaws in global trade especially in agriculture, it is not to open markets in developing countries in order for developed countries to have access for their subsidized products to developing countries. We say the Round should correct the structural flaws and distortions in the system, and there should be fair trade, not only free trade. The USA say 'we want market access and only if we get it the way we want it can we correct the structural flaws.' There is no equity in that argument. (Kamal Nath, quoted at http://www.globalissues.org/TradeRelated/FreeTrade/dohacollapse.asp. 28 July 2006)**

Fair and ethical trade

The issues of fair and ethical trade have received much attention in recent years and there have been serious attempts, particularly by charities and pressure groups, to influence the working conditions and remuneration of workers in poor countries (Hughes, 2001). For example, in September 1999 *The Independent*, a UK-based newspaper, launched its 'Global Sweatshop' campaign to increase awareness about the sale of 'sweatshop' goods in British high-street stores. The newspaper revealed that 13,000 workers in 32 garment factories on the US-administered Northern Mariana Islands in the western Pacific were producing 'designer' shirts and other garments for well-known retailers in the USA under appalling working conditions, long hours of work and poor wages (The Independent, 1999). Another report by the charity Christian Aid revealed that plantation workers growing bananas in Costa Rica commonly received only 5.5 per cent of the average price of a banana, while on tea plantations workers' wages accounted for just 7 per cent of the final price (Christian Aid, 1996).

There has been a strong call for retailers to report annually on their codes of conduct, to make compulsory the country-of-origin labelling on all imported clothes, and to introduce an 'ethical trade kitemark' funded by retailers but independently monitored, indicating acceptable standards of workers' pay and working conditions (The Independent, 1999).

Such fair trade 'kitemarking' is more advanced in relation to food crops, where a distinctive label is used to denote commodities which have been produced by workers receiving a fair wage and working under acceptable conditions. Producers are encouraged to

> **continuously improve working conditions and product quality, to increase the environmental stability of their activities and to invest in the development of their organizations and the welfare of their producers/workers.**
>
> **(Fairtrade Foundation, 2002: 1)**

The first Fairtrade Label was created in the Netherlands in 1988, and by 2002 there were 17 Fairtrade Labelling Organizations, including related initiatives in the UK, USA, Japan, Netherlands, Germany and Switzerland. The UK Fairtrade Foundation was established in 1992, with support from development agencies such as Christian Aid, CAFOD, New Consumer, Oxfam, Traidcraft Exchange and the World Development Movement. In 1997 Fairtrade Labelling Organizations International was established, which is an association of 20 national labelling initiatives that promote and market the Fairtrade label in their countries. FLO is the worldwide standard setting and certification body for labelled Fairtrade (see www.fairtrade.net).

In 2001 there was a 21 per cent increase in UK sales of Fairtrade marked commodities, such as coffee, tea, chocolate, honey, sugar, orange juice, bananas and mangoes. In 2005 the entire range of UK high street retailer Marks & Spencer's coffee and tea, totalling 38 lines, switched to Fairtrade in a move which is estimated to increase the value of all Fairtrade instant and ground coffee sold in UK supermarkets by 18 per cent, and that of Fairtrade tea by approximately 30 per cent. Some 350 certified Fairtrade producer groups operate in 36 producer countries selling to hundreds of Fairtrade registered importers, licensees and retailers in 17 countries. Fairtrade estimates that some 500,000 farmers and workers in Latin America, Caribbean, Africa and Asia benefit from the better deal that the Fairtrade Mark guarantees (Fairtrade Foundation, 2002).

Recent trends in world trade

Two important trends in world trade have developed during the 1980s and 1990s. First, with the end of the Cold War and the collapse of Soviet communism in 1991, trade between Eastern Europe and the Western capitalist countries has accelerated. Western investment is playing an important role in the restructuring of the former Communist bloc countries, and some would argue that these countries are receiving both the attention and the funding which the world's poorest countries urgently require. In the early 1990s, over 400 agreements were signed between Western businesses and bodies in the newly democratised Czechoslovakia, Poland and Hungary, much of this investment going to the major cities and heavily industrialised regions.

A second and very significant development in world trade has been the increasing power and participation of the export-oriented newly industrialising countries (NICs), such as Hong Kong, Malaysia, South Korea, Singapore and Taiwan, in addition to the already powerful Japan (see also Chapter 4). Industrial employment in South Korea increased by 77 per cent between 1974 and 1983, and by 75 per cent in Malaysia during the same period.

> **Trade, trade and more trade was what propelled the so-called Pacific Rim states out of agrarian destitution or post-World War II destruction and decline into world economic prominence.**
>
> **(Aikman, 1986: 10)**

The growth of international trade in the countries of the western Pacific Rim has been remarkable (Figure 8.2). From 1982 to 1988 the growth in export volume from East Asia (even excluding Japan) was over 12 per cent per annum, a rate almost double that of South Asia, three times that of the Middle East, North Africa and Latin America, and about six times higher than in sub-Saharan Africa (Hodder, 1992: 67). The western Pacific Rim's share of world trade increased

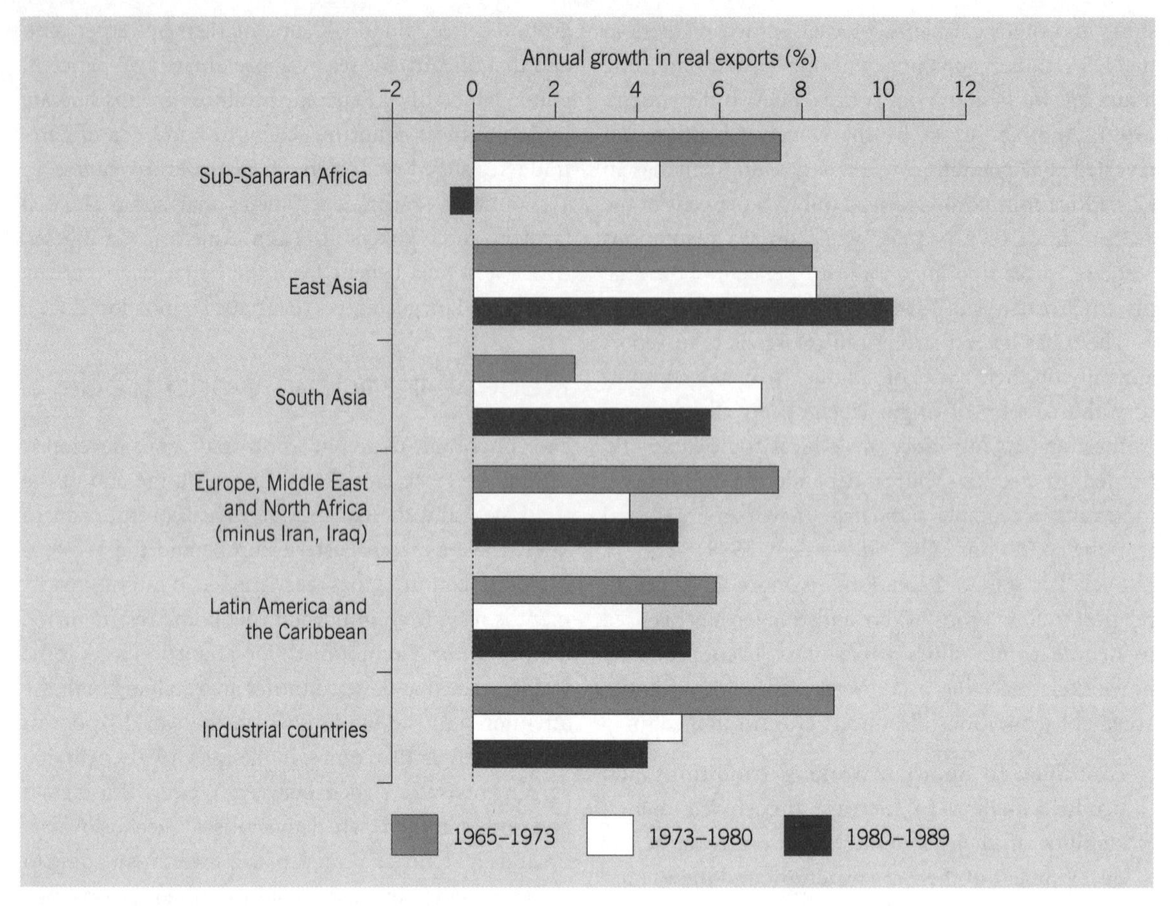

Figure 8.2 Estimated percentage annual growth in real exports for selected regions, 1965–1989

Source: Adapted from Hodder (1992)

from 14.3 per cent in 1971 to 22.8 per cent in 1984. This reflects the success of export-led growth strategies and the readiness of the peoples of these countries to undertake programmes of rapid structural adjustment, the development of new products and the exploitation of new markets (Dicken, 1993).

However, in 1997 and 1998 many of the Asian NICs suffered a serious economic downturn which had widespread implications both domestically and internationally. In South Korea, for example, the crisis was precipitated by a growing concern in 1997 about heavily indebted and overextended companies with weak profitability that were borrowing heavily to finance long-term investment. Banks were also borrowing heavily, such that in November 1997 the USA made it difficult for three large South Korean banks to continue borrowing from abroad to lend at home. Despite the South Korean government trying to restore foreign and domestic confidence in the country, the problems continued and the stock market fell sharply, while external liabilities were estimated to be as high as US$200 billion. Negotiations took place with the IMF and a rescue package of US$20 billion was announced, with a further US$40 billion coming from the World Bank, the Asian Development Bank, Japan, the USA and other lenders. World reaction to this was mixed. The South Korean currency was allowed to float freely from mid December 1997 and some confidence was only restored when the IMF brought forward part of its assistance package. However, South Korea's economy bounced back in 1999–2000, such that GDP grew by 10.9 per cent in 1999 and 8.6 per cent in 2000, representing the fastest growth rates since 1987. In the following year, despite recession in South Korea's important export markets of USA and Japan, economic performance at home was strong. GDP in 2001 grew by 3.3 per cent, while Taiwan and Singapore experienced sharp falls in output (Economist Intelligence Unit, 2002b).

The western Pacific Rim countries conduct most trade with the USA and Canada, representing some

Plate 8.6 Container port, Kowloon, Hong Kong
(photo: Tony Binns)

15 per cent of world trade. International trade as a whole is dominated by Japan and the USA, with Japan supplying 30 per cent of the US car market. In most of Southeast Asia, except Singapore, there is still evidence of the colonial pattern of trade, in which countries export raw materials and primary products and import most of their manufactured goods. The reverse is true in Japan, where 55 per cent of its imports are crude materials and primary products and 99 per cent of its exports are manufactured goods. In the case of Hong Kong and South Korea, a clear majority of both imports and exports are manufactured goods (Hodder, 1992; Plate 8.6).

There is considerable potential for tapping into the vast Chinese market, and Southeast Asia's 'overseas Chinese' have established a close network of business links with the Chinese in Taiwan, Singapore and Hong Kong, and they are well placed to take full advantage of potential business opportunities. Hong Kong has provided about 80 per cent of investment in southern China's Guangdong Province, which has experienced massive industrial development (Box 8.6), while Taiwan has invested heavily in Fujian Province, opposite Taiwan on China's eastern coast.

Whereas in the 1960s and 1970s the USA was the dominant world trading power, in the 1990s a 'multipolar' system developed, with power concentrated in three blocs: North America, Europe and, increasingly, the Pacific Rim. Trade relations are being transformed by such features as increased flows of foreign investment, the globalisation of production under the auspices of transnational corporations and trade liberalisation in developing countries. Institutional structures have also changed, such as the customs union between Brazil, Argentina and Uruguay created in 1995, and the Asia-Pacific Economic Cooperation (APEC), established in 1993, which could lead to links between the Pacific Rim states (Japan, China, South Korea, Malaysia, Philippines and Thailand) and the North American Free Trade Agreement (NAFTA), comprising the USA, Mexico and Canada (Watkins, 1995: 113).

Transnational corporations

It is often assumed that trade is an activity which is conducted between countries, each of which controls its own economic destiny. However, as noted in Chapter 4, world trade flows are in reality dominated by incredibly powerful transnational corporations. Dicken has examined the nature of TNCs and concludes that

> because these big companies are often based in a single country, though they operate in at least two countries, including the firm's home country, they are now usually termed 'transnational' rather than 'multinational'. All multinational corporations are transnational corporations, but not all transnational corporations are multinational corporations.
>
> (Dicken, 1992: 47)

The role of transnational corporations in world trade should not be underestimated, indeed they are vital actors in the global economic and trading system.

BOX 8.6

Migration and economic development in China

The reforms introduced since 1979 under Deng Xiaoping's rule have had a major impact on population movements within China. Although migration did occur before 1979, it was generally involuntary and much more centrally controlled than in recent years. Resettlement programmes took place, most notably from the densely populated provinces of the east to the sparsely populated western regions. In the case of Xinjiang Province in the far north-west of China, there was a deliberate government policy to change its ethnic balance. Xinjiang has many minority groups, but in-migration of Han Chinese from the east increased the proportion of Han within the population from under 10 per cent in 1949 to 40 per cent in 1982. Furthermore, many Han were appointed to important administrative and political posts in Xinjiang, so Beijing is able to maintain strong central control over the region.

Another wave of migrations occurred during the period of the first five-year plan (1953–1957), when millions of peasants moved into towns looking for jobs during a phase of intensive reconstruction and industrialisation after the Second World War and the subsequent civil war. As a result, between 1949 and 1957 China's urban population increased by 60 per cent, whereas the rural population grew by only 13 per cent (Jowett, 1990).

During the Great Leap Forward (1958–1960), despite a strong emphasis on rural industrialisation, this initiative was thwarted by widespread famine in 1959–1961, leading to 'surplus' population being moved back into the countryside. Later, during the Cultural Revolution of the 1960s and 1970s, the Chinese government imposed strict controls on rural–urban migration, and urban youths were sent to work in the countryside in the so-called rustication programme. During this period the country's largest cities scarcely grew through migration; for example, there was a relatively small net gain of 350,000 migrants in Tianjin during the 30 years between 1950 and 1980, and China's largest city, Shanghai, actually experienced a net loss of 1 million people through out-migration during the same period.

The reforms of 1979, however, had a major effect on population movements within China. Probably the most important reform was the replacement of people's communes with individual farming units under the 'household responsibility system'. This new approach to agricultural production led to the collapse of collective farming and gave rise to abundant rural surplus labour, due to a substantial increase in the efficiency and productivity of the agricultural sector. Controls on internal migration were relaxed, such that surplus rural labour could move freely without the need for permanent registration. The late 1970s and early 1980s were characterised by increasing migration, and it was estimated that in Shanghai alone there was a net in-migration in 1979 of 264,800 (Jowett, 1990).

A second significant element in the reforms of 1979 was the establishment of the first four special economic zones (SEZs). Located on China's south-eastern coast, close to Hong Kong and Taiwan, and with a series of tax inducements, these areas were designed to attract foreign investment and 'joint ventures' between Chinese and overseas companies, to manufacture export goods which would generate foreign exchange. The SEZs were also seen as 'social and economic laboratories, in which foreign technological and managerial skills might be observed and adopted' (Phillips and Yeh, 1990: 236).

Two of the first SEZs, Zhuhai and Shenzhen, are located in the Pearl River Delta zone of Guangdong Province. Shenzhen, the largest SEZ (327.5 square kilometres) is situated adjacent to the Hong Kong border and has grown spectacularly during the 1980s and 1990s, from being just a small rural town to an ultramodern city with over a million people, less than an hour's travel from the 'throbbing heart of capitalism' in Kowloon and Hong Kong Island. China's southern coast has become an innovative capitalist periphery for the new international division of labour and capital flowing into the region from all over the world (see Figures 8.3 and 8.4).

Female migration to southern China

These two factors – the changes in the rural production system and the creation of the SEZs – have provided an important stimulus for the massive increase in migration which China has experienced in

BOX 8.6 (continued)

Figure 8.3 Province-level administrative divisions of China (Xinjiang, Tibet and Inner Mongolia are autonomous regions)
Source: Adapted from So (1997)

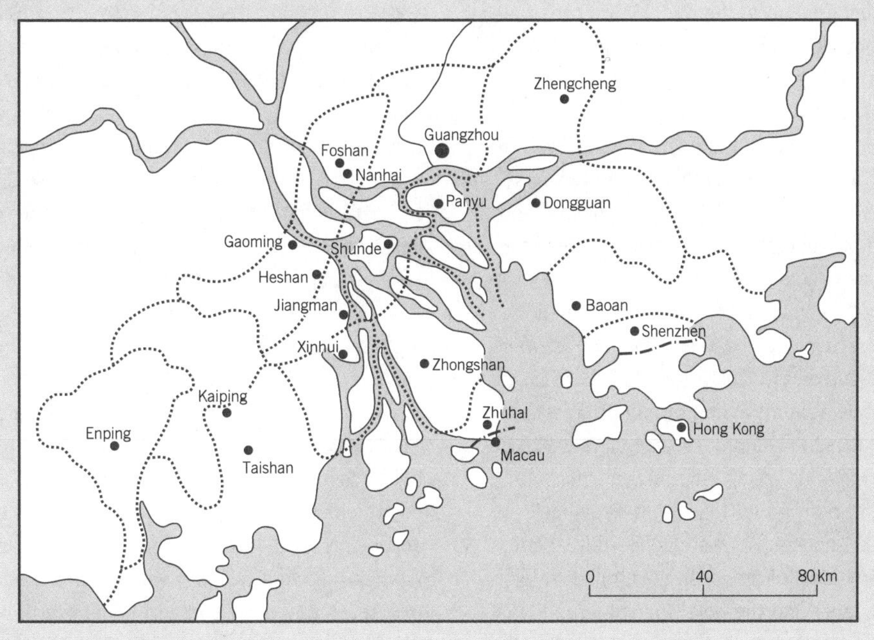

Figure 8.4 The major cities in Guangdong Province
Source: Adapted from So (1997)

BOX 8.6 (continued)

recent years. The 1990 population census revealed there were 525 million population movements during 1982–1985, 740 million during 1985–1987 and 660 million during 1987–1990. Whereas in many developing countries migration is usually dominated by males, China's 1990 census indicates that 56 per cent of migrants were male and 44 per cent female. However, if intraprovincial migration alone is considered, females were in the majority at 66 per cent, due largely to migration for marriage, since brides commonly move to their husband's home. Yet aggregate data mask striking differences at the local level and, in fact, females dominate migration from some counties of the vast, and predominantly rural, Sichuan Province (China's most populous province), to the growing industries of Guangdong Province, particularly in the Pearl River Delta region (Davin, 1996). Regulations relating to labour migration were further relaxed from 1984, and from July 1985 peasants were allowed to be temporary residents in urban areas, applying for six-monthly permits through their work units. As a result, in Guangdong Province the 'temporary resident' population rose from 280,000 in 1982 to 3.3 million in 1990, representing nearly a 12-fold increase (So, 1997).

A study undertaken in the mid 1990s of female migrants from the rural areas of Sichuan to the industries of Guangdong found that only 12 per cent of household heads were strongly against the migration of their female members (So, 1997). The decision for a household member to migrate was taken by the entire family, but migration was seen as important in generating cash income to pay for such items as education, marriage, consumer goods, farming inputs, house building and maintenance. Migration was also seen as a 'risk aversion' strategy to diversify sources of income (So, 1997: 161).

The considerable attraction of migration is summed up in a popular Chinese phrase, *dongnanxibeizhong, facaidaoGuangdong* – 'east, south, west, north or central; to get rich, go to Guangdong Province'. However, rural households were concerned about the social problems of cities, such as prostitution, robbery and rape, and the potential vulnerability of their women in a strange environment. In relation to the income of rural households, the economic cost of migration can be considerable, with very long journeys by bus and train. A typical journey from Sichuan to Guangdong would take up to three days and cost more than 100 yuan, a considerable financial outlay for poor rural households; 1 yuan = US$0.12 (12 cents) in 1996.

Food and accommodation in Guangdong are provided by the factory, although the employer may charge for certain services. It appears that female migration had little effect on the rural household's productivity. On the contrary, the income from migrants' remittances far outweighed the impact of the loss of labour. Furthermore, the status of migrant women improved after they engaged in wage labour, and young migrants felt they had 'grown up' by learning new skills, experiencing a different environment and earning their own wage before getting married. The great majority of female migrants (79 per cent) were between the ages of 16 and 24, and a similar proportion were unmarried (So, 1997: 174, 189).

In the late 1990s the Chinese government started to regulate the flow of labour migration by coordinating efforts between the migrant-sending and migrant-receiving provinces, in order to relieve pressure on the transport system and reduce social problems. Coordinating offices were established from 1991 to regulate the flow of labour from Sichuan and other provinces to Guangdong, and factories recruit migrant workers through local labour bureaux. Potential migrants find out about job prospects through labour offices or industrial enterprises, which may even contact their villages. Well over half (57.8 per cent) of female migrants interviewed in 1994 found out about job opportunities through families and friends, and many already had contacts in Guangdong, who played a key role in helping migrants adjust to the new ways of living and working (So, 1997: 180).

Employment was mainly in producing electrical goods, toys, clothing and shoes, many of the products being destined for the export market. Working conditions were difficult and the work was manual and highly repetitive, with many industrial accidents. Women migrants typically worked for 70 hours or more in a seven-day week and additional overtime at a rate of 1 or 2 yuan per hour was common. Wages averaged 10–15 yuan per day, with a typical monthly income of between 250 and 450 yuan, including bonuses.

BOX 8.6 (continued)

Factories then often deducted as much as 100 yuan for provision of meals and accommodation (usually in dormitories), leaving a monthly net income of 150–350 yuan. The migrants were seen by the indigenous population as poor peasant workers and in a much inferior position compared with local residents, who invariably earned higher wages and held more responsible positions in the factories. Upward mobility for migrants was therefore very difficult, and workers coped with the poor conditions through a mutual support network known as the *tongxiang* system (people from the same village or county).

Migrants retain strong links with their villages by exchanging letters. However, the Chinese New Year (spring) holiday provides a valuable opportunity for migrants to return to their villages, taking money, consumer goods and much information about life and work in Guangdong. Between 70 and 90 per cent of Sichuan migrants return home at least once a year, usually for the spring holiday, which lasts about 60 days around the New Year period (Davin, 1996). However, transport costs increase at this time and there is much competition for places.

Typical remittances from migrants varied between 100 and 200 yuan per month (1200–2400 yuan per year), which is quite considerable in relation to the average annual net income in 1993 of only 698 yuan for a Sichuan peasant. An estimated five billion yuan is remitted each year to Sichuan province from migrant workers, a substantial proportion of whom were working in southern coastal cities, predominantly in Guangdong (So, 1997: 237).

The advantages and problems of migration on such a massive scale within China are highly complex, and the extent to which these factors are reflected in the fortunes of individuals and their rural households needs further careful investigation. However, one is inclined to agree with Davin's conclusion that

> **migration in the form it takes at present in China has the potential to return human and financial resources to the villages, and thus helps prevent the gap [between the poor countryside and the prosperous urban areas] becoming even wider.**
>
> **(Davin, 1996: 665)**

Many TNCs have their origins in the colonial period (Chapter 2). Walter Rodney (1972: 182) traces the development of the large TNC, Unilever, as what he calls 'a major beneficiary of African exploitation'. Originally founded in 1885 by William H. Lever, the firm of Lever made soap from palm oil imported from West Africa. Large concessions were also obtained in the Belgian Congo and, through a series of mergers, the company (renamed Lever Brothers) gained a foothold in every colony in West Africa. Lever bought the Niger Company in 1920, and then in 1929 further company takeovers led to the establishment of the new United Africa Company (UAC).

Yet more mergers took place with Dutch soap and margarine companies, and in 1930 Unilever Ltd (registered in Britain) and Unilever NV (registered in Holland) were formed, with the UAC subsidiary supplying the oils and fats. With further takeovers the organisation grew from strength to strength. As Unilever's information division commented, 'Unilever's centre of gravity lies in Europe, but far and away its largest member (the UAC) is almost wholly dependent for its livelihood . . . on the well-being of West Africa' (quoted in Rodney, 1972: 182).

In the early twenty-first century Unilever is still a major operator on the world commercial stage, with a total turnover in 2006 of €39,642 million and an operating profit of €5405 million. Unilever's stated mission is 'to add vitality to life – we meet everyday needs for nutrition, hygiene, and personal care with brands that help people feel good, look good and get more out of life' (Unilever, 2006a; see Chapter 4).

Until the 1960s most TNCs were of either US or UK origin, but in recent years Japanese, German and other companies have become important on the global scene. In the future, it is likely that companies based in the BRIC countries will become increasingly significant (see Key idea box on BRICs). Although TNCs are by no means homogeneous, they typically have their headquarters or strategic base in one country, but with a

Key idea

BRICs and the global economy

Economists at Goldman Sachs, a global investment banking and management firm, were probably the first to use the term 'BRICs' in 2003, to refer to Brazil, Russia, India and China – a group of rapidly developing economies which are becoming progressively more important in international investment trends (Wilson and Purushothaman, 2003). Goldman Sachs suggested that in less than 40 years the BRICs economies could become larger than those of the world's six most developed countries, possibly leading to a significant shift in the global balance of power. It is predicted that by 2050, of the current six largest economies only the USA and Japan will remain in the top six. Just as Germany and Japan experienced massive economic growth after 1945, so it is argued that the BRIC countries are capable of achieving similar success.

India has the potential to show the fastest growth over the next 30 and 50 years, with an estimated average annual growth rate of about 5 per cent. China's growth is likely to slow down to an average of 3.5 per cent, very similar to that of Brazil. Meanwhile, Russia's growth projections are hampered by its shrinking population, but by 2050 it will have the highest per capita GDP in the BRIC group, and comparable to other countries in the G6. Russia's economy is expected to overtake Italy in 2018 and the UK in 2027.

Before 2010 the number of people in the BRIC countries with an annual income over a threshold of US$3000 will more than double, leading to a massive increase in the size of the middle class in these nations and a corresponding increase in demand for higher priced consumer goods. However, by 2050 individuals in the BRIC countries are still likely to be on average poorer than in the G6 economies, though Russia should catch up with the G6 during this period. In China, however, per capita income in 2050 is likely to be approximately where the developed economies are now (Wilson and Purushothaman, 2003).

The Goldman Sachs thesis has generated much debate, for example in possibly underestimating the level of economic growth in China and India. Meanwhile, Brazil's economic potential has long been recognised, but so far it has failed to achieve investor expectations. Other factors such as human rights issues, civil unrest, poor governance, disease and terrorism could have an effect on the economic progress of one or more of the BRIC countries during the period up to 2050.

China's strengthening relations with Africa illustrate the global activity of just one of the BRIC countries. In the first 10 months of 2005, trade between China and African nations increased by 39 per cent to US$32.17 billion (£18 billion). During that period China's exports to Africa totalled US$15.25 billion, while China's imports from Africa were US$16.92 billion. China is investing heavily in construction projects in Africa, and particularly in oil exploration in countries like Sudan, in order to meet its rapidly growing petroleum needs. China has scrapped tariffs on 190 kinds of imported goods from 28 of the least developed African countries. In the first 10 months of 2005 alone, Chinese companies invested a total of US$175 million in African countries and Africa has become significantly more important for China as a source of raw materials to supply its growing manufacturing sector (BBC News, http://newsvote.bbc.co.uk/mpapps/pagetools/print/news.bbc.co.uk/2/hi/business/4587, 6 Jan 2006).

variety of production sites and subsidiary operations in other countries (Plate 8.7). TNCs also generally have a good amount of geographical flexibility, enabling them to shift resources and operations from one global location to another as production factors change, in order to seek new competitive advantages.

The relocation of operations in developing countries may often be because labour costs are cheaper, there is generally less militancy among labour unions, and fewer health and safety restrictions in the workplace. Abundant, low-cost and largely illiterate workforces, with little industrial tradition, are very attractive. Nike, for example, is a large, disaggregated TNC which subcontracts out production of its footwear and apparel to 700 factories around the world (Chapter 4). The company originally produced goods in the USA, but when production costs there rose, it subcontracted production to Taiwan and South Korea and then to other

Plate 8.7 TNC headquarters: Lonrho building, Nairobi, Kenya
(photo: Tony Binns)

developing countries. In 2006/07, Nike had net profits of US$1.05 billion, and a total income of US$12 billion, higher than the GNP of all except six sub-Saharan African countries (Nike, 2007; World Bank, 2007d).

Foreign direct investment (FDI) is when one firm invests in another (for example, in an overseas subsidiary) with the intention of gaining some control in that firm's operations (Dicken, 1998: 69). In order to attract TNC investment, governments frequently offer tax breaks, lax regulations, low minimum wages, cheap rent and, if necessary, military assistance to crush any labour unrest. Governments establish export-processing zones (EPZs) with the intention of attracting foreign investors, who then stay and hopefully make the development permanent (Gwynne, 2002). The level of FDI is now nine times that of international development aid and about 100 TNCs account for all FDI in developing countries (Madeley, 1999; World Bank, 2007d). By the end of the 1990s 38 per cent of global FDI went to developing countries, but one-third of this went to China alone, making that country the single largest host developing country of inward investment, with at least 18 million people employed in 124 EPZs (Hoogvelt, 2001; see also Chapter 4).

Schneider and Frey (1985) and Clayton and Potter (1996) found that other factors, such as the size of the home market, price and exchange rate stability, and political and institutional stability, were also important considerations in TNC overseas investment decisions. The power and influence of TNCs continue to grow, facilitated both by governments withdrawing controls on foreign investment, and thus encouraging the greater mobility of capital, and also through government support of WTO trade rules, which limit the rights of governments to control TNC activities.

TNCs have been hailed by some as the new development agents, as they can provide assistance to developing countries in the shape of economic, technical and managerial resources. Furthermore, the World Bank believes that 'private sector investment is the most important source of growth in developing economies' (World Bank, in Madeley, 1999: 24). In contrast, dependency theorists argue that TNCs represent 'core' countries exploiting 'peripheral' countries, and that TNC investment leads to the international division of labour, foreign indebtedness, capital monopolisation and economic impoverishment (Bury, 2001). Curtis is particularly critical of TNCs and the effects of their operations, suggesting that

> **Under TNC-led globalisation, the evolution of the world economy continues to be driven by expanding inequalities in wealth both between and within most countries. Income differentials are generally widening, skewing wealth distribution towards the middle classes and the rich. The result is that markets, products and services are increasingly focused on supplying the needs of these powerful and dominant consumers.**
>
> (Curtis, 2001b: 115)

In reality, the impact of a TNC on a developing country will partly depend on the nature of the employment generated. The low-skill production work that Nike provides has particular negative implications for development. Employment will not necessarily be beneficial for workers in terms of pay, working conditions or skills enhancement; the low-tech nature of the work does not hold much scope for useful technology transfer; nor will it bring many local linkages, since few local suppliers are used; trade advantages will be moderate, as the country is mainly being used as an export platform and the government may provide tax incentives; positive impacts may be unsustainable, since investment is often insecure and short-term, and there is frequently a lack of provision for social services, such as education and healthcare. Much of the literature has evaluated the impact of TNCs internationally and at the level of the state, but there is also an urgent need to focus on individual households as a key unit of analysis (Bury, 2001).

Critical reflection

TNCs and corporate social responsibility

As a result of adverse media attention about the activities of certain TNCs in developing countries, most major firms have been through a process of reconceptualising their relationship with broader society. Many firms now employ corporate social responsibility (CSR) managers, whose tasks include promoting their firm's role as corporate citizens, and assisting in the publication of annual reports detailing the firm's contributions to the societies within which they operate (Jones *et al.*, 2007; Seyfang, 2002).

In the case of UK retail firms (such as Tesco, Sainsbury's and Marks & Spencer), much attention is given to ensuring that working conditions at the site of production meet acceptable standards, for example that farms in developing countries do not use child labour, that their pay levels are in line with national legislation and that proper health and safety equipment is provided. International codes of practice are used to confirm that appropriate standards are being met.

For some companies these approaches represent a defensive role, as they attempt to protect themselves from external criticism which could be potentially damaging to their business operations. Other companies are more proactive and seek to promote responsible practice as a central component of their corporate brand image, as illustrated by this statement by Marks & Spencer's Chief Executive in the firm's 2006 CSR Report:

> **We believe that being a responsible business is the right thing to do, but we also believe that it makes good business sense. Put simply, it helps us to attract shoppers to our stores, recruit and retain the best people, form better partnerships with our suppliers and create greater value for our shareholders. (http://www2.marksandspencer.com/thecompany/investorrelations/downloads/2006/complete_csr_report.pdf, 27 February 2007)**

Why do you think large companies are showing an interest in Corporate Social Responsibility (CSR)? Examine some company policy statements to see what reasons are motivating companies on CSR.

TNCs and the globalisation of fresh food

During the 1990s, with trade liberalisation and the associated change in the global regulatory network, improvements in transportation technology and changing consumer demand resulted in an increasingly integrated global food production system dominated by TNCs. Overproduction of staple crops in the European Union under Fordist production systems and favourable government subsidies through the Common Agricultural Policy (CAP) have meant that, during the 1980s, trade in cereals and sugar, as well as tropical beverages from developing countries, has declined (Dixon, 1990).

One element of the global food trade that has shown a spectacular increase in recent years is the export of high-value crops, such as fresh fruit, vegetables and cut flowers. Between 1989 and 1997, the value of exports of fresh vegetables from sub-Saharan Africa to the EU increased by 150 per cent (Dolan and Humphrey, 2000). In 1989 the trade in these items comprised 5 per cent of global commodity trade and was equivalent in volume to trade in crude petroleum (Jaffee, 1994; Watts, 1996). Developing countries contributed one-third by value to this lucrative trade, twice the value of

Plate 8.8 Rose-growing for export, south of Nairobi, Kenya
(photo: Tony Binns)

their traditional agricultural exports of cocoa, coffee, cotton, sugar, tea and tobacco.

In 1990 24 low- and middle-income countries, mainly in Asia and Latin America, exported annually in excess of US$500 million worth of high-value, fresh horticultural products. The main producers were Chile, Argentina, Brazil and Uruguay in South America, and Malaysia and Thailand in Asia. Elsewhere, in order to maintain their foreign exchange earnings, as well as to diversify their economies, some African countries have been giving more attention to the production and export of high-value horticultural produce, most notably in Egypt, Kenya, post-apartheid South Africa, Zambia and Zimbabwe. Africa has been part of the global food market for centuries, with efficient and well-integrated marketing chains developing from the late nineteenth century to move cash crops such as tropical beverages, sugar, cotton and tobacco from African producers to consumers in Europe. However, these traditional marketing chains are not suitable for the export of highly perishable items such as fruit, vegetables and cut flowers (Plate 8.8). New chains have therefore evolved which, perhaps more than anything, reflect the considerable power of the large European retailers in responding to changing consumer demands (Box 8.7).

Until the beginning of the twentieth century, urban populations in Europe and the temperate regions of North America could only eat fresh produce seasonally and had to rely on canned and, later, frozen foods. The major change came with bananas, a tropical fruit that could withstand a long transportation link between producer and consumer, provided the temperature could be controlled. Early experiments in the banana trade began in the 1870s, when nationally based British, French and US specialist firms produced bananas in their tropical colonies, or 'semi-colonies' in the case of the USA, for consumption in Europe and the USA. As the banana industry grew, the US firms became extensively involved in the internal politics of states such as Cuba and the Central American 'banana republics' (Friedland, 1994).

Three firms involved in the early production and trade of bananas are Dole, Chiquita and Del Monte Tropical Products. All three firms also have major stakes in food labelling and transportation, with refrigerated cargo ships. Dole is a US-based transnational, known until 1991 as Castle & Cooke. It originally began as a merchant firm in the Hawaiian Islands and then became involved in food processing, real estate and fresh fruit and vegetable activities. Chiquita is also originally US-based and was formerly known as United Brands, and before that as the United Fruit Company. The Del Monte Fresh Produce Company, originally a US-based company, was known as Del Monte Tropical Products until late 1992.

Although bananas were important in the early history of these companies, they have, like other TNCs,

BOX 8.7

The globalisation of food: horticultural exports from Kenya

Trade in fresh fruit, vegetables and flowers from sub-Saharan Africa to the EU increased dramatically during the 1980s and 1990s (Barrett *et al.*, 1999; Dolan and Humphrey, 2000; Hughes, 2001). The European consumer now demands high-quality fresh commodities throughout the year. With a flight time of about 9 hours from Europe to Kenya, major supermarket chains have established links and organised production to ensure these items can be on their shelves in less than 24 hours after harvest.

Exports of fresh horticultural produce from Kenya have grown steadily since independence, such that in 2000 they accounted for 16 per cent of total export earnings and were the second most important export after tea (Economist Intelligence Unit, 2001). The value of horticultural exports rose more than five-fold from Ksh 3780 million (about US$66 million) in 1991 to Ksh 21,216 million (about US$360 million) in 2000, and to Ksh 28,200 million (about US$480 million) in 2004.

Over the last two decades the horticultural industry has undergone dramatic change, coinciding with economic liberalisation. Huge private investments have been made, particularly by the country's ten largest producers in the cut flower and pre-packaged vegetable sectors. This has been in response to increased demand from Europe, and especially to attract and keep lucrative supply arrangements with large UK supermarkets. Kenya is a major supplier of green beans, mange-touts, avocados, mangos and cut flowers, as well as a significant range of Asian vegetables. Green beans are the most important vegetable crop exported, although quantities declined in 1995–1996. However, this decline has been compensated by adding value to green bean exports through sorting and packaging in Kenya. The country faces increasingly stiff competition from other producers, such as Egypt, for green beans, South Africa for avocados and mangos, and Israel for avocados and cut flowers.

The ongoing troubles in Zimbabwe, leading to serious disruption of that country's horticultural exports, have provided a gap in the market which Kenya, Zambia and other African countries are aiming to fill. The growth of the Kenyan cut flower sector has been spectacular, and in 1996 Kenya overtook Israel to become the leading supplier of cut flowers to the Dutch auctions. In 1995, the tonnage of cut flower exports exceeded that of vegetables for the first time (Barrett *et al.*, 1997, 1999; Hughes, 2001). Kenya exports most of its fruit and vegetables to the United Kingdom and France, whereas cut flowers are destined for the Netherlands and Germany, as well as the United Kingdom.

The growth of the Kenyan horticulture industry has been due in no small measure to government support since the late 1960s. The Horticultural Crops Development Authority (HCDA) was formed in 1967, through which state policy and support for the sector have been channelled. The government has generally restricted itself to the role of facilitator and has not interfered with market mechanisms or pricing policy, a role which has been endorsed by a series of structural adjustment programmes since 1979. *Sessional Paper 1, 1986* specifically emphasised that agricultural growth was to be achieved by higher productivity, the expansion of high-value crops (such as fruit and vegetables) and improved export competitiveness.

The promotion of the horticultural industry is seen as a partnership between government departments, quasi-parastatal organisations, such as the Export Promotion Council, and the main growers, such as Sulmac – Brooke Bond's flower-growing subsidiary and the country's largest producer of cut flowers – and Homegrown, the largest producer of vegetables. Meanwhile, under structural adjustment, the influence of agricultural marketing boards has been greatly reduced. Duty exemptions have been particularly helpful, for example from 1991 the exemption of imported packaging materials used in the industry, and from 1994 the exemption of fertilisers, tools and greenhouse sheeting.

Two distinct marketing chains can be identified in the Kenyan export horticulture industry. One chain, which has developed since the 1960s, involves mainly small and medium-sized growers and in the mid 1990s accounted for 31 per cent of Kenya's horticultural exports. Small farmers may either supply medium growers, who then sell on to exporters or, alternatively, sell to intermediaries and agents who then sell to

BOX 8.7 (continued)

exporters. There is some criticism at various points in this chain about the quality and reliability of produce supply. This chain supplies large quantities of vegetables for the Asian market and strong links have developed between Asian exporters in Kenya and Asian importers and retailers in the United Kingdom. Importers of Asian vegetables are keen to have a wide variety of produce, but in smaller quantities than the supermarkets. Consumers are concerned with flavour and value for money rather than presentation or packaging. Produce for this market is imported and sold loose, not in pre-packs, and it does not have to meet strict supermarket specifications.

In sharp contrast, a fully integrated chain, which has developed since 1990, links Kenya's largest producers to major companies in the EU. Virtually all their vegetable produce is sold under contract to supermarket chains, mostly in pre-packs, which are processed in packing stations where standards exceed EU requirements. The packs use approved materials, and are bar-coded and priced in Kenya as directed by the supermarkets, which supply the pricing and other stickers. Many flowers are also sold to supermarket chains, and bouquets are made up in pack-houses on the farm. The rest of the flowers are sent in bulk to Dutch flower auctions, where they constitute over 25 per cent of all flowers sold. The major producers have all invested heavily in EU-standard pack-stations, refrigerated trucks and cold stores.

The large UK supermarket chains are at the top of the power hierarchy in this business, but shoulder few of the risks until produce actually reaches their shelves. Supermarkets depend on UK importers, through their associated exporting companies in Kenya, for getting produce out of Kenya and into the United Kingdom. The requirements of the UK Food Safety Act (1990) have had a major effect on production and marketing, since the Act calls for 'due diligence', requiring importers to know exactly where and how the crops were produced (including fertilisers and pesticides used) and there must be documentation to prove it. Traceability is now as crucial in the horticultural trade as quality, reliability and price, and logistically this favours dealing with a few large commercial farmers who can maintain strict standards and detailed records, rather than with many smallholder producers among whom there could be much variability. UK consumers are highly sensitive to issues concerning toxic chemicals or the perceived exploitation of local labour.

Two factors are absolutely crucial in exporting horticultural produce: the freight space and the cold chain. The larger exporters have more control than smaller exporters over freight space, because they can negotiate guaranteed space on aircraft and either fill it with their own produce or sell it on. Kenya's largest horticultural exporter has a pre-booked arrangement for cargo space on British Airways' nightly airfreight service to London. Maintenance of the cold chain is also vital in dealing with such highly perishable goods in a tropical climate. Exporters with their own dedicated cold storage facilities, at Jomo Kenyatta International Airport in Nairobi, run much less risk of breaking the cold chain than if they have to rely on using the general cold-store facilities.

If Kenya's export-oriented horticulture industry is to expand further these constraints need to be addressed, particularly in relation to the possible incorporation of more small-scale producers into the trade. In addition, the improvement of road infrastructure and the expansion of airport facilities are necessary. In the late 1990s a new international airport at Eldoret was completed, which could both reduce the pressures on Nairobi airport and also create potential for exporting more produce from the country's north-western region. Other possible measures might include the establishment of an agency to monitor controls and standards for all export crops, including banning sales of chemicals not permitted under EU regulations. Codes of conduct between growers, exporters and freight agents might also be enforced by trade associations, which also work to promote the industry.

But in the context of a poor developing country, such as Kenya, surely a key question is the extent to which poverty can be reduced among small producers engaged in export-oriented production. If horticultural production is to have a meaningful impact on poverty alleviation in Kenya then more attention needs to be given to small-scale producers – perhaps coordinated within producer groups that help them gain access to export markets – as well as controlling quality, post-harvest handling and marketing techniques.

diversified considerably since the Second World War. Chiquita bought seven lettuce-producing firms in California in 1969 and integrated them into a single subsidiary, Interharvest, which dominated US lettuce production and distribution. The Dole Food Company emerged in its modern form in 1961, when Castle & Cooke acquired the Dole Company, a pineapple producer. It expanded into bananas from 1964, as it bought an increasing share of the Standard Fruit Company, a banana producer for the North American market. In 1967, when Dole owned 87 per cent of Standard, Standard supplied 31 per cent of North American banana requirements. Ten years later, in 1977, Dole followed Chiquita into lettuce production by purchasing Bud Antle, the second largest lettuce producer in the USA. Subsequently, it was from the Bud Antle base that Dole expanded into a wide variety of other commodities.

During the 1980s Chiquita, Del Monte and Dole all expanded substantially into global sourcing and distribution based on their banana operations. The recipe for success of these TNCs has involved: (1) attracting capital from investors to make the initial purchases; (2) continuously generating new capital and demonstrating good profit levels; and (3) making good acquisitions, which have to fit into an overall strategy. Acquisitions should ideally be clustered geographically rather than be spread all over the world, and should be fully consolidated before venturing into new areas (Friedland, 1994).

Critical reflection

Should we be reducing 'food miles'?

There have been massive changes in the production and supply of food in the post-war period and particularly in the last 20 years. With globalisation of the food industry and the setting up of supermarket regional distribution centres, food is travelling greater distances by air, road and rail to reach the consumer – a trend which has led to some concern about 'food miles'. Steadily increasing food miles from the farm to the consumer have led to increased carbon dioxide emissions, air pollution, congestion, accidents and noise. Air transport increased by 140 per cent between 1992 and 2002. Although air travel only represents 0.1 per cent of total vehicle distance travelled, it contributes 11 per cent to greenhouse gas emissions. It is estimated that CO_2 equivalent emissions from all food transport increased by 12 per cent between 1992 and 2002.

In reflecting on the issue of food miles we might ask ourselves: Are organic bananas really worth the cost of the jet fuel that carried them from the West Indies? Does an apple grown a few hundred miles away taste better than one grown 3000 miles away? Is it better to support the local green bean producer than farmers in developing countries such as Kenya? What would happen to the economies of the world's poorest countries if food exports to richer countries were cut back? And how would richer, but distant, countries such as New Zealand manage such changes given the key importance of exported meat, fruit and milk products to the national economy?

The food miles issue is not straightforward, as an article in *The Economist* suggests:

> **Obviously it makes sense to choose a product that has been grown locally over an identical product shipped in from afar. But such direct comparisons are rare. And it turns out that the apparently straightforward approach of minimising the 'food miles' associated with your weekly groceries does not, in fact, always result in the smallest possible environmental impact. The term 'food mile' is itself misleading. A mile travelled by a large truck full of groceries is not the same as a mile travelled by a sport-utility vehicle carrying a bag of salad. It is more helpful to think about food-vehicle miles (ie, the number of miles travelled by vehicles carrying food) and food-tonne miles (which take the tonnage being carried into account).**
>
> **(© The Economist Newspaper Limited, London, 7 December 2006)**

(see also DEFRA, 2005).

What are your views about the 'food miles' issue? Choose one supermarket item that you eat regularly and that has been transported over a long distance. Examine the feasibility and implications of sourcing a similar product more locally.

Developing countries and the debt crisis

The external debt of low- and middle-income countries has soared from US$658 billion in 1980, to US$1375 billion in 1988, to US$1945 billion in 1994, to US$2560 billion in 1999 and to US$3039 billion in 2005 (World Bank, 2002a; 2007d). The origins of the debt crisis are complex, but undoubtedly major factors were the long-term effects of rising oil prices in the 1970s, compounded by developed countries adopting monetarist policies in the late 1970s following the 'second oil shock' of 1979, which forced up interest rates on debt repayments (Corbridge, 2002b). Added to this was the collapse of commodity prices in the early 1980s, such that in 1993 prices were 32 per cent lower than in 1980; and in relation to the price of manufactured goods they were 55 per cent lower than in 1960. As a result, there was a sharp deterioration in the terms of trade affecting developing countries (ICPQL, 1996). Facing massive debt repayments, the IMF and World Bank, which were created in part to transfer the savings of surplus countries to deficit countries, then imposed structural adjustment programmes on these countries, which required deep cuts in public spending, often with little concern for local circumstances and human welfare (see Chapters 3 and 7). As Watkins observes,

> **In Latin America, the epicentre of the debt crisis, average incomes fell by 10 per cent in the 1980s and investment declined from 23 per cent to 16 per cent of national income, causing widespread unemployment and poverty.**
>
> **(Watkins, 1995: 174)**

The debt crisis occurred suddenly in the early 1980s, with the financial collapse of Mexico in August 1982, and affected other middle-income developing countries which were heavily dependent on commercial lending, particularly Brazil and Argentina. The poorest countries were also badly hit, but since commercial banks had been reluctant to lend to them, most of their borrowing has been through public sector aid programmes. During the 1980s, the question of rescheduling the massive debts of certain developing countries became a major issue, since some were unable to repay the interest on the sums borrowed, let alone reduce the basic sum.

As we saw earlier, the newly industrialising countries of East and Southeast Asia did not experience such problems due to the relative buoyancy of their economies, although in 1997 and 1998 many Asian NICs registered a serious downturn in their economies. Countries such as South Korea, while borrowing heavily, have been able to service their debt due to a high level of exports. During the 1990s private and often highly speculative capital flows, mainly in the form of direct foreign investment, benefited China and middle-income countries such as Argentina, Malaysia, Mexico and Thailand. Furthermore, in 1989 the Brady Plan assisted middle-income countries by recognising that commercial debt could be reduced with IMF and World Bank support and by extending repayment periods.

However, the world's poorest countries, particularly those in sub-Saharan Africa, continue to suffer from a huge debt crisis, and are heavily dependent on official aid flows for their financial survival. As the flow of aid declines in real terms, trade and debt reform assume a greater importance. Between 1980 and 1999 sub-Saharan Africa's debt more than tripled to around US$216 billion and, although considerably less than that of Latin America (US$813 billion), the region's debt increased from the equivalent of 28 per cent of its GNP to 72 per cent, compared with 40 per cent for Latin America. Between 1985 and 1992, Africa disbursed US$81.6 billion in debt payments, diverting government funds from vital expenditure on education, health and other urgent priorities (Oxfam, 1993: 13). A number of countries have had to reschedule their debts, which has contributed to the steady build-up of arrears. Between 1989 and 1991 the official creditors cancelled some US$10 billion worth of debt to sub-Saharan countries, but the debt problem remains severe.

So what can and should be done about Africa's plight? As Oxfam observed,

> **It is difficult to avoid being struck by the contrast between the urgency with which Western governments have responded to the financial problems of Eastern Europe and Russia, and their neglect, for more than a decade, of Africa's far deeper problems.**
>
> **(Oxfam, 1993: 17)**

Oxfam advocates a fresh approach on the part of Northern governments, arguing that Africa's problem is not temporary, but rather a serious matter of bankruptcy which must be recognised 'in placing debtors' *ability to pay* above the claims of creditors'. Furthermore, Oxfam called on the industrialised countries to 'agree to the cancellation of between 90 and 100 per cent of *all* non-concessional debt' (Oxfam, 1993: 17).

Oxfam is also highly critical of the IMF, stating that it is

> **not an instrument for providing long-term concessional development finance; and it is governed by apparently immutable orthodoxies entirely inappropriate to African conditions. The time has come therefore either fundamentally to reform the IMF, or to extricate it from Africa. In either case, measures to write off obligations due to it from low-income African countries are long overdue.**
>
> **(Oxfam, 1993: 17)**

It seems that the recovery of Africa will depend on substantial investment of foreign capital, since in 2005 sub-Saharan African countries together only received about 1.7 per cent of worldwide foreign direct investment, compared with the 8 per cent received by China alone. With the lack of private investment, Africa is becoming increasingly dependent on government and multilateral agency development assistance, such that aid accounts for 80 per cent of all financial resource flows into the region, and for 5 per cent of the region's entire GDP. Just as Africa was shown to be off the e-map in Chapter 5, so it is also virtually off the investment map.

There is no doubt that some real progress has been made in the late 1990s and early twenty-first century in alleviating the debt crisis in Africa and elsewhere among the world's poorest countries. The call from Oxfam and other influential charities and NGOs for urgent attention to be given to the debt problem was taken forward by targeted anti-debt pressure groups, most notably 'Jubilee 2000', and with pressure also from the world's most powerful leaders, the World Bank and the International Monetary Fund launched the significant HIPC initiative in 1996 (Box 8.8).

Jubilee 2000 proved to be particularly successful, and originated when a number of major development-focused charities decided in 1996 to petition political leaders of the rich countries to 'cancel the unpayable debts of the poorest countries by the year 2000 under a fair and transparent process' (Jubilee 2000, 2002: 2). The petition grew to become a major international campaign, such that by the end of the campaign some 24 million signatures had been collected for the Jubilee 2000 petition, the first ever global petition. The campaign brought the issue of Third World debt to ordinary people across the world, through similar campaigns in 60 countries. The decision to introduce the Enhanced Initiative for HIPC countries in 1999 was undoubtedly a sign of the power and influence of Jubilee 2000 (Roodman, 2001).

Worldwide concern about the debt crisis in poor countries by no means ended in 2000. Jubilee Research was established as a successor to Jubilee 2000, and in April 2001 the Jubilee Movement International was established at a conference in Mali, resolving 'to lift nations and their peoples out of foreign debt bondage; and to struggle for global economic and social justice' (Jubilee 2000, 2002: 1).

Aid to developing countries

In 1970, through Resolution 2626 on the International Development Strategy for the Second United Nations Development Decade, the UN General Assembly set out for the first time agreed targets for finance resource transfers and flows of overseas development assistance – aid. The UN urged developed countries to achieve an allocation level of 0.70 per cent of their gross national product in overseas aid by 1975. As we have already seen, this figure was subsequently emphasised in the Brandt Report in 1980, and was reaffirmed at numerous world gatherings such as the Earth Summit in Rio de Janeiro in 1992, the Conference on Population and Development in Cairo in 1994 and the World Summit on Social Development in Copenhagen in 1995.

However, there has unfortunately been little progress on this, and in some cases governments of developed countries which are members of the Development Assistance Committee (DAC) of the Organisation for Economic Co-operation and Development (OECD) have actually reduced their aid budgets substantially. For example, the United Kingdom's allocation of overseas development assistance reached an all-time high in 1979 at 0.51 per cent of Gross National Income (GNI), but then declined steadily to an all-time low of 0.27 per cent in 1990. There was a slight increase to 0.32 per cent in 1991, but at the time of the Labour Party's general election victory in May 1997 the UK aid budget had fallen again to 0.26 per cent. Since the incoming Labour government pledged not to raise key taxes for two years, there seemed little prospect of an increase in the UK aid budget before the year 1999. However, in 2000 there was a significant increase in UK Official Development Assistance to 0.32 per cent of

BOX 8.8

The Heavily Indebted Poor Countries (HIPC) Debt Relief Initiative

The Heavily Indebted Poor Countries (HIPC) Debt Relief Initiative was launched by the World Bank and the IMF in 1996 and represented a recognition of the significant effort needed to bring about a once and for all reduction in the debts of some of the world's poorest countries. According to the World Bank,

> **It was the first comprehensive approach to reduce the external debt of the world's poorest, most heavily indebted countries, and represented an important step forward in placing debt relief within an overall framework of poverty reduction.**
>
> **(World Bank, 2001b: 1)**

The experience of some of the first countries to qualify for HIPC debt relief led the UK government to push for a thorough review of the initiative, which started in January 1999.

> **The [UK] Government believed it would have to be redesigned to ensure that it provided poor countries with a permanent solution to their debt problems and freed up resources to tackle poverty.**
>
> **(Department for International Development, 2000b: 2)**

This was supported by the leaders of the Group of Seven (G7) at their summit in Cologne, Germany, in July 1999, and in September 1999 the World Bank and IMF approved an 'enhanced initiative' as an integral part of the new poverty reduction strategy. This strategy was designed to link external support to domestically formulated, results-based poverty strategies and also to improve relations between the World Bank, IMF and recipient countries.

The value of the new HIPC framework is over US$50 billion, compared with US$12.5 billion under the original HIPC Initiative. The World Bank suggests that, 'the net-present value of public debt in the 33 countries likely to qualify (approximately US$90 billion) would be reduced by about half after HIPC and traditional debt relief' (World Bank, 2001b: 4).

To be eligible for relief, a country must:

- Be very poor – as defined by the World Bank and IMF;
- Have an unsustainable debt burden – with a stock of debt which is more than 150 per cent of exports in terms of present value;
- Have a Debt Sustainability analysis prepared by the World Bank/IMF together with local officials;
- Pursue good policies – macro-economic, structural and social policies consistent with poverty reduction and sustained growth (World Bank, 2001a).

In response to problems with the original initiative, the enhanced HIPC Initiative aims to provide:

- Faster debt relief, beginning immediately or soon after the 'decision point';
- Stronger links between debt relief and poverty reduction strategies, formulated with civil society participation;
- Deeper and broader debt relief, with a reduction of US$50 billion in external debt servicing. This should reduce by more than two-thirds the outstanding debt in more than 30 countries.

By February 2006, some 33 countries had reached their 'decision point' under the enhanced HIPC Initiative, and six countries had reached the 'completion point' under the original HIPC Initiative. The 33 countries were receiving debt relief amounting to some US$38.2 billion (WB-IEG, 2006). Table 8.3 shows the countries identified as HIPCs and the stage they had reached by February 2006 in terms of planned debt relief. It should be noted that a large proportion of the countries listed are in sub-Saharan Africa.

In June 2001 Bolivia became the second HIPC country to reach completion point after Uganda, the latter being the first country to receive approval in May 2000. The conditions which Bolivia had to meet in order to receive assistance were: first, continued implementation of strong macro-economic and structural policies; second, the establishment of a fully defined Poverty Reduction Strategy Paper (PRSP); and, third, confirmation of participation in the enhanced HIPC framework from Bolivia's other creditors.

Bolivia met virtually all of the financial requirements in its programme with the IMF and, in addition, reformed its customs and internal revenue agencies; introduced a new tax procedure code to

▶

BOX 8.8 (continued)

Table 8.3 Status of countries under the Enhanced HIPC Initiative, February 2006

	Post-completion point	At decision point	Pre-decision point	Potentially eligible for HIPC
Early (before July 2002)	Bolivia, Burkina Faso, Mauritania, Mozambique, Tanzania, Uganda **(6)**	Benin, Cameroon, Chad, Ethiopia, Gambia, Ghana, Guinea, Guinea-Bissau, Guyana, Honduras, Madagascar, Malawi, Mali, Nicaragua, Niger, Rwanda, São Tomé and Principe, Senegal, Sierra Leone, Zambia **(20)**	Burundi, Central African Republic, Comoros, Côte d'Ivoire, Democratic Republic of Congo, Lao PDR, Liberia, Myanmar, Republic of Congo, Somalia, Sudan, Togo **(12)**	
Late (after July 2002)	Benin, Ethiopia, Ghana, Guyana, Honduras, Madagascar, Mali, Nicaragua, Niger, Rwanda, Senegal, Zambia **(12)**	Burundi, Democratic Republic of Congo **(2)**		Bangladesh, Bhutan, Eritrea, Haiti, Kyrgyz Republic, Nepal, Sri Lanka, Tonga **(8)**
Total (as of February 2006)	Benin, Bolivia, Burkina Faso, Ethiopia, Ghana, Guyana, Honduras, Madagascar, Mali, Mauritania, Mozambique, Nicaragua, Niger, Rwanda, Senegal, Tanzania, Uganda, Zambia **(18)**	Burundi, Cameroon, Chad, Democratic Republic of Congo, Gambia, Guinea, Guinea-Bissau, Malawi, Republic of Congo, São Tomé and Principe, Sierra Leone **(11)**	Central African Republic, Comoros, Côte d'Ivoire, Lao PDR, Liberia, Myanmar, Somalia, Sudan, Togo **(9)**	Bangladesh, Bhutan, Eritrea, Haiti, Kyrgyz Republic, Nepal, Sri Lanka, Tonga **(8)**

Source: WB-IEG (2006), p. 4, Table 1.1 © International Bank for Reconstruction and Development/The World Bank

strengthen tax administration; and launched a new financial management information system to increase transparency and accountability in the administration of public expenditure. Including assistance provided under the original 1996 Initiative, Bolivia received debt service relief of over US$2 billion. As a result of HIPC assistance, Bolivia's total external debt has been reduced by 50 per cent, and annual debt service payments in the ten years from 2001 have been reduced by some US$120 million per year.

Sierra Leone, which has emerged from over a decade of civil war and internal turmoil, was only admitted to the Enhanced HIPC Debt Relief Initiative in March 2002. The country's government formulated a detailed plan for the transparent and accountable use of funds received, with emphasis placed on education, health and rural development.

HIPC has been effective in channelling additional resources to qualifying countries, the number of which is increasing, with eight more low-income countries identified as being potentially eligible. Transfers to HIPC countries have increased from US$8.8 billion in 1999 to US$17.5 billion in 2004, while transfers to other low-income countries grew by only a third. HIPC has already reduced $19 billion of debt in 18 countries. The requirement for countries to implement a poverty reduction strategy at the same time as receiving debt relief has been an important and beneficial outcome of the programme.

GNI and an even greater increase to 0.48 in 2005 (Figure 8.5). This figure is expected to eventually reach 0.7 per cent of GNI by 2013.

The US record has been disappointing. Despite giving the largest total amount of overseas aid in 2005, some US$27.4 billion in 2005, the proportion of GNI given as aid to developing countries by the USA was only 0.22. Although this represents a significant increase on the 0.13 per cent figure for 2002, this is still the lowest figure among OECD countries, with the

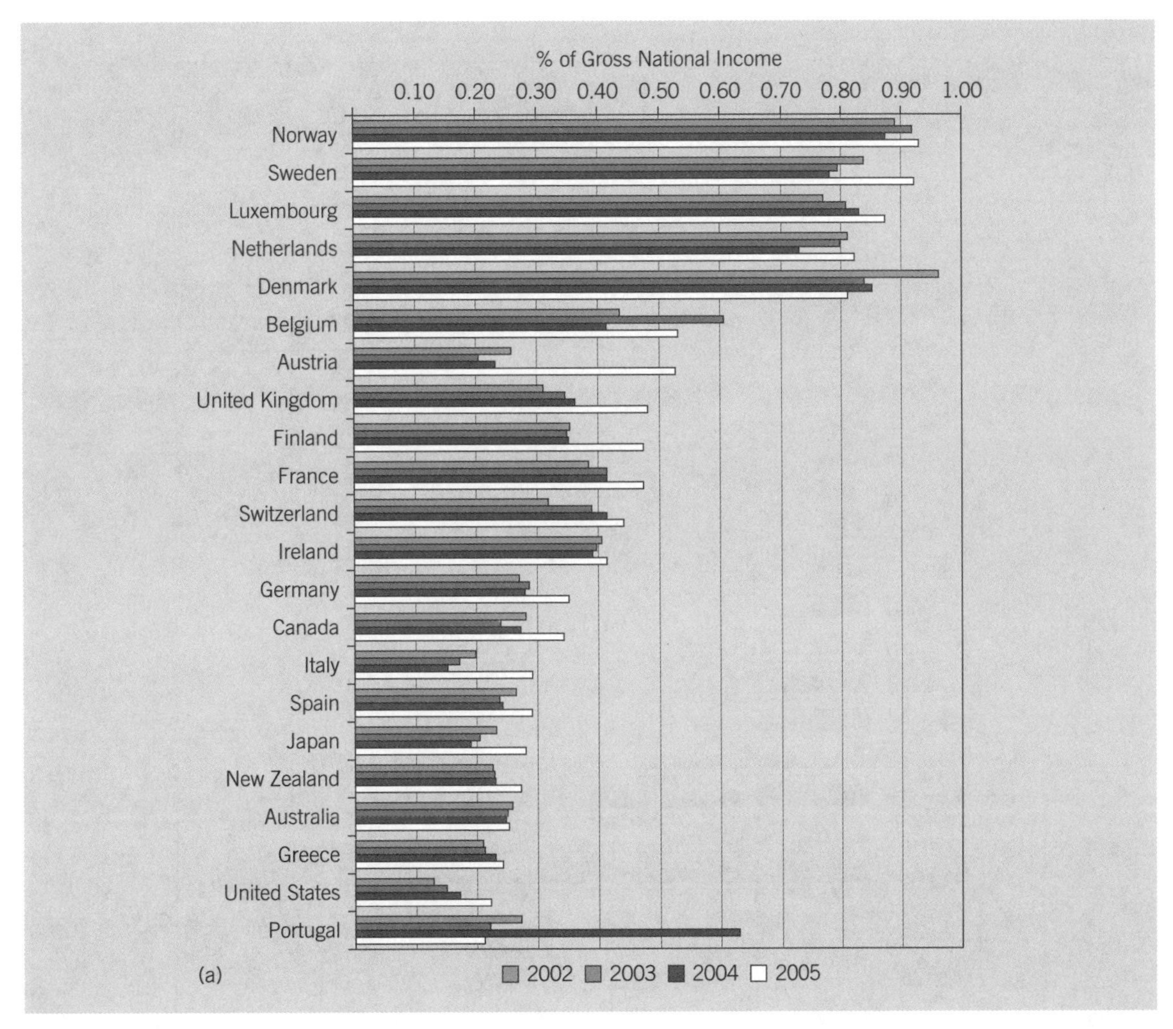

Figure 8.5 Net official development assistance (ODA) to developing countries, 2002–2005: (a) by donor country (% GNI)
Source: OECD (2006)

exception of Portugal. In 2005, only five countries – Denmark, the Netherlands, Luxembourg, Norway and Sweden – had reached, and indeed exceeded the 0.70 target. Norway, with 0.93 per cent of GNI, was ahead of other countries, and has shown a steadily increasing commitment to development assistance since 1975.

In addition to the quantity, it is important to consider the nature of the aid and the reasons for giving it. Whether overseas development assistance takes the form of short-term disaster relief, longer term development aid, food aid or military aid, each has many complex ramifications and implications, both in relation to the successful alleviation of poverty in developing countries and also in relation to the priorities, and frequently ulterior motives, of donor countries. During the 1980s and early 1990s, Britain tied a higher proportion of its aid than most donors, and 74 per cent of bilateral aid in 1991 was tied to the purchase of British goods and services (German and Randel, 1993). The construction of the controversial Pergau Dam in Malaysia in the early 1990s, costing US$350 million, was the largest single project ever financed under the UK aid programme. However, critics argued that it was an expensive and inefficient source of power, and a waste of money for both Malaysian consumers and British taxpayers. Nevertheless, the project went ahead, because it was linked to large quantities of British exports to Malaysia, including over US$1 billion of arms exports.

From 1 April 2001 all UK development assistance became fully untied. It is argued that

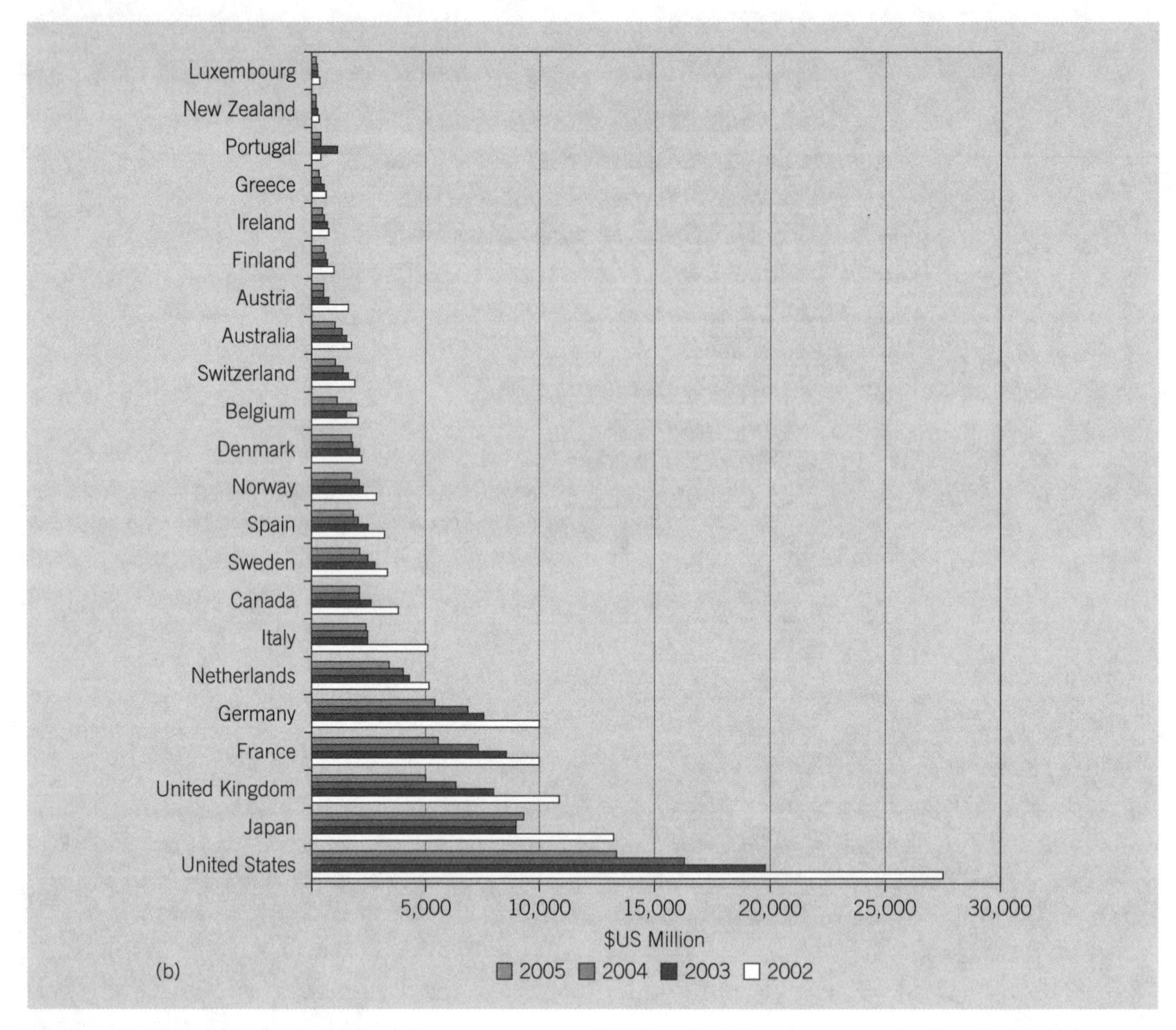

Figure 8.5 (continued) (b) by donor country ($US million)
Source: OECD (2006)

> **Tying aid protects donor exporters from international competition for contracts. This leads to higher costs and lower quality, and consequently reduces the impact on poverty. The World Bank estimates that tying aid reduces its effective value by as much as 25 per cent.**
>
> **(Department for International Development, 2001b: 1)**

The UK government is trying to encourage other governments to untie their aid, and on 1 January 2002 the OECD's 'Recommendation on Untying Official Development Assistance to Least Developed Countries' came into effect, representing an important step forward.

A report written in 1993 by German and Randel, and reflecting on the policies of the Reagan and Bush administrations in the USA, noted that US aid was generally directed to two broad categories of countries: countries of strategic importance, and poor but 'politically correct' countries. Egypt and Israel, both of strategic importance, respectively received 32.1 and 8.3 per cent of aid in 1990–91. During the civil war in El Salvador, Central American countries received about US$1 billion of bilateral aid from the USA each year. One-third of US bilateral aid in 1991 was spent on food aid. Interestingly, the primary stated purpose of US aid is to 'advance US interests by helping co-operating countries to expand their economies and the opportunities they offer their people' (German and Randel, 1993: 59). Poverty alleviation was not a specifically stated goal of US aid, though it is an element of certain programmes.

Since the 1990 *World Bank Report on Poverty*, some governments have made a stronger commitment to poverty alleviation, and the DAC is keen to monitor donor performance in this area. NGOs are concerned that too much aid is used to promote exports and subsidise domestic industry rather than to benefit the poorest people in developing countries. Even Norway, with a significant proportion of Gross National Income allocated to aid, and with a strong poverty alleviation focus, was keen in the early 1990s to expand Norwegian commercial interests. With Prime Minister Gro Harlem Brundtland's high-profile involvement on the world development stage, environmental issues became a particularly important element in Norwegian development aid.

In the case of another Scandinavian country, Finland's overseas assistance budget reached 0.66 per cent of GNI in 1991. But dramatic cuts were then made due to an economic recession at home, and since 1996 it has been in the 0.31–0.33 per cent range. China is the biggest recipient of Finland's development assistance and, despite China's human rights record, the country is perceived as having a vast potential for commercial exploitation.

In its White Paper on International Development, published in December 2000, the UK government stated its aim to focus development assistance on 'systematic poverty reduction':

> **In 1999/2000 we spent 74 per cent of the UK's development assistance in low income countries, up from 68 per cent in 1996/97. We will further increase the focus on low income countries over the next three years . . . [and increase] our capacity to provide more resources to countries that implement pro-poor policies by establishing policy and performance funds for Africa, Asia and for the multilateral development institutions.**
>
> **(Department for International Development, 2000a: 87)**

Types of aid

Food aid continues to be significant in commodity flows from developed to developing countries, and although most food aid is provided bilaterally on a government-to-government basis, the creation of the World Food Programme (WFP) in 1961 added a significant multilateral dimension, such that it is now the main provider of international food aid for development and disaster relief. Food aid is a controversial form of development assistance, since political and economic motives may be significant in sustaining flows and determining their direction (Shaw and Clay, 1993). Food aid may be divided into three types:

- *Programme food aid* is usually given as a grant or a soft loan on a government-to-government basis to fill the gap between the demand and supply of food from domestic production and any commercial imports. Such aid may reduce the amount of foreign exchange a country needs to spend on buying imports, and if the food is sold it provides additional local currency for the government.
- *Project food aid* is primarily aimed at satisfying the nutritional needs of the poor, mainly in rural areas, and is given as a grant, with the food aid closely targeted. Although the WFP is the main provider, other government and NGO bodies may also be involved. WFP is also involved in a variety of rural projects concerned with health centres for mother-and-child care, primary education and training, and food-for-work programmes.
- *Emergency food aid* is a response to sudden disasters such as drought, floods and pest attack, as well as civil war. Emergency food aid is provided both bilaterally and multilaterally, mainly by the WFP. The 1980s crisis in Ethiopia, the severe floods in Bangladesh and the refugee crisis in Africa's Great Lakes region in the 1990s all triggered large quantities of emergency food aid. In the period 1987–1991 Bangladesh dominated the recipient countries, followed by Pakistan, India and Tunisia (Shaw and Clay, 1993).

The world food aid system is highly complex and diverse, with the USA being the largest single provider; some of the other countries involved are Canada, Australia, Japan, Norway and Sweden. The European Union operates a union-wide programme, in addition to separate national programmes.

Much has been written about the merits and problems of food aid. Although most commentators would agree with the importance of food aid as part of a disaster-relief package, there is more concern about longer term food aid, which might affect local production and disrupt food-marketing systems. It is suggested that food aid can lower local food prices, encourage governments to neglect the drive to food self-sufficiency, create a dependency mentality and change eating habits. However, although potential problems in moving to food self-sufficiency are recognised, 'the widespread professional view of practitioners and economists [is] that disincentive effects are avoidable' (Shaw and Clay, 1993: 15).

Military assistance

With the end of the Cold War, it has been suggested that some of the so-called peace dividend might be allocated to overseas development assistance aimed at poverty alleviation, rather than on military expenditure. It is both ironical and deeply disturbing, however, that although military expenditure within developed countries has fallen, the sale of arms and military hardware to the poorer countries of the world is still big business. In 1987 the USA gave US$5.4 billion in worldwide military assistance, and the former Soviet Union gave US$13.5 billion. These figures declined to US$3.4 billion (USA) and zero (former Soviet Union) by 1993, but there was still a total military assistance of US$4.6 billion, some 74 per cent of this coming from the USA. The United Nations Development Programme (UNDP) argued that

> **military assistance to the Third World formed one cornerstone of the cold war . . . and also had commercial motives, helping sustain the output of the arms industry by subsidizing exports and unloading outdated weaponry.**
>
> **(UNDP, 1994: 53)**

Military assistance has many damaging effects for poor countries. Even after conflicts are resolved, large quantities of weaponry within developing countries pose a continuing threat to internal stability, and considerably strengthen the army and its ability to seize power.

UNDP advocates the phasing out of military assistance and tighter controls imposed on the arms trade. In 1994, 86 per cent of conventional weapons exported to developing countries came from (in descending order): the former Soviet Union, the USA, France, China and the United Kingdom, all permanent members of the UN Security Council. Two-thirds of these arms were sold to ten developing countries, including Afghanistan, India and Pakistan.

A comprehensive policy for arms production and sales is urgently needed, with special emphasis placed on cutbacks in the production of chemical weapons and landmines (UNDP, 1994). In Angola and Cambodia it is estimated that millions of landmines have been planted, causing continual suffering among local populations. The landmine issue came to the fore in 1997, and many governments, including the UK government, have now agreed to ban their sale overseas.

Aid: quantity and quality

In conclusion, there needs to be an improvement in the quality as well as the quantity of overseas development assistance if poverty alleviation, leading ultimately to poverty eradication, is to be achieved throughout the world. There have been proposals to make development assistance obligatory, perhaps by introducing a form of international tax on the rich countries (Watkins, 1995). Another proposal is a tax on international currency transactions, which would deter vast flows of speculative capital, estimated at US$1 trillion per day, which are destabilising economies in both the developed world and the developing world. The so-called 'Tobin Tax' is considered in some detail in Chapter 4. National and international action would be needed to bring financial markets under more effective control. Some reform of the IMF is also necessary to help developing countries with serious foreign exchange shortages to increase their reserves without resorting to deflationary measures or constraining growth by cutting essential imports (Watkins, 1995).

Aid should focus on the poorest people in the poorest countries, and the nature of future development assistance must place greater emphasis on key aspects such as health and education, rather than emphasising the potential for exports from donor countries. There is a need to discuss aid and development priorities with local communities, and also to reduce the costs of delivering aid through an army of relatively well-paid expatriate consultants and developers (Oxfam, 1993). Perhaps most important of all, development assistance should concentrate more on achieving a sustainable improvement in the quality of life.

There are so many questions relating to aid, and much detailed evidence is available to support the many different perspectives in the ongoing debate. It is impossible to cover all aspects here. But, we might ask, does aid actually work? Why, for example, are so many sub-Saharan African countries worse off now than they were at independence in the 1960s, despite having vast amounts of development assistance pumped into them? It is a sobering thought that the number of Africans living below the poverty line actually increased from an estimated 217 million in 1987 to 291 million in 1998, and primary school enrolment fell by 1 per cent (Reality of Aid, 2002). A survey undertaken in Zambia in 2003 found that 75.8 per cent of the population was living on less than US$1 a day (World Bank, 2006d). These trends have occurred at the same time as the developed world has experienced an information technology revolution and steadily improving standards of living.

The issue of conditionality is of great significance. As we have seen in the case of the HIPC Initiative, the World Bank and IMF expect recipient countries to fulfil a number of conditions before they will be assisted in relieving their debt burden. UNICEF reports on another case where, in 1998,

> **the IMF, the World Bank and other international agencies loaned Indonesia more than US$50 billion. But with the loans came stringent restrictions . . . [such that] the IMF-imposed austerity measures exacerbated the mushrooming social crisis . . . Between 1997 and 1998, according to the World Bank, the number of Indonesians living in poverty doubled.**
>
> **(UNICEF, 2000: 36)**

In many cases the consultants who undertake the research which might subsequently lead to the formulation of aid packages are expatriates, who during their commonly all-too-brief field visits fail to understand local situations and are conditioned by donor country approaches and priorities. As the Reality of Aid report comments, with reference to the experience of NGOs in Uganda, 'even when countries have developed their own national strategies for addressing poverty, donors insist on additional processes which undermine the very ownership and accountability that donors are claiming to promote' (Reality of Aid, 2002: 9).

Perhaps understandably, evidence of good governance has now become a prerequisite for the receipt of most overseas development assistance. As the UK Government's White Paper comments,

> **Effective governments are needed to build the legal, institutional and regulatory framework without which market reforms can go badly wrong, at great cost – particularly to the poor . . . Effective governments are also needed to put in place good social policies . . . [to] ensure the provision of key public services.**
>
> **(Department for International Development, 2000a: 24–5)**

But does good governance as a prerequisite for the receipt of aid necessarily lead to development and poverty alleviation? Furthermore, the UK White Paper speaks of

> **Making political institutions work for poor people [which] means helping to strengthen the voices of the poor and helping them to realise their human rights. It means empowering them to take their own decisions, rather than being the passive objects of choices made on their behalf.**
>
> **(Department for International Development, 2000a: 27)**

These may be admirable objectives, but is there any concrete field-based evidence to indicate that this is actually happening, or is it more likely that the wealthy, educated and politically astute elites will continue to dictate policy and the implementation of projects funded by aid from overseas? These and many more questions are in themselves a strong motivation for further in-depth research.

Conclusion

As we have seen, there are a great many different movements and flows between the developed and developing world and also within and between particular countries. It is quite impossible to catalogue all such movements and flows, simply because of their diversity and complexity. It is hoped that, through the use of real examples, this chapter has clarified some of the key concepts and important issues which need to be monitored in the future. Furthermore, it is impossible to fully appreciate the character and underlying causes of different geographies of development without an understanding of the ways in which people and places are connected through movements of people themselves, as well as flows of commodities, finance and knowledge.

Key points

- Producers in poor countries have very little control over the prices they receive for their commodities and world price fluctuations can have a significant effect on their livelihoods.
- Migrants' remittances can have a significant effect on the quality of life in their home areas and countries.
- Tourism makes valuable contributions to the economies of some developing countries, but the negative effects of having large numbers of international visitors need to be fully appreciated in planning for the expansion of tourism.
- We live in an interdependent, but unequal world. Many poor countries are heavily dependent on producing just a single commodity for world markets and are therefore vulnerable to changes in the terms of trade.
- World trade policy and institutions are dominated by the world's richest countries. Developing countries should be given a larger role in decision making, and fair trading policies should be strengthened.
- Transnational corporations play an important role in world trade. Criticism of some poor working practices have encouraged TNCs to show a greater interest in corporate social responsibility.
- The so-called BRIC countries (Brazil, Russia, India and China) are playing a significantly greater role in world trade and investment, such that in less than 40 years their economies are predicted to become larger than many of the world's richest countries.
- Although some progress is being made in reducing the crippling debt in the world's poorest countries, there is still much to be achieved.
- It is not just the amount of overseas development assistance (aid) that richer countries give to poorer countries that is important, but also the nature of that assistance.

Further reading

Curtis, M. (2001) *Trade for Life: Making Trade Work for Poor People*. London: Christian Aid.

A thought-provoking study of world trade and its role in poverty alleviation.

Department for International Development (2000) *Eliminating World Poverty: Making Globalisation Work for the Poor*, White Paper on International Development, London: DfID.

The UK government's official statement on the importance of eliminating poverty within the broader context of globalisation.

Dicken, P. (2003) *Global Shift: Reshaping the Global Economic Map in the 21st Century*, 4th edn. London: Sage.

A key text which examines how economic globalisation arises from the dynamic interplay between transnational corporations as prime actors and nation states as regulators, facilitated by processes of technological change.

Hoogvelt, A. (2001) *Globalization and the Postcolonial World: The New Political Economy of Development*, 2nd edn. Basingstoke: Palgrave.

An important and thought-provoking text on the ramifications of globalisation.

Jenkins, R., Pearson, R. and Seyfang, G. (eds) (2002) *Corporate Responsibility and Labour Rights: Codes of Conduct in the Global Economy*. London: Earthscan.

A collection of papers examining issues relating to the operating activities of companies in light of corporate social responsibility.

Madeley, J. (1999) *Big Business, Poor Peoples: The Impact of Transnational Corporations on the World's Poor*. London: Zed Press.

A fascinating study of how poor people are affected by the operations of transnational companies.

Roberts, J.T. and Hite, A.B. (eds) (2007) *The Globalization and Development Reader*. Oxford: Blackwell.

An excellent collection of papers, which were originally published elsewhere, but have now been assembled in this collection. Includes classic works of Marx, Engels, Weber and Rostow, as well as many more recent contributions on the theme of globalisation.

Websites

http://www.cafedirect.co.uk/, Cafedirect, 2007
This website provides a useful insight into the operations of a successful Fairtrade organisation involved in the purchase and marketing of tea and coffee from some 37 producer organisations in developing countries.

http://www.ico.org/history.asp, International Coffee Organization, 2007
Includes useful details and statistics on world coffee production and trade.

http://www.CarnegieEndowment.org, Carnegie Endowment for International Peace, 2007
This website contains some helpful papers on world trade.

http://www.commissionforafrica.org/, Commission for Africa, 2007
This website has details on the establishment of Tony Blair's Commission for Africa in 2004 and includes a copy of the Commission's Report published in March 2005.

www.fairtrade.org.uk, Fairtrade Foundation 2007

www.fairtrade.net/, Fairtrade Labelling Organizations International, 2007
Two wide-ranging websites on the Fairtrade movement, with statistics and case studies relating to Fairtrade.

http://www.hipc-cbp.org/, Heavily Indebted Poor Countries, Capacity Building Programme (HIPC CBP), 2007
The official website of HIPC CBP, concerned with building the capacity of HIPC governments to conduct debt strategy analysis and negotiate debt relief and new financing.

http://www.jubileeresearch.org/, Jubilee Research, 2007
Jubilee Research, successor to Jubilee 2000, UK, is part of the Global and National Economics programme at the New Economics Foundation in London. The website has up-to-date news and data about debt and international financial reform.

http://www.oxfam.org/en/policy/, Oxfam, 2007
Contains detailed statements on Oxfam policies and reports on particular countries and projects where Oxfam is involved.

www.propoortourism.org.uk, Pro-poor Tourism Partnership, 2007
Includes a range of research reports and studies that focus on how tourism can contribute more to poverty reduction.

http://www.unhcr.org/cgi-bin/texis/vtx/home, UNHCR **(United Nations High Commission for Refugees) 2007**
The key international organisation concerned with the condition and relief of refugees.

http://web.worldbank.org/WBSITE/EXTERNAL/TOPICS/EXTDEBTDEPT/, World Bank, 2007
World Bank website with information on economic policy and debt issues.

Discussion Topics

- Select a single commodity which is produced in a poor country and is traded internationally. Examine the significance of the commodity in the producing country's economy, and suggest how improved terms of trade might lead to a fairer return for both the producing country and those engaged in production.
- How might governments in developing countries ensure that local people gain greater benefit from the development of international tourism?
- What do you feel are the implications of the greater involvement of the BRIC countries in world trade and investment?
- Investigate the significance of debt with reference to selected developing countries. Suggest some possible strategies for overcoming the debt burden.
- Examine the levels of aid given by different countries and the progress in achieving the 0.70 per cent target of Gross National Income for aid to developing countries. What types of aid do you feel should receive priority in the future?

Chapter 9

Urban spaces

As explored in previous parts of the book, particularly Chapters 1, 3 and 5, through time urbanisation has been regarded as being linked directly to industrialisation and overall development. This chapter considers this relationship, at different scales, from the global down to the level of the individual city. It also considers changing views concerning this association.

The chapter starts with the global setting in stressing the contemporary relation existing between rapid urbanisation and poor nations. This is manifest in the locus of the largest and fastest growing cities to be found in the world today. Moving down a scale, the regional role of cites is considered, in respect of the generic approaches that can be used in efforts to change urban circumstances by urban and regional planners and policy makers. The suggestion that the urban and the rural are closely interlinked and that new forms of urban–rural relations are characterising cities in the developing world in the twenty-first century is then explored. Attention is then directed at individual urban areas, in respect of their structure, and the important role the informal sector plays in providing both jobs and homes. To conclude, the pressing nature of environment–urbanisation issues is discussed, with a strong accent on the need to promote more sustainable forms of urbanisation. Specifically the chapter:

- Shows how in the twenty-first century rapid urbanisation is characteristic of many developing countries. Indeed, it has been occurring at rates that exceed those that were experienced in developed countries in their heydays;
- Explores the occurrence of urban primacy and unequal development and emphasises how these often have to be examined and understood at the regional rather then the national scales;
- Reviews the generic strategies that can be employed by states as part of urban and regional planning, these approaches being referred to as 'national urban development strategies';
- Considers the relations between urban–rural areas and looks at the argument that new forms of urban–rural relations and zones can be recognised in respect of developing nations;
- Reviews the various models that have been developed in order to generalise about the internal structure of cities in the developing world. It is clear that cities in the developing world are not following a single unified path, or one that is converging in a simple way on cities found in the developed world;
- Stresses the ways in which the informal and self-help sectors have provided both homes and jobs for the majority poor in developing world cities;
- Reviews the links between urbanisation and environment in respect of the Brown Agenda.

Urbanisation and development: an overview

Over recent history it has generally been assumed that urbanisation – defined as an increase in the proportion of a given population that is to be found living in urban spaces – goes hand in hand with the process of 'development'.

Through time, since the emergence of the first cities some 6000–9000 years ago (Pacione, 2005; Potter and Lloyd-Evans, 1998), it has been assumed that urbanisation, industrialisation and development occur together as joint processes. For this reason it can be argued that urbanisation is one of the most significant processes affecting societies in the late twentieth century and beyond (Devas and Rakodi, 1993; Drakakis-Smith, 2000; Gilbert and Gugler, 1992; Lloyd-Evans and Potter, 2008; Potter 1992a, 2000; Potter and Lloyd-Evans, 1998; Satterthwaite, 2008).

As stressed in Chapter 3, dualistic conceptualisations regard the development process as endeavouring to change what are regarded as traditional, rural, agrarian-based societies into so-called modern, urban–industrial ones, following the model provided by European nations. Hence urbanisation and industrialisation are conflated as essentially synonymous processes.

However, today it is the countries of the developing world that are experiencing the fastest rates of urbanisation and, as this chapter will serve to demonstrate only too clearly, their urban proportions are increasing much more rapidly than those of European countries at their fastest.

Such rapid growth is shown in the world map reproduced as Figure 9.1. As this historic map shows, during the 1980s through to the end of the 1990s the global South was characterised by annual growth rates of urban population well in excess of 2 per cent per annum. Indeed, Figure 9.1 shows that considerable tracts of both Africa and Asia were, for the last two decades of the twentieth century, characterised by urban growth rates higher than 4, 6 or even 8 per cent.

This trend is very different from the past, largely because it has involved a change whereby the current rapid growth of large cities is now firmly associated with poor countries. In rich countries, many central cities are declining in size due to inner-city decay and counter-urbanisation, whereby people are moving from the cities to the suburbs and, indeed, the rural areas. This gives rise to the linked idea that increasing population densities are often associated with cities in the developing world, whereas often these are declining in the core cities in the developed world.

In the mid-1970s Dwyer (1975: 13) observed that 'in all probability we have reached the end of an era of association of urbanisation with Western style industrialisation and socio-economic characteristics' (Plate 9.1).

This is best illustrated by the disparity which characterises the relationship between levels of urbanisation and industrialisation in the Global South. In 1970, the non-communist, less-developed countries, taken as

Plate 9.1 Part of the commercial centre of Tijuana, Mexico, with peripheral squatter settlements in the distance
(photo: Rob Potter)

Figure 9.1 Average annual rate of urban growth 1980–1998 by country
Source: Adapted from *Third World Cities*, 2e, Drakakis-Smith, D., Copyright (© 2000), Routledge. Reproduced by permission of Taylor & Francis Books UK.

Critical reflection

Urban definitions and characteristics

Urbanisation may be defined as the proportion of the population of a nation or region that is to be found living in towns and cities, and is generally represented as a percentage. But, of course, the measure depends on a prior categorisation of what exactly constitutes an urban settlement. In fact, each and every nation exercises its own judgement in determining urban status. Indeed, different criteria and thresholds are used, involving variations on total population size, density of population, predominant economic activity and even legal definition. Bearing this in mind, what do you normally expect of a town or a city? Do you think in terms of a given threshold size of population, or are there key facilities that you expect to be present and which therefore define urbanity in your eyes? Are you aware of the definitions of urbanity that are used by countries around the globe?

a whole, showed a level of urbanisation of 21 per cent, whereas only 10 per cent of the active population was employed in manufacturing, yielding an excess of urbanisation over industrialisation of 110 per cent (Bairoch, 1975; see Critical reflection on Urban definitions).

By contrast, Europe in the 1930s showed a 32 per cent level of urbanisation, but at this point some 22 per cent of the active population were engaged in manufacturing. Thus, the excess of urbanisation over industrialisation was appreciably lower, at around 45 per cent.

As we know from the consideration of global shifts provided in Chapter 4, the industrial expansion that has occurred in developing countries is being concentrated in just a few nations, mainly in Asia, with rapid recent growth in China and India.

In contrast, the majority of poor developing nations now account for a reduced proportion of total world manufacturing output. Thus, for many poor nations, urbanisation currently has little to do with industrialisation, but rather is linked with the creation of jobs in the service sector. This trend will be fully considered later in this chapter.

Urbanisation in the contemporary developing world

A focus on cities

The very rapid growth of towns and cities that has been occurring in the developing world over the last 40 to 50 years can be illustrated in a number of different ways.

One is the increase that has occurred in cities that have a population of a million or more persons in the tropical world. Statistics show that in the 1920s, 24 of the world's cities had more than 1 million inhabitants (Table 9.1). By the early 1980s, the number of such cities had increased to 198. But more significant, during each decade between these two dates the average latitude of these cities had moved steadily toward the Equator.

This figure had changed from 44°30′ in the 1920s, to 34°7′ in the 1980s (Table 9.1). Thus, million-population cities became increasingly associated with the tropical regions of the world (Mountjoy, 1976). In 1950, there were 31 cities of a million or more inhabitants in developing countries, but by 1985 there were 146. It is estimated that by 2025 there will be as many as 486 cities of a million or more in developing countries (Harris, 1989).

The distribution of million cities in the world is shown in Figure 9.2. The map shows metropolitan

Table 9.1 The world distribution of 'million cities': cities with more than 1 million inhabitants, 1920s–1980s

Date	Number of million cities	Mean latitude north or south of Equator	Mean population (millions)	Percentage of world population living in million cities
Early 1920s	24	44°30′	2.14	2.86
Early 1940s	41	39°20′	2.25	4.00
Early 1960s	113	35°44′	2.39	8.71
Early 1980s	198	34°07′	2.58	11.36

Source: Potter (1992a)

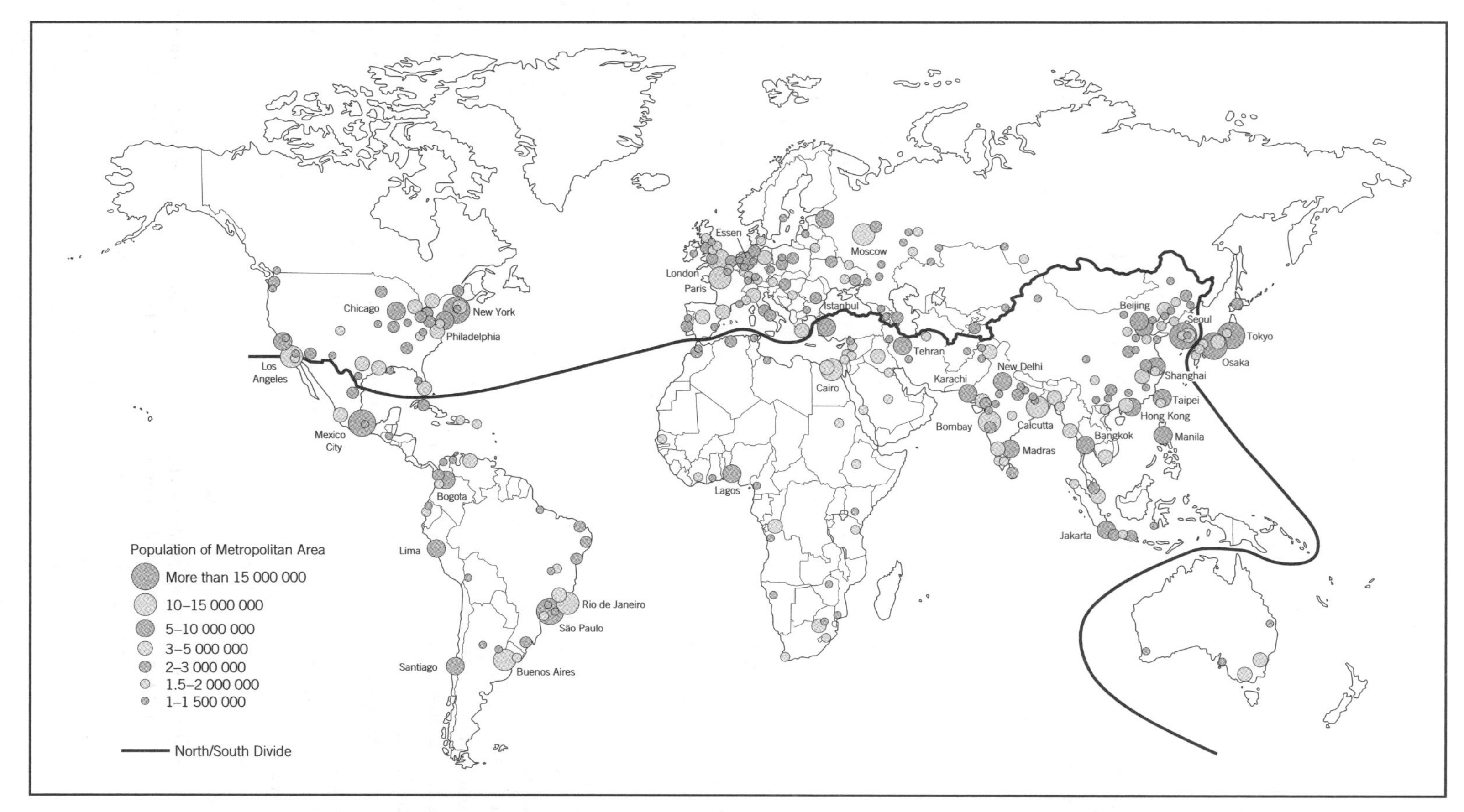

Figure 9.2 Cities with more than 1 million inhabitants

Table 9.2 The largest cities in the world, 1950 and 2005

1950			2005		
Rank	City	Population (millions)	Rank	City	Population (millions)
1	New York	12.3	1	Tokyo	35.2
2	London	10.4	2	Mexico City	19.4
3	Rhine–Ruhr	6.9	3	New York	18.7
4	Tokyo	6.7	4	São Paulo	18.3
5	Shanghai	5.8	5	Mumbai	18.2
6	Paris	5.5	6	Delhi	15.1
7	Buenos Aires	5.3	7	Shanghai	14.5
8	Chicago	4.9	8	Kolkata	14.3
9	Moscow	4.8	9	Jakarta	13.2
10	Calcutta	4.6	10	Buenos Aires	12.4
11	Los Angeles	4.0	11	Dhaka	12.4
12	Osaka	3.8	12	Los Angeles	12.3
13	Milan	3.6	13	Karachi	11.6
14	Bombay	3.0	14	Rio de Janeiro	11.5
15	Mexico City	3.0	15	Osaka-Kobe	11.3

Source: UN (1989, 2005b)

areas with more than 1, 1.5, 2, 3, 5, 10 and 15 million populations. Outside of western Europe and north-eastern America, the largest concentration of million cities occurs in Asia, and particularly in China and India, and there are also concentrations in South and Central America. The substantial number of million plus cities south of the North–South global divide is plain to see from the figure.

The trend towards rapid urbanisation in the developing world is also shown by the league table of the largest urban places in the world. This is given for the years 1950 and 2005 in Table 9.2. In 1950 the three largest cities in the world, New York, London and the Rhine–Ruhr conurbation, were all located in developed countries and seven out of the ten largest world cities were located in the developed world. By 2005, out of the 10 largest cities listed in Table 9.2, eight can be described as being in the developing world, and 11 out of the 15 largest world cities in 2005 were in the 'South'. The largest developing world cities in 2005 included Mexico City (19.4 million), São Paulo (18.3 million), Mumbai (18.2 million), Delhi (15.5 million), Shanghai (14.5 million), Kolkata (14.3 million), Jakarta (13.2 million) and Buenos Aires (12.4 million).

The world's fastest growing cities during the period 1985–2000 are depicted as part of Figure 9.3, along with broad levels of urbanisation by continental division. The map also indicates the commonly accepted geographical definition of the Global South or developing world (see also Chapter 1).

It is noticeable that the fastest growing cities are all located to the south of the line that divides the rich 'North' from the poor 'South'. Thus, large cities such as Mexico City, São Paulo, Lagos, Cairo, Delhi, Bangkok, Manila and Jakarta grew rapidly between 1985 and 2000, as did smaller agglomerations such as Zibo, Surabaya, Pune, Bangalore, Kinshasa, Casablanca, Caracas and Porto Alegre. The salient point is that these cities will continue to grow rapidly into the future. Figure 9.3 gives a very clear indication of the contemporary link between rapid urbanisation and urban growth on the one hand and the less-developed world on the other.

In fact, cities are growing so fast in the Third World that in certain areas they are merging together to form large linked or 'compound' urban regions, a point which is elaborated later in this chapter. These regions are sometimes referred to as megalopolitan systems and although initially associated with the developed world, at least eight of them can now be recognised in the Third World (Gottmann, 1978; Potter and Lloyd-Evans, 1998).

These large urban systems are based on Third World urban regions such as Mexico City, São Paulo, Lagos and Cairo. The terms *mega-city*, *super-city*, *giant city*, *conurbation* are also used to indicate these large, sprawling and complex urban regions.

Since the early 1990s the expression 'extended metropolitan region (EMR)' has increasingly been used to describe new urban forms, especially in the

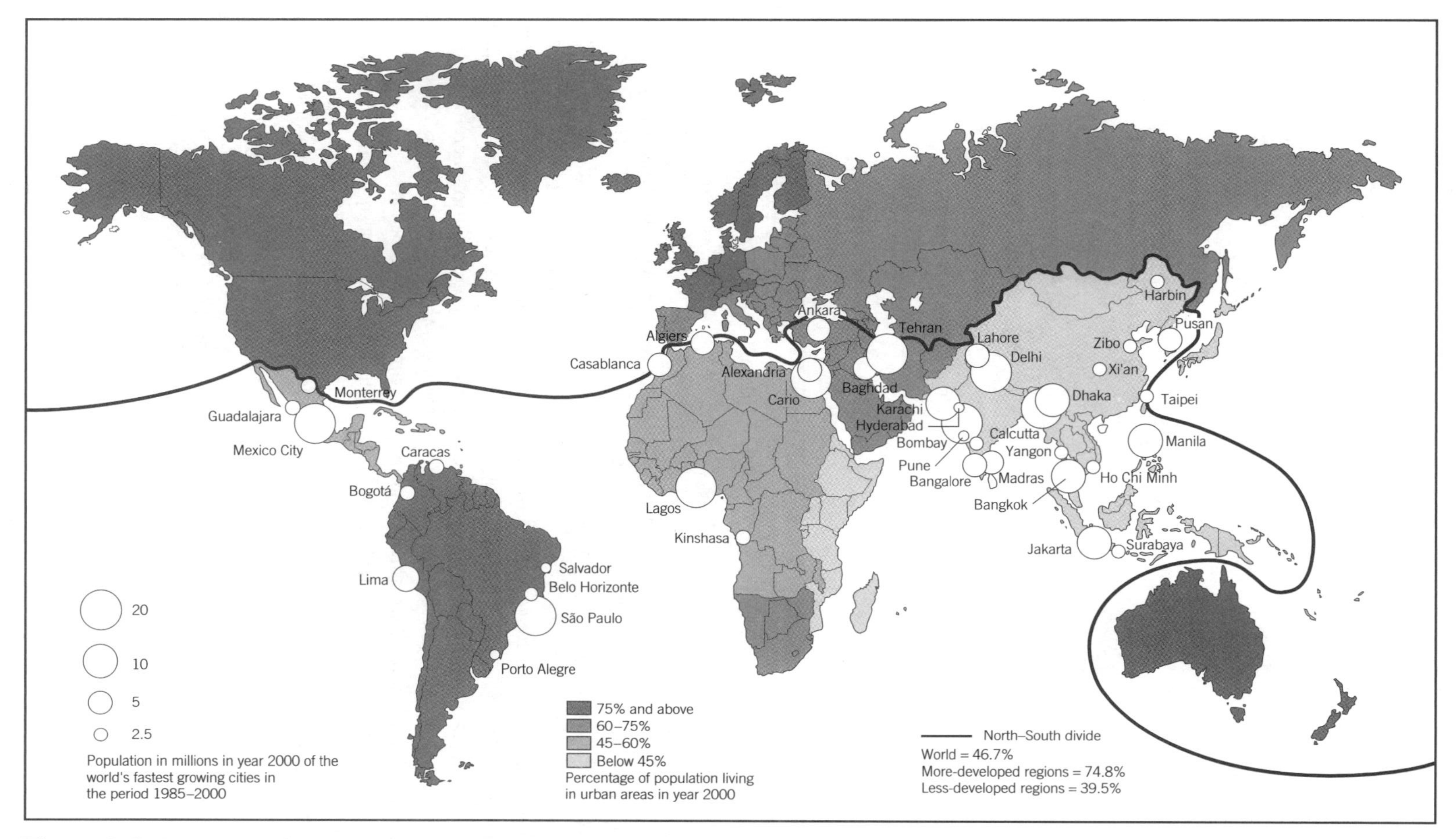

Figure 9.3 Aspects of contemporary world urbanisation

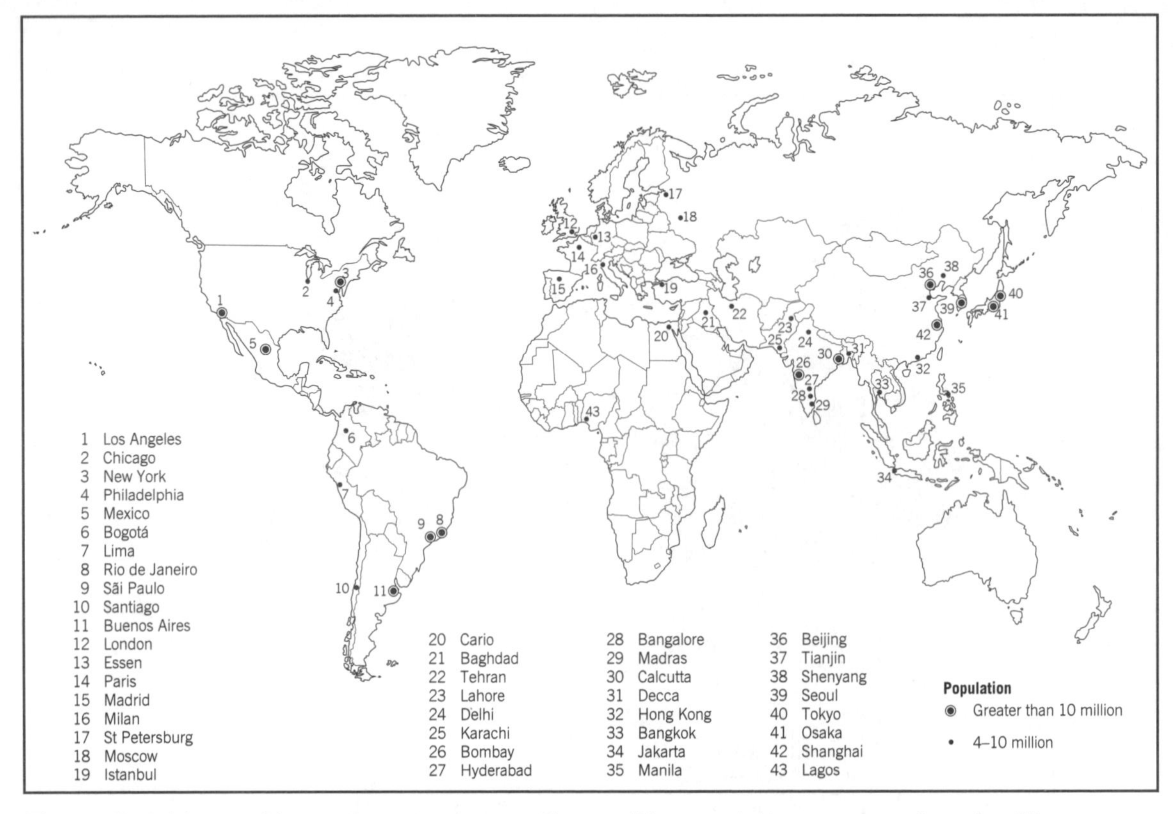

Figure 9.4 Mega-cities: urban agglomerations with populations exceeding 4 million
Source: Adapted from Gugler (1996)

Asian context. The nature of these large urbanised tracts of land is examined later in the chapter when the focus is placed on the changing form of urban and rural linkages in developing countries.

Yet other analysts refer to the concept of the world city, as already reviewed in Chapter 4. This term relates primarily to the international functional importance of certain contemporary cities, irrespective of their size, whereas *mega-cities* are generally defined as those with populations in excess of 8 million (Gilbert, 1996; Oberai, 1993).

Gugler (1996) mapped 43 urban agglomerations with more than 4 million inhabitants, of which around 29 are located in less-developed countries, as shown in Figure 9.4, and again the concentration of such mega-cities in India and China is highly apparent.

A focus on levels of urbanisation

Figure 9.5 summarises the overall increase in the level of urbanisation between the years 1950 and 2025 for the more-developed and less-developed regions of the world.

The clearest feature depicted in the graph is that, although it has flattened out for the developed world since 1965, the urbanisation curve has increased very sharply in the less-developed world. By 2000, 74.8 per cent of the population of more-developed areas was urban, but this was true of only 39.5 per cent of those living in less-developed areas.

The situation is also summarised by the data in Table 9.3. At the global scale, a good deal has been made of the assertion that from 2000 half the world's

Table 9.3 Percentage of total population living in urban areas, 1970–2025

Date	World	Less-developed regions	More-developed regions
1970	37.2	25.5	66.6
2000	46.7	39.5	74.8
2025	60.5	56.9	79.0

Source: UN (1989); further reading: UNCHS (1996)

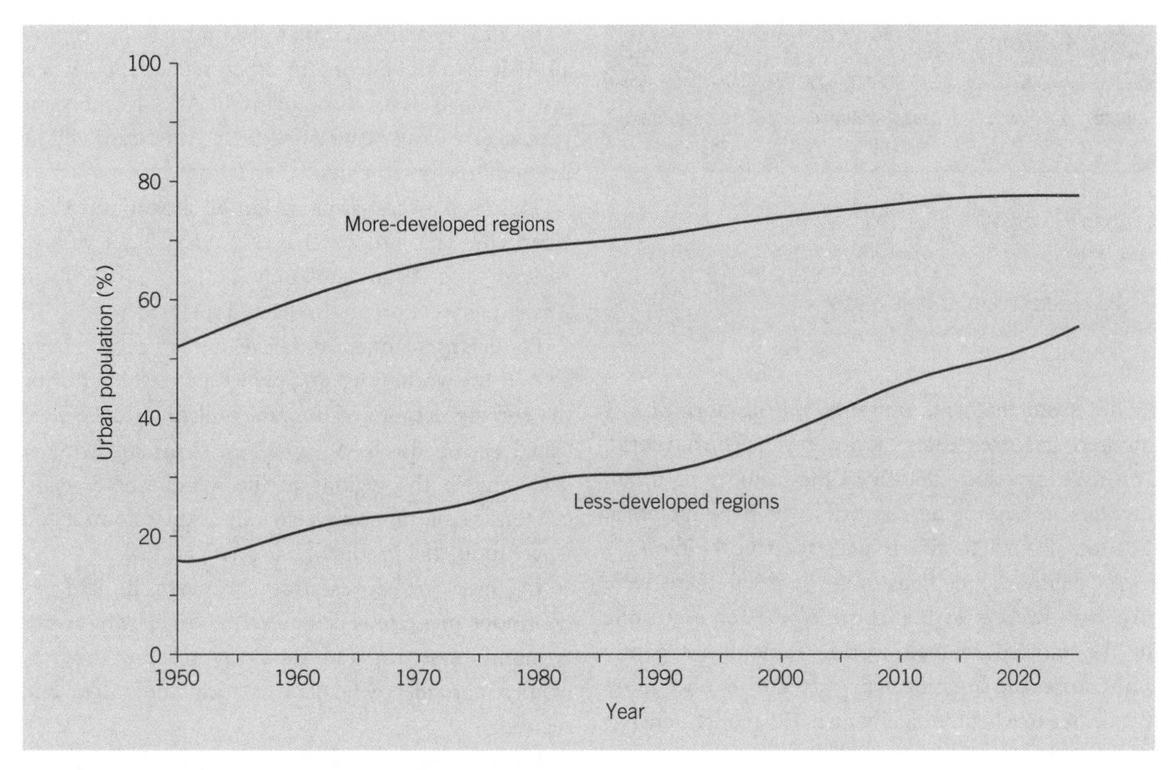

Figure 9.5 The proportion of population residing in more- and less-developed regions, 1950–2025
Source: Adapted from UN (1989)

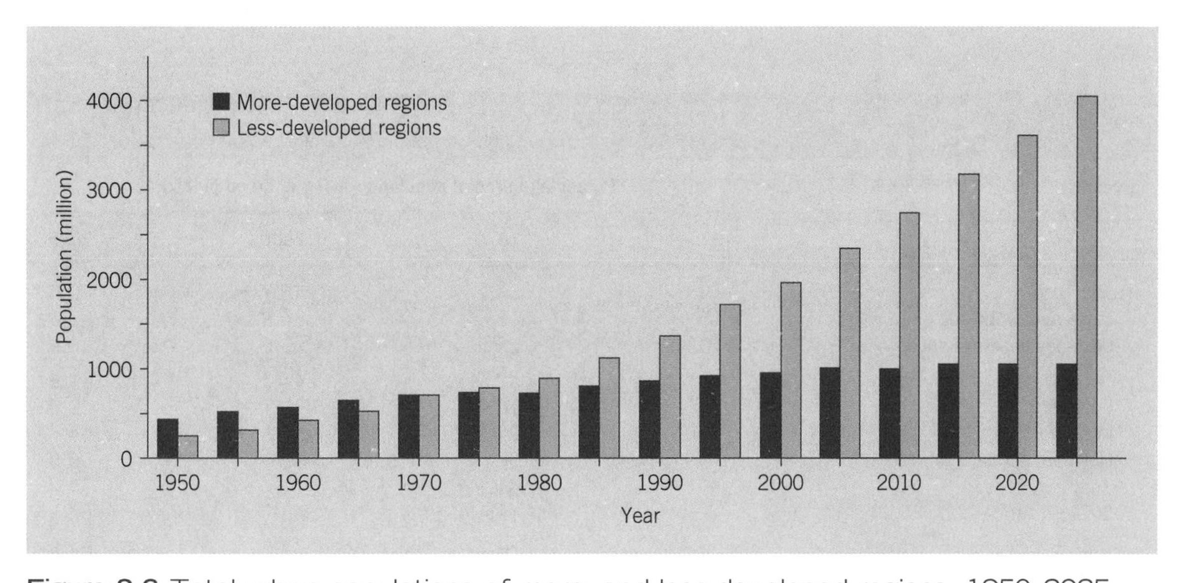

Figure 9.6 Total urban populations of more- and less-developed regions, 1950–2025

total population was to be found living in towns and cities (see, for example, Clark, 1996). However, the official data suggested that this tipping point is only being crossed around the mid to late 2000s, as implied by Table 9.3 (see also Lloyd-Evans and Potter, 2008; Satterthwaite, 2008).

However, the percentage figures tell us only one limited part of the story, and it is the total numbers involved that are truly awe-inspiring. This is illustrated in Figure 9.6. Up to 1970 the absolute number of city dwellers was larger in the more-developed regions than in the less-developed areas of the world.

Table 9.4 Number of people living in urban areas, 1970–2025 (millions)

Date	World	Less-developed regions	More-developed regions
1970	1374	675	698
2000	2916	1971	944
2025	5118	4050	1068

Source: UN (1989); UNCHS (1996)

But from that date onwards, the number of city dwellers has risen dramatically in the Third World. During the period 1950–2025 the number of urban dwellers in developing countries will have increased 14 times, from 300 million to a staggering 4 billion.

As Figure 9.6 demonstrates, by 2000 there were two city dwellers in the Third World for every one in the more-developed world. Table 9.4 cogently summarises the situation. This fact stresses once more the degree to which urban living in the modern era has come to be associated with the poorer countries of the globe.

These data can be disaggregated by major world region, as shown in Table 9.5.

By 2025 it is believed that just under 60 per cent of all Africans will be living in urban settlements, as will just over half of all those living in Asia. In the same year, nearly 85 per cent of all Latin Americans will be living in towns and cities.

This is all a very far cry from the situation in 1920, when UN data indicate that less than 10 per cent of Africans and Asians and only 22 per cent of Latin Americans were urban dwellers (Table 9.5).

Politicians, planners and development experts from all over the world must grapple with these facts during the coming decades of the new millennium. As illustrated above, this very rapid rise in urban living is occurring in the regions of the world where socio-economic conditions are generally at their poorest and where industrial production is relatively low.

In these areas, resources are very limited, so enormous pressure is being exerted on existing socio-economic systems, and especially on the children, women and men who live in such poor areas and regions.

Indeed, dealing with the challenges that are presented by these fundamental changes represents one of the major tasks faced by planners and politicians in the twenty-first century.

Table 9.5 Percentage of population living in urban areas by major continental region, 1920–2025

Region	Percentage of total population living in urban places			
	1920	1970	2000	2025
World	19	37.2	46.7	60.5
More-developed regions	40	66.6	74.8	79.0
Less-developed regions	10	25.5	39.5	56.9
Africa	7	22.9	41.3	57.8
East Africa	–	10.3	30.1	48.0
Middle Africa	–	24.8	47.6	64.7
Northern Africa	–	36.0	49.9	65.3
Southern Africa	–	44.1	61.7	74.2
Western Africa	–	19.6	40.7	58.9
Latin America	22	57.3	77.2	84.8
Caribbean	–	45.7	65.5	75.5
Central America	–	54.0	71.1	80.5
South America	–	60.0	81.0	87.5
Asia	9	23.9	35.0	53.0
Eastern Asia	–	26.9	32.6	49.0
Southeast Asia	–	20.2	35.5	54.3
Southern Asia	–	19.5	33.8	52.6
Western Asia	–	43.2	63.9	76.3

Source: UN (1988); UNCHS (1996)

Causes and consequences of rapid urbanisation in the Third World

Introduction

As already noted, urbanisation can be defined as the process which leads a higher proportion of the total population of an area to live in towns and cities (Potter and Lloyd-Evans, 1998).

Urbanisation is thus a relative measure, recording the percentage of the total population of a nation or a region that is to be found in towns and cities. This should not be confused with the absolute growth of urban areas and urban populations. These are best described by the term *urban growth*.

Throughout the Third World, people are migrating from the rural areas to towns and cities. It is suggested that often about half the growth of cities reflects rural-to-urban migration. For example, during the 1960s the World Bank estimated that migrants as a percentage of total population increase amounted to 50 per cent in Caracas, 52 per cent in Bombay, 54 per cent in Djakarta, 50 per cent in Nairobi and 68 per cent in São Paulo.

In the Philippines in the 1970s the inmigration rate to cities was 1.9 per cent per year, out of an urban population growth per annum of 3.9 per cent. For Brazil during the same decade inmigration accounted for 2.2 per cent per annum out of a total rate of 4.4 per cent per annum (Devas and Rakodi, 1993; UN, 1989).

But why are migrants currently travelling to cities in the Third World in such large numbers? This mainly stems from the widespread existence of poverty, unemployment and deprivation in the rural areas of many developing countries.

Data show that where jobs do exist rates of pay are higher in urban areas, and also that average incomes increase with city size in a progressive manner (Hoch, 1972).

In addition, social and health facilities are better in the principal towns and cities (Phillips, 1990), although access to them is a major problem for the poor in society (Potter and Lloyd-Evans, 1998; Chapter 5).

The informal sector

But just as important is the fact that, with so many people and activities around in cities, it is literally possible to make a living for yourself doing what are known as *informal* sector jobs.

The informal sector is also called the 'tertiary refuge sector'; it is made up of jobs such as street hawking, shoe shining, ice-cream or 'snow cone' vending, car washing, taxi driving and many others (Plates 9.2, 9.3 and 9.4; Lloyd-Evans and Potter, 2008; McGee, 1979; Portes *et al.*, 1991; Santos, 1979).

Such activities subscribe to the notion that 'even to eat crumbs you have to be sitting at the table' (Jones and Eyles, 1977). Later on in this chapter we examine in detail the structure and role of the informal sector.

Plate 9.2 Motorised rickshaw driver, Delhi
(photo: Rob Potter)

Plate 9.3 Informal sector furniture production, Castries, St Lucia
(photo: Rob Potter)

Plate 9.4 Female hawkers in Georgetown, Guyana
(photo: Rob Potter)

Critical reflection

The Informal Sector

As already noted, Plates 9.2, 9.3 and 9.4 show various activities in the informal sector of several cities. Specifically, these show a taxi driver, furniture maker and street-side retailers. In what ways do these activities appear to be essentially similar to those you might encounter in your home locality, or in towns and cities that are well known to you? And in what principal ways do they appear to differ from similar types of activities in other places that are well known to you? What advantages do informal sector activities have over more formal ones? And what challenges might informal sector activities pose for local policy makers and environmentalists? As we shall see later in this chapter, houses are also frequently built within the informal sector. Are you aware of the construction of what may be regarded as informal sector homes in your locality – in the past as well as currently? Or, alternatively, are you aware of informal sector homes constructed in other localities that are well-known to you?

Urban bias in development

The perceived advantages of living in large urban places are of great importance. This is because in the

Key idea

Urban bias in development

A strong argument is mounted by some commentators that development, due to its western origins and orientation, has in the past always favoured the urban over the rural. It is argued that this has been the result of explicit development polities that have emphasised urban industrialisation, and is thereby also reflected in the operation of day-to-day processes in society.

Michael Lipton has been a leading figure in this arena, and since the 1970s has championed the role of small-scale agriculture in development (Harriss, 2006). In a book published in 1977 under the title *Why Poor People Stay Poor: A Study of Urban Bias in World Development*, Lipton stressed what he referred to as 'urban bias' in development (see also Jones and Corbridge, 2008)

Through time, as noted in Chapter 3 in particular, the presumption has been that developing nations should follow a path to industrialisation starting with import-substituting industrialisation. Thus, Lipton argued that agriculture tends to be neglected in the allocation of public investment, whereas the aim of policy should be to reduce the riskiness of agriculture (see Harriss, 2005). It is argued that urban bias in the past has acted against what would be more efficient and equitable ways of using public resources in the agriculture sector, rather than in the urban-industrial sector.

Lipton argued that the urban classes, both rich and poor, have a vested interest in forming a powerful political alliance in the maintenance of urban bias. They have an interest in keeping the prices of agricultural goods cheap and in the transfer of productive resources out of agriculture and into urban-based activities. In this way, urban bias is maintained.

Critical reflection

Bearing in mind the diverse range of theories and strategies of development reviewed in Chapter 3, how do you respond to the thesis of urban bias? Which approaches to development theory and practice reviewed in Chapter 3 seem to support the thesis of urban bias, and which appear to run counter to it? Is it right to argue that states get the urban–rural space-economies that they sign up for in the first place? For example, Table 9.11 shows the idealised settlement policies and patterns that can be followed by states. Can you bring the experience of particular nations with which you may be familiar to bear on the issue of urban bias in development?

past, where they did exist, factories, roads, infrastructure and other facilities focused on the major urban areas in most developing countries. Similarly, efforts to plan the urban system in many nations have likewise stressed the salience of urban areas

Sometimes this so-called 'urban bias' in development has been attributed to the outcome of rational–Western forms of development planning, as discussed in Chapter 3 (Jones and Corbridge, 2008; Lipton, 1977). This key concept is elaborated in the Key idea box.

The demographics: birth and death rates and the cycle of urbanisation

Another very important factor has been enhanced medical facilities in the post-war period, while birth rates have remained at traditionally high levels. Thus, Third World cities are growing by natural increase as well as by migration, although the proportions have varied from region to region.

This was not true of cities in Britain during the Industrial Revolution. Such areas tended to be far less healthy than the surrounding countryside, and were thereby regarded as deathtraps. Quite simply, health conditions are much better in the urban areas of developing nations, where the large teaching hospitals, clinics and the majority of doctors are to be found.

This theme was elaborated in Chapter 5. Third World populations are growing at an average of 2 per cent per annum, so urbanisation is working on much larger base populations.

These sorts of demographic features of contemporary urbanisation are shown if the demographic transition and what is referred to as 'the cycle of urbanisation' are juxtaposed (Figure 9.7).

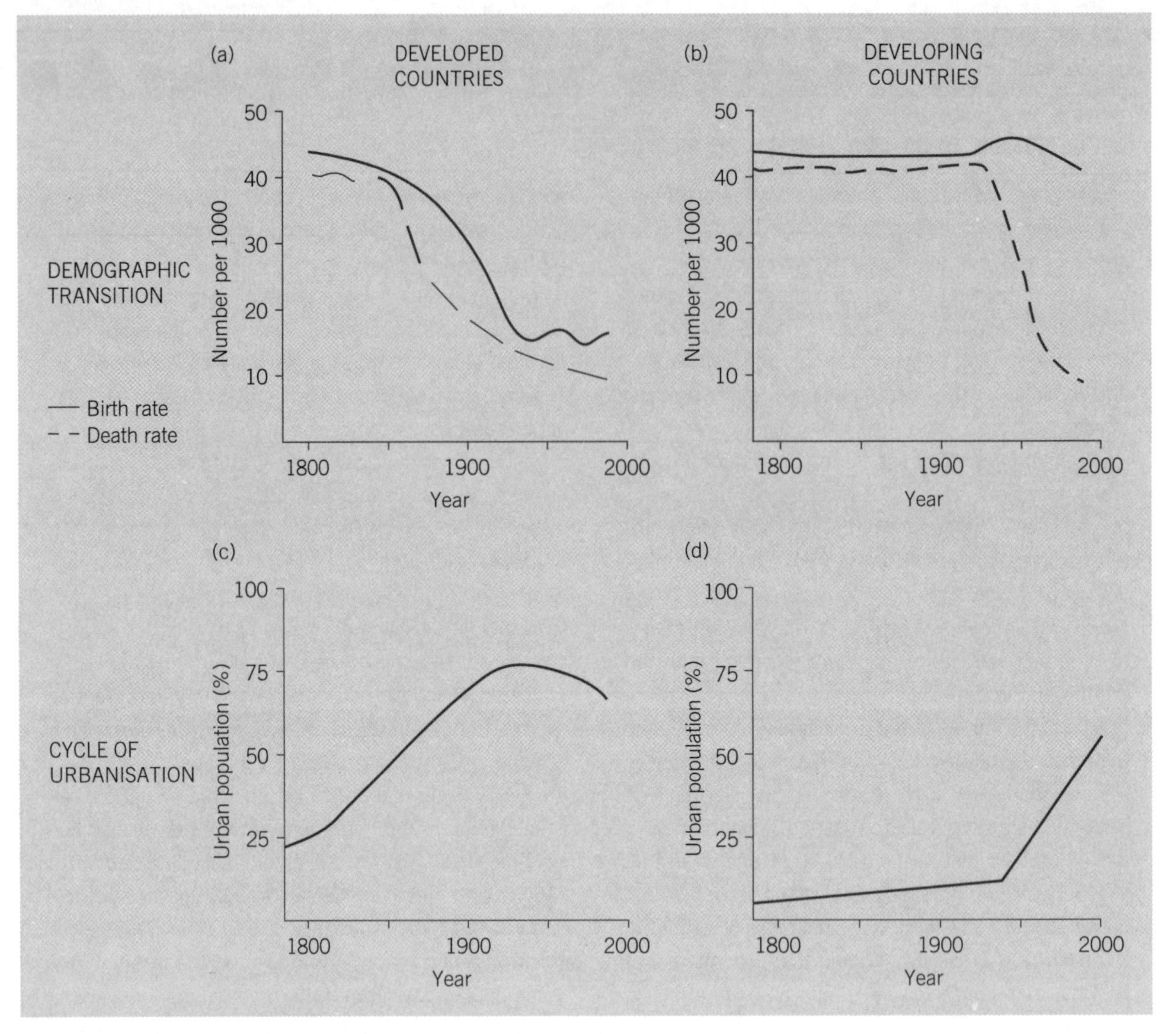

Figure 9.7 The cycle of urbanisation and the demographic transition for developed and developing countries
Source: Adapted from Potter (1985, 1992a)

The countries of the developed world experienced a gradual process of demographic change, as explained in Chapter 5. The rapid urbanisation that occurred in Western Europe and the USA during the late nineteenth and early twentieth centuries was also associated with the rise of the factory system and industrialisation.

Figure 9.7a and c show just how gradual this has been. Birth and death rates both fell relatively gradually from 1800 onwards. In less-developed countries the birth rate has generally continued at traditional levels of 40–45 per 1000 of the population. But since around 1950 crude death rates have often fallen very dramatically, leading to very rapid rates of total population increase (Figure 9.7b). This is frequently known as the telescoping of the demographic transition by developing countries.

Figure 9.7 also shows that, in a similar fashion, urbanisation is occurring at a much accelerated rate in Third World countries. The gradual increase in urbanisation which occurred in the more-developed world is described as the 'cycle of urbanisation' mentioned above. This takes the form of an attenuated or squashed S-shape curve (Figure 9.7c).

However, urbanisation is occurring much more rapidly in Third World countries. The very rapid rise in the urban proportion (Figure 9.7d) occurs at the same time as the massive spurt in population.

Real growth statistics for several nations are graphed in Figure 9.8. The urban proportion for England and Wales increased gradually from around 25 per cent in 1800 to approximately 80 per cent in 1975. The swiftest rise came in the period 1811–1851, and the rate of increase dropped somewhat after that.

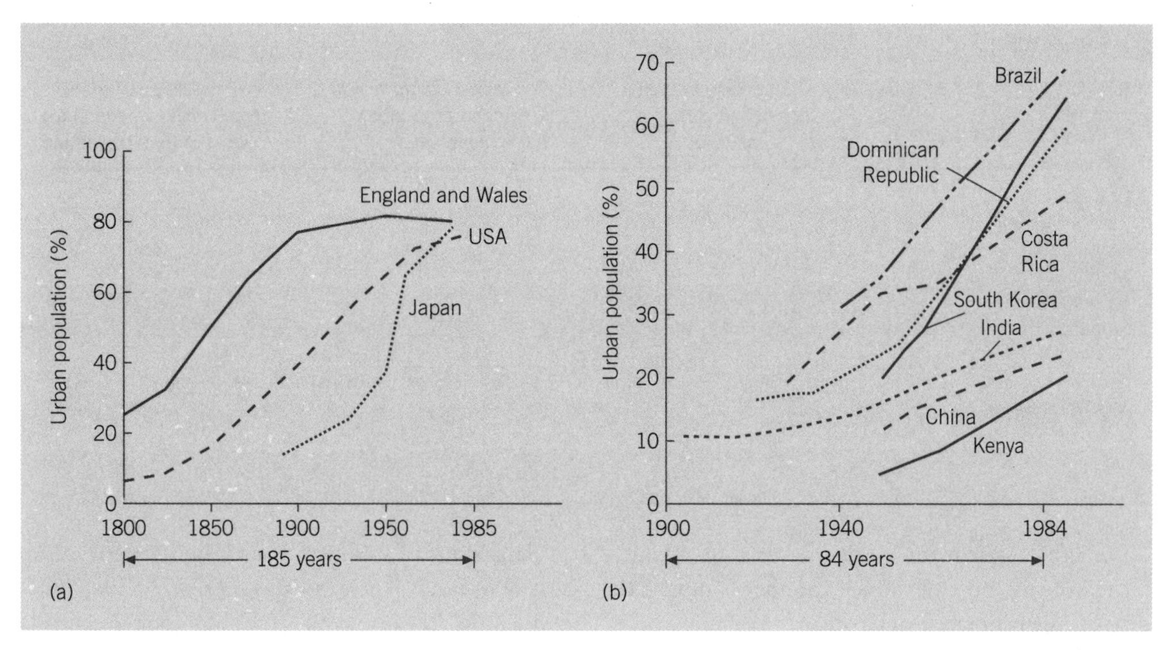

Figure 9.8 Examples of urbanisation curves for developed and developing countries
Source: Adapted from Potter (1992a)

In comparison, countries such as Brazil, Egypt, South Korea and India have shown very rapid rates of urbanisation in the relatively short period since 1945.

Urbanisation as modernisation and dependency

As noted in Chapter 3, some people also point to the fact that various organisations and governments in Third World countries have tended to follow the example set by rich industrial countries. Initially this was due to colonial rule and influence, but even since independence such domination has frequently occurred in the form of imperialism.

This is as true today as in the past, in terms of emulation, where ways of doing things and lifestyles from overseas are valued over and above those from the home country. This is seen as being intensified by the import of foreign goods and the activities of multinational companies, as discussed in Chapter 4.

The arrival of large numbers of tourists from wealthier countries can also give rise to what are described as 'demonstration effects'. Together, these two influences are seen as giving rise to the continuing dependency of many Third World countries on those of the First World (see Chapter 3 for a full discussion of dependency).

It can be argued that too much reliance has in the past been placed on overseas goods, and on ways of doing things that have been derived from abroad. This has also come to be associated with an overconcentration in urban areas, and industry not agriculture has been stressed as the natural path to development (Chapters 1 and 3) – the same conclusion as reached as part of the thesis of urban bias. In other words, development has been based on exogenous rather than endogenous factors (see Chapter 3).

City systems and development: questions of urban primacy, regional inequalities and unequal development

Later in this chapter, socio-economic conditions within Third World cities are examined in detail.

But before that we focus attention on some of the most important discussions about the sets of towns and cities which make up the so-called 'urban system' of nations and regions. Indeed, the idea of an urban system can just as easily be applied at the continental and global scales.

It has frequently been argued that urban primacy – the eminence of one or more cities – is characteristic of Third World urban systems.

Table 9.6 An overview of urban development in the Third World

Region	Percentage of total population classified as urban	Capital city as a percentage of urban population	Percentage of urban population living in cities over 1 million population
Sub-Saharan Africa	29	33	34
East Asia/Pacific	29	12	37
South Asia	25	8	38
Middle East/North Africa	55	26	41
Latin America	73	24	46
World	42	15	38

Sources: Dickenson *et al.* (1996); World Bank (1994b) © International Bank for Reconstruction and Development/ The World Bank

An urban system can be defined as a set of interrelated towns and cities which together comprise the urban settlement fabric of an area.

Table 9.6 shows that for the world as a whole, in the mid-1990s, 15 per cent of the urban population was to be found living in capital cities; the fraction is considerably higher for several of the world's developing regions. The proportion of the total urban population living in the capital was as high as 33 per cent for sub-Saharan Africa, and approximately 25 per cent for Latin America, the Middle East and North Africa.

But despite this broad association, when we turn to the level of urban primacy recorded in individual countries, the issue becomes more complex.

In a pioneering paper, Berry (1961) showed that, at the level of nation states, there is no clear statistical relationship between a nation's city size distribution and either its level of urbanisation or economic development, as measured by GNP per capita.

In the face of these negative findings, Berry speculated that a whole complex of forces serve to influence relative city size distributions. In particular, it was posited that if a few strong forces operate, a primate distribution will be the outcome (Figure 9.9a).

It was argued that fewer forces are likely to influence the urban situation in the case of a smaller country, a shorter history of urbanisation, a simpler economic and political life, and a poorer overall degree of socio-economic development.

The opposites of these cases suggest that a country will develop a range of specialised cities performing a variety of functions. The smooth distribution of cities that results can be called a rank–size distribution (Figure 9.9b) or a log-normal city size distribution (Figure 9.9c).

Berry's line of argument was taken up afterwards in several research papers, notably those by Mehta (1964) and Linsky (1965), both of whom took essentially the same approach, correlating a number of variables against degree of urban primacy for a sample of nations. The results of this approach are summarised in Table 9.7.

Linsky pre-specified the predicted relationships between urban primacy and six variables. He suggested that primacy was positively associated with the degree of export-orientation of the nation, the proportion of the workforce employed in agriculture and the overall rate of population growth. The areal extent of dense population and per capita income levels were envisaged as being negatively correlated with primacy. Finally, somewhat curiously perhaps, an open verdict was initially pronounced on the association between primacy and former colonial status.

The empirical analysis showed that all the hypothesised associations between the variables and urban primacy were as expected. In addition, former colonial status was positively related to levels of urban primacy. This might be expected, for colonial status involves a strong coastal–mercantile orientation, and the attendant urban polarisation that goes with this, as mapped into the mercantile and plantopolis models of settlement evolution in developing countries (Chapter 3).

Significantly, the strongest relationship was the negative correlation recorded between urban primacy and the size of countries ($q = -0.37$ in Table 9.7).

The other statistically significant relationship was the positive one existing between urban primacy and the overall rate of population growth ($q = +0.33$).

Thus, Linsky's work was significant in confirming that although urban primacy is characteristic of small

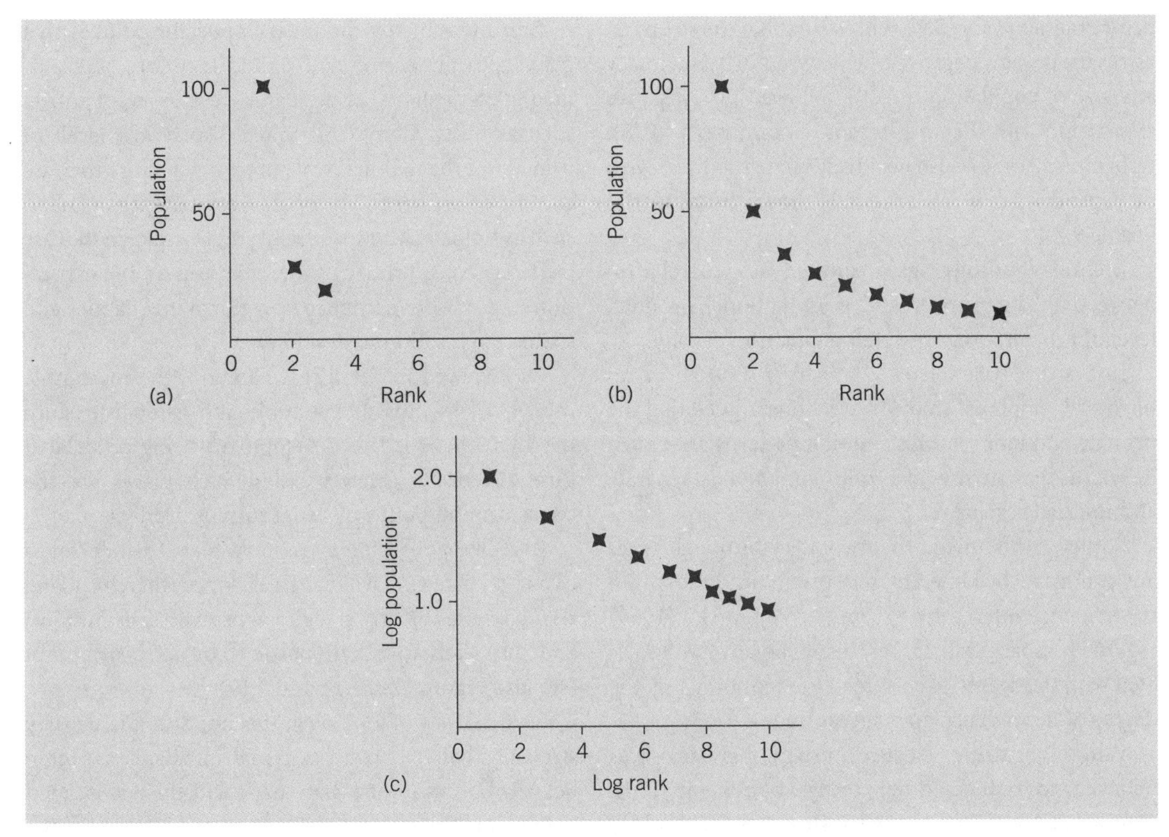

Figure 9.9 Urban settlement distributions: (a) primate, (b) rank–size, (c) log-normal

Table 9.7 The relationships between urban primacy and a selection of other variables

Variable	Expected relation with primacy	Correlation
A. Linsky (1965)		
Areal extent of dense population	Negative	−0.37[a]
Per capita income	Negative	−0.22
Export orientation	Positive	+0.22
Colonial history	?	+0.21
Number in agriculture	Positive	+0.14
Rate of population growth	Positive	+0.33[a]
B. Mehta (1964)		
Gross national product	Negative	−0.08
Level of urbanisation	Negative	−0.12
Overall population density	Negative	+0.02
Export dependency on raw materials	Positive	+0.19
Area of country	Negative	−0.28
Size of population of country	Negative	−0.29

[a] Correlation statistically significant.
Source: Potter and Lloyd-Evans (1998)

nations which have low per capita incomes, a high dependence on exports, a former colonial status, an agricultural economy and a fast rate of population growth, it is certainly not precluded elsewhere. For example, contemporary Thailand exhibits few of these features, yet it presents extreme urban primacy.

At virtually the same juncture, essentially similar conclusions were being arrived at by Mehta (1964), a demographer (Table 9.7), who found that the strongest association was a negative one between urban primacy and size of population (–0.29), followed by a negative association with the areal extent of countries (–0.28). Urban primacy was also positively related to the degree of dependency on raw materials shown by countries (Table 9.7).

Mehta's findings again showed no correlation between urban primacy and gross national product, level of urbanisation or overall population density.

Just a few years later, Vapnarsky (1969) in an historical–empirical study of Argentina argued that the primate and rank–size distribution patterns are not to be seen as the extremes of a continuum, that is, separate and mutually exclusive.

Rather, he argued, the two distributional types are produced by different sets of circumstances. On the evidence offered by Argentina, Vapnarsky (1969) regarded urban primacy as being positively associated with the degree of closure of the economy (i.e. the degree of dependence on overseas trade).

With increasing closure, urban primacy was believed to reduce, other things being equal. In contrast, Vapnarsky saw the distribution pattern as being affected by the level of interdependence existing in a country (i.e. the extent to which its various regions are interlinked by virtue of flows of people, goods, capital, etc.).

It was believed that as internal interdependence increases, the smooth or log-normal pattern will progressively be approached. This gives rise to the argument that various admixtures of the primate distribution and the log-normal distribution can be recognised.

The basic divisions are shown in Table 9.8. The classic primate distribution is seen as the outcome of low closure together with low interdependence; this may prevail for longer in small countries than elsewhere. The classic log-normal distribution and the classic rank–size distribution result from high closure together with high interdependence (Table 9.8).

But undoubtedly the most important point is that primacy really occurs and should therefore be examined at the regional scale. Large developing countries such as India, China and Brazil display low levels of primacy at the national scale precisely because they are comprised of several primate regional urban areas, such as Kolkata, Mumbai, Delhi and Chennai in the case of India. Thus, primacy should be seen as one expression of the wider existence of regional inequalities and spatial polarisation in development.

As reviewed in detail in Chapter 3, the mercantile model shows how global trade and capitalism since the 1400s have led to development being articulated through coastal gateway cities. Such places are the concentration points of social surplus product.

It is likely in a large territory that a whole series of coastal gateways will have developed, thereby giving rise to a series of strong regional primate distributions; and this leads to a national log-normal distribution (see also Figures 3.8, 3.11 and 3.12).

Another way of looking at this important issue is by stressing that it is really regional imbalances – often between an array of urban areas and their associated regions on the one hand, and the rural areas on the other – that are most pronounced. Here the evidence shows quite clearly that the regional inequalities that characterise developing countries are generally far more marked than those which exist in the case of more-developed nations.

For example, if the gross regional products of the richest and poorest regions of nations are compared then developed countries typically show ratios between 1.5 and 3.0 (Table 9.9). Examples are France, where the richest region is 2.09 times better off than the poorest, the United Kingdom (1.43) and the Netherlands (1.56). With respect to developing nations, the ratio between richest and poorest region may be as high as nine or ten times, as Table 9.9 shows for Brazil (10.14) and Iran (10.07).

Perhaps a clearer way of looking at the nature of the differences which currently characterise rural and

Table 9.8 Vapnarsky's cross-classification of city size distributions

High closure, low interdependence	No clear pattern
Low closure, low interdependence	Primate distribution
Low closure, high interdependence	Primate city superimposed on a rank–size pattern
High closure, high interdependence	Rank–size distribution

Source: Based on *Economic Development and Cultural Change*, 17, 1969, pp. 584–95 © The University of Chicago Press, (Vapnarsky, C.A.)

Table 9.9 Regional inequalities in a sample of Third World and developed countries

Country	Ratio of gross regional product between the richest and poorest regions
Developed countries	
France	2.09
Italy	2.20
Japan	2.92
Netherlands	1.56
United Kingdom	1.43
Developing countries	
Brazil	10.14
India	2.24
Iran	10.07
Thailand	6.34
Venezuela	5.72

Reading: Renaud (1981)

urban areas is to consider the proportions of the rural, urban and total populations that are designated as living in poverty.

These data are shown for a range of developing nations in Table 9.10. For Brazil, 73 per cent of the rural populace are classified as living in poverty, but the proportion is only 38 per cent for urban areas. Even for Rwanda – the nation in Table 9.10 that displays the highest national level of poverty – the incidence of poverty is given as 30 per cent of the urban population and 90 per cent of the rural population (UNCHS, 1996).

However, as migration occurs, it has to be recognised that urban poverty is growing faster than rural poverty in some nations.

Evidence suggests to many commentators that differences between regions of this sort are showing relatively little sign of decreasing over time and with development (see Drakakis-Smith, 2000; Friedmann and Weaver, 1979; Gilbert and Goodman, 1976; Gilbert and Gugler, 1992; Potter and Lloyd-Evans, 1998; Stöhr and Taylor, 1981).

Table 9.10 Urban–rural incidence of poverty for a range of developing countries

	Population in poverty, 1980–1990		
	% total population	% rural population	% urban population
Bolivia	60	86	30
Botswana	43	55	30
Brazil	47	73	38
Burundi	84	85	55
Dominican Republic	55	70	45
Ghana	42	54	20
Guatemala	71	74	66
Haiti	76	80	65
Honduras	37	55	14
India	40	42	33
Kenya	52	55	10
Malaysia	16	22	8
Mexico	30	51	23
Morocco	37	45	28
Mozambique	59	65	40
Nepal	60	61	51
Nigeria	40	51	21
Panama	42	65	21
Papua New Guinea	73	75	10
Peru	32	75	13
Philippines	54	64	40
Rwanda	85	90	30
St Kitts and Nevis	46	50	40
Tanzania	58	60	10
Thailand	30	34	17
Venezuela	31	58	28

Reading: UNCHS (1996)

This is despite the fact that the classic view is that regional inequalities increase at first during the early stages of economic growth, but decrease with time thereafter. This U-shaped patterning of inequality over time was first recognised for European countries by Williamson (1965), and others have taken the same essentially laissez-faire stance (Alonso, 1968, 1971; Mera, 1973, 1975, 1978).

However, Gilbert and Goodman (1976) argue that for every Third World country that can be cited as showing a tendency towards regional equality, another can be found that is showing increasing disequilibrium. Indeed, Gilbert and Gugler (1992) and Potter and Lloyd-Evans (1998) argue that for a number of reasons it seems that regional income and welfare convergence is likely to be weak and slow in developing nations (see also Rapley, 2001, for a wider review of the 'convergence debate' in relation to incomes).

This is in no small measure due to the magnitude of the differentials which now exist in Third World countries as well as the vast base populations that are involved.

A more contemporary view of the issues of primacy and inequality can be derived if we go back to the approach presented in Chapter 4, where it is argued that cities are the key points in the dual processes of global convergence and divergence.

The use of the word 'convergence' is somewhat different to the use in the previous paragraph. Following this approach, urban areas in Third World countries are seen as the points of introduction and diffusion of global norms and patterns of consumption (convergence), as discussed in Chapter 1 when dealing with globalisation.

At the same time, cities in the Third World are regarded as the spatial localities at which production, capital and decision making are increasingly being concentrated (divergence). The divergence processes, it is argued, are increasingly coming to be controlled by a relatively small band of large transnational companies and concerns.

It was in this light that Armstrong and McGee (1985: 41) described Third World cities as simultaneously acting as both 'theatres of accumulation' and 'centres of diffusion'. At this scale we are talking about what is happening to the global urban system.

It certainly has to be acknowledged that simple theoretical frameworks and conceptualisations – where urban systems are seen as developing from a primate distribution to a log-normal distribution and a central-place hierarchical structure – are highly unrealistic in the contemporary world.

Such formulations can also be regarded as Eurocentric and linear in conception. Urban systems are better understood as being affected by a larger number of complex forces in the modern and postmodern times in which we live. And the global processes of convergence–divergence offer an interesting framework for exploring the multiplicity of meanings involved in the process of Third World development and change.

Urban and regional planning in Third World countries

The foregoing discussion demonstrates that the crucial point is whether there is faith that the free market will lead to the spread of growth and the equalisation (or convergence) of production, incomes and welfare throughout the national space.

Referring back to Chapter 3, it was Myrdal, in contrast to Hirschman, who believed that such equalisation or 'polarisation reversal' would not occur spontaneously.

Friedmann (1966) had explicitly taken up this theme in his four-stage core–periphery model (refer back to Figure 3.8). Specifically, Friedmann believed that the transition from the second stage of the model, associated with a single strong national core, to the third stage, witnessing the emergence of peripheral subcores, would only come about in developing societies as the result of direct state intervention.

This debate is reflected just as clearly in the literature of the 1970s in what amounted to a debate about the efficacy of the growth of large cities.

In the 1970s, several economists and regional economists followed the neo-classical approach, basically arguing that any attempt to retard the spontaneous growth of large cities would, by definition, be counterproductive because it would serve to retard national rates of economic growth (Alonso, 1971; Hoch, 1972; Richardson, 1973, 1976).

However, in a series of exchanges with Richardson, Gilbert (1976, 1977) took issue with the laissez-faire doctrine of unrestrained growth. From a primarily economic viewpoint, Gilbert (1977) questioned the assumption that higher productivity in big cities is brought about via economies of scale.

Gilbert argued that such growth is essentially achieved at the expense of productivity elsewhere. It was maintained that if 'infrastructure of the same quality were provided in medium-sized centres, the productivity in these centres would rise' (Gilbert, 1976: 29). This is tantamount to saying that capitalist development has promoted an essentially circular argument. It has enhanced the productivity of large cities by concentrating investment in them, and has subsequently argued that only large cities are productive (Gilbert, 2008; Potter and Lloyd-Evans, 1998).

Hence development theory ends up with 'modernisation surfaces' and the equation of urbanisation, modernity and development as virtually synonymous.

Thus, the degree to which the state should become involved in regulating urban growth, and redirecting it at the regional and national scales, is the crucial development planning issue (Gilbert and Gugler, 1992; Potter and Lloyd-Evans, 1998: Chapter 4).

Many social commentators have inferred, or stated overtly, that only socialist states have seriously endeavoured to reduce urban and rural imbalances in national change and development. For example, avowedly anti-urban policies were implemented in South Vietnam between 1975 and 1980, and policies of zero urban growth were followed in China periodically from the late 1950s. But it is Cuba since the socialist revolution in 1959 that is frequently cited as the best example of redressing urban–rural imbalance (Case study 9.1).

It is vital to recognise that urban and regional planning policies must be based on economic, political, even moral and ethical considerations, not just on economic foundations.

Indeed, although Richardson had argued from a strongly pro-large-city standpoint, in reviewing national urban development strategies in the early 1980s he noted for the first time that the key goals were the same as societal goals in general, and that such strategies need to be highly country-specific (Richardson, 1981). In other words, there is no panacea or general solution to urban and regional problems.

Case study 9.1

Cuba: urban and regional planning in a revolutionary state

Cuba, the largest of the Caribbean islands, was discovered by Columbus in 1492 at the dawn of the mercantile period. With the exception of a brief spell of British rule in 1762, Spain retained its colonial power over Cuba until defeated in the Spanish–American War of 1898. This represented the start of a period during which the island was dominated by the United States of America, first militarily and then economically, after independence in 1902.

During the first half of the twentieth century the country was governed by a series of dictators, the last one being Fulgencio Batista whose corrupt regime had ruled the country from 1933. After a two-year guerrilla campaign, law student Fidel Castro and his followers ousted Batista from power in 1959. It is generally accepted that the leaders of the revolution were not initially communists, but fervent nationalists who were opposed to the corruption and inequalities that had existed before. But the antagonistic stance taken by the United States after the revolution resulted in the Cubans increasingly turning to the Soviet Union. Before the revolution, Havana, the capital, was a classic primate city (see Figure 9.10). Most of the wealth and activities of the country were concentrated there. However, it was also characterised by shanty towns, poverty, gambling and vice. By 1953 the Greater Havana area had grown to 1.2 million people, containing 21 per cent of the country's total population. At this time 75 per cent of all industry was found in Havana and 80 per cent of the nation's exports passed through the port, serving to stress the dependent relation of the country to the United States. Most of the country's health care facilities, schools, colleges and cultural organisations were also situated in and around Havana.

Castro and his followers regarded the city as representing capitalist (American) interests and overprivilege. From around 1963, Havana was increasingly discriminated against. Its physical fabric was left to decay so as to make it less attractive to potential rural migrants. Two key policies were adopted: the decentralisation of people and activities from Havana; and the reduction of the striking differences which had come to exist between the urban and rural areas of the nation.

▶

Case study 9.1 (continued)

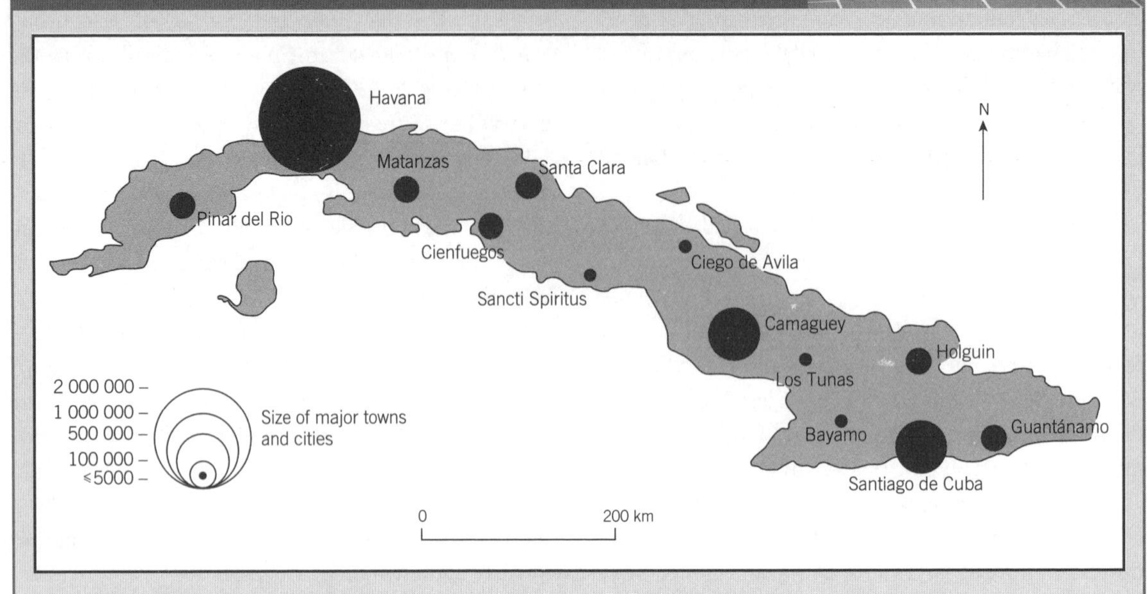

Figure 9.10 The principal towns and cities of Cuba

Thus, since 1959, promoting a more even geographical pattern of development has been the express aim of the state. The growth of provincial towns having populations between 20,000 and 200,000 has been encouraged. At the next level down, the regrouping of villages into rural new towns (*comunidades*) has occurred. Each rural new town has been developed with its own food and clothing stores, nurseries, primary schools, small clinic, social centre, bookshop and cafes. By 1982 some 360 *comunidades* had been created. Control has also been exercised over migration, with ministerial permission being required in order to move to a job in Havana.

Most important, massive efforts have been made to develop primary, secondary and tertiary health care facilities throughout the country. Treatment at the centres is free. Primary health care is available throughout Cuba, whereas secondary and tertiary facilities are located in towns and cities. There have also been great improvements in education. In 1971 only 7 out of 478 secondary schools were to be found in rural areas; by 1979 this had changed to 633 rural schools out of a total of 1318. All students are expected to work in agriculture at some stage in an effort to reduce elitist attitudes and values.

Today Havana has nearly reached the 2 million mark and is certainly much larger than Santiago (351,000) and Camaguey (252,000), the second and third largest cities (see Figure 9.10). Some 69 per cent of the total population of Cuba is to be found living in urban settlements.

Cuba has done much to reduce the differences between town and country, although critics of the Marxist approach which has been followed suggest that the same could have been achieved without the state apparatus that controls all sectors of the economy. They also argue that much unemployment is disguised, that rural–urban differences still exist and that elite privileges have re-emerged.

Whatever the ultimate judgement, there can be no doubting that Cuba's 10 million inhabitants are now highly dependent on external controlling factors. In 1983 the total Soviet assistance to Cuba was US$4100 million, consisting of US$1000 million in development aid and US$3100 million in trade subsidies. Thus, Cuba has remained a dependent state, even though this is now a socialist rather than a capitalist form of dependency.

Source: Potter (1992a)

Table 9.11 Richardson's categorisation of national urban development strategies

Concentrated urbanisation
1. Free market or do nothing
2. Polycentric development of the primate city
3. 'Leap-frog' decentralisation within the primate city

Deconcentration and decentralisation
4. Development corridors and axes
5. Growth poles and growth centres
6. 'Countermagnets'
7. Secondary cities
8. Provincial capitals
9. Regional centres and hierarchy
10. Small service centres and rural development

Source: After 'National urban development strategies in developing countries', *Urban Studies*, Vol. 18, pp.267–83; permission from Taylor & Francis, www.tandf.co.uk/journals (Richardson, H.W. 1981)

Richardson's paper was particularly useful in itemising the wide range of policies which can be employed in attempts to decentralise people, jobs and social infrastructure away from primate cities and congested core regions.

The variety of policy reactions is shown in Table 9.11. These range from three policies of continued concentrated urbanisation to seven representing genuine interregional deconcentration and decentralisation.

The first policy of concentrated urbanisation is the laissez-faire policy of letting the market take its course. If, however, problems of congestion and imbalance are recognised in the primate city, efforts may be made to decentralise, though merely within the core region. Thus, a polycentric pattern of growth on the edge of the primate city, or a form of leap-frog decentralisation to the edge of the existing core, may be envisaged (Table 9.11).

Strategies of genuine deconcentration can be categorised into seven generic types, as shown in Table 9.11. Development corridors or axes can be designated, leading from the core region, and growth can be focused upon them.

Alternatively, growth may be channelled into what are regarded as dynamic growth poles or growth centres. A variation on essentially the same theme sees the strengthening of a few distant major nodes as countermagnets.

Other forms of decentralisation can be created by the promotion of a limited number of secondary or intermediate cities, or the establishment of provincial state and departmental capitals.

Yet another variant involves the promotion of regional metropolises and an associated hierarchy of urban places. At the far end of the spectrum, a dispersed policy of small service centres and associated rural development throughout the periphery may be pursued.

Of course, these strategies are not mutually exclusive and several of them are very similar; various elements of these strategies can be combined into any number of hybrid forms. Examples of the ways in which Cuba and Nigeria have applied national urban development strategies like these are provided in Case studies 9.1 and 9.2.

In conclusion, it is re-emphasised that arguments about urban and regional systems planning cannot sensibly be based on economic reasoning alone.

As demonstrated by Case studies 9.1 and 9.2, strategic social, political and ideological issues are just as important in the equation. This is also apparent when it is stressed that decentralisation and deconcentration are relative concepts and policy instruments.

The choices are socio-political and moral, so once again we encounter a classic position where it must be accepted that there are many urban and regional geographies of the future which may be promoted by the state or other responsible agencies. These circumstances are reflected in the current policy arena. Thus, although much of the literature over the past 10–15 years has stressed the importance of bottom-up and grassroots approaches to national planning, the World Bank and the United Nations Development Programme have been returning to the argument that urban growth and large cities are the keys to development and change (UNDP, 1991; World Bank, 1991).

This argument is strongly based on the success of the Asian newly industrialising countries (NICs), and on what is regarded as the overall failure of rural-based development programmes.

The policies of the New Right, involving deregulation, privatisation, the rolling back of the state, export-based programmes of industrialisation, structural adjustment programmes (SAPs) and poverty reduction strategies (PRSs) are all signifiers of what some have called the new urban management programme of the World Bank or the 'Washington consensus' (Drakakis-Smith, 2000; Gould, 1992; Harris, 1992; T. McGee, 1994; Potter, 2000; Potter, and Lloyd-Evans, 1998: Chapter 10; Rojas, 1995; Yeung, 1995).

Case study 9.2

Nigeria: urban and regional planning in a top-down context

Nigeria is the largest nation in Africa and currently has a population of 94 million. In 1471 the Portuguese were the first Europeans to visit what is today the Nigerian coast, and they were followed by visitors from other European countries. British colonial rule dated from 1900. From the colonial era to the present, policies have tended to be top-down or from above, and development has been concentrated into a limited number of areas. Planning strategies, since their introduction in 1946, have been essentially market-oriented, concentrating on the production of agricultural crops for export and import substitution industrialisation. Investment and industrial plants have focused on the cities.

Today the 12 major cities of Nigeria account for nearly 77 per cent of all industrial establishments in the country and 87 per cent of the total industrial employment. However, in 1985 only 23 per cent of the population lived in towns and cities. Of the total employment in manufacturing, 76 per cent is to be found along the coastal belt (see Figure 9.11). The capital, Lagos, which currently has a population nearing 2 million, accounts for well over 50 per cent of the nation's industrial wages, nearly 60 per cent of its gross output, 49 per cent of all industrial employment and 38 per cent of total industrial plants. Within each of the states making up the country, services and jobs are also strongly concentrated into the state capital. For example, in the north of the country, in Kano State, the Kano metropolitan area contained 71 of the 73 industries that were operating in 1971, and 11 of the 12 banks.

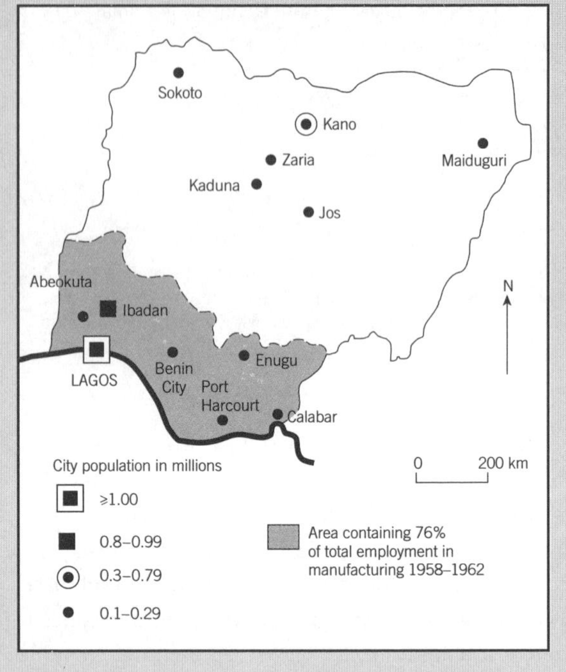

Figure 9.11 The principal cities and manufacturing zone of Nigeria

In such circumstances it is perhaps not surprising that rural-to-urban migration has been very strong and the main cities have grown extremely quickly. For example, during the period 1952–1963, Port Harcourt grew at the exceptional rate of 10.5 per cent per annum, whereas Lagos and Kano increased their populations at 8.6 and 7.6 per cent per annum respectively. A long search for oil proved to be successful in the mid 1950s, and by 1963 oil accounted for 3 per cent of government revenues. In 1982 oil represented 90 per cent of the value of the country's exports. However, many people maintain that the oil monies have been used inefficiently, leading to massive imports of expensive foreign goods.

Although many agree that the Nigerian economy has grown, others maintain that it has not developed. They suggest that the majority of the population are not better off and that deep regional inequalities still characterise the country. These critics claim there have been relatively few trickle-down effects of growth from the urban areas to the rural areas. Too much emphasis has been placed on sectoral growth – the promotion of different areas of the economy, such as industry – but little regard has been paid to the geographical consequences. Agriculture has been neglected, the drift from the land to the cities has not been reduced, and the country remains strongly dependent on the nations of the West.

Source: Potter (1992a)

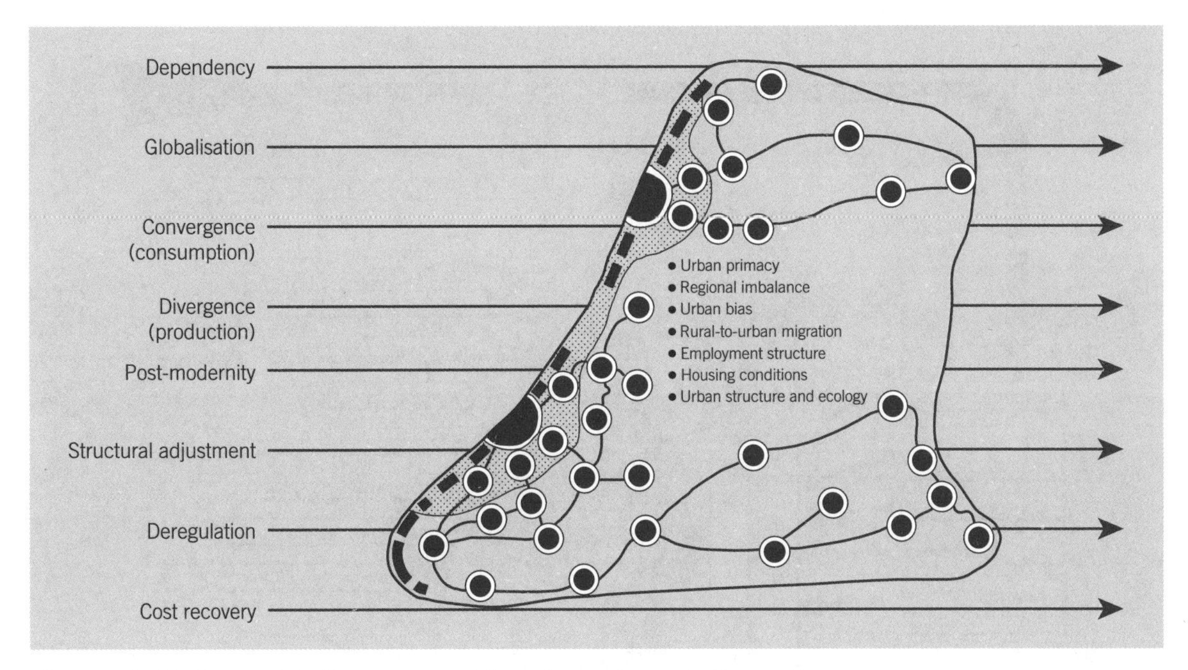

Figure 9.12 Current influences on the character of city systems
Source: Adapted from Potter (1995a)

Some of the forces affecting urban systems in the Third World are summarised in Figure 9.12. They provide the contemporary context in which aspects of urban–rural imbalances and urban primacy need now to be studied.

Whatever the balance of policies followed in the global urban and regional arena, massive urban growth and development are inevitable.

Some evidence does now exist to suggest that the rate of growth of major cities may have lessened in Latin America and the Middle East in the 1980s, partly due to the effects of austerity packages; but on the other hand, secondary and intermediate cities appear to be growing more rapidly than ever before (Gilbert, 1993; Portes *et al.*, 1997).

Thus, it is inevitable that continued urban growth and urbanisation will be the order of the day for the foreseeable future in developing societies.

Rural–urban interrelations in developing countries

Just as Chapter 3 has shown that in the past too much of a distinction has been drawn between the categories 'developed' and 'developing', along with 'core' and 'periphery', the same can be said about the terms 'urban' and 'rural'.

Although definitional exercises involve the recognition of a gradual transition between rural and urban settlements (that is the existence of a rural–urban continuum); the politico-administrative need to define hamlets, villages, towns and cities has involved the imposition of clear thresholds and boundaries.

It seems that the same sort of categorical thinking has come to affect the identification of rural and urban zones in most other respects as well. In other words, all too frequently, writers have implied that the urban and the rural are essentially discrete and separate entities, in both physical and functional terms. The aim of the present section is to show that in respect of urban policy and urban management, this is a very simplistic and essentially unhelpful view.

Work over the last decade or so has stressed how urban and rural areas are closely interrelated. This theme is fleshed out under the three headings: (i) rural–urban interaction in developing countries, (ii) the nature of peri-urban zones and (iii) extended metropolitan regions. Several of the themes raised here are picked up and are developed in Chapter 10, which deals with rural landscapes and development issues.

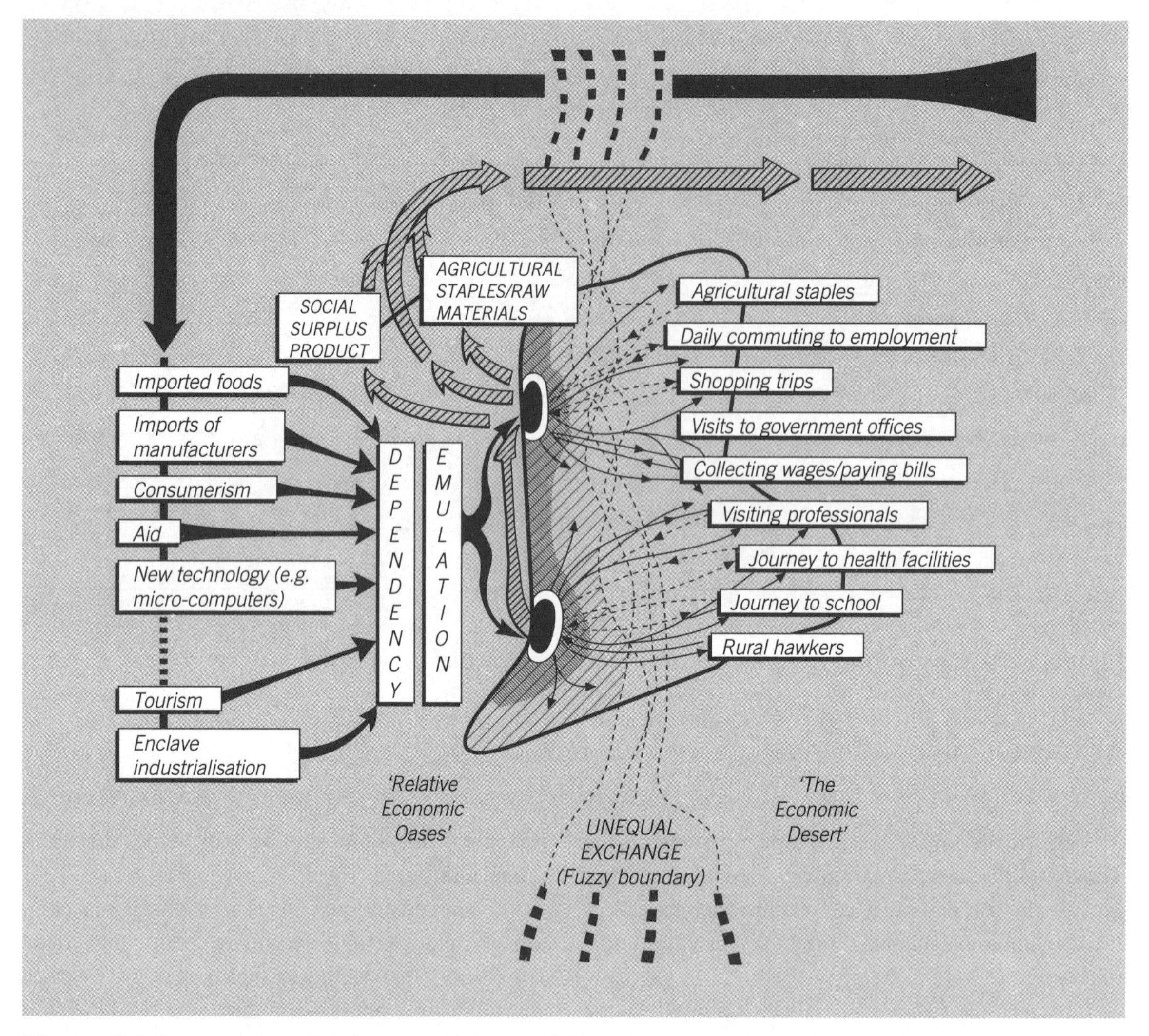

Figure 9.13 Aspects of urban–rural interrelation in an hypothetical developing country
Source: Based on Potter (1989, 2000)

Rural–urban interaction in developing countries

Some of the first efforts to contest dichotomous thinking about the urban and the rural occurred in the 1980s. Several works focused attention on the fact that strong and complex interactions occur between rural and urban areas, and that these underpin many aspects of the development equation (see Dixon, 1987; Potter and Unwin, 1987).

Just like the argument about development and underdevelopment being opposite faces of the same coin (that is mutual or linked conditions), so it can be stressed that the differences between urban and rural at their extremes are maintained by the strong functional interlinkages which exist between them on a day-to-day basis.

This is suggested in diagrammatic fashion in Figure 9.13. For example, much of the cheap produce sold on urban stalls and in urban markets is picked and transported by rural dwellers who then transport such goods to town. Some of these will be part-time rural agriculturists that also act as urban higglers.

Via processes of unequal exchange and urban bias, the low support prices of agricultural products serve to reduce the costs of city life. Other urban–rural flows involve trips to shop, to visit offices, and to health and education facilities. All of these movements serve to etch out what are major differences between rural and urban zones.

And yet, it is these strong flows which generate such wide and marked differences in the first place. Work on the complex nature of urban–rural differences continues at the present time.

The nature of peri-urban zones

Another rural–urban topic has commanded a mounting volume of attention among geographers and others interested in development studies over the last ten years or so.

This has principally involved recognising the social and economic salience of the areas that exist at the very edges of cities in the developing world. These tracts of land are frequently referred to as 'peri-urban zones', and represent areas of active assimilation of agricultural lands into the urban network.

As such, these zones show a mixture of land uses. The recent growth of urban population is reflected in the existence of large peripheral squatter settlements and shanty towns.

At the same time, these very areas are often the locales for intensive agricultural production in order to supply the food needs of the rapidly burgeoning urban populations (Lynch, 2002). Such urban-based agriculture is common in many cities of the developing world. Case study 9.3 look at the peri-urban zones of Kano, Nigeria (see also Plate 9.5, showing the Kano close-settled zone, Nigeria).

The research that is being done on such urban agriculture is examining the advantages and limitations of such production. The cons are perhaps far less obvious than the pros. They include the potentially worrying possibility of the use of polluted groundwater for intensive agricultural production. This point is returned to later in this chapter.

Extended metropolitan regions (EMRs)

Recently, some analysts have suggested that what amount to new rural–urban complexes are developing and that the nature of the urbanisation process and urban structure are thereby changing fundamentally. This idea has been presented by Terry McGee (1991, 1995) (see Key thinker box) among others, in the context of Asian cities.

Key thinker

Terry McGee

Terry McGee's contribution to development studies has been summarised by Lea (2005). Terry McGee was born in Cambridge in the Northern island of New Zealand in the heart of the Waikato agricultural area. After a teaching diploma he took a degree specialising in geography at the Victoria University of Wellington, New Zealand.

His first academic appointment was to a lectureship in Geography at the University of Malaya in Kuala Lumpar in 1959 and he remained in this position for six years, during which time he undertook a PhD on Malay migration to Kuala Lumpur City and travelled extensively through Southeast Asia. This led directly to the publication of the book *The Southeast Asian City: A Social Geography of the Primate Cities of Southeast Asia* in 1967. In this book, McGee saw urbanisation in the region as being characterised by conspicuous consumption and the extraction of surplus from the countryside (see Chapter 3). He employed the term 'pseudo-urbanisation' to imply that urbanisation was acting as a brake on effective and progressive development. In the 1970s, McGee's research focused strongly on the contribution made by the informal sector in developing world cities.

Urban–rural differences became a topic of particular concern to Terry McGee during the 1980s and this gave rise to the identification of what have come to be known as Extended Metropolitan Regions (EMRs) – areas within which new urban–rural forms are to be found mixed together. In the mid 1980s, with Warwick Armstrong, McGee published *Theatres of Accumulation*, which proved to be influential, not least in stressing the relevance of the concepts of convergence and divergence in relation to patterns and processes of urbanisation in developing nations.

Plate 9.5 Traditional Hausa village in the Kano close-settled zone, northern Nigeria
(photo: Tony Binns)

Case study 9.3

Urban and peri-urban agriculture in Kano, Nigeria

Nigeria, with a total population exceeding 120 million, is Africa's most populous country. The country had an average annual population growth rate of 2.9 per cent between 1990 and 1997, whereas the annual growth rate of the urban population was approaching 5 per cent, leading to a projected 55 per cent of the total population being urban-based by 2015 (UNDP, 1998). As in many developing countries, there is an increasing demand for fresh foodstuffs in Nigeria's cities, and large quantities of food are now being produced within the urban and peri-urban areas.

The city of Kano, with an estimated population of 2–3 million, dominates northern Nigeria. Located in the semi-arid savanna belt, with an annual average rainfall between 1961 and 1991 of under 700 mm, and a long dry season from late September to May, the region experiences considerable variation in both the amount and frequency of rainfall from one year to the next. Dry season cultivation is dependent upon irrigation, and low-lying areas in river valleys and depressions, where the water table is close to the surface (known locally as fadamas), are valuable locations for such cultivation. The construction in the last 20 years of a number of dams and associated irrigation schemes in Kano State, together with the sinking of wells and boreholes, has resulted generally in more water being available for dry season cultivation. However, where facilities such as abattoirs and tanneries discharge their effluence into rivers and drains, pollution of water sources can be a serious problem, particularly during the dry season when rainfall that might dilute and flush out toxic elements is absent (Lewcock, 1995).

A survey undertaken in 1996 discovered considerable amounts of fruit and vegetables being produced in and around Kano, within 10 km of the walls of the old city (Figure 9.14), and mainly located near major routes. Fruit and vegetables were also piled high along the roadside, waiting for collection and passing trade. The production, transporting and marketing of such fresh produce is a significant income-generating business, as well as satisfying the basic needs of the urban population. While wealthy households and businessmen see fruit trees as a form of investment, 'resource-poor' cultivators grow mainly vegetables,

Case study 9.3 (continued)

with some fruit, for home consumption and sale. As Figure 9.14 shows, many plots are actually located in built-up areas, with limited amounts of cultivation even within the walls of urban family compounds, in some cases undertaken by women who, under Islamic tradition, are in seclusion. Plots in the built-up area are typically small, ranging from 0.01 to 0.40 hectares, while in the peri-urban area they are generally larger, between 0.1 and 2 hectares. Most growers outside compounds are men between the ages of 30 and 70, with little, if any, formal education. Traditional tools such as hoes and cutlasses predominate, though some farmers use water pumps and apply chemical fertilisers and pesticides, when available, as well as manure and compost. The most common vegetable crops grown are spinach, maize, okra, lettuce, onion, tomato, carrot, sorrel, pepper and sugar cane, while the main fruits are mango, guava, cashew, orange and pomegranate.

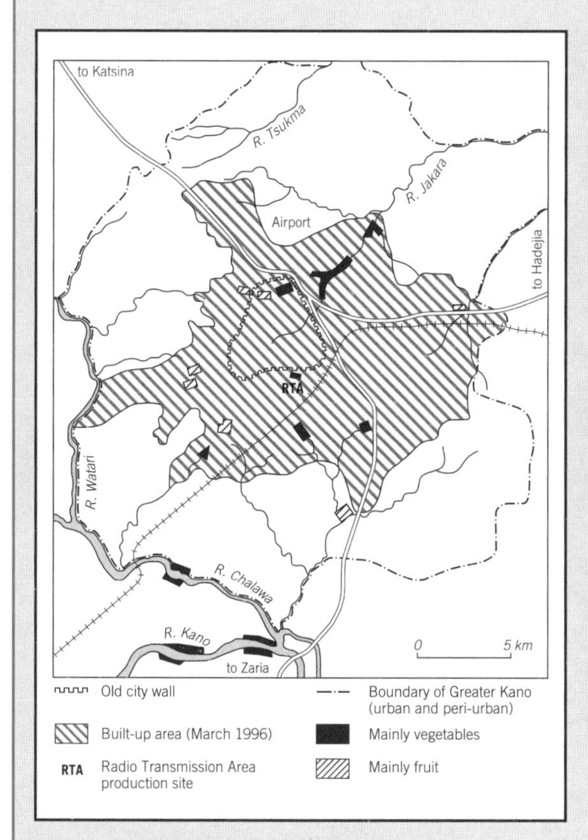

Figure 9.14 Main horticultural production sites, Kano

Source: Adapted from Bayero University Kano Survey map

A particularly important area of vegetable production in Kano lies underneath the transmission masts of the Federal Aviation Authority (FAA), on the southern side of the old city and just across the main road from the city walls (RTA on Figure 9.14). The site covers an area roughly 1 by 0.5 km, and draws its main water supply from a drain leading from the old city (Binns and Fereday, 1996). This area was opened up to cultivation in the early 1980s, when the last civilian government under President Shagari gave permission under its 'Green Revolution' initiative that all vacant public lands within urban areas could be used for cultivation without charge. However, this permission has never been formalised, so tenure of such land is by no means secure. The Federal Aviation Authority is generally satisfied that farming activities have improved the condition of the site and there seems to have been little negative impact on the environment. The area is divided into two sections; the first, located alongside and as far as 200 metres to the west of the main drain, uses irrigation for the year-round cultivation of vegetables and grains, such as lettuce, spinach, okra, maize and rice, while the other section of the site grows rainfed staple crops, such as sorghum and millet during the wet season.

Prospective cultivators must first seek permission from the Aviation Authority's officers and land is allocated on a 'first-come, first-served' basis. There is stiff competition for plots, and those acquired more recently are generally smaller than those occupied earlier. Plot size in the irrigated area is between 0.01 and 0.4 hectares, while plots located in the rainfed cultivation section are generally two to four times larger. During the dry season there is considerable competition for water, such that farmers may even irrigate at night. Buckets are mainly used to raise water and only one of the farmers interviewed in February 1996 had a petrol pump. Those farmers with plots situated away from the drain have in some cases sunk shallow wells to tap rather less polluted groundwater, though FAA officials were concerned about possible damage to underground transmission cables.

▶

Case study 9.3 (continued)

Plate 9.6 Boy with donkey conveying urban waste to peri-urban fields around Kano, northern Nigeria
(photo: Tony Binns)

Most farmers use a combination of family and hired labour. There are three cropping seasons for vegetables: the dry season with full irrigation lasts from November to March; a transition season starts in April and continues into the rainfed cropping season in June; wet season cropping usually begins in early June and continues until late September. Some chemical fertiliser is used, but farmers experienced difficulty in acquiring it, many using ash, household refuse and animal manure instead to maintain soil fertility, although there is also often some difficulty in acquiring these items, as shown in Plate 9.6 (Lewcock, 1995). Farmers generally received little, if any, advice from extension workers and a 'self-help' credit association collapsed because of the lack of transparency in the way its leaders managed the funds. Over 75 per cent of those interviewed on the FAA site said that they relied on horticultural production to supplement earnings from their main occupation and that horticultural production was more lucrative than growing rainfed staples.

Vegetables and fruit were generally head-loaded or transported by bicycle to a local market on the southern edge of the production site. However, in some cases crops were sold directly to local consumers or to market traders and middlemen. Sometimes, an entire plot of maize or carrots, for example, will be sold to a trader who first visits the farm to negotiate a price and then arranges for the harvest and transport of the produce to the larger city markets. The largest quantities of vegetable crops are sold in the dry season, when growing conditions for crops such as tomato and pepper are more favourable.

The most common marketing chain goes from the producer to the major wholesaler, to the lesser wholesaler, the retailer and, finally, to the consumer. However, some farmers sell produce such as leafy vegetables directly to local consumers or street traders. Given the perishability of fruit and vegetables, ease of movement between production sites and the main fruit and vegetable markets is important. Transportation seems to be the responsibility of whoever owns the crop at the time of transport. Large (about 9 tonne) trucks, minibuses, taxis, motor cycles, bicycles and sometimes donkeys are the main forms of transport.

Some markets sell a wide range of fresh products and manufactured goods, while others are more specialised, such as Yan Kaba market which sells vegetables, Yan Lemo fruit and Bachirawa market

Case study 9.3 (continued)

onions. Market traders identified the lack of cold storage facilities, the high cost of transport and communication difficulties between producers and buying agents as the main problems in dealing with fruit and vegetables.

Urban agriculture in Kano is a significant form of land use, employment and food supply. In terms of the future sustainability of urban and peri-urban agriculture, Kano respondents identified two important concerns; first, the heavily polluted nature of much of the water used for irrigation and, second, the uncertainties surrounding security of tenure in many crop production locations (Lynch *et al.*, 2001). These and other issues need careful consideration, but there is divided opinion among urban authorities on the merits of encouraging the further development of urban and peri-urban agriculture. Urban planners are concerned about such issues as land-use conflicts and possible disease transmission when irrigated areas are closely juxtaposed to dwellings.

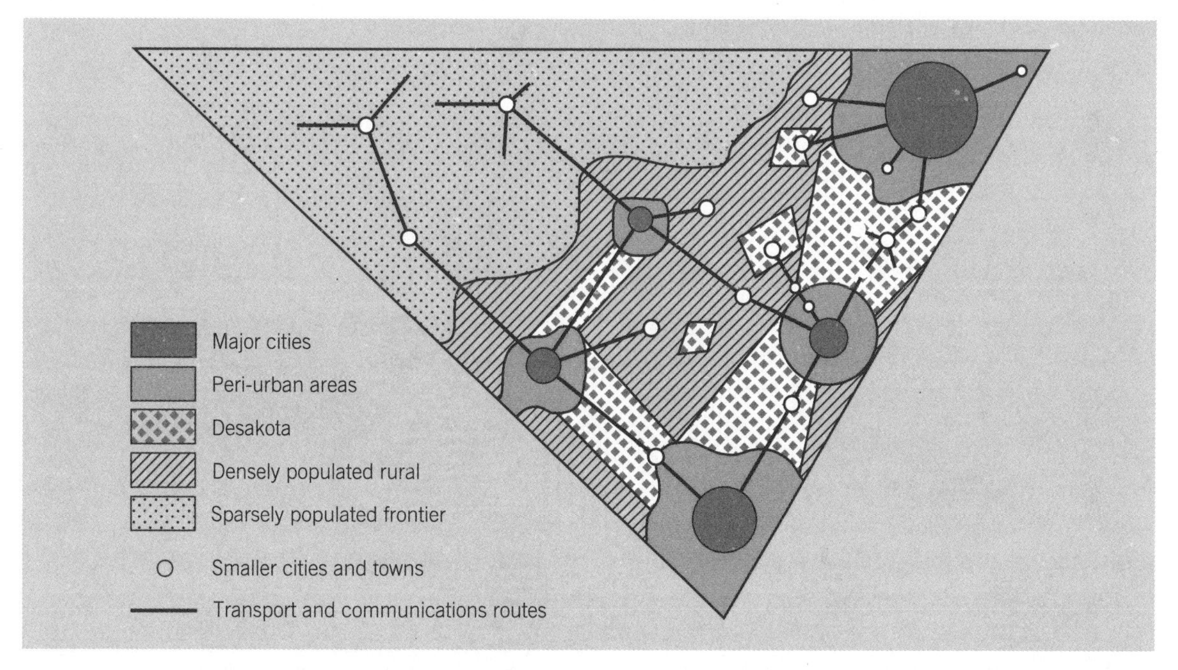

Figure 9.15 A simplified depiction of desakota regions within a national space-economy
Source: Adapted from McGee (1991)

Stated simply, the approach stresses the fusion of urban and rural, as cities extend along corridors of transport and communications (see Figure 9.15). McGee (1995) referred to regions of extended urban activity surrounding the core cities of many Asian countries. McGee originally called these kotadesasi, and later desakota (Indonesian for city-village). Such forms are now referred to in more generic terms as 'extended metropolitan regions'.

The early diagram used by McGee to explain the nature of such EMRs is reproduced here as Figure 9.15. Five principal zones were identified as making up a hypothetical Asian country. These consisted of:

(ii) first, the major cities, which generally amount to one or two large urban places in the Asian context;

(ii) these are surrounded by peri-urban zones, which are defined as those within daily commuting reach of the city;

(iii) the *desakota* areas are identified beyond the peri-urban zones, and witness the intense mixing of agricultural and non-agricultural activities, along major national corridors. Many are associated with intensive wet rice cultivation;

(iv) on the national periphery are the densely inhabited rural areas;

(v) and these in turn are surrounded by the less populated frontier agricultural regions.

It is envisaged that the process of urban expansion has bypassed small towns and villages, which have thus become embedded in the expanding metropolitan region. These may later start to experience *in situ* changes both to their functions and their occupational profiles.

Thus, urban and rural type settlements become juxtaposed and enmeshed within one variegated area, and localities with high agricultural employment counts exist within what ostensibly appears to be a major urban zone. Throughout, the overall population density is high.

McGee identified a range of desakota zones in Asia, and these are shown in Figure 9.16. In so doing, three different types of desakota region were recognised, according to their processes of formation and development over time.

The first were those associated with rapid rural–urban shifts, but where agriculture has remained important, such as in Japan and South Korea.

The second type were identified as being based on rapid changes in economic activity, that is the movement to secondary economic activities. The Taipei–Kaohsiung corridor in Taiwan was given as an example, where the proportion of the workforce engaged in agriculture declined from 56 to 20 per cent between 1956 and 1980. The Bangkok–Central Plains region of Thailand and the main coastal cities of China were given as further instances of this type of development (see Figure 9.16).

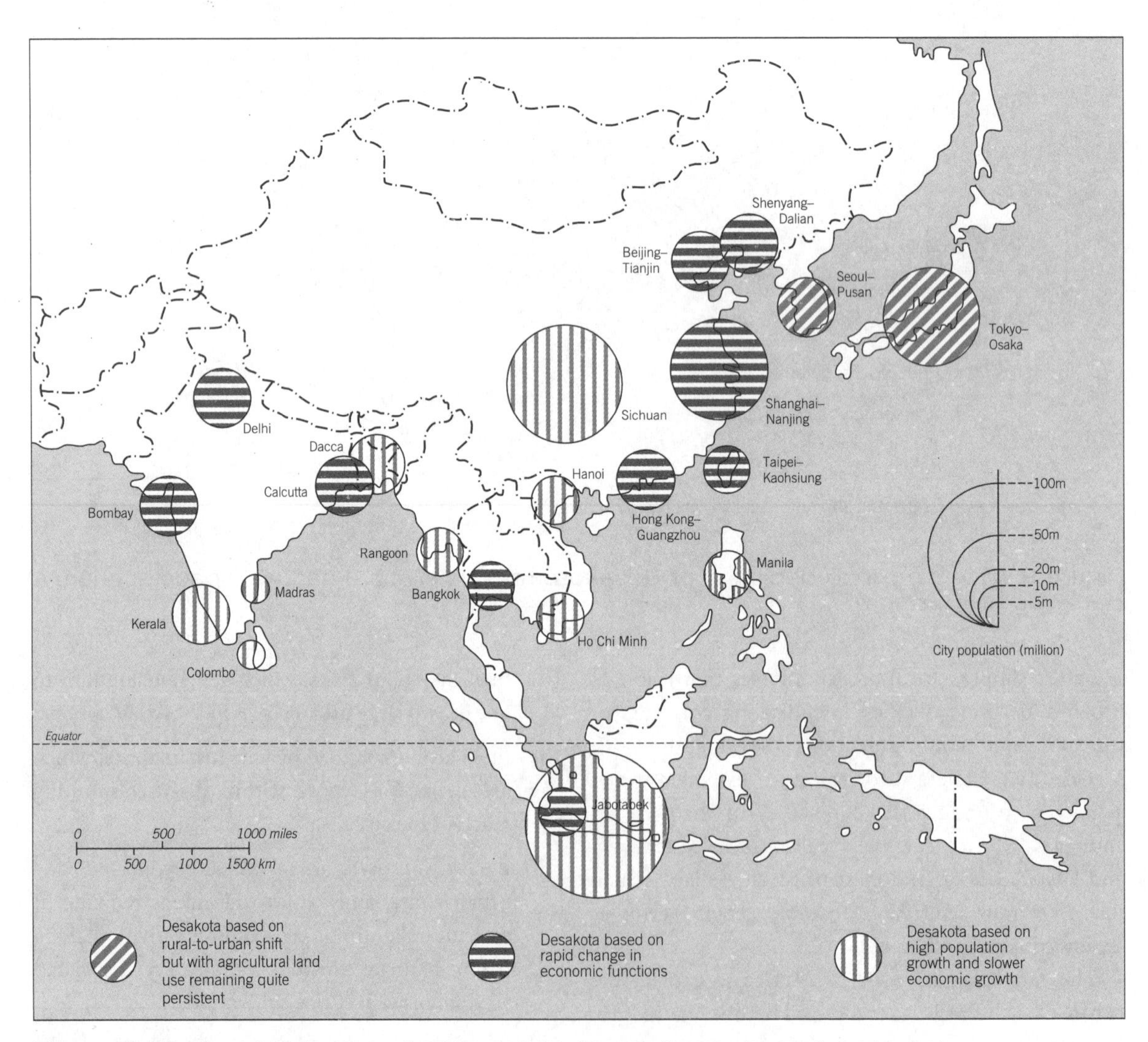

Figure 9.16 The distribution of different types of desakota regions in Asia
Source: Adapted from McGee (1991)

Third, urban zones characterised by high population growth but slower rates of economic change were identified, resulting in underemployment and self-employment in unpaid family work and enterprises. Such trends were identified in respect of Kerala and Tamil Nadu in India.

Another major feature of the EMR is the formation of multiple nuclei, as major functions decentralise into special zones for business, recreation, finance, production, entertainment and tourism. As a prime example of this, Drakakis-Smith (2000) used the case of Hanoi, Vietnam, as shown in Figure 9.17. With growth, the

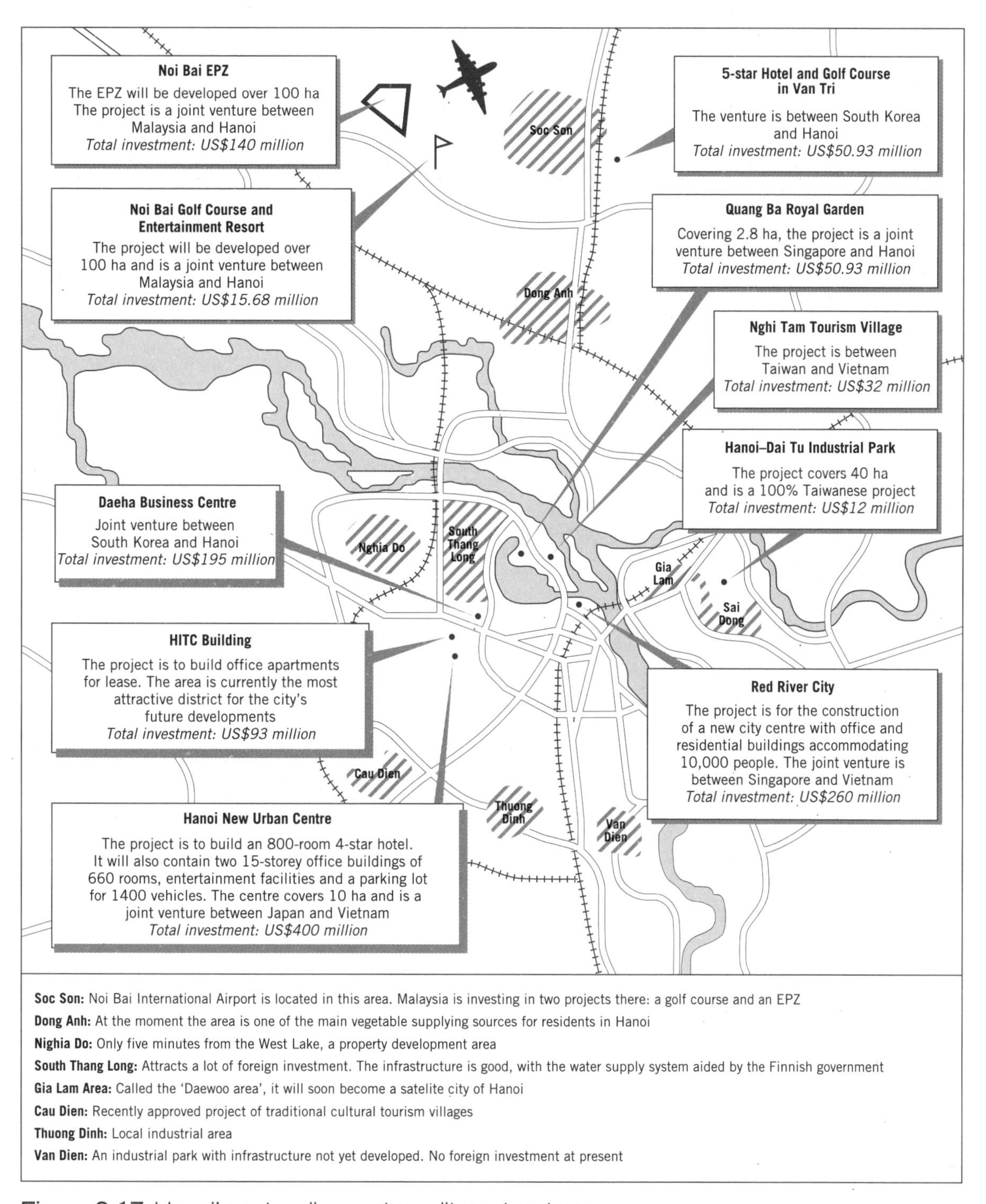

Figure 9.17 Hanoi's extending metropolitan structure
Source: Adapted from Drakakis-Smith (2000). Reproduced by permission of Taylor & Francis Books UK.

wider city of Hanoi is showing marked signs of regionalisation. Notably, several of these developments reflect the input of international finance. These include the development of the Noi Bai export processing zone (EPZ), and the Noi Bai golf course and entertainment resort, both joint ventures between Malaysia and Hanoi. The figure enumerates a number of other such developments, which are giving rise to a multiple-centred urban zone.

To a considerable extent, the emergence of EMRs can be recognised as the developing world equivalent of what the geographer Jean Gottmann (1961) had much earlier referred to as 'megalopolis' in the context of the developed world.

The term 'megalopolis' was first applied to the continuously urbanised north-east seaboard of the USA, stretching from Boston in the north to Washington in the south (the so-called 'Boshwash' megalopolis), and later to other major linked metropolitan regions. These included Brazil based on Rio de Janeiro and São Paulo, and China, based on Shanghai in the developing world (Gottmann, 1957).

In more recent work, Rimmer (1991) has considered EMR surrounding the Pacific Ocean, and including major American urban systems such as the Pacific Northwest corridor (Vancouver to Seattle), plus the Californian corridor (based on San Francisco, Los Angeles and San Diego), as well as the East Asian and Southeast Asian corridors (see Figure 9.18). In most respects these can be seen as evolving megalopolitan forms in the way Gottmann originally envisaged.

Also in the 1960s, the Greek scholar Doxiadis (1967) suggested that such urban complexes would coalesce into chain-like forms. It was ventured that these urban lineaments would eventually connect all of the urban cores in South America, Africa and Asia.

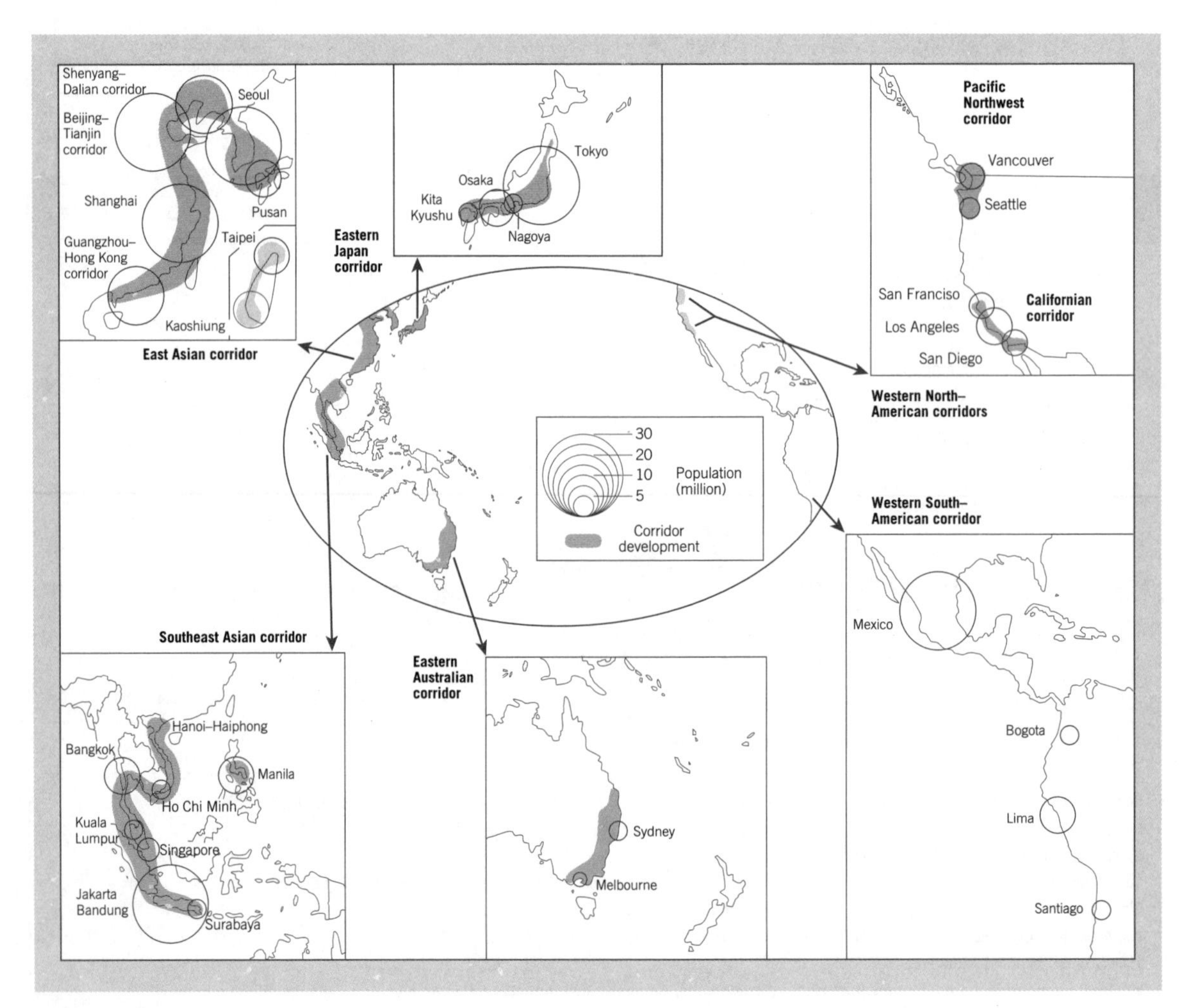

Figure 9.18 The distribution of extended metropolitan regions on the margins of the Pacific
Source: Adapted from Rimmer (1991)

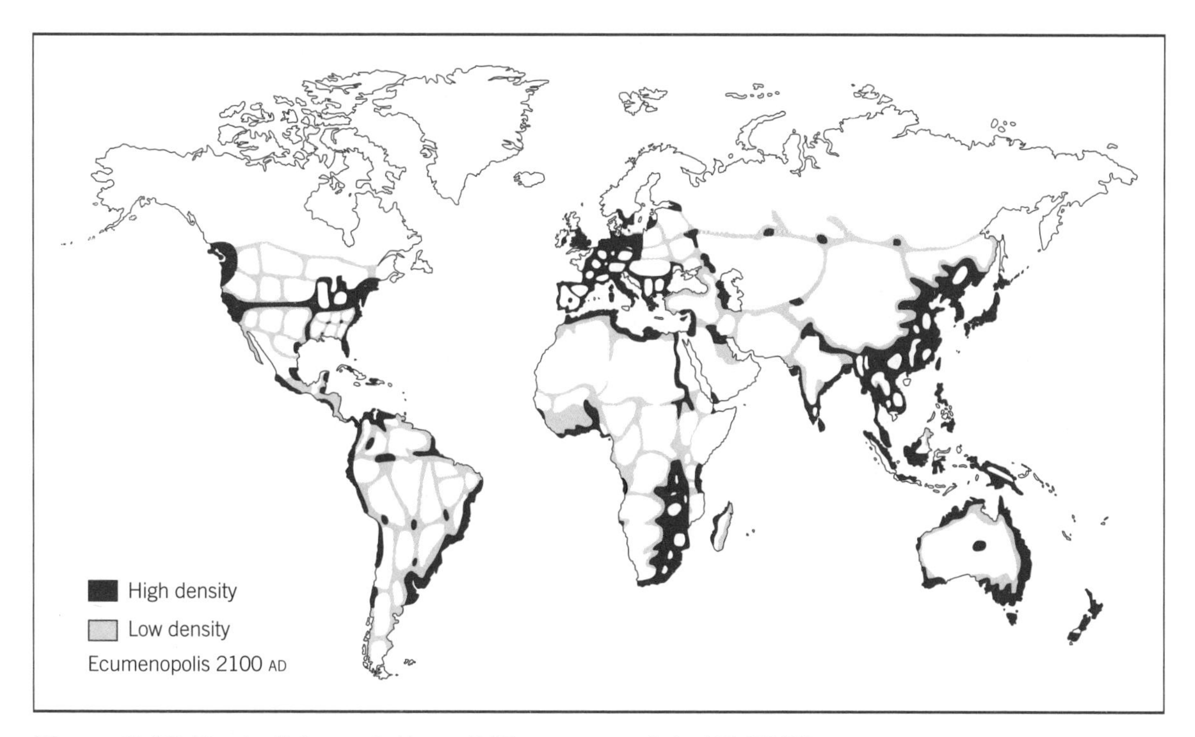

Figure 9.19 Doxiadis' prediction of 'Ecumenopolis', AD 2100

The term 'ecumenopolis' or 'world city' was employed to describe the development of this functionally integrated urban whole by the year 2050, with a global population of 9600 million (see Figure 9.19).

Doxiadis (1967) referred to this as the 'inevitable city of the future', and the development of extended metropolitan regions in the developed and developing worlds can be seen as the major ingredient, and one which will witness the progressive functional integration and physical interdigitation of the urban and the rural in the twenty-first century.

In this respect, it is notable that several writers have referred to the evolution of essentially similar flexible urban forms in much smaller urban zones and in other regions of the developing world.

Such developments have often been related to the evolution of major tourist resorts. Thus, Potter (2000, 2001c) and Sahr (1998) have described what Potter referred to as the 'mini-metropolitan area' of Castries, the capital of the Caribbean island of St Lucia.

Since the 1980s, Castries Harbour has been developed as a major cruise ship port of call. The city shows all of the usual zonations, including actively growing squatter areas. To the north of the city proper, along the coast, the major hotels of St Lucia are located (Potter, 2001c). In the 1970s, a major new marina was developed at Rodney Bay.

Gradually in the intervening years major supermarket developments have occurred, along with places of entertainment and recreation. High-income residential enclaves have developed in two clearly demarcated and separate locations.

At the same time, wedging in and out of these locations are traditional agricultural and rural activities, including the growing of bananas and ground provisions.

Notably, all of this has happened in a linear or ribbon fashion, along the major road transport artery which extends from the capital north along the coast (see Potter, 2001c).

The example serves to demonstrate a major reality concerning twenty-first century processes of urbanisation: that much contemporary urban development occurs alongside the development of tourism and recreationally oriented activities.

Inside Third World cities

Understanding the processes and the patterns

Although many towns and cities in developing countries exhibit the signs of urban growth discussed earlier in this chapter, their individual characteristics vary enormously according to a wide range of factors.

Such factors specifically include cultural context, the legacy of colonialism and the role cities and nations play in the broader regional and global economies, as shown in Chapters 2 and 4.

Regardless of this, many scholars have sought to identify common denominators in the complex process by constructing models of urban development. For example, Anthony King (1976, 1990) has presented theories on the nature of the colonial city.

However, because of their variety few scholars have attempted to construct a model for all Third World cities, so the most relevant studies have investigated city formation in particular regions such as Southeast Asia (McGee, 1967), post-apartheid South Africa (Simon, 1992a) and contemporary China (Yeh and Wu, 1995).

In respect of China, Yeh and Wu (1995) drew attention to an essentially concentric form in Nanjing, with the occurrence of peripheral industrial zones, as shown in Figure 9.20.

Stressing the regional diversity of city forms, Potter and Lloyd-Evans (1998) drew on other authors in presenting idealised models of the African (Figure 9.21a), Southeast Asian (Figure 9.21b), South Asian colonial (Figure 9.21c) and Latin American (Figure 9.21d) cities.

Although each of these models is used to attempt to explain changing processes and their impact on the city, their use often seems to restrict the scope of the discussion, giving pre-eminence to function and form, when cities are, above all, agglomerations of people, many of whom have migrated there in search of work.

As noted previously, few find the well-paid jobs they seek and face enormous problems in meeting the basic needs of both themselves and their families. Thus, issues of the socio-economic and environmental conditions faced by urban residents are more pertinent to any understanding of the contemporary urban condition.

In order to understand fully the range of problems faced by urban managers and the processes which give rise to such issues, we need a more comprehensive, more flexible and less ideographic conceptual approach. Such a perspective started to emerge in the 1990s and is concerned with sustainable urbanisation (Burgess *et al.*, 1997; Pugh, 1996; Satterthwaite, 1999, 2008).

This must not be confused with sustained growth, in which the city is seen to have a pivotal role in initiating and maintaining national economic growth. Economic

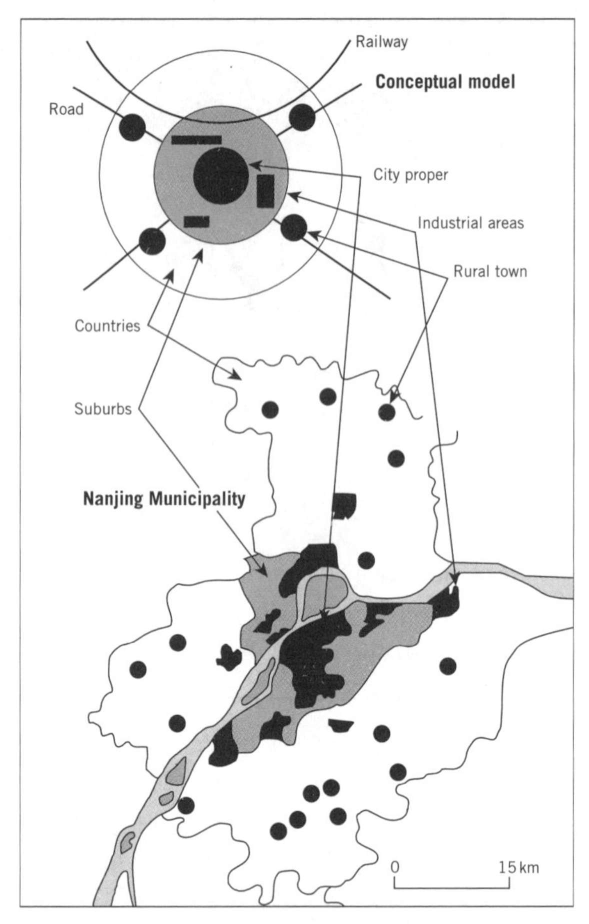

Figure 9.20 Spatial structure of Chinese cities: conceptual model and a case study of Nanjing Municipality
Source: Adapted from Yeh and Wu (1995)

growth is a vital component of sustainable urbanisation, but it is only one in the array of interlinked processes which make up the contemporary city.

Sustainable urbanisation can and should constitute an important goal for any urban management team, irrespective of the level and nature of economic development. Figure 9.22 indicates some of the major components of sustainable urban development.

Drakakis-Smith (1995, 1996, 2000) has given these issues fuller consideration – and, more importantly, the ways in which they interlink, illustrating the complexity of many urban problems faced in towns and cities.

Thus the economic dimension of urban development is not just related to the role of the city in the national economy but also to the repercussions of the economy on the residents of the city in terms of employment, incomes and poverty at the household

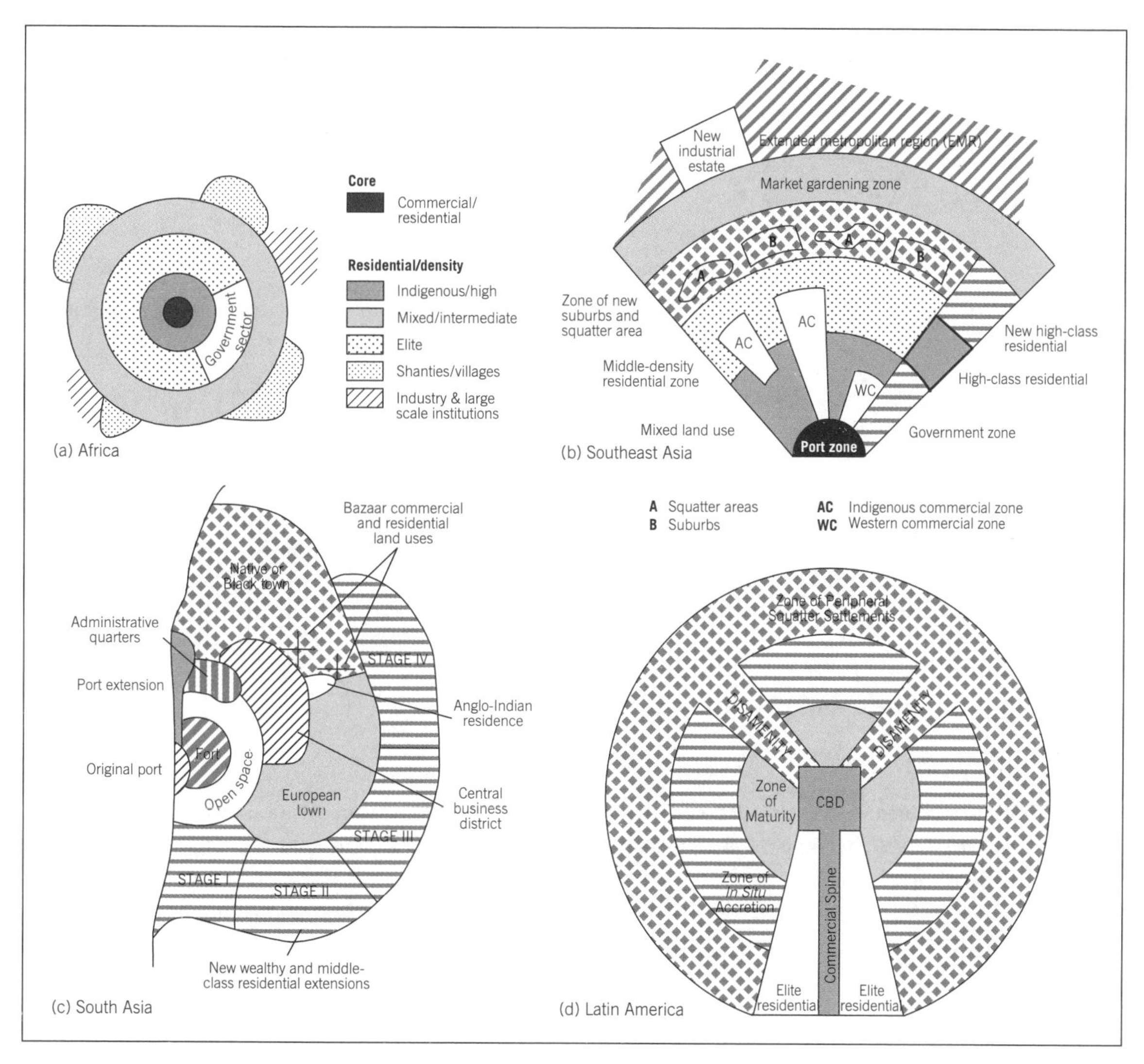

Figure 9.21 Various regional models of urban form and structure
Source: Adapted from Potter and Lloyd-Evans (1998)

level, as well as its impact on the urban environment and social issues, such as workers' rights.

The main components comprise demographic factors, economic factors, social, political and environmental factors.

The management of urban development in order to achieve sustainable rather than just sustained growth is clearly a complex task. Nevertheless, many of the issues raised by such an approach have been carefully researched over the past two decades, though separately rather than in an interlinked manner.

Moreover, urban management for sustainable urban development requires a new set of attitudes towards the objectives of intervention. So, in addition to seeking to create and sustain economic growth, there must be other, equally important priorities. These might be enumerated as including:

- The pursuit of equity and social justice
- The satisfaction of basic needs
- The recognition of social and ethnic self-determination and human rights
- Environmental awareness and integrity
- Appreciation of the interlinkages across space and time

Whether these principles do, or could, form a basis for urban management will be discussed towards the end of this chapter. The following sections pick up on some of the main components of sustainable

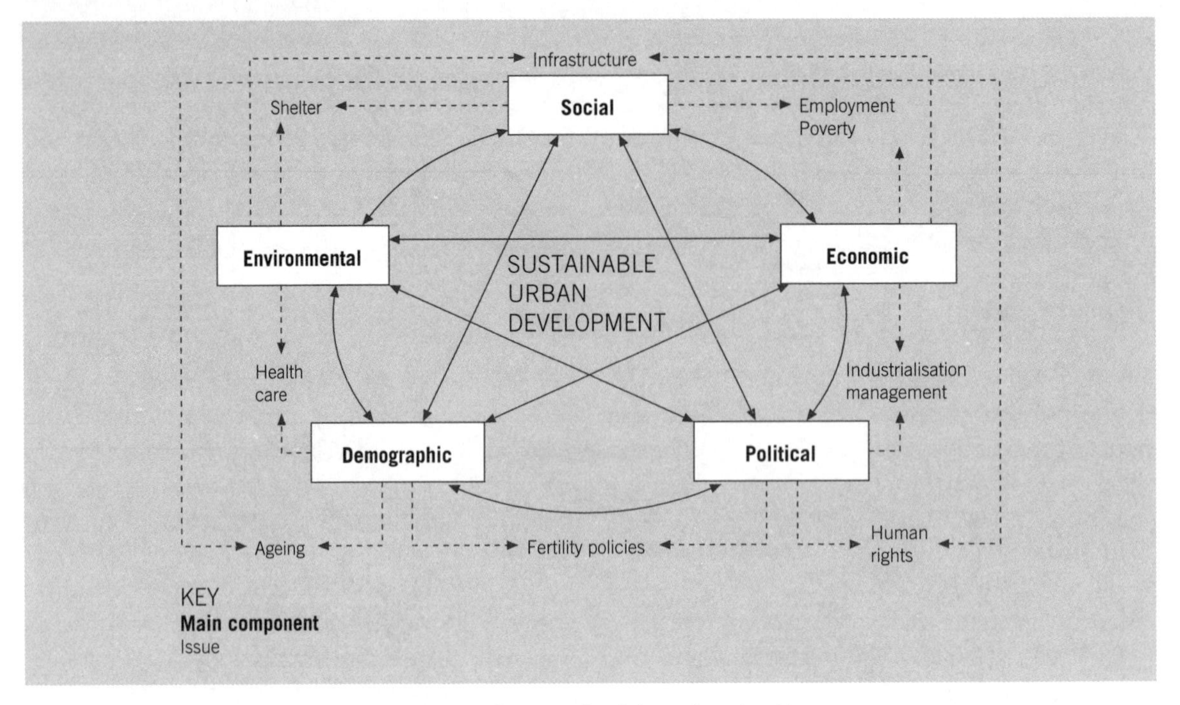

Figure 9.22 The main components of sustainable urbanisation

urbanisation as outlined in Figure 9.22 and discuss some of the main issues that might relate to appropriate policy measures.

Demographic factors

Some of the main demographic issues involved in urban development have already been alluded to in the first half of this chapter.

Indeed, urban growth by migration has probably been one of the most researched topics in urban studies. Many of the studies have attempted to construct predictive models, most of which have concurred with Todaro (1994) that much migration is motivated by perceived differences in economic opportunities.

However, there is immense variation in the relationship between urban migration and economic circumstances. For example, large-scale rural poverty in Bangladesh has induced nowhere near the scale of migration to its cities as it has in southern Africa.

History, culture and even the nature of communications and transport all have a role to play in the decision to migrate. Certainly the last of these factors has induced considerable change in the nature of both internal and international migration.

With regard to internal migration, in most countries it is almost as cheap and easy to access the major cities as it is to move to smaller local centres; consequently, the smaller local centres are often bypassed in the rural–urban migration process.

This has refocused attention onto the role that small and intermediate towns can and should play in the development process (Aeroe, 1992; Baker and Pedersen, 1992). Simon (1992b) states that no such role has yet emerged; and unfortunately the attention of most researchers has been diverted away from small towns per se and towards the phenomenon of mega-urbanisation.

In the previous section, the emergence of extended metropolitan regions was considered, whereby vast compound urban–rural zones are the outcome. Some writers refer to mega-urbanisation as the process which leads to rapid urban expansion along major lines of communication, enveloping villages and villagers *in situ*, and creating multi-nodal settlements, where rural and urban become blurred and indistinguishable, obfuscating the precise nature of movements between the two (McGee, 1989; McGee and Robinson, 1995; Potter and Unwin, 1995).

As explained, some see this as an inevitable trend which needs to be managed, rather than prevented, in the process implying that 'hard decisions [will have to be taken] against fostering small town development and rural industrialisation' (McGee and Greenberg, 1992: 7).

One of the reasons why mega-urbanisation is occurring in countries with rapid economic growth is the expansion of international urban migration.

In Pacific-Asia the waves of economic growth which have rippled from Japan, through the four tigers to the industrialising ASEAN (Association of South East Asian Nations) states have been followed by streams of job-seekers moving in search of work in manufacturing and tertiary activities.

It is estimated that such cross-border urban migration amounts to some 2 million people in the region (Dixon and Drakakis-Smith, 1997). The result is an extremely complex situation in which multiple nationalities move between the various countries, particularly in Southeast Asia.

In Africa there is much less cross-border migration specifically to cities. Indeed, such are the socio-economic circumstances pertaining in much of Africa (see Chapter 4, etc.), that in some countries not only has rural–urban migration slowed considerably, it is occasionally being accompanied by reverse migration to rural areas (Mijere and Chilivumbo, 1987).

However, there is more to the demographic aspects of sustainable urbanisation than the nature and management of migration, not the least of which is the two-way relationship between urbanisation and fertility.

For example, living in the city not only raises the cost of rearing children but also increases access to family planning programmes, and yet cities in developing countries have overwhelmingly young populations (Boyden and Holden, 1991), with cities as geographically distinct as Bogotá, Delhi and Jakarta all having half their populations aged 15 years or less.

On the other hand, in some of the cities of Pacific-Asia ageing populations, and their impact on the labour force and social welfare (Graham, 1995), provide different but equally pressing issues.

Other demographically linked issues related to sustainable urbanisation could encompass household composition and the roles of women in generating income and meeting basic needs, or ethnicity and the ways in which migration has created more complex and tense situations in the competition for limited urban resources (Box 9.1).

BOX 9.1

Urban migration and ethnicity

The fact that migration to cities is being drawn from increasingly extensive geographical areas has often resulted in a broader diversification of ethnic groups. In most countries this ethnic complexity is spontaneous as people from economically, geographically and ethnically marginal regions are drawn to capital cities. Almost two-thirds of the migrants to Bangkok are from the Lao-dominated north-east of Thailand. But in Malaysia, increased urban ethnic diversity has been the consequence of deliberate government policies designed to increase Malay participation in urban economic activities (Eyre and Dwyer, 1996).

Although some might argue this process has occurred without increasing ethnic tension, it was in fact prompted by ethnic tensions in the first place. In other cities where ethnic mixing has accompanied migratory growth, tensions have increased markedly; for example, in many African cities where national politics reflect tribal antagonisms.

This growing urban ethnic diversification and its consequences have been exacerbated by the increased internationalisation of labour movements. In Southeast Asia this has produced a particularly complex pattern of movement. Singapore was the initial magnet for migrants from Malaysia, Indonesia and the Philippines, for factory work and domestic work. More recently, construction labour has come from Thailand and India. And as it has developed its own economy Malaysia has recruited both legal and illegal workers from Thailand and, more particularly, Indonesia. There are now an estimated 1 million Indonesians working in Malaysia, most of whom are illegal, and local resentment at narrowing access to jobs has increased substantially. Meanwhile, illegal workers from the transitional socialist economies of Southeast Asia are also beginning to flow across weakly policed borders into the regional capitals of Thailand, which have lost migrants to Bangkok.

This increasingly complex pattern of ethnodevelopment is threatening urban sustainability in a variety of ways. Although it may provide a larger labour pool for economic growth, it is also leading to growing ethnic and class antagonism and exploitation. In Singapore, for example, there are increasing reports in local newspapers of crude racism, especially against domestic workers (Teo and Ooi, 1996).

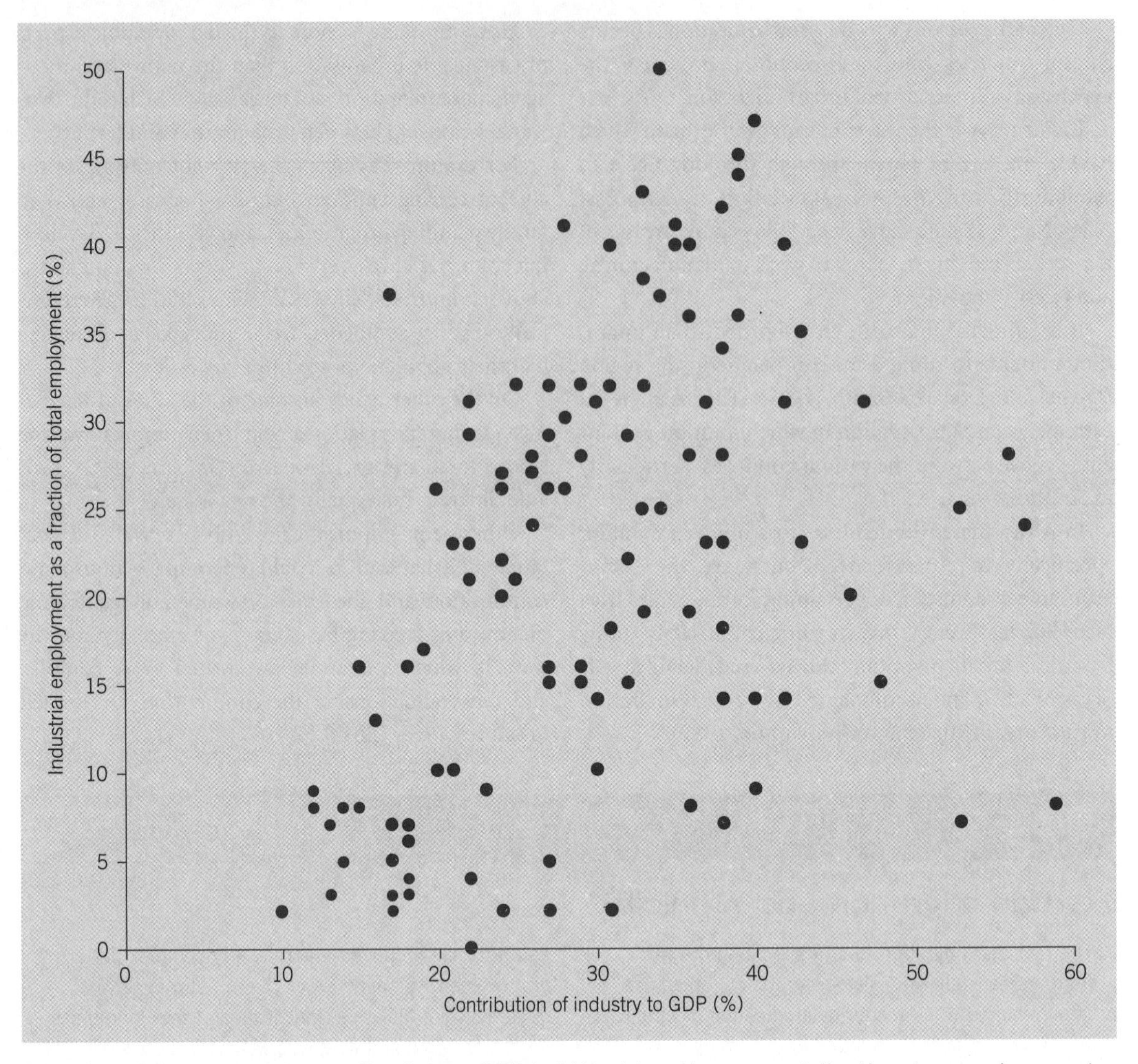

Figure 9.23 Industrial contributions of Third World nations: contribution to employment versus contribution to GDP

Economic matters

Urban-based industrial growth fuelled by transnational corporation (TNC) investment has been the focus for many development policies in Third World countries, as discussed in Chapters 1, 3 and 4.

However, it is well known that TNC capital has been very selective in where it chooses to invest, and the countries which have experienced urban-based industrialisation have been relatively few (Dicken, 1992, 2007): Mexico, Brazil, the four tigers and some members of ASEAN (see Chapter 4).

But even in these countries the factories are not large, and throughout the Third World the contribution of industry to GDP is far greater than its contribution to the labour force in terms of jobs (Figure 9.23). Indeed, even in the four Asian tigers, most firms tend to be small to medium-sized, a prime factor in promoting industrial flexibility.

In many countries, therefore, those migrants who come in search of work in the city are not usually incorporated into the formal sector, particularly given the downsizing that has occurred in government employment as a result of structural adjustment.

Most find work in what has come to be known as the informal sector, as already mentioned earlier in this chapter (see Plates 9.7, 9.8 and 9.9). This is difficult to define adequately and an informal activity may possess any one or more of a number of features allegedly

Plate 9.7 Rubbish collection in Hanoi, Vietnam
(photo: David Smith)

Plate 9.8 Street barber in Kashgar, Kashi, western China
(photo: Tony Binns)

typical of it: semi-legal, small-scale, family-oriented, traditional technology.

The main characteristics of formal (upper circuit) and informal (lower circuit) activities are shown in Table 9.12. Certainly, there is considerable overlap between the two sectors (Santos, 1979).

For example, many trishaw owners work in the formal sector and rent their machines out to the riders, whereas domestic outworkers are often essential to small businesses in helping them absorb fluctuations in demand. In Plate 9.10, an elephant is shown in a low-income housing area of Delhi, India. This animal is rented out by one of the better off residents of the informal settlement, for participation at formal social events in the city.

Perhaps the easiest way to look at the informal sector is in terms of a continuum of activities, as shown in

Plate 9.9 Street traders in central Johannesburg, South Africa
(photo: Tony Binns)

Table 9.12 Characteristics of the two circuits of the urban economy in Third World countries

	Upper circuit (formal)	Lower circuit (informal)
Technology	Capital-intensive	Labour-intensive
Organisation	Bureaucratic	Primitive
Capital	Abundant	Limited
Labour	Limited	Abundant
Regular wages	Prevalent	Exceptional
Inventories	Large quantity and/or high quality	Small quantity and poor quality
Prices	Generally fixed	Negotiable (haggling)
Credit	Banks and institutions	Personal and non-institutional
Profit margin	Small per unit, but large turnover and considerable in aggregate	Large per unit, but small turnover
Relations with customers	Impersonal and/or on paper	Direct, personalised
Fixed costs	Substantial	Negligible
Advertising	Necessary	None
Re-use of goods	None (waste)	Frequent
Overhead capital	Essential	Not essential
Government aid	Extensive	None or almost none
Direct dependence on foreign countries	Considerable	Small or none

Source: Santos (1979)

Figure 9.24. This figure is derived from Potter and Lloyd-Evans (1998).

At its extreme, the informal sector is associated with reproduction and domestic labour, and is unremunerated or unpaid. It then links to subsistence production (for example, in agriculture and household work) and, in turn, to petty commodity production by small-scale operators. Finally, one moves to waged employment in the full capitalist system of the formal sector.

For many years the activities of the informal sector, although clearly useful, have been anathema to urban management as they mar the modernising image it is trying to create in order to attract investment (Chapter 3).

Plate 9.10 Elephant used for ceremonial functions by one of the residents of a low-income settlement in Delhi, India
(photo: Rob Potter)

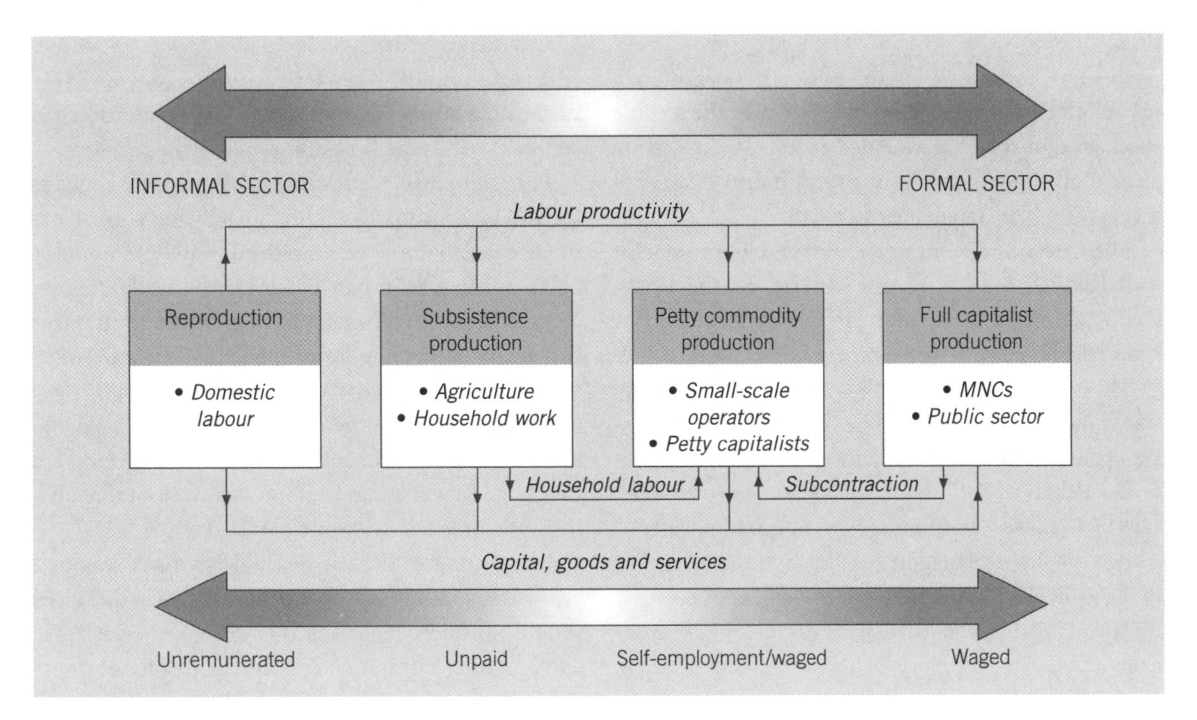

Figure 9.24 The informal–formal sector continuum
Source: Adapted from Potter and Lloyd-Evans (1998)

Many regulations have sought to keep the informal sector in check, despite the 'unconventional wisdom' of the 1970s, which saw support for the informal sector as a way of improving basic needs provision (Richards and Thomson, 1984).

Yet few urban governments were as enthusiastic as the experts and relatively little was achieved, particularly in the field of improved employment opportunities. However, the 1990s and 2000s are witnessing a revival of interest in the informal sector, particularly in the wake of the reduced employment opportunities and reduced incomes which have followed structural adjustment programmes, particularly in Africa (Gibbon, 1995).

The reaction was to introduce a social component to adjustment and latterly to move to what are styled poverty reduction strategies. As poverty was due to limited job access, and as the informal sector seemed capable of creating employment, it was reasoned that removing some of the constraints on the informal sector (deregulation) would help to expand work opportunities and absorb those in poverty (Urban Foundation, 1993).

This approach has not yet been the success it was hoped. There is a limit to the capacity of the informal sector to involute and create employment and income.

Moreover, very few small informal firms have the capacity to upgrade and 'formalise' on their own without assistance from the state. Furthermore, the alleged deregulation created undesirable knock-on effects as employers were given freedom to exploit their workers further (Wilson, 1994), and to pollute without fear of prosecution.

In short, as Parnwell and Turner (1998) note, the urban informal sector does not equate with the flexible specialisation that has emerged in the West, despite structural similarities; it is much more a survival mechanism than an engine of growth.

The consequence of these urban labour market problems has been increasing poverty. By the 1990s, conflicting views were coming from the global institutions concerned with development.

The World Bank (1993) was alleging that poverty was being rapidly eliminated by the spread of the market economy. On the other hand, many other institutions, such as UNDP (1991) and even the World Bank (1991) itself, were noting a growth in urban poverty.

Part of the explanation for this contradiction lies in the difficulties of defining poverty and consequently the quality of the data, particularly for making comparisons (Drakakis-Smith, 1996; Rakodi, 1995; Wratten, 1995).

But despite doubts about the data, in general terms it is possible to discern a steady shift in the distribution of poverty associated with the growth of urban populations. Table 9.13 therefore reveals that in countries with higher levels of urbanisation there are now more urban poor than rural poor, in terms of absolute totals.

Bringing the discussion around to poverty further humanises the debate on urban economies and employment. We must not think of labour simply as an input into the development process; it comprises many different groups of people, and some of their specific problems and needs often overlap with those of other groups.

Women have been incorporated into the urban labour market in many different ways that depend on local economic and social conditions. However, as McIlwaine (1997) has noted, this enhanced value in the workplace has not always reduced gender inequalities in society in general or within the household.

The changing links between gender and urban economic growth need therefore to be followed through other dimensions of urban sustainability, such as those related to basic needs provision and human rights.

It is clear in the early part of the twenty-first century that these realities are forcing development agencies to focus on the need for poverty reduction. Thus, structural adjustment programmes have all but disappeared from the World Bank's literature. They are being replaced with Poverty Reduction Programmes, as elaborated in Chapters 3 and 7 (Simon, 2008).

Children too form an identifiable group whose specific needs must be taken into account in any review of sustainable urban development. The value of child labour is well recognised and exploited by employers.

For example, it has been estimated that in Thailand the child labour force is approximately the same size as the female labour force. One-third of these 1.5 million children work in urban factories where they receive about half the adult minimum wage. Although it is true that in many developing countries children are often important income earners in the family (Clifford, 1994; Gilbert, 1994) and are often proud of their household role, there are ways in which their conditions of work and their life as a whole can be improved without threatening household survival strategies (Lefevre, 1995).

But as long as the use of child labour is seen as a 'comparative advantage' which 'humanitarian measures' should not threaten (Silvers, 1995: 38), the sustained and sustainable improvement in the quality of life of this segment of the labour force will not occur.

It must not be thought, however, that those who find themselves disadvantaged in the labour market are passive acceptors of their fate.

Low-income households display a wide range of coping mechanisms (Rakodi, 1995), some of which are indicated in Table 9.14 (see Plate 9.11 showing subcontracted assembly work in a low-income area of Delhi, India).

Table 9.13 Absolute poverty in urban and rural areas (percentage below poverty line)

	Urban areas	Rural areas	Percentage urban	Ratio of rural poor to urban poor[a]
Africa				
Botswana	30.0	64.0	29	4.8
Egypt	34.0	33.7	47	1.1
Ivory Coast	30.0	26.0	41	1.3
Morocco	28.0	32.0	49	1.2
Mozambique	40.0	70.0	28	4.5
Tunisia	7.3	5.7	55	0.7
Uganda	25.0	33.0	11	10.3
Asia				
Bangladesh	58.2	41.3	17	3.4
China	0.4	11.5	60	18.9
India	37.1	38.7	27	2.8
Indonesia	20.1	16.4	31	1.8
Malaysia	8.3	22.4	44	3.2
Nepal	19.2	43.1	10	21.0
Pakistan	25.0	31.0	33	2.5
Philippines	40.0	54.1	43	1.8
South Korea	4.6	4.4	73	0.3
Sri Lanka	27.6	45.7	22	6.1
Latin America				
Argentina	14.6	19.7	87	0.2
Brazil	37.7	65.9	76	0.6
Colombia	44.5	40.2	71	0.4
Costa Rica	11.6	32.7	48	3.2
Guatemala	61.4	85.4	40	2.1
Haiti	65.0	80.0	29	3.1
Honduras	73.0	80.2	45	1.3
Mexico	30.2	50.5	73	0.6
Panama	29.7	51.9	54	1.3
Peru	44.5	63.8	71	0.6
Uraguay	19.3	28.7	86	0.2
Venezuela	24.8	42.2	85	0.3

[a] In absolute totals, less than 1.0 indicates more urban poor than rural poor.
Sources: Drakakis-Smith (1996), UNCHS (1996), World Bank (1994b)

Table 9.14 Urban household strategies for coping with worsening poverty

Changing household composition
- Migration
- Increasing household size in order to maximise earning opportunities
- Not increasing household size through fertility controls

Consumption controls
- Reducing consumption
- Buying cheaper items
- Withdrawing children from school
- Delaying medical treatment
- Postponing maintenance or repairs to property or equipment
- Limiting social contacts, including visits to rural areas

Increasing assets
- More household members into workforce
- Starting enterprises where possible
- Increased subsistence activity such as growing food or gathering fuel
- Increased scavenging
- Increased sub-letting of rooms and/or shacks

Source: Rakodi (1995)

Plate 9.11 Subcontracted assembly work in an informal sector settlement, Delhi, India (photo: Rob Potter)

Not all strategies are available to all households, depending on individual and local circumstances, but the ways in which poor families sustain themselves in the city ought to be the basis on which policy responses are formulated. These strategies will be discussed in the final section of this chapter.

Meeting basic needs and human rights

As noted in Chapter 3, the concept of 'basic needs' is a rather fluid one, with many of the early approaches in the 1970s encompassing what would now be regarded as environmental issues, together with employment and poverty.

At present, the term basic needs tends to refer to those core areas of personal and household needs, such as education, health care and housing, in which it is possible for both state and community to make a contribution towards improving the situation (Chapter 3).

Indeed, such partnerships became almost the norm in the 1970s and 1980s as international agencies attempted to encourage an improved response to a deteriorating urban situation. Before the emergence of what became known as the 'basic needs approach', most governments had neither the funds nor the inclination to invest in the social overheads of welfare programmes, preferring instead to use their resources to encourage and sustain economic growth, as stressed in Part I of this book.

The late 1980s and 1990s saw a return to this position as debt crises and imposed structural adjustment have forced governments to withdraw from social welfare programmes of all kinds, shifting responsibilities on to the 'market' or the poor themselves. However, as noted above, poverty alleviation and so-called poverty eradication have become major themes in the early 2000s.

To illustrate the changing circumstances affecting access to, and provision of, basic needs, the remainder of this section will focus on housing, but the issues raised are often common to other basic needs.

Housing poverty (Pugh, 1996) has been well researched since the 1960s when John Turner (1967, 1982) and William Mangin (1967) first drew attention to the positive qualities of squatter settlements (illegal settlements) and shanty towns (poorly built settlements).

Essentially they argued that such housing should be seen as a solution to the housing shortage, rather than as a problem. Turner (1967) in particular argued that if security of tenure was provided, and if real incomes are rising, then self-help housing will improve year by year and little by little.

Thus, self-help housing was not seen as leading to slums, but rather to building sites. It was argued that the role of the state is to help the poor to help

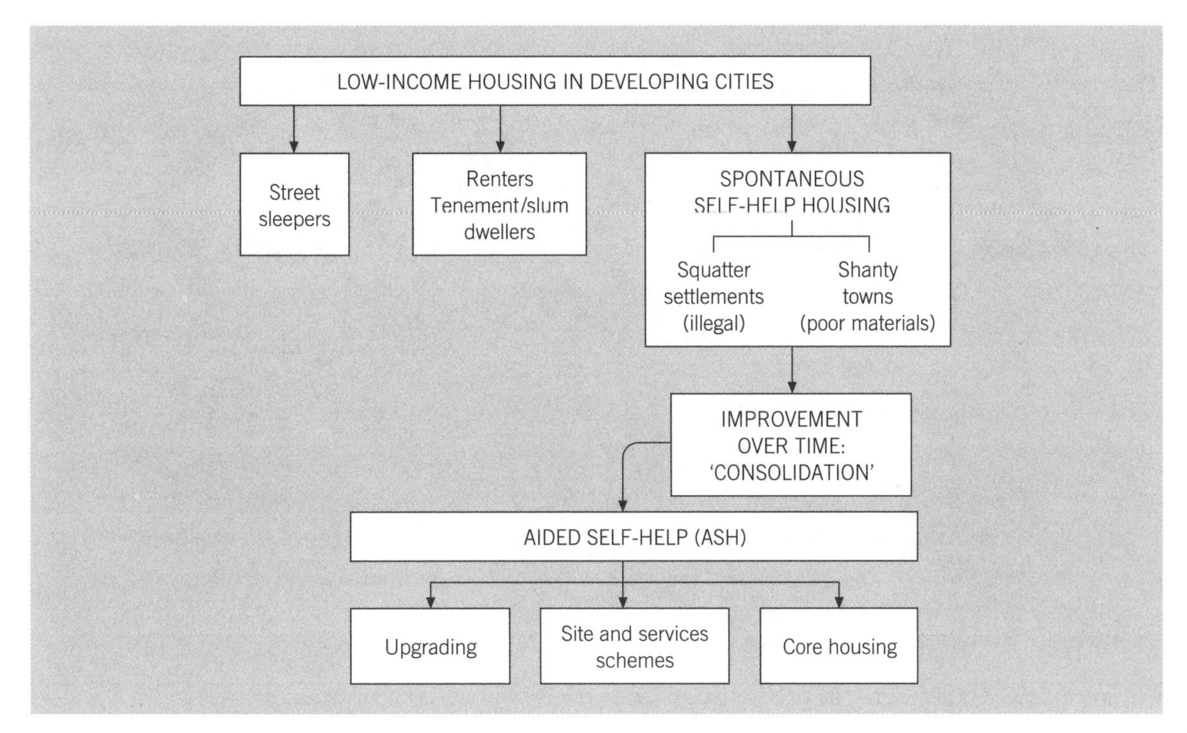

Figure 9.25 Different types of low-income housing in developing world cities
Source: Adapted from Potter (1992a)

Plate 9.12 Low-cost housing at Cradock, Eastern Cape Province, South Africa
(photo: Tony Binns)

themselves, by promoting aided self-help schemes (ASH; see Figure 9.25).

Aided self-help consists of upgrading, that is improving existing areas, developing site and service schemes, where new lands are opened up. Sometimes the core of a new house is also provided on the site (see Plate 9.12). Both Mangin and Turner showed how squatter settlements often house residents who have a job and who are trying to make their own way in the urban economy.

Thus, movements within and to the city have been well explored, and show that rural migrants often

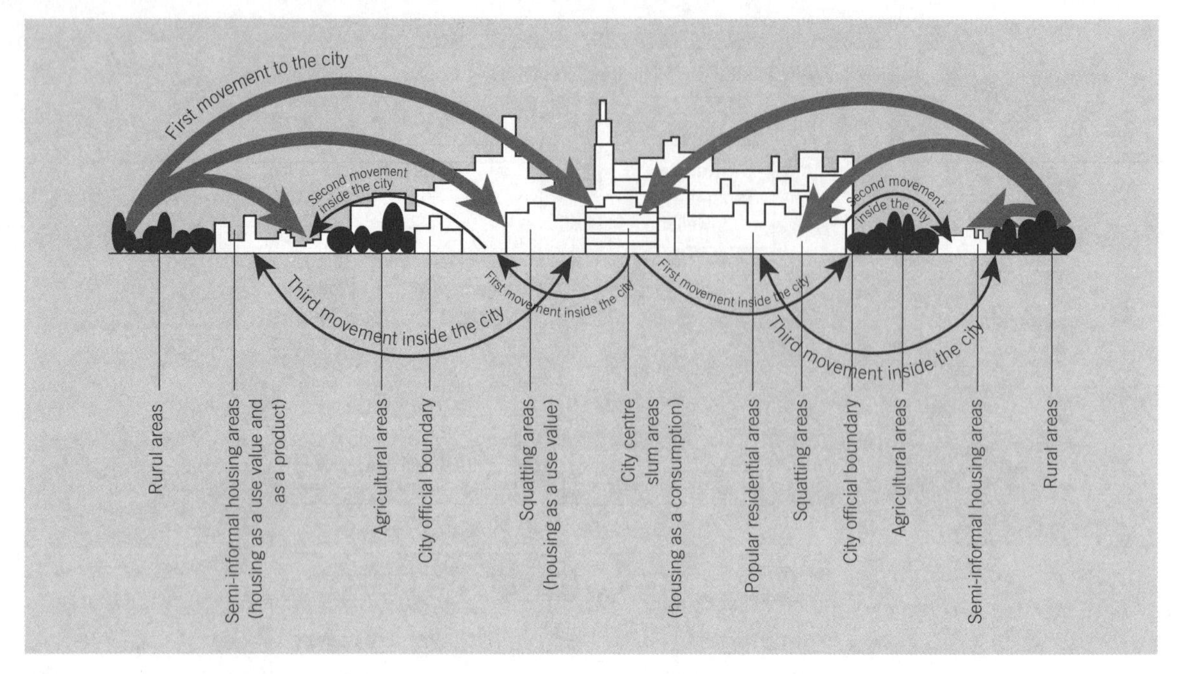

Figure 9.26 Mobility of the urban poor to and within the city of Alexandria
Source: Adapted from Soliman (1996)

locate in rental housing in the city and only move out to squatter settlements once they have a solid foothold in the urban economy (see Figure 9.26).

In spite of almost four decades of admittedly varied responses to shelter needs, the problems seem to be as widespread as ever. It is generally believed that at least 20 per cent, and perhaps even as high as 50 per cent, of the world's population lack decent housing.

Even in Pacific-Asia, where economic successes have occurred, it was estimated that in 2000 some 60 per cent of the region's urban population was living in slum or squatter settlements (Pinches, 1994).

Although much of the statistical information on housing poverty is unreliable and incompatible, it is useful in illustrating trends over time and between regions.

In some cities, renting is far more usual and acceptable than ownership, yet the policy responses of much of the last 30 years have been based on the assumption that tenure security through ownership is the fundamental desire of most low-income populations.

Certainly, the discussions about shelter itself have evolved into a sort of dualism. On the one hand are the debates about the role of shelter provision in the development process as a whole, a debate which in recent years has increasingly been conducted at the global rather than the national level, with the main international development agencies dominating the discussion, the funding and hence the policy.

Increasingly distinct from these events are the national and urban debates about programmes and projects for the real world.

In recent years this has tended to focus upon the practical ways the various stakeholders involved can help improve access by the poor to better housing. As Ward and Macoloo (1992) observe, these two sides of the housing debate have been moving further apart; less and less is practical planning informed by the conceptual debates on the role of housing in sustainable urbanisation, and vice versa.

As noted above, for many years the debate on practical responses to housing poverty has revolved around aided self-help programmes, in which the energies and ambitions of the poor themselves are combined with tenure, material and land inputs from the state to produce developments which are largely self-built, but which have the support and approval of the state.

Although many low-income households benefited substantially from such schemes, they were nevertheless subject to considerable criticisms (Plates 9.13 and 9.14). Pinches (1994: 118), in particular, claimed that aided self-help schemes 'served the narrow economic interest of states, elites and international agencies' by offering cheap solutions to demands for housing,

Plate 9.13 Self-help housing in Caracas, Venezuela, soon consolidates
(photos: Rob Potter)

Plate 9.14 'Shack farming': backyard shack housing in Harare, Zimbabwe
(photo: David Smith)

Plate 9.15 Low-income settlements in Cape Town, South Africa
(photo: Rob Potter)

Plate 9.16 Low-income dwellings in Cape Town, South Africa
(photo: Rob Potter)

containing restive populations and formalising part of the informal sector (Plates 9.15 and 9.16).

However, since the 1990s enthusiasm for these approaches has diminished substantially, partly because there was an enforced retreat of the state from welfare programmes under structural adjustment, partly because the scale of the housing problem was not reducing substantially and partly because funds from the international agencies dried up.

As urban populations have continued to grow, so housing poverty has remained an important issue related to sustainable urbanisation.

In some cities, particularly in Africa, this has meant a resurgence of squatter settlements; in others, market forces have produced a rapid expansion of renting and sharing (Gilbert, 1992). The growing research interest in these phenomena has indicated a wide range of types and circumstances.

In Latin America, for example, Gilbert (1992) argues that most landlords operate on a small scale and are not exploitative; but in many African cities there is widespread exploitation of shack tenants in gardens or yards attached to formal housing (Plate 9.14), conditions are cramped and there are grossly inadequate washing and toilet facilities (Auret, 1995; Grant, 1995). Essentially, this witnesses the privatisation of housing, the benefits of which usually filter upwards through a hierarchy of landlords and owners.

State responses to the continued housing crisis have been strongly influenced by the neo-liberal trends in international development and have shifted away from the more direct subsidies of aided self-help to the formulation of partnerships between the national and local states, together with a variety of local agencies, such as non-governmental and community-based organisations (NGOs and CBOs).

The focus for these partnerships is on facilitating the access of households to land or credit through the removal of existing constraints – helping the poor to help themselves. However, the poorest and most needy are often not capable of the organised and sustained collective action required to improve their housing, health care or education.

Enablement programmes are often used by the authorities as an excuse to abandon many of their social responsibilities, privatising them to the NGOs and CBOs. Since the 1990s, therefore, the pursuit of adequate shelter as a human right has proceeded in theory rather than in practice.

Reliance on market forces is simply resulting in a mounting but hidden problem that is liable to create social and political tensions for many years to come. This situation has many resonances of the modes of production argument presented in Chapter 3 (see also Potter, 1994; Potter and Conway, 1997).

The brown agenda

For many people, urban sustainability in developing countries equates only to environmental issues. Moreover, this environmental agenda tends to prioritise those issues which are of greatest concern to the West, e.g. global warming and the rapid use of finite resources.

The result has been a Western-led series of programmes in many developing countries, programmes which do not correlate with, or respond to, many of the real concerns and priorities of the residents of cities in the developing world.

Indeed, the environmental problems facing such cities vary enormously according to the local combination of contributing factors. In broad terms, these encompass the following:

- The nature of the urbanisation process itself – the rate, scale and degree of concentration in growth;
- The ecosystem within which the settlement is located;
- The level and nature of the development process, which affects the ability of the family and the state to respond to problems;
- The development priorities of the state.

Within the development process, urban environmental problems usually emanate from two principal sources.

The first is environmentally irresponsible or poorly managed economic development (see Figure 9.27). Despite the arguments of development economists, the market has responded poorly to environmental problems created by the philosophy of 'grow now and clear up later', unless coerced by the enforcement of regulatory legislation. In general, such controls have been weak, often because urban managers themselves are frequent beneficiaries of uncontrolled development.

The second major contributor is poverty and vulnerability, which forces low-income households to survive as best they can, leaving the environment to look after itself (Figure 9.27). This does not mean that the poor are unaware of the environmental consequences of their actions, rather that they have other priorities. The implication is that improvements to the urban environment must be strongly linked to poverty alleviation as well as the regulation of industry. This is another area where poverty reduction/amelioration have become of great salience.

This combination of contributory factors to the brown agenda also varies spatially and in terms of scale – from the household scale to the regional and global scales. In general, the concerns of the household, workplace or community are more immediate and relate primarily to health and to equality of access to basic services (Table 9.15). At the regional and global levels the problems are more long term in nature and are linked to the impact of resource use on future generations – the major concerns of the West. Between these sets of concerns lies the city itself, combining all these issues in a complex situation that requires careful management to ensure sustainable urbanisation.

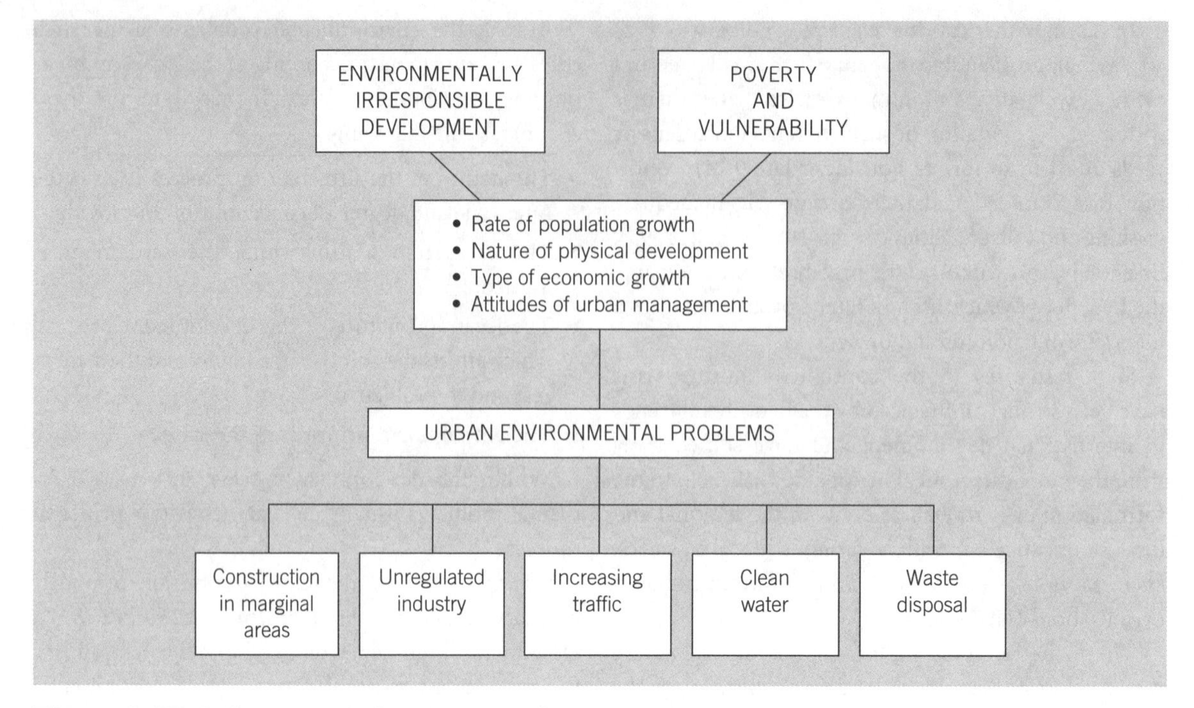

Figure 9.27 A framework for the consideration of urban environmental problems in developing cities

Source: Adapted from *Third World Cities*, 2e, Drakakis-Smith, D., Copyright (© 2000), Routledge. Reproduced by permission of Taylor & Francis Books UK.

Table 9.15 Spatial dimensions of the brown agenda

Principal service infrastructure	Problem issues
Household/workplace	
Shelter	Substandard housing
Water provision	Lack of water, expensive
Toilets	No sanitation
Solid waste	No storage
Ventilation	Air pollution
Community	
Piped water	Inadequate reticulation
Sewerage system	Human waste pollution
Drainage	Flooding
Waste collection	Dumping
Streets (safety)	Congestion, noise
City	
Industry	Accidents, hazards, air pollution
Transport	Congestion, noise, air pollution
Waste treatment	Inadequate, seepage
Landfill	Unmonitored, toxic, seepage
Energy	Unequal access
Geomorphology	Natural hazards
Region	
Ecology	Pollution, deforestation, degradation
Water sources	Pollution, overuse
Energy sources	Overextended, pollution

Source: Bartone *et al.* (1994) © International Bank for Reconstruction and Development/The World Bank

Table 9.16 Water charges in selected cities[a]

	Average piped tariff (US$ M^3)	Private vendor tariff (US$ M^3)	Private/public
Jakarta	0.363	1.848	5.1
Bandung	0.268	6.161	23.0
Manila	0.232	1.873	8.1
Calcutta	0.049	2.099	42.8
Madras	0.046	0.875	19.0
Karachi	0.047	1.747	37.2
Ho Chi Minh City	0.045	1.511	33.6

[a] Data from various sources

Household and community matters: health, water and waste

It is estimated that some 600 million urban residents live in conditions that continually threaten their health. For most families, simply trying to feed themselves takes up most of their income, so there is little money for shelter or health care.

Many are forced to live in squatter settlements or tenements that exhibit a range of environmental problems. The second United Nations Centre for Human Settlements (Habitat) Report (UNCHS, 1996) highlighted four particular problems: water, sewerage, overcrowding and air pollution.

Of all basic needs, access to clean water is probably the most important (see Figure 9.27), and yet some 170 million urban residents lack access to potable water near (not in) their homes (World Bank, 1992).

For example, in Indonesia only one-third of the urban population has access to safe drinking water. Moreover, those with such access are usually the better off; the poor, who can least afford it, are forced to buy their water from vendors at much higher prices (Table 9.16).

Little wonder that the poor are often forced to resort to contaminated water with disastrous consequences for their health. With rising populations some cities have been forced to overexploit their aquifer resources, so cities such as Bangkok and Mexico City have experienced widespread subsidence.

Even where, as in Amman, the capital of Jordan, water supply connections exist to virtually every dwelling, water is rationed to one or two supply periods a week and issues of social inequity mean that the poor have to spend more time and proportionately more of their income to get adequate water, especially in the summer months (see Case study 9.4) (Potter *et al.* 2007a, 2007b).

Case study 9.4

Urban water supply issues – the example of Amman, Jordan

Amman, the capital city of Jordan, is one of the ten most water scarce nations in the world (see Potter, Barham and Darmame, 2007; Potter, Darmame and Nortcliff, 2007; Potter *et al.*, 2007). In 2004, the total water consumption for the city area was 105 million cubic metres and local resources are insufficient to meet this. In its *National Water Master Plan* 2004, the Jordanian government stressed that the first priority is to meet the basic needs of the people. Indeed, as the population of the city has grown, various strategies have been implemented, most notably the transfer of waters from the Jordan Valley, from distant reservoirs and aquifers and the recycling of wastewater.

Today, Amman receives around 50 per cent of its water from the Jordan Valley. Water is pumped from –225 m in the Jordan Valley to a modern treatment plant at Zai, which is located to the north-west of the city at an altitude of 1035 m. The remaining water demands of the city are met from the Al-Mafraq well, the Azraq aquifer (some 70 km east of Amman), and from Qatrana, Swaqa and Wala to the south of the

▶

Case study 9.4 (continued)

city. Looking to the future, providing the city with adequate water is a priority for the government. One of the major projects to achieve this is the Disi Project. This involves the proposed construction of a 325 km pipeline from the Disi aquifer that lies on Jordan's border with Saudi Arabia. This will provide the city with around 100 million cubic metres per year for the next 100 years at an estimated base capital cost of $US 600 million.

Unlike many cities in the developing world, 98 per cent of households in Amman are connected to the water supply network. However, since 1987 the supply of water to households has been rationed. For most parts of the city, water is supplied on just one or two days of the week, and the problem for households is one of storage. Wealthy families have been able to invest in large underground storage tanks or cisterns, whilst less wealthy families have been dependent on the ubiquitous 2 cubic metre rooftop storage tank. For such consumers, purchasing a second tank is likely to be very costly, amounting to around 90 JD.

The rationing of the urban water supply system in Amman reflected not just the relative scarcity of water, but also the generally dilapidated physical state of the network. Until 1999, 54 per cent of the water entering the city's distribution system was classified as 'unaccounted for', with half of this being lost through leakage. This situation reflected the fact that over time, extensions to the network have not generally been planned and have consisted of small diameter pipes. Over the years, operators have generally responded to problems of water pressure by increasing pump size rather than by reinforcing the network, thereby increasing overall pressures within the system. The remaining 'unaccounted for water' has been due to inadequate billing, lax payment collection and the illegal use of water, which in 2004 amounted to over 30,000 instances. The Water Authority of Jordan calculates that on average an illegal user of water consumes two to three times more water than a legal subscriber.

In order to meet the demand, especially in the dry summer months, various sub-markets for urban water have developed, such as private water tankers (see Plate 9.17), water bottled from private wells and distilled mineral water derived from small reverse osmosis machines. In this context, household income and family size are vitally important variables. The cost of purchased water, storage tanks, pipework and filters are prohibitive for poor households in the eastern and southern areas of Amman. This is one of the reasons for the low average domestic water consumption of 94 litres per head per day in the city. Not surprisingly, the social polarity that characterises Amman is, therefore, also reflected in patterns of water consumption within the city (Potter, Barham and Darmame, 2007).

In respect of management, Amman's water supply system was placed in the hands of the private sector in February 1999. At this time, a four-year contract was granted to ONDEO, the commercial arm of Suez Environmental, of which Lyonnaise des Eaux, France is a leading subsidiary. A local company known as LEMA was created, owned 75 per cent by Suez Environmental and 25 per cent by Arabtech Jardaneh (Jordan) and Montgomery Watson (UK). LEMA operated with an operational investment fund of US$25 million for urgent maintenance and repairs. The contract was extended twice and continued through to December 2006.

LEMA operated over an area of 3000 sq km, supplying 2 million people and managing 350,000 accounts. It is generally acknowledged that LEMA's major contribution has been in improved billing and debt collection, customer service in general, and in the regulation of rationing. For instance, in winter 2006 continuous supply was introduced to 15.8 per cent of LEMA's customers, and it is clear that some technical sources feel that the entire system should move toward continuous supply both for technical and supply reasons. However, it seems equally clear that at the present time government does not feel this is a step in the right direction – or at least, one they wish to follow (see Potter, Darmame and Barham, 2007).

After much debate during the period 2005–2006, the era of privatisation came to an end in January 2007, and the management of Amman's water was placed in the hands of a new 'public company' named *Meyahona* ('Our Water'). This is owned by the Water Authority of Jordan (WAJ), but will be

Case study 9.4 (continued)

Plate 9.17 A water tanker in its depot on the north-east fringe of Greater Amman (photo: Rob Potter)

run on the lines of a private company. This is exactly the model that has been in operation in the second city, Aqaba, since 2004 and is being presented by the Ministry of Water and Irrigation as a crucial alternative to private sector involvement in the water sector in Jordan. In this sense, sources in the Ministry stress that while the water system of Amman may no longer be privatised, it will remain commercialised.

Closely linked to water provision is the problem of waste removal (Figure 9.27) through sewerage systems (Pernia, 1992).

Again, in most developing countries because of increasing populations this situation is worsening; during the 1980s alone the number of urban residents without access to adequate sanitation increased by 25 per cent (World Bank, 1992).

Human waste, therefore, often lies untreated around the household, increasing health risks, and is eventually washed into waterways, lakes or seas, polluting aquifers and poisoning aquatic resources (Stren *et al.*, 1992).

The health problems created by poor water and sanitary conditions are often exacerbated by poor diets and by overcrowding and poor ventilation, which intensifies the transfer of respiratory infections, especially where biomass fuels are used.

Those who are more involved in domestic activities (women and children) are therefore more prone to tuberculosis or bronchitis, still major killers in the cities of the Third World (Satterthwaite, 1997).

The city environment

The problems experienced in and around the household can be compounded by city-wide issues that often reflect the particular setting of the settlement.

For example, many cities are located in hazard-vulnerable zones (Figure 9.27) and it is usually the poor who are forced to live in the most marginal areas, such as steep slopes or flood-prone lowlands (Main and Williams, 1994).

All too often the impact of natural disasters is intensified by poor urban management in allowing such areas to be settled without providing adequate safeguards. In Rio de Janeiro in 1988 the floods and landslides which followed torrential rain were partially caused by neglected, blocked or inadequate drainage systems in the favelas (World Bank, 1993).

In the same way, Case study 9.5 demonstrates how the increased incidence of landslides and mudslides in Caracas, Venezuela has been attributed to the growth of informal rancho areas (Jimenez-Dias, 1994; Potter, 1996).

Humans also contribute to these problems as a result of inadequate supervision of economic growth; governments are reluctant to enforce what few regulatory controls they have for fear of discouraging investment.

As a result, industrial air and water pollution from uncontrolled discharges increasingly contaminate Third World cities. For example, industrial discharges have increased 12-fold over the last 14 years in the newer Asian industrialising countries of Thailand, Indonesia and the Philippines.

The most infamous example of such pollution remains the Union Carbide plant in Bhopal, India, where in 1984 poisonous gases killed some 3300 people and seriously injured another 150,000 (Figure 9.28). Most were from poor households living adjacent to the plant (Gupta, 1988).

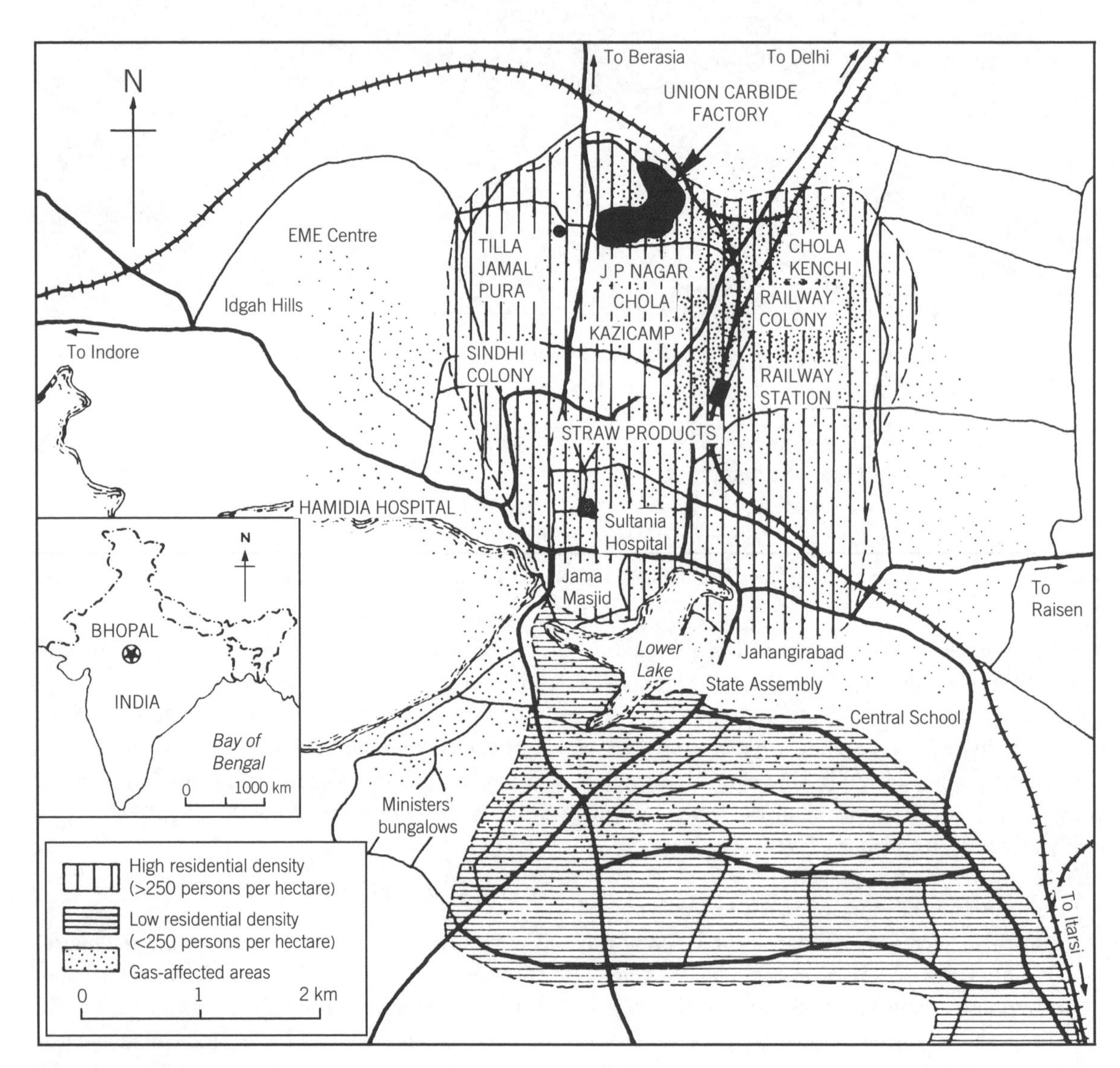

Figure 9.28 Bhopal: the gas escape at Union Carbide
Source: Adapted from Gupta (1988)

The disastrous explosion occurred in a storage tank containing methyl isocyanate gas at the plant. The Bhopal plant was an unprofitable operation, which Shrivastava (1992: 3) suggests was for the most part 'ignored by the top Union Carbide officials'.

This, combined with the unregulated development of two large slum areas across from the plant, these housing several thousand residents, gave rise to the preconditions for a major disaster. Many are still suffering from serious health problems stemming from the incident.

Increasing vehicle ownership and extensive use of fossil fuels are also contributing to air pollution in developing countries. In Bangkok, 26 million workdays are lost annually through respiratory problems.

Figures suggest that the incidence of lung cancer in Chinese cities is up to seven times greater than in the country as a whole. However, we must put this into global perspective, since many East European cities have much worse urban air pollution levels and the three leading global producers of carbon emissions are the USA, Canada and Australia.

Case study 9.5

Urbanisation and environment: the case of Caracas, Venezuela

Caracas is the primate capital city of Venezuela, and as such exhibits many of the features discussed earlier in this chapter. It may be recalled that in developing his ideas about core–periphery relations in transitional societies, John Friedmann (1966) specifically used Caracas and the rest of Venezuela as his case study (Chapter 3). The city grew very rapidly indeed following the development of the oil industry in the early part of the twentieth century. In 1950 it housed a population of just over 500,000; but by 1981 the figure had increased to just over 2 million.

One of the most conspicuous features of Caracas is its location in a very narrow east-to-west valley (see Plate 9.13 (bottom photograph) and Figures 9.29a and 9.29b). With rapid urban growth, the sites for new development have become increasingly scarce.

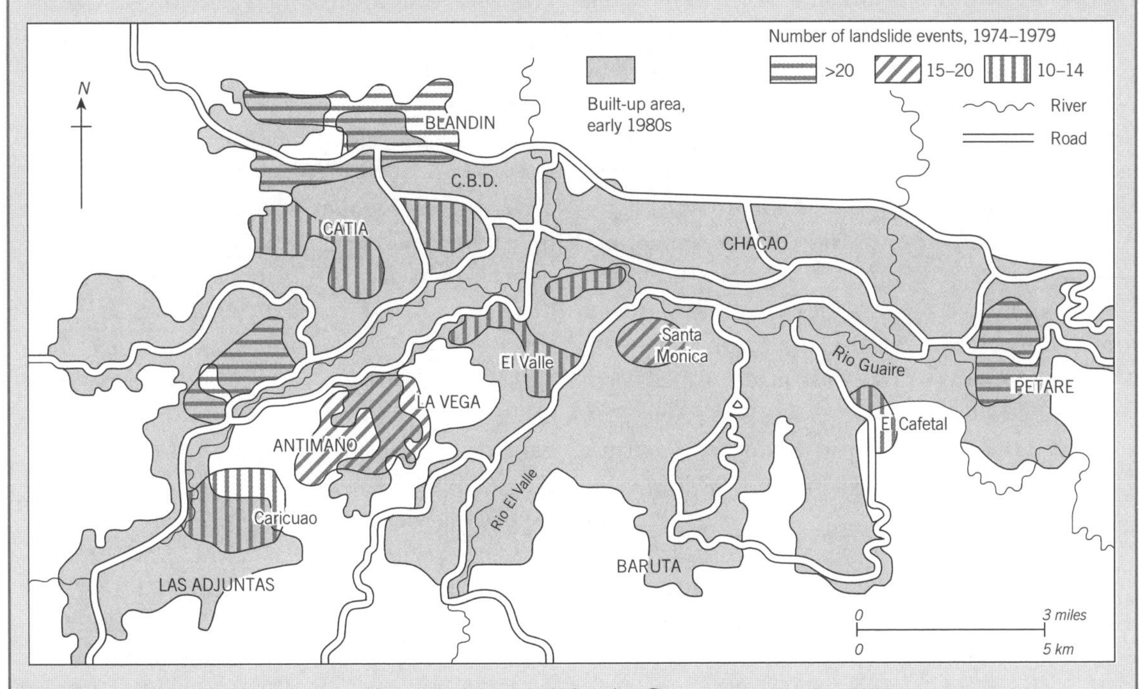

Figure 9.29a Main areas of landslide activity in Caracas

Case study 9.5 (continued)

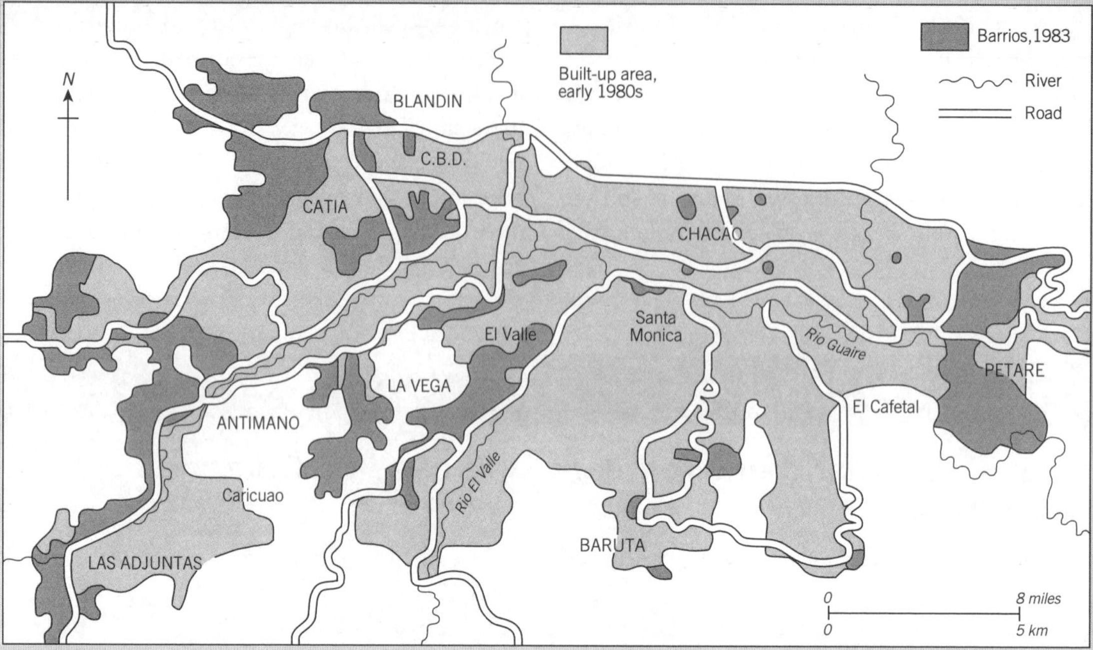

Figure 9.29b Principal barrio areas in Caracas

Like most cities in the developing world, since the 1950s a very high proportion of the growth of the city has been accounted for by self-help low-income settlements. These are referred to locally as barrios. Within these areas, many individual houses may start as relatively poor dwellings, but the majority undergo reasonably rapid improvement, upgrading and consolidation. By the mid 1980s, over 61 per cent of all homes in Caracas were self-built informal sector dwellings.

Given the geography of the city, it was inevitable that many of these new informal dwellings were to be found located on steep slopes. Jimenez-Dias (1994) records that by the middle of the 1980s 67 per cent of the total area occupied by barrios were on land which was unstable enough to justify the legal eviction of the residents.

Records for Caracas suggest that before 1950 landslides were comparatively rare events. Although it is always possible that some were not recorded, only 12 were documented between 1800 and 1949. But the records then show that by the 1950s and 1960s landslides were occurring at the rate of one a year. Thereafter, research shows that failures increased dramatically, from an average of 25 per year in the 1970s to 35 per year during the first half of the 1980s (Jimenez-Dias, 1994).

However, the most important fact is that up to the 1960s the majority of slope failures recorded in Caracas were associated with the occurrence of earthquakes as the initiating mechanism. However, from 1970 onwards, slope failures and mass movements became associated with heavy rainfall events rather than seismic activity. Spatially it is noticeable that slope failures have tended to occur in the barrio areas. This is shown very clearly in the maps of the city in Figures 9.29a and b. The areas with significant landslide events between 1974 and 1979 (Figure 9.29a) correspond almost exactly with the main barrio areas of the city (Figure 9.29b). As a recent example of the outcome, when tropical storm Bret hit Caracas in August 1993, over 150 are believed to have been killed and thousands more lost their homes as a result of the major mudslides which occurred in the hilly low-income settlements.

Solid waste disposal compounds these problems for most cities (Plate 9.7). City-wide collection services are a rarity, and where they exist they are often confined to the wealthier districts.

Sometimes private garbage removal services do exist, not necessarily provided by companies but by groups of people traditionally associated with such activities.

Again, however, the fees necessary for such services restrict them to those households that can afford them. In poor districts the rubbish is simply dumped in open spaces, from where it is occasionally removed, usually for uncontrolled incineration.

Ironically, it is the poor themselves who often recycle solid waste, saving and selling bottles, cans or paper (Plate 9.7 see also Plates 1.6a–d). Sometimes families even live on the city garbage dumps, as in the famous Smokey Mountain site in Manila, where some 20,000 scavengers live and work.

Regional impacts

The regional impact of cities has now spread far beyond their immediate hinterland. Food, fuel and material goods are drawn into cities from all over the nation and the world, affecting the lives of many.

The area that the city affects by its waste output has also grown – its so-called footprint. Indeed, the wealthier the individual or the city, the more distance they can afford to put between themselves and their waste. This is most clearly illustrated by the export of toxic and noxious wastes from developed countries to the more poverty-stricken parts of the Third World.

It is useful to divide the regional impact of cities into two zones: (1) the immediate peri-urban area around the city where the urban footprint looms large and heavy; (2) the broader region beyond this.

Peri-urban zones exhibit two broad areas of concern (Satterthwaite, 1997):

- *Unplanned and uncontrolled urban sprawl.* Often in the form of squatter settlements and illegal small-scale industries beyond the city boundaries, these areas can also contain large-scale municipal uses such as power stations or sports stadia.
- *Liquid waste disposal.* Untreated sewage and industrial effluent enter rivers, lakes or aquifers, making peri-urban areas concentrations of intense contamination. This problem was briefly discussed earlier in this chapter.

The impact of these processes is worsened by the fact that the peri-urban area is often a zone of major importance for recent migrants, offering land for shelter and agriculture and perhaps for woodfuel. The destruction and pollution of the peri-urban area is thus viewed rather differently by urban and peri-urban residents.

Further afield in terms of regional impact, the urban footprint is becoming more marked every year, breaking the traditional links between the city and its hinterland.

Thus the immediate regions around African capital cities are now producing exotic vegetables, flowers and other crops for export rather than to feed the city population, thereby exacerbating the nutritional problems of the poor (Smith, 1998).

On the other hand, meeting the energy needs of the city has often created other environmental problems such as denudation of biomass or expansion of coal mining and associated waste tipping.

There are many other examples of regional urban footprints; the growing demand for cement and bricks lead, respectively, to quarrying and pollution, or to loss of valuable soil resources.

Water demands too can affect regions way beyond the city, particularly for cities in semi-arid areas. Bulawayo in Zimbabwe is seeking funds to draw on the waters of the Zambesi. If this succeeds, what will be the impact on the fragile ecosystems of this part of southern Africa? The Three Gorges Dam on the Yangtse is also partly designed to provide energy for urban industry but will have enormous environmental effects on the 600 km of the new lake between Chongqing and Ichang. As noted previously, Amman, Jordan is bringing water some 325 km from the Disi aquifer in southern Jordan (Potter, Darmame and Nortcliff, 2007).

Final comments: urban management for sustainable urbanisation

The above discussion of the brown agenda reveals some of the main issues related to urban management for sustainable urbanisation, such as false priorities, lack of adequate legislation, self-interest, poor knowledge and training, and a susceptibility to external influences, both benign and malign.

Lack of public awareness of appropriate policy responses and the fact that protest is usually fragmented means the authorities discover that opposition is easy to contain. Indeed, most pressure for change in urban management towards greater responsiveness to problems of sustainability has come from external sources.

Both the World Bank (1991) and the United Nations (UNDP, 1991) became increasingly nervous about potentially restive urban populations in the early 1990s and they have recommended new management strategies, as noted earlier in this chapter.

Such approaches have mirrored the emergence of neo-liberal development strategies in general (see Chapters 3 and 4), with an emphasis on market-led solutions and limited interference by the state, with many urban management policies being substantially shaped by national governments in conjunction with their external advisers.

Municipal governments thus have to work within limits set by agencies beyond their control, although even within the city there are often more specific management problems too. Many would claim that overall the shift to the market economy has simply resulted in the transfer of responsibilities from the state to the poor, largely by removing the constraints on letting them help themselves.

Certainly the poor have taken advantage of such moves in the development of their coping mechanisms but these are essentially small in scale and focused on the household. Collective and larger scale responses are difficult without proper knowledge, training and funding. Recycling waste is possible for the poor, constructing a sewerage system is not.

Increasingly, intermediaries have become involved in order to 'improve local capacity'. The private sector, the embodiment of the market, has been slow to respond to the needs of sustainability.

Although there are some examples of the private sector meeting some basic needs in large Latin American and Pacific-Asian cities (Gilbert, 1992), they are far from universally replicable and, unsurprisingly, tend not to affect those in greatest need of assistance. Indeed, the role of the private sector in water vending has increased the exploitation of the poor and has to be countermanded by the state (Choguill, 1994).

Despite early pessimism, community participation and cooperation emerged on a substantial scale, perhaps due to growing competition for resources by increasingly diverse ethnic and social groups.

In this context, the role of NGOs and CBOs as facilitators of urban community development has been equivocal. Many NGOs are themselves large global organisations driven by Western agendas and funding sources. Indeed, many NGOs have been criticised for assisting in the privatisation of resource provision and the retreat of the state. In this context the underprivileged have little option but to engage in their own protests and civil action for improved access to urban resources.

For some analysts, such as Escobar (1995; see also Chapter 1), these movements could form the basis of new development strategies, but often they are deliberately limited in their objective and they are consciously non-political. If and when they achieve some of their aims, such social movements tend to fade away.

There is no doubt that the appropriate level at which to tackle problems related to urban sustainability is the local state, the city itself. At present, however, most but not all urban management is poorly informed, poorly motivated and poorly organised. A decentralisation of power, funding and responsibilities from the national to the local state would be a start, but it must also be accompanied by greater democratisation at the level of the city itself.

Those most affected by the inadequacies and inequalities of unsustainable urban development – in short, those whose coping mechanisms sustain the unsustainable – must become part of the process of policy formulation and enactment.

At present, there is little sign of this occurring on a wide scale, and with urban populations and poverty continuing to grow, the problems of urban sustainability for a whole range of different cities within the Third World are likely to get worse, rather than better.

Key points

- The chapter has illustrated the key importance of the contemporary processes of urbanisation and urban growth in the poorer nations of the world.
- It is these nations that are now showing some of the highest rates of urbanisation ever recorded. Further, they now account for some of the largest cities in the world.
- Levels of regional inequality and regional urban primacy are frequently high in developing countries.
- Urban and regional planning is needed to ameliorate some of the worst effects of spatially polarised and unequal development. There is a range of national urban development strategies that can be employed by the state machinery.
- The linkages existing between urban and rural components of the national space are frequently appreciably stronger and more complex than has previously been conceptualised. This is witnessed in the recognition of extended metropolitan regions (EMRs) and 'compound urban regions'.
- Cities in the developing world are showing a variety of forms as they grow and develop. They are not universally converging on the norms associated with 'Western' cities.
- The informal sector has played a vital role in the provision of both homes and jobs for the poor in the cities of the developing world.
- The brown agenda serves to stress the environmental impacts of urbanisation, with emphasis on the need for sustainable principles of urbanisation.

Further reading

Desai, V. and Potter, R.B. (eds) (2008) *The Companion to Development Studies*. London: Arnold, Part 5: Urbanization, 241–72.

An accessible source book that brings together over 100 key essays dealing with all aspects of the field of development studies, and Part 5 deals with urban processes.

Drakakis-Smith, D. (2000) *The Third World City*, 2nd edn. London: Routledge.

A good introductory account on cities in the developing world.

Hardoy, J.E., Mitlin, D. and Satterthwaite, D. (2001) *Environmental Problems in an Urbanizing World*. London: Earthscan Publications.

The environmental aspects of rapid urban growth and urbanisation are the focus of this detailed text.

Pacione, M. (2005) *Urban Geography*, London: Routledge.

The chief merit of this general introduction to urban geography is that it covers cities both in the less developed as well as the more developed regions of the world.

Potter, R.B. and Lloyd-Evans, S. (1998) *The City in the Developing World*. London: Pearson.

A detailed overview of the role of cities and urbanisation in the development process.

United Nations (2005) *World Urbanization Prospects Report, 2005*. New York: United Nations.

A good source of basic statistics concerning urbanisation in the current global context.

Websites

www.un.org

Website of the United Nations, which includes access to urban-oriented reports such as the United Nations World Urbanization Prospects 2005 revised report cited here.

www.unchs.org

This is the website for UN-HABITAT, the United Nations Human Settlement Programme, the mission of which is to promote sustainable development. The accent is very much on issues of shelter and the provision of legal titles to land, and on urban governance.

www.iied.org

IIED is the International Institute for Environment and Development. The Human Settlements section contains features on urban poverty, urban environmental issues, and rural–urban linkages.

Discussion topics

- 'Rapid urbanisation, slow everything else.' Examine this statement in relation to urbanisation in developing countries.
- Examine the view that only socialist nations have seriously endeavoured to curb the growth of large cities.
- Examine the evidence suggesting that cities in the developing world are not converging on a global norm in respect of their urban structure.
- Assess the extent to which the term 'self-help' aids our understanding of both housing and employment in developing societies.
- 'Environmental problems in developing world cities require good governance.' Discuss.

Chapter 10

Rural spaces

This chapter examines the ways in which the strategies, policies and processes of development examined in earlier sections and chapters impact in rural spaces of the developing world. Whilst the world officially became predominantly an urban one in 2007 with over 50 per cent of the world's population living in towns and cities classified as urban, this chapter details the continued significance of rural development challenges for developing countries in particular, for some of the poorest groups of people worldwide and for securing the conservation of globally valued resource functions and landscapes, for example.

This chapter:

- Details the core patterns of rural living in the developing world and a number of core differences in basic development opportunities between rural and urban areas;
- Identifies the legacy of colonial impacts in structuring the nature of the rural development challenge such as through patterns and processes of landholding;
- Investigates the notion of rural livelihoods systems and in particular how this has raised understanding of rural development processes as encompassing more than agricultural development;
- Outlines a number of principal systems of livelihood within rural areas and a range of sources for change that increasingly extend beyond the local level;
- Overviews the major components of rural development policy and practice through the post-independence era and considers in some detail the characteristics of and optimism for a 'new paradigm' in rural development through the 1990s;
- The final section examines the way in which rural spaces are currently considered in development theory and practice and in particular the moves towards more participatory, locally relevant and sustainable initiatives.

Introduction

Although the world's population is becoming increasingly urbanised, in many parts of the developing world it is the rural areas which will continue to accommodate the majority of people for the foreseeable future. Not only will these rural areas have to deliver food and incomes to expanding populations in the next decades, but they will also be expected to safeguard many aspects of the 'global commons' such as biodiversity (Morse and Stocking, 1995). However, these areas are the locus for some of the most insecure livelihoods anywhere in

Table 10.1 Broad processes and trends of the rural South

- Increased diversification of occupations and livelihoods
- More common and pronounced occupational multiplicity
- A shift in the balance of household income from farm to non-farm
- A de-linking of livelihoods and poverty from land (and from farming)
- Livelihoods are increasingly delocalised as lives become more mobile
- A growing role of remittances within rural household incomes
- Rising average age of farmers
- Cultural and social changes are operating in new ways to modify livelihoods

Source: Compiled from Rigg (2006)

the world, with three-quarters of the world's poor estimated to live in rural areas (IFAD, 2001).

While occupations and livelihoods in rural areas are diversifying, as highlighted in Table 10.1 and as investigated in this chapter, the trajectories of change within farming and agricultural production and issues of control over fundamental resources such as land remain key to the development of rural areas (World Bank, 2001a). However, the mounting evidence (particularly from Asia and Latin America) that lives in developing regions are becoming more mobile and livelihoods less localised (see Chapter 8) demands new approaches in rural development planning and interventions beyond the redistribution of rural resources and/or the reinvigoration of agricultural production (Rigg, 2006).

Rural spaces in development thinking

Eurocentric models

Early models of development (as explored in Chapter 3) often emphasised the importance of agricultural innovation and improved rural productivity for the release of capital and surplus labour, which could then be used in emerging urban and industrial activities. Based largely on the historical experience of Western Europe, the interdependence of the rural-agricultural and urban-industrial sectors and the transformation of a country's economy from 'one that is dominantly rural and agricultural to one that is dominantly urban, industrial, and service-oriented in composition' (Mellor, 1990: 70), were central to the concept of development itself. For example, in his classic text *The Theory of Economic Growth*, Arthur Lewis (1955: 433) argues that:

> **industrialisation is dependent upon agricultural improvement: it is not profitable to produce a growing volume of manufactures unless agricultural production is growing simultaneously. This is also why industrial and agricultural revolutions always go together, and why economies in which agriculture is stagnant do not show industrial development.**

Rigg (2001: 19) has recently noted that 'on the face of it' this 'classical agrarian question' remains 'highly pertinent' in the developing world where:

> **the transition to capitalism is an ongoing project and, notwithstanding far-reaching structural change in some countries, agriculture remains an important economic sector employing the largest proportion of the workforce. Thus we have, on the one hand, a completed historical process in the developed North, while in the South the process is contemporary and continuing. In the North, obstacles to accumulation were overcome and capitalist transformation and industrialisation ensured. In the South, significant obstacles remain.**

The colonial encounter in much of the developing world was centred principally on the raw materials, labour supply and opportunities which the rural areas offered for both export production and satisfying the food requirements of growing urban populations. Post-independence, national development plans frequently attached great importance to rural and agricultural development, and substantial domestic and foreign financial resources have been directed to rural areas. However, official development assistance (and assistance from the IFIs) to agriculture, for example, has generally fallen in recent decades (IFAD, 2001).

Urban bias

One criticism levelled against governments both during and since the colonial period has been the suggested state and process of 'urban bias' in development that

Key idea

The bias of rural development planning and practice

In 1983 Robert Chambers put forward a number of reasons why rural development interventions were failing to have the impacts that they were planned to have. He suggested that key national decision makers and development institutions more widely had inadequate knowledge about rural dwellers and their needs due to a number of 'biases' in research and practice. Fundamentally, rural communities and their people are often regarded as being 'out of the way' and 'off the beaten track' as far as development planners are concerned:

> **In Third World countries as elsewhere, academics, bureaucrats, foreigners and journalists are all drawn to towns or based in them. All are victims, though usually willing victims, of the urban trap.**
>
> **(Chambers, 1983: 7)**

Further biases in the investigation process may also contribute to an incomplete and inaccurate understanding of rural needs such as 'dry season bias' that results from only visiting rural areas during the dry season, when travel is usually easier. This is despite the fact that in tropical countries with well-defined wet and dry seasons, it is during the rainy season that most crops are grown, people have to work for long hours in the fields, and disease and malnutrition are more common. 'Tarmac bias' refers to the reality that many visitors to rural areas travel only on good roads and rarely venture into remote areas, thus failing to make contact with what are often the poorest communities. A further factor in generating inaccurate perceptions is 'person bias', where visitors only speak to influential community leaders, who are invariably men. The views of women and 'ordinary' community members are therefore rarely heard (Chambers, 1983).

Chambers' subsequent texts (1993, 1997) have been important in continuing to foster the importance of understanding local people's priorities first in rural development and continue to be influential currently. As Bebbington (1999: 2021) suggests, misperceptions concerning the 'way people get by and get things done' in rural areas continue to compromise rural development efforts in developing countries.

they have fostered, in which urban areas have been consistently favoured relative to rural areas (Lipton, 1977). The suggestion has been that in reality national politicians, keen to maintain a hold on power, have generally been much more concerned to keep their urban populations contented, since these communities are invariably better educated, more articulate, organised in trade unions and other groupings, and therefore likely to be a much greater potential threat to economic and political stability than the less educated and less well-organised rural poor.

However, as seen in Chapter 7, the state's capacity to intervene in all arenas of development was reduced through the 1980s and 1990s such as through pressures of mounting debt and requirements to meet the demands of the international financial institutions. Civil society organisations have also emerged to contest the inadequacies of the state, including around fundamental rural concerns over land and control of rural resources leading to challenges to the processes of urban bias.

A further constraint on rural development interventions in the past has been the suggested bias within rural development planning and practice itself as highlighted in the Key idea box.

'Old' and 'new' challenges in rural development

An important factor in prompting a re-evaluation of rural areas in development thinking and practice has been the search for new patterns and processes of 'sustainable development', as identified in Chapter 6. In short, if globally valued environmental resources, landscapes and resource functions are to be conserved, it will depend on meeting the development needs of some of the poorest groups worldwide, often based in rural areas of the developing world. Towards meeting these 'new' challenges, the work over the last decade that has witnessed important shifts in the way that indigenous skills and capacities are considered in the explanation of rural resource outcomes and environmental transformations is particularly relevant (Batterbury and Warren, 2001; Elliott, 2002; Leach and Mearns, 1996). In addition, understanding of the diversity and flexibility of rural livelihoods within the developing world (emphasised throughout this chapter) are features that have been regularly neglected within past development efforts, but form the basis for many of the more successful and sustainable rural development initiatives into the twenty-first century.

However, further new challenges in rural development are emerging as the global processes identified in Chapter 4 have profound impacts on local development and environmental outcomes. For example, the prospects for local small-scale producers (or indeed Southern-based commercial agricultural enterprises) to thrive under current World Trade Organization rulings has been the focus of challenges to globalisation and neo-liberalism launched by international non-governmental organisations like Oxfam (2002) as considered in Chapter 7. Others point to how the options for governments in the South to influence rural development agendas more specifically are widely becoming increasingly circumscribed (see Rigg, 2001). New challenges under globalisation processes may also come from within the 'developing' world. The Critical reflection box considers the demand from China for minerals (and oil in particular) and the impact on initiatives towards democracy on the African continent specifically.

Critical reflection

China's thrust to sub-Saharan Africa

The emergence of China on the world stage is being felt around the world (Flavin and Gardner, 2006) but its impact may be particularly significant in the world's poorest region, sub-Saharan Africa (SSA). Kaplinsky *et al.* (2006) have researched the opportunities and threats of China's recent rapid growth in trade with the region. Critically, they identify that it is China's strategic search for raw materials and energy sources that are the primary drivers for investment and aid (rather than for final markets or low-cost production platforms, for example). Similarly, Taylor (2006) reviews the burgeoning and unprecedented trade links between China and Africa that increased in value from $2 billion in 1999 to $39.7 billion by 2005 and is predicted to reach $100 billion per year within the next five years. Taylor refers to a current 'oil safari' driven by China's need in the short term to secure oil supplies and in the long term to position itself as a global player in the oil market.

Whilst some SSA economies are undoubtedly gaining through the growing demand for these commodity exports to China and the associated increase in global commodity prices, there is concern regarding China's current investment in countries with known problems of government transparency and with human rights abuses. For example, Taylor (2006) suggests an explicit stance being taken by China that emphasises state sovereignty and non-interference in domestic affairs. Kaplinsky *et al.* (2006) identify that China has actively forged closer links with 'fragile states' including the Sudan, Angola and the Democratic Republic of the Congo within the energy and resource sectors that is undermining attempts by Western donors and NGOs, for example, to promote more transparent and better corporate and environmental governance in these countries; 'poorly handled, a resource-boom can easily become a resource-curse' (Kaplinsky *et al.*, 2006: ii). They warn of 'very severe problems of economic management' and the 'inherently unequalising' nature of capital-intensive and concentrated ownership of mineral and hard commodity production.

Consider the benefits and costs of China's interest in the continent of Africa. You could investigate the impact on world prices for commodities and patterns of development in China itself as possible benefits. But you could also consider the outcomes of primary resource developments historically (considered in Chapter 6) and research China's programme of official development assistance in particular African countries, for example.

Table 10.2 Population and employment in selected countries

	Population, (2005[a] millions)	Rural population as percentage of total population, 2004[b]	Agriculture as percentage of GDP, 2005[a]	Percentage of economically active in agriculture 2004[b]
Low-income economies				
Haiti	9	62	28	60
India	1095	71	19	58
Nigeria	132	52	24	30
Middle-income economies				
Indonesia	221	53	14	46
Egypt	74	58	14	31
Venezuela	27	11	5	7
Upper-middle-income economies				
Brazil	186	16	10	15
South Africa	45	42	3	8
Malaysia	25	35	9	16
High-income economies				
Republic of Korea	48	20	4	8
United Kingdom	60	11	1	2
USA	296	19	1	2

Source: [a] World Bank (2007d); [b] FAO (2005–06)

The persistence of rural poverty

In addition to the new challenges in rural development identified above, the rural areas of the developing world continue to host the major and persistent 'old' problems in development – of hunger and poverty (Sen, 2000), for example. An estimated 20 per cent of the population of the developing world, equating to some 785 million people, were chronically undernourished in 1990 (Robinson, 2004). A broad sense of the reality of rural living in the developing world is shown in the following tables. For example, in Table 10.2, it is clear that although there are some variations in the overall pattern, the proportion of the total population living in rural areas is generally higher in lower income countries than in more wealthy countries. The importance of the agricultural sector within overall national production is also evident.

Table 10.3 confirms that poverty (on the basis of national poverty lines) is higher in rural areas than urban within the developing world. Table 10.4 highlights the disparities in access to the basic service of improved drinking water between rural and urban areas in selected countries. Such disparities have a major bearing on the development opportunities in rural areas, most overtly through impacting on the health of the population. The higher infant mortality rates found in rural areas compared with urban centres, as shown in Table 10.5, are a further expression of the generally more impoverished conditions characteristic of rural areas in developing countries.

Table 10.3 The challenge of poverty: the percentage of population below national poverty lines

Economy	Rural	Urban
Bangladesh (1995–1996)	39.8	14.3
India (1994)	36.7	30.5
Indonesia (1998)	22.0	17.8
Morocco (1998–1999)	27.2	12.0
Nigeria (1992–1993)	36.4	30.4
Peru (1997)	64.7	40.4
Philippines (1997)	51.2	22.5
Sri Lanka (1990–1991)	38.1	28.4
Thailand (1992)	15.5	10.2
Tunisia (1990)	21.6	8.9

Source: World Bank (2001a) © International Bank for Reconstruction and Development/The World Bank

Table 10.4 Rural-urban gaps in access to improved drinking water; percentage of population served, 2000[a]–2004[b]

Country	Rural		Urban	
	2000	2004	2000	2004
Afghanistan	11	31	19	63
Bangladesh	97	72	99	82
Brazil	53	57	95	96
Chad	26	43	31	41
China	66	67	94	93
Ghana	62	64	91	88
India	79	83	95	95
Kenya	42	46	88	83
Mexico	69	87	95	100
Nicaragua	59	63	91	90
Pakistan	87	89	95	96
Peru	62	65	87	89
Philippines	79	82	91	87
Rwanda	40	69	60	92
Sierra Leone	46	46	75	75
South Africa	73	73	99	99
Sri Lanka	70	74	98	98
Thailand	81	100	95	99
Tunisia	58	82	92	85
Venezuela	70	70	85	90
Zambia	48	40	88	89

Reading: [a] http://unstats.un.org/unsd;
Source: [b] WHO/UNICEF, 2006

Table 10.5 Infant mortality rate and urban-rural residence for selected countries

Country	Infant mortality rate	
	Rural	Urban
Sub-Saharan Africa		
Ghana	87	67
Ivory Coast	121	70
Kenya	59	57
Asia		
India	105	57
Indonesia	74	57
Philippines	55	42
Thailand	43	28
Latin America		
Guatemala	85	65
Mexico	79	29
Panama	28	22
Peru	101	54

Reading: Allen and Thomas (1992)

Note: IMR is the number of deaths in first year of life per 1,000 live births.

De-agrarianisation in rural areas

The longstanding view amongst development theorists and practitioners has been of rural areas of developing countries (particularly the African continent) dominated by agrarian livelihoods (Bryceson, 2002; Rigg, 2006). This view has underpinned many of the rural development strategies considered below, towards redistributing land and reinvigorating agriculture, for example, as a means to overcoming poverty and fostering development. However, this view is now changing as processes of livelihood diversification and 'de-agrarianisation' are better understood. These key concepts are defined in Box 10.1.

Whilst the value of remittances from family members beyond the rural area have been a long standing interest for those engaged in rural development theory and practice, documentation of the significance of 'non-farm' activities on behalf of those within rural localities is only recently being understood. The Key idea box (p. 450) on rural life highlights how rural people are 'not just farmers'.

BOX 10.1

Key concepts in understanding rural change

Rural livelihood diversification: the process by which rural households construct an increasingly diverse portfolio of activities and assets in order to survive and to improve their standard of living.

(Ellis, 2000b: 15)

De-agrarianisation: the long-term process of occupational adjustment, income-earning reorientation, social identification and spatial relocation of rural dwellers away from strictly agricultural-based modes of livelihood.

(Bryceson, 2002: 726)

While there are some methodological problems associated with comparing studies over time, there is mounting research evidence that suggests the significance of non-farm income to overall livelihoods in rural areas is rising. For example, Reardon *et al.* (2001) examined data for 11 country case studies in Latin America, and identified that rural non-farm incomes (defined as self-employment and wage employment in manufacture and services) constituted on average 40 per cent of rural income in the 1990s, from typically between 25 and 30 per cent of total income in the 1960s and 1970s. The Deagrarianisation and Rural Employment (DARE) research programme covering six countries in Africa found non-agricultural activities accounting for between 60 per cent and 80 per cent of household income in 1996–1998, whereas studies conducted in similar contexts in the mid 1980s had suggested figures of around 40 per cent (Bryceson, 2002). Research in India suggests that the share of non-farm incomes in total rural incomes rose from 19 per cent to 48 per cent in a period of research stretching from 1971 to 1999 (Rigg, 2006).

As the patterns of change in rural livelihoods are revealed, research is now focusing on understanding the processes of de-agrarianisation and the social and economic implications of diversification. Both Rigg (2006) and Bryceson (2002), for example, refer to the 'squeezing' of agriculture under market liberalisation as a key factor that has eroded the profitability and returns to smallholder agricultural production in particular. It can be considered that the same broad processes of globalisation have also been part of the creation of new non-farm opportunities within and beyond the rural sector. Associated with de-agrarianisation are deep-rooted social change, with new divisions of labour and decision making, increased wealth differences and changing desires on behalf of parents and the next generation, for example.

These new patterns and processes of rural livelihood challenge planners and practitioners to find new means for supporting sustainable futures in rural areas. It is likely, for example, that new measures for targeting those in need will have to be found taking into account non-agrarian assets since as Rigg (2006: 192) states, 'it is becoming increasingly difficult, often impossible, to "read off" poverty on the basis of the usual markers'.

Bryceson (2002) urges new interventions prioritising the youth and children towards equipping them with the kinds of skills to make realistic assessments of the prospects of agrarian and non-agrarian options in an era of rapid change in markets. Most fundamentally, these processes and impacts of change are geographically very diverse such that rural development initiatives need to be tailored to the specific local context and circumstance.

Plate 10.1 Transportation of vegetables, Kolkata suburbs
(photo: J. Elliott)

Key idea

There is more to rural life than agriculture

Livelihoods in the Rural South do, in many places and for many households – perhaps even in most places and for most households – continue to depend on small-holder agricultural production. (Rigg, 2006: 181)

However, it is increasingly appreciated that there is more to rural life than agriculture and rural people are not just farmers (Scoones, 1996).

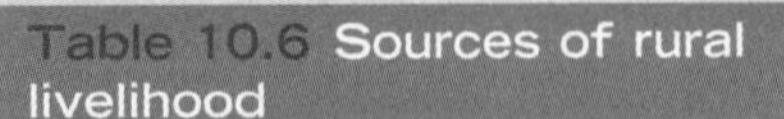
Table 10.6 Sources of rural livelihood

- Home gardening – the exploitation of small, local micro-environments
- Common property resources – access to fuel, fodders, fauna, medicines, etc. through fishing, hunting, gathering, grazing and mining
- Processing, hawking, vending and marketing
- Share-rearing of livestock – the lending of livestock for herding in exchange for rights to some products including offspring
- Transporting goods
- Mutual help – small loans from saving groups or borrowing from relatives and neighbours
- Contract outwork
- Casual labour or piecework
- Specialised occupations such as tailors, blacksmiths, carpenters, sex workers
- Domestic service
- Child labour – domestic work at home in collecting fuel and fodder, herding, etc. and working away in factories, shops or other people's houses
- Craft work – basket making, carving, etc.
- Selling assets – labour, children
- Family splitting – putting children out to other families or family members
- Migration for seasonal work
- Remittances from family members employed away
- Food for work and public works relief projects
- Begging
- Theft

Source: Compiled from Chambers (1997)

Gathering 8%
Crop output 25%
Livestock 15%
Farm wage 10%
Non-farm wage 12%
Non-farm self-employment 15%
Remittances 15%

Figure 10.1 The average rural livelihood in SSA

Source: Compiled from Ellis (2000b). By permission of Oxford University Press, Inc.

Table 10.6 identifies a range of ways in which people may compile the stocks and flows of food and cash to meet their needs in rural areas. Evidently, many go beyond the direct production of agricultural goods, or what Ellis (2000) refers to as 'own-account agriculture'.

The idea of a 'livelihood portfolio' is used to encompass the varied ways in which people secure an income (as the most visible outcomes of the processes by which a livelihood is constructed) within rural areas. Figure 10.1 represents a portfolio for an 'average' rural livelihood within sub-Saharan Africa based on recent research (Ellis, 2000). While own-account agricultural activities are evidently highly significant (delivering 40 per cent of the total income portfolio), 18 per cent is secured through 'off-farm' activities (that may include income gained through working for other people on neighbouring farms as well as gathering activities). A further 42 per cent of income of typical households comes through 'non-farm' activities; including remittances from outside the rural area but also self-employment locally in the services and manufacturing activities illustrated in Figure 10.1.

Agrarian structures and landholding in rural areas

Agriculture remains fundamental to economy and society in the rural areas of the developing world. In turn, patterns of landholding in agrarian economies are key determinants of 'agrarian structure', which concerns the different ways in which land and labour are combined in varying forms of production, as well as the social relations, such as class, which 'structure' the processes of production and reproduction.

Fundamentally, an analysis of agrarian structure attempts to answer the questions: 'Who owns what, does what, gets what and what do they do with it?' (Bernstein, 1992c: 24). Not only do patterns of landownership provide a context in which land and labour are combined in the production process, but land is also often the primary means through which rural inhabitants of the developing world define their personal, social and political identities. As such, land ownership is a major correlate of political and social prestige in these areas. Furthermore, since food is the major product of the land, such patterns have clear implications for the relative and absolute well-being of the population (Ghose, 1983; Plate 10.2).

Agrarian structures are neither static nor uniform across space. They reflect varied historical experience in rural areas, encompassing factors such as environment, culture and political economy. While subsequent sections here consider core patterns on a regional basis, it should be recognised that the fortunes of particular groups of rural people are also often influenced by processes operating beyond the specific rural and even wider regional setting. For example, colonialism and the penetration of capitalism have had a significant impact on agrarian structures across many parts of the developing world. Such processes have taken diverse forms, were variously imposed, taken up or rejected over time and space, and have left distinctive legacies within specific contemporary agrarian structures. However, common legacies can be identified such as the considerable inequality in landholding that exists in some countries, as shown in Table 10.7. However, the dynamics of agrarian structure is exhibited in the case of Zimbabwe, for example, where many white farmers have been (often violently) dispossessed of their land in the last five years.

Table 10.7 Land distribution in selected countries

- Zimbabwe: Some 70,000 whites (0.5 per cent of the population) own 70 per cent of the land; 4000 whites own nearly one-third of the farmland
- South Africa: Blacks, who account for 75 per cent of the population, occupy 15 per cent of the land
- Namibia: Some 4000 whites (less than 1 per cent of the population) own 44 per cent of the territory
- Brazil: Just 3 per cent of the population owns two-thirds of the land
- India: Some 9 per cent of the farm population owns 44 per cent of the agricultural land
- USA: Only 16 per cent of farmers control 56 per cent of all the land

Source: From Halweil (2002)

Plate 10.2 Banana cutting in Dominica, West Indies
(photo: Philip Wolmuth, Panos Pictures)

Latin America

While land is relatively abundant in Latin America, inequality in landownership is marked and an important factor in the persistence of poverty across the region (Department for International Development, 2002). Land is increasingly concentrated in the hands of a small minority, within what is known as the latifundio system. This system includes very large plantation estates, cattle ranches and the haciendas persisting from the colonial era. The majority of the population of these countries have access to very small family farms, or minifundios, and/or are subject to exploitative landlord–tenant systems.

Typically, the share of total agricultural land accommodated by the minifundios is very small. In

Guatemala, for example, farms under 5 hectares constituted 90 per cent of all farms in 1979, but occupied only 16 per cent of the land area. In Brazil in 1980 these figures were 37 per cent and 1 per cent, respectively (Bernstein, 1992b). The Inter-American Development Bank (IADB) has recently estimated that over 60 per cent of the poor in Latin America is based in rural areas, and half of them have very limited or no access to productive resources including land (Inter-American Development Bank, 1998).

The polarisation of landholding in contemporary Latin America is deeply rooted in the colonial history of the region, although subsequent policies in the agricultural sector and processes of commercialisation during the nineteenth and twentieth centuries further reinforced and extended the latifundio system. In the seventeenth century, Spanish colonists were given rights to expropriated land, originally in return for military service, and were able to levy tribute from indigenous communities in the form of labour or goods. With the capital acquired, large landed estates, or haciendas, were established, on which peasants typically cultivated land allocated to them in return for paying rent to the landlord in cash or a share of their crops, or alternatively worked on the landlord's farm in return for a small subsistence plot.

The expansion of this system depended on acquiring further labour and appropriating land from Indians, a process which was often unpopular and resisted by indigenous communities. The scale and nature of expansion varied according to local factors such as population density, the strength of peasant organisations, the extent of land fragmentation and the demand for cash crop production. For those people who resisted such incorporation, the alternative minifundio system often proved inadequate for subsistence production, since plots were often too small or located in areas which were marginal for agricultural production. Many small farmers had no alternative but to seek out other sources of income, often through wage employment. Historically, large-scale agribusiness has dominated agricultural production in Latin America and is a situation substantially maintained by the contemporary policy environment (Department for International Development, 2002).

Asia

Landholding in Asia is heavily concentrated, as in Latin America, but due to land scarcity the size of farms is much smaller and the cultivation of land is more decentralised through systems of tenancy and sharecropping. The significance of access to land and tenancy in determining human welfare is shown in Table 10.8, where it can be seen that landless households in the Philippines and Indonesia own less than the average size of landholding, or are tenants, and are among the poorest households.

Much of Asia has a relatively recent colonial history, and the legacy of that period is evident in present agrarian structures, particularly with respect to the introduction and reinforcement of private property ownership. Before British rule in India, for example, property rights had rested with the community, with village heads allocating individual rights to land, supervising the management of common resources, such as meadows and rangelands, and collecting taxes in kind.

Under Crown rule from 1868, a system of private property rights in land and taxation payable in cash was introduced. For example, in the Zamindari system

Table 10.8 Characterising the poor in Indonesia and the Philippines

	Indonesia	Philippines
Rural-based	✓	✓
Dependent on farming	✓	✓
Concentrated in peripheral regions	✓	✓
Smaller than average landholdings	✓	✓
More likely to be tenants or landless agricultural wage labourers	✓	✓
Larger than average family	✓	✓
Young household head		✓
Household head with a low level of education	✓	✓
Low level of access to social services such as health and education facilities	✓	✓
High level of underemployment		✓

Source: Rigg (1997)

of landholding, which was particularly widespread in northern India, de facto landlords were created as intermediaries between the tenants (the real landowners) and the British Administration. These 'landlords' were required to pass on to the government treasury an agreed percentage of land rents collected from tenants. In return, they were effectively given private property rights through their assigned privilege to extract rents from tenants at whatever level they thought was feasible (Shariff, 1987).

This system of 'parasitic landlordism' (Bernstein, 1992a) enabled the British to control land in the colony and to extract rents from the peasants who worked it without intervening directly in the production process. The system was parasitic in the sense that rents were not reinvested to enhance the productivity of farming and, unlike the communal land system, landlords had no duties to the tenant, such as providing assistance in times of hardship. Although peasant cash crop production grew, indebtedness and landlessness also rose and peasants found themselves subordinated to a newly created class of overlords – the landlords and moneylenders.

At India's independence in 1947, 50 per cent of the land area was held by approximately 4 per cent of the rural population. Meanwhile, some 27 per cent of the population were landless and a further 53 per cent had farms of less than 5 acres (Ghose, 1983).

Africa

In broad terms, African agrarian structures are more widely characterised by communal systems of landownership, a lower concentration of landholding and less widespread tenancy and leasing arrangements than in Asia or Latin America. At a continental scale, and in comparison with other major world regions, land is relatively abundant in Africa, and it was labour rather than land shortages which was generally perceived throughout history to be the more critical constraint on agricultural development.

Colonialism occurred much later in Africa than elsewhere and, generally speaking, peasant farmers were neither incorporated into a system of private property rights nor had their land expropriated. However, there are some exceptions to this rule, particularly in French colonial countries such as Cameroon, Madagascar, Guinea and Ivory Coast, where large-scale European settlement occurred (Plate 10.3). The so-called White Highlands in British-ruled Kenya are another example of this.

In southern Africa, the expropriation of land for European agriculture forced African farmers into 'reserve' areas. These reserves were often more marginal for cultivation, and together with policies such as compulsory labour recruitment, the active blocking of opportunities for cash cropping by African farmers and

Plate 10.3 Harvest of pineapples on cash crop plantation, Dabou, Ivory Coast
(photo: Ron Giling, Panos Pictures)

Plate 10.4 Communal/commercial boundary, Marondera district, Zimbabwe
(source: Surveyor General's Office, government of Zimbabwe)

many more indirect means such as policies on soil conservation, African labour was released into European agriculture, mining and infrastructural development projects (Elliott, 1991).

The legacies of settler colonialism are evident within contemporary agrarian structures in Southern Africa. In Zimbabwe, for example, on independence in 1980 approximately one-half of the country's agricultural land was in private freehold ownership of around 7000 European farmers. The majority of the rural population was resident in the former reserve or 'communal' areas, with usufruct rights to land held in trust by the community (Elliott, 1995). Plate 10.4 is an extract from an aerial photo taken at the boundary between an 'African' and a 'European' farming area in the 1980s. The contrast in land use and population density between the two sectors is very evident.

Colonial interests in increasing cash crop production in Africa were generally undertaken without fundamental change to the system of landholding. Although peasant producers remained in control of their lands, the 'colonial triad' of taxation, export commodity production and monetisation impacted significantly on the relations of production throughout the continent, as shown in Figure 10.2.

On the basis of work within Hausa communities in northern Nigeria, both Scott (1976) and Watts (1984) have suggested that although some groups during this period were able to benefit from increased agricultural

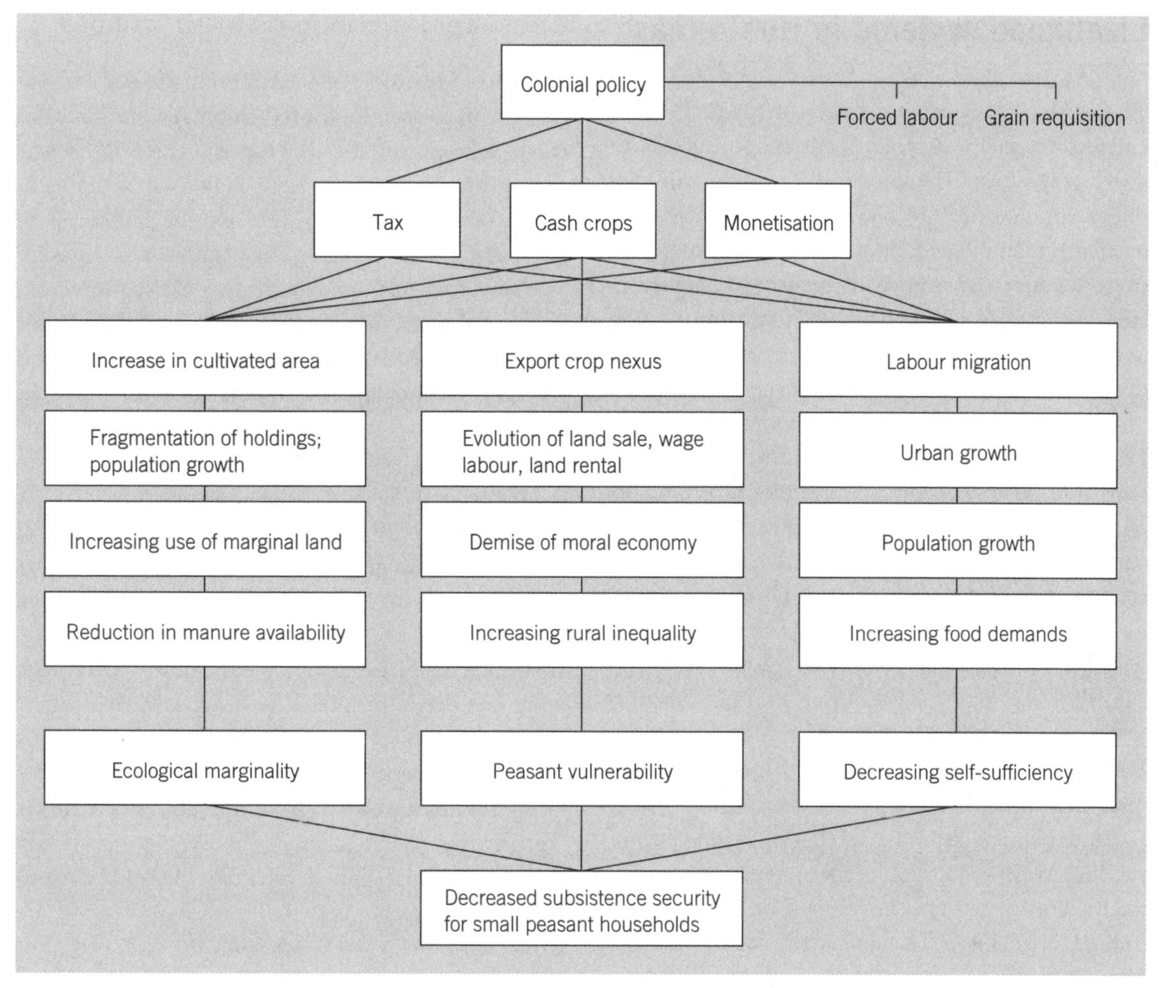

Figure 10.2 The effects of colonial penetration on food security in northern Nigeria
Source: Adapted from Watts (1984)

production and sales, or engaged in trading and moneylending enterprises, many more people became increasingly vulnerable to longstanding hazards such as drought and market fluctuations. Land and labour became goods to be bought and sold, and there was a decline in the 'moral economy' – the reciprocal sharing of tasks and goods – thereby removing options and flexibility in times of hardship.

The legacy of colonial policies on agrarian structures in Africa is therefore extremely varied. Rarely in Africa today is there a neat distinction between imposed systems of individual freehold ownership and customary African tenure, as has often been suggested. Rather, as Siddle and Swindell (1990: 72) proposed,

> **diverse and parallel systems of tenure and rights to farm, which include communal usufructary systems, loaning, pledging, different forms of labour renting associated with squatters and share contractors, fixed rents, leasing freehold purchase and land nationalization. These different forms of landholding and farming are rooted in different relations of production underwritten by religion, kinship and political authority, as well as varying with ecological circumstance.**

There is further evidence that land in Africa is not always plentiful. Many recent case studies indicate that access to land – notably the best quality land, land surrounding urban areas and land in areas of population expansion – is becoming difficult for some groups to obtain, particularly women, pastoralists and poorer households (Bryceson, 2002: Toulmin and Quan, 2000).

Livelihood systems in rural areas

This section overviews a number of key and persistent features of the systems for livelihood across the developing world. The intention is to draw out some of the common features, but the danger is clearly of overgeneralisation and, indeed, the most important characteristic (and source of security as argued by commentators like Chambers, 1997; Mortimore, 1998; and see also Box 10.1 below) is their diversity and dynamism. As an example of the diversity of systems, research in 1999 in one small community in the Peruvian Amazon identified 12 distinct farming systems and 39 ways of combining the 12 production types among the 46 households surveyed (Brookfield and Stocking, 1999). Similarly, the dynamism of indigenous food production systems is illustrated through longitudinal research into household coping strategies, which has shown how even the concept of what constitutes 'food' may change during times of hardship, with communities utilising their considerable environmental knowledge to seek out wild foods to supplement their diets (Mortimore, 1989).

Rural households in the developing world generally produce a high proportion of their subsistence requirements directly, certainly in comparison to households in the more developed regions where food production and consumption have become separated. However, non-agricultural income-generating activities can play a greater or lesser role in securing livelihood needs for particular rural households and at certain times. There is also evidence that diversification of rural livelihoods and de-agrarianisation is increasing over time within rural regions, as discussed above.

Agricultural production

In the space available here, some generalisation of the diverse agricultural systems in the developing world is unavoidable. Table 10.9 contrasts traditional subsistence production with modern, commercial agriculture. This type of analysis should not discount the significance of cash crop production within small-scale livelihood systems, as discussed below. Indeed, the same crop may be grown as a food staple, and any surplus sold subsequently for cash, such as in the case of rice in Asia or maize in many parts of sub-Saharan Africa.

Subsistence concerns in agriculture

Table 10.6 identifies the need to minimise risk as a feature of small-scale food production systems, and this can be a key factor affecting the willingness of farmers to innovate. Innovations such as new crops and production methods like irrigation and the application of fertilisers and pesticides are frequently costly, unavailable and unreliable, and require a major departure from well-tested methods, which are often carefully adapted to the particularities of the environment. Farmers with low-level technology and limited financial resources cannot afford to take such risks. Yet many 'outside' commentators have been slow to appreciate the rationality of indigenous cultivation practices.

Indigenous techniques, such as the burning of debris after clearing and the intercropping of a number of crop types in the same plot, for example, were once strongly criticised by Western observers as being environmentally destructive, but are now viewed much more positively, including as being environmentally more sustainable (Conway, 1997; Hill, 1972). Intercropping is a valuable risk-aversion strategy; in simple terms, if one crop suffers from drought or pest attack, there will be others to supply household food needs. In addition, the fertility of the soil is likely to be maintained for a longer period if a variety of crops are grown, and total crop yields from intercropped plots are actually often greater than those where a single crop is grown (Richards, 1985, 1986).

Population expansion as a source of change

An important factor affecting the nature of land use and farming systems is the size and density of population. Traditionally, in areas of sparse population within the tropics, extensive farming systems such as shifting cultivation and rotational bush fallow have been commonly practised (Ruthenberg, 1976).

Both types of system depend on long fallow periods, sometimes of 20 years or more; to restore soil fertility after relatively short periods of cultivation, perhaps only one or two years of fallow are allocated. However, fallowing practices are often complex and can vary over relatively short distances due to factors such as population density, the physical environment, land tenure arrangements and the management practices of individual farmers (Gleave, 1996).

Table 10.9 Differences between subsistence and commercial farmers

	Traditional, subsistence	Modern, commercial
Proportion of output sold off the farm	Low	High
Destination of foods	Local direct consumption and some processed locally	High proportion processed and to food manufactures
Origin of inputs		
Power	Draught animals, human labour	Petroleum, electricity
Plant nutrients	Legumes, ash, bones, manure	Chemical fertilisers
Pest control	Crop rotations, intercropping	Insecticides, fungicides, break crops
Weed control	Rotations, hoeing, use of plough	Herbicides
Implements and tools	Hoe, plough, sickle, scythe	Machinery, often self-propolled combine harvesters
Seed	From own harvest	Purchased from seed merchants
Livestock feeds	Grass and fodder crops grown on farm or common land	Purchased from compound feed mixers
Economic aims	Prime aim to provide family food	Profit maximisation
	Land and labour main inputs; few capital inputs	Capital and land major inputs; labour a declining input
	Diversity of crops grown	Specialised production
	Aims at maximising gross output and yield per acre	Aims at maximising output per head and minimising production costs
	Prime aim avoidance of risk; reluctant to innovate	Innovation

Source: Grigg (1995)

Plate 10.5 Intercropping, central Zimbabwe
(photo: J. Elliott)

Typically within such fallowing systems, households clear and cultivate land around settlements, moving from one plot to another when falling yields indicate declining soil fertility. Whereas settlements are usually fixed under rotational bush fallowing systems, under true shifting cultivation, settlements will also move when all the land in the surrounding area has been cultivated. Plots under such fallowing systems are often irregularly shaped and unfenced clearings in the forest or bush, and their size varies according to the availability of land and labour.

Much has been written on population pressure leading to a reduction in fallow periods and environmental degradation, but in reality there is a complex relationship between population size and land use (see Chapters 5 and 6). Ester Boserup has been prominent in suggesting that population pressure frequently induces rural communities to become more efficient, adapting their environment, engaging in more intensive production and, in time, increasing productivity (Boserup, 1993).

A study of Machakos District in Kenya examining 60 years of change has given much support to the Boserup hypothesis, revealing progressive intensification of land use, increased productivity and careful environmental management in the face of steadily growing population (Tiffen *et al.*, 1994). With increased population and land intensification, plot boundaries generally become more clearly defined, often with fences, trees or bushes to demarcate one household's land from another's. As fallow periods become shorter, fertility can no longer be maintained without additions to the soil, which usually take the form of animal manure or compost made from kitchen waste and other organic material. Sometimes crop rotations and successions are used to maintain fertility, with root crops such as sweet potatoes or cassava being planted after a harvest of grain crops such as rice, maize, millet or sorghum. Another response to increasing population pressure is to cultivate hitherto unused land, perhaps by constructing terraces or irrigation systems. Plot size may decline even further, and more intensive, even permanent cultivation develops.

In low-technology agricultural systems the most important input is human labour. Where most family labour is unpaid, the size and composition of the household labour force, together with access to communal labour, can have a crucial impact on agricultural productivity and the nutritional status of both individuals and households (Moock, 1986; Mortimore, 1998). The availability of farm labour can be a problem where there is a predominance of either very young or old members whose contribution to farm work is generally less effective than that of able-bodied young adults.

A further factor affecting the quality of household labour is ill-health. Debilitating diseases such as malaria, onchocerciasis (river blindness) and schistosomiasis (bilharzia) can have a marked effect on labour availability and efficiency. Since the 1980s the spread of HIV/AIDS has had a significant effect on certain communities in rural Africa, depriving households of valuable labour resources (Chapter 5).

The degree of monetisation can also be an important factor in determining household labour supply and performance even in subsistence production. This will determine the extent to which households require cash income to purchase inputs, and therefore they must sell their own labour in other activities and/or pay for labour at certain times. All such factors can influence the ability of households to undertake productive tasks at the required point in the cultivation cycle, or to fulfil essential 'reproductive' tasks such as fuelwood or water collection.

Capital is one factor of production that is frequently in short supply among rural cultivators in developing countries. For many producers, traditional tools such as hoes, cutlasses and axes are little changed from those used by their ancestors, but they are often well adapted to the local environment and the nature of the work undertaken. In contrast, more sophisticated tools and machinery, such as tractors and ploughs, may be quite inappropriate and they also require fuel and regular maintenance.

Simple irrigation technologies, such as shadoofs and other water-lifting devices, are used to raise water from wells, streams and rivers onto neighbouring fields. There is often little incentive to invest time and effort into improving land under extensive rotation systems with only short periods of cultivation. However, with more permanent farming systems, capital investment is common in such forms as terracing, soil erosion control, irrigation infrastructure, fencing, farm buildings and tree planting. In the absence of mechanised implements on many farms, such improvements can represent hours of work by the household or community.

The commercialisation of agricultural production

In contrast to small-scale agricultural production for largely subsistence purposes, commercial or plantation agriculture, as shown in Table 10.9, is characterised by labour-intensive production methods, high investments in technologies, a limited range of crops grown, expatriate ownership and management, often by a multinational company, and a hierarchical relationship between management and labour force. Plantations are frequently 'total institutions', with workers living on the plantations and being heavily dependent on them for their livelihood (Laing and Pigott, 1987; Tiffen and Mortimore, 1990).

As already seen, it was the colonial encounter which introduced many new cash crops, primarily directed at export, and forced the commodification of land and labour in rural areas of the developing world. Such crops included cotton, sugar, coffee, cocoa, tea, sisal and groundnuts.

The earliest plantations were probably established on the Canary Islands, although they became more widespread in parts of the Americas from about 1550. Coffee became an important plantation crop in Brazil from the late eighteenth century, with cocoa joining it in the nineteenth century. In West Africa, whereas the French colonialists established plantations for growing coffee, cocoa and bananas in Ivory Coast and bananas in Guinea, the Americans grew rubber in Liberia, and from 1884 the Germans set up rubber, oil palm and banana plantations in Cameroon. Meanwhile, in East Africa White-settler farmers established estates in countries such as Kenya to grow tea, coffee, pyrethrum, tobacco, cotton and sugar. The rapid development of the car industry in the late nineteenth century prompted the widespread establishment of rubber estates by Europeans, particularly in peninsular Malaysia.

The specific impacts of plantation agricultural production on the wider economy and society in any particular location can be complex; this was discussed in Chapter 7 with reference to enhanced cash crop production as an element of structural adjustment programmes. Young summarises it as follows:

> **Expansion of cash crop production may be beneficial for food security if land ownership is relatively equitable and if exports are diverse, but clearly a rush to promote cash crops for export may just intensify food insecurity, increase economic and political marginalisation, and increase the likelihood of environmental catastrophe.**
>
> **(Young, 1996: 72)**

But cash crops, whether destined for domestic consumption or export, are by no means always grown on estates and plantations. Indeed, in the case of Ghana, Hill showed in 1963 that it was mainly small farmers who planned and developed cocoa production in that country from the early 1900s, reinvesting their wealth in the expansion of cocoa lands and also in infrastructure, such as roads and bridges.

A major concern over cash crop cultivation on smallholder farms relates to problems of food security and the nutritional status of households that may be neglecting food production to cultivate crops for the market. The work of Longhurst (1988), for example, suggested that the children of cash crop farmers do not enjoy better nutritional status than children of non-growers; studies of cocoa production in Nigeria and Mexico, sisal in Brazil, sugar cane in Kenya and coffee in Papua New Guinea all revealed negative effects of cash crop production on family food consumption and/or nutritional status.

In determining the nutritional effects of cash crop production among smallholder farmers, a critical factor is identified as being the extent to which women control production and marketing. Although the picture remains complex and varied between and within rural communities, it is apparent that whoever controls these aspects has the greatest influence on how cash is spent. Increased women's wealth is therefore likely to translate into improved household nutrition, particularly children's nutrition.

Pastoralism

According to Sandford (1983: 1), pastoralists are people who 'derive most of their income or sustenance from keeping domestic livestock in conditions where most of the feed that their livestock eat is natural forage rather than cultivated fodders and pastures' (Plate 10.6).

Different degrees of mobility

A distinction is often made between settled ranching and various forms of livestock herding that involve migration; these migratory forms range from pure nomadism, through semi-nomadism to transhumance,

Plate 10.6 Cattle being taken to Ballyera market, Niger
(photo: Mark Edwards, Still Pictures)

involving seasonal movements from permanent settlements (Galaty and Johnson, 1990). Rural livelihood systems based on livestock production are most commonly found in arid and semi-arid areas, where crop cultivation is restricted by water availability. Pastoralists utilise marginal environments where resources are widely dispersed and land may be too dry, rocky or steep for cultivation.

It was estimated in the early 1980s that some 30–40 million people were engaged in 'animal-based' economies in the world's arid and semi-arid areas, of which '50–60 per cent were found in Africa, 25–30 per cent in Asia, 15 per cent in all of America, and less than 1 per cent in Australia' (Sandford, 1983: 2). The most important countries in terms of numbers of pastoralists are Sudan, Somalia, Chad, Ethiopia, Kenya, Mali, Mauritania, Mongolia, India and China. Some of the main pastoral groups in the developing world include the Masai, Sumburu and Karamojong of East Africa, the Fulani and Tuareg in the savanna–sahel zone of West Africa, the Bedouin of the Middle East and the Kazakhs and Mongols of Central Asia.

Whereas many rural households may own some cattle or small livestock, such as chickens, sheep and goats, to provide draught power and supplement diets, livestock are central to pastoralist production, as is a degree of mobility within their lifestyles. Among pastoral communities, cattle, camels, sheep and goats provide the primary sustenance base for the household, providing milk, yoghurt and meat at certain times, although meat may not be eaten on a regular basis. Livestock are also the source of income, capital and savings. In addition, social status is often linked closely to the size and quality of animal herds. A completely nomadic lifestyle is now quite rare, and it is more common for pastoralists to make seasonal migrations, returning to a permanent base which may be occupied throughout the year by women and children. The precise timing and extent of migrations may be governed by factors such as altitude, the farming calendar, the prevalence of tsetse infestation, and the progression and severity of the dry season (Stock, 1995).

Increased diversification on behalf of pastoralists

Pastoralists, like cultivators, have been frequently misunderstood by academic writers and government officials, who criticise herders for keeping livestock purely for prestige purposes, for allowing overgrazing of rangeland, for transmitting human and animal diseases, for avoiding tax payments and for disregarding international boundaries. Another common misunderstanding is that pastoralists do not grow crops. In fact, a large proportion of pastoralists engage in cultivation to a greater or lesser extent, and some groups attach equal importance to farming and are therefore more accurately described as 'agro-pastoralists'.

As an illustration of this diversity, the Fulani of northern Nigeria are frequently subdivided into four groups (Binns, 1994b). The Bororo can be considered 'true nomads' in that they shift camp regularly and move with their herds in search of pasture and water. Semi-nomadic Fulani have permanent homesteads with adjacent farms, where a variety of crops are grown. The elders stay in the homesteads for most of the year, to be joined by the returning young herdsmen at the start of the rainy season when the fields must be prepared for planting. A third group, the semi-settled Fulani, regard cultivation and livestock rearing as equally important. They have fixed settlements and migrate over shorter distances, mainly at the height of the dry season. A fourth group, the settled stock owners, graze or corral their herds close to the villages, and young men take animals out daily into the surrounding area. These stock owners often supplement their income by tending the livestock owned by wealthy town-based people, who increasingly invest in livestock as a symbol of wealth and status. All four groups of Fulani show a common attachment to their animals and are respected for their skills in rearing and managing livestock (Binns, 1994a: 105).

Recent research in East Africa (Little *et al.*, 2001) has shown that livestock herders in the region are increasingly pursuing non-pastoral income strategies to meet consumption needs and to guard against climatic, disease and economic shocks. While diversification in terms of changing herd composition or varying the level of trading activities have long been strategies for these herders, diversification strategies as identified in Table 10.10 are relatively recent, dating most widely from the 1980s. They also increasingly involve engagement with the market, even in very remote areas.

Time series data for pastoral groups in Kenya suggest that dependence on livestock income decreased from approximately 75 per cent to around 50 per cent of household incomes for some herders across the 1980s and the number of household members engaged in wage labour doubled. Across the wider region, it has been found that herders are owning fewer head of livestock, and in many cases these are below the thresholds considered necessary for pastoral self-sufficiency, such that diversification strategies as shown in Table 10.10 have become a necessity. 'What is surprising is how much of this has occurred since the 1970s, and how rapidly it has happened' (Little *et al.*, 2001: 422).

Table 10.10 Types of pastoral diversification

- Trading – of milk, firewood, animals or other products
- Wage employment – within and outside area
- Retail shop activities
- Rental property ownership and sales
- Gathering and selling wild products
- Farming – for subsistence and cash incomes

Source: Little *et al.* (2001)

Table 10.11 Fishing and aquaculture worldwide

Region	Inland and marine fisheries production (thousand metric tones) 2000–2002		Exports of fish products, 2002 (thousand metric tones)	Number of fishers, 2000
	Capture	Aquaculture		
Asia	44,189	33,275	19,051	28,890,352
Europe	15,773	2,064	19,356	855,333
Middle East and North Africa	3,049	526	1,355	746,955
Sub-Saharan Africa	5,160	63	1,862	1,995,694
North America	6,072	629	6,346	303,784
Central America and Caribbean	1,990	147	1,525	446,390
South America	16,315	869	5,232	784,051
Oceania	1,104	122	1,794	85,324
World	93,650	37,694	56,520	34,501,411
More developed countries	27,917	3,641	28,159	1,467,401
Less developed countries	65,694	34,060	28,378	32,640,482

Source: World Resources Institute (2005)

The causes of pastoral diversification are multi-faceted, may not always relate to managing risk, and as with agricultural diversification considered below, not all members of the community will have similar diversification opportunities. Little *et al.* (2001), for example, have found considerable variation, including along wealth and gender lines, within communities. For poorest herders, unskilled waged labour and petty trade are the most common non-pastoral option, whereas wealthier herders tend towards trading, business and opportunities for higher income depending on more skilled waged labour.

In addition, they identified that sedentarisation provides increased earning opportunities, particularly for low-income women, such as in petty trade, handicrafts, beer brewing and local wage employment. In contrast, wealthier women herders were found to be more likely to rely on income from livestock and milk sales rather than on other revenue sources. Diversification opportunities such as shop ownership, retail business and labour migration remain strategies largely undertaken by male members of these households.

Rethinking pastoral-based systems of livelihood

It is increasingly acknowledged that many previous interventions in pastoral development, most notably particular programmes for sedentarisation and stocking rate adjustments, have been limited by what has been termed an 'equilibrium view' of range management (Scoones, 1995). For example, the concept of 'carrying capacity' has been used widely to indicate the human and animal population limit that cannot be exceeded without setting in motion the process of land degradation.

This concept, however, is largely based on relatively sophisticated beef ranching systems in places such as Australia and the USA; whereas in the dryland areas of Africa and Asia, equilibrium conditions just do not apply. These areas are characterised instead by unpredictable climates and considerable spatial and temporal heterogeneity in terms of rainfall, for example, but also in the complex dynamic interactions between species and indeed human activities, i.e. non-equilibrium dynamics may apply much of the time in drylands, such that conventional range management based on equilibrium concepts and the adjustments of animal numbers become inappropriate. As Homewood and Rogers (1987) suggest:

The unpredictable nature of the environment is the central factor affecting not only attempts to measure productivity and population density in semiarid rangelands, but also the applicability of the concept of carrying capacity that is used to link the two. Irregular rainfall falling at unpredictable places and times results in unpredictable primary production. Measures of grassland productivity may be accurate but are only valid for a particular place and time and do not support extrapolation to a wider or longer term.

(Quoted in Mortimore, 1998: 73)

Similarly, Scoones (1995: 26) reflects on the contrasting approaches of conventional range management and indigenous strategies:

Stocking rate [numbers/area/time] adjustments have always concentrated on the changing of animal numbers, rather than seeking management options that manipulated area or time. Pastoralists operating in the non-equilibrium environments of the drylands use the range of strategies, but flexible movement and spatial and temporal adjustments are the key to success.

Where the traditional flexibility and mobility of pastoral systems have been constrained by government policies directed at sedentarisation, or the construction of large dams and irrigation systems across traditional grazing areas and migratory routes, tensions between pastoralists and cultivators have regularly arisen (Adams, 2001; Binns and Mortimore, 1989).

The Tribal Grazing Lands Policy (TGLP), implemented in Botswana since the mid 1970s, centred on fenced, borehole-based ranches and sedentarised livestock production. The policy aims to reduce environmental pressure and foster social and economic development in the Kalahari. However, it has been much criticised, including for its suggested role in aggravating desertification, and for enhancing rural poverty and social inequality.

Recent research by Thomas *et al.* (2000) has looked explicitly at the interplay of the environmental and social outcomes of the TGLP and they suggest that the impacts of the policy have not been entirely negative, nor has it specifically caused severe degradation of the resource base. What they do assert, however, is that

future policies that impact on livestock production and residents in rural areas need to be underpinned with a clearer knowledge of resource-use practices

prior to policy implementation, a stronger and better implemented social component, and *a clear understanding of the nature of likely environmental impacts, including their relationship to natural environmental variability.*

(Thomas *et al.*, 2000: 340, emphasis added)

A further strong influence on previous interventions in pastoral development has been the idea of the so-called 'tragedy of the commons'. In a situation where grazing land is held communally, but livestock are owned by individuals or households, it is argued that individuals have an incentive to graze as many animals as they can without concern for the sustainable management of the communal grazing resources. As a result, it is suggested that overstocking leads to overgrazing, which in turn can lead to land degradation and even desertification (Hardin, 1968).

Sandford (1983: 119), however, is concerned about the assumption that the 'tragedy of the commons' scenario is widespread, whereas in reality there is much variation from one region and pastoral group to another:

Not only are the argument's assumptions often not fulfilled in practice in particular cases, but its importance has been exaggerated to the point where it sometimes appears as the only factor to be considered in deciding on land-tenure policy in pastoral areas; and a number of unjustifiable conclusions have been drawn from it.

There is evidence that many pastoral communities do in fact control the intensity of grazing on communally owned land, and in arid and semi-arid regions a move to individual ownership of land would probably cut across the need for flexibility and mobility, both natural adaptive responses to environments with inadequate and unreliable water and grazing resources. However, as Scoones (1996) suggests, the 'science of ecological calamity' that stresses the damaging potentials of livestock grazing and the threats of degradation and desertification remains strong in underpinning public policy on rangeland and livestock development. Hence, many government policies continue to emphasise the need to control livestock numbers and grazing movement.

Fishing

It is estimated that fish provide the primary source of protein for 1 billion people, mostly in the developing countries, and particularly among those living in rural areas (World Resources Institute, 2000). However, its importance as an economic activity, source of employment and as a contribution to diets and household nutrition is frequently undervalued.

A complex but important source of livelihood

Fishing activities may be broadly classified by location into marine fishing and inland (freshwater fishing). 'Aquaculture' refers to the farming of aquatic organisms, i.e. encompasses interventions in the rearing processes to enhance production such as through regular feeding, stocking and protection from predators (Plate 10.7). Aquaculture has a long and successful history in China and parts of Southeast Asia where complex systems for the rearing of not just fish but also ducks and geese make a substantial contribution to income but also nutrition. Table 10.11 (p. 461) highlights the importance of fishing and aquaculture in the economies and livelihoods of regions of the developing world in particular.

Plate 10.7 Mollusc collection, Kolkata surrounds
(photo: J. Elliott)

Plate 10.8 Fishing boats landing their catches, Dunga Beach, Kisumu, Kenya
(photo: Tony Binns)

Fishing activities are also classified regularly by method into artisanal (or traditional) and modern. In reality, however, there is a vast and complex array of different fishing methods, ranging from line and spear fishing to basket traps, and a great variety of nets (Plate 10.8). For example, in the West African region alone there is a great diversity of fishing activities, with the Fante and Ewe fishermen of Ghana using mainly wooden sea-going boats and beach-seine nets off the Atlantic coast, whereas the Bozo and Somono peoples use canoes, basket traps and seine, gill and cast nets in the inland Niger Delta of Mali (Binns, 1992; see also FAO/DFID, 2003).

Fishermen typically have a good knowledge of the breeding cycles and movements of fish, and they often migrate in search of better catches. The Somono fishermen of Mali migrate in the early part of the rainy season following the *Alestes* migration, and also in the dry season when water levels in the inland Niger Delta fall (Sundstrom, 1972). Considerable numbers of Fante fishermen have moved westwards from Ghana and have established communities on the Gambian coast, from where they catch mainly shark and ray, which are then dried, salted and bagged before being sent by ship to Ghana, where there is considerable demand.

The large numbers of people in the sub-Saharan region of Africa involved in fishing was shown in Table 10.11. These statistics are compiled by the Food and Agriculture Organization on the basis of annual questionnaires submitted to the national reporting offices of member countries and include those employed in both commercial and subsistence fishing. Whilst clearly such official statistics have their limitations, it is certainly known that many more people in the rural South are also employed in the subsequent marketing and processing of fish – many of them women – who buy fish from fishermen, and sell them in the local marketplace or to long-distance traders. Further numbers are employed in the processing of fish where, given the often hot climates and absence of refrigeration, fish are often fried or smoked over wood or charcoal to preserve them (Geheb and Binns, 1997).

Stock resources under pressure

Fish catches worldwide are under increasing pressure as demand for fish globally has risen with many stocks in decline (World Resources Institute, 2005). The impact of EU trawling off the coast of West Africa, for example, has been a recent cause for concern (The Ecologist, 2003) as well as a focus for community-based interventions on behalf of multinational and bilateral donors (FAO/DFID, 2003). European Union trawlers pay minimal licence fees to poor coastal nations to catch their fish and sell it in Europe. Once among the richest fishing grounds in the world, fish stocks off the coast of West Africa have fallen by as much as 80 per cent (Pearce, 2002). As a result of the increased fishing by commercial trawlers, fish stocks in the shallow inshore waters of this coast where artisanal fishers have historically plied their trade, have dropped by more than half over the period 1985 to 1990 (World Resources Institute, 2000). The small, yet important, Kenyan sector of Lake Victoria is also experiencing a variety of pressures on stocks and livelihoods, as seen in Box 10.2.

BOX 10.2

Farming and fishing on the Kenyan shores of Lake Victoria

Although Kenya controls only 4100 square kilometres of Lake Victoria, 6 per cent of its surface area, the lake produces a massive 85 per cent of Kenya's fish supply, valued at US$80 million in 1994 (see Figure 10.3; Geheb, 1995; Geheb and Binns, 1997). In recent years, however, the ecology and economy of the lake have come under serious threat, and fishermen have been forced to devote more of their time to other activities, such as farming, in an effort to satisfy household food requirements and supplement income.

The people living around the Kenyan shores of Lake Victoria are mainly of the Luo ethnic group, whose

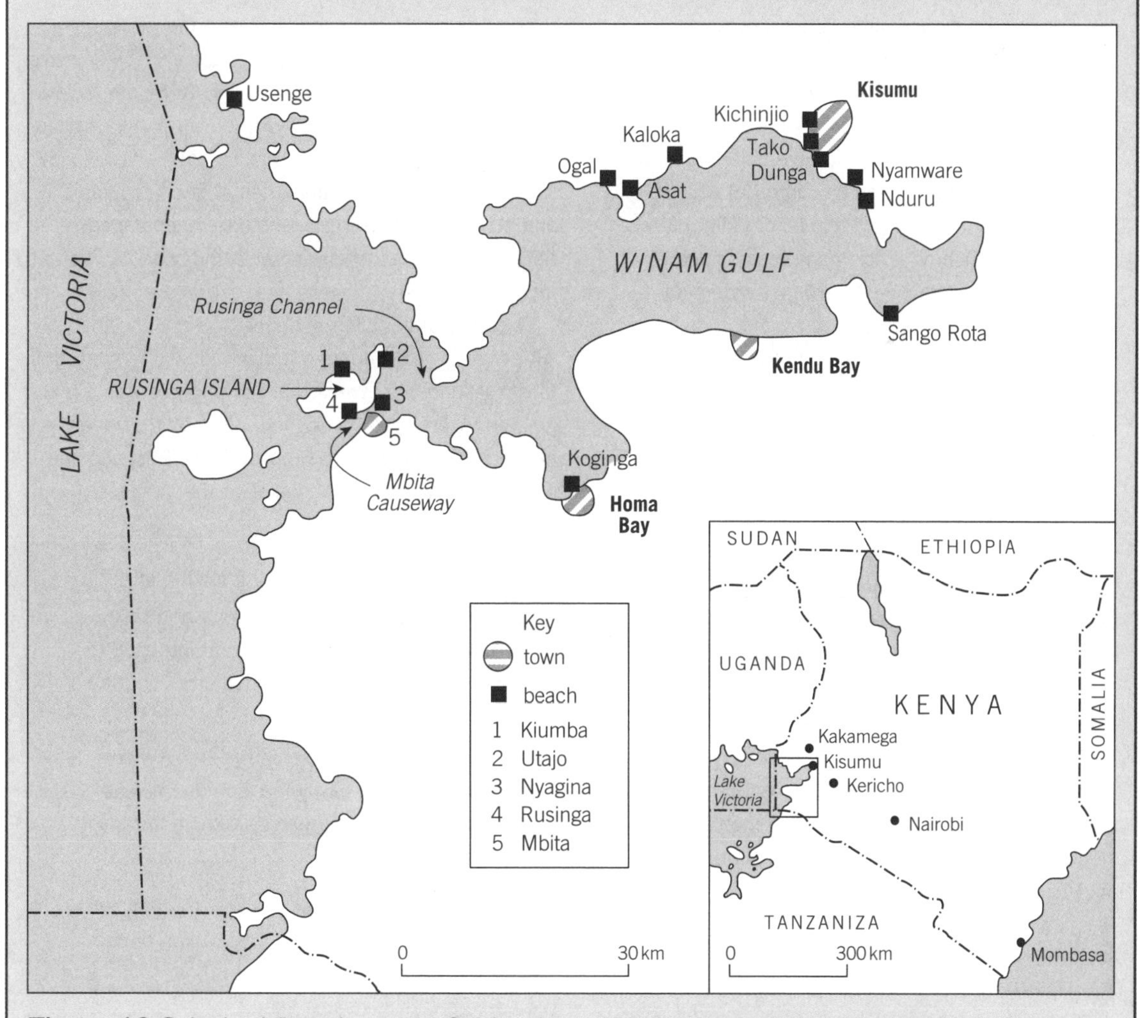

Figure 10.3 Lake Victoria: main fishing beaches in the Kenyan sector
Source: Adapted from Geheb and Binns (1997) in *African Affairs*, 1997, 96, pp.73–93 © The Royal African Society, by permission of The Royal African Society, http://afraf.oxfordjournals.org/

BOX 10.2 (continued)

ancestors migrated south from Sudan, arriving in what is now Kenya in the late fifteenth or early sixteenth centuries. The livelihood system of Luo lakeside communities around the Winam Gulf has traditionally involved three important elements: fishing, farming and livestock herding. Each of these elements varies in importance at different times during an annual cycle, whereas longer term variations depend upon such factors as the availability of fish or periods of drought which affect farming.

In the pre-colonial period, Luo artisanal fishing activities were characterised by the use of low-technology gears, such as reed traps, papyrus beach seines and sisal long-lines, which served to control the level of exploitation of fish resources. Furthermore, lakeside Luo clans and sub-clans each had well-defined territories, extending into the lake as far as their boats could travel. Territorial rules were strictly enforced, forbidding members of one clan from fishing in their neighbours' waters. The pre-colonial period was therefore characterised by control of access through territorial rules and regulations that closely controlled the fishing 'effort', and by a shared responsibility for the clan's territorial waters and maintenance of fish reserves.

In 1901 the railway from Mombasa and Nairobi reached Kisumu on the shores of Lake Victoria, opening up a vast potential market for fish. An acceleration of the fishing effort was prompted by the introduction of hut tax in 1900 and poll tax in 1910, so the fishermen had to earn money from something. Further expansion was enabled by the introduction of the flax gill-net in 1905 and the beach seine soon afterwards. Between 1949 and 1953 the number of fishermen increased from 35,000 to around 60,000. During this same period, however, there was a steady decrease in overall catch tonnage, such that the British colonial authorities introduced regulatory measures, including restrictions on net mesh-size. As the fishing intensity continued to increase, a number of exotic fish species were introduced. In 1954 the Nile perch, a voracious predator, was introduced into the Ugandan sector of the lake, together with six exotic tilapia species. The introduction of the Nile perch had a drastic effect on the ecology of the lake, not least in reducing the species diversity with the loss of between 200 and 300 Haplochromis species.

Following Kenya's independence in 1963 there was a further rapid growth in fishing activity, but costs of hired labour, boats and nets escalated. The fishery has increasingly attracted outside interests with higher investment potential, such that in the 1990s there are many absentee boat owners, many of whom are Nairobi-based business people and politicians. Probably as many as 50 per cent of the boats in the Kenyan sector of Lake Victoria are now owned by people who do not themselves work on them. The local communities have therefore lost much of the control they once had over the fishery, and fish yields have suffered.

Pollution levels are also rising due to fertiliser and pesticide use on neighbouring agricultural land, which is contributing to eutrophication and the increased incidence of oxygen-using algal blooms. In 1997 parts of the Winam Gulf, particularly close to Kisumu, were suffering badly from water hyacinth, a floating mass of vegetation which makes any fishing activities virtually impossible. It is the combination of such economic and ecological pressures which have encouraged fishermen to invest their energies and seek additional income elsewhere.

A survey undertaken in 1994–1995 found that 94 per cent of fishermen interviewed also farmed, and most respondents agreed that farming is becoming a greater necessity for fishermen, particularly since the early 1990s (Geheb and Binns, 1997). Respondents gave three main reasons to explain the increased attention to farming:

1. fishing yields and incomes are declining;
2. after almost four years of irregular and poor rainfall, many fishermen were turning back to the land after better crop yields and rising food prices in 1994;
3. farming is generally regarded as being more reliable and more easily monitored, whereas fishing is increasingly perceived as a 'hit and miss' activity.

Many Luo feel that the land is likely to offer better security than the lake, given the problem of declining fish yields, the greater incidence of boat thefts and the rising prices of fishing gear.

This illustrates the adaptability of rural communities and the importance of diversification

BOX 10.2 (continued)

into more than one activity in an unpredictable environment and resource base under pressure. Future development strategies will need to consider ways to ensure the sustainability of both land and water resources and their associated food production systems, if nutritional security is to be assured for lakeside communities. The essence of a solution may well be to build upon the diversification of activities which have long been an integral feature of Luo communities. With regard to achieving a sustainable future for the lake, there is an urgent need to improve the efficiency of regulations controlling the exploitation of fish stocks. Possibly the most appropriate solution is to strengthen traditional Luo institutions, which seemed to provide effective controls during the pre-colonial period.

Simultaneously, the various fishing cooperatives located on key beaches around the Winam Gulf might also be strengthened to provide investment loans, security for boats and gear, and reliable marketing links. What is clear, however, is that well-informed action is urgently needed, both to preserve the ecology of the lake and to protect the livelihoods of many thousands of people.

In addition, mangrove destruction (as identified in Chapter 6) has serious implications for the productivity of tropical fisheries since mangroves provide the spawning grounds, in sheltered coastal areas and river estuaries, for a wide range of fish species. Dam construction, and in particular the reduction in flooding downstream, can also seriously affect fish populations and the economic returns from fishing. For example, dams can act as a complete barrier to fish movement, contribute to high fish mortality as fish become stranded in deoxygenated pools and can cause a failure of fish to spawn through altered salinity.

In south-west India, the depletion of forests resulted in a shortage of wood for building traditional fishing boats. Artisanal fishing was also being threatened by more sophisticated fishing methods such as bottom trawling and purse seining. In 1981, the UK-based Intermediate Technology Group in cooperation with local organisations such as the South Indian Federation of Fishermen Societies and the Fishermen's Welfare Society collaborated to produce a new type of boat that was light, stable, powered by sail and oars and has a life of six to ten years. However, under the pressure of increasing competition from trawlers, fish catches in inshore waters declined and many artisanal fishermen considered using outboard motors on the new-style boats. By 1984, the collaboration led to new and stronger plywood boats with outboard motors being built. The use of motors increased the fishing range and led to enhanced yields. Several boatyards have since been established in Kerala State and, having acquired the necessary constructional skills, some workers have set up their own small businesses to build similar plywood boats (Kurien, 1988).

Forestry

Trees, woodlands and forests are multipurpose resources that provide varied functions in global society (as discussed in Chapter 6) and differing roles in rural livelihoods across the developing world. At all scales, patterns of forest resource use and management reflect the diverse ecologies of forests over space, the numerous different combinations of forest type, the changing value placed on particular forest products and services, and the many interventions over time which have aimed to secure those resources for human development.

Forests and woodlands in livelihoods of the developing world

The tropical regions of the world are host to over 60 per cent of the world's remaining forests as seen in Chapter 6. Such forests have been the locus of indigenous livelihoods built upon a close association with the forest and dating back thousands of years. It is estimated that as many as 15–20 million people dependent on hunting, gathering and shifting cultivation techniques lived in the Amazon Basin in the sixteenth century (Mather and Chapman, 1995). Indeed, shifting cultivation systems, by definition, depend on the manipulation of the dynamics of forest ecologies:

> **All tropical regions provide examples of the way in which farmers, as part of their traditional systems of shifting cultivation, incorporate trees from the original forest stand into their fields, for ground protection or for useful products.**
>
> **(Weidelt, 1993: 39)**

Plate 10.9 Collecting firewood, Eastern Cape, South Africa
(photo: Tony Binns)

For many more rural people in the developing world generally, it is the resources of open woodlands and scrub vegetation which play a particularly significant role in farming systems and livelihoods. The generally high dependence of communities on biomass sources of energy, as noted in Chapter 6, in itself demands a close association with local woodland ecologies to secure fuelwood for cooking, heating and lighting (Plate 10.9). In addition to firewood, woodlands may also provide a whole host of products, including:

> **building timber; wood for kraal fences, tools, transport and construction (boats, scotchcarts, sledges, etc.); edible leaves, pods, nuts and fruits; honey; natural fibres; fodder; medicines; utensils; and a whole range of other items.**
>
> **(Munslow *et al.*, 1988: 45)**

Table 10.12 illustrates the varied uses of wood in rural livelihoods based on recent research in Zimbabwe. The table also highlights the different value that such uses have within the overall system, such as in providing inputs into productive activities or raising cash through sales.

Securing such woodland products often requires extensive local environmental knowledge and varied degrees of management. For example, in arid zones, particularly where populations are sparse and livestock herding is the dominant livelihood system, woodland management may be based on the selective harvesting of natural vegetation, such as is characteristic of large parts of sub-Saharan Africa. A knowledge of local ecologies and the regenerative capacity of the natural vegetation is used to secure fodder for livestock, and to provide shade and rubbing poles for pest control. At the other end of a continuum of management, in areas of high population density and plentiful rainfall, trees of various indigenous and exotic species are planted and managed carefully and intensively for their varied products. In many parts of Southeast Asia multistorey, home or kitchen gardens are host to varied tree and plant species, often with distinct horizontal levels: food crops at ground level, coffee bushes and medicinal plants in the next zone, through fruit, fuel and fodder species at higher levels (Christanty, 1986).

Approaches to forestry development

Many previous approaches to forest management have been associated with the separation of people from forests and excluding people from accessing forest resources, such as through the establishment of forest plantations for industrial purposes, or through the creation of forest reserves, particularly for watershed management (see Fairhead and Leach, 1998, for policy examples in West Africa). Table 10.13 contrasts this conventional, or what Cline-Cole (1996: 126) terms outsider forestry, with indigenous or insider forestry.

Table 10.12 The multiple economic uses of wood in Zimbabwe

The multiple use of wood	Consumption	Durable	Production input	Asset formation	Sale
Timber (for commercial use, carvings)		*	*		*
Firewood (cooking, heat, light, beer brewing, brick burning)	*		*		*
Construction wood (huts, granaries, livestock pens, field fencing)		*	*	*	*
Agricultural implements (carts, yokes, hoes, axe handles, ploughs)		*	*		*
Furniture (wardrobes, beds, tables, chairs, stools, shelving, etc.)		*		*	*
Household utensils (cook sticks, mortars, pestles, plates, etc.)		*	*		*
Musical instruments (*mbiras*, *marimba*, *drums*, guitars)		*			*
Hunting implements (knobkerries, bows, arrows, fishing rods, etc.)		*	*		
Rope from bark (roofing, binding, whips, baskets, mats, nets)			*		*

Source: Cavendish (2000)

Table 10.13 Characteristics of outsider and insider forestry compared

Outsider forestry	Insider forestry
Introduced	Indigenous
Focused on production and protection functions of forestry	Integrated social, cultural, health and economic values of forestry and woodlands
Uses powerful environmental conservation imperatives to enforce legislation and management	Multipurpose species prioritised
Forestry and agriculture as separate endeavours	Forest management integral to agricultural activities and vice versa
Nature and culture conceived as separate and shapes interventions in forestry	Environmental change not separate from human history

Source: Compiled from Cline-Cole (1996)

In Peru and India, this 'scientific perspective' towards forestry management has led to large-scale afforestation programmes. These programmes may meet industrial timber demands, but they are often at the expense of local use of grazing and agricultural lands and the loss of varied 'non-timber forestry products' so valued by local communities (Dankelman and Davidson, 1988).

In Latin America, the importance of local environmental knowledge in enabling the sustainable management of forests is evidenced by the deforestation that has occurred as people are forced into the forests, by the increasing concentration of land ownership to secure resources for agriculture, but without the necessary knowledge or interest in forest ecologies and dynamics (Lohmann, 1993). Similarly, state-sponsored programmes of transmigration in Indonesia, although alleviating land pressure in the more populated islands, have led in some cases to deleterious impacts on forest environments in receiving areas, as migrants have lacked sufficient skills for forest management (Secrett, 1986).

An improved understanding of the role of trees in farming systems of the developing world and the value of indigenous skills and institutions is now being used to modify conventional forest management practices. The term *agroforestry* refers to the deliberate growth and management of woody perennials in conjunction with annual crops and/or animals. Although the concept is not entirely new (certainly to local people), agroforestry experienced a resurgence of interest amongst development practitioners in the 1980s, as the relative failure of many large agriculture and forestry initiatives started to become evident, for example. Table 10.14 shows aspects of the extent and diversity of agroforestry systems in the developing world.

In addition to the investments in agroforestry, programmes of 'social forestry' are now increasingly recognised as a promising approach to forestry development. The concept of social forestry may encompass a broad spectrum of activities, with the common aim of managing more explicitly the varied social values of trees in community development. Programmes are

Table 10.14 Examples of agroforestry systems in the developing world

Major system	Subsystem and practices	South Pacific examples	Southeast Asian examples
AGROSILVICULTURAL SYSTEMS	Improved fallow (in shifting cultivation areas)		Forest villages of Thailand; various fruit trees and plantation crops used as 'fallow' species in Indonesia
	The Taungya system	(e.g. Taro with *Cedrella* and *Anthocephalus* trees)	Widely practised; forest villages of Thailand are improved forms
	Tree gardens	Involving fruit trees	Dominated by fruit trees
	Hedgerow intercropping (alley cropping)	Extensive use of *Sesbania grandiflora*, *Leucaena leucocephala* and *Calliandra callothyrsus*	
	Multipurpose trees and shrubs on farmlands	Mainly fruit or nut trees (e.g. *Canarium*, *Pometia Barringtonia*, *Pandanus*, *Artocarpus altilis*)	Dominated by fruit trees; also *Acacia mearnsii* cropping system, Indonesia
	Crop combination with plantation crops	Plantation crops and other multipurpose trees (e.g. *Casuarina* and coffee in the highlands of PNG; also *Cliricidia* and *Leucaena* with cacao)	Plantation crops and fruit trees; smalholder systems of crop combinations with plantation crops; plantation crops with spice trees
	Agroforestry fuelwood production	Multipurpose fuelwood trees around settlements	Several examples in different ecological regions
	Shelterbelts, windbreaks, soil conservation hedges	*Casuarina oligodon* in the highlands as shelterbelts and soil improvers	Terrace stabilisation in steep slopes
SILVOPASTORAL SYSTEMS	Protein bank (cut-and-carry fodder production)	Rare	Very common, especially in highlands
	Living fence of fodder trees and hedges	Occasional	*Leucaena*, *Calliandra*, etc., used extensively
	Trees and shrubs on pastures	Cattle under coconuts, pines and *Eucalyptus deglupta*	Grazing under coconuts and other plantations
AGROSILVOPASTORAL SYSTEMS	Woody hedges for browse, mulch, green manure, soil conservation, etc.	Various forms; *Casuarina oligodon* widely used to provide mulch and compost	Various forms
	Home gardens (involving a large number of herbaceous and woody plants with or without animals)	Several types of home gardens and kitchen gardens	Very common; Java home gardens often quoted as examples; involving several fruit trees
OTHER SYSTEMS	Agrosilvofishery (aquaforestry)		Silviculture in mangrove areas; trees on bunds of fish-breeding ponds e.g. swidden farming
	Various forms of shifting cultivation	Common	
	Agriculture with trees	Common	Common

Continues

South Asian examples	Middle East and Mediterranean examples	East and Central African examples	West African examples	American Tropics examples
Improvements to shifting cultivation; several approaches (e.g. in the north-eastern parts of India)		Improvements of shifting cultivation (e.g. gum gardens of the Sudan)	*Acioa barterii*, *Anthonontha macrophylla*, *Gliricidia sepium*, etc., tried as fallow species	Several forms
Several forms, several names		The 'Shamba' system	Several forms	Several forms
In all ecological regions	The Dehesa system; 'Parc arboree'			e.g. 'Paraiso Woodlot' of Paraguay
Several experimental approaches (e.g. conservation farming in Sri Lanka)		The corridor system of Zaire	Experimental systems on alley croping with *Leucaena* and other woody perennial species	Experimental
Several forms both in lowlands and highlands (e.g. hill farming in Nepal; 'Khejri'-based system in the dry parts of India)	The oasis system; crop combinations with the Carob tree; the Dehesa system; irrigated systems; olive trees and cereals	Various forms; the Chagga system of Tanzania highlands; the Nyabisindu system of Rwanda	*Acacia albida*-based food production systems in dry areas; *Butyrospermum* and *Parkia* systems; 'Parc arboree'	Various forms in all ecological regions
Integrated production systems in smallholdings; shade trees in plantations; other crop mixtures including various tree spices	Irrigated systems; olive trees and cereals	Integrated production; shade trees in commercial plantations; mixed systems in the highlands	Plantation crop mixtures; smallholder production systems	Plantation crop mixtures; shade trees in commercial plantations; mixed systems in smallholdings; spice trees; babassu palm-based systems
Various forms including some forms of social forestry		Various forms	Common in the dry regions	Several forms in the dry regions
Use of *Casuarina* spp. as shelterbelts; several windbreaks	Tree spices for errosion control	The Nyabisindu system of Rwanda	Various forms	Live fences, windbreaks especially in highlands
Multipurpose fodder trees on or around farmlands especially in highlands		Very common	Very common	Very common
Sesbania, Euphorbia, Syzigium, etc., common		Very common in all ecological regions		Very common in the highlands
Several tree species used very widely	Very common in the dry regions; the Dehesa system	The *Acacia* dominated system in the arid parts of Kenya, Somalia and Ethiopia	Cattle under oil palm; cattle and sheep under coconut	Common in humid as well as dry regions (e.g. grazing under plantation crops in Brazil)
Various forms especially in lowlands		Common; variants of the 'Shamba' system	Very common	Especially in hilly regions
Common in all ecological regions; usually involving fruit trees	The oasis system	Various forms (e.g. the Chagga homegardens; the Nyabisindu system)	Compound farms of humid lowlands	Very common in the thickly populated areas
Occasional				
Very common; various names		Very common	Very common in the lowlands	Very common in all ecological regions
Common	Common	Common	Common	

Source: Gholz (1987)

seen particularly as a means of addressing women's concerns for food production and energy management, and for potentially meeting the needs of the poorest households, which have been left out of forest management programmes in the past.

Approaches to agricultural and rural development

Strategies designed to promote development in rural areas and among rural peoples have taken many forms. Since agriculture provides both a direct and indirect source of income in the rural sector, development efforts have focused frequently, though not exclusively, on raising agricultural production. Furthermore, since land is a fundamental means for agricultural production, many rural development programmes have looked to redistribute land resources, particularly where these may be 'underutilised', towards greater social equality and economic efficiency. This section will consider some key ideas and practices relating to rural and agricultural development in developing countries.

Rural development initiatives may be undertaken for different reasons and at different times by various agencies or institutions, including community organisations and government departments, as identified in Chapter 7. Poverty alleviation, the desire to secure political patronage or to effect social transformation, are just some of many possible motivations for interventions in the development of rural areas. As was considered earlier in this chapter, any intervention in the prevailing agrarian structures has implications not only for the material conditions of rural life, but also for personal and social identity, for power relations within and between households and for traditions and customs associated with land, and its use and significance within society (Shipton and Goteen, 1992). Such complexities have, unfortunately, often been overlooked in rural development strategies and may actually be an important factor in explaining the continuing prevalence of widespread rural poverty in the developing world.

The specific nature of rural development strategies has also been shaped by prevailing development ideologies. For example, in the 1950s and 1960s it was generally assumed that the benefits of urban, industry-led development would spontaneously 'trickle-down' without active planning in the rural sector. Where specific agricultural development schemes were introduced, they were often characterised by 'top-down' planning, inadequate environmental knowledge and heavy mechanisation. Such an example was the ill-fated East African Groundnut Scheme in the 1950s, which involved ploughing up vast areas of semi-arid Tanganyika (now Tanzania) using inappropriate machinery to grow groundnuts to supply Britain with vegetable oils (Binns, 1994a: 97–98).

In contrast, during the 1970s many governments felt that substantial state intervention was required to achieve greater equity and poverty alleviation in rural areas. Organisations such as the World Bank stressed the need for 'integrated rural development' schemes which, in addition to raising agricultural productivity and improving nutrition, also emphasised the importance of improving rural health care, education, transport and marketing. But in reality, many so-called integrated schemes actually showed little evidence of integration, such that rural education and health care were often neglected in favour of raising agricultural output through the use of high-yielding varieties, fertilisers, pesticides and mechanisation (Airey *et al.*, 1979; Binns and Funnell, 1983). Also during the 1970s, and following the publication of Schumacher's influential book *Small is Beautiful*, a lengthy debate ensued concerning the issue of transferring technology from rich to poor countries and the need for 'appropriate' technology (Schumacher, 1974).

By the 1980s, however, as considered in Chapter 3, development ideologies reflected a concern for accountability and efficiency, which was translated into limiting the use of state resources. International donors and governments themselves began to look for alternative institutions to deliver services and foster development in rural areas, particularly within the private and non-governmental sectors, as discussed in Chapter 7. NGOs were considered as frequently more successful in promoting local participation, democracy and empowerment, as well as being more cost effective. In consequence, rural development projects during the 1980s were generally smaller, involving a better understanding of rural communities; they were also more responsive to the perceived aspirations and constraints of local farmers.

Through the 1990s, the reconceptualisation of the role of public, private and civic organisations in development generally, as considered in Chapters 3 and 7, was also evidenced within rural development interventions. For example, an increased role for locally derived and managed policy choices was seen in the expansion of 'community-based natural resource management' initiatives throughout Asia and Africa (Agrawal and Gibson, 1999; Elliott, 2002). However, by the turn of the twenty-first century the potentially 'over-idealistic' notions of community and indigenous institutions, for example, had been replaced with more nuanced understandings of communities and ecologies, which are directing further changes in rural development interventions. These are discussed in the final section of this chapter.

Land reform

> **At one time or another, but especially since 1960, virtually every country in the world has passed land reform laws . . . Yet in spite of decades of land reform activities, land ownership remains extremely skewed, concentration of land ownership is almost universally increasing, the mass of landless is growing rapidly, and the extent of rural poverty and malnutrition has reached horrendous proportions.**
>
> **(de Janvry, 1984: 263)**

This quotation alludes to the very widespread and longstanding recognition of the linkages between access to land and the prospects for human development. While land reform is a relatively recent phenomenon certainly in parts of Africa where it has been undertaken most overtly since independence, land reforms have a very substantial history such as illustrated in the enclosure movements of Western Europe through the eighteenth and nineteenth centuries. In some parts of Latin America, land reforms go back almost a century (Robinson, 2004).

Land reform programmes may take many different forms, but generally involve tenancy reforms and/or changes in the distribution and scale of land ownership. In practice, therefore, land reform may involve such measures as the elimination of certain kinds of rent or cropping arrangements, the creation of new kinds of farm, such as cooperatives or state farms, or the expropriation and redistribution of lands, including through the implementation of ceilings on ownership and resettlement programmes. The objectives of land reform may combine social, political and economic intentions, including enhancing social stability, increasing political participation and patronage, widening economic opportunity and promoting more efficient use of land and labour.

The political significance of land reform is confirmed by the number of countries which initiated programmes following their gaining independence or other major political events. In cases such as post-revolutionary Ethiopia in 1974, or China in 1958, the radical nature of land reform is confirmed by the new forms of societal organisation created. In Cuba, land reform was high on the agenda of the revolutionaries in the late 1950s, as shown in Box 10.3. Under colonial rule in Africa, the dispossession of lands and oppression of the agricultural workforce were less prevalent than in Latin America or Asia. Radical land reform programmes following independence were most likely in settler-dominated regimes such as Zimbabwe (Box 10.3).

Programmes of land reform have also been undertaken with more reformist intentions, such as to prevent the accumulation of large landholdings or to enhance security of tenure. In South Korea, during a major programme of land reform initiated after the Japanese colonial period in 1949, those landlords holding more than 3 hectares had to turn over lands to tenants in return for compensation from the state. In Taiwan, during a programme of land reform in the 1950s, rents were cut to a maximum of 37.5 per cent of crop value from a figure in excess of 70 per cent (Barke and O'Hare, 1991). Land reform in the Philippines was effected through the conversion of share tenancies to fixed rents in the 1970s. In India, the exploitative tenancy arrangements inherited from the British colonial period were identified within the first five-year plan of 1950 as fundamental barriers to raising agricultural production and as the source of much social injustice in the rural sector. A four-fold programme of land reform was subsequently introduced in 1951 (Bernstein *et al.*, 1992).

The specific impacts of land reform are hard to identify in many cases since they often arise from diverse programmes and are frequently implemented in phases and in conjunction with wider agrarian reform measures. Many such programmes remain incomplete, and there is frequently a lack of data monitoring their performance. They often demand considerable financial investment, but this may not be forthcoming, particularly as international backing for land reform has declined. For example, large amounts of aid from the USA are generally considered to have been a critical factor in the relative success of South Korea's land reform programme, whereas underfunding is deemed a principal reason for the failure of land redistribution efforts in the Philippines (Dixon, 1990).

BOX 10.3

Some experiences of land reform

Zimbabwe

On independence in 1980 the new majority government in Zimbabwe quickly committed itself to a programme of land reform. The objectives included overcoming the dualist agricultural history of the country, satisfying the political demands of the peasantry and fostering the proposed socialist transformation of society. The resettlement programme aimed to provide 162,000 landless families and persons displaced by the war with land in newly serviced resettlement villages on former European farms. Several 'models' for resettlement were implemented, including cooperative farms run on a collective basis, and the more widespread, individual family farms with common grazing resources (Elliott, 1995).

However, the Lancaster House Agreement between Britain and Zimbabwe served to protect private property interests in the country for a further ten years and was an important factor in limiting the pace of land reform. By the beginning of the 1990s, only 52,000 families had been moved to 3.3 million hectares of land, and since the government had no control as to where lands became available for purchase, scheme areas tended to be in fragmented rather than contiguous blocks, and they were often located in more marginal areas. The cooperative model for resettlement proved particularly limited, with uptake on those schemes only 42 per cent of planned capacity by 1991 (Elliott, 1995). There were also wider problems of management, infrastructural developments and finance, with the British government refusing to provide aid to this model, for example.

In 1992 legislation was passed to enable the compulsory purchase of lands for the resettlement programme and in 1997 a large number of privately owned 'commercial' farms were designated for acquisition 'in an attempt to answer the clamour for land reform and shore up the waning popularity of ZANU (PF) – the former liberation movement party' (Wolmer *et al.*, 2004: 91). Many of these designations were successfully challenged in court, but in 2000 a wave of farm occupations occurred (led by members of the War Veterans Association and with the tacit support of Mugabe's government). While these farm occupations are the basis for the official 'Fast Track' land reform programme in Zimbabwe currently, they also heralded a new era of political violence, economic decline and collapse of the rule of law in the country.

However, escalating costs, population increase, changing class interests and mounting political unrest are among a variety of factors that restrict the impact of land reform on Zimbabwe's economy and society.

Cuba

At the end of the revolution in Cuba in 1959, 28 sugar companies controlled 20 per cent of the farmlands of the country and 40 cattle ranches controlled a further 10 per cent of the cultivated area (MacEwan, 1982: 162). Approximately 70 per cent of the land area was held by fewer than 10 per cent of the farming community. Agrarian reform laws passed in 1959 and 1963, together with other social and political reforms, are widely accepted to have destroyed these inherited forms of rural inequality and to have provided the basis for Cuba's economic transformation into the 1990s.

The main elements of land reform were the staged expropriation of large farming units to state control and the enlargement of the small-scale private sector through legislation which gave landownership to all tenants, sharecroppers and squatters cultivating up to 67 hectares at the time of the revolution. The state farm sector provided the basis for socialist agricultural development, and the large units enabled the effective provision of services in health and education. With a greater number of peasant farmers after the revolution, the new government also undertook measures to organise the role of the enlarged private agricultural sector within the economy. For example, new institutions were created to provide credit, to initiate the formation of cooperatives and to purchase products from the peasantry.

Corruption and considerable vested interests in land among senior politicians and businessmen have certainly compromised the outcomes of land reform in many African countries (Toulmin and Quan, 2000). Political conflict and upheaval are regularly generated by land reform programmes, not solely between the landed elite and popular interests, but also between such groups as cultivators and pastoralists, as seen in Eritrea where there is no provision within the current Land Proclamation to protect the rights of pastoralist claims to land (Fullerton-Joireman, 1996).

Land reform is evidently not in itself a panacea for rural development, and it is now generally recognised that such reforms need to be undertaken in conjunction with other measures, such as ensuring effective social institutions at the local level and raising land productivity. Furthermore, particular attention is required to examine the socio-political, economic and environmental contexts in countries and regions where land reform programmes are being introduced as a component of future rural development strategies.

Agricultural intensification

In contrast to programmes of land reform that have looked to intervene directly in the agrarian structure within a country, many more rural development strategies have focused on transforming the way in which agricultural production occurs, typically through various packages of introduced technology and assisting the move from labour-intensive farming to more capital-intensive production. In the 1960s, a number of breakthroughs in the development of high-yielding varieties (HYVs) of grains such as rice and wheat were made in research institutes in the Philippines and in Mexico (largely with overseas public finance). The extensive transfer of these technologies (along with the fertilisers, pesticides, irrigation and machinery required) to many poorer countries of the developing world is widely referred to as the green revolution, as discussed in the Key idea box. The largest impacts of these developments were in Asia, where more than 75 per cent of the wheat plantings and 30 per cent of the rice plantings by the early 1980s were HYVs (Barke and O'Hare, 1991: 107).

Part of the attraction for planners of the green revolution lay in the assumed scale-neutrality of the technologies and the power of the market to encourage and disseminate improvements in well-being. It was assumed that the biochemical technologies of seeds and fertilisers would be equally viable at all scales of operation, whether on small or large farms. It was therefore thought that the yields and incomes of all farmers could be enhanced without raising rural inequalities. In practice, however, the green revolution had very uneven regional and social impacts.

In India the green revolution is considered to have been responsible for delivering national self-sufficiency in food grains by the late 1970s, but as Bernstein (1992a) pointed out, per capita grain production

Key idea

The future source of world food supply

Increased food production can be achieved in one of two ways; by extending the area under production or through raising the intensity of production. Since the area of land available for cultivation globally is fundamentally limited and world population numbers are increasing, further increases in global food production will have to come through increasing the intensity of production. In recent years, the debate concerning how this will be achieved has shifted from a 'Green' to a 'Gene' revolution (Atkins and Bowler, 2001).

Green revolution – a series of phases of the development of high-yielding varieties of crops via breakthroughs in plant breeding and the cross-breeding of one crop with another, the development of associated technologies allied to the cultivation of these and the transfer of these packages to further and further areas of the world.

Gene revolution – the development of new organisms through the artificial introduction of alien genetic material into existing plants and crops. Genetically modified crops are produced that can fix their own nitrogen, require less pesticide and will yield in very dry conditions, for example. Encompasses research and development financed by major biotechnology companies of the world.

actually fell in 11 of the 15 major states of India between 1960 and 1985, with success being correlated strongly with the distribution of irrigation to secure multiple cropping. Social inequality was also aggravated in circumstances where rising landlessness forced increasing numbers of women into wage employment to ensure household survival.

However, due to mechanisation, some employment opportunities for women in the traditional areas of harvesting and grain processing declined. In sharp contrast, for women in households which owned land, their burden in agriculture was often increased with double cropping. Research has shown that investment in technology at this income level tends to be in areas which save male labour time, such as tractors for land clearance, rather than women's time taken up in tasks such as weeding (Pearson, 1992). Table 10.15 summarises the gendered impacts of rural development strategies based on technocratic packages, including within the green revolution.

This era of the green revolution also had limited impacts in South America and Africa. As noted by Dixon (1990: 92), 'the crops that have formed the mainstay of the green revolution, rice and wheat, are simply not grown by large numbers of third world farmers'. In much of Africa, food production is dominated by rainfed cultivation of coarse grains such as maize, millet and sorghum. Research into improved varieties of these crops has generally been more limited and the impact more confined to larger scale, export-oriented, irrigated production (Dixon, 1990).

There have been efforts to develop a second green revolution, more suited to the vast dry land areas of the developing world and to the resource-poor conditions of many farmers who live there. For example, conventional plant breeding schemes in the 1980s developed high-yielding strains of millet, sorghum, cowpeas and cassava, which achieved some success in raising production in parts of India and Africa, without requiring increased farm inputs (Barke and O'Hare, 1991). However, many consider that further increases in yields through conventional plant breeding methods are now limited and it is through biotechnology, encompassing genetic engineering and also many more diverse and complex processes and technologies, that the most profound changes are predicted (see Key idea box on the future source of food). Biotechnology is believed to offer the potential for raising food security in the developing world, as well as delivering many other human goals globally, ranging from detoxification of hazardous wastes to a possible cure for HIV/AIDS (Conway, 1997; Morse, 1995). However, in Chapter 6 some of the ways in which genetically modified crops are considered to threaten, rather than improve, the opportunities for small-scale farmers in the developing world were highlighted.

Irrigation

Irrigation has played an important role in the development of human civilisation and in enabling rural production in the developing world (Chapter 6). Rural development interventions based on irrigation technologies may be undertaken with varied objectives, such as to control flooding or to increase agricultural production through the timely application of water, which can lead to extended cropping seasons or double cropping.

The extension of irrigation in the developing world occurred most rapidly from the 1960s, and it was closely associated with the ideology of development as modernisation and the era of the green revolution in agricultural development as discussed above and in Chapter 6. Irrigation worldwide has been achieved mainly through the construction of large-scale reticulation systems using big dams and barrages on permanent rivers (Heathcote, 1983). Large-scale dam construction provided a means for subsidising industrial production and, where associated with irrigation, a technocratic route to rural development in developing nations by increasing agricultural production.

South Asia and Southeast Asia account for over half the developing world's irrigated area, with China, India and Indonesia all having a significant proportion of their agricultural land under some form of irrigation (Table 10.16; Plate 10.10). The spread of irrigation in Africa, however, has been much slower and is concentrated in countries such as Sudan and Egypt, where agriculture is virtually wholly under irrigation.

One of the world's largest irrigation developments is the Gezira scheme in eastern Sudan, covering almost 1 million hectares and involving 100,000 tenants (Stock, 1995). Project planning for the Gezira started in the early twentieth century and was developed in stages from the 1920s using water from the Blue Nile. Before this development, small farmers used the banks of the Nile for the cultivation of date palms, vegetables and a type of millet known as dura, which is the staple food crop. Water was raised from the river using ox-powered saskias.

Table 10.15 Rural development: technology and its impacts on women

Property ownership	Employment	Decision making	Status	Level of living and nutrition	Education
GREEN REVOLUTION: NEW SEEDS, BREEDS AND AGROCHEMICALS					
May lose usufruct rights as land is used more intensively. Land owned by women is often physically marginal and not suitable for optimum applications of new inputs	Women exclude themselves from use of chemicals because of threat to their reproductive role. New crops may not need traditional labour inputs of women. Women generally displaced from the better paid, permanent jobs	Decline. Training in new methods in agriculture limited to men. Use of new technology and crops generally subsumed by men. Women farmers equally innovative when given opportunity	Increase in family income may allow women to concentrate on reproductive activities. In patriarchal society this increases status of male head of household	New crops may be less acceptable in family diet and nutritionally inferior because of chemicals	Increase in additional disposable income of family may be used for children's education
MECHANISATION					
Women operate smaller farms in general and so may not find it economic to invest in new implements	Women usually excluded from use of mechanical equipment. Women farmers have difficulty obtaining male labourers	Decline	Decline because of reduced role on farm and downgrading of female skills	New implements not used for subsistence production	Growth of interest in mechanical training but limited to males
COMMERCIALISATION OF AGRICULTURE AND CHANGES IN CROP PATTERNS					
Female-operated farms tend to concentrate on subsistence crops and crops for local market. Tend to remain at small scale	Decline because technical inputs substituted for female labour	Decline because less involved in major crop production activity	Decline	Decline because cash crops take over land traditionally used for subsistence production by women. Males allocate more income to developing enterprise and for personal gratification than to family maintenance	Increased time available for education
POST-HARVEST TECHNOLOGY					
New equipment owned by men	Women's traditional food processing skills no longer in demand. May employ young women in unskilled jobs in agro-industries	Decline because ownership of equipment and skills passed to men	Decline because female skill downgraded	Decline because loss of women's independent income from food processing activities. New product may be nutritionally inferior. Women deprived of use of waste products for animal feed and so lose important part of traditional family diet	

Source: Compiled from Momsen (1991, 2004)

Table 10.16 Irrigated land as percentage of cropland, 2003

Country	Irrigated land as a proportion of arable land (%)
Nigeria	1
India	33
China	35
Indonesia	13
Zimbabwe	5
Philippines	15
Mexico	23
Brazil	4

Source: FAO Statistical Yearbook 2005–06

The British colonial concern in the Gezira scheme was to expand cotton production, and a centrally planned rotation of cotton, wheat, sorghum, groundnuts and fallow was introduced onto the scheme. In the early years, the Gezira programme was quite successful in producing export and food crops, certainly in relation to other similar schemes, such as the inland Niger Delta programme in Mali. However, cropping intensities on Gezira fell in the 1970s and 1980s, as maintenance of irrigation structures and the clearance of silt from channels became less effective, leading to water-supply problems and the withdrawal of some lands from cultivation, often in mid season (Fadl, 1990). Furthermore, problems of pesticide pollution and salinisation threaten the future productivity of the scheme (Stock, 1995).

Programmes for the expansion and rehabilitation of irrigation in Southeast Asia have been closely related to the perceived opportunities provided by the green revolution. The Muda irrigation scheme in north-west Malaysia was completed in 1974 with World Bank support, and aimed to increase rice production in the region by upgrading canal, drain and farm road density, together with fertiliser subsidies and price supports. Although rice production did rise, income distribution between groups involved in the programme had worsened by the end of the 1980s. The number of tenants declined as landowners perceived the productivity gains to be made and demanded their lands back, although increased mechanisation reduced the demand for hired labour, and the poorest tenants and small owner-operators experienced losses in income (Ghee, 1989).

With respect to irrigation schemes in sub-Saharan Africa, it has been suggested that 'the larger the projects are, and the higher the level of their technology, the poorer is their performance' (*The Courier*, 1990: 84). Large-scale irrigation schemes are certainly costly to construct, maintain and operate. The three largest schemes in Nigeria had a development cost of approximately US$25,000 per hectare of land developed (Stock, 1995: 168). Frequently, costs fall to certain individuals who are not compensated fully, for example

Plate 10.10 Farmer irrigating his fields in Guilin, southern China (photo: Tony Binns)

when their lands are submerged through the construction of dams. There are also various unquantified social costs for farmers involved in these schemes, such as their loss of autonomy. Large irrigation schemes are usually operated by semi-state institutions in conjunction with peasant organisations, but farmers are required to adhere to specific crops and cropping calendars, schedules of maintenance, and to use designated sources of credit and marketing outlets.

Further hidden costs of large-scale irrigation schemes fall on downstream farmers who operate small-scale irrigation techniques, based on a sensitive knowledge of local flows and ecologies. Pastoralists may lose access to valuable grazing lands (Binns and Mortimore, 1989). Fish catches may also decline downstream of large dams and irrigation schemes (Adams, 1985). The environmental costs of irrigation developments and the associated dams include the disturbance of water regimes in rivers and floodplains, the loss of habitat and, frequently, a decline in water quality and health (New Internationalist, 1995).

Funding for large irrigation schemes (as with large dam construction, discussed in Chapter 6) is declining, in part due to their poor performance up to now. Dam construction is becoming increasingly privatised in the developing world and it is suggested that 'only in Asia are governments determined to continue to build dams, with or without international aid' (Pearce, 1997: 353). The Narmada dams in India are going ahead despite the withdrawal of World Bank funding, for example, and in 2003 the world's biggest dam to date, the Three Gorges in China, was completed, largely with internal finance. However, it seems likely that future irrigation projects will be even more expensive, if only because the most easily constructed projects are already built. By the mid-1990s more than half of World Bank spending on irrigation was going towards the rehabilitation and upgrading of old irrigation schemes rather than new schemes (World Bank, 1996b).

Since the mid 1980s, many development institutions have been increasingly involved in small-scale and less formal irrigation. Small-scale systems based on indigenous practices such as 'flood cropping', stream diversion and simple lift irrigation are now widely held to offer a more acceptable strategy for agricultural development (Binns and Funnell, 1989; Kimmage, 1991).

However, although smaller may be cheaper, easier and potentially more sustainable, Adams and Anderson warned in 1988 that small-scale schemes may also fall prey to the same basic problems as large-scale systems. In particular, they pointed to the general failure within irrigation schemes to frame development initiatives in the reality of historical experience and 'the operation of the prevailing social systems and economic networks that sustain local production' (Adams and Anderson, 1988: 535) that remains true today.

Investing in non-farm activities in rural areas

While obtaining sufficient food and income to satisfy domestic subsistence requirements remains the key livelihood objective for the majority of rural households in developing countries, it was seen in the Key idea box on page 450 that in many regions, this increasingly means pursuing opportunities beyond agriculture. Box 10.4 considers some of the benefits (and costs) of this de-agrarianisation. In recent years, theorists and practitioners have appreciated widely that in many rural communities there is a wealth of knowledge, skills and investment in activities such as textile production, baking, craft making (Plate 10.11), wood- and metalworking, pottery and various forms of construction, for example. Furthermore, in the light of the general failure of large-scale industrialisation policies to alleviate underemployment and poverty in many countries, an interest emerged (particularly among NGOs but also amongst the IFIs) in fostering rural industry as a strategy towards rural development and as a means for supplying urban and/or tourist markets.

Rural industries have a longstanding presence in China that has not been widely seen in many developing countries. This pattern can be understood as due to the widespread richness and nature of minerals and ores in China but also as a result of the country's past development strategies. For example, a central feature of the Great Leap Forward between 1958 and 1961 was the mobilisation of rural peasants 'to convert surplus labour into capital with the aim of producing more iron and steel and of creating greater agricultural and light industrial production' (Endicott, 1988: 51). All communities were encouraged to establish 'five small industries': agricultural machinery, coal mining, hydroelectricity, chemical fertilisers and building materials, reflecting Mao's pronouncement that 'the peasants can become workers right where they are' (Endicott, 1988: 87).

Unfortunately, with the diversion of labour into industrial growth, agricultural production suffered, and in the winter of 1960 there was widespread starvation

Plate 10.11 Paper-making in Guizhou, southern China
(photo: Tony Binns)

BOX 10.4

The benefits and costs of de-agrarianisation

Research into the dynamics of rural livelihood systems in Indonesia, including the introduction of technological packages within agriculture, has suggested that increasing and diversified opportunities for employment and income in the non-farm economy have enabled the poor to increase their standard of living, in some cases compensating for the predicted land concentration and impoverishment of some groups. It was found that off-farm, local opportunities in informal activities such as trading, home-based shops, food manufacturing and taxi services, but also employment (part-time, seasonal or full-time) in factory work and the public sector, were growing in importance. The location of large factory enterprises in rural areas of Java and the central and northern regions of Thailand enabled people to remain in rural areas and commute daily for employment. As Rigg comments,

> **In large part, the emergence of occupational diversity has put off the displacement of the rural poor from their land and has, perversely, helped to sustain 'traditional' rural economies.**
>
> **(Rigg, 1997: 189)**

Often, however, it is the younger members of the household who engage in alternative activities. This may impact in economic terms, as the most productive labour is lost to agriculture; but perhaps most significant are its social and cultural effects. Rigg (1997: 241) observes:

> **The slack season is a period for courting, festivals and fairs, religious devotions, and the telling of tales. It is the season when households and villages collectively renew their bonds and identities. If young men and women are absent from such festivities, this creates the conditions in which the activities either become meaningless, or their meanings and roles change.**

In spite of such observations, there seems no doubt that the diversification of rural economies can have some benefits as an element of rural development strategies, perhaps most particularly in reducing the dependence of households and communities on the success of agricultural production. For example, where crop yields are affected by low rainfall, pests or disease, income generated by non-farm activities can reduce the vulnerability of rural households and enable them to survive until the next harvest. In fact, such activities have often been an important part of the repertoire of coping strategies which communities have developed to ensure their survival from one year to the next.

across China. A decade later, the Walking on Two Legs initiative emphasised a more balanced approach to rural development; it involved 'being open to the new and the modern while not overlooking the tried and trusted' (Endicott, 1988: 81).

Elsewhere in the developing world, the significance of rural industries relative to urban industries can be considerable in particular countries. For example, in Sierra Leone it was estimated that 86 per cent of total industrial sector employment and 95 per cent of industrial establishments in the late 1980s were located in rural areas. Textile and clothing production accounted for 53 per cent of such employment. Rural enterprises were typically small, family-owned and family-run (Chuta and Liedholm, 1990).

However, it has been widely recognised for some time that many people in rural areas frequently lack funds and equipment to increase productivity and to innovate in enterprises beyond agriculture. In addition, it was understood that in many cases, particularly amongst the poorest groups, people were often deeply in debt to local business people and moneylenders. In the 1950s it was envisaged that the state should step in to provide an alternative to moneylenders, but although substantial credit was provided particularly to the agricultural sector through state-run banks and cooperatives, the impacts are considered to have been generally limited and of doubtful benefit (Sinha, 1998).

Through the 1980s and 1990s, a host of innovative credit initiatives through what are termed microfinance institutions (MFIs) were developed to overcome some of the problems encountered by formal lending institutions to deliver credit to the poor. In short, these initiatives made small loans to poor borrowers organised into groups and used joint liability structures and peer monitoring to overcome the previous problems. Case study 10.1 details perhaps the best-known example of a MFI, the Grameen Bank in Bangladesh. Over its history, this bank has lent more than £2.9 billion and has achieved a recovery rate from its loans of over 98 per cent (Huggler, 2006). It has been particularly successful in delivering credit to the

Case study 10.1

The Grameen Bank and poverty alleviation in rural Bangladesh

The Grameen Bank originated in 1976 as an experimental research project led by economist Professor Muhammad Yunus at Chittagong University. The initial aim of the project was to tackle the serious problem of indebtedness among the rural poor in Bangladesh through providing loans to households which owned less than 0.5 acres of land.

The word Grameen means 'village' and the project was concerned with assessing whether if the poor received financial help at reasonable terms and conditions, they could then generate productive self-employment without external assistance. Professor Yunus managed to persuade a local branch of a commercial bank to provide credit at an interest rate of 13 per cent a year, provided he could guarantee recovery of the loans.

This pilot project was successful and in June 1979 the Grameen Bank Project was launched in five districts of Bangladesh, with support from rural branches of commercial banks and the agricultural development bank, and with financial assistance from the state bank of Bangladesh and the International Fund for Agricultural Development. Within a year, 24 branches had been set up, though expansion was constrained by some reluctance from participating banks. A government ordinance in September 1983 transformed the project into the Grameen Bank, a specialised financial institution for the rural poor, in which government has a 60 per cent share and the borrowers 40 per cent.

As Todd (1996: 7) comments:

> The 'essential Grameen' . . . is an exclusive focus on the poor, with preference to poor women, simple loan procedures administered in the village, small loans repaid weekly and used for any income-generating activity chosen by the woman herself, collective responsibility through groups, bolstered by compulsory group savings, strict credit discipline and close supervision through weekly meetings and home visits.

▶

Case study 10.1 (continued)

Each branch of the Grameen Bank covers between 15 and 20 villages, and members of households owning less than 0.5 acres of cultivated land, or with assets with a value equivalent to less than 1.0 acres of medium-quality land, are eligible to receive a loan. The loan and interest must be repaid in 52 weekly instalments. The bank's default rate of less than 3 per cent is most impressive when compared with default rates of 40–60 per cent in other rural credit programmes in developing countries. Since many poor rural dwellers are illiterate, bank workers often visit their homes to assist with loan arrangements. One of the guiding mottos of the bank is: 'Take the bank to the people, not the people to the bank.'

In order to receive loans, the poor have to organise themselves into groups and associations and must be prepared to interact with each other. Several groups from each village constitute a 'centre' and weekly meetings of all centre members are held with the bank representative in attendance. All business is conducted in a transparent manner in front of the members, and an important incentive for regular repayment is the assurance of a new and bigger loan at the end of the repayment cycle.

The Grameen Bank is no longer merely an agency for disbursing loans, but it has generated a wider 'culture' of development, involving many other aspects of life among the rural poor in Bangladesh. In 1984, a social development programme called Sixteen Decisions was introduced to promote discipline, unity, hard work and improve living standards.

The success of the Grameen Bank indicates the importance of credit as a starting point for the introduction of a wide-ranging economic and social development programme and poverty alleviation. A study undertaken in 1985 found that those people who had joined the bank had about 50 per cent higher income than target groups in villages where there were no bank groups (Hossain, 1988).

The Sixteen Decisions of the Grameen Bank

1. The four principles of Grameen Bank – discipline, unity, courage and hard work – we shall follow and advance in all walks of our lives
2. We shall bring prosperity to our families
3. We shall not live in dilapidated houses. We shall repair our houses and work towards constructing new houses as soon as possible
4. We shall grow vegetables all the year round. We shall eat plenty of them and sell the surplus
5. During the planting seasons, we shall plant as many seedlings as possible
6. We shall plan to keep our families small. We shall minimise our expenditures. We shall look after our health
7. We shall educate our children and ensure that they can earn enough to pay for their education
8. We shall always keep our children and the environment clean
9. We shall build and use pit latrines
10. We shall drink tubewell water. If it is not available, we shall boil water or use alum
11. We shall not take any dowry in our sons' weddings, neither shall we give any dowry in our daughters' weddings. We shall keep the centre free from the curse of dowry. We shall not practise child marriage
12. We shall not inflict any injustice on anyone, neither shall we allow anyone to do so
13. For higher income we shall collectively undertake bigger investments
14. We shall always be ready to help each other. If anyone is in difficulty, we shall all help
15. If we come to know of any breach of discipline in any centre, we shall all go there and help restore discipline
16. We shall introduce physical exercise in all our centres. We shall take part in all social activities collectively

Note: Formulated in a national workshop of 100 female centre chiefs in March 1984, these sixteen decisions might be called the social development constitution of Grameen Bank. All Grameen Bank members are expected to carry out these decisions.

Source: Hossain (1988)

poorest rural households and in enabling the expansion of non-agricultural rural enterprises. This case has received widespread acclaim, including the award of the Nobel Peace Prize in 2006 to Muhammad Yunus, the key architect of the approach, and has formed the basis of further micro-credit programmes in many more countries.

In 1997, the first International Micro-credit Summit was held in Washington and reported that over 1000 MFIs at that date reached over 13 million people (75 per cent of whom were women) in over 100 countries (Wheat, 2000). Microfinance had become central to projects funded by donors including the Department for International Development and NGOs such as Oxfam throughout countries of the developing world as a means for addressing poverty alleviation and empowerment objectives. Significantly such projects remained consistent with the dominant development objectives of the time – market-led growth as discussed in Chapters 3 and 4.

However, the Grameen Bank and MFIs more widely are not entirely without criticisms or reservations. For example, the majority of borrowers are women (although the Grameen programme was not specifically designed as a project targeted at women). Proponents of the model point to how women's involvement in micro-credit programmes not only raises women's involvement in productive economic activities, but also raises their status within families and communities, suggesting the empowering effects of their involvement. Women are also regularly found to be more reliable as borrowers and research has shown that the increased income to the family is most likely to be spent on improvements to the general welfare of the family and particularly children (Pearson, 2000).

There is substantial dispute concerning the empowerment potential of micro-credit programmes (Kabeer, 2001; Mayoux, 2001) and it is recognised that the outcomes can be highly varied according to the complexities and specificities of gender relations within the household, the community and wider society. Research within the gender and development (GAD) approach (see also Chapter 7) is now exposing some of the problems surrounding women combining income-generating activities alongside other domestic and productive work, for example, and of the possible conflicts within households as women's access to financial resources increases.

It is apparent that in some contexts women's access to credit does not increase their control over economic activities, the credit in fact facilitates activities controlled by men, and women may just have the responsibility of repayment. Similarly, such research has investigated the impact of access to credit among women on intra-household gender relations, on the quality of spousal relations, the authority and assertiveness of women in their domestic roles and the levels of domestic violence, for example (Kabeer, 2001). The research reveals quite divergent experiences in different contexts, confirming that gender relations are an important factor in shaping the outcomes of development initiatives such as promoted by the Grameen Bank.

A new paradigm in rural development?

Through the 1990s, substantial progress towards an improved understanding of social and environmental changes in rural areas of the developing world occurred that was considered a significant source for optimism in terms of future rural development interventions (Elliott, 2002). Shepherd (1998), for example, suggested a 'new paradigm' in rural development, many elements of which have been identified in this chapter. In particular, a different ideology of rural development was promoted, based on what has been termed 'last' rather than 'first' priorities; in short, those of the small, impoverished farmers rather than external researchers and practitioners in rural development; and on 'reversing' the centralising and simplifiying tendencies of many past rural development efforts (Chambers, 1983, 1993).

Many commentators would agree that the work of Robert Chambers has been central in promoting this new ideology of rural development (see the Key thinker box). For example, in his book *Challenging the Professions: Frontiers for Rural Development*, Chambers (1993: 110) identified the fundamental limitations of previous ideologies as being:

> **a planner's core, centre–outwards, top–down view of rural development. They start with the economies, not the people; with the macro and not the micro; with the view from the office, not the view from the field. And in consequence, their prescriptions tend to be uniform, standard and for universal application.**

The brief overview within this chapter of the green revolution, for example, indicated that a standard

Key thinker

Robert Chambers

Robert Chambers (1932–) has worked as both a practitioner in development (including as a District Officer in Kenya, 1958–1962) and as an academic (particularly at the Institute of Development Studies at the University of Sussex from 1972) and is widely recognised to have had a profound impact on the ideologies, strategies and practice of development. As Parnwell (2006: 77) suggests,

> **his thumbprint is everywhere one looks in the modern development field: bottom-up development, participatory development, sustainable livelihoods, the redefinition of poverty, the strengthening of civil society, and the entire ethos of appropriate development.**

Since the early 1980s his burgeoning writings in articles and books have detailed why and how the most marginalised, historically excluded and most powerless groups in society should and could be put at the heart of the processes of research, policy formulation, decision making and project implementation in development. He identified and popularised notions of 'Farmer First', for example and 'challenged' the professions (of academia and within state and donor structures) to change their values and the ways and with whom they work to deliver the 'reversals' required.

package was indeed required by farmers, including HYV seeds, fertilisers, pesticides and irrigation systems. The technologies were generally developed on experimental research stations, not on farms, and required substantial government investment to ensure subsequent adoption at the local level. The package was frequently not modified sufficiently for successful implementation in the very contrasting regional and local environments within the developing world. Furthermore, there continue to be concerns as to the sustainability of such energy-intensive forms of agricultural production (Chapters 6 and 7).

By definition, within this new paradigm for rural development there are no blueprints for development interventions; instead, poor rural people provide 'starting points which are at once dispersed, diverse and complicating' (Chambers, 1993: 120). Guiding principles in this approach to rural development included emphasis on providing 'baskets of choices' rather than uniform packages, in combining traditional and Western science in the generation and dissemination of technologies, on participatory methods of enquiry and action, and on empowering communities through appropriate interactions with accountable and democratic state and non-governmental organisations (Chambers, 1993, 1997; Chambers *et al.*, 1989).

The 'reversals' required in research and development confirm that rural poverty alleviation, the health of ecosystems and the sustainability of rural livelihoods depend on much more than just raising agricultural production and, indeed, it was important to remove the agricultural bias that has characterised many rural development policies in the past (Reardon *et al.*, 2001).

An increasing number of rural development programmes now demonstrate greater empathy with the needs of poor rural households and eschew some of the principles of 'farmer first'. Indeed, the spread of this new paradigm, and the search for more participatory means for research and development in particular, has been referred to as the new 'tyranny' in development (Cooke and Kothari, 2001).

Critical consideration 'from within' is also underway with ongoing examination of the technical limitations of the approaches, ethical concerns and in the development of new concepts and styles of investigation, for example, in an attempt to move 'beyond farmer first' and to further the original work and evident successes (Guijt and Shah, 1998; Scoones and Thompson, 1994). The Key idea box highlights how such work has been extended within what is called the 'sustainable livelihoods framework' (SLF) and the ways in which this is now being used by a number of development institutions in the understanding of rural livelihoods and for identifying more effective policy interventions.

Key idea

The sustainable livelihoods framework (SLF)

The 'sustainable livelihoods framework' is an analytical framework that was developed in the 1990s and owes much to the earlier work of Robert Chambers and Gordon Conway at the Institute of Development Studies. It has been most well developed in relation to rural livelihoods, but is also increasingly applied in urban situations (see Rakodi, 2002).

A livelihood comprises:

> **the capabilities, assets (stores, resources, claims and access) and activities required for a means of living: a livelihood is sustainable which can cope with and recover from stress and shocks, maintain or enhance its capabilities and assets, and provide sustainable livelihood opportunities for the next generation; and which contributes net benefits to other livelihoods at the local and global levels in the long and short term.**
>
> **(Chambers and Conway, 1992: 7–8).**

The framework aims to help in the understanding and analysis of the livelihoods of poor people and to assist in the identification of appropriate entry points and sequencing of more effective development policy and interventions. It is a simplification of real life and does not attempt to represent that complex reality directly, but rather to ease the identification of the main factors affecting people's livelihoods and the typical relationships between them. It is

> **essentially people-centred and aims to explain, in a necessarily abstract and simplified way, the relationships between people, their livelihoods, and their environments, (macro) policies, and all kinds of institutions.**
>
> **(Neefjes, 2000: 82)**

Substantial development and application of the framework has been funded by the UK Department for International Development (DfID), and has evolved through the research and practical activities of DfID, UNDP, CARE and Oxfam through the 1990s (Carney *et al.*, 1999). It is proving a useful tool in raising understanding of local livelihoods, for planning new development activities and in evaluating the impact of existing interventions.

The 'asset pentagon' is at the core of the SLF, as shown in Figure 10.4. Five asset categories or types of capital that people may or may not have as a basis for pursuing their livelihoods, are identified. More fully, these capitals are:

1 Human capital: the skills, knowledge, ability to work and good health. It is both a means of achieving livelihood outcomes and an end in itself; overcoming a lack of education, for example, can be a primary livelihood objective.

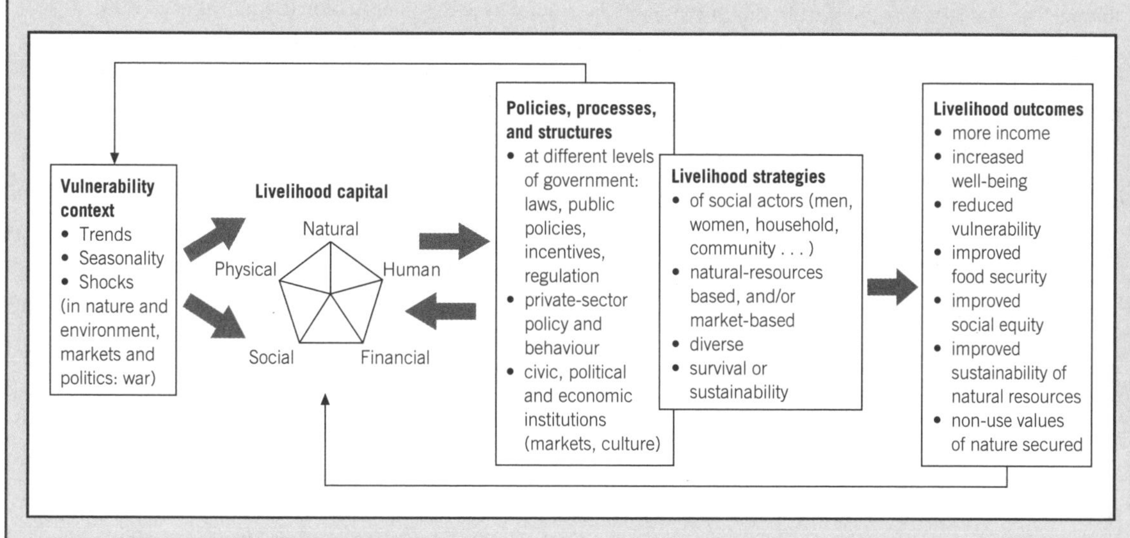

Figure 10.4 The framework
Source: Neefjes (2000)

Key idea (continued)

2 Social capital: in the context of the SLF is taken to mean the social resources (networks, membership of groups, relationships of trust, access to wider institutions of society) upon which people draw in pursuit of their livelihood objectives. Social resources may be enhanced through networks that increase people's ability to work together and to access wider institutions such as political or civic bodies. Relationships of trust and exchange also provide the basis for many informal safety nets among the poor.

3 Natural capital: the natural resource stocks from which resource flows and services useful for livelihoods are derived. These may include assets used directly in production, such as land or trees, and less tangible public goods such as biodiversity or the atmosphere.

4 Physical capital: the basic infrastructure and producer goods needed to enable people to meet their basic needs and function more productively. These typically include secure shelter, affordable energy, adequate water and sanitation and access to transport and information.

5 Financial capital: the financial resources that are available to people (savings, credit, remittances or pensions) that provide different livelihood options.

A large part of people's asset status, however, is the external environment in which people operate, termed the 'vulnerability context' in the SLF. Fire or flood can (and do) devastate the livelihoods of the poor. The term vulnerability is used to draw attention to how the poorest groups often have very little control over many of the factors that are responsible, directly or indirectly, for their poverty. Fire and flood would be considered natural 'shocks' within the framework in that they are largely unpredictable. Other shocks may be economic, conflict scenarios and shocks relating to human, livestock or crop health.

Changes to the vulnerability context can also occur through 'trends', the source of which may similarly lie in the economy, natural resources, in governance or population issues, but are generally more predictable than shocks. 'Seasonality' in terms of prices, production, health and employment opportunities, is certainly one of the biggest and most enduring sources of hardship and a strong determinant of vulnerability among the poor, whose margins may be very small.

The arrows within the framework as shown in Figure 10.4 are used to suggest that policies, processes and structures, at various levels and spheres, can influence this vulnerability context. Government fiscal or health policy, for example, can shape particularly the non-natural trends and shocks. Furthermore, the arrows are used to confirm that there is a need to think beyond people's assets per se in any context, to consider how these are transformed into 'livelihood outcomes' via varied processes and structures.

'Livelihood strategy' refers to the range and combination of activities and choices people undertake to achieve their livelihood goals. The outcomes of those strategies are diverse. A number of categories of outcome are listed in the diagram, and give a sense of what motivates people to behave as they do. These can assist agencies in considering how likely people are to respond to new initiatives, and can also be used to identify performance indicators in project monitoring.

Table 10.17 displays the basic differences in origins of the sustainable livelihood approach and how it is being used in the activities of a number of organisations. It is apparent that the sustainable livelihoods approach is both sufficiently comprehensive to frame large-scale challenges such as poverty elimination, for example, and sufficiently flexible to address very context-specific constraints such as within households (Carney *et al.*, 1999).

For Oxfam, the framework can accommodate a way of thinking about environmental change that combines with its concerns for globalising markets, gender inequality and the need to strengthen people's participation in the development process. Economic, social, institutional and ecological sustainability are very much to the fore in its corporate aim to help secure rights to sustainable livelihood.

For DfID, the sustainability requirement as in the original definition is less prominent in its work, such that sustainable livelihoods is a broad approach to achieving the institutional goal of poverty elimination in poor countries, rather than sustainability being a goal in its own right. Two types of activity dominate DfID's implementation; direct support for asset building and changing the functioning of the structures and processes that influence the strategies that are open to poor people.

Key idea (continued)

Table 10.17 The origins and the use of the SLF in different organisations

Agency	Origins of SL approach	When introduced	Change from what . . .	Status of SL within the agency	Current uses	Types of activity	Strengths emphasised	Core ideas/organising principles
CARE	CARE Long Range Strategic Plan as central programme thrust	1994	Primarily a sectoral focus	Primary organisation-wide framework for programming	Relief through development Urban and rural	Livelihood protection Livelihood promotion Livelihood provisioning	Comprehensive yet flexible Improves sectoral coordination Increases multiplier effects	Household livelihood security People-centred
DFID	White Paper commitment to supporting policies and actions that promote SL Overall aim of poverty elimination	1998	Resource focused activity (within former natural resource division) Sectoral focus	Support from the top but still associated with rural side	Started rural, now more interest from urban side Various uses through development project cycle	Various to meet international development targets (including poverty elimination) Link to rights and sector approach	Builds upon existing experience and lessons Offers a practical way forward in a complex environment	People-centred Multilevel Partnership Various types of sustainability Dynamic Poverty focused
Oxfam	Need to link envt. change with poverty issues Strategic planning exercise looking for unifying concepts	1993	Primary environmental care	One approach for achieving poverty eradication One of five strategic change objectives	Across development emergency and advocacy Mostly rural Used for strategic planning purposes, seldom at field level	Strategic planning activities	Participatory analysis Enables links to social and human rights approaches	People-centred Multilevel Partnership Various types of sustainability Dynamic
UNDP	Parts of the overall Sustainable Human Development agenda	1995	Partly a reaction against economic, employment focused initiatives	One of five corporate mandates An approach for achieving sustainable human development	Rural and urban Country programme planning Small and micro enterprise activity	A conceptual and programming framework	Links micro–macro Integrates poverty, environment and governance issues Gets the most out of communities and donors	Adaptive strategies Conditioning factors (shocks and stresses that affect asset use)

Source: From Carney *et al*. (1999)

Key idea (continued)

UNDP has a corporate mandate centred on sustainable human development and the sustainable livelihood approach is used to bring together the issues of poverty, governance and environment within that mandate. Like DfID, UNDP seeks a non-sectoral entry point for development activities, and its focus has been the strengths of the adaptive strategies that people employ in their livelihoods. In particular, UNDP is working to find technological options that help people improve the productivity of their assets and eliminate poverty.

For CARE, as an NGO that has historically worked with the poorest groups in basic needs activities, the focus is less on the transforming structures and processes and macro–micro links, but rather with household livelihood security and local matters of community mobilisation and local institutional strengthening.

Plate 10.12 Brick making, in rural India
(photo: J. Elliott)

Conclusion

There remains much to be learnt concerning the nature and dynamics of rural livelihoods in the developing world, as suggested in the opening sections of this chapter. The sustainable livelihoods framework aims to accommodate a richness of potential livelihood goals in rural areas that have been underestimated in previous rural development approaches and have been limited within sectoral approaches to rural challenges. The SLF also goes some way towards embracing the varied importance for different people in different places of a wide range of assets in constructing and changing their livelihoods. It therefore has the potential to be used to take future public investments in the rural sector into areas beyond agriculture or natural-resource-based interventions that was identified as a significant 'new' challenge for rural development in the opening sections of this chapter.

The notion of assets and capitals developed with the framework is proving useful for considering not just how people build their livelihoods materially, but also experientially. As Bebbington (1999: 2022–3) argues, assets are not simply resources that people use in building livelihoods, but they also give meaning to a person's world and give the capability to engage with others to change relationships and transactions; 'assets are thus as much implicated in empowerment and change, as they are in survival and "getting by"'.

While earlier 'farmer first' approaches could be suggested to have overemphasised indigenous skills and capabilities, the SLF very explicitly links these to the external context in which people operate and which influences livelihood outcomes. As Scoones (1996: 3) identified, 'the social and economic worlds that influence local-level decisions go well beyond the farm gate'. In addition, 'sustainability' within the framework aims to encompass consideration of the implications of livelihood strategies for future natural capital that is evidently a profound concern for (rural and urban) policy and practice worldwide, as considered in Chapters 6 and 9, but also provides space to identify whether such strategies enhance or compromise people's capabilities to change or renegotiate the various political, economic and social relationships that shape the way they interact with the environment – an essential feature for rural development interventions as considered in this chapter.

It seems as if considerable progress has been made in understanding and responding to the needs of rural households and communities. However, despite some success stories, the experience on the ground in terms of outcomes continues to lag behind the 'theory'. In particular, sub-Saharan Africa still has a long, steep hill to climb out of rural poverty. It is to be hoped that, in time, a greater use of participatory methodologies and the introduction of more appropriate rural development strategies will be translated into a genuine improvement in rural livelihoods. But the reality is still too often one of urban bias and political expediency, such that the rural poor continue to be well down the 'development agenda' in many developing countries.

Key points

- While the world's population is increasingly an urban one, the development of rural areas of the developing world remain key to understanding and resolving global challenges of poverty and resource conservation, for example.
- In recent years, there has been an identified de-linking of rural livelihoods (and poverty) from land and from agricultural production specifically that presents new challenges for rural development planning and practice.
- Rural livelihoods are diverse and dynamic and increasingly shaped by processes of change in wider regional, national and international arenas. They remain in large measure closely related to the natural resource functions and ecosystems of specific locales.
- Substantial investment has been made in both interventions to radically change access to productive assets such as land and to transform the way in which production is carried out towards rural development.
- Future rural development interventions need to be closely linked to the particular local context and circumstances and substantially developed through the participation and empowerment of the most marginal and poorest groups.

Further reading

Binns, T. (ed.) (1995) *People and Environment in Africa*. Chichester: Wiley – comprises a large number of short accounts of key resource and development issues organised on a regional basis.

Chambers, R. (1997) *Whose Reality Counts? Putting the First Last*. London: IT Publications – a thorough examination of how and why the rural poor are so often 'left out' of rural development.

Conway, G.R. (1997) *The Doubly Green Revolution: Food for all in the Twenty First Century*. London: Penguin – a readable book that explores the challenge of enhancing world food production but in ways that conserve natural resources and the environment.

Ellis, F. (2000) *Rural Livelihoods and Diversity in Developing Countries*. Oxford: OUP – explores the many factors underpinning the patterns and processes of livelihood diversification going on in rural areas of the developing world.

Mortimore, M. (1998) *Roots in the African Dust: Sustaining the Drylands*. Cambridge: Cambridge University Press – another important text based on long-term research particularly in West Africa that explores the challenges and opportunities of environmental change from the perspective of smallholder households.

Robinson, G. (2004) *Geographies of Agriculture: Globalisation, Restructuring and Sustainability*. Harlow: Pearson Education – a wide-ranging and thorough review of world agriculture and the processes of change underway.

Websites

www.livelihoods.org
Home page of the sustainable livelihoods programme hosted by the UK Department for International Development and the Institute of Development Studies, Sussex. Provides lots of practical guidance, distance learning materials and guidance sheets, for example towards promoting sustainable livelihoods to eliminate poverty.

www.id21.org
A reporting service funded by DfID that aims to bring UK-based development research findings and policy recommendations to policy makers and development practitioners worldwide. Provides easy access to one-page summaries of research highlights, for example.

www.iied.org
Site for the International Institute for Environment and Development which is an independent non-profit organisation promoting sustainable patterns of development. A good source for recent research in areas of pastoral development and land reform, for example, and of tools and methodologies towards more participatory development interventions.

www.fao.org
Site of the Food and Agriculture Organisation of the United Nations. Responsible for sharing policy experience, undertaking field projects, disseminating information and facilitating meetings of UN members. Site hosts substantial statistical databases.

Discussion topics

- Review the evidence given in the chapter to suggest that there are no 'blueprints' or 'standard packages' for rural development.
- Illustrate the suggestion (Mortimore, 1998) that there is nothing simple about survival in the drylands.
- Why is it hard for 'outsiders' to understand the realities of rural living in the developing world?
- In the Key idea box on page 475, the prospects of both the green and gene revolutions for enhancing world supply were considered. Widen your research to consider whether the problems of world hunger are due to issues of supply and food availability or factors of distribution and access to food.
- Select a particular developing country. What are the prospects for smallholder agriculture in this country? Find out as much as you can concerning its economy, its environmental conditions and its policies on agriculture, for example. Sources like the World Resources Institute, the Food and Agriculture Organisation and the World Bank are useful starting points. Recent academic research can also be sourced through the web-links above.

Conclusion

Development is undoubtedly a complex topic, but we hope that *Geographies of Development* has illuminated some of the key concepts, patterns and processes involved, as well as presenting case studies which illustrate and clarify particular issues. We have endeavoured to 'demystify' development, although in reaching this section of the book readers will be all too aware that development remains a highly contested topic and we encourage you to engage in further reading and debate concerning the different perspectives and questions raised in this text.

As we stand back from the material presented in *Geographies of Development*, we should perhaps ask ourselves once again, what is development all about? You will be aware by now that there are many possible answers to this question, but essentially, development is about a process of change leading hopefully to an improvement in the quality of life among individuals and groups living on this planet (see particularly Chapters 1 and 3). We have seen also how the processes of development can be examined at many different scales from the local to the regional, national, international and global.

This book has focused mainly on the world's poorest countries and peoples, but development can also be examined in the context of our own communities. Dudley Seers (1969, 1972), a pioneer in charting a clear direction for development studies, saw development as primarily about reducing poverty, inequality and unemployment, and fulfilling individual and community ambitions. Such key diagnostic features can be identified and evaluated in all communities since, as we can all too easily observe for ourselves, even the world's wealthiest countries, such as those in Western Europe and North America, have their share of poor and marginalised people, and in some countries there are staggering inequalities both between and among different groups and different regions. As Chapter 4 showed all too clearly, gross inequalities at the global scale have been rising progressively since the 1970s, and this global reality is increasingly being etched out at the national and regional scales, under the influence of post-1980s 'rampant' neo-liberalism. Development should begin at home, but, as we have argued in this book, it is essential that we understand more distant places where peoples and environments are often very different from our own, notwithstanding that globalisation is bringing us ever closer – especially for those who can afford new technologies and lifestyles.

Throughout *Geographies of Development* we have repeatedly alluded to the 'plural' nature of development, with different disciplines having their own distinctive focus and methodology for interpreting issues concerning development. Politics, international relations, economics, anthropology, sociology and history, among others, have particular contributions to make in understanding development patterns and processes.

Similarly, within geography, there is a longstanding interest in how people interact with environments, as well as a concern for appreciating development issues at different scales. Perhaps geography, more than any other social science discipline, is concerned with adopting an holistic approach to the study of development.

But the links between practitioners in these various disciplines are themselves complex, and development studies draws upon the different perspectives and methodologies offered by a wide range of social sciences, as well as other disciplines such as agricultural science, ecology and health science.

Many of the issues considered in development studies are truly global and transnational in reach, ranging from climate change to poverty, to global trade and aid and migration, to the transmission and impact of diseases such as malaria and HIV/AIDS (Chapters 5 and 8). But at the local level we may be just as interested in issues such as community-based development and the reduction of social and personal inequality within particular communities. All these issues can be considered in terms of the challenges and opportunities for more sustainable patterns and processes of development in the future (see particularly Chapters 6 and 7).

In studying development at different scales and in different locations, we need to ask ourselves whether development is actually occurring – is development working? This question was raised directly in Chapter 1, specifically in relation to the arguments of the 'anti-development' lobby. Further, in Part II of this book we considered 'development in practice' and examined various aspects concerning people, environments, resources, institutions and communities. In Chapter 7, the Millennium Development Goals were identified as one illustration of an emerging consensus on 'what development is about' and, for example, there have been recent moves towards identifying measurable targets for progress in development and for monitoring and accounting for international aid.

In some parts of the world the development record is lamentable. Many countries in sub-Saharan Africa, for example, are now relatively worse off than they were over 40 years ago when they gained independence in the 1960s. The high levels of poverty that characterise sub-Saharan Africa as a whole were emphasised in Chapter 1.

One of the world's poorest countries, Sierra Leone, had an average life expectancy of 44 years in 1974, yet over 30 years later it was still only 41 years. In Zimbabwe male life expectancy in 1980 was 56 years, but this is now below 40. In some southern African countries, HIV/AIDS has had a devastating impact on a range of development indicators including life expectancy. The reasons behind these seemingly 'no development' or 'regressive' scenarios are both complex and variable, through space and time.

The Western media (as well as some academic writing) seem to quite relentlessly portray negative images of Africa, inferring that people are the victims of the environment and, in turn, that the environment is a victim of human malpractice (Binns, 1997; Milner-Smith and Potter, 1995). This is a heavily distorted and oversimplified view of things, which fails to appreciate that despite widespread poverty, people within these countries who are actually experiencing such environmental change frequently have an impressive understanding of the characteristics and capabilities of their environments, as important research by scholars such as Ester Boserup (1965), Polly Hill (1963, 1970, 1972) and Robert Chambers (1983, 1993, 1997, 2005) has revealed. Some of the biases, constraints and inherently political factors that serve to sustain such images of African environments and people were considered very explicitly in Chapters 7 and 10. Chapter 7, for example, revealed that, despite substantial change in the way that the key institutions in development are working, external factors such as trade and international finance are still often seen as fundamental to overcoming poverty, rather than considered as part of the dynamic of creating poverty, thereby enabling 'blame' and 'responsibility' to be placed on governments and society of the South.

It is essential that all of us engaged in development theory and practice try to understand why specific development initiatives succeed and why others fail. It is certainly possible to identify common characteristics and patterns that underpin both successes and failures and it is important to learn from these. A number of principles for projects exhibiting greater sustainability, for example, were considered in Chapter 7, including engaging with and empowering local community organisations. Projects regularly fail for similar reasons, such as through inadequate inputs, a lack of human capacity, or through inadequate understanding of new market opportunities.

Good practice must be understood, but so too the context in which that success is achieved. For example, the often cited case of the Grameen Bank in Bangladesh, in providing much-needed credit for community-based development projects, was highlighted in Chapter 10. That 'Grameen model' has been extended and developed widely in other countries. However, it has also been shown that simply because something succeeds in one particular social and environmental context, it does not necessarily mean that it will succeed in a very different context. A 'blueprint approach' to development often fails because development initiatives are not carefully attuned to local environments, local cultures and the needs and aspirations of local people. Evidently, whilst it is important to understand the common factors and features that shape the nature of development in particular contexts, it is also essential to learn and change through the very processes of development research, planning and implementation – quite different challenges to a blueprint approach.

There is also a need to be aware of the structural–political argument that world patterns of development have been so unequal for so many centuries, that all 'development projects' and 'schemes' can be considered as attempts merely to paper over the geo-political realities of structural poverty and inequality. For those of a radical orientation, it is the global system that is wrong – or rather the lack of a global set of solutions to what are global problems that is needed (see Chapters 1 and 4). In Chapter 4, for example, we noted that some form of global taxation, such as a 'Tobin tax' could generate sums of finance large enough to pay for key Millennium Development Goals, such as the establishment of universal primary education.

In the decade that has elapsed since publication of the First Edition of *Geographies of Development*, some clear and measurable targets for development have emerged, such as those embodied in the Millennium Development Goals (MDGs) (see Chapters 1 and 7). The eight Goals, if they are to be achieved by the target date of 2015, could lead to very significant improvements in the quality of life of millions of people. The MDGs have received much publicity and considerable political support internationally. Critically, they identify the significance of global responsibilities and partnerships. They offer the potential to move beyond a view that developing countries need to be responsible for getting their 'own house in order' and the role of the developed countries is to support such efforts. Chapter 7, for example, highlighted the ongoing challenges for some of the key global institutions, as well as governments of developed countries, to further change the ways in which they operate and how they work with other organisations towards achieving these goals.

There is a consensus that achieving success in the MDGs will depend on improvements in governance locally within the developing world, as well as on scaling up commitments in the level of overseas development assistance (ODA) from the world's richer countries. But, as concluded in Chapter 1, there also remains serious concern (despite evident progress in some regions with many of the MDGs) as to the prospects for achieving the targets by 2015. As we have seen, the challenges of 'good governance' are multifaceted (Chapter 7) and encompass much more than overcoming corruption within government, for example. There remains also a long way to go before many countries achieve the 0.7 per cent of GNP target for ODA to developing countries that was set back in 1960 (Chapter 8). Once again, those on the political right and those on the political left take very different views as to the likelihood of policies working in the absence of major global reforms.

The context for development has changed a great deal in the last decade. For example, the pace of globalisation can be considered to have quickened with the progressive sophistication of information technology (Chapter 4). Powerful North American and European-based transnational companies continue to locate more of their activity in certain developing countries and very rapid processes of economic development are being seen in China and India specifically (Chapters 6 and 7). Meanwhile, in British and American supermarkets, produce from developing countries is available throughout the year, whether it is bananas from the Caribbean and Central America, green beans from Kenya, or orchids from Thailand. But the essential challenge remains the same: to continue to question critically the benefits and costs to particular groups of people and to places both distant and less further afield. A commitment to the geographies of development includes pursuing an understanding of what economic benefits from our own consumption falls to producers through such trade, and to a consideration of how different consumer, industry, governmental and multilateral responses to climate change may impact, for example, on the future long-distance transportation of such commodities (Chapters 1, 4, 6 and 7).

It is salutary to remember that despite the extension of more sophisticated communications technologies and the increasing connectivity between developed and developing countries, there are still millions of people in the world's poorest regions who are without a supply of fresh drinking water, let alone the electricity to power a computer (Chapters 4, 7 and 9). Meanwhile, satellite-based mobile phone technologies are having a significant impact in many developing countries, where land-based communications systems are often in a parlous state.

Another important change in the context of development seen over the last decade has been the emergence of the so-called BRIC countries (Brazil, Russia, India and China), including their rapid economic successes and current power to shape resource use and new trade linkages. As we saw in Chapters 6 and 8, China's trade with the African continent has burgeoned in the last decade and very considerable amounts of Chinese investment are now going into resource extraction and construction projects in many countries within Africa that are raising new questions regarding the prospects for economic stability, sovereign control, sustainability and democracy.

We firmly believe that geography has an important and distinctive role to play in understanding development, and we trust that this Third Edition of *Geographies of Development* provides good evidence of this. There has been no shortage of academic debate concerning the commitment to understanding development issues, and Potter (1993, 2001), Gilbert (1987) and others, for example, have criticised their fellow geographers for a lack of attention towards developing countries (Binns, 2007). We strongly suspect that there are similar ongoing debates in other social science disciplines, where, as in Geography, there is significantly more research and publishing that focuses on developed, rather than developing, countries. But we would strongly agree with Smith (1994: 366) that there is a genuine moral and professional motivation for being concerned about improving the quality of life for the world's poorest and most marginalised people; that 'We owe distant others far more than we give them'.

We also believe that a greater understanding of the world and its peoples is an essential endeavour. As Johnston succinctly puts it:

> **World understanding is fundamental to world peace and ultimately to world survival. Ignorance leads to the development of stereotypes, negative reactions to other peoples and cultures which breed hostility. Geography must be used to break down those barriers of ignorance.**
>
> **(Johnston, 1984: 458)**

We hope that *Geographies of Development* can assist in this evidently necessary reduction in stereotypical, negative and pejorative thinking about other peoples, cultures and lands.

Bibliography

Actionaid (1994) *Kyuso Rural Development Area Plan and Budget*. Nairobi: Actionaid

Actionaid (1995) *Listening to Smaller Voices: Children in an Environment of Change*. Chard: Actionaid

Actionaid International (2004) *Money Talks: How Aid Conditions Continue to Drive Utility Privatisation in Poor Countries*. London: Actionaid International

Adams, W.M. (1985) The downstream impacts of dam construction: a case study from Nigeria. *Transactions of the Institute of British Geographers*, 10, 292–302

Adams, W.M. (1990) *Green Development: Environment and Sustainability in the Third World*. London: Routledge

Adams, W.M. (1996) Irrigation, erosion and famine: visions of environmental change in Marakwet, Kenya, in Leach, M. and Mearns, R. (eds) *The Lie of the Land: Challenging Received Wisdom on the African Environment*. Oxford: International African Institute

Adams, W.M. (2001) *Green Development: Environment and Sustainability in the Third World*, 2nd edn. London: Routledge

Adams, W.M. and Anderson, D.M. (1988) Irrigation before development: indigenous and induced change in agricultural water management in East Africa. *African Affairs*, 87, 519–35

Adger, W.N., Huq, S., Brown, K., Conway, D. and Hulme, M. (2003) Adaptation to climate change in the developing world. *Progress in Development Studies*, 3(3), 179–95

Aeroe, A. (1992) The role of small towns in regional development in Southeast Africa, in Baker, J. and Pedersen, P.O. (eds) *The Rural–Urban Interface in Africa*. Uppsala: Nordic Institute for African Studies, 51–65

African Union (2002) Transition from the OAU to the African Union, www.au2002.gov.za/docs/background/oau_to_au.htm

Ageing and Development (2002) News and analysis of issues affecting the lives of older people, www.helpage.org/publications

Agrawal, A. and Gibson, C.C. (1999) Enchantment and disenchantment: the role of community in natural resource conservation. *World Development*, 27(4), 629–49

Ahmend, K. (2002) British arms sales to Africa soar. *The Observer*, 3 February

Aikman, D. (1986) *Pacific Rim: Area of Change, Area of Opportunity*. Boston, MA: Little Brown

Ainger, K. (2004) The scramble for Africa. *New Internationalist*, 367, 8–12

Airey, A., Binns, T. and Mitchell, P.K. (1979) To integrate or . . . ? Agricultural development in Sierra Leone. *IDS Bulletin*, 10(4), 20–7

Allen, C. (1979) *Tales from the Dark Continent*. London: BBC/Andre Deutsch

Allen, C. (1999) Warfare, endemic violence and state collapse in Africa. *Review of African Political Economy*, 81, 367–84

Allen, J. (1995) Global worlds, in Allen, J. and Massey, D. (eds) *Geographical Worlds*. Oxford: Oxford University Press and Open University, 105–44

Allen, J. and Hamnett, C. (1995) Uneven worlds, in Allen, J. and Hamnett, C. (eds) *A Shrinking World*. Oxford: Oxford University Press, 233–54

Allen, T. and Thomas, A. (eds) (1992) *Poverty and Development in the 1990s*. Oxford: Oxford University Press

Allen, T. and Thomas, A. (eds) (2000) *Poverty and Development into the 21st Century*. Oxford: Oxford University Press

Alonso, W. (1968) Urban and regional imbalances in economic development. *Economic Development and Cultural Change*, 17, 1–14

Alonso, W. (1971) The economics of urban size. *Papers of the Regional Science Association*, 26, 67–83

Amnesty International (2002a) Children devastated by war: Afghanistan's lost generations, www.amnesty.org

Amnesty International (2002b) Sierra Leone: childhood – a casualty of conflict, www.amnesty.org

AND Cartographic Publishers (1997, 1999) *Political Atlas of the World*. Abingdon, Oxfordshire: Helicon Publishing

Anderson, A. (ed.) (1990) *Alternatives to Deforestation: Steps Towards Sustainable Use of the Amazon Rainforest*. New York: Columbia University Press

Apter, D. (1987) *Rethinking Development: Modernization, Dependency and Postmodern Politics*. Newbury Park, CA: Sage

Armstrong, W. and McGee, T.G. (1985) *Theatres of Accumulation: Studies in Asian and Latin American Urbanization*. London: Methuen

Arndt, C. and Lewis, J.D. (2000) *The Macro Implications of HIV/AIDS in South Africa: A Preliminary Assessment*. Muldersdrift: Trade and Industrial Policy Secretariat

Arnwell, N.W., Livermore, M.J.L., Kovats, S., Levey, P.E., Nicholls, R., Parry, M.J. and Gaffin, S.R. (2002) Climate and socio-economic scenarios for global-scale climate change impact assessments: characterising the SRES storylines. *Global Environmental Change*, 14, 3–20

Ashcroft, B., Griffiths, G. and Tiffin, H. (1998) *Key Concepts in Post-Colonial Studies*. London: Routledge

Ashley, C. and Roe, D. (2002) Making tourism work for the poor: strategies and challenges in southern Africa. *Development Southern Africa*, 19, 61–82

Atkins, P.J. and Bowler, I.R. (2001) *Food and society: Economy, Culture, Geography*. London: Arnold

Augelli, J.P. and West, R.C. (1976) *Middle America: Its Land and Peoples*. Englewood Cliffs, NJ: Prentice Hall

Auret, D. (1995) *Urban Housing: A National Crisis*. Gweru: Mambo Press

Austin-Broos, D.J. (1995) Gay nights and Kingston Town: representations of Kingston, Jamaica, in Watson, S. and Gibson, K. (eds) *Postmodern Cities and Spaces*. Oxford: Blackwell, 149–64

Auty, R. (1979) World within worlds. *Area*, 11, 232–35

Auty, R. (1993) *Sustaining Development in Mineral Economies: The Resource-Curse Thesis*. London: Routledge

Auty, R. (1994) *Patterns of Development*. London: Methuen

Azam, J.P. and Gubert, F. (2006) Migrants' remittances and the household in Africa: a review of evidence. *Journal of African Economies*, 15, AERC Supplement 2, 426–62

Baez, A.L. (1996) Learning from experience in the Monteverde Cloud Forest, Costa Rica, in Price, M.F. (ed.) *People and Tourism in Fragile Environments*. Chichester: John Wiley, 109–22

Bairoch, P. (1975) *The Economic Development of the Third World Since 1900*. London: Methuen

Baker, J. and Pedersen, P.O. (eds) (1992) *The Rural–Urban Interface in Africa*. Uppsala: Nordic Institute for African Studies

Baran, P. (1957) *Political Economy of Growth*. Monthly Review Press: New York

Baran, P. (1973) *The Political Economy of Growth*. Harmondsworth: Penguin

Baran, P. and Sweezy, P. (1968) *Monopoly Capitalism*. Harmondsworth: Penguin

Baran, P. and Sweezy, P. (1998) *Monopoly Capital*. Harmondsworth: Penguin

Barff, R. and Austen, J. (1993) 'It's gotta be da shoes': domestic manufacturing, international subcontracting, and the production of athletic footwear. *Environment and Planning A*, 25, 1103–14

Barke, M. and O'Hare, G. (1991) *The Third World*, 2nd edn. Harlow: Oliver & Boyd

Barnett, T. (2002a) HIV/AIDS impact studies II: some progress evident. *Progress in Development Studies*, 2, 219–25

Barnett, T. (2002b) The social and economic impacts of HIV/AIDS on development, ch. 8.3 in Desai, V. and Potter, R.B. (eds) *The Companion to Development Studies*. London: Arnold, 391–5

Barnett, T., Whiteside, A. and Desmond, D. (2001) The social impact of HIV/AIDS in poor countries: a review of studies and lessons. *Progress in Development Studies*, 1, 151–70

Barratt-Brown, M. (1974) *The Economics of Imperialism*. Harmondsworth: Penguin

Barrett, H. and Browne, A. (1995) Gender, environment and development in Sub-Saharan Africa, in Binns, T. (ed.) *People and Environment in Africa*. Chichester: John Wiley, 31–8

Barrett, H.R., Binns, T., Browne, A.W., Ilbery, B.W. and Jackson, G.H. (1997) Prospects for horticultural exports under trade liberalisation in adjusting African economies. Unpublished report to the Overseas Development Administration, London

Barrett, H.R., Browne, A.W., Ilbery, B.W. and Binns, T. (1999) Globalisation and the changing networks of food supply: the importation of fresh horticultural produce from Kenya into the UK.

Transactions of the Institute of British Geographers, NS 24(2), 159–74

Barrow, C. (1987) *Water Resources and Agricultural Development in the Tropics*. London: Longman

Barrow, C.J. (1995) *Developing the Environment: Problems and Management*. London: Longman

Bartone, C. *et al.* (1994) *Towards Environmental Strategies for Cities*, Urban Management Policy Paper 18, 'Strategic Options for Managing the Urban Environment'. Washington, DC: World Bank

Bassett, T.J. (1993) The land question and agricultural transformation in sub-Saharan Africa, in Bassett, T.J. and Crummey, D.E. (eds) *Land in African Agrarian Systems*. Madison, WI: University of Wisconsin Press, 3–34

Batley, R. (2002) The changing role of the state in development, ch. 2.16 in Desai, V. and Potter, R.B. (eds) *The Companion to Development Studies*. London: Arnold, 135–9

Batterbury, S. and Warren, A. (2001) The African Sahel 25 years after the great drought: assessing progress and moving towards new agendas and approaches. *Global Environmental Change*, 11, 1–8

Bauer, P.T. (1975) Western guilt and Third World poverty. *Quadrant*, 20(4), 13–22

Bauer, P.T. (1976) *Dissent on Development*. London: Weidenfeld & Nicolson

Bebbington, A. (1999) Capitals and capabilities: a framework for analysing peasant viability, rural livelihoods and poverty. *World Development*, 27(12), 2021–44

Becker, C.M. and Morrison, A.R. (1997) Public policy and rural–urban migration, in Gugler, J. (ed.) *Cities in the Developing World*. Oxford: Oxford University Press, 88–105

Beckford, G. (1972) *Persistent Poverty: Underdevelopment in Plantation Economies of the Third World*. New York: Oxford University Press

Bek, D., Binns, T., Nel, E. and Ellison, B. (2006) Achieving grassroots transformation in post-apartheid South Africa. *International Journal of Development Issues*, 5(2), 65–94

Bell, M. (1980) Imperialism: an introduction, in Peet, R. (ed.) *An Introduction to Marxist Theories of Underdevelopment*, Monograph HG14, RSPACS. Canberra: Australian National University, 39–50

Bernstein, H. (1992a) Agrarian structures and change: India, in Bernstein, H., Crow, B. and Johnson, H. (eds) *Rural Livelihoods: Crises and Responses*. Oxford: Oxford University Press, 51–64

Bernstein, H. (1992b) Agrarian structures and change: Latin America, in Bernstein, H., Crow, B. and Johnson, H. (eds) *Rural Livelihoods: Crises and Responses*. Oxford: Oxford University Press, 27–50

Bernstein, H. (1992c) Poverty and the poor, in Bernstein, H., Crow, B. and Johnson, H. (eds) *Rural Livelihoods: Crises and Responses*. Oxford: Oxford University Press, 13–26

Bernstein, H., Crow, B. and Johnson, H. (eds) (1992) *Rural Livelihoods: Crises and Responses*. Oxford: Oxford University Press

Berry, B.J.L. (1961) City size distributions and economic development. *Economic Development and Cultural Change*, 9, 573–87

Berry, B.J.L. (1972) Hierarchical diffusion: the basis of development filtering and spread in a system of growth centres, in Hansen, N.M. (ed.) *Growth Centres in Regional Economic Development*. New York: Free Press

Bhabha, H. (1994) *The Location of Culture*. London: Routledge

Bigg, T. (ed.) (2004) *Survival for a Small Planet: The Sustainable Development Agenda*. London: Earthscan/IIED

Binns, T. (1992) Traditional agriculture, pastoralism and fishing, in Gleave, M.B. (ed.) *Tropical African Development*. London: Longman, 153–91

Binns, T. (1994a) *Tropical Africa*. London: Routledge

Binns, T. (1994b) Ghana: West Africa's latest success story? *Teaching Geography*, 19(4), 147–53

Binns, T. (1995a) Geography in development: development in geography. *Geography*, 80(4), 303–22

Binns, T. (ed.) (1995b) *People and Environment in Africa*. Chichester: John Wiley

Binns, T. (1997) People, environment and development in Africa. *South African Geographical Journal*, 79(1), 13–18

Binns, T. (2007) Marginal lands, marginal geographies. *Progress in Human Geography*, 31(5), 587–91

Binns, T. and Fereday, N. (1996) Feeding Africa's urban poor: urban and peri-urban horticulture in Kano, Nigeria. *Geography*, 81, 380–4

Binns, T. and Funnell, D.C. (1983) Geography and integrated rural development. *Geografiska Annaler B*, 65(1), 57–63

Binns, T. and Funnell, D.C. (1989) Irrigation and rural development in Morocco. *Land Use Policy*, 6(1), 43–52

Binns, T. and Lynch, K. (1998) Feeding Africa's growing cities into the 21st century: the potential of urban agriculture. *Journal of International Development*, 10, 777–93

Binns, T. and Maconachie, R. (2006) Post-conflict reconstruction and sustainable development: diamonds, agriculture and rural livelihoods in Sierra Leone. *International Journal of Cultural, Economic and Social Sustainability*, 2, 205–16

Binns, T. and Mortimore, M. (1989) Ecology, time and development in Kano State, Nigeria, in Swindell, K., Baba, J.M. and Mortimore, M.J. (eds) *Inequality and Development: Case Studies from the Third World*. London: Macmillan, 359–80

Binns, T. and Nel, E.L. (2002) Tourism as a local development strategy in South Africa. *Geographical Journal*, 168(3), 235–47

Binns, T. and Robinson, R. (2002) Sustaining democracy in the 'new' South Africa. *Geography*, 87(1), 25–37

Biswas, A.K. (1992) Water for Third World development. *Water Resources Development*, 8(1), 3–9

Biswas, A.K. (1993) Management of international waters. *International Journal of Water Resources Development*, 9(2), 167–89

Biswas, A.K. (2004) From Mar del Plata to Kyoto: an analysis of global water policy dialogue. *Global Environmental Change*, 14, 81–8

Biswas, A.K. and Biswas, A. (1985) The global environment. *Resources Policy*, 11(1), 25–42

Black, R. (1996) Refugees and environmental change: the case of the forest region of Guinea. Unpublished Project CFCE Report No. 2, University of Sussex, Brighton

Black, R. (1997) Refugees, land cover, and environmental change in the Senegal River Valley. *Geojournal*, 41(1), 55–67

Black, R. and White, H. (eds) (2004) *Targeting Development: Critical Perspectives on the Millennium Development Goals*. Abingdon, London and New York: Routledge

Blaikie, P. (1985) *The Political Economy of Soil Erosion in Developing Countries*. London: Longman

Blaikie, P. (2000) Development, post-, anti-, and populist: a critical review. *Environment and Planning A*, 32, 1033–50

Blaikie, P. (2002) Vulnerability and disasters, in Desai, V. and Potter, R.B. (eds) *The Companion to Development Studies*. London: Arnold, 298–305

Blaikie, P. and Brookfield, H. (eds) (1987) *Land Degradation and Society*. London: Methuen

Blaut, J. (1993) *The Colonizers' Model of the World*. London: Guildford

Blouet, B.W. and Blouet, O.M. (2002) *Latin America and the Caribbean: A Systematic and Regional Survey*. New York: John Wiley

Blunt, A. (1994) *Travel, Gender and Imperialism: Mary Kingsley and West Africa*. New York: Guilford

Blunt, A. and McEwan, C. (eds) (2002) *Postcolonial Geographies*. London: Continuum

Blunt, A. and Wills, J. (2000) *Dissident Geographies*. London: Prentice Hall

Bogert, C. (1995) Midlife crisis. *Newsweek*, 30 October, 12–18

Bongaarts, J. (1994) Demographic transition, in Eblen, R.A. and Eblen, W.R. (eds) *Encyclopedia of the Environment*. Boston MA: Houghton-Mifflin, 132

Bongaarts, J. (1995) Global and regional population projections to 2025, in Islam, N. (ed.) *Population and Food in the Early Twenty-First Century: Meeting Future Food Demand of an Increasing Population*. Washington, DC: International Food Policy Research Institute, 7–16

Booth, D. (1985) Marxism and development sociology: interpreting the impasse. *World Development*, 13, 761–87

Booth, D. (1993) Development research: from impasse to new agenda, in Schurmann, F. (ed.) *Beyond the Impasse: New Directions in Development Theory*. London: Zed Books

Booth, K. (1997) Exporting ethics in place of arms. *The Times Higher Education Supplement*, 7 November, 118

Borchert, J.R. (1967) American metropolitan evolution. *Geographical Review*, 57, 301–23

Boserup, E. (1965) *The Conditions of Agricultural Growth: The Economics of Agricultural Change Under Population Pressure*. London: Allen & Unwin

Boserup, E. (1993) *The Conditions of Agricultural Growth*. London: Earthscan (first published in 1965)

Botes, L.J. (1996) Promoting community participation in development initiatives. Paper presented to the Development Studies Association Annual Conference, University of Reading

Bourdieu, F. (1998) The essence of neoliberalism. *Le Monde Diplomatique*, December 1998

Boyden, J. and Holden, P. (1991) *Children of the Cities*. London: Zed Books

Brac de la Perriere, R.A. and Seuret, F. (2000) *Brave New Seeds: The Threat of GM Crops to Farmers*. London: Zed Books

Brandt, W. (1980) *North–South: A Programme For Survival*. London: Pan

Brandt, W. (1983) *Common Crisis. North–South: Co-operation for World Recovery*. London: Pan

Branford, S. (2002) Sow resistant. *The Guardian*, 17 April

Brazier, C. (1994) Winds of change. *New Internationalist*, 262, 4–7

Brierley, J. (1989) A review of development strategies and programmes of the People's Revolutionary Government in Grenada, 1979–83. *Geographical Journal*, 151, 40–52

Brierley, J.S. (1985a) Idle land in Grenada: a review of its causes and the PRG's approach to reducing the problem. *Canadian Geographer*, 29, 298–309

Brierley, J.S. (1985b) The agricultural strategies and programmes of the People's Revolutionary Government in Grenada, 1979–1983, in *Conference of Latin American Geographers Yearbook*, 55–61

Brimblecombe, P. (2000) Urban air pollution and public health, in O'Riordan, T. (ed.) *Environmental Science for Environmental Management*. Harlow: Pearson Education, 399–416

Brohman, J. (1996) *Popular Development: Rethinking the Theory and Practice of Development*. Oxford: Blackwell

Brookfield, H. (1975) *Interdependent Development*. London: Methuen

Brookfield, H. (1978) Third World Development. *Progress in Human Geography*, 2(1), 121–32

Brookfield, H. and Stocking, M. (1999) Agrodiversity: definition, description and design. *Global Environmental Change*, 9, 77–80

Brown, D.L. and Fox, J. (2001) Transnational civil society coalitions and the World Bank: lessons from project and policy influence campaigns, in Edwards, M. and Gaventa, J. (eds) *Global Citizen Action*. London: Earthscan, 43–58

Brown, L.R. (1996a) *The Potential Impact of AIDS on Population and Economic Growth Rates*, Food, Agriculture and the Environment, Discussion Paper 15. Washington DC: International Food Policy Research Institute

Brown, L.R. (ed.) (1996b) *Vital Signs, 1996/1997: The Trends That are Shaping our Future*. London: Earthscan

Browne, A.W. and Barrett, H.R. (1995) Children and AIDS in Africa, African Studies Centre Paper 2. Coventry: Coventry University

Bruce, J.W. (1998) *Country Profiles of Tenure: Africa, 1996*. Land Tenure Center, University of Wisconsin-Madison

Brundtland Commission (1987) *Our Common Future*. Oxford: Oxford University Press

Bruno, K. (2005) Bluewash. *New Internationalist*, 375, 26–7

Bryant, R.L. and Bailey, S. (1997) *Third World Political Ecology*. London: Routledge

Bryceson, D.F. (2002) The scramble in Africa: reorienting rural livelihoods. *World Development*, 30(5), 725–39

Buchanan, K. (1964) Profiles of the Third World. *Pacific Viewpoint*, 5(2), 97–126

Buckley, R. (1994) *NAFTA and GATT: The Impact of Free Trade*, Understanding global issues series, 94/2. Cheltenham: Understanding Global Issues Ltd.

Buckley, R. (1995) *The United Nations: Overseeing the New World Order*, Understanding global issues series 93/6. Cheltenham: Understanding Global Issues Ltd.

Buckley, R. (ed.) (1996) *Fairer Global Trade: The Challenge for the WTO*, Understanding global issues series 96/6. Cheltenham: Understanding Global Issues Ltd.

Budds, J. and McGranahan, G. (2003) Are the debates on water privatisation missing the point? *Environment and Urbanisation*, 15(2), 87–113

Burgess, R. (1990) The state and self-help building in Pereira, Colombia. Unpublished PhD thesis, University of London

Burgess, R. (1992) Helping some to help themselves: Third World housing policies and development strategies, in Mathéy, K. (ed.) *Beyond Self-Help Housing*. London: Mansell, 75–91

Burgess, R., Carmona, K. and Kolstree, T.C. (1997) *The Challenge of Sustainable Cities*. London: Zed Books

Burns, J.P. (1999) The Hong Kong civil service in transition. *Journal of Contemporary China*, 8(20), 67–87

Bury, J. (2001) Corporations and capitals: a framework for evaluating the impacts of transnational corporations in developing countries. *Journal of Corporate Citizenship*, 1(1), 75–91

Buvinic, M. (1993) The feminisation of poverty? Research and policy needs, Paper presented at *ILS Symposium on Poverty: New Approaches to Analysis and Policy*, Geneva, 22–24 November

Campbell, D.J., Zinyama, L.M. and Matiza, T. (1991) Coping with food deficits in rural Zimbabwe: the sequential adoption of indigenous strategies. *Research in Rural Sociology and Development*, 5, 73–85

Carbon Trust (2006) *The Carbon Trust Three Stage Approach to Developing a Robust Offsetting Strategy*. London: Carbon Trust, www.carbontrust.co.uk/publications

Cardoso, F.H. (1969) *Dependency and Development in Latin America*. Los Angeles: University of California Press

Carney, D., Drinkwater, M., Rusinow, T., Neefjes, K., Wanmali, S. and Singh, N. (1999) *Livelihood Approaches Compared*. London: Department for International Development

Cardoso, F.H. (1976) The consumption of dependency theory in the United States, in *Proceedings of the Third Scandinavian Research Conference on Latin America*, Bergen

Cashin, P.C., Liang, H. and McDermott, C.J. (1999) Do commodity price shocks last too long for stabilization schemes to work? *Finance and Development*, 36(3), 40–43

Castells, M. (1977) *The Urban Question: A Marxist Approach*. London: Edward Arnold

Castells, M. (1978) Urban social movements and the struggle for democracy. *International Journal of Urban and Regional Research*, 1, 133–46

Castells, M. (1983) *The City and the Grassroots*. London: Edward Arnold

Castells, M. (1996) *The Rise of the Network Society*. Oxford: Blackwell

Cater, E. (1992) Must tourism destroy its resource base? in Mannion, A.M. and Bowlby, S.R. (eds) *Environmental Issues in the 1990s*. London: John Wiley, 309–24

Cavendish, W. (2000) Empirical regularities in the poverty–environment relationship of rural households: evidence from Zimbabwe. *World Development*, 28(11), 1979–2000

Chambers, R. (1983) *Rural Development: Putting the Last First*. London: Longman

Chambers, R. (1993) *Challenging the Professions: Frontiers for Rural Development*. London: Intermediate Technology Publications

Chambers, R. (1997) *Whose Reality Counts*? London: Intermediate Technology Publications

Chambers, R. (2005) *Ideas for Development*. London: Earthscan

Chambers, R. and Conway, G. (1992) Sustainable rural livelihoods: practical concepts for the twenty-first century. *IDS Discussion Paper 296*. Brighton: IDS

Chambers, R., Pacey, A. and Thrupp, L.A. (1989) *Farmer First*. London: Intermediate Technology Publications

Chandra, R. (1992) *Industrialization and Development in the Third World*. London and New York: Routledge

Chant, S. (1996) *Gender, Uneven Development and Housing*. New York: UNDP

Chatterjee, P. (1994) Riders of the apocalypse. *New Internationalist*, 262, 10–11

Choguill, C. (1994) Crisis, chaos, crunch: planning for urban growth in the developing world. *Urban Studies*, 31, 935–45

Christaller, W. (1933) Die zentralen Onte in Suddeutschland. Doctoral thesis translated by Baskin, C.W. (1966) *Central Places in Southern Germany*. Englewood Cliffs, NJ: Prentice Hall

Christanty, L. (1986) Traditional agroforestry in West Java: the *pekarangan* (home garden) and *kebuntalun* (annual perennial rotation) cropping systems, in Marten, G.G. (ed.) *Traditional Agroforestry in Southeast Asia: A Human Ecology Perspective*. Boulder, CO: Westview, 132–58

Christian Aid (1996) *The Global Supermarket*. London: Christian Aid

Christian Aid (2004) *Fuelling Poverty*. London: Christian Aid

Chuta, E. and Liedholm, C. (1990) Rural small-scale industry: empirical evidence and policy issues, in Eicher, C.K. and Staatz, J.M. (eds) *Agricultural Development in the Third World*, 2nd edn. Baltimore, MD: Johns Hopkins University Press, 327–41

Clapham, C. (1985) *Third World Politics*. London: Croom Helm

Clark, D. (1996) *Urban World–Global City*. London: Routledge

Clark, D. (2006) *The Elgar Companion to Development Studies*. Cheltenham: Edward Elgar

Clarke, C. (2002) The Latin American structuralists, ch. 2.7 in Desai, V. and Potter, R.B. (eds) *The*

Companion to Development Studies. London: Arnold, 92–6

Clarke, K. (2001) ICT: What does it all mean? *Developments (DfID)*, 16, 5–9

Clayton, A. and Potter, R.B. (1996) Industrial development and foreign direct investment in Barbados. *Geography*, 81, 176–80

Clayton, K. (1995) The threat of global warming, in O'Riordan, T. (ed.) *Environmental Science for Environmental Management*. London: Longman, 110–31

Cliff, A.D. and Smallman-Raynor, M.R. (1992) The AIDS pandemic: global geographical patterns and local spatial processes. *Geographical Journal*, 158(2), 182–98

Clifford, M. (1994) Social engineers. *Far Eastern Economic Review*, 14 April, 56–58

Cline-Cole, R. (1996) Dryland forestry: manufacturing forests and farming trees in Nigeria, in Leach, M. and Mearns, R. (eds) *The Lie of the Land*. Oxford: James Currey

Clinton, B. (2001) The struggle for the soul of the 21st century, The Richard Dimbleby Lecture 2001, www.bbc.co.uk/arts/news-comment/dimbleby

Cochrane, A. (1995) Global worlds and worlds of difference, in Anderson, J., Brook, C. and Cochrane, A. (eds) *A Global World?* Oxford: Oxford University Press and Open University, 249–80

Colchester, M. (1991) Guatemala: the clamour for land and the fate of the forests. *The Ecologist*, 21(4), 177–85

Colchester, M. and Lohmann, L. (eds) (1993) *The Struggle for Land and the Fate of the Forests*. London: Zed Books

Commodityexpert (1999) World coffee production estimates detailed, 5 August 1999, www.commodityexpert.com/Archive/Analysis/99 08 05wrdprod2.htm

Concise Oxford Dictionary (1999) Oxford: Oxford University Press

Conroy, C. and Litvinoff, M. (1988) *The Greening of Aid: Sustainable Livelihoods in Practice*. London: Earthscan

Conway, D. and Heynen, N. (2002) Classical dependency theories: from ECLA to Andre Gunder Frank, ch. 2.8 in Desai, V. and Potter, R.B. (2002) *The Companion to Development Studies*, London: Arnold

Conway, D. and Heynen, N. (2006) *Globalization's Contradictions: Geographies of Discipline, Destruction and Transformation*. London and New York: Routledge

Conway, D. and Potter, R.B. (2007) Caribbean transnational return migrants as agents of change. *Geography Compass*, I, 25–45

Conway, G. (1997) *The Doubly Green Revolution: Food for all in the 21st Century*. London: Penguin

Conway, G. and Barbier, E. (1995) Pricing policy and sustainability in Indonesia, in Kirkby, J., O'Keefe, P. and Timberlake, L. (eds) *The Earthscan Reader in Sustainable Development*. London: Earthscan, 151–7

Conyers, D. (1982) *An Introduction to Social Planning in the Third World*. Chichester: Wiley

Cooke, P. (1990) Modern urban theory in question. *Transactions of the Institute of British Geographers, New Series*, 15, 331–43

Cooke, B. and Kothari, U. (eds) (2001) *Participation: The New Tyranny?* London: Zed Books

Corbridge, S. (1986) *Capitalist World Development*. London: Macmillan

Corbridge, S. (1992) Third World development. *Progress in Human Geography*, 16(54), 584–95

Corbridge, S. (1993a) Colonialism, post-colonialism and the Third World, in Taylor, P. (ed.) *Political Geography of the Twentieth Century*. London: Belhaven, 173–205

Corbridge, S. (1993b) Marxisms, modernities and moralities: development praxis and the claims of distant strangers. *Environment and Planning D*, 11, 449–72

Corbridge, S. (ed.) (1995) *Development Studies: A Reader*. London: Edward Arnold

Corbridge, S. (1997) Beneath the pavement only soil: the poverty of post-development. *Journal of Development Studies*, 33, 138–48

Corbridge, S. (2002a) Development as freedom: the spaces of Amartya Sen. *Progress in Development Studies*, 2, 183–217

Corbridge, S. (2002b) Third World debt, in Desai, V. and Potter, R.B. (eds) *The Companion to Development Studies*. London: Arnold, 477–80

Cornia, G.A., Jolly, R. and Stewart, F. (eds) (1987) *Adjustment with a Human Face: Vol. 1, Protecting the Vulnerable and Promoting Growth*. Oxford: Clarendon

Courtenay, P.P. (ed.) (1994) *Geography and Development*. Melbourne: Longman Cheshire

Cowen, M.P. and Shenton, R. (1995) The invention of development, in Crush, J. (ed.) *Power of Development*. London: Routledge, 27–43

Cowen, M.P. and Shenton, R.W. (1996) *Doctrines of Development*. London: Routledge

Craig, G. and Mayo, M. (eds) (1995) *Community empowerment: A Reader in Participation and Development*. London: Zed Books

Craig, D. and Porter, D. (2006) *Development beyond Neo-liberalism? Governance, Poverty Reduction and Political Economy*. London: Routledge

Crang, P. (2000) Worlds of consumption, ch. 4 in Daniels, P., Bradshaw, M., Shaw, D. and Sidaway, J. (eds) *Human Geography: Issues for the Twenty-First Century*. London and New York: Prentice Hall, 399–426

Crehan, K. (1992) Rural households: making a living, in Bernstein, H., Crow, B. and Johnson, H. (eds) *Rural Livelihoods: Crises and Responses*. Oxford: Oxford University Press, 87–112

Crook, C. (1991) Two pillars of wisdom. *The Economist*, 12 October, 3–4

Crook, N. (1997) *Principles of Population and Development*. Oxford: Oxford University Press

Crow, B. (1992) Rural livelihoods: action from above, in Bernstein, H., Crow, B. and Johnson, H. (eds) *Rural Livelihoods: Crises and Responses*. Oxford: Oxford University Press, 251–74

Crush, J. (1995a) Imagining development, in Crush, J. (ed.) *Power of Development*. London: Routledge, 1–26

Crush, J. (ed.) (1995b) *Power of Development*. London: Routledge

Curtis, M. (2001a) What's wrong with international trade rules? *Christian Aid News*, 14(Autumn), 12–13

Curtis, M. (2001b) *Trade for Life: Making Trade Work for Poor People*. London: Christian Aid

Cuthbert, A. (1995) Under the volcano: postmodern space in Hong Kong, in Watson, S. and Gibson, K. (eds) *Postmodern Cities and Space*. Oxford: Blackwell, 138–48

Dandekar, H.C. (1997) Changing migration strategies in Deccan Maharashtra, India, 1885–1990, in Gugler, J. (ed.) *Cities in the Developing World*. Oxford: Oxford University Press, 48–61

Daniel, M.L. (2000) The demographic impact of HIV/AIDS in sub-Saharan Africa. *Geography*, 85(1), 46–55

Dankelman, I. and Davidson, J. (1988) *Women and Environment in the Third World: Alliance for the Future*. London: Earthscan

Dann, G. and Potter, R.B. (2001) Supplanting the planters: new plantations for old in Barbados. *International Journal of Tourism and Hospitality Research*, 2, 51–84

Dann, G.M.S. and Potter, R.B. (1994) Tourism and postmodernity in a Caribbean setting. *Cahiers du Tourisme, Series C*, 185, 1–45

Dann, G.M.S. and Potter, R.B. (1997) Tourism in Barbados: rejuvenation or decline?, in Lockhart, D.G. and Drakakis-Smith, D. (eds) *Island Tourism: Trends and Prospects*. London: Mansell, 205–28

Datta, G. and Meerman, J. (1980) *Household Income and Household Income Per Capita in Welfare Comparisons*, World Bank Staff Working Paper 378. Washington DC: World Bank

Davin, D. (1996) Migration and rural women in China: a look at the gendered impact of large-scale migration. *Journal of International Development*, 8(5), 655–65

de Albuquerque, K. (1996) Computer technologies and the Caribbean. *Caribbean Week*, 8, 32–33

Debray, R. (1974) *A Critique of Arms*. Paris: Seuil

DEFRA (UK Government) (2005) The validity of food miles as an indicator of sustainable development, statistics.defra.gov.uk/esg/reports/foodmiles/execsumm.pdf

de Janvry, A. (1984) The role of land reform in economic development, in Eicher, C. and Staatz, J.M. (eds) *Agricultural Development in the Third World*. Baltimore MD: Johns Hopkins University Press, 262–77

Denny, C. (2001) For richer – and for poorer. *The Guardian*, 23 January

Department for International Development (1997) *White Paper on Eliminating World Poverty: A Challenge for the Twenty First Century*. London: Government Stationery Office

Department for International Development (1999) *Working in Partnership with the United Nations*, Institutional Strategy Paper. London: DfID

Department for International Development (2000a) *Eliminating World Poverty: Making Globalisation Work for the Poor*, White Paper on International Development. London: DfID

Department for International Development (2000b) *Debt Relief for Poverty Reduction*, background briefing, September. London: DfID

Department for International Development (2000c) *The Crisis in Ethiopia*, background briefing, April. London: DfID

Department for International Development (2001a) *Addressing the Water Crisis, Strategies for Achieving the International Development Targets*. London: DfID

Department for International Development (2001b) *Untying Aid*, background briefing, September. London: DfID

Department for International Development (2002) *Better Livelihoods for Poor People: The Role of Land Policy*, Issues Paper Discussion Draft, Rural Livelihoods Department. London: DfID

Desai, V. (2002) Role of non-government organizations (NGOs), ch. 10.6 in Desai, V. and Potter, R.B. (eds) *The Companion to Development Studies*. London: Arnold, 495–9

Desai, V. and Potter, R.B. (eds) (2006) *Doing Development Research*. London, Thousand Oaks and New Delhi: Sage Publications

Desai, V. and Potter, R.B. (eds) (2008) *The Companion to Development Studies*, 2nd edn. London: Hodder-Arnold and New York: Oxford University Press

Devas, N. and Rakodi, C. (eds) (1993) *Managing Fast Growing Cities: New Approaches to Urban Planning and Management in the Developing World*. Harlow: Longman

Devereux, S. and Maxwell, S. (2001) *Food Security in Sub-Saharan Africa*. London: ITDG Publishing

Dey, J. (1981) Gambian women: unequal partners in rice development projects? *Journal of Development Studies*, 17(3), 109–22

Dicken, P. (1992) *Global Shift: The Internationalization of Economic Activity*, 2nd edn. London: Paul Chapman

Dicken, P. (1993) The growth economies of Pacific Asia in their changing global context, in Dixon, C. and Drakakis-Smith, D. (eds) *Economic and Social Development in Pacific Asia*. London: Routledge, 22–42

Dicken, P. (1998) *Global Shift: Transforming the World Economy*, 3rd edn. London: Paul Chapman

Dicken, P. (2007) *Global Shift: Mapping the Changing Contours of the World Economy*, 5th edn. London, Thousand Oaks and New Delhi: Sage

Dickenson, J.P. (1994) Manufacturing industry in Latin America and the case of Brazil, in Courtenay, P.P. (ed.) *Geography and Development*. Melbourne: Longman Cheshire, 165–191

Dickenson, J., Gould, B., Clarke, C., Mather, C., Prothero, M., Siddle, D., Smith, C. and Thomas-Hope, E. (1996) *A Geography of the Third World*, 2nd edn. London: Routledge

Dixon, C. (ed.) (1987) *Rural–Urban Interaction in the Third World*. London: Developing Areas Research Group

Dixon, C. (1990) *Rural Development in the Third World*. London: Routledge

Dixon, C. (1998) *Thailand*. London: Routledge

Dixon, C. and Drakakis-Smith, D. (eds) (1997) *Uneven Development in Southeast Asia*. Aldershot: Ashgate

Dixon, C. and Heffernan, M. (eds) (1991) *Colonialism and Development in the Contemporary World*. London: Mansell

Dodds, F. (2002) Reforming the international institutions, in Dodds, F. (ed.) *Earth Summit 2002: A New Deal*. London: Earthscan, 291–314

Dodds, K. (2008) The Third World, developing countries, the South, poor countries, ch. 1.1 in Desai, V. and Potter, R.B. (eds) *The Companion to Development Studies*, 2nd edn. London: Hodder-Arnold and New York: Oxford University Press, 3–7

Dolan, C. and Humphrey, J. (2000) Governance and trade in fresh vegetables: the impact of UK supermarkets on the African horticulture industry. *Journal of Development Studies*, 37(2), 147–76

Dolman, P. (2000) Biodiversity and ethics, in O'Riordan, T. (ed.) (2000) *Environmental Science for Environmental Management*. Harlow: Pearson Education, 119–48

Donaghue, M.T. and Barff, R. (1990) Nike just did it: international subcontracting, flexibility and athletic footwear production. *Regional Studies*, 24, 537–52

Dooge, J.C.I. (1992) *An Agenda for Science for Environment and Development in a Changing World*. Cambridge: Cambridge University Press

Dorian, J.P., Franssen, H.T. and Simbeck, D.R. (2006) Global challenges in energy. *Energy Policy*, 34, 1984–91

Dos Santos, T. (1970) The structure of dependency. *American Economic Review*, 60, 125–158

Dos Santos, T. (1977) Dependence relations and political development in Latin America: some considerations. *Ibero-Americana*, 7, 245–59

Dowdeswell, E. (undated) Editorial. *Our Planet*, 6(5), 2

Doxiadis, C.A. (1967) Developments toward ecumenopolis: the Great Lakes megalopolis. *Ekistics*, 22, 14–31

Doxiadis, C.A. and Papaioannou, J.G. (1974) *Ecumenopolis: The Inevitable City of the Future*. New York: Norton

Drakakis-Smith, D. (1981) *Urbanization, Housing and the Development Process*. London: Croom Helm

Drakakis-Smith, D. (1983) Advance Australia fair: internal colonialism in the Antipodes, in Drakakis-Smith, D. and Wyn Williams, S. (eds) *Internal Colonialism: Essays Around a Theme*, Developing Areas Research Group, Monograph 3. London: Institute of British Geographers, 81–103

Drakakis-Smith, D. (1987) *The Third World City*. London: Methuen

Drakakis-Smith, D. (1989) Urban social movements and the built environment. *Antipode*, 21(3), 207–31

Drakakis-Smith, D. (1990) Food for thought or thought about food: urban food distribution systems in the Third World, in Potter, R.B. and Salau, A.T. (eds) *Cities and Development*. London: Mansell, 100–120

Drakakis-Smith, D. (1991) Colonial urbanization in Africa and Asia: a structural review. *Cambria*, 16, 123–50

Drakakis-Smith, D. (1992) *Pacific Asia*. London: Routledge

Drakakis-Smith, D. (1995) Third World cities: sustainable urban development I. *Urban Studies*, 32, 659–77

Drakakis-Smith, D. (1996) Third World cities: sustainable urban development II. *Urban Studies*, 33, 673–701

Drakakis-Smith, D. (1997) Third World cities: sustainable urban development III. *Urban Studies*, 34(5/6), 797–823

Drakakis-Smith, D. (2000) *Third World Cities*, 2nd edn. London: Routledge

Drakakis-Smith, D. and Dixon, C. (1997) Sustainable urbanisation in Vietnam. *Geoforum*, 28(1), 21–38

Drakakis-Smith, D., Doherty, J. and Thrift, N. (1987) What is a socialist developing country? *Geography*, 72(4), 333–5

Drakakis-Smith, D., Graham, E., Teo, P. and Ling, O.G. (1993) Singapore: reversing the demographic transition to meet labour needs. *Scottish Geographical Magazine*, 109, 152–63

Driver, F. (1992) Geography's empire: histories of geographical knowledge. *Environment and Planning D: Society and Space*, 10, 23–40

Duncan, J.S., Johnson, N.C. and Schein, R.H. (eds) (2004) *A Companion to Cultural Geography*. Oxford: Blackwell

Dunn, S. and Flavin, C. (2002) Moving the climate change agenda forward, in Worldwatch Institute *State of the World 2002: Progress Towards a Sustainable Society*. London: Earthscan, 24–50

Dwyer, D.J. (1975) *People and Housing in Third World Cities*. London: Longman

Dwyer, D.J. (1977) Economic development: development for whom? *Geography*, 62(4), 325–34

Earth Summit (2002) Earth Summit 2002, briefing paper, www.earthsummit2002.org

Economist Intelligence Unit (1996a) *Country Profile: Côte d'Ivoire*. London: EIU

Economist Intelligence Unit (1996b) *Country Profile: Kenya*. London: EIU

Economist Intelligence Unit (2001) *Kenya: Country Profile, 2001*. London: EIU

Economist Intelligence Unit (2002a) *Côte d'Ivoire and Mali: Country Profile, 2002*. London: EIU

Economist Intelligence Unit (2002b) *South Korea, North Korea: Country Profile, 2002*. London: EIU

Eden, M.J. and Parry, J. (eds) (1996) *Land Degradation in the Tropics: Environment and Policy Issues*. London: Mansell

Edge, G. and Tovey, K. (1995) Energy: hard choices ahead, in O'Riordan, T. (ed.) *Environmental Science for Environmental Management*. London: Longman, 317–34

Edwards, M. (2001a) The rise and rise of civil society. *Developments: The International Development Magazine*, 14(2nd quarter), 5–7

Edwards, M. (2001b) Introduction, in Edwards, M. and Gaventa, J. (eds) *Global Citizen Action*. London: Earthscan, 1–14

Edwards, M. and Gaventa, J. (eds) (2001) *Global Citizen Action*. London: Earthscan

Edwards, M. and Hulme, D. (eds) (1992) *Making A Difference: NGOs and Development in a Changing World*. London: Earthscan

Edwards, M. and Hulme, D. (eds) (1995) *Nongovernmental Organisations – Performance and Accountability: Beyond the Magic Bullet*. London: Earthscan

Ehrlich, P.R. (1968) *The Population Bomb*. New York: Ballantine Books

Eicher, C.K. and Staatz, J.M. (eds) (1990) *Agricultural Development in the Third World*, 2nd edn. Baltimore, MD: Johns Hopkins University Press

Elliott, J.A. (1990) The mechanical conservation of soil in Zimbabwe, in Cosgrove, D. and Petts, G. (eds) *Water, Engineering and Landscape*. London: Belhaven, 115–28

Elliott, J.A. (1991) Environmental degradation, soil conservation and the colonial and post-colonial state in Rhodesia/Zimbabwe, in Dixon, C. and Heffernan, M. (eds) *Colonialism and Development in the Contemporary World*. London: Mansell, 72–91

Elliott, J.A. (1994) *An Introduction to Sustainable Development: The Developing World*. London: Routledge

Elliott, J.A. (1995) Government policies and the population–environment interface: land reform and distribution in Zimbabwe, in Binns, T. (ed.) *People and Environment in Africa*. Chichester: John Wiley, 225–30

Elliott, J.A. (1999) *An Introduction to Sustainable Development*, 2nd edn. London: Routledge

Elliott, J.A. (2002) Towards sustainable rural resource management in sub-Saharan Africa. *Geography*, 87(3), 197–204

Elliott, J.A. (2006) *An Introduction to Sustainable Development*, 3rd edn. Abingdon: Routledge

Elliott, J.A. (2008, in press) Development: sustainable development, in Kitchen, R. and Thrift, N. (eds) *International Encylopedia of Human Geography*. Oxford: Elsevier

Elliott, L. (2000) A setback for Global Megabucks plc. *The Guardian*, 11 December, p. 27

Elliott, L. (2001) Brown must push harder for G7 change. *The Guardian*, 19 November, 23

Ellis, F. (2000a) The determinants of rural livelihoods diversification in developing countries. *Journal of Agricultural Economics*, 24(3), 281–97

Ellis, F. (2000b) *Rural Livelihoods and Diversity in Developing Countries*. Oxford: Oxford University Press

Elsom, D. (1995) *Male Bias in the Development Process*, 2nd edn. Manchester: Manchester University Press

Elsom, D. (1996) *Smog Alert: Managing Urban Air Quality*. London: Earthscan

Emel, J., Bridge, G. and Krueger, R. (2002) The earth as input: resources, in Johnston, R.J., Taylor, P.J. and Watts, M. (eds) *Geographies of Global Change: Remapping the World*, 2nd edn. London: Blackwell, 377–90

Endicott, S. (1988) *Red Earth: Revolution in a Sichuan Village*. London: I.B. Tauris

Engler, M. (2005) Human development. *New Internationalist*, 375, 30–31

Escobar, A. (1995) *Encountering Development*. Princeton NJ: Princeton University Press

Estes, R. (1984) World social progress, 1969–1979. *Social Development Issues*, 8, 8–28

Esteva, G. (1992) Development, in Sachs, W. (ed.) *The Development Dictionary*. London: Zed Books, 6–25

Evans, R. (1993) Reforming the union. *Geographical Magazine*, February, 24–27

Evans, R. (2001) Uganda: winning one battle in the long war against AIDS. *The Courier*, 188(September–October), 27–30

Eyre, J. and Dwyer, D.J. (1996) Ethnicity and uneven development in Malaysia, in Dwyer, D.J. and Drakakis-Smith, D. (eds) *Ethnicity and Development*. London: John Wiley, 181–94

Fadl, O.A.A. (1990) Gezira: the largest irrigation scheme in Africa. *The Courier*, November/December, 91–95

Fage, J.D. (1995) *A History of Africa*, 3rd edn. London: Routledge

Fairhead, J. (2004) Achieving sustainability in Africa, in Black, R. and White, H. (eds) *Targeting Development: Critical Perspectives on the Millennium Development Goals*. Abingdon: Routledge, 292–306

Fairhead, J. and Leach, M. (1995) Local agro-ecological management and forest–savanna transitions: the case of Kissidougou, Guinea, in Binns, T. (ed.) *People and Environment in Africa*. Chichester: John Wiley, 163–70

Fairhead, J. and Leach, M. (eds) (1998) *Reframing Deforestation: Global Analysis and Local Realities: Studies in West Africa*. London: Routledge

Fairtrade Foundation (2002) Guide to the fairtrade mark, www.fairtrade.org.uk/guide.htm

Fairtrade (2007) *About Fairtrade*, www.fairtrade.org/uk/about_sales.htm

Feeney, G. and Wang, F. (1993) Parity progression and birth intervals in China: the influence of policy in hastening fertility decline. *Population and Development Review*, 19(1), 61–100

Ferguson, J. (1990) *Grenada: Revolution in Reverse*. London: Latin American Bureau

Fik, T.J. (2000) *The Geography of Economic Development: Regional Changes, Global Challenges*, 2nd edn. Boston MA: McGraw Hill

Financial Times (2001) Leaders in denial as graves fill up, South Africa survey. *Financial Times*, 26 November

Flavin, C. and Gardner, G. (2006) China, India and the New World Order, in *State of the World: The Challenge of Global Sustainability*. London: Earthscan and WorldWatch Institute, 3–23

Foley, G. (1991) *Global Warming: Who is Taking the Heat?* London: Panos

Food and Agriculture Organization (1987) *Consultation on Irrigation in Africa*. Irrigation and Drainage Paper 42, Rome: FAO

Food and Agriculture Organization (2001) *Global Forest Resources Assessment 2000*, FAO Forestry Paper 140. Rome: FAO

Food and Agriculture Organization (FAO) (2005) *Global Forest Resources Assessment 2005*. Rome: FAO (UN)

Food and Agriculture Organization (FAO) (2005–06) *Statistical Yearbook*, available online www.fao.org/statistics/yearbook/vol1_1_1/pdf

Food and Agriculture Organization (FAO)/DFID (2003) *Grassroots Potential Unleashed: Good News from West African Fishing Communities*. Rome: FAO

Ford, L.H. (1999) Social movements and the globalisation of environmental governance. *IDS Bulletin*, 30(3), 68–74

Fox, J.A. (2000) The World Bank Inspection Panel: lessons from the first five years. *Global Governance*, 6(11), 279–318

Frank, A.G. (1966) The development of underdevelopment. *Monthly Review*, September, 17–30

Frank, A.G. (1967) *Capitalism and Underdevelopment in Latin America*. New York: Monthly Review Press

Frank, A.G. (1980) North–South and East–West paradoxes in the Brandt Report. *Third World Quarterly*, 2(4), 669–80

French, H. (2002) Reshaping global governance, in Worldwatch Institute *State of the World 2002*, 174–98

Friedland, W.H. (1994) The global fresh fruit and vegetable system: an industrial organization analysis, in McMichael, P. (ed.) *The Global Restructuring of Agro-food Systems*. Ithaca, NY: Cornell University Press, 173–89

Friedman, M. (1962) *Capitalism and Freedom*. Chicago, IL: University of Chicago Press

Friedmann, J. (1966) *Regional Development Policy: A Case Study of Venezuela*. Cambridge, MA: MIT Press

Friedmann, J. (1986) The world city hypothesis. *Development and Change*, 17, 69–83

Friedmann, J. (1992) *Empowerment: The Politics of Alternative Development*. Oxford: Blackwell

Friedmann, J. (1995) Where we stand: a decade of world city research, in Knox, P.L. and Taylor, P.J. (eds) *World Cities in a World-System*. Cambridge: Cambridge University Press, 21–37

Friedmann, J. and Weaver, C. (1979) *Territory and Function: the Evolution of Regional Planning*. London: Edward Arnold

Friedmann, J. and Wulff, G. (1982) World city formation: an agenda for research and action. *International Journal of Urban and Regional Research*, 6, 309–43

Fukuyama, F. (2001) Social capital, civil society and development. *Third World Quarterly*, 22, 7–20

Fullerton-Joireman, S. (1996) The minefield of land reform: comments on the Eritrean Land Proclamation. *African Affairs*, 95, 269–85

Furniss, C. (2006) The hungry dragon and the dark continent. *Geographical Magazine*, 78(12), 53–61

Furtado, C. (1964) *Development and Underdevelopment*. Berkeley, CA: University of California Press

Furtado, C. (1965) *Diagnosis of the Brazilian Crisis*. Berkeley, CA: University of California Press

Furtado, C. (1969) *Economic Development in Latin America*. Cambridge: Cambridge University Press

Galaty, J.G. and Johnson, D.L. (1990) Introduction: pastoral systems in global perspective, in Galaty, J.G. and Johnson, D.L. (eds) *The World of Pastoralism*. London: Belhaven, 1–31

Gale, D.J. and Goodrich, J.N. (eds) (1993) *Tourism Marketing and Management in the Caribbean*. London: Routledge

Gandhi, L. (1998) *Postcolonial Theory: A Critical Introduction*. Edinburgh: Edinburgh University Press

Gasper, D. (2004) *The Ethics of Development: From Economism to Human Development*. Edinburgh: Edinburgh University Press

Geheb, K. (1995) Exploring people–environment relationships: the changing nature of the small-scale fishery in the Kenyan sector of Lake Victoria, in Binns, T. (ed.) *People and Environment in Africa*. Chichester: John Wiley, 91–101

Geheb, K. and Binns, T. (1997) 'Fishing farmers' or 'farming fishermen'? The quest for household income and nutritional security on the Kenyan shores of Lake Victoria. *African Affairs*, 96, 73–93

George, S. (2002) Global citizens movement. *New Internationalist*, 343, 7

German, T. and Randel, J. (eds) (1993) *The Reality of Aid*. London: Actionaid

Getis, A., Getis, J. and Fellman, J. (1994) *Introduction to Geography*, 4th edn. Dubuque, IA: William C. Brown

Ghee, L.T. (1989) Reconstituting the peasantry: changes in landholding structure in the Muda irrigation scheme, in Hart, G., Turton, A. and White, B. (eds) *Agrarian Transformations: Local Processes and the State in Southeast Asia*. Berkeley, CA: University of California Press, 193–212

Gholz, H.L. (ed.) (1987) *Agroforestry: Realities, Possibilities and Potentials*. Dordrecht: Martinus Nijhoff

Ghose, A.K. (ed.) (1983) *Agrarian Reform in Contemporary Developing Countries*. London: Croom Helm

Gibbon, D. (ed.) (1995) *Structural Adjustment and the Working Poor in Zimbabwe*. Uppsala: Nordic Institute for African Studies

Gibbs, C., Fumo, C. and Kuby, T. (1999) *Nongovernmental Organisations in World Bank-Supported Projects: A review*, Operations Evaluation Department. Washington, DC: World Bank

Gilbert, A.G. (1976) The arguments for very large cities reconsidered. *Urban Studies*, 13, 27–34

Gilbert, A.G. (1977) The argument for very large cities reconsidered: a reply. *Urban Studies*, 14, 225–7

Gilbert, A. (1987) Research policy and review, No.15. From little Englanders to big Englanders: thoughts on the relevance of relevant research. *Environment and Planning A*, 19, 143–51

Gilbert, A.G. (1992) Third World cities: housing, infrastructure and servicing. *Urban Studies*, 29, 435–60

Gilbert, A.G. (1993) Third World cities: the changing national settlement system. *Urban Studies*, 30, 721–40

Gilbert, A.G. (1994) Third World cities: poverty, employment, gender roles and the environment during a time of restructuring. *Urban Studies*, 31, 605–33

Gilbert, A.G. (1996) *The Mega-City in Latin America*. Tokyo: United Nations University Press

Gilbert, A. (2002) The new international division of labour, ch. 4.2 in Desai, V. and Potter, R.B. (eds) *The Companion to Development Studies*. London: Arnold, 186–91

Gilbert, A.G. and Goodman, D.E. (1976) Regional income disparities and economic development, in Gilbert, A.G. (ed.) *Development Planning and Spatial Structure*. Chichester: John Wiley

Gilbert, A.G. and Gugler, J. (1982) *Cities, Poverty and Development: Urbanization in the Third World*. Oxford: Oxford University Press

Gilbert, A.G. and Gugler, J. (1992) *Cities, Poverty and Development: Urbanization in the Third World*, 2nd edn. Oxford: Oxford University Press

Girvan, N. (1973) The development of dependency economics in the Caribbean and Latin America: review and comparison. *Social and Economic Studies*, 22, 1–33

Gleave, M.B. (1996) The length of the fallow period in tropical fallow farming systems: a discussion with evidence from Sierra Leone. *Geographical Journal*, 162(1), 14–24

Global Eye (2002) Focus on population: Kerala, South India, www.globaleye.org.uk

Goodrich, R. (2001) *Sustainable Rural Livelihoods: A Summary of Research in Mali and Ethiopia*. Brighton: IDS

Gottmann, J. (1957) Megalopolis, or the urbanization of the north-eastern seaboard. *Economic Geography*, 33, 189–200

Gottmann, J. (1961) *Megalopolis: The Urbanization of the North-East Seaboard of the United States*. Oxford: Oxford University Press

Gottmann, J. (1978) Megalopolitan systems around the world, in Bourne, L.S. and Symmons, J.W. (eds) *Systems of Cities*. Oxford: Oxford University Press, 53–60

Goudie, A. (1990) *The Human Impact on the Natural Environment*, 3rd edn. London: Blackwell

Gould, P. (1969) The structure of space preferences in Tanzania. *Area*, 1, 29–35

Gould, P. (1970) Tanzania, 1920–63: the spatial impress of the modernisation process. *World Politics*, 22, 149–70

Gould, P. and White, R. (1974) *Mental Maps*. Harmondsworth: Penguin

Gould, W.T.S. (1992) Urban development and the World Bank. *Third World Planning Review*, 14, iii–vi

Gould, W.T.S. (1993) *People and Education in the Third World*. Harlow: Longman

Gould, W.T.S. and Prothero, R.M. (1975) Space and time in African population mobility, in Kosinski,

L.A. and Prothero, R.M. (eds) *People on the Move*. London: Methuen, 39–49

Graham, E. (1995) Singapore in the 1990s: can population policies reverse the demographic transition? *Applied Geography*, 15, 219–32

Grainger, A. (1993) *Controlling Tropical Deforestation*. London: Earthscan

Grant, M.C. (1995) Movement patterns and the intermediate sized city. *Habitat International*, 19, 357–70

Greater Anatolia Project (2007) www.gap.cov.tr

Greenpeace (2006) *A Guide to the Climate Negotiations in Nairobi*, www.greenpeace.org/international/press/reports/guide-to-Nairobi

Gregson, S., Garnett, G.P. and Anderson, R.M. (1994) Assessing the potential impact of the HIV-1 epidemic on orphanhood and the demographic structures of populations in sub-Saharan Africa. *Population Studies*, 48, 435–58

Greig, A., Hulme, D. and Turner, M. (2007) *Challenging Global Inequality: Development Theory and Practice in the 21st Century*. Basingstoke: Palgrave Macmillan

Griffin, K. (1980) Economic development in a changing world. Annual Lecture of the Development Studies Association, University of Swansea

Griffiths, I.L.L. (1993) *The Atlas of African Affairs*, 2nd edn. London: Routledge

Griffiths, I.L. (1995) *The African Inheritance*. London: Routledge

Grigg, D. (1995) *An Introduction to Agricultural Geography*, 2nd edn. London: Routledge

Grummer-Strawn, L., Hughes, M., Khan, L.K. and Martorell, R. (2000a) Obesity in women from developing countries. *European Journal of Clinical Nutrition*, 54, 247–52

Grummer-Strawn, L., Hughes, M., Khan, L.K. and Martorell, R. (2000b) Overweight and obesity in preschool children from developing countries. *International Journal of Obesity*, 24, 959–67

Gugler, J. (ed.) (1996) *The Urban Transformation of the Developing World*. Oxford: Oxford University Press

Gugler, J. (1997) Over-urbanization reconsidered, in Gugler, J. (ed.) *Cities in the Developing World*. Oxford: Oxford University Press, 114–23

Guha, R. (ed.) (1982) *Subaltern Studies I: Writings on South Asian History and Society*. Delhi: Oxford University Press

Guijt, I. and Shah, M. (eds) (1998) *The Myth of the Community: Gender Issues in Participatory Development*. London: IT Publications

Gupta, A. (1988) *Ecology and Development in the Third World*. London: Methuen

Gutkind, P.C.W. (1969) Tradition, migration, urbanization, modernity and unemployment in Africa: the roots of instability. *Canadian Journal of African Studies*, 3, 343–65

Gwin, C. (1995) A comparative assessment, in Ul-Haq, M., Jolly, R. Streeten, P. and Haq, K. (1995) *The UN and the Bretton Woods Institutions: New Challenges for the Twenty-First Century*. Basingstoke: Macmillan, 95–116

Gwynne, R.N. (2002) Export processing and free trade zones, in Desai, V. and Potter, R.B. (eds) *The Companion to Development Studies*. London: Arnold, 201–6

Haar, L.N. and Haar, L. (2006) Policy making under uncertainty: commentary upon the European Union Emissions Trading Scheme, *Energy Policy*, 34, 2615–29

Habitat (1996) *An Urbanising World: Global Report on Human Settlements, UN Centre for Human Settlements*. Oxford: Oxford University Press

Haddad, L. (1992) Introduction, in *Understanding How Resources are Allocated Within Households*. Washington, DC: International Food Policy Research Institute

Haffajee, F. (2001) AIDS in South Africa – bold steps in a discouraging climate. *The Courier*, 188(September–October), 46–7

Hagerstrand, T. (1953) *Innovationsforloppet ur Korologisk Synpunkt*. Lund: University of Lund

Haggett, P. (1990) *Geography: A Modern Synthesis*. London: Harper & Row

Haggett, P. (1990b) *The Geographer's Art*. Basil Blackwell

Hall, P. (1982) *Urban and Regional Planning*, 3rd edn. London: George Allen & Unwin

Hall, S. (1995) New cultures for old, in Massey, D. and Jess, P. (eds) *A Place in the World?* Oxford: Oxford University Press and Open University, 175–213

Hall, S. and Gieben, B. (1992) *Foundations of Modernity*. Cambridge: Polity

Halweil, B. (2002) Farming in the public interest, in World Institute *State of the World 2002*. London: Earthscan, 51–74

Hancock, G. (1997) Transmigration in Indonesia: how millions are uprooted, in Rahnema, M. and Bawtree, V. (eds) *The Post-Development Reader*.

London: Zed Books, 234–43 (Reprinted from Hancock, G. 1989, *Lords of Poverty*. London: Macmillan)

Hansen, N.M. (1981) Development from above: the centre-down development paradigm, in Stöhr, W.B. and Taylor, D.R.F. (eds) *Development from Above or Below? The Dialectics of Regional Development in Developing Countries*. Chichester: John Wiley

Harden, B. (1993) *Africa: Dispatches from a Fragile Continent*. London: HarperCollins

Hardin, G. (1968) The tragedy of the commons. *Science*, 162, 1243–8

Hardoy, J.E., Cairncross, S. and Satterthwaite, D. (eds) (1990) *The Poor Die Young: Housing and Health in Third World Cities*. London: Earthscan

Harmsen, R. (1995) The Uruguay Round: a boon for the world economy. *Finance and Development*, March, 24–26

Harris, J. (2001) The second 'Great Transformation'? Capitalism at the end of the twentieth century, in Allen, T. and Thomas, A. (eds) *Poverty and Development into the 21st Century*. Oxford: Oxford University Press, 325–42

Harris, N. (1989) Aid and urbanization. *Cities*, 6, 174–85

Harris, N. (1992) *Cities in the 1990s: The Challenge for Developing Countries*. London: UCL Press

Harrison, D. (ed.) (1992) *Tourism and the Less Developed Countries*. London: Belhaven

Harrison, P. and Palmer, R. (1986) *News Out of Africa: Biafra to Band Aid*. London: Hilary Shipman

Harriss, B. and Crow, B. (1992) Twentieth century free trade reform: food market deregulation in subSaharan Africa and south Asia, in Wuyts, M., Mackintosh, M. and Hewitt, T. (eds) *Development Policy and Public Action*. Oxford: Oxford University Press, 199–227

Harriss, J. (2006) Michael Lipton, in Simon, D. (ed.) *Fifty Key Thinkers on Development*. London and New York: Routledge, 149–54

Harriss, J. and Harriss, B. (1979) Development studies. *Progress in Human Geography*, 3(4), 577–82

Harvey, D. (1973) *Social Justice and the City*. London: Edward Arnold

Harvey, D. (1989) *The Condition of Postmodernity*. Oxford: Blackwell

Healey, P. (1997) *Collaborative Planning: Shaping Places in Fragmented Societies*. London: Macmillan

Healey, P. (1998) Building institutional capacity through collaborative approaches to urban planning. *Environment and Planning A*, 30, 1531–46

Healey, P. (1999) Deconstructing communicative planning theory: a reply to Tewdwr-Jones and Allmendinger. *Environment and Planning A*, 31, 1129–35

Heathcote, R.L. (1983) *The Arid Lands: Their Use and Abuse*. London: Longman

Henning, R.O. (1941) The furrow makers of Kenya. *Geographical Magazine*, 12, 268–79

Hentati, A. (undated) Taking effective action. *Our Planet*, 6(5), 5–7

Hettne, B. (1990) *Development Theory and the Three Worlds*. London: Longman

Hettne, B. (1995) *Development Theory and the Three Worlds*, 2nd edn. Harlow: Longman

Hewitt, T., Johnson, H. and Wield, D. (eds) (1992) *Industrialization and Development*. Oxford: Oxford University Press and Open University

Hiebert, M. (1993) Long shot? *Far Eastern Economic Review*, 14 October, 58

Hildyard, N. (1994) The big brother bank. *Geographical*, June, 26–8

Hill, P. (1963) *The Migrant Cocoa-Farmers of Southern Ghana: A Study in Rural Capitalism*. Cambridge: Cambridge University Press

Hill, P. (1972) *Rural Hausa: A Village and a Setting*. Cambridge: Cambridge University Press

Hill, P. (1970) *Studies in Rural Capitalism in West Africa*. Cambridge: Cambridge University Press

Hill, P. (1986) *Development Economics on Trial: The Anthropological Case for a Prosecution*. Cambridge: Cambridge University Press

Hill, R., O'Keefe, P. and Snape, C. (1995) Energy planning, in Kirkby, J., O'Keefe, P. and Timberlake, L. (eds) *The Earthscan Reader in Sustainable Development*. London: Earthscan, 78–101

Hirschman, A.O. (1958) *The Strategy of Economic Development*. New Haven, CT: Yale University Press

Hoch, I. (1972) Income and city size. *Urban Studies*, 9, 299–328

Hodder, R. (1992) *The West Pacific Rim*. London: Belhaven

Holdern, J. and Pachauri, R.K. (1992) 'Energy', p. 111 in Dooge (ed) *An Agenda of Science for Environment and Development into the 21st Century*. Cambridge: Cambridge University Press

Homewood, K. and Rogers, W.A. (1987) Pastoralism, conservation and the overgrazing controversy, in Anderson, D. and Grove, R. (eds) *Conservation in*

Africa: People, Policies and Practice. Cambridge: Cambridge University Press, 111–28

Hoogvelt, A. (2001) *Globalization and the Postcolonial World: The New Political Economy of Development*, 2nd edn. Basingstoke: Palgrave

Hopkins, A.G. (1973) *An Economic History of West Africa*. London: Longman

Horvath, R. (1988) National development paths 1965–1987: measuring a metaphor. Paper presented to the International Geographic Congress, Sydney University

Hossain, M. (1988) *Credit for Alleviation of Rural Poverty: The Grameen Bank in Bangladesh*, Research Report 65. Washington DC: International Food Policy Research Institute

Hoyle, B.S. (1979) African socialism and urban development: the relocation of the Tanzanian capital. *Tijdschrift voor Economische en Sociale Geografie*, 70, 207–16

Hoyle, B.S. (1993) The 'tyranny' of distance – transport and the development process, in Courtney, P.P. (ed.) *Geography and Development*. Melbourne: Longman Cheshire, 117–43

Hudson, B. (1989) The Commonwealth Eastern Caribbean, in Potter, R.B. (ed.) *Urbanization, Planning and Development in the Caribbean*. London and New York: Mansell

Hudson, B. (1991) Physical planning in the Grenada Revolution: achievement and legacy. *Third World Planning Review*, 13, 179–90

Hudson, J.C. (1969) Diffusion in a central place system. *Geographical Analysis*, 1, 45–58

Huggler, J. (2006) The banker who changed the world. *The Independent*, 14 October, 38–9

Hughes, A. (2001) Global commodity networks, ethical trade and governmentality: organizing business responsibility in the Kenyan cut flower industry. *Transactions of the Institute of British Geographers*, NS 26, 390–406

Hughes, J.M.R. (1992) Use and abuse of wetlands, in Mannion, A.M. and Bowlby, S.R. (eds) *Environmental Issues in the 1990s*. London: John Wiley, 211–26

Hunt, D. (1984) *The Impending Crisis in Kenya: The Case for Land Reform*. London: Gower

Huntington, E. (1945) *Mainsprings of Civilisation*. New York: John Wiley

Hutton, W. (1993) Gatt's principles have been corrupted by free market nihilism. *The Guardian*, 16 November

ICO (2007) Total production of exporting countries, crop years 2001/02 to 2006/07, 22 May, www.ico.org/prices/po.htm

ICPQL (Independent Commission on Population and Quality of Life) (1996) *Caring for the Future*. Oxford: Oxford University Press

Ignatieff, M. (1995) Fall of a blue empire. *The Guardian*, 17 October

Iliffe, J. (1995) *Africans: The History of a Continent*. Cambridge: Cambridge University Press

ING Barings (2000) *Economic Impact of AIDS in South Africa: A Dark Cloud on the Horizon*. Johannesburg: ING Barings

Inter-American Development Bank (1998) *The Path Out of Poverty: The Inter-America Development Bank's Approach to Reducing Poverty*. Washington, DC: IADB

International Business Leaders Forum (2002) *Business and Human Rights: A Geography of Corporate Risk*. IBLF/Amnesty International

International Energy Agency (2004) *World Energy Outlook 2004*. Paris: IEA

International Food Policy Research Institute (1995a) *A 2020 Vision for Food, Agriculture, and the Environment in Latin America: A Synthesis*. Washington, DC: IFPRI

International Food Policy Research Institute (1995b) *A 2020 Vision for Food, Agriculture, and the Environment in South Asia: A Synthesis*. Washington, DC: IFPRI

International Fund for Agricultural Development (2001) *Rural Poverty Report: The Challenge of Ending Rural Poverty*. Oxford: Oxford University Press

International Monetary Fund (1998) 'Cameroon Statistical Appendix', *IMF Staff Country Report*, no. 98/17. Washington, DC: IMF

International Monetary Fund (2006) *World Economic Outlook 2006*. Washington, DC: IMF

International Rivers Network (2007) China's Rivers at risk, www.irn.org/programs/china

IPCC Intergovernmental Panel on Climatic Change (1990) *Climate Change: The IPCC Assessment*. Cambridge: Cambridge University Press

IPCC (2001) *Climate Change 2001: Synthesis Report*, Robert T. Watson and the Core Writing Team (eds). Cambridge: Cambridge University Press

IPCC (2007) *Climate Change 2007: The Physical Science Basis. Summary for Policymakers*. Geneva: IPCC

Jaffee, S. (1994) *Exporting High Value Food Commodities*. Washington, DC: World Bank

Jain, P.S. (1996) Managing credit for the rural poor: lessons from the Grameen Bank. *World Development*, 24(1), 79–89

Jamal, V. and Weeks, J. (1994) *Africa Misunderstood: Or Whatever Happened to the Rural–Urban Gap?* Basingstoke: Macmillan

James, H. (1998) From grandmotherliness to governance: the evolution of IMF conditionality. *Finance and Development*, December, 44–7

Jameson, F. (1984) Postmodernism, or the cultural logic of late capitalism. *New Left Review*, 146, 53–92

Janelle, D.G. (1969) Spatial reorganization: a model and a concept. *Annals of the Association of American Geographers*, 59, 348–64

Janelle, D.G. (1973) Measuring human extensibility in a shrinking world. *Journals of Geography*, 72, 8–15

Jenkins, R. (1987) *Transnational Corporations and Uneven Development*. London: Methuen

Jenkins, R. (1992) Industrialization and the global economy, in Hewitt, T., Johnson, H. and Wield, D. (eds) *Industrialization and Development*. Oxford: Oxford University Press in association with the Open University

Jimenez-Diaz, V. (1994) The incidence and causes of slope failures in the barrios of Caracas, Venezuela, in Main, H. and Williams, S.W. (eds) *Environment and Housing in Third World Cities*. Chichester: Wiley

Joekes, S., Leach, M. and Green, C. (eds) (1995) Gender relations and environmental change. *IDS Bulletin*, 26, 1

Johnson, B.L.C. (1983) *India: Resources and Development*, 2nd edn. London: Heinemann

Johnston, R. (1984) The world is our oyster. *Transactions of the Institute of British Geographers*, NS 9(4), 443–59

Johnston, R.J. (1996) *Nature, State and Economy: A Political Economy of the Environment*, 2nd edn. Chichester: John Wiley

Jones, G. and Corbridge, S. (2008) Urban bias, ch. 5.2 in Desai, V. and Potter, R.B. (eds) *The Companion to Development Studies*, 2nd edn. London: Hodder-Arnold and New York: Oxford University Press, 243–7

Jones, E. and Eyles, J. (1977) *An Introduction to Social Geography*. Oxford: Oxford University Press

Jones, G. and Hollier, G. (1997) *Resources, Society and Environmental Management*. London: Paul Chapman

Jones, J.P., Natter, W. and Schatzki, T.R. (1993) *Postmodern Contentions: Epochs, Politics, Space*. London: Guildford Press

Jones, I., Pollit, M. and Bek, D. (2007) *Multinationals in their Communities: A Social Capital Approach to Corporate Citizenship Projects*. London: Palgrave

Jordan, A. and Brown, K. (1997) The international dimensions of sustainable development: Rio reconsidered, in Auty, R.M. and Brown, K. (eds) *Approaches to Sustainable Development*. London: Pinter, 270–95

Jowett, J. (1990) People: demographic patterns and policies, in Cannon, T. and Jenkins, A. (eds) *The Geography of Contemporary China: The Impact of Deng Xiaoping's Decade*. London: Routledge, 102–32

Jubilee 2000 (2002) About 'Jubilee Research', successor to Jubilee 2000, UK, www.jubilee2000uk.org

Kaarsholm, P. (ed.) (1995) *From Post-Traditional to Post-Modern? Interpreting the Meaning of Modernity in Third World Urban Societies*, Occasional Paper 14, International Development Studies, Roskilde University

Kabbani, R. (1986) *Imperial Fictions*. London: Pandora

Kabeer, N. (1992) Beyond the threshold: intrahousehold relations and policy perspectives, in *Understanding How Resources are Allocated Within Households*. Washington DC: International Food Policy Research Institute, 51–52

Kabeer, N. (2001) Conflicts over credit: re-evaluating the empowerment potential of loans to women in rural Bangladesh. *World Development*, 29(1), 63–84

Kaplinsky, R., McCormick, D. and Morris, M. (2006) *The Impact of China on Sub-Saharan Africa*. Sussex: Institute of Development Studies Asian Drivers Programme and London: DFID, www.ids.ac.uk/ids/global/Asiandriverpdfs/DFIDAgendaPaper06.pdf

Kats, G. (1992) Achieving sustainability in energy use in developing countries, in Holmberg, J. (ed.) *Policies for a Small Planet*. London: Earthscan, 258–89

Keeling, D.J. (1995) Transport and the world city paradigm, in Knox, P.L. and Taylor, P.J. (eds) *World Cities in a World-System*. Cambridge: Cambridge University Press, 115–31

Kelly, M. and Granich, S. (1995) Global warming and development, in Morse, S. and Stocking, M. (eds)

People and Environment. London: UCL Press, 69–107

Kennedy, E. and Bouis, H.E. (1993) *Linkages Between Agriculture and Nutrition: Implications for Policy and Research*. Washington DC: International Food Policy Research Institute

Kennes, W. (1990) The European Community and food security. *IDS Bulletin*, 21(3), 67–71

Kiely, R. (1999a) Globalisation, (post)-modernity and the Third World, in Kiely, R. and Marfleet, P. (eds) *Globalisation and the Third World*. London: Routledge, 1–22

Kiely, R. (1999b) Transnational companies, global capital and the Third World, ch. 2 in Kiely, R. and Marfleet, P. (eds) *Globalisation and the Third World*. London: Routledge, 45–66

Kiely, R. (1999c) The last refuge of the noble savage? A critical assessment of post-development theory. *The European Journal of Development Research*, 11, 30–55

Kiely, R. (2002) Global shift: industrialization and development, ch. 4.1 in Desai, V. and Potter, R.B. (eds) *The Companion to Development Studies*. London: Arnold, 183–6

Killick, A. (1990) Whither development economics? *Economics*, 26(2), 62–69

Killick, T. (1995) Structural adjustment and poverty alleviation: an interpretative survey. *Development and Change*, 26, 305–31

Kimbrell, A. (1998) Why biotechnology and high-tech agriculture cannot feed the world. *The Ecologist*, 28(5), 294–8

Kimmage, K. (1991) Small-scale irrigation initiatives in Nigeria: the problems of equity and sustainability. *Applied Geography*, 11(5), 5–20

King, A. (1976) *Colonial Urban Development*. London: Routledge & Kegan Paul

King, A. (1990) *Urbanism, Colonialism and the World Economy*. London: Routledge

Kirkby, J., O'Keefe, P. and Timberlake, L. (eds) (1995) *The Earthscan Reader in Sustainable Development*. London: Earthscan

Kirton, C.D. (1988) Public policy and private capital in the transition to socialism: Grenada 1979–85. *Social and Economic Studies*, 37, 125–50

Klak, T. (2008) World-systems theory: cores, peripheries and semi-peripheries, ch. 2.8 in Desai, V. and Potter, R.B. (eds) *The Companion to Development Studies*, 2nd edn. London: Hodder-Arnold and New York: Oxford University Press, 101–7

Klein, N. (2001) Between McWorld and Jihad. *The Guardian Weekend*, 27 October, 30–32

Knight, J.B. (1972) Rural–urban income comparisons and migration in Ghana. *Bulletin of the Oxford University Institute of Economics and Statistics*, 34(2), 199–229

Knox, P. and Marston, S. (2001) *Places and Regions in Global Context: Human Geography*. Englewood Cliffs, NJ: Prentice Hall

Knox, P.L. and Taylor, P.J. (eds) (1995) *World Cities in a World-System*. Cambridge: Cambridge University Press

Komin, S. (1991) Social dimensions of industrialization in Thailand. *Regional Development Dialogue*, 12, 115–37

Korten, D. (1990a) *Voluntary Organisations and the Challenge of Sustainable Development*, Briefing Paper 15, Australia Development Studies Network, Australian National University, Canberra

Korten, D.C. (1990b) *Getting to the Twenty-First Century: Voluntary Action and the Global Agenda*. Connecticut: Kumarian Press

Kothari, U. (ed.) (2005) *A Radical History of Development Studies: Individuals, Institutions and Ideologies*. London and New York: Zed Books and Cape Town: David Philip

Kuhn, T. (1962) *The Structure of Scientific Revolutions*. Chicagom IL: University of Chicago Press

Kurien, J. (1988) Kerala Fishing-Boat Project, South India, in Conroy, C. and Litvinoff, M. (eds) *The Greening of Aid*. London: Earthscan, 108–12

La Chard, L.W. (1906) Some recent impressions of northern Nigeria. *The Geographical Teacher*, 3, 191–201

Laing, R. and Pigott, M. (1987) Meeting the health and housing needs of plantation workers. *IDS Bulletin*, 18(2), 23–29

Lasuen, J.R. (1973) Urbanisation and development – the temporal interaction between geographical and sectoral clusters. *Urban Studies*, 10, 163–88

Lea, J.P. (2006) Terence Gary McGee, in Simon, D. (ed.) *Fifty Key Thinkers on Development*. London and New York: Routledge, 176–80

Leach, M. (1991) Locating gendered experience: an anthropologist's view from a Sierra Leonean village. *IDS Bulletin*, 22(1), 44–50

Leach, M. and Mearns, R. (eds) (1996) *The Lie of the Land: Challenging Received Wisdom on the African Environment*. Oxford: James Currey

Lean, G. (2002) World will ratify protocol that Bush wants to destroy. *The Independent*, 4 September

Lee, J. and Bulloch, J. (1990) Spirit of war moves on Mid-East waters. *The Independent on Sunday*, 13 May, 13

Leeming, F. (1993) *The Changing Geography of China*. Oxford: Blackwell

Lefevre, A. (1995) *Islam, Human Rights and Child Labour*. Copenhagen: Nordic Institute of Asian Studies

Leftwich, A. (1993) Governance, democracy and development in the Third World. *Third World Quarterly*, 14(3), 605–24

Leinbach, T.R. (1972) The spread of modernization in Malaya: 1895–1969. *Tijdschrift voor Economische en Sociale Geografie*, 63, 262–77

Lester, A., Nel, E. and Binns, T. (2000) *South Africa Past, Present and Future*. Harlow: Longman

Lewcock, C. (1995) Farmer use of urban waste in Kano. *Habitat International*, 19, 225–34

Lewis, W.A. (1950) The industrialisation of the British West Indies. *Caribbean Economic Review*, 2, 1–61

Lewis, W.A. (1955) *The Theory of Economic Growth*. London: George Allen & Unwin

Leys, C. (1996) *The Rise and Fall of Development Theory*. London: James Currey

Leyshon, A. (1995) Annihilating space? The speed-up of communications, in Allen, J. and Hamnett, C. (eds) *A Shrinking World*? Oxford: Oxford University Press and the Open University, 11–54

Lin, G.C.S. (1997) *Red Capitalism in South China: Growth and Development of the Pearl River Delta*. Vancouver: University of British Columbia Press

Linsky, A.S. (1965) Some generalizations concerning primate cities. *Annals of the Association of American Geographers*, 55, 506–13

Lipton, M. (1977) *Why Poor People Stay Poor: Urban Bias in World Development*. London: Temple Smith

Little, P.D., Smith, K., Cellarius, B.A., Coppock, D.L. and Barrett, C.B. (2001) Diversification and risk management amongst East African Herders. *Development and Change*, 32, 401–33

Littlejohns, M. and Silber, L. (1997) UN prepares to bite the bullet on reforms. *The Financial Times*, 16 September

Livingstone, D. (1993) *The Geographical Tradition: Episodes in the History of a Contested Enterprise*. Oxford: Blackwell

Lloyd-Evans, S. and Potter, R.B. (1996) Environmental impacts of urban development and the urban informal sector in the Caribbean, in Eden, M.J. and Parry, J. (eds) *Land Degradation in the Tropics*. London: Mansell, 245–60

Lloyd-Evans, S. and Potter, R.B. (2008) Third World cities, in Kitchen, R. and Thrift, N. (eds) *International Encyclopedia of Human Geography*. Oxford: Elsevier

Lockhart, D. (1993) Tourism to Fiji: crumbs off a rich man's table? *Geography*, 78(3), 318–23

Lohmann, L. (1993) Against the myths, in Colchester, M. and Lohmann, L. (eds) *The Struggle for Land and the Fate of the Forests*. London: Zed Books, 16–34

Longhurst, R. (1988) Cash crops and food security. *IDS Bulletin*, 19(2), 28–36

Lösch, A. (1940) *Die räumliche Ordnung der Wirtschaft*, Jena, translated by Woglom, W.H. and Stolpen, W.F., 1954, *The Economics of Location*. New Haven, CT: Yale University Press

Lonsdale, J. and Berman, B. (1979) Coping with the contradictions: The development of the colonial state in Kenya, 1895–1914. *Journal of African History*, 20(4), 487–505

loveLife/Henry J. Kaiser Family Foundation (2001) Impending Catastrophe Revisited: An Update on the HIV/AIDS Epidemic in South Africa, Parklands: loveLife, www.lovelife.org.za

Lowder, S. (1986) *Inside Third World Cities*. Beckenham: Croom Helm

Lowenthal, D. (1960) *West Indian Societies*. Oxford: Oxford University Press

Lucas, C. (2001a) *Stopping the Great Food Swap – Relocalising Europe's Food Supply*. London: The Green Party

Lucas, C. (2001b) The crazy logic of the continental food swap. *The Independent*, 25 March, 15

Lugard, F.J.D. (1965) *The Dual Mandate in British Tropical Africa*. London: Frank Cass

Lundqvist, J. (1981) Tanzania: socialist ideology, bureaucratic reality, and development from below, in Stöhr, W.B. and Taylor, D.R. (eds) *Development from Above or Below*? Chichester: John Wiley, 329–49

Lynch, K. (2002) Urban agriculture, ch. 5.6 in Desai, V. and Potter, R.B. (eds) *The Companion to Development Studies*. London: Arnold, 268–72

Lynch, K., Binns, T. and Olofin, E.A. (2001) Urban agriculture under threat; the land security question in Kano, Nigeria. *Cities*, 18, 159–71

MacAskill, E. (2000) Britain's ethical foreign policy: keeping the Hawk jets in action. *The Guardian*, 20 January

McAslan, E. (2002) Social capital and development, ch. 2.17 in Desai, V. and Potter, R.B. (eds) *The Companion to Development Studies*. London: Arnold, 139–43

MacCannell, D. (1976) *The Tourist: A New Theory of the Leisure Class*. New York: Schocken

McCormick, J. (1995) *The Global Environment Movement*, 2nd edn. Chichester: John Wiley

McCully, P. (2001) *Silenced Rivers: The Ecology and Politics of Large Dams*, 2nd edn. London: Zed Books

McElroy, J.L. and Albuquerque, K. (1986) The tourism demonstration effect on the Caribbean. *Journal of Travel Research*, 25, 31–4

MacEwan, A. (1982) Revolution, agrarian reform and economic transformation in Cuba, in Jones, S., Joshi, P.C. and Murmis, M. (eds) *Rural Poverty and Agrarian Reform*. New Delhi: Allied Publishers, 162–82

McEwan, C. (2002) Postcolonialism, in Desai, V. and Potter, R.B. (eds) *The Companion to Development Studies*. London: Arnold, 127–31

McGee, T. (1979) Conservation and dissolution in the Third World city: the 'shanty town' as an element of conservation. *Development and Change*, 10, 1–22

McGee, T. (1994) The future of urbanisation in developing countries: the case of Indonesia. *Third World Planning Review*, 16, iii–xii

McGee, T.G. (1967) *The Southeast Asian City: A Social Geography of the Primate Cities of Southeast Asia*. London: Bell

McGee, T.G. (1989) 'Urbanisasi' or Kotadesasi: evolving patterns of urbanisation in Asia, in Costa, F.J. (ed.) *Urbanization in Asia*. Honolulu, HI: University of Hawaii Press

McGee, T.G. (1991) The emergence of desakota regions in Asia: expanding a hypothesis, ch. 1 in Ginsburg, N., Koppell, B. and McGee, T.G. (eds) *The Extended Metropolis: Settlement Transition in Asia*. Honolulu, HI: University of Hawaii Press, 3–25

McGee, T.G. (1995) Eurocentralism and geography, in Crush, J. (ed.) *Power of Development*. London: Routledge, 192–207

McGee, T.G. (1997) The problem of identifying elephants: globalization and the multiplicities of development. Paper presented at the Lectures in Human Geography Series, University of St Andrews

McGee, T.G. and Greenberg, L. (1992) The emergence of extended metropolitan regions in ASEAN. *ASEAN Economic Bulletin*, 1(6), 5–12

McGee, T.G. and Robinson, I. (eds) (1995) *The Mega-Urban Regions of Southeast Asia*. Vancouver: UBC Press

McGinn, A.P. (2002) Reducing our toxic burden, in Worldwatch Institute, *State of the World 2002: Progress Towards a Sustainable Society*. London: Earthscan, 75–100

McIlwaine, C. (1997) Fringes or frontiers? Gender and export-oriented development in the Philippines, in Dixon, C. and Drakakis-Smith, D. (eds) *Uneven Development in Southeast Asia*. Aldershot: Ashgate, 100–23

Mackenzie, F. (1992) Development from within? The struggle to survive, in Taylor, D.R. and Mackenzie, F. (eds) *Development from Within: Survival in Rural Africa*. London: Routledge, 1–33

Mackintosh, M. (1992) Questioning the state, in Wuyts, M., Mackintosh, M. and Hewitt, T. (eds) *Development Policy and Public Action*. Oxford: Oxford University Press, 61–89

MacLeod, S. and McGee, T. (1990) The last frontier: the emergence of the industrial palate in Hong Kong, in Drakakis-Smith, D. (ed.) *Economic Growth and Urbanization in Developing Areas*. London: Routledge

McLennan, A. and Ngomas, W.Y. (2004) Quality governance for sustainable development? *Progress in Development Studies*, 4(4), 279–93

McLeod, J. (2000) *Beginning Postcolonialism*. Manchester: Manchester University Press

McLuhan, M. (1962) *The Gutenburg Galaxy: The Making of Typographic Man*, London: Routledge and Kegan Paul

McMichael, P. (2000) *Development and Social Change: a Global Perspective*, 2nd edn. London: Sage Publications

Maddox, B. (2002) GM food labelling row could mutate into trade war. *The Times*, 24 May

Madeley, J. (1999) *Big Business, Poor Peoples: The Impact of Trans-National Corporations on the World's Poor*. London: Zed Books

Madeley, J. (2000) An astonishing week in Seattle. *Developments*, 9, 6–9

Madeley, J. (2001) WTO members agree new trade round. *Developments*, Fourth Quarter, 26/27

Madeley, J. (2002) *Food For All: The Need For a New Agriculture*. London: Zed Books

Mail and Guardian (2001a) Drug giants back down. *Mail and Guardian*. Johannesburg, 20–25 April

Mail and Guardian (2001b) A disastrous reign. *Mail and Guardian*. Johannesburg, 26 April–3 May

Main, H. and Williams, S.W. (eds) (1994) *Environment and Housing in Third World Cities*. London: John Wiley

Makuch, Z. (1996) The World Trade Organisation and the General Agreement on Tariffs and Trade, in Werksman, J. (ed.) *Greening International Institutions*. London: Earthscan, 94–116

Malena, C. (2000) Beneficiaries, mercenaries, missionaries and revolutionaries: unpacking NGO involvement in World Bank financed project. *IDS Bulletin*, 31(3), 19–34

Maltby, E. (1986) *Waterlogged Wealth: Why Waste the World's Wet Places*? London: Earthscan

Mangin, W. (1967) Latin American squatter settlements: a problem and a solution. *Latin American Research Reviews*, 2, 65–98

Mannion, A.M. and Bowlby, S.R. (eds) (1992) *Environmental Issues in the 1990s*. London: John Wiley

Manzo, K. (1995) Black consciousness and the quest for counter-modernist development, in Crush, J. (ed.) *Power of Development*. London: Routledge, 228–52

Marshall, D. (2002) The New World group of dependency scholars: reflections on a Caribbean Avant-garde movement, ch. 2.9 in Desai, V. and Potter, R.B. (eds) *The Companion to Development Studies*. London: Arnold, 102–7

Martorell, R. (2001) Obesity – an emerging health and nutrition issue in developing countries, in Pinstrup-Andersen, P. and Pandya-Lorch, R. (eds) *The Unfinished Agenda: Perspectives on Overcoming Hunger, Poverty and Environmental Degradation*. Washington, DC: International Food Policy Research Institute, 49–53

Masselos, J. (1995) Postmodern Bombay: fractured discourses, in Watson, S. and Gibson, K. (eds) *Postmodern Cities and Spaces*. Oxford: Blackwell, 200–215

Massey, D. (1991) A global sense of place. *Marxism Today*, June, 24–29

Massey, D. (1995) Imaging the world, in Allen, J. and Massey, D. (eds) *Geographical Worlds*. London: Oxford University Press, 5–52

Massey, D. and Jess, P. (1995) *A Place in the World? Places, Cultures and Globalization*. Oxford: Oxford University Press and the Open University

Mather, A.S. and Chapman, K. (1995) *Environmental Resources*. London: Longman

Matthews, E. (2001) Understanding the Forest Resources Assessment 2000. *World Resources Briefing No. 1*, www.pdf.wri.org/fra2000.pdf

Maxwell, S. (1988) National food security planning: first thoughts from Sudan, unpublished paper presented to Workshop on Food Security in the Sudan, Institute of Development Studies, Sussex, 3–5 October

Maxwell, S. (1996) Food security: a post-modern perspective. *Food Policy*, 21(2), 155–70

Maxwell, S. (2004) Heaven or hubris: reflections on the 'New Poverty Agenda', in Black, R. and White, H. (eds) *Targeting Development: Critical Perspectives on the Millennium Development Goals*. Abingdon: Routledge, 25–46

Mayoux, L. (2001) Tackling the down side: social capital, women's empowerment and micro-finance in Cameroon. *Development and Change*, 32, 435–64

Mayhew, S. (1997) *A Dictionary of Geography*, 2nd edn. Oxford: Oxford University Press, 122

Meadows, D.H., Meadows, D.L., Randers, J. and Behrens, W.W. (1972) *The Limits to Growth*. London: Pan

Mehmet, O. (1995) *Westernising the Third World*. London: Routledge

Mehmet, O. (1999) *Westernizing the Third World*, 2nd edn. London: Routledge

Mehta, S.K. (1964) Some demographic and economic correlates of primate cities: a case for revaluation. *Demography*, 1, 136–47

Mehta, L., Leach, M., Newell, P., Scoones, I., Sivaramakrishnan, K. and Way, S. (1999) Exploring understandings of institutions and uncertainty: new directions in natural resource management, IDS Discussion Paper 372. Sussex: Institute of Development Studies

Meier, G.M. and Baldwin, R.E. (1957) *Economic Development: Theory, History, Policy*. New York: John Wiley

Meillassoux, C. (1972) From reproduction to production. *Economy and Society*, 1, 93–105

Meillassoux, C. (1978) The social organization of the peasantry: the economic basis of kinship, in Seddon, D. (ed.) *Relations of Production: Marxist Approaches to Economic Anthropology*. London: Frank Cass, 159–70

Mellor, J.W. (1990) Agriculture on the road to industrialization, in Eicher, C.K. and Staatz, J.M. (eds) *Agricultural Development in the Third World*, 2nd edn. Baltimore, MD: Johns Hopkins University Press, 70–88

Menzel, M. (2006) Walt William Rostow, in Simon, D. (ed.) *Fifty Key Thinkers on Development*. London and New York: Routledge, 211–17

Mera, K. (1973) On the urban agglomeration and economic efficiency. *Economic Development and Cultural Change*, 21, 309–24

Mera, K. (1975) *Income Distribution and Regional Development*. Tokyo: University of Tokyo Press

Mera, K. (1978) The changing pattern of population distribution in Japan and its implications for developing countries, in Lo, F.C. and Salih, K. (eds) *Growth Pole Strategies and Regional Development Policy*. Oxford: Pergamon

Mercer, C. (2002) NGOs, civil society and democratization: a critical review of the literature. *Progress in Development Studies*, 2, 5–22

Merriam, A. (1988) What does 'Third World' mean?, in Norwine, J. and Gonzalez, A. (eds) *The Third World: States of Mind and Being*. London: Unwin-Hyman

Merrick, T. (1986) World population in transition. *Population Bulletin*, 41(2), 1–51

Messkoub, M. (1992) Deprivation and structural adjustment, in Wuyts, M., Mackintosh, M. and Hewitt, T. (eds) *Development Policy and Public Action*. Oxford: Oxford University Press, 175–98

Middleton, N., O'Keefe, P. and Moyo, S. (1993) *Tears of the Crocodile: From Rio to Reality in the Developing World*. London: Pluto Press

Mijere, N. and Chilivumbo, A. (1987) Rural urban migration and urbanization in Zambia during the colonial and post-colonial periods, in Kaliperi, E. (ed.) *Population, Growth and Environmental Degradation in Southern Africa*. New York: Reinner

Milich, L. and Varady, R.G. (1998) Managing transboundary resources: lessons from river-basin accords. *Environment*, 40(8), 10–13

Millennium Ecosystem Assessment (2005) *Ecosystems and Well-being: Biodiversity Synthesis*. Washington, DC: World Resources Institute

Miller, D. (1992) The young and the restless in Trinidad: a case of the local and the global in mass consumption, in Silverstone, R. and Hirsch, E. (eds) *Consuming Technology*. London: Routledge, 163–82

Miller, D. (1994) *Modernity: An Ethnographic Approach: Dualism and Mass Consumption in Trinidad*. Oxford: Berg

Milner-Smith, R. and Potter, R.B. (1995) *Public Knowledge of Attitudes Towards the Third World*. CEDAR Research Paper 13, Royal Holloway College, University of London

Mitchell, R.E. and Reid, D.G. (2001) Community integration: island tourism in Peru. *Annals of Tourism Research*, 28, 113–39

MoBbrucker, H. (1997) Amerindian migration in Peru and Mexico, in Gugler, J. (ed.) *Cities in the Developing World*. Oxford: Oxford University Press, 74–87

Mohan, G. (1996) SAPs and Development in West Africa. *Geography*, 81(4), 364–8

Mohan, G. (2002) Participatory development, in Desai, V. and Potter, R.B. (eds) *The Companion to Development Studies*. London: Arnold, 49–54

Mohan, G. and Stokke, K. (2000) Participatory development and empowerment: the dangers of localism. *Third World Quarterly*, 21, 247–68

Mohan, G., Brown, E., Milward, B. and Zack-Williams, A.B. (2000) *Structural Adjustment: Theory, Practice and Impacts*. London: Routledge

Momsen, J.H. (1991) *Women and Development in the Third World*. London: Routledge

Momsen, J.H. (2004) *Gender and Development*. London: Routledge

Moock, J.L. (ed.) (1986) *Understanding Africa's Rural Households and Farming Systems*. London: Westview Press

Morrissey, D. (1999) An ageing world. *The Courier*, 176, 38–9

Morrissey, O. (2001) Does aid increase growth? *Progress in Development Studies*, 1, 37–50

Morse, S. (1995) Biotechnology: a servant of development? in Morse, S. and Stocking, M. (eds)

People and Environment. London: UCL Press, 131–55

Morse, S. and Stocking, M. (eds) (1995) *People and Environment*. London: UCL Press

Mortimore, M.J. (1989) *Adapting to Drought: Farmers, Famines and Desertification in West Africa*. Cambridge: Cambridge University Press

Mortimore, M. (1998) *Roots in the African Dust: Sustaining the Drylands*. Cambridge: Cambridge University Press

Mountjoy, A.B. (1976) Urbanization, the squatter and development in the Third World. *Tijdschrift voor Economische en Sociale Geografie*, 67, 130–7

Mountjoy, A.B. (1980) Worlds without end. *Third World Quarterly*, 2(4), 753–57

Moyo, S. (1995) *The Land Question in Zimbabwe*. Harare: Sapes Books

Munslow, B. and Ekoko, F. (1995) Is democracy necessary for sustainable development? *Democratisation*, 2, 158–78

Munslow, B., Katerere, Y., Ferf, A. and O'Keefe, P. (1988) *The Fuelwood Trap: A Study of the SADCC Region*. London: Earthscan

Murphy, D.F. and Mathew, D. (2001) Nike and global labour practices: a case study prepared for the New Academy of Business Innovation Network for Socially Responsible Business, www.new-academy.ac.uk/nike/nike-report.pdf

Murray, M. (1995) The value of biodiversity, in Kirkby, J., O'Keefe, P. and Timberlake, L. (eds) *The Earthscan Reader in Sustainable Development*. London: Earthscan, 17–29

Murray, W.E. (2006) *Geographies of Globalization*. London and New York: Routledge

Myint, H. (1964) *The Economics of Developing Countries*. London: Hutchinson

Myrdal, G. (1957) *Economic Theory and Underdeveloped Areas*. London: Duckworth

Naim, M. (2000) Fads and fashion in economic reforms: Washington consensus or Washington confusion? *Third World Quarterly*, 21(3), 505–28

Najam, A. (ed.) (2003) *Environment, Development and Human Security: Perspectives from South Asia*. Lanham, MD: University Press of America

Narayan, D., Patel, R., Schafft, K., Rademacher, A. and Koche-Schulte, S. (2000) *Voices of the Poor: Can Anyone Hear Us*? New York: Oxford University Press

Nederveen Pieterse, J. (2000) After post-development. *Third World Quarterly*, 21, 175–91

Neefjes, K. (2000) *Environments and Livelihoods: Strategies for Sustainability*. Oxford: Oxfam

Nelson, N. and Wright, S. (eds) (1995) *Power and Participatory Development: Theory and Practice*. London: IT Publications

Nelson, P. (2000) Whose civil society? Whose governance? Decision making and practice in the new agenda at the Inter-American Development Bank and the World Bank. *Global Governance*, 6, 405–31

Nelson, P.J. (2002) The World Bank and NGOs, in Desai, V. and Potter, R.B. (eds) *The Companion to Development Studies*. London: Arnold, 499–504

NEPAD (2001) New Partnership for Africa's Development, October, www.nepad.org

NEPAD (2002) Declaration on democracy, political, economic and corporate governance, 18 June, New Partnership for Africa's Development, www.nepad.org

Neumayer, E. (2001) *Greening Trade and Investment: Environmental Protection Without Protectionism*. London: Earthscan

New Internationalist (1995) Flood of protest. *New Internationalist*, Issue 273

New Internationalist (2001a) World Trade Organization: shrink it or sink it, Issue 334

New Internationalist (2001b) Faces of global resistance: we are everywhere, Issue 338

New Internationalist (2006) *CO2nned: Carbon Offsets Stripped Bare*, 391(July)

Newsweek (1995) The UN turns. *Newsweek*, 30 October

Nike (2007) Financial statement, www.nike.com/nikebiz/nikebiz.jhtml?page=16

Noonan, T. (1996) In the rough. *Far Eastern Economic Review*, 25 January, 38–9

Norwine, J. and Gonzalez, A. (1988) Introduction, in Norwine, J. and Gonzalez, A. (eds) *The Third World: States of Mind and Being*. London: Unwin-Hyman, 1–6

Oberai, A.S. (1993) *Population Growth, Employment and Poverty in Third World Mega-Cities: Analytical and Policy Issues*. Basingstoke: Macmillan and New York: St Martin's Press

O'Brien, R. (1991) *Global Financial Integration: the End of Geography*. London: Pinter

O'Connor, A. (1976) Third World or one world. *Area*, 8, 269–71

O'Connor, A. (1983) *The African City*. London: Hutchinson

O'Connor, A. (1991) *Poverty in Africa: A Geographical Approach*. London: Belhaven

O'Hare, G. (2002a) Climate change and the temple of sustainable development. *Geography*, 87(3), 234–46

O'Hare, G. (2000b) Reviewing the uncertainties in climate change science. *Area*, 32(4), 357–68

Ohlsson, L. (ed.) (1995) *Hydropolitics*. London: Zed Books

Oliver, R. and Fage, J.D. (1966) *A Short History of Africa*. Harmondsworth: Penguin

Olthof, W. (1995) Wildlife resources and local development: experiences from Zimbabwe's Campfire programme, in van de Breemer, J.P.M., Drijver, C.A. and Venema, L.B. (eds) *Local Resource Management in Africa*. Chichester: John Wiley, 111–28

O'Riordan, T. (ed.) (1995) *Environmental Science for Environmental Management*. London: Longman

O'Riordan, T. (2000a) Climate change, in O'Riordan, T. (ed.) *Environmental Science for Environmental Management*. Harlow: Pearson Education, 171–211

O'Riordan, T. (ed.) (2000b) *Environmental Science for Environmental Management*. Harlow: Pearson Education

O'Riordan, T. (2001) *Globalism, Localism and Identity*. London: Earthscan

O'Riordan, T. and Jordan, A. (2000) Managing the global commons, in O'Riordan, T. (ed.) (2000) *Environmental Science for Environmental Management*. Harlow: Pearson Education, 485–511

O'Tuathail, G. (1994) Critical geopolitics and development theory: intensifying the dialogue. *Transactions of the Institute of British Geographers, New Series*, 19, 228–38

OECD (Organisation for Economic Cooperation and Development) (2006) *Development Cooperation Report, 2005*. Paris: OECD

Oxfam (1984) *Behind the Weather: Lessons to be Learned. Drought and Famine in Ethiopia*. Oxford: Oxfam

Oxfam (1993) *Africa: Make or Break. Action for Recovery*. Oxford: Oxfam

Oxfam (1994) *The Coffee Chain Game*. Oxford: Oxfam

Oxfam (2002) Rigged rules and double standards: trade, globalisation and the fight against poverty, www.maketradefair.com

Pachai, B. (ed.) (1973) *Livingstone: Man of Africa. Memorial Essays 1873–1973*. Harlow: Longman

Pacione, M. (2005) *Urban Geography: A Global Perspective*, 2nd edn. London and New York: Routledge

Panayiotopoulos, P. and Capps, G. (2001) *World Development: An Introduction*. London: Pluto Press

Parfitt, T. (2002) *The End of Development: Modernity, Post-Modernity and Development*. London: Pluto Press

Parnwell, M. (1994) Rural industrialisation and sustainable development in Thailand. *Quarterly Environment Journal*, 2, 24–29

Parnwell, M. (2006) Robert Chambers, in Simon, D. (ed.) *50 Key Thinkers on Development*. London: Routledge, 73–7

Parnwell, M. and Turner, S. (1998) Sustaining the unsustainable: city and society in Southeast Asia. *Third World Planning Review*, 20, 147–164

Parry, M. (1990) *Climate Change and World Agriculture*. London: Earthscan

Patullo, P. (1996) *Last Resort? Tourism in the Caribbean*. London: Mansell and the Latin American Bureau

Pearce, D. (1995) *Blueprint 4: Capturing Global Environmental Value*. London: Earthscan

Pearce, F. (1993) How green is your golf? *New Scientist*, 25 September, 30–5

Pearce, F. (1997) The biggest dam in the world, in Owen, L. and Unwin, T. (eds) *Environmental Management: Readings and Case Studies*. Oxford: Blackwell, 349–54

Pearce, F. (2002) Heading into troubled waters. *The Independent*, 31 August

Pearlstein, S. (1998) Tropical and landlocked make a poor combination. *Washington Post*, 23 April, C1, C12

Pearson, R. (1992) Gender matters in development, in Allen, T. and Thomas, A. (eds) *Poverty and Development in the 1990s*. Oxford: Oxford University Press, 291–313

Pearson, R. (2000) Rethinking gender matters in development, in Allen, T. and Thomas, A. (eds) *Poverty and Development into the Twenty First Century*. Oxford: Oxford University Press, 383–402

Pedersen, P.O. (1970) Innovation diffusion within and between national urban systems. *Geographical Analysis*, 2, 203–54

Peet, R. and Watts, M. (eds) (1996) *Liberation Ecologies: Environment, Development, Social Movements*. London: Routledge

Pelling, M. (2002) The Rio Earth Summit, ch. 6.3 in Desai, V. and Potter, R.B. (eds) *The Companion to Development Studies*. London: Arnold, 284–9

Pepper, D. (1996) *Modern Environmentalism: An Introduction*. London: Routledge

Perloff, H.S. and Wingo, L. (1961) Natural resource endowment and regional economic growth, in Spengler, J.J. (ed.) *Natural Resources and Economic Growth*. Washington DC: Resources for the Future

Pernia, E. (1992) Southeast Asia, in R. Stren (ed.) *Sustainable Cities: Urbanization and the Environment in International Perspective*. Oxford: Westview Press, 233–58

Perroux, F. (1950) Economic space: theory and applications. *Quarterly Journal of Economics*, 64, 89–104

Perroux, F. (1955) Note sur la notion de 'pôle de croissance'. *Économie Appliquée*, 1(2), 307–20

Phillips, D.R. (1990) *Health and Health Care in the Third World*. Harlow: Longman

Phillips, D.R. and Yeh, G.O. (1990) Foreign investment and trade: impact on spatial structure of the economy, in Cannon, T. and Jenkins, A. (eds) *The Geography of Contemporary China*. London: Routledge, 224–48

Pinches, M. (1994) Urbanisation in Asia: development, contradiction and conflict, in Jayasuriya, L. and Lee, M. (eds) *Social Dimensions of Development*. Sydney: Paradigm Press

Pinstrup-Andersen, P. (1994) *World Food Trends and Future Food Security*. Washington, DC: International Food Policy Research Institute

Pletsch, C. (1981) The three worlds or the division of social scientific labour 1950–1975. *Comparative Studies in Society and History*, 23, 565–90

Pleumarom, A. (1992) Course and effect: golf tourism in Thailand. *The Ecologist*, 22(3), 104–10

Porteous, D. (1995) in Crush, J. (ed.) *Power of Development*. London: Routledge

Porter, D.J. (1995) Scenes from childhood, in Crush, J. (ed.) *Power of Development*. London: Routledge, 63–86

Porter, G. (1996) SAPs and road transport deterioration in West Africa. *Geography*, 81(4), 368–71

Porter, P.W. and Sheppard, E.S. (1998) *A World of Difference: Society, Nature, Development*. London: The Guilford Press

Portes, A., Castells, M. and Benton, L. (eds) (1991) *The Informal Economy: Studies in Advanced and Less Developed Countries*. Baltimore, MD: Johns Hopkins University Press

Portes, A., Dore-Cabral, C. and Landolt, P. (1997) *The Urban Caribbean: Transition of the New Global Economy*. Baltimore, MD: Johns Hopkins University Press

Postel, S. (2000) Redesigning irrigated agriculture, in Brown, L.R. (ed.) *State of the World 2000*. London: Earthscan, 39–58

Potter, D. (2000) Democratisation, 'good governance' and development, in Allen, T. and Thomas, A. (eds) *Poverty and Development into the 21st Century*. Oxford: Oxford University Press, 365–82

Potter, R.B. (1981) Industrial development and urban planning in Barbados. *Geography*, 66, 225–8

Potter, R.B. (1983) Tourism and development: the case of Barbados, West Indies. *Geography*, 68, 46–50

Potter, R.B. (1985) *Urbanisation and Planning in the Third World: Spatial Perceptions and Public Participation*. London: Croom Helm and New York: St Martin's Press

Potter, R.B. (1989) Rural–urban interaction in Barbados and the southern Caribbean, in Potter, R.B. and Unwin, T. (eds) *Urban–Rural Interaction in Developing Countries*. London and New York: Routledge, 257–93

Potter, R.B. (1990) Cities, convergence, divergence and Third World development, in Potter, R.B. and Salau, A.T. (eds) *Cities and Development in the Third World*. London: Mansell

Potter, R.B. (1992a) *Urbanisation in the Third World*. Oxford: Oxford University Press

Potter, R.B. (1992b) *Housing Conditions in Barbados: A Geographical Analysis*. Mona, Kingston, Jamaica: Institute of Social and Economic Research, University of the West Indies

Potter, R.B. (1993a) Little England and little geography: reflections on Third World teaching and research. *Area*, 25, 291–4

Potter, R.B. (1993b) Basic needs and development in the small island states of the Eastern Caribbean, in Lockhart, D. and Drakakis-Smith, D. (eds) *Small Island Development*. London: Routledge

Potter, R.B. (1993c) Urbanization in the Caribbean and trends of global convergence–divergence. *Geographical Journal*, 159, 1–21

Potter, R.B. (1994) *Low-Income Housing and the State in the Eastern Caribbean*. Barbados: University of the West Indies Press

Potter, R.B. (1995a) Urbanisation and development in the Caribbean. *Geography*, 80, 334–41

Potter, R.B. (1995b) Whither the real Barbados? *Caribbean Week*, 7(4), 64–7

Potter, R.B. (1996) Environmental impacts of urban-industrial development in the tropics: an overview, in Eden, M. and Parry, J.T. (eds) *Land Degradation in the Tropics*. London: Pinter

Potter, R.B. (1997) Third World urbanisation in a global context. *Geography Review*, 10, 2–6

Potter, R.B. (1998) From plantopolis to mini-metropolis in the eastern Caribbean, ch. 3 in McGregor, D., Barker, D. and Lloyd-Evans, S. (eds) *Resource Sustainability and Caribbean Development*. Barbados: University of the West Indies Press, 51–68

Potter, R.B. (2000) *The Urban Caribbean in an Era of Global Change*. Aldershot: Ashgate

Potter, R.B. (2001a) Geography and development: core and periphery? *Area*, 33, 422–7

Potter, R.B. (2001b) Progress, development and change. *Progress in Development Studies*, 1, 1–4

Potter, R.B. (2001c) Urban Castries, St Lucia revisited: global forces and local responses. *Geography*, 86, 329–36

Potter, R.B. (2002a) Geography and development: core and periphery? A reply. *Area*, 34, 213–14

Potter, R.B. (2002b) Making progress in development studies. *Progress in Development Studies*, 2, 1–3

Potter, R.B. (2003) The environment of development. *Progress in Development Studies*, 3, 1–4

Potter, R.B. (2008a) Global convergence, divergence and development, ch. 4.3 in Desai, V. and Potter, R.B. (eds) *The Companion to Development Studies*, 2nd edn. London: Hodder-Arnold and New York: Oxford University Press, 192–6

Potter, R.B. (2008b) World cities and development, ch. 5.3 in Desai, V. and Potter, R.B. (eds) *The Companion to Development Studies*, 2nd edn. London: Hodder-Arnold and New York: Oxford University Press, 247–52

Potter, R.B. and Conway, D. (eds) (1997) *Self-Help Housing, the Poor and the State in the Caribbean*. Knoxville, TN: Tennessee University Press and Barbados: University of the West Indies Press

Potter, R.B. and Dann, G.M.S. (1994) Some observations concerning postmodernity and sustainable development in the Caribbean. *Caribbean Geography*, 5, 92–107

Potter, R.B. and Dann, G. (1996) Globalization, postmodernity and development in the Commonwealth Caribbean, in Yeung, Yue-man (ed.) *Global Change and the Commonwealth*. Hong Kong: Hong Kong Institute of Asia-Pacific Studies, Chinese University of Hong Kong, 103–29

Potter, R.B. and Lloyd-Evans, S. (1998) *The City in the Developing World*. Harlow: Longman

Potter, R.B. and Lloyd-Evans, S. (2008) Development: the Brandt Commission, in Kitchen, R. and Thrift, N. (eds) *International Encyclopedia of Human Geography*. Oxford: Elsevier

Potter, R.B. and Philips, J. (2004) The rejuvenation of tourism in Barbados, 1993–2003: reflections on the Butler model. *Geography*, 89, 240–47

Potter, R.B. and Phillips, J. (2006a) 'Mad dogs and transnational migrants?' Bajan-Brit second-generation migrants and accusations of madness. *Annals of the Association of American Geographers*, 96, 586–600

Potter, R.B. and Phillips, J. (2006b) Both black and symbolically white: the Bajan-Brit return migrants as post-colonial hybrid. *Ethnic and Racial Studies*, 29, 901–27

Potter, R.B. and Phillips, J. (2008) 'The past is still right here in the present': second-generation Bajan-Brit transnational migrants' views on issues relating to race and colour class, *Environment and Planning D: Society and Space* (in press)

Potter, R.B. and Pugh, J. (2001) Planning without plans and the neo-liberal state: the case of St Lucia, West Indies. *Third World Planning Review*, 23, 323–40

Potter, R.B. and Unwin, T. (1987) *Urban–Rural Interaction in Developing Countries: Essays in Honor of Alan B Mountjoy*. London and New York: Routledge

Potter, R.B. and Unwin, T. (1988) Developing areas research in British geography. *Area*, 20, 121–6

Potter, R.B. and Unwin, T. (eds) (1992) *Teaching the Geography of Developing Areas*. Monograph 7, Developing Areas Research Group. London: Institute of British Geographers

Potter, R.B. and Unwin, T. (1995) Urban–rural interaction: physical form and political process in the Third World. *Cities*, 12, 67–73

Potter, R.B. and Welch, B. (1996) Indigenization and development in the Caribbean. *Caribbean Week*, 8, 13–4

Potter, R.B., Barham, N. and Darmame, K. (2007) The polarised social and residential structure of the city of Amman: a contemporary view using GIS data. *Bulletin of the Council for British Research in the Levant*, 2, 48–52

Potter, R.B., Darmame, K. and Nortcliff, S. (2007) The provision of water under conditions of 'water stress', privatisation and de-privatisation in Amman. *Bulletin of the Council for British Research in the Levant*, 2, 52–4

Potter, R.B., Barham, N., Darmame, K. and Nortcliff, S. (2007) An introduction to the urban geography of Amman, Jordan. *Reading Geographical Paper*, 182

Potts, D. (1995) Shall we go home? Increasing urban poverty in African cities and migration processes. *Geographical Journal*, 161(3), 245–64

Power, M. (2002) Enlightenment and the era of modernity, ch. 2.2 in Desai, V. and Potter, R.B. (eds) *The Companion to Development Studies*. London: Arnold, 65–70

Power, M. (2003) *Rethinking Development Geographies*. London and New York: Routledge

Prebisch, R. (1950) *The Economic Development of Latin America*. New York: United Nations

Pred, A. (1977) *City-Systems in Advanced Economies*. London: Hutchinson

Pred, A.R. (1973) The growth and development of systems of cities in advanced economies, in Pred, A. and Törnqvist, G. (eds) *Systems of Cities and Information Flows: Two Essays*. Lund: University of Lund, 9–82

Preston, D. (1987) Population mobility and the creation of new landscapes, in Preston, D. (ed.) *Latin American Development: Geographical Perspectives*. London: Longman, 229–59

Preston, P.W. (1985) *New Trends in Development Theory*. London: Routledge

Preston, P.W. (1987) *Making Sense of Development: An Introduction to Classical and Contemporary Theories of Development and their Application to Southeast Asia*. London: Routledge

Preston, P.W. (1996) *Development Theory: An Introduction*. Oxford: Blackwell

Pro-Poor Tourism (2002) How is PPT different from other forms of 'alternative' tourism? www.propoortourism.org.uk/ppt_vs_alternative.html

Prothero, R.M. (1959) *Migrant Labour from Sokoto Province, Northern Nigeria*. Kaduna: Government Printer, 46

Prothero, R.M. (1994) Forced movements of population and health hazards in tropical Africa. *International Journal of Epidemiology*, 23(4), 657–64

Prothero, R.M. (1996) Migration and AIDS in West Africa. *Geography*, 81(4), 374–7

Pryer, J. (1987) Production and reproduction of malnutrition in an urban slum in Khulna, Bangladesh, in Momsen, J.H. and Townsend, J. (eds) *Geography of Gender in the Third World*. London: Hutchinson, 131–49

Pugh, C. (ed.) (1996) *Sustainability, the Environment and Urbanization*. London: Earthscan

Pugh, J. (2002) Local Agenda 21 and the Third World, ch. 6.4 in Desai, V. and Potter, R.B. (eds) *The Companion to Development Studies*. London: Arnold, 289–93

Pugh, J. and Potter, R.B. (2000) Rolling back the state and physical development planning: the case of Barbados. *Singapore Journal of Tropical Geography*, 21, 175–91

Pugh, J. and Potter, R.B. (eds) (2003) *Participatory Planning in the Caribbean: Lessons from Practice*. Aldershot, UK and Burlington, VT: Ashgate

Purcell, M. and Brown, J.C. (2005) Against the local trap: scale and study of environment and development. *Progress in Development Studies*, 5, 279–97

Putnam, R. (1993) The prosperous community: social capital and public life. *American Prospect*, 13, 35–42

Pye-Smith, C. and Feyerarbend, G.B. (1995) What next? in Kirkby, J., O'Keefe, P. and Timberlake, L. (eds) *The Earthscan Reader in Sustainable Development*. London: Earthscan, 303–9

Quader, M.A. (2000) Ruralopolis: the spatial organization and residential land economy of high-density rural regions in South Asia. *Urban Studies*, 37, 1583–600

Raath, J. (1997) Mugabe wants aid to seize white land. *The Times*, 20 October

Rakodi, C. (1995) Poverty lines or household strategies? *Habitat International*, 19(4), 407–26

Rakodi, C. (2002) A livelihood approach – conceptual issues and definitions, in Rakodi, C. and Lloyd-Jones, T. (eds) *Urban Livelihoods: A People-Centred Approach to Reducing Poverty*. London: Earthscan

Rakodi, C. and Lloyd-Jones, T. (eds) (2002) *Urban Livelihoods: A People-Centred Approach to Reducing Poverty*. London: Earthscan

Ransom, D. (2001) A world turned upside down. *New Internationalist*, 334(May), 26–8

Ransom, D. (2005) Upside down: the United Nations at 60. *New Internationalist*, 375, 9–12

Ransom, D. (2005b) The big charity bonanza, *New Internationalist*, 383, 2–5

Rapley, J. (1996) *Understanding Development: Theory and Practice in the Third World*. London: University College of London Press

Rapley, J. (2001) Convergence: myths and realities. *Progress in Development Studies*, 1, 295–308

Reading, A.J., Thompson, R.D. and Millington, A.C. (1995) *Humid Tropical Environments*. Oxford: Blackwell

Reality of Aid (2002) An independent review of poverty reduction and development assistance, www.devint.org/realityof aid

Reardon, T. (1997) Using evidence of household income diversification to inform study of the rural nonfarm labor market in Africa. *World Development*, 25(5), 735–47

Reardon, T., Berdegue, J. and Escobar, G. (2001) Rural nonfarm employment and incomes in Latin America: overview and policy implications. *World Development*, 29(3), 395–409

Redclift, M. (1987) *Sustainable Development: Exploring the Contradictions*. London: Methuen

Redclift, M. (1997) Sustainable development: needs, values and rights, in Owen, L. and Unwin, T. (eds) *Environmental Management: Readings and Case Studies*. Oxford: Blackwell, 438–50

Reed, D. (ed.) (1996) *Structural Adjustment: The Environment and Sustainable Development*. London: Earthscan

Rees, J. (1990) *Natural Resources: Allocation, Economics and Policy*, 2nd edn. London: Methuen

Renaud, B. (1981) *National Urbanization Policy in Developing Countries*. Oxford: Oxford University Press for the World Bank

Renner, M. (2002) Breaking the link between resources and repression, in Worldwatch Institute, *State of the World 2002: Progress Towards a Sustainable Society*. London: Earthscan, 149–72

Rich, B. (1994) *Mortgaging the Earth: The World Bank, Environmental Impoverishment and the Crisis of Development*. London: Earthscan

Richards, P. (1985) *Indigenous Agricultural Revolution*. London: Hutchinson

Richards, P. (1986) *Coping with Hunger: Hazard and Experiment in an African Rice-Farming System*. London: George Allen & Unwin

Richards, P. and Thomson, A. (1984) *Basic Needs and the Urban Poor*. London: Croom Helm

Richardson, H.W. (1973) *The Economics of Urban Size*. Farnborough: Saxon House

Richardson, H.W. (1976) The argument for very large cities reconsidered: a comment. *Urban Studies*, 13, 307–10

Richardson, H.W. (1977) City size and national spatial strategies in developing countries. World Bank Staff Working Paper 252

Richardson, H.W. (1980) Polarization reversal in developing countries. *Papers of the Regional Science Association*, 45, 67–85

Richardson, H.W. (1981) National urban development strategies in developing countries. *Urban Studies*, 18, 267–83

Riddell, J.B. (1970) *The Spatial Dynamics of Modernization in Sierra Leone: Structure, Diffusion and Response*. Evanston, IL: Northwestern University Press

Riddell, J.B. (1978) The migration to the cities of West Africa: some policy considerations. *Journal of Modern African Studies*, 16(2), 241–60

Rigg, J. (1997) *Southeast Asia*. London: Routledge

Rigg, J. (2001) *More Than the Soil: Rural Change in Southeast Asia*. Harlow: Pearson Education

Rigg, J. (2002) The Asian crisis, ch. 1.6 in Desai, V. and Potter, R.B. (eds) *The Companion to Development Studies*. London: Arnold, 27–32

Rigg, J. (2006) Land, farming, livelihoods, and poverty: rethinking the links in the rural South. *World Development*, 34(1), 180–202

Rigg, J. (2007) *An Everyday Geography of the Global South*. Abingdon: Routledge

Rigg, J. (2008) The Millennium Development Goals, ch. 1.7 in Desai, V. and Potter, R.B. (eds) *The Companion to Development Studies*, 2nd edn. London: Hodder-Arnold and New York: Oxford University Press, 37–40

Righter, R. (1995) *Utopia Lost: the United Nations and the World Order*. New York: Twentieth Century Fund Press

Riley, S. (1988) Structural adjustment and the new urban poor: the case of Freetown. Paper presented at the Workshop on the New Urban Poor in Africa, School of Oriental and African Studies, London, May 1988

Rimmer, P.J. (1991) International transport and communications interactions between Pacific Asia's world cities, in Lo, F.-C. and Yeung, Y.-M. (eds) *Emerging World Cities in Asia*. Tokyo: United Nations University Press, 48–97

Robb, C. (1998) PPAs: a review of the World Bank's experience, in Holland, J. and Blackburn, J. (eds) *Whose Voice: Participatory Research and Policy Change*. London: IT Publications

Robins, K. (1989) Global times. *Marxism Today*, December 1989, 20–27

Robins, K. (1995) The new spaces of global media, in Knox, P.C. and Taylor, P.J. (eds) *World Cities in a World-System*. Cambridge: Cambridge University Press, 248–62

Robinson, G. (2004) *Geographies of Agriculture: Globalisation, Restructuring and Sustainability*. Harlow: Pearson Education

Robson, B.T. (1973) *Urban Growth: An Approach*. London: Methuen

Robson, E. (1996) Working girls and boys: children's contributions to household survival in West Africa. *Geography*, 81(4), 43–7

Rodney, W. (1972) *How Europe Underdeveloped Africa*. Washington, DC: Howard University Press

Rojas, E. (1989) Human settlements of the Eastern Caribbean: development problems and policy options. *Cities*, 6, 243–58

Rojas, E. (1995) Commentary: government–market interactions in urban development policy. *Cities*, 12, 399–400

Roodman, D.M. (2001) *Still Waiting for the Jubilee: Pragmatic Solutions for the Third World Debt Crisis*, Worldwatch Paper 155. Washington, DC: Worldwatch Institute

Rostow, W.W. (1960) *The Stages of Economic Growth: A Non-communist Manifesto*. Cambridge: Cambridge University Press

Routledge, P. (1995) Resisting and reshaping the modern: social movements and the development process, in Johnston, R.J. *et al.* (eds) *Geographies of Global Change*. London: Blackwell, 263–79

Routledge, P. (2002) Resisting and reshaping destructive development: social movements and globalising networks, in Johnston, R.J., Taylor, P.J. and Watts, M.J. (eds) *Geographies of Global Change: Remapping the World*, 2nd edn. Oxford: Blackwell, 310–27

Rowley, C. (1978) *The Destruction of Aboriginal Society*. Ringwood: Penguin

Ruthenberg, H. (1976) *Farming Systems in the Tropics*, 2nd edn. Oxford: Clarendon Press

Sachs, W. (1992) *The Development Dictionary*. London: Zed Books

Sachs, J. (ed.) (2005) *Investing in Development: A Practical Plan to Achieve the MDGs. Overview*. New York: UN

Safier, M. (1969) Towards the definition of patterns in the distribution of economic development over East Africa. *East African Geographical Review*, 7, 1–13

Sahnoun, M. (1994) Flashlights over Mogadishu. *New Internationalist*, 262, 9–11

Sahr, W.D. (1998) Micro-metropolis in the eastern Caribbean: the example of St Lucia, in McGregor, D., Lloyd-Evans, S. and Barker, D. (eds) *Resources, Sustainability and Caribbean Development*. Barbados: University of the West Indies Press

Said, E. (1978) *Orientalism*. New York: Vintage

Said, E. (1979) *Orientalism*. New York: Village Books

Said, E. (1993) *Culture and Imperialism*. London: Chatto

Sanchez-Rodriguez, R. (2006) Fernando Henrique Cardoso, in Simon, D. (ed.) *Fifty Key Thinkers on Development*. London and New York: Routledge, 61–6

Sandbrook, R. (1999) Institutions for global environmental change. *Global Environmental Change*, 9, 171–4

Sandford, S. (1983) *Management of Pastoral Development in the Third World*. Chichester: John Wiley

Santos, M. (1979) *The Shared Space: the Two Circuits of the Urban Economy in Underdeveloped Countries*. London: Methuen

Sapsford, D. (2008) Smith, Ricardo and the world market place, 1776–2007: back to the future, in Desai, V. and Potter, R.B. (eds) *The Companion to Development Studies*, 2nd edn. London: Hodder-Arnold and New York: Oxford University Press, 75–81

Sardar, Z. (1998) *Postmodernism and the Other: The New Imperialism of Western Culture*. London and Chicago, IL: Pluto Press

Sartre, J.P. (1964, 2001) *Colonialism and Neo-colonialism*. English translation (2001), London: Routledge

Sassen, S. (1991) *The Global City*. Princeton, NJ: Princeton University Press

Sassen, S. (2002) *Global Networks, Linked Cities*. New York: Routledge

Satterthwaite, D. (1997) Sustainable cities or cities that contribute to sustainable development. *Urban Studies*, 35

Satterthwaite, D. (ed.) (1999) *The Earthscan Reader in Sustainable Cities*. London: Earthscan

Satterthwaite, D. (2008) Urbanization in low- and middle-income nations, ch. 5.1 in Desai, V. and Potter, R.B. (eds) *The Companion to Development Studies*, 2nd edn. London: Hodder-Arnold and New York: Oxford University Press, 237–43

Save the Children Fund (1995) *Towards a Children's Agenda: New Challenges for Social Development*. London: SCF

Schech, S. and Haggis, J. (2000) *Culture and Development: A Critical Introduction*. Oxford: Blackwell

Schneider, F. and Frey, B. (1985) Economic and political determinants of foreign direct investment. *World Development*, 13(2), 167–75

Schultz, T.W. (1953) *The Economic Organization of Agriculture*. New York: McGraw-Hill

Schumacher, E.F. (1974) *Small is Beautiful*. London: Abacus

Schumpeter, J.A. (1911) *Die Theorie des Wirtschaftlichen Entwicklung*. Leipzig

Schumpeter, J.A. (1934) *The Theory of Economic Development*. Cambridge, MA: Harvard University Press

Schuurman, F. (ed.) (1993) *Beyond the Impasse: New Directions in Development Theory*. London: Zed

Schuurman, F. (2000) Paradigms lost, paradigms regained? Development studies in the twenty-first century. *Third World Quarterly*, 21, 7–20

Schuurman, F.J. (2001) *Globalization and Development Studies: Challenges for the 21st Century*. London, Thousand Oaks and New Delhi: Sage Publications

Schuurman, F.J. (2008) The impasse in development studies, ch. 1.3 in Desai, V. and Potter, R.B. (eds) *The Companion to Development Studies*, 2nd edn. London: Hodder-Arnold and New York: Oxford University Press, 12–15

Scoones, I. (1995) Policies for pastoralists: new directions for pastoral development in Africa, in Binns, T. (ed.) *People and Environment in Africa*. Chichester: John Wiley, 23–30

Scoones, I. (1996) Range management science and policy: politics, polemics and pasture in Southern Africa, in Leach, M. and Mearns, R. (eds) *The Lie of the Land: Challenging Received Wisdom on the African Environment*. Oxford: International African Institute/James Currey, 34–53

Scoones, I. and Thompson, J. (1994) *Beyond Farmer First*. London: Intermediate Technology Productions

Scott, J.C. (1976) *The Moral Economy of the Peasant: Rebellion and Subsistence in South-East Asia*. New Haven, CT: Yale University Press

Secrett, C. (1986) The environmental impact of transmigration. *The Ecologist*, 16(2/3), 77–89

Seers, D. (1969) The meaning of development. *International Development Review*, 11(4), 2–6

Seers, D. (1972) What are we trying to measure? *Journal of Development Studies*, 8(3), 21–36

Seers, D. (1979) The new meaning of development, in Lehmann, D. (ed.) *Development Theory: Four Critical Studies*. London: Frank Cass, 25–30

Seitz, J.L. (2000, 2002) *Global Issues: An Introduction*, 2nd edn. Oxford: Blackwell

Sen, A. (1981) *Poverty and Famines*. Oxford: Clarendon Press

Sen, A. (2000) *Development as Freedom: Human Capability and Global Need*. New York: Anchor Books

Sengupta, K. (2002) Atlas maps investment in a world of abuses. *The Independent*, 13 February, 11

Seyfang, G. (ed.) (2002) Corporate Responsibility and Labour Rights: Codes of Conduct in the Global Economy. London: Earthscan

Shah, P. (1994) Participatory watershed management in India: the experience of the Aga Khan Rural Support Programme, in Scoones, I. and Thompson, J. (eds) *Beyond Farmer First: Rural People's Knowledge, Agricultural Research and Extension Practice*. London: Intermediate Technology Publications

Shankland, A. (1991) The devil's design. *New Internationalist*, 219, 11–13

Shariff, I. (1987) Agricultural development and land tenure in India. *Land Use Policy*, 4(3), 321–30

Sharp, R. (1992) Organising for change: people-power and the role of institutions, in Holmberg, J. (ed.) *Policies for a Small Planet*. London: Earthscan/IIED, 39–65

Sharpley, R. (2002) Tourism management: rural tourism and the challenge of tourism diversification: the case of Cyprus. *Tourism Management*, 23, 233–44

Shaw, J. and Clay, E. (eds) (1993) *World Food Aid: Experiences of Recipients and Donors*. London: James Currey

Shepherd, A. (1998) *Sustainable Rural Development*. Basingstoke: Macmillan

Shibusawa, M., Ahmad, Z.H. and Bridges, B. (1992) *Pacific Asia in the 1990s*. London: Routledge

Shipton, P. and Goteen, M. (1992) Understanding African land-holding: power, wealth and meaning. *Africa*, 62(3), 307–25

Shiva, V. (2000) *Stolen Harvest: The Hijacking of Global Food Supply*. London: Zed Books

Short, C. (2000) Speech to the Seattle Assembly of the World Trade Organization. *Developments*, 9, 10–11

Shrivastava, X. (1992) *Bhopal*. London: Paul Chapman

Sidaway, J.D. (1990) Post-Fordism, post-modernity and the Third World. *Area*, 22, 301–3

Sidaway, J. (2007) Spaces of postdevelopment. *Progress in Human Geography*, 31, 345–61

Sidaway, J.D. (2008) Post-development, ch. 1.4 in Desai, V. and Potter, R.B. (eds) *The Companion to Development Studies*, 2nd edn. London: Hodder-Arnold and New York: Oxford University Press, 16–20

Siddle, D. and Swindell, K. (1990) *Rural Change in Tropical Africa*. Oxford: Blackwell

Silvers, J. (1995) Death of a slave. *Sunday Times*, 10 October, 36–41

Simms, A. (2002) Disasters waiting to happen. *The Guardian*, 19 June

Simon, D. (1992a) *Cities, Capital and Development: African Cities in the World Economy*. London: Belhaven

Simon, D. (1992b) Conceptualizing small towns in African development, in Baker, J. and Pedersen, P.O. (eds) *The Rural–Urban Interface in Africa*. Uppsala: Nordic Institute for African Studies, 29–50

Simon, D. (1993) The world city hypothesis: reflections from the periphery. CEDAR Research Paper 7, Royal Holloway College, University of London

Simon, D. (1998) Rethinking (post)modernism, postcolonialism, and posttraditonalism: north–south perspectives. *Environment and Planning D, Society and Space*, 16, 219–45

Simon, D. (ed.) (2006) *Fifty Key Thinkers on Development*. London and New York: Routledge

Simon, D. (2007) Beyond antidevelopment; discourses, convergences, practices. *Singapore Journal of Tropical Geography*, 28, 205–18

Simon, D. (2008) Neoliberalism, structural adjustment and poverty reduction strategies, ch. 2.5 in Desai, V. and Potter, R.B. (eds) *The Companion to Development Studies*, 2nd edn. London: Hodder-Arnold and New York: Oxford University Press, 86–91

Simon, J.L. (1981) *The Ultimate Resource*. London: Martin Robertson

Singh, R.P.B. (2006) Mohandas (Mahatma) Gandhi, in Simon, D. (ed.) *Fifty Key Thinkers on Development*. London and New York: Routledge, 106–10

Sinha, S. (1998) Introduction and overview. *IDS Bulletin*, 29(4), 1–10

Singer, H. (1980) The Brandt Report: a north-western point of view. *Third World Quarterly*, 2(4), 694–700

Slater, D. (1992a) On the borders of social theory: learning from other regions. *Environment and Planning D*, 10, 307–27

Slater, D. (1992b) Theories of development and politics of the post-modern: exploring a border zone. *Development and Change*, 23, 283–319

Slater, D. (1993) The geopolitical imagination and the enframing of development theory. *Transactions of the Institute of British Geographers, New Series*, 18, 419–37

Smith, A. (2002) Translocals, critical area studies and geography's others, or why 'development' should not be geography's organizing framework: a response to Potter. *Area*, 34, 210–13

Smith, D. (2000) *Moral Geographies: Ethics in a World of Difference*. Edinburgh: Edinburgh University Press

Smith, D.M. (2008) Responsibility to distant others, ch. 2.14 in Desai, V. and Potter, R.B. (eds) *The Companion to Development Studies*, 2nd edn. London: Hodder-Arnold and New York: Oxford University Press, 129–32

Smith, D.W. (1994) On professional responsibility to distant others. *Area*, 26, 359–67

Smith, D.W. (1998) Urban food systems and the poor in developing countries, *Transactions of the Institute of British Geographers*, 23, 207–19

Smith, N. (2000) Global Seattle. *Environment and Planning: Society and Space*, 18, 1–5

So, Chin-Hung (1997) Economic development, state control and labour migration of women in China. Unpublished PhD thesis, University of Sussex, Brighton

Soil Association (2002) *Seeds of Doubt: Experiences of North American Farmers of Genetically Modified Crops*. London: Soil Association

Soil Association (2005) *Organic Market Report, 2005*. Bristol: Soil Association

Soja, E.W. (1968) *The Geography of Modernization in Kenya: A Spatial Analysis of Social, Economic and Political Change*. Syracuse, NY: Syracuse University Press

Soja, E.W. (1974) The geography of modernization: paths, patterns, and processes of spatial change in developing countries, in Bruner, R. and Brewer, G. (eds) *A Policy Approach to the Study of Political Development and Change*. New York: Free Press

Soja, E.W. (1989) *Postmodern Geographies: The Reassertion of Space in Critical Social Theory*. London: Verso

Soliman, A.M. (1996) Legitimising informal housing: accommodating low income groups in Alexandria, Egypt. *Environment and Urbanisation*, 1, 183–194

Soussan, J. (1988) *Primary Resources in the Third World*. London: Routledge

Sparr, P. (1994) *Mortgaging Women's Lives: Feminist Critiques of Structural Adjustment*. London: Zed Books

Spivak, G.C. (1993) Can the subaltern speak?, in Williams, P. and Chrisman, L. (eds) *Colonial Discourse and Postcolonial Theory*. London: Prentice Hall

Stern, N. (2007) *The Economics of Climate Change: The Stern Review*. Cambridge: Cambridge University Press

Stewart, C. (1995) One more river to cross. *New Internationalist*, 273, 16–17

Stock, R. (1995) *Africa South of the Sahara: A Geographical Interpretation*. New York: Guildford Press

Stocking, M. (1987) Measuring land degradation, in Blaikie, P. and Brookfield, H. (eds) *Land Degradation and Society*. London: Methuen, 49–64

Stocking, M. (1995) Soil erosion and land degradation, in O'Riordan, T. (ed.) *Environmental Science for Environmental Management*. London: Longman, 223–43

Stocking, M. (2000) Soil erosion and land degradation, in O'Riordan, T. (ed.) *Environmental Science for Environmental Management*. Harlow: Pearson Education, 287–321

Stöhr, W.B. (1981) Development from below: the bottom-up and periphery-inward development paradigm, in Stöhr, W.B. and Taylor, D.R.F. (eds) *Development from Above or Below*? Chichester: John Wiley, 39–72

Stöhr, W.B. and Taylor, D.R.F. (1981) *Development from Above or Below? The Dialectics of Regional Planning in Developing Countries*. Chichester: John Wiley

Streeten, P. (1995) *Thinking About Development*. Cambridge: Cambridge University Press

Stren, R., White, R. and Whitney, J. (eds) (1992) *Sustainable Cities: Urbanization and the Environment in International Perspective*. Oxford: Westview Press

Stycos, J.M. (1971) Family planning and American goals, in Chaplin, D. (ed.) *Population Policies and Growth in Latin America*. Lexington, KY: Heath, 111–31

Sunday Times (2002) The biggest show in town. *Sunday Times*, Johannesburg, 7 July, 17

Sundstrom, L. (1972) Ecology and symbiosis: Niger water fold. *Studia Ethnographica Uppsalensia*, No. 25. Uppsala: Uppsala University Press

Sutton, K. and Zaimeche, S.E. (2002) The collapse of state socialism in the socialist Third World, ch. 1.5 in Desai, V. and Potter, R.B. (eds) *The Companion to Development Studies*. London: Arnold, 20–26

Taaffe, E.J., Morrill, R.L. and Gould, P.R. (1963) Transport expansion in underdeveloped countries: a comparative analysis. *Geographical Review*, 53, 503–29

Tasker, R. (1995) Tee masters. *Far Eastern Economic Review*, 5 January

Tata, R. and Schultz, R. (1988) World variations in human welfare: a new index of development status. *Annals of the Association of American Geographers*, 78(4), 580–92

Taylor, A. (2003) Trading with the environment, in Bingham, N., Blowers, A. and Belshaw, C. (eds) *Contested Environments*. Chichester: Wiley

Taylor, D.R. and Mackenzie, F. (eds) (1992) *Development from Within: Survival in Rural Africa*. London: Routledge

Taylor, I. (2006) China's oil diplomacy in Africa. *International Affairs*, 82(5), 937–59

Taylor, P. (1985) *Political Geography*. London: Longman

Taylor, P.J. (1986) The world-systems project, in Johnston, R.J. and Taylor, P.J. (eds) *A World in Crisis? Geographical Perspectives*. Oxford: Basil Blackwell, 333–54

Teo, P. and Ooi, G.L. (1996) Ethnic differences and public policy in Singapore, in Dwyer, D.J. and Drakakis-Smith, D. (eds) *Ethnicity and Geography*. London: John Wiley, 249–70

Tewdwr-Jones, M. and Allmendinger, P. (1998) Deconstructing communicative rationality: a critique of Habermasian collaborative planning. *Environment and Planning A*, 30, 1975–89

The Courier (1990) Irrigation. *The Courier*, 124, 64–95

The Courier (1996) Country report – Kenya. *The Courier*, 157, 19–36

The Ecologist (2003) Atlantic dawn. *The Ecologist*, 33(3), 1 April, 18–19

The Economist (2006) Voting with your trolley. *The Economist*, 7 December, www.economist.com/business/displaystory.cfm?story_id=8380592

The Guardian (2001) G8 leaders survive the siege of Genoa. *The Guardian*, 28 July 2, 9

The Independent (1998) The population bomb defused. *The Independent*, 12 January

The Independent (1999) Shirts for the fashionable, at a price paid in human misery. *The Independent*, 24 September, 3

The Independent (2002) Mr Blair's visit will not heal Africa's scars, but it is better than ignoring them. *The Independent*, 6 February, 11

Thirlwall, A.P. (1999) *Growth and Development: With Special Reference to Developing Economies*, 6th edn. London: Macmillan

Thirlwall, A.P. (2002) Development as economic growth, ch. 1.9 in Desai, V. and Potter, R.B. (eds) *The Companion to Development Studies*. London: Arnold, 41–4

Thomas, A. (1992) Non-governmental organisations and the limits to empowerment, in Wuyts, M., Mackintosh, M. and Hewitt, T. (eds) *Development Policy and Public Action*. Oxford: Oxford University Press, 117–46

Thomas, A. (2000) Development as practice in a liberal capitalist world. *Journal of International Development*, 12, 773–87

Thomas, A. (2001) NGOs and their influence on environmental policies in Africa, in Thomas, A. *et al.* (2001) *Environmental Policies and NGO Influence*. 1–22

Thomas, A. and Allen, T. (2000) Agencies of development, in Allen, T. and Thomas, A. (eds) *Poverty and Development in the 21st Century*. Oxford: Oxford University Press, 189–216

Thomas, C.Y. (1989) *The Poor and the Powerless: Economic Policy and Change in the Caribbean*. London: Latin American Bureau

Thomas, D.H.L. (1996) Fisheries, tenure and mobility in a West African floodplain. *Geography*, 81(4), 35–40

Thomas, D.S.G. (1993) Storm in a teacup? Understanding desertification. *Geographical Journal*, 159(3), 318–31

Thomas, D.S.G., Sporton, D. and Perkins, J. (2000) The environmental impact of livestock ranches in the Kalahari, Botswana: natural resource use, ecological change and human response in a dynamic dryland system. *Land Degradation and Development*, 11, 327–41

Thomas, G.A. (1991) The gentrification of paradise: St John's, Antigua. *Urban Geography*, 12, 469–87

Thompson, M. and Warburton, M. (1985) Uncertainty on a Himalayan scale. *Mountain Research and Development*, 5, 115–35

Thompson, R.D. (1992) The changing atmosphere and its impact on Planet Earth, in Mannion, A.M. and Bowlby, S.R. (eds) *Environmental Issues in the 1990s*. London: John Wiley, 61–78

Thrift, N. and Forbes, D. (1986) *The Price of War: Urbanisation in Vietnam 1954–1986*. London: Allen & Unwin

Tickell, O. (2000) Carbon trading is the burning issue for pollution talks. *The Independent*, 15 December

Tiffen, M. and Mortimore, M. (1990) *Theory and Practice in Plantation Agriculture: An Economic Review*. London: Overseas Development Institute

Tiffen, M., Mortimore, M.J. and Gichuki, F. (1994) *More People, Less Erosion: Environmental Recovery in Kenya*. Chichester: John Wiley

Todaro, M. (1994) *Economic Development*. Harlow: Longman

Todd, H. (ed.) (1996) *Cloning Grameen Bank: Replicating a Poverty Reduction Model in India, Nepal and Vietnam*. London: Intermediate Technology Publications

Toffler, A. (1970) *Future Shock*. London: Bodley Head

Tordoff, W. (1992) The impact of ideology or development in the Third World. *Journal of International Development*, 4(1), 41–53

Toulmin, C. (2001) *Lessons From the Theatre: Should This be the Final Curtain Call for the Convention to Combat Desertification*? WSSD Opinion Series. London: IIED

Toulmin, C. and Quan, J. (eds) (2000) *Evolving Land Rights: Policy and Tenure in Africa*. London: IIED

Tourism Concern (2002) Briefing on ecotourism, www.tourismconcern.org.uk

Toye, J. (1987) *Dilemmas of Development*. Oxford: Blackwell

Traisawasdichai, M. (1995) Chasing the little white ball. *New Internationalist*, 263, 16–17

Turner, J.R. (1967) Barriers and channels for housing development in modernizing countries. *Journal of the American Institute of Planners*, 33, 167–81

Turner, J.R. (1982) Issues in self-help and self-managed housing, ch. 4 in Ward, P. (ed.) *Self-Help Housing: A Critique*. London: Mansell, 99–113

Turner, M. and Hulme, D. (1997) *Administration and Development: Making the State Work*. Basingstoke: Macmillan

Turner, J.L. and Zhi Lu (2006) Building a green civil society in China, in WorldWatch Institute, *The State of the World, 2006: The Challenge of Global Sustainability*. London: Earthscan, 152–70

Tussie, D. and Tuozzo, M.F. (2001) Opportunities and constraints for civil society participation in multilateral lending operations: lessons from Latin America, in Edwards, M. and Gaventa, J. (eds) *Global Citizen Action*. London: Earthscan, 105–17

Uitto, J.I. (2004) Multi-country cooperation around shared waters: role of monitoring and evaluation. *Global Environmental Change*, 14, 5–14

ul-Haq, M. (1994) The new deal. *New Internationalist*, 262, 20–23

ul-Haq, M., Jolly, R., Streeten, P. and Haq, K. (1995) *The UN and the Bretton Woods Institutions: New Challenges for the Twenty-First Century*. Basingstoke: Macmillan

UN (1988) *World Population Trends and Policies, 1987: Monitoring Report*. New York: United Nations Department of International Economic and Social Affairs

UN (1989) *Prospects for World Urbanization 1988*. New York: United Nations

UN (1993) *The Global Partnership for Environment and Development: A Guide to Agenda 21*. New York: United Nations

UN (1998) *Kyoto Protocol to the UNFCCC*. New York: United Nations

UN (2000a) *World Population Prospects: The 1998 Revision, Volume III: Analytical Report*. New York: United Nations

UN (2000b) *We the Peoples: The Role of the United Nations in the Twenty First Century*. New York: UN Department of Public Information

UN (2000c) *United Nations Millennium Declaration*. New York: United Nations

UN (2001) *Road Map Towards the Implementation of the UN Millennium Declaration*, Report of the Secretary General. New York: United Nations

UN (2005a) *World Population Prospects: The 2004 Revision, Highlights*. Department of Economic and Social Affairs, Population Division. New York: United Nations, www.un.org/esa/population/publications/WPP2004/2004Highlights_finalrevised.pdf

UN (2005b) *World Urbanization Prospects 2005*. New York: United Nations

UN. United Nations Environmental Programme (2005c): Millennium Development Project, www.environmenttimes.net/article.cfm?pageID=196&group.

UN (2006a) Fact Sheet for the International Day of Peace, www.un.org/events/peaceday/2006/factsheet.html

UN (2006b) Millennium Development Goals Report, www.un.org/millenniumgoals

UN (2007a) *World Population Prospects: The 2006 Revision, Highlights*. Department of Economic and Social Affairs, Population Division. New York: United Nations, www.un.org/esa/population/publications/wpp2006/wpp2006.htm

UN (2007b) *World Population Prospects: The 2006 Revision Database*, online database, United Nations Population Division, http://esa.un.org/unpp

UN (2007c) The Millennium Development Goals Report 2007, mdg2007.pdf, www.un.org/millenniumgoals (see graph on p. 6, entitled 'Extreme poverty is beginning to fall in sub-Saharan Africa')

UNAIDS (1998a) *Joint United Nations Programme on HIV/AIDS and World Health Organization*. Geneva: UNAIDS

UNAIDS (1998b) *AIDS and the Military*. Geneva: UNAIDS, www.unaids.org/publications/documents/sectors/military/militarypre.pdf

UNAIDS (2001) *Joint United Nations Programme on HIV/AIDS: AIDS Epidemic Update*. Geneva: UNAIDS, www.unaids.org

UNAIDS (2006) AIDS epidemic update, December 2006. Geneva: UNAIDS, www.unaids.org/en/HIV_data/epi2006

UNCHS (1996) *An Urbanizing World: Global Report on Human Settlements, 1996*, United Nations Centre for Human Settlements (Habitat). Oxford: Oxford University Press

UNCTAD (2001) *World Investment Report*, United Nations Conference on Trade and Development. New York: United Nations

UNCTAD (2005) Trends in world commodity trade, enhancing Africa's competitiveness and generating development gains. Presented to AU Extraordinary Conference of Ministers of Trade on African Commodities Arusha, Tanzania, 21–24 November 2005, http://r0.unctad.org/infocomm/comm_docs/sources.asp

UNDP (1991) *Cities, People and Poverty*, United Nations Development Programme. New York: UNDP

UNDP (1993) *Human Development Report, 1993*, United Nations Development Programme. Oxford: Oxford University Press

UNDP (1994) *Human Development Report, 1994*, United Nations Development Programme. Oxford: Oxford University Press

UNDP (1996) *Human Development Report, 1996*, United Nations Development Programme. Oxford: Oxford University Press

UNDP (1997) *Human Development Report, 1997*, United Nations Development Programme. Oxford: Oxford University Press

UNDP (1998) *Human Development Report, 1998*, United Nations Development Programme. New York: Oxford University Press

UNDP (2000) *Human Development Report, 2000*, United Nations Development Programme. Oxford: Oxford University Press

UNDP (2001) *Human Development Report, 2001: Promoting Linkages*, United Nations Development Programme. Oxford: Oxford University Press

UNDP (2003) *Making Global Trade Work for People*. London: Earthscan

UNDP (2006) *Human Development Report 2006*. New York: UNDP, http://hdr.undp.org/hdr2006/statistics/build_your_table/default.cfm#

UNEP (1995) *Global Biodiversity Assessment*, United Nations Environment Programme. Cambridge: Cambridge University Press

UNEP (1997) *World Atlas of Desertification*, 2nd edn, United Nations Environment Programme. London: John Wiley

UNEP (2000) *Global Environment Outlook 2000*. London: Earthscan

UNEP (2002) *Global Environment Outlook 3*. London: Earthscan

UNEP (2005) Millennium Development Project. www.environmenttimes.net/article.cfm?pageID=196&group

UNEP/WHO (1993) *City Air Quality Trends*, Vol. 2, United Nations Environment Programme/World Health Organisation. Nairobi: UNEP

UNESCO (undated) World Water Assessment Programme, About the programme, www.unesco.org/water/wwap/description/index.shtml

UNESCO-UIS (2006) *Global Education Digest 2006*. Montreal: UNESCO-UIS, www.uis.unesco.org/TEMPLATE/pdf/ged/2006/GED2006.pdf

UNHCR (2002) Afghanistan – the long road home, www.unhcr.ch/cgi-bin/texis/ytx/afghan

UNHCR (2006) *2005 Global Refugee Trends*, www.unhcr.org/statistics

UNICEF (1997) *The State of the World's Children*, UN Children's Fund. Oxford: Oxford University Press

UNICEF (1998) *The Impact of Conflict on Children in Afghanistan*, United Nations Children's Fund. Geneva: UNICEF

UNICEF (2000) *The State of the World's Children 2000*, United Nations Children's Fund. New York/Geneva: UNICEF

UNICEF (2001) *The State of the World's Children 2001*, United Nations Children's Fund. Geneva: UNICEF, www.unicef.org

UNICEF (2006) *State of the World's Children 2007*. New York: United Nations Children's Fund, www.unicef.org/sowc07/docs/sowc07.pdf

Unilever (2002) Financial highlights, www. unilever.com, 31 August

Unilever (2006a) Annual report and accounts 2006: Financial review, http://unilever.com/ourcompany/investorcentre/annual_reports/annual_report_Form.asp

Unilever (2006b) Financial highlights, www.unilever.com

United States Department of Energy (1994) *Energy Use and Carbon Emissions: Some International Comparisons*. Washington DC: Energy Information Administration

UNRISD (United Nations Research Institute for Social Development) (1995) *States of Disarray: the Social Effects of Globalisation*. Geneva: UNRISD

Unwin, T. and de Bastion, G. (2008) Information and communication technologies for development, ch. 1.12 in Desai, V. and Potter, R.B. (eds) *The Companion to Development Studies*, 2nd edn. London: Hodder-Arnold and New York: Oxford University Press, 54–8

Unwin, T. and Potter, R.B. (1992) Undergraduate and postgraduate teaching on the geography of the Third World. *Area*, 24, 56–62

Urbach, J. (2007) Development goes wireless. *Journal of the Institute of Economic Affairs*, 27, 20–28

Urban Foundation (1993) *Managing Urban Poverty*. Johannesburg: Urban Foundation

Urry, J. (1990) *The Tourist Gaze*. London: Sage

US Census Bureau (1994) *Trends and Patterns of HIV/AIDS Infection in Selected Developing Countries*. Country Profiles, Research Note 15, Health Studies Branch. Washington, DC: Center for International Research

US Census Bureau (2002), International Data Base, www.census.gov/ipc/www/idbnew.html

Usbourne, D. (2001) Annan is honoured for bringing new life to the UN. *The Independent*, 13 October

Vance, J.E. (1970) *The Merchant's World: The Geography of Wholesaling*. Englewood Cliffs, NJ: Prentice Hall

van der Gaag, N. (1997) Gene dream. *New Internationalist*, 293, 7–10

Van Rooy, A. (2002) Strengthening civil society in developing countries, in Desai, V. and Potter, R.B. (eds) *The Companion to Development Studies*. London: Arnold, 489–95

Vapnarsky, C.A. (1969) On rank-size distributions of cities: an ecological approach. *Economic Development and Cultural Change*, 17, 584–95

Vesiland, P.J. (1993) Water: the Middle East's critical resource. *National Geographic*, 183(5), 38–71

Vidal, J. (2002) New Delhi opens door to GM crops. *The Guardian*, 27 March

Vivian, J.M. (1992) Foundations for sustainable development: participation, empowerment and local resource management, in Ghai, D. and Vivian, J.M. (eds) *Grassroots Environmental Action: Peoples' Participation in Sustainable Development*. London: Routledge, 50–80

Vivian, J. (1995) How safe are safety nets? In Vivian, J. (ed.) *Adjustment and Social Sector Restructuring*. London: Frank Cass

von Moltke (1994) The World Trade Organisation: its implications for sustainable development. *Journal of Environment and Development*, 3(1), 43–57

Wallerstein, I. (1974) *The Modern World System I*. New York: Academic Press

Wallerstein, I. (1979) *The Capitalist World Economy*. Cambridge: Cambridge University Press

Wallerstein, I. (1980) *The Modern World System II*. New York: Academic Press

War on Want (2004) *Profiting from Poverty: Privatisation Consultants, DFID and Public Services*. London: War on Want

War on Want (2007) *Growing Pains: The Human Cost of Cut Flowers in British Supermarkets*. London: War on Want

Ward, P. and Macoloo, C. (1992) Articulation theory and self-help housing practice in the 1990s. *International Journal of Urban and Regional Research*, 16, 60–80

Watkins, K. (1995) *The Oxfam Poverty Report*. Oxford: Oxfam

Watson, K. (ed.) (1982) *Education in the Third World*. London: Croom Helm

Watson, M. and Potter, R.B. (2001) *Low-Cost Housing in Barbados: Evolution or Social Revolution?* Barbados: University of the West Indies Press

Watson, R.T. (ed.) (2001) *Climate Change 2001, Synthesis Report*. Cambridge: Cambridge University Press

Watters, R.F. and McGee, T.G. (eds) (1997) *Asia–Pacific: New Geographies of the Pacific Rim*. London: Hurst

Watts, M. (1984) The demise of the moral economy: food and famine in a Sudano-Sahelian region in historical perspective, in Scott, E. (ed.) *Life Before the Drought*. Boston, MA: Allen & Unwin, 124–48

Watts, M. (1996) Development in the global agrofood system and late twentieth-century development (or Kautsky reduxe). *Progress in Human Geography*, 20(2), 230–45

Watts, M.J. (2004) Violent environments: petroleum conflict and the political ecology of rule in the Niger delta, Nigeria, in Peet, R. and Watts, M. (2004) *Liberation Ecologies: Environment, Development, Social Movements*, 2nd edn. London: Routledge, 273–98

Watts, M. (2006) Andre Gunder Frank, in Simon, D. (ed.) *Fifty Key Thinkers on Development*. London and New York: Routledge, 90–95

Watts, M. and McCarthy, J. (1997) Nature as artifice, nature as artefact: development, environment and modernity in the late twentieth century. Paper presented in the Lectures in Human Geography Series, University of St Andrews

WCED (1987) *Our Common Future*, World Commission on Environment and Development. Oxford: Oxford University Press

Weaver, D.B. (1998) *Ecotourism in the Less Developed World*. Wallingford: CAB International

Webb, D. (1997) *HIV and AIDS in Africa*. London: Pluto Press

Weidelt, H.J. (1993) Agroforestry systems in the tropics – recent developments and results of research. *Applied Geography and Development*, 41, 39–50

Wellard, K. and Copestake, J. (1993) *Non-governmental Organizations and the State in Africa*. London: Routledge

Wen, Y.-K. and Sengupta, J. (eds) (1991) *Increasing the International Competitiveness of Exports from Caribbean Countries*. Washington DC: World Bank

Werksman, J. (1995) Greening Bretton Woods, in Kirkby, J., O'Keefe, P. and Timberlake, L. (eds) *The Earthscan Reader in Sustainable Development*. London: Earthscan, 274–87

Werksman, J. (ed.) (1996) *Greening International Institutions*. London: Earthscan

Wheat, S. (1993) Playing around with nature. *Geographical Magazine*, LXV(8), 10–14

Wheat, S. (2000) A path out of poverty?, *The Courier*, 183, 60–62

Whitaker's (1999) *Whitaker's Almanack 2000*, 132nd edn. London: The Stationery Office

Whitman, J. (2002) The role of the United Nations in developing countries, in Desai, V. and Potter, R.B. (eds) *The Companion to Development Studies*. London: Arnold, 466–70

WHO/UNICEF (2000) *Global Water Supply and Sanitation Assessment 2000*, Joint Report, World Health Organization and UNICEF. Geneva and New York: WHO/UNICEF

Wignaraja, P. (1993) Rethinking development and democracy, in Wignaraja, P. (ed.) *New Social Movements in the South*. London: Zed Books, 4–35

Williams, I. (1995) *The UN for Beginners*. London: Writers and Readers

Williams, M. (1994) Making golf greener. *Far Eastern Economic Review*, May, 40–41

Williams, M. (1995) Role of the multilateral agencies after the Earth Summit, in ul-Haq, M., Jolly, R. Streeten, P. and Haq, K. (1995) *The UN and the Bretton Woods Institutions: New Challenges for the Twenty-First Century*. Basingstoke: Macmillan, 210–38

Williams, M. and Ford, L. (1999) The World Trade Organization, social movements and global environmental management. *Environmental Politics*, 8(1), 268–89

Williams, P. and Chrisman, L. (eds) (1993) *Colonial Discourse and Postcolonial Theory*. London: Prentice Hall

Williams, S. (2007) Comparative study of cut roses for the British market produced in Kenya and the Netherlands, Precis Report for World Flowers, Cranfield University, www.world-flowers.co.uk/12news/news4.html

Williamson, J.G. (1965) Regional inequality and the process of national development: a description of the patterns. *Economic Development and Cultural Change*, 13, 3–45

Wills, J. (2002) Political economy III: neoliberal chickens, Seattle and geography. *Progress in Human Geography*, 26, 90–100

Wilson, D. and Purushothaman, R. (2003) *Dreaming with BRICs: The Path to 2050*. Goldman Sachs, Global Economics Paper No. 99

Wilson, F. (1994) Reflections on the present predicament of the Mexican garment industry, in Pedersen, P. (ed.) *Flexible Specialization*. London: IT Publications, 147–58

Wolf, A.T., Natharius, J.A., Danielson, J.J., Ward, B.S. and Pender, J.K. (1999) International river basins of the world. *Water Resources Development*, 15(4), 387–427

Wolfe-Phillips, L. (1987) Why Third World – origins, definitions and usage. *Third World Quarterly*, 9(4), 1311–9

Wolmer, W., Chaumba, J. and Scoones, I. (2004) Wildlife management and land reform in southeastern Zimbabwe: a compatible pairing or a contradiction in terms? *Geoforum*, 35, 87–98

Wolpe, H. (1975) The theory of internal colonialism, in Oxaal, J., Barnett, T. and Booth, D. (eds) *Beyond the Sociology of Development*. London: Routledge & Kegan Paul, 229–52

Womankind Worldwide (undated) How to challenge a colossus: engaging with the World Bank and the International Monetary Fund, www.womankind.org.uk

Women and Geography Study Group (1997) *Feminist Geographies: Explorations in Diversity and Difference*. London: Longman

Woods, N. (2000) The challenge of good governance for the IMF and the World Bank themselves. *World Development*, 28(5), 823–41

World Bank (1986) *Poverty and Hunger: Issues and Options for Food Security in Developing Countries*. Washington, DC: World Bank

World Bank (1988) *World Development Report*. Washington, DC: World Bank

World Bank (1989) *Sub-Saharan Africa: From Crisis to Sustainable Growth: A Long Term Perspective Study*. Washington, DC: World Bank

World Bank (1990) *World Development Report, 1990*. Oxford: Oxford University Press

World Bank (1991) *Urban Policy and Economic Development: An Agenda for the 1990s*. Washington, DC: World Bank

World Bank (1992) *World Development Report*. Washington, DC: World Bank

World Bank (1993) *East Asian Miracle*. Washington, DC: World Bank

World Bank (1994a) *World Bank and the Environment: Fiscal 1993*. Washington, DC: World Bank

World Bank (1994b) *World Development Report, 1994*. Oxford: Oxford University Press

World Bank (1995) *World Development Report, 1995*. Oxford: Oxford University Press

World Bank (1996a) *World Development Report, 1996*. Oxford: Oxford University Press

World Bank (1996b) *A Review of World Bank Experience in Irrigation*. Washington, DC: World Bank Operations Evaluation Department

World Bank (1997a) *World Development Report: The State in a Changing World*. Oxford: Oxford University Press

World Bank (1997b) *The Impact of Environmental Assessment: a Review of World Bank Experience*, Environment Department. Washington, DC: World Bank

World Bank (1999) Safe motherhood and the World Bank: lessons from 10 years of experience, www.worldbank.org/html/extdr/hnp/population/tenyears/text.pdf

World Bank (2000a) *Operational Manual February 2000 GP14.70 Good Practices Involving Nongovernmental Organisations in Bank-supported Activities*. Washington, DC: World Bank

World Bank (2000b) *Third Environmental Assessment Review (FY96-00)* Environment Department. Washington: World Bank

World Bank (2001a) *World Development Report 2000/2001*. Oxford: Oxford University Press

World Bank (2001b) The HIPC debt initiative, December 2001, www.worldbank.org/hipc

World Bank (2001c) *Environment Matters*. Washington, DC: World Bank

World Bank (2001d) *Poverty Reduction and the World Bank: Progress in Operationalizing the World Development Report 2000/01*. Washington, DC: World Bank

World Bank (2001e) *Making Sustainable Commitments: An Environment Strategy for the World Bank, Summary, December 2001*. Washington, DC: World Bank

World Bank (2002a) *World Development Report 2002*. Oxford: Oxford University Press, Oxford

World Bank (2002b) World Bank to commit $500 million more to fight HIV/AIDS in Africa, News Release No. 2002/197/HD, wbln0018.worldbank.org/news/pressrelease.nsf/673fa6c5a2d50a67852565e2006

World Bank (2002c) *Environment Matters*. Washington, DC: World Bank

World Bank (2002d) *Africa Issue Briefs*, www.worldbank.org/news, 10 December

World Bank (2002e) *African Development Indicators 2002*. Oxford: Oxford University Press

World Bank (2003a) *World Development Report*. Oxford: Oxford University Press

World Bank (2003b) *Environment Matters at the World Bank*. Washington, DC: World Bank

World Bank (2004) *The Millennium Development Goals for Health: Rising to the Challenges*. Washington, DC: World Bank

World Bank (2005a) *Third Environmental Assessment Review*. Washington, DC: World Bank Environment Department

World Bank (2005b) *World Development Report 2006: Equity and Development*. New York: World Bank and Oxford University Press

World Bank (2005c) *World Development Indicators 2005*. Washington, DC: World Bank

World Bank (2005d) *Annual Report 2005.* Washington, DC: World Bank

World Bank (2006) *World Development Indicators 2006.* Washington, DC: World Bank

World Bank (2006a) *Global Monitoring Report 2006.* Washington, DC: World Bank

World Bank (2006b) *Annual Report 2006.* Washington, DC: World Bank

World Bank (2006c) *Environment Matters at the World Bank.* Washington, DC: World Bank

World Bank (2006d) *Africa Development Indicators, 2006.* Washington, DC: World Bank

World Bank (2006e) *World Development Indicators, 2006.* Washington, DC: World Bank http://devdata.worldbank.org/wdi2006/contents/cover.htm

World Bank (2007a) *World Development Report: Development and the Next Generation.* Washington, DC: World Bank

World Bank (2007b) *Global Development Finance, 2007.* Washington, DC: World Bank

World Bank (2007c) *A Decade of Measuring the Quality of Governance.* Washington, DC: World Bank

World Bank (2007d) *World Development Indicators,* http://ddp-ext.worldbank.org/ext/DDPQQ/member.do?method=getMembers&userid=1&queryId=135

World Bank (undated) *World Bank–Civil Society Collaboration – Progress Report for Fiscal Years 2000 and 2001,* Social Development Report. Washington, DC: World Bank

World Bank/IMF (2004) *Poverty Reduction Strategy Papers: Progress in Implementation.* Washington, DC: World Bank

World Bank Independent Evaluation Group (WB-IEG) (2006) *Debt Relief for the Poorest: An Evaluation Update of the HIPC Initiative.* Washington, DC: World Bank

World Commission on Dams (2000) *Dams and Development: A New Framework for Decision-Making,* The Report of the World Commission on Dams. London: Earthscan

World Development Movement (2006) *Pipe Dreams: The Failure of the Private Sector to Invest in Water Services in Developing Countries.* London: World Development Movement

World Health Organization (1994) *The Current Global Situation of the HIV/AIDS Pandemic.* Geneva: WHO

World Health Organization (1998) *Obesity – Preventing and Managing the Global Epidemic.* Geneva: WHO

World Health Organization (1999) Reduction of maternal mortality: a joint WHO/UNFPA/UNICEF/World Bank statement, www.unfpa.org/news/ pressroom/1999/maternal.htm

World Health Organization (2002) *World Health Report 2002: Reducing Risks and Promoting Healthy Life.* Geneva: WHO

World Health Organization (2006a) Obesity and overweight, online fact sheet, www.who.int/mediacentre/factsheets/fs311/en/index.html

World Health Organization (2006b) *WHO Global InfoBase Online,* www.who.int/ncd_surveillance/infobase/web/InfoBaseCommon

World Health Organization (2007) Epidemiological fact sheets on HIV/AIDS and sexually transmitted infections, www.who.int/globalatlas/default.asp

World Health Organization/UNICEF (2006) Meeting the MDG Drinking Water and Sanitation Target. Geneva: WHO

World Resources Institute (1990) *World Resources 1990–1991.* Oxford: Oxford University Press

World Resources Institute (1992) *World Resources 1992–1993.* Oxford: Oxford University Press

World Resources Institute (1994) *World Resources 1994–1995.* Oxford: Oxford University Press

World Resources Institute (1996) *World Resources 1996–1997.* Oxford: Oxford University Press

World Resources Institute (1998) *World Resources 1998–99: Environment and Health.* Oxford: Oxford University Press

World Resources Institute (2000) *World Resources: 2000–2001: People and Ecosystems.* Oxford: Oxford University Press

World Resources Institute (2003) *World Resources, 2002–4.* Washington, DC: World Resources Institute

World Resources Institute (2004) *Earth Trends,* www.wri.org/earthtrends

World Resources Institute (2005) *World Resources, 2005: The Wealth of the Poor.* Washington, DC: World Resources Institute

World Resources Institute (2005b) *Earth Trends, 2005.* Washington, DC: World Resources Institute

World Tourism Organization (2002) www.world-tourism.org/market_research/facts&figures

World Tourism Organization (2006) World Tourism Highlights: 2006 edition, www.unwto.org/facts/menu.html

World Trade Organization (2000) Seven common misunderstandings about the WTO, in Lechner, F.J. and Boli, J. (eds) *The Globalization Reader*. Oxford: Blackwell, 236–9

Worldwatch Institute (2002) *State of the World 2002: Progress Towards a Sustainable Society*. London: Earthscan

Worldwatch Institute (2006) *The State of the World, 2006: The Challenge of Global Sustainability*. London: Earthscan

Worsley, P. (1964) *The Third World*. London: Weidenfeld & Nicolson

Worsley, P. (1979) How many worlds? *Third World Quarterly*, 1(2), 100–108

Wratten, E. (1995) Conceptualizing urban poverty. *Environment and Urbanization*, 7(1), 11–37

Wuyts, M., Mackintosh, M. and Hewitt, T. (eds) (1992) *Development Policy and Public Action*. Oxford: Oxford University Press

Yeh, A.G.O. and Wu, F.L. (1995) Internal structure of Chinese cities in the midst of economic reform. *Urban Geography*, 16(6), 521–54

Yeung, Y.-M. (1995) Commentary: urbanization and the NPE: an Asia-Pacific perspective. *Cities*, 12, 409–11

Young, E.M. (1996) *World Hunger*. London: Routledge

Zack-Williams, A.B. (2001) No democracy, no development: reflections on democracy and development in Africa. *Review of African Political Economy*, 88, 213–23

Zimmerman, E.W. (1951) *World Resources and Industries*. New York: Harper & Row

Zook, M.A. (2005) *The Geography of the Internet Industry*. Oxford: Blackwell

Zulkifli (2007) Speech by Mr Masagos Zulkifli, Senior Parliamentary Secretary, Ministry of Education, at the 2007 Trim & Fit Award Ceremony on Monday, 19 March 2007, 1500 hrs at MOE Edutorium, www.moe.gov.sg/speeches/2007/sp20070319.htm

Index

KEY TEXTS FROM PEARSON EDUCATION

The Developing Areas Research Group Series

1999 | 296pp
ISBN 13: 9780582414174
ISBN 10: 0582414172

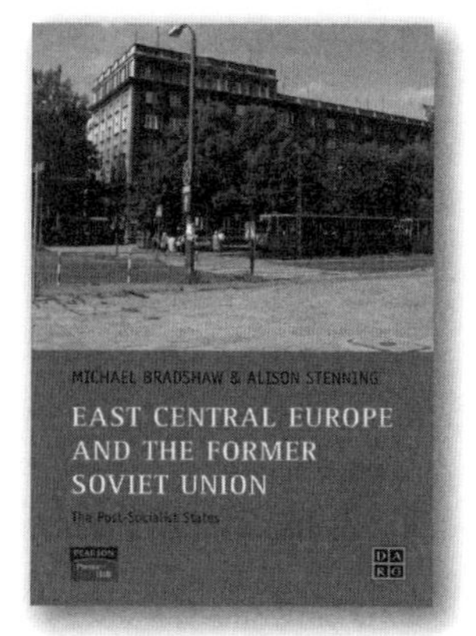

2004 | 288pp
ISBN 13: 9780130182524
ISBN 10: 013182524

2005 | 296pp
ISBN 13: 9780130259493
ISBN 10: 0130259497

2002 | 304pp
ISBN 13: 9780582404854
ISBN 10: 0582404851

2004 | 408pp
ISBN 13: 9780130264688
ISBN 10: 0130264687

2003 | 304pp
ISBN 13: 9780130259479
ISBN 10: 0130259470

OTHER KEY TEXTS

2007 | 376pp
ISBN 13: 9780131293168
ISBN 10: 0131293168

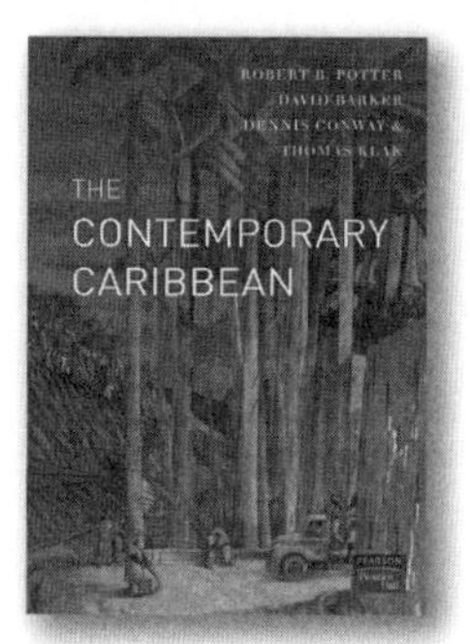

2004 | 528pp
ISBN 13: 9780582418530
ISBN 10: 0582418534

1998 | 264pp
ISBN 13: 9780582357419
ISBN 10: 0582357411

2000 | 392pp
ISBN 13: 9780582356269
ISBN 10: 0582356261

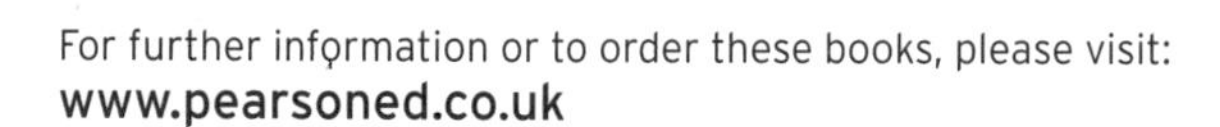
For further information or to order these books, please visit:
www.pearsoned.co.uk